Lecture Notes in Artificial Intelligence   4874

Edited by J. G. Carbonell and J. Siekmann

Subseries of Lecture Notes in Computer Science

José Neves   Manuel Filipe Santos
José Manuel Machado (Eds.)

Progress in Artificial Intelligence

13th Portuguese Conference
on Aritficial Intelligence, EPIA 2007
Workshops: GAIW, AIASTS, ALEA, AMITA, BAOSW
BI, CMBSB, IROBOT, MASTA, STCS, and TEMA
Guimarães, Portugal, December 3-7, 2007
Proceedings

 Springer

Series Editors

Jaime G. Carbonell, Carnegie Mellon University, Pittsburgh, PA, USA
Jörg Siekmann, University of Saarland, Saarbrücken, Germany

Volume Editors

José Neves
Manuel Filipe Santos
José Manuel Machado
Universidade do Minho
Departamento de Informática
4710-057 Braga, Portugal
E-mail: {jneves, mfs, jmac}@di.uminho.pt

Library of Congress Control Number: 2007939905

CR Subject Classification (1998): I.2, H.2, F.1, H.3, D.1.6

LNCS Sublibrary: SL 7 – Artificial Intelligence

ISSN   0302-9743
ISBN-10   3-540-77000-3 Springer Berlin Heidelberg New York
ISBN-13   978-3-540-77000-8 Springer Berlin Heidelberg New York

Springer is a part of Springer Science+Business Media

springer.com

© Springer-Verlag Berlin Heidelberg 2007
Printed in Germany

Typesetting: Camera-ready by author, data conversion by Scientific Publishing Services, Chennai, India
Printed on acid-free paper   SPIN: 12197330   06/3180   5 4 3 2 1 0

Preface

The 13th Portuguese Conference on Artificial Intelligence, EPIA 2007, took place in the old city of Guimarães and was sponsored by the University of Minho and APPIA, the Portuguese Association for Artificial Intelligence. The city of Guimarães is classified as a UNESCO World Cultural Heritage, being located in the North of Portugal, in the Minho Region, approximately 350 Km north of the capital, Lisbon, and about 50 Km from the second largest city, the city of Oporto. Guimarães has its origin in times previous to the foundation of the Portuguese nationality, the place where Portugal was born in the twelveth century. It is proudly referred to as the Cradle of the Nation.

EPIA was firstly held in the city of Oporto, in 1985. Its purpose was to promote the research in artificial intelligence (AI) and the scientific exchange among AI researchers, practitioners, scientists and engineers. The conference is devoted to all areas of artificial intelligence and covers both theoretical and foundational issues as well as applications. As in previous editions, the program was prearranged in terms of workshops dedicated to specific themes of AI, invited lectures, tutorials and sessions, all selected according to the highest standards. The program of the conference included for the first time a Doctoral Symposium.

A total of 11 workshops were considered, following the guidelines of the Organizing and Advisory Committees. The list includes workshops held in the previous edition, such as GAIW, BAOSW07, IROBOT07, MASTA07, TEMA07, and ALEA, rearrangements of previous ones, such as CMBSB, and novel ones such as AIASTS, AmITA, BI07 and STCS.

In this edition a total of 210 contributions were received from 29 countries, namely, Australia, Belgium, Brazil, Canada, China, Czech Republic, Denmark, France, Germany, Greece, Hungary, India, Iran, Italy, Jamaica, Lithuania, Mexico, Netherlands, Pakistan, Portugal, Romania, Russia, Spain, South Africa, UK, USA and Venezuela. Each paper was refereed by three independent reviewers in a double-blind process. From the submitted contributions, 58 were selected for publication in this volume as full papers. The acceptance rate was 27.6%, ensuring the high quality and diversity of topics and perspectives. Geographically, the contributions are distributed as follows: Belgium (1); Brazil (2); Czech Republic (1); Spain (1); China (2); France (1); Germany (4); Lithuania (1); Portugal (41); Romania (1); UK (2); and South Africa(1).

In addition to APPIA, the main sponsor of the conference, our special thanks are due to the American Association for Artificial Intelligence (AAAI); the Association for Computing Machinery (ACM); the European Coordinating Committee for Artificial Intelligence (ECCAI); and the Institute of Electrical and Electronics Engineers, Inc. (IEEE-SMC). We wish to thank the members of all committees, specifically the Advisory Committee, the Program Committee and the Local Organization. We would also like to thank all the Chairs of the

Workshops, the authors, the referees and the organizers of the previous edition for their contribution and efforts which made EPIA 2007 possible. The final thanks go to Springer for their assistance in producing this book.

The present volume includes the full papers organized in chapters, and serialised in the form of workshops.

December 2007   José Neves
                                                                 Manuel Santos
                                                                 José Machado

EPIA 2007 Conference Organization

Program and Conference Co-chairs

José Maia Neves	Universidade do Minho, Portugal
Manuel Filipe Santos	Universidade do Minho, Portugal
José Manuel Machado	Universidade do Minho, Portugal

Advisory Committee

Amílcar Cardoso	Universidade de Coimbra, Portugal
António Porto	Universidade Nova de Lisboa, Portugal
Arlindo Oliveira	Universidade Técnica de Lisboa, Portugal
Carlos Bento	Universidade de Coimbra, Portugal
Carlos Ramos	Instituto Politécnico do Porto, Portugal
Ernesto Costa	Universidade de Coimbra, Portugal
Ernesto Morgado	Universidade Técnica de Lisboa, Portugal
Eugénio Oliveira	Universidade do Porto, Portugal
Gabriel Lopes	Universidade Nova de Lisboa, Portugal
Gaël Dias	Universidade da Beira Interior, Portugal
Hélder Coelho	Universidade de Lisboa, Portugal
João Pavão Martins	Universidade Técnica de Lisboa, Portugal
Luís Damas	Universidade do Porto, Portugal
Luís Moniz Pereira	Universidade Nova de Lisboa, Portugal
Miguel Filgueiras	Universidade do Porto, Portugal
Pavel Brazdil	Universidade do Porto, Portugal
Pedro Barahona	Universidade Nova de Lisboa, Portugal
Pedro Henriques	Universidade do Minho, Portugal
Salvador Abreu	Universidade de Évora, Portugal

Workshop Chairs

GAIW 2007

José Neves	Universidade do Minho, Portugal
Manuel Santos	Universidade do Minho, Portugal
José Machado	Universidade do Minho, Portugal

AIASTS 2007

Rosaldo Rossetti	Universidade do Porto, Portugal
José Telhada	Universidade do Minho, Portugal
Ronghui Liu	University of Leeds, UK
Elisabete Arsénio	Laboratório Nacional de Engenharia Civil, Portugal

ALEA 2007

Luís Correia	Universidade de Lisboa, Portugal
Agostinho Rosa	Universidade Técnica de Lisboa, Portugal

AMITA 2007

Carlos Ramos	Instituto Politécnico do Porto, Portugal
Fariba Sadri	Imperial College of Science, UK

BAOSW 2007

H. Sofia Pinto	Instituto Superior Técnico, Portugal
Andreia Malucelli	Pontifícia Universidade Católica do Paraná, Brazil
Fred Freitas	Universidade Federal de Pernambuco, Brazil
Philipp Cimiano	University of Karlsruhe, Germany

BI 2007

Paulo Cortez	Universidade do Minho, Portugal
Robert Stahlbock	University of Hamburg, Germany

CMBSB 2007

Rui Camacho	Universidade do Porto, Portugal
Miguel Rocha	Universidade do Minho, Portugal

IROBOT 2007

Luís Paulo Reis	Universidade do Porto, Portugal
Nuno Lau	Universidade de Aveiro, Portugal
Cesar Analide	Universidade do Minho, Portugal

MASTA 2007

Luís Paulo Reis	Universidade do Porto, Portugal
João Balsa	Universidade de Lisboa, Portugal
Paulo Novais	Universidade do Minho, Portugal
Eugénio Oliveira	Universidade do Porto, Portugal

STCS 2007

Francisco Azevedo	Universidade Nova de Lisboa, Portugal
Inês Lynce	Universidade Técnica de Lisboa, Portugal
Vasco Manquinho	Universidade Técnica de Lisboa, Portugal

TEMA 2007

Joaquim Silva	Universidade Nova de Lisboa, Portugal
José Gabriel Lopes	Universidade Nova de Lisboa, Portugal
Gaël Dias	Universidade da Beira Interior, Portugal
Vitor Rocio	Universidade Aberta, Portugal

Local Organization Committee

António Abelha
Cesar Analide
Hélder Quintela
Paulo Cortez
Paulo Novais
Pedro Gago
Víctor Alves

Doctoral Symposium Committee

Cesar Analide
Paulo Novais
Pedro Henriques

Referees

Adam Kilgarriff	Ana Figueiredo	Arlindo Oliveira
Adriano Moreira	Ana Paiva	Armando Sousa
Agachai Sumalee	Ana Rocha	Armando Tachella
Alan Winfield	André Carvalho	Artur Lugmayr
Aldo Gangemi	Andrea Omicini	Barbara Smith
Alexander Kleiner	Andreia Malucelli	Beatriz De la Iglesia
Alexandre Agustini	Anna Costa	Benedita Malheiro
Aline Villavicencio	Ansgar Bredenfeld	Boris Motik
Alípio Jorge	António Abelha	Carla Gomes
Álvaro Seco	António Costa	Carlos Bento
Amílcar Cardoso	António Moreira	Carlos Carreto
Anália Lourenço	António Porto	Carlos Gershenson
Ana Bazzan	Antonio Sanfilippo	Carlos Ramos

Carlos Soares
Carsten Sinz
Catherine Havasi
Cesar Analide
Clive Best
Colin Chen
Cristián F.-Sepulveda
Cristiano Castelfranchi
Chris Welty
Christel Vrain
Christoph Tempich
Christopher Brewster
Daniel Le Berre
Daniel Oberle
Daniel Polani
Daniel Shapiro
David Hales
Eduardo Silva
Elaine Holmes
Elena Simperl
Elisabete Arsénio
Enrico Pagello
Enrique Pelaez
Eric de la Clergerie
Ernesto Costa
Ernesto Morgado
Eugénio Ferreira
Eva Iglesias
Fariba Sadri
Fazel Famili
Fei-Yue Wang
Felip Manyà
Fernando Diaz-Gomez
Fernando Lobo
Fernando Mouta
Fernando Pereira
Fernando Ribeiro
Florentino Riverola
Frédérique Segond
Francesca Lisa
Francisco Couto
Francisco Pereira
Franziska Klügl
Gaël Dias
Giovanna Serugendo

Giuseppe Vizzari
Goreti Marreiros
Gregory Grefenstette
Guy Theraulaz
Hélder Coelho
Hans-Dieter Burkhard
Harith Alani
Harry Timmermans
Hasan Jamil
Henry Lieberman
Hongbo Liu
Hussein Dia
Inês Dutra
Inman Harvey
Irene Rodrigues
Isabel Praça
Isabel Rocha
J. Gabriel Lopes
J. Norberto Pires
Jacky Baltes
Jaime Sichmann
Javier Ruiz-del-Solar
Jianhua Ma
João Balsa
João Gama
João Leite
João Marques-Silva
João Martins
João Silva
Joaquim Silva
John Davies
John-Jules Meyer
Jonas Almeida
Jordi Sabater
Jorg Muller
Jorge Dias
Jorge Louçã
Jorge Santos
Jorge Simão
Jorge Vieira
José Alferes
José Costa
José Iria
José Machado
José Mendes

José Mendez-Reboredo
José Neves
José Pereira
José Telhada
Juan Augusto
Juan Corchado
Juan Merelo
Juergen Branke
Julian Padget
Junji Nishino
Kenji Araki
Kiran Patil
Klaas Dellschaft
Klaus Fischer
Larry Birnbaum
Leandro de Castro
Leo Obrst
Ljiljana Stojanovic
Logbing Cao
Luc Dehaspe
Luca Gambardella
Lucas Bordeaux
Luís Antunes
Luís Camarinha-Matos
Luís Correia
Luís Damas
Luís Lopes
Luís Moniz Pereira
Luís Moniz
Luís Nunes
Luís Reis
Luís Rocha
M. Isabel Ribeiro
M. João Viamonte
M. Sameiro Carvalho
Manuel Barata
Manuel Delgado
Manuel Ferro
Manuel Santos
Marc Schonenauer
Marcelo Finger
Marco Dorigo
Maria Nunes
Maria Vargas-Vera
Mariano Lopez

Maribel Santos	Paulo Gomes	Spela Vintar
Marie-France Sagot	Paulo Leitão	Stefan Lessmann
Marie-Laure Reinberger	Paulo Novais	Stefano Borgo
Marie-Pierre Gleizes	Paulo Quaresma	Stephane Lallich
Mário Silva	Paulo Urbano	Stephen Balakirsky
Mark d'Inverno	Pavel Brazdil	Steven Prestwich
Mark Wallace	Pedro Barahona	Tim Kovacs
Marko Grobelnik	Pedro Henriques	Timothy M. Ebbels
Martin Riedmiller	Pedro Marrion	Vítor Costa
Maybin Muyeba	Pedro Meseguer	Vasilis Aggelis
Michael ONeil	Peter Eggenberger	Veska Noncheva
Michel Klein	Peter Geczy	Victor Alves
Miguel Calejo	Peter Haase	Victor Rocio
Miguel Filgueiras	Peter Todd	Vijay Kumar
Miguel Rocha	Philippe Lenca	Virginia Dignum
Miguel-Angel Sicilia	Pierre Zweigenbaum	Vojtech Svatek
Mikhail Prokopenko	Raphael Volz	Walter Daelemans
Mohammed Zaki	Raul Rojas	Wojtek Jamroga
Nikos Vlassis	Reinaldo Bianchi	Wolfgang Banzhaf
Nuno Fonseca	Renata Vieira	Wolfgang Jank
Nuno Gomes	Richard Vaughan	Wolfram-M. Lippe
Nuno Lau	Rolf Pfeifer	Wong Man-leung
Nuno Silva	Ronghui Liu	Xiaoping Chen
Oliver Kullmann	Rosa Vicari	Yin Ding
Oliver Obst	Rosaldo Rossetti	Ying Tan
Olivier Roussel	Rui José	Yonghong Peng
Orcar Corcho	Rui Mendes	York Sure
Orlando Belo	Salvador Abreu	Yves Demazeau
Pablo Otero	Santiago Schnell	Zary Segall
Paolo Petta	Sascha Ossowski	Zbigniew Michalewicz
Patrick Meyer	Sherief Abdallah	Zeynep Kiziltan
Paul Guyot	Shuming Tang	Zita Vale
Paul Valckenaers	Siegfried Handschuh	
Paulo Cortez	Sophia Ananiadou	

Organizing Institutions

DI-UM - Departamento de Informática, Universidade do Minho
DSI-UM - Departamento de Sistemas de Informação, Universidade do Minho
CCTC - Centro de Ciências e Tecnologias da Comunicação, Universidade do
    Minho
ALGORITMI - Centro Algoritmi, Universidade do Minho
APPIA - Associação Portuguesa para a Inteligência Artificial

In Collaboration with

IEEE - Institute of Electrical and Electronics Engineers, Inc.
ECCAI - European Coordinating Committee for Artificial Intelligence
AAAI - Association for the Advancement of Artificial Intelligence
ACM - Association for Computing Machinery

Table of Contents

Chapter 1 - Second General Artificial Intelligence Workshop (GAIW 2007)

Towards Tractable Local Closed World Reasoning for the Semantic
Web 3
  Matthias Knorr, José Júlio Alferes, and Pascal Hitzler

Optimal Brain Surgeon for General Dynamic Neural Networks 15
  Christian Endisch, Christoph Hackl, and Dierk Schröder

Answer-Set Programming Based Dynamic User Modeling for
Recommender Systems 29
  João Leite and Manoela Ilić

Application of Logic Wrappers to Hierarchical Data Extraction from
HTML 43
  Amelia Bădică, Costin Bădică, and Elvira Popescu

Relaxing Feature Selection in Spam Filtering by Using Case-Based
Reasoning Systems 53
  *J.R. Méndez, F. Fdez-Riverola, D. Glez-Peña, F. Díaz, and
  J.M. Corchado*

Gödel and Computability 63
  Luís Moniz Pereira

Prospective Logic Agents 73
  Luís Moniz Pereira and Gonçalo Lopes

An Iterative Process for Building Learning Curves and Predicting
Relative Performance of Classifiers 87
  Rui Leite and Pavel Brazdil

Modelling Morality with Prospective Logic 99
  Luís Moniz Pereira and Ari Saptawijaya

Change Detection in Learning Histograms from Data Streams 112
  Raquel Sebastião and João Gama

Real-Time Intelligent Decision Support System for Bridges Structures
Behavior Prediction 124
  Hélder Quintela, Manuel Filipe Santos, and Paulo Cortez

Semi-fuzzy Splitting in Online Divisive-Agglomerative Clustering 133
  Pedro Pereira Rodrigues and João Gama

On the Use of Rough Sets for User Authentication Via Keystroke
Dynamics   145
    Kenneth Revett, Sérgio Tenreiro de Magalhães, and
    Henrique M.D. Santos

The Halt Condition in Genetic Programming   160
    José Neves, José Machado, Cesar Analide, António Abelha, and
    Luis Brito

Two Puzzles Concerning Measures of Uncertainty and the Positive
Boolean Connectives   170
    Gregory Wheeler

Chapter 2 - First Workshop on AI Applications for Sustainable Transportation Systems (AIASTS 2007)

Nonlinear Models for Determining Mode Choice: Accuracy is not
Always the Optimal Goal   183
    Elke Moons, Geert Wets, and Marc Aerts

Adaptation in Games with Many Co-evolving Agents   195
    Ana L.C. Bazzan, Franziska Klügl, and Kai Nagel

Chapter 3 - Third Workshop on Artificial Life and Evolutionary Algorithms (ALEA 2007)

Symmetry at the Genotypic Level and the Simple Inversion Operator . . .   209
    Cristian Munteanu and Agostinho Rosa

A Genetic Programming Approach to the Generation of
Hyper-Heuristics for the Uncapacitated Examination Timetabling
Problem   223
    Nelishia Pillay and Wolfgang Banzhaf

Asynchronous Stochastic Dynamics and the Spatial Prisoner's Dilemma
Game   235
    Carlos Grilo and Luís Correia

Improving Evolutionary Algorithms with Scouting   247
    Konstantinos Bousmalis, Gillian M. Hayes, and
    Jeffrey O. Pfaffmann

Stochastic Barycenters and Beta Distribution for Gaussian Particle
Swarms   259
    Rui Mendes and James Kennedy

Exploiting Second Order Information in Computational Multi-objective
Evolutionary Optimization   271
    Pradyumn Kumar Shukla

Chapter 4 - First Workshop on Ambient Intelligence Technologies and Applications (AMITA 2007)

Ambient Intelligence – A State of the Art from Artificial Intelligence
Perspective.. 285
  Carlos Ramos

Ubiquitous Ambient Intelligence in a Flight Decision Assistance
System 296
  Nuno Gomes, Carlos Ramos, Cristiano Pereira, and Francisco Nunes

Argumentation-Based Decision Making in Ambient Intelligence
Environments 309
  Goreti Marreiros, Ricardo Santos, Paulo Novais, José Machado,
  Carlos Ramos, José Neves, and José Bula-Cruz

Intelligent Mixed Reality for the Creation of Ambient Assisted
Living 323
  Ricardo Costa, José Neves, Paulo Novais, José Machado,
  Luís Lima, and Carlos Alberto

Medical Imaging Environment – A Multi-Agent System for a Computer
Clustering Based Multi-display 332
  Victor Alves, Filipe Marreiros, Luís Nelas, Mourylise Heymer, and
  José Neves

Chapter 5 - Second Workshop on Building and Applying Ontologies for the Semantic Web (BAOSW 2007)

Partial and Dynamic Ontology Mapping Model in Dialogs of Agents.... 347
  Ademir Roberto Freddo, Robison Cris Brito,
  Gustavo Gimenez-Lugo, and Cesar Augusto Tacla

Using Ontologies for Software Development Knowledge Reuse 357
  Bruno Antunes, Nuno Seco, and Paulo Gomes

Chapter 6 - First Workshop on Business Intelligence (BI 2007)

Analysis of the Day-of-the-Week Anomaly for the Case of Emerging
Stock Market 371
  Virgilijus Sakalauskas and Dalia Kriksciuniene

A Metamorphosis Algorithm for the Optimization of a Multi-node
OLAP System 383
  Jorge Loureiro and Orlando Belo

Experiments for the Number of Clusters in K-Means   395
    Mark Ming-Tso Chiang and Boris Mirkin

A Network Algorithm to Discover Sequential Patterns   406
    Luís Cavique

Adaptive Decision Support for Intensive Care......................   415
    Pedro Gago, Álvaro Silva, and Manuel Filipe Santos

A Tool for Interactive Subgroup Discovery Using Distribution Rules   426
    Joel P. Lucas, Alípio M. Jorge, Fernando Pereira,
    Ana M. Pernas, and Amauri A. Machado

Quantitative Evaluation of Clusterings for Marketing Applications: A
Web Portal Case Study   437
    Carmen Rebelo, Pedro Quelhas Brito, Carlos Soares,
    Alípio Jorge, and Rui Brandão

Resource-Bounded Fraud Detection   449
    Luis Torgo

Chapter 7 - First Workshop on Computational Methods in Bioinformatics and Systems Biology (CMBSB 2007)

System Stability Via Stepping Optimal Control: Theory and
Applications..   463
    Binhua Tang, Li He, Sushing Chen, and Bairong Shen

Evaluating Simulated Annealing Algorithms in the Optimization of
Bacterial Strains   473
    Miguel Rocha, Rui Mendes, Paulo Maia, José P. Pinto,
    Isabel Rocha, and Eugénio C. Ferreira

Feature Extraction from Tumor Gene Expression Profiles Using DCT
and DFT   485
    Shulin Wang, Huowang Chen, Shutao Li, and Dingxing Zhang

Chapter 8 - Second Workshop on Intelligent Robotics (IROBOT 2007)

An Omnidirectional Vision System for Soccer Robots   499
    António J.R. Neves, Gustavo A. Corrente, and Armando J. Pinho

Generalization and Transfer Learning in Noise-Affected Robot
Navigation Tasks   508
    Lutz Frommberger

Heuristic Q-Learning Soccer Players: A New Reinforcement Learning
Approach to RoboCup Simulation   520
  Luiz A. Celiberto Jr., Jackson Matsuura, and Reinaldo A.C. Bianchi

Human Robot Interaction Based on Bayesian Analysis of Human
Movements   530
  Jörg Rett and Jorge Dias

Understanding Dynamic Agent's Reasoning   542
  Nuno Lau, Luís Paulo Reis, and João Certo

Chapter 9 - Fourth Workshop on Multi-agent Systems: Theory and Applications (MASTA 2007)

Convergence of Independent Adaptive Learners   555
  Francisco S. Melo and Manuel C. Lopes

Multi-agent Learning: How to Interact to Improve Collective Results ...   568
  Pedro Rafael and João Pedro Neto

A Basis for an Exchange Value-Based Operational Notion of Morality
for Multiagent Systems   580
  Antônio Carlos da Rocha Costa and Graçaliz Pereira Dimuro

Intelligent Farmer Agent for Multi-agent Ecological Simulations
Optimization   593
  *Filipe Cruz, António Pereira, Pedro Valente, Pedro Duarte, and
  Luís Paulo Reis*

Tax Compliance Through MABS: The Case of Indirect Taxes   605
  Luis Antunes, João Balsa, and Helder Coelho

Chapter 10 - First Workshop on Search Techniques for Constraint Satisfaction (STCS 2007)

Efficient and Tight Upper Bounds for Haplotype Inference by Pure
Parsimony Using Delayed Haplotype Selection   621
  João Marques-Silva, Inês Lynce, Ana Graça, and Arlindo L. Oliveira

GRASPER: A Framework for Graph Constraint Satisfaction
Problems   633
  Ruben Viegas and Francisco Azevedo

Chapter 11 - Second Workshop on Text Mining and Applications (TEMA 2007)

Text Segmentation Using Context Overlap   647
  Radim Řehůřek

Automatic Extraction of Definitions in Portuguese: A Rule-Based
Approach   659
    Rosa Del Gaudio and António Branco

N-Grams and Morphological Normalization in Text Classification: A
Comparison on a Croatian-English Parallel Corpus   671
    *Artur Šilić, Jean-Hugues Chauchat, Bojana Dalbelo Bašić, and
    Annie Morin*

Detection of Strange and Wrong Automatic Part-of-Speech Tagging   683
    Vitor Rocio, Joaquim Silva, and Gabriel Lopes

New Techniques for Relevant Word Ranking and Extraction   691
    João Ventura and Joaquim Ferreira da Silva

Author Index   703

Chapter 1 - Second General Artificial Intelligence Workshop (GAIW 2007)

Towards Tractable Local Closed World Reasoning for the Semantic Web*

Matthias Knorr[1], José Júlio Alferes[1], and Pascal Hitzler[2]

[1] CENTRIA, Universidade Nova de Lisboa, Portugal
[2] AIFB, Universität Karlsruhe, Germany

Abstract. Recently, the logics of minimal knowledge and negation as failure MKNF [12] was used to introduce hybrid MKNF knowledge bases [14], a powerful formalism for combining open and closed world reasoning for the Semantic Web. We present an extension based on a new three-valued framework including an alternating fixpoint, the well-founded MKNF model. This approach, the well-founded MKNF semantics, derives its name from the very close relation to the corresponding semantics known from logic programming. We show that the well-founded MKNF model is the least model among all (three-valued) MKNF models, thus soundly approximating also the two-valued MKNF models from [14]. Furthermore, its computation yields better complexity results (up to polynomial) than the original semantics where models usually have to be guessed.

1 Introduction

Joining the open-world semantics of DLs with the closed-world semantics featured by (nonmonotonic) logic programming (LP) is one of the major open research questions in Description Logics (DL) research. Indeed, adding rules, in LP style, on top of the DL-based ontology layer has been recognized as an important task for the success of the Semantic Web (cf. the Rule Interchange Format working group of the W3C[1]). Combining LP rules and DLs, however, is a non-trivial task since these two formalisms are based on different assumptions: the former is nonmonotonic, relying on the closed world assumption, while the latter is based on first-order logic under the open world assumption.

Several proposals have been made for dealing with knowledge bases (KBs) which contain DL and LP statements (see e.g. [2,3,4,9,14,15]), but apart from [4], they all rely on the stable models semantics (SMS) for logic programs [6]. We claim that the well-founded semantics (WFS) [17], though being closely related to SMS (see e.g. [5]), is often the better choice. Indeed, in applications dealing with a large amount of information like the Semantic Web, the polynomial worst-case complexity of WFS is preferable

---

* This research was partly funded by the European Commission within the 6th Framework Programme projects REWERSE number 506779 (cf. http://rewerse.net/). Pascal Hitzler is supported by the German Federal Ministry of Education and Research (BMBF) under the SmartWeb project (grant 01 IMD01 B), and by the Deutsche Forschungsgemeinschaft (DFG) under the ReaSem project.

[1] http://www.w3.org/2005/rules/

to the NP-hard SMS. Furthermore, the WFS is defined for all programs and allows to answer queries by consulting only the relevant part of a program whereas SMS is neither relevant nor always defined.

While the approach in [4] is based on a loose coupling between DL and LP, others are tightly integrated. The most advanced of these approaches currently appears to be that of hybrid MKNF knowledge bases [14] which is based on the logic of Minimal Knowledge and Negation as Failure (MKNF) [12]. Its advantage lies in a seamless integration of DL and LP within one logical framework while retaining decidability due to the restriction to DL-safe rules.

In this paper[2], we define a three-valued semantics for hybrid MKNF knowledge bases, and a well-founded semantics, restricted to nondisjunctive MKNF rules, whose only model is the least three-valued one wrt. derivable knowledge. It compares to the semantics of [14] as the WFS does to the SMS of LP, viz.:

- the well-founded semantics is a sound approximation of the semantics of [14];
- the computational complexity is strictly lower;
- the semantics retains the property of [14] of being faithful, but now wrt. the WFS, i.e. when the DL part is empty, it coincides with the WFS of LPs.

2   Preliminaries

MKNF notions. We start by recalling the syntax of MKNF formulas from [14]. A *first-order atom* $P(t_1, \ldots, t_n)$ is an MKNF formula where P is a predicate and the t_i are first-order terms[3]. If φ is an MKNF formula then $\neg\varphi$, $\exists x : \varphi$, $\mathbf{K}\,\varphi$ and $\mathbf{not}\,\varphi$ are MKNF formulas and likewise $\varphi_1 \wedge \varphi_2$ for MKNF formulas φ_1, φ_2. The symbols $\vee$, $\subset$, $\equiv$, and $\forall$ are abbreviations for the usual boolean combinations of the previously introduced syntax. Substituting the free variables x_i in φ by terms t_i is denoted $\varphi[t_1/x_1, \ldots, t_n/x_n]$. Given a (first-order) formula φ, $\mathbf{K}\,\varphi$ is called a *modal* **K**-*atom* and $\mathbf{not}\,\varphi$ a *modal* **not**-*atom*. If a modal atom does not occur in scope of a modal operator in an MKNF formula then it is *strict*. An MKNF formula φ without any free variables is a *sentence* and *ground* if it does not contain variables at all. It is *modally closed* if all modal operators (**K** and **not**) are applied in φ only to sentences and *positive* if it does not contain the operator **not**; φ is *subjective* if all first-order atoms of φ occur within the scope of a modal operator and *objective* if there are no modal operators at all in φ; φ is *flat* if it is subjective and all occurrences of modal atoms in φ are strict.

Apart from the constants occurring in the formulas, the signature contains a countably infinite supply of constants not occurring in the formulas. The Herbrand Universe of such a signature is also denoted $\triangle$. The signature contains the equality predicate $\approx$ which is interpreted as an equivalence relation on $\triangle$. An *MKNF structure* is a triple (I, M, N) where I is an Herbrand first-order interpretation over $\triangle$ and M and N are nonempty sets of Herbrand first-order interpretations over $\triangle$. MKNF structures (I, M, N) define satisfiability of MKNF sentences as follows:

---

[2] Preliminary work on this subject was presented in [11].

[3] We consider function-free first-order logic, so terms are either constants or variables.

$$(I, M, N) \models p(t_1, \ldots, t_n) \text{ iff } p(t_1, \ldots, t_n) \in I$$
$$(I, M, N) \models \neg\varphi \qquad \text{iff } (I, M, N) \not\models \varphi$$
$$(I, M, N) \models \varphi_1 \wedge \varphi_2 \qquad \text{iff } (I, M, N) \models \varphi_1 \text{ and } (I, M, N) \models \varphi_2$$
$$(I, M, N) \models \exists x : \varphi \qquad \text{iff } (I, M, N) \models \varphi[\alpha/x] \text{ for some } \alpha \in \Delta$$
$$(I, M, N) \models \mathbf{K}\,\varphi \qquad \text{iff } (J, M, N) \models \varphi \text{ for all } J \in M$$
$$(I, M, N) \models \mathbf{not}\,\varphi \qquad \text{iff } (J, M, N) \not\models \varphi \text{ for some } J \in N$$

An *MKNF interpretation* M is a nonempty set of Herbrand first-order interpretations over[4] Δ and *models* a closed MKNF formula φ, i.e. $M \models \varphi$, if $(I, M, M) \models \varphi$ for each $I \in M$. An MKNF interpretation M is an *MKNF model* of a closed MKNF formula φ if (1) M models φ and (2) for each MKNF interpretation M' such that $M' \supset M$ we have $(I', M', M) \not\models \varphi$ for some $I' \in M'$.

Hybrid MKNF Knowledge Bases. Quoting from [14], the approach of hybrid MKNF knowledge bases is applicable to any first-order fragment $\mathcal{DL}$ satisfying these conditions: (i) each knowledge base $\mathcal{O} \in \mathcal{DL}$ can be translated into a formula $\pi(\mathcal{O})$ of function-free first-order logic with equality, (ii) it supports *A-Boxes*-assertions of the form $P(a_1, \ldots, a_n)$ for P a predicate and a_i constants of $\mathcal{DL}$ and (iii) satisfiability checking and instance checking (i.e. checking entailment of the form $\mathcal{O} \models P(a_1, \ldots, a_n)$) are decidable[5].

We recall MKNF rules and hybrid MKNF knowledge bases from [14]. For the rational behind these and the following notions we also refer to [13].

Definition 2.1. *Let $\mathcal{O}$ be a description logic knowledge base. A first-order function-free atom $P(t_1, \ldots, t_n)$ over Σ such that P is $\approx$ or it occurs in $\mathcal{O}$ is called a* DL-atom; *all other atoms are called non-DL-atoms. An MKNF rule r has the following form where H_i, A_i, and B_i are first-order function free atoms:*

$$\mathbf{K}\,H_1 \vee \ldots \vee \mathbf{K}\,H_l \leftarrow \mathbf{K}\,A_1, \ldots, \mathbf{K}\,A_n, \mathbf{not}\,B_1, \ldots, \mathbf{not}\,B_m \qquad (1)$$

The sets $\{\mathbf{K}\,H_i\}$, $\{\mathbf{K}\,A_i\}$, and $\{\mathbf{not}\,B_i\}$ are called the rule head, *the* positive body, *and the* negative body, *respectively. A rule is* nondisjunctive *if $l = 1$; r is* positive *if $m = 0$; r is a* fact *if $n = m = 0$. A* program *is a finite set of MKNF rules. A hybrid MKNF knowledge base $\mathcal{K}$ is a pair $(\mathcal{O}, \mathcal{P})$ and $\mathcal{K}$ is* nondisjunctive *if all rules in $\mathcal{P}$ are nondisjunctive.*

The semantics of an MKNF knowledge base is obtained by translating it into an MKNF formula ([14]).

Definition 2.2. *Let $\mathcal{K} = (\mathcal{O}, \mathcal{P})$ be a hybrid MKNF knowledge base. We extend π to r, $\mathcal{P}$, and $\mathcal{K}$ as follows, where x is the vector of the free variables of r.*

$$\pi(r) = \forall x : (\mathbf{K}\,H_1 \vee \ldots \vee \mathbf{K}\,H_l \subset \mathbf{K}\,A_1, \ldots, \mathbf{K}\,A_n, \mathbf{not}\,B_1, \ldots, \mathbf{not}\,B_m)$$

$$\pi(\mathcal{P}) = \bigwedge_{r \in \mathcal{P}} \pi(r) \qquad \pi(\mathcal{K}) = \mathbf{K}\,\pi(\mathcal{O}) \wedge \pi(\mathcal{P})$$

---

[4] Due to the domain set Δ, the considered interpretations are in general infinite.

[5] For more details on DL notation we refer to [1].

An MKNF rule r is *DL-safe* if every variable in r occurs in at least one non-DL-atom $\mathbf{K}\, B$ occurring in the body of r. A hybrid MKNF knowledge base $\mathcal{K}$ is *DL-safe* if all its rules are DL-safe. Given a hybrid MKNF knowledge base $\mathcal{K} = (\mathcal{O}, \mathcal{P})$, the *ground instantiation of* $\mathcal{K}$ is the KB $\mathcal{K}_G = (\mathcal{O}, \mathcal{P}_G)$ where $\mathcal{P}_G$ is obtained by replacing in each rule of $\mathcal{P}$ all variables with constants from $\mathcal{K}$ in all possible ways. It was shown in [13], for a DL-safe hybrid KB $\mathcal{K}$ and a ground MKNF formula ψ, that $\mathcal{K} \models \psi$ if and only if $\mathcal{K}_G \models \psi$.

3   Three-Valued MKNF Semantics

Satisfiability as defined before allows modal atoms only to be either true or false in a given MKNF structure. We extend the framework with a third truth value $\mathbf{u}$, denoting undefined, to be assigned to modal atoms only, while first-order atoms remain two-valued due to being interpreted solely in one first-order interpretation. Thus, MKNF sentences are evaluated in MKNF structures with respect to the set $\{\mathbf{t}, \mathbf{u}, \mathbf{f}\}$ of truth values with the order $\mathbf{f} < \mathbf{u} < \mathbf{t}$ where the operator max (resp. min) chooses the greatest (resp. least) element with respect to this ordering:

$$- (I, M, N)(p(t_1, \ldots, t_n)) = \begin{cases} \mathbf{t} & \text{iff } p(t_1, \ldots, t_n) \in I \\ \mathbf{f} & \text{iff } p(t_1, \ldots, t_n) \notin I \end{cases}$$

$$- (I, M, N)(\neg\varphi) = \begin{cases} \mathbf{t} & \text{iff } (I, M, N)(\varphi) = \mathbf{f} \\ \mathbf{u} & \text{iff } (I, M, N)(\varphi) = \mathbf{u} \\ \mathbf{f} & \text{iff } (I, M, N)(\varphi) = \mathbf{t} \end{cases}$$

$$- (I, M, N)(\varphi_1 \wedge \varphi_2) = \min\{(I, M, N)(\varphi_1), (I, M, N)(\varphi_2)\}$$

$$- (I, M, N)(\varphi_1 \supset \varphi_2) = \mathbf{t} \text{ iff } (I, M, N)(\varphi_2) \geq (I, M, N)(\varphi_1) \text{ and } \mathbf{f} \text{ otherwise}$$

$$- (I, M, N)(\exists x : \varphi) = \max\{(I, M, N)(\varphi[\alpha/x]) \mid \alpha \in \Delta\}$$

$$- (I, M, N)(\mathbf{K}\,\varphi) = \begin{cases} \mathbf{t} & \text{iff } (J, M, N)(\varphi) = \mathbf{t} \text{ for all } J \in M \\ \mathbf{f} & \text{iff } (J, M, N)(\varphi) = \mathbf{f} \text{ for some } J \in N \\ \mathbf{u} & \text{otherwise} \end{cases}$$

$$- (I, M, N)(\mathbf{not}\,\varphi) = \begin{cases} \mathbf{t} & \text{iff } (J, M, N)(\varphi) = \mathbf{f} \text{ for some } J \in N \\ \mathbf{f} & \text{iff } (J, M, N)(\varphi) = \mathbf{t} \text{ for all } J \in M \\ \mathbf{u} & \text{otherwise} \end{cases}$$

To avoid having modal atoms which are true and false at the same time, we restrict MKNF structures to consistent ones.

Definition 3.1. *An MKNF structure* (I, M, N) *is called* consistent *if, for all MKNF formulas* φ *over some given signature, it is not the case that* $(J, M, N)(\varphi) = \mathbf{t}$ *for all* $J \in M$ *and* $(J, M, N)(\varphi) = \mathbf{f}$ *for some* $J \in N$.

First of all, this evaluation is not really a purely three-valued one since first-order atoms are evaluated like in the two-valued case. In fact, a pure description logic knowledge base is only two-valued and it can easily be seen that it is evaluated in exactly the same way as in the scheme presented in the previous section. This is desired in particular in the case when the knowledge base consists just of the DL part. The third truth value can thus only be rooted in the rules part of the knowledge base. So, the main difference to the previous two-valued scheme consists of two pieces:

1. Implications are no longer interpreted classically: $\mathbf{u} \leftarrow \mathbf{u}$ is true while the classical boolean correspondence is $\mathbf{u} \vee \neg\mathbf{u}$, respectively $\neg(\neg\mathbf{u} \wedge \mathbf{u})$, which is undefined. The reason for this change is that rules in this way can only be true or false, similarly to logic programming, even in the case of three-valued semantics.
2. While in the two-valued framework M is used solely for interpreting modal $\mathbf{K}$-atoms and N only for the evaluation of modal $\mathbf{not}$-atoms, the three-valued evaluation applies symmetrically both sets to each case.

The second point needs further explanations. In the two-valued scheme, $\mathbf{K}\varphi$ is true in a given MKNF structure (I, M, N) if it holds in all Herbrand interpretations occurring in M and false otherwise, and in case of $\mathbf{not}$ exactly the other way around wrt. N. However, the truth space is thus fully defined leaving no gap for undefined modal atoms. One could change the evaluation such that e.g. $\mathbf{K}\varphi$ is true in M if φ is true in all $J \in M$, false in M if φ is false in all $J \in M$, and undefined otherwise. Then $\mathbf{not}\,\varphi$ would only be true if it is false in all models in N and we no longer have a negation different from the classical one. Thus, we separate truth and falsity in the sense that whenever a modal atom $\mathbf{K}\varphi$ is not true in M then it is either false or undefined. The other set, N, then allows to obtain whether $\mathbf{K}\varphi$ is false, namely just in case $\mathbf{not}\,\varphi$ is true[6]. We only have to be careful regarding consistency: we do not want structures which evaluate modal atoms to true and false at the same time and thus also not that $\mathbf{K}\varphi$ and $\mathbf{not}\,\varphi$ are true with respect to the same MKNF structure. The last case might in fact occur in the two-valued evaluation but does not do any harm there since the explicit connection between $\mathbf{K}\varphi$ and $\mathbf{not}\,\varphi$ is not present in the evaluation, and these inconsistencies are afterwards inhibited in MKNF interpretations.

Obviously, MKNF interpretations are not suitable to represent three truth values. For this purpose, we introduce interpretation pairs.

Definition 3.2. *An* interpretation pair (M, N) *consists of two MKNF interpretations M, N and models a closed MKNF formula φ, written $(M, N) \models \varphi$, if and only if $(I, M, N)(\varphi) = \mathbf{t}$ for each $I \in M$. We call (M, N) consistent if (I, M, N) is consistent for any $I \in M$ and φ is consistent if there exists an interpretation pair modeling it.*

M contains all interpretations which model only truth while N models everything which is true or undefined. Evidently, just as in the two-valued case, anything not being modeled in N is false.

We now introduce a preference relation on pairs in a straightforward way.

Definition 3.3. *Given a closed MKNF formula φ, a (consistent) interpretation pair (M, N) is a* (three-valued) MKNF model *for φ if* (1) $(I, M, N)(\varphi) = \mathbf{t}$ *for all $I \in M$ and* (2) *for each MKNF interpretation M' with $M' \supset M$ we have $(I', M', N)(\varphi) = \mathbf{f}$ for some $I' \in M'$.*

The idea is, having fixed the evaluation in N, i.e. the modal $\mathbf{K}$-atoms which are false (and thus also the modal $\mathbf{not}$-atoms which are true), to maximize the set which evaluates modal $\mathbf{K}$-atoms to true, thus only incorporating all the minimally necessary knowledge into M. In this sense, we remain in a logic of minimal knowledge. As a side-effect,

---

[6] This concurs with the idea that $\mathbf{not}$ is meant to represent $\neg\mathbf{K}$.

we also minimize the falsity of modal **not**-atoms, which is justified by the relation of **K** and ¬**not** . This feature is not contained in the MKNF semantics, but not necessary in the two-valued case anyway. We nevertheless obtain that any (two-valued) MKNF model M corresponds exactly to a three-valued one.

Proposition 3.1. *Given a closed MKNF formula* φ, *if* M *is an MKNF model of* φ *then* (M, M) *is a three-valued MKNF model of* φ.

Example 3.1. Consider the following knowledge base $\mathcal{K}$ containing just two rules.

$$\mathbf{K}\, p \leftarrow \mathbf{not}\, q$$
$$\mathbf{K}\, q \leftarrow \mathbf{not}\, p$$

The MKNF models of $\mathcal{K}$ are $\{\{p\}, \{p, q\}\}$ and $\{\{q\}, \{p, q\}\}$, i.e. $\mathbf{K}\, p$ and $\mathbf{not}\, q$ are true in the first model, and $\mathbf{K}\, q$ and $\mathbf{not}\, p$ are true in the second one.

We thus obtain two three-valued MKNF models: $(\{\{p\}, \{p, q\}\}, \{\{p\}, \{p, q\}\})$ and $(\{\{q\}, \{p, q\}\}, \{\{q\}, \{p, q\}\})$. Besides that, any interpretation pair which maps $\mathbf{K}\, p$, $\mathbf{K}\, q$, $\mathbf{not}\, p$ and $\mathbf{not}\, q$ to undefined is also a three-valued model. Among those, only $(\{\emptyset, \{p\}, \{q\}, \{p, q\}\}, \{\{p, q\}\})$ is an MKNF model while e.g. $(\{\emptyset, \{p, q\}\}, \{\{p, q\}\})$ is not. Finally, the pair which maps both, p and q, to true is a model but not MKNF either since $(\{\{p, q\}\}, \{\{p, q\}\})$ is also dominated by $(\{\emptyset, \{p\}, \{q\}, \{p, q\}\}, \{\{p, q\}\})$

There is one alternative idea for defining three-valued structures. We can represent a first-order interpretation by the set of all atoms which are true and the set of all negated atoms which are false. Thus, in the previous example we would obtain sets consisting of p, q, $\neg p$ and $\neg q$ where e.g. $\{p, \neg q\}$ instead of $\{p\}$ represents that p is true and q is false. This results in an MKNF model $\{\{p, \neg q\}\}$ which represents the knowledge $\mathbf{K}\, p$ and $\mathbf{K}\, \neg q$, respectively $\mathbf{not}\, \neg p$ and $\mathbf{not}\, q$. Unfortunately, for the model where $\mathbf{K}\, p$ and $\mathbf{K}\, q$ are undefined we obtain $(\{\{p, \neg q\}, \{p, q\}, \{\neg p, \neg q\}, \{\neg p, q\}\}, \{\{p, q, \neg p, \neg q\}\})$ as the representation. This is not very useful since it forces us to include inconsistent interpretations into interpretation pairs to state e.g. that neither $\mathbf{K}\, \neg p$ nor $\mathbf{K}\, p$ hold.

4   Three-Valued MKNF Models and Partitions

As shown in [13], since MKNF models are in general infinite, they can better be represented via a 1st-order formula whose models are exactly contained in the considered MKNF model. The idea is to provide a partition (T, F) of true and false modal atoms which uniquely defines φ. The 1st-order formula is then obtained from T as the objective knowledge contained in the modal atoms. We extend this idea to partial partitions where modal atoms which neither occur in T nor in F are supposed to be undefined. To obtain a specific partial partition we apply a technique known from LP: stable models ([6]) for normal logic programs correspond one-to-one to MKNF models of programs of MKNF rules (see [12]) and the well-founded model ([17]) for normal logic programs can be computed by an alternating fixpoint of the operator used to define stable models ([16]).

We proceed similarly: we define an operator providing a stable condition for nondisjunctive hybrid MKNF knowledge bases and use it to obtain an alternating fixpoint, the well-founded semantics. We thus start by adapting some notions from [13] formalizing partitions and related concepts.

Definition 4.1. *Let σ be a flat modally closed MKNF formula. The* set of **K**-atoms of σ, *written* $\mathsf{KA}(\sigma)$, *is the smallest set that contains (i) all* **K**-atoms occurring in σ, *and (ii) a modal atom* **K** ξ *for each modal atom* **not** ξ *occurring in σ.*

For a subset S of $\mathsf{KA}(\sigma)$, the objective knowledge *of S is the formula* $\mathrm{ob}_S = \bigcup_{\mathbf{K} \xi \in S} \xi$. *A (partial) partition (T, F) of $\mathsf{KA}(\sigma)$ is* consistent *if* $\mathrm{ob}_T \not\models \xi$ *for each* **K** $\xi \in F$.

We now connect interpretation pairs and partial partitions of modal **K**-atoms similarly to the way it was done in [13].

Definition 4.2. *We say that a partial partition (T, F) of $\mathsf{KA}(\sigma)$ is* induced by *a consistent interpretation pair (M, N) if (1) whenever* **K** $\xi \in T$ *then* $M \models \mathbf{K} \xi$ *and* $N \models \mathbf{K} \xi$, *(2) whenever* **K** $\xi \in F$ *then* $N \not\models \mathbf{K} \xi$, *and (3) whenever* **K** $\xi \in T$ *or* **K** $\xi \in F$ *then it is not the case that* $M \not\models \mathbf{K} \xi$ *and* $N \models \mathbf{K} \xi$.

The only case not dealt with in this definition is the one where M models **K** ξ and N does not model **K** ξ, i.e. N modeling **not** ξ. But this cannot occur since interpretation pairs are restricted to consistent ones.

Based on this relation we can show that the objective knowledge derived from the partition which is induced by a three-valued MKNF model yields again that model.

Proposition 4.1. *Let σ be a flat modally closed MKNF formula, (M, N) an MKNF model of σ and (T, F) a partition of $\mathsf{KA}(\sigma)$ induced by (M, N). Then (M, N) is equal to the interpretation pair (M', N') where $M' = \{I \mid I \models \mathrm{ob}_T\}$ and $N' = \{I \mid I \models \mathrm{ob}_{\mathsf{KA}(\sigma) \setminus F}\}$.*

Proof. Let I be an interpretation in M. Since (M, N) induces the partition (T, F) for each **K** $\xi \in T$ we have $M \models \mathbf{K} \xi$ and thus $I \models \xi$. Hence, $I \models \mathrm{ob}_T$ which shows $M \subseteq M'$. Likewise, for each **K** $\xi \notin F$ we have $N \models \mathbf{K} \xi$ and so for each $I \in N$ it holds that $I \models \xi$. Then $I \models \mathrm{ob}_{\mathsf{KA}(\sigma) \setminus F}$ which also shows $N \subseteq N'$.

Conversely, consider at first any I' in N'. We know for all $I' \in N'$ that $I' \models \mathrm{ob}_{\mathsf{KA}(\sigma) \setminus F}$, i.e. $I' \models \bigcup_{\mathbf{K} \xi \in \mathsf{KA}(\sigma) \setminus F} \xi$. Thus $I' \models \xi$ holds for all **K** ξ occurring in T and for all **K** ξ that neither occur in T nor in F. Since the partition was induced by (M, N) we obtain in both cases that $N \models \mathbf{K} \xi$, i.e. for all $I \in N$ we have $I \models \xi$ for all **K** $\xi \in \mathsf{KA}(\sigma) \setminus F$. We conclude that $N = N'$. For showing that also $M = M'$ we assume that $M' \setminus M$ is not empty but contains an interpretation I'. Then, for each $K\xi \in T$ we obtain $(I', M', N)(\mathbf{K} \xi) = \mathbf{t}$ just as we have $(I, M, N)(\mathbf{K} \xi) = \mathbf{t}$ for all $I \in M$. Likewise, for each $K\xi \in T$ we have $(I', M', N)(\mathbf{not} \xi) = \mathbf{f}$ for any $I' \in M'$ and $(I, M, N)(\mathbf{not} \xi) = \mathbf{f}$ for any $I \in M$. We also know for each $K\xi \in F$ that $(I', M', N)(\mathbf{not} \xi) = \mathbf{t}$ and $(I, M, N)(\mathbf{not} \xi) = \mathbf{t}$, and $(I', M', N)(\mathbf{K} \xi) = \mathbf{f}$ and $(I, M, N)(\mathbf{K} \xi) = \mathbf{f}$ for any I, I' since N remains the same. For the same reason, and since augmenting M does not alter the undefinedness of a modal atom, all modal atoms which are undefined in (I, M, N) are also undefined in (I, M', N) The truth value of a flat σ in a structure (I', M', N) for some $I' \in M'$ is completely defined by the truth values of the modal atoms and since they are all identical to the ones in (I, M, N) for all $I \in M$ we have that $(I', M', N)(\sigma) = \mathbf{t}$. This contradicts the assumption that M is a three-valued MKNF model of σ.

This result is used later on to show that the partition obtained from the alternating fixpoint yields in fact a three-valued MKNF model. For that, we at first adapt the notions about partitions to hybrid MKNF knowledge bases like in [14].

Consider a hybrid MKNF knowledge base $\mathcal{K} = (\mathcal{O}, \mathcal{P})$. Note that $\mathbf{K}\,\pi(\mathcal{O})$ occurs in $\mathsf{KA}(\sigma)$ and must be true in any model of $\mathcal{K}$. The set of the remaining modal $\mathbf{K}$-atoms is denoted $\mathsf{KA}(\mathcal{K}) = \mathsf{KA}(\sigma) \setminus \{\mathbf{K}\,\pi(\mathcal{O})\}$. Furthermore, for a set of modal atoms S, S_{DL} is the subset of DL-atoms of S, and $\widehat{S} = \{\xi \mid \mathbf{K}\,\xi \in S\}$. These changes allow to rewrite the objective knowledge in the following way where S is a subset of $\mathsf{KA}(\mathcal{K})$: $\mathrm{ob}_{\mathcal{K},S} = \mathcal{O} \cup \bigcup_{\mathbf{K}\,\xi \in S} \xi$.

We now adapt the monotonic operator $T_{\mathcal{K}}$ from [14] which allows to draw conclusions from positive hybrid MKNF knowledge bases.

Definition 4.3. *For $\mathcal{K}$ a positive nondisjunctive DL-safe hybrid MKNF knowledge base, $R_{\mathcal{K}}$, $D_{\mathcal{K}}$, and $T_{\mathcal{K}}$ are defined on the subsets of $\mathsf{KA}(\mathcal{K})$ as follows:*

$$R_{\mathcal{K}}(S) = S \cup \{\mathbf{K}\,H \mid \mathcal{K} \text{ contains a rule of the form (1) such that } \mathbf{K}\,A_i \in S$$
$$\text{for each } 1 \le i \le n\}$$
$$D_{\mathcal{K}}(S) = \{\mathbf{K}\,\xi \mid \mathbf{K}\,\xi \in \mathsf{KA}(\mathcal{K}) \text{ and } \mathcal{O} \cup \widehat{S}_{DL} \models \xi\} \cup$$
$$\{\mathbf{K}\,Q(b_1, \ldots, b_n) \mid \mathbf{K}\,Q(a_1, \ldots, a_n) \in S \setminus S_{DL},\ \mathbf{K}\,Q(b_1, \ldots, b_n) \in \mathsf{KA}(\mathcal{K}),$$
$$\text{and } \mathcal{O} \cup \widehat{S}_{DL} \models a_i \approx b_i \text{ for } 1 \le i \le n\}$$
$$T_{\mathcal{K}}(S) = R_{\mathcal{K}}(S) \cup D_{\mathcal{K}}(S)$$

The difference to the operator $D_{\mathcal{K}}$ in [14] is that given e.g. only $a \approx b$ and $\mathbf{K}\,Q(a)$ we do not derive $\mathbf{K}\,Q(b)$ explicitly but only as a consequence of $\mathrm{ob}_{\mathcal{K},P}$.

A transformation for nondisjunctive hybrid MKNF knowledge is defined turning them into positive ones, thus allowing the application of the operator $T_{\mathcal{K}}$.

Definition 4.4. *Let $\mathcal{K}_G = (\mathcal{O}, \mathcal{P}_G)$ be a ground nondisjunctive DL-safe hybrid MKNF knowledge base and $S \subseteq \mathsf{KA}(\mathcal{K})$. The MKNF transform $\mathcal{K}_G/S = (\mathcal{O}, \mathcal{P}_G/S)$ is obtained by $\mathcal{P}_G/S$ containing all rules $\mathbf{K}\,H \leftarrow \mathbf{K}\,A_1, \ldots, \mathbf{K}\,A_n$ for which there exists a rule $\mathbf{K}\,H \leftarrow \mathbf{K}\,A_1, \ldots, \mathbf{K}\,A_n, \text{not } B_1, \ldots, \text{not } B_m$ in $\mathcal{P}_G$ with $\mathbf{K}\,B_j \notin S$ for all $1 \le j \le m$.*

Now an antitonic operator can be defined using the fixpoint of $T_{\mathcal{K}}$.

Definition 4.5. *Let $\mathcal{K} = (\mathcal{O}, \mathcal{P})$ be a nondisjunctive DL-safe hybrid MKNF knowledge base and $S \subseteq \mathsf{KA}(\mathcal{K})$. We define:*

$$\Gamma_{\mathcal{K}}(S) = T_{\mathcal{K}_G/S} \uparrow \omega$$

Applying $\Gamma_{\mathcal{K}}(S)$ twice is a monotonic operation yielding a least fixpoint by the Knaster-Tarski theorem (and dually a greatest one) and can be iterated as follows: $\Gamma_{\mathcal{K}}^2 \uparrow 0 = \emptyset$, $\Gamma_{\mathcal{K}}^2 \uparrow (n+1) = \Gamma_{\mathcal{K}}^2(\Gamma_{\mathcal{K}}^2 \uparrow n)$, and $\Gamma_{\mathcal{K}}^2 \uparrow \omega = \bigcup \Gamma_{\mathcal{K}}^2 \uparrow i$, and dually $\Gamma_{\mathcal{K}}^2 \downarrow 0 = \mathsf{KA}(\mathcal{K})$, $\Gamma_{\mathcal{K}}^2 \downarrow (n+1) = \Gamma_{\mathcal{K}}^2(\Gamma_{\mathcal{K}}^2 \downarrow n)$, and $\Gamma_{\mathcal{K}}^2 \downarrow \omega = \bigcap \Gamma_{\mathcal{K}}^2 \downarrow i$.

These two fixpoints define the well-founded partition.

Definition 4.6. *Let* $\mathcal{K} = (\mathcal{O}, \mathcal{P})$ *be a nondisjunctive DL-safe hybrid MKNF knowledge base and let* $\mathbf{P}_\mathcal{K}, \mathbf{N}_\mathcal{K} \subseteq \mathrm{KA}(\mathcal{K})$ *with* $\mathbf{P}_\mathcal{K}$ *being the least fixpoint of* $\Gamma^2_\mathcal{K}$ *and* $\mathbf{N}_\mathcal{K}$ *the greatest fixpoint. Then* $(P_W, N_W) = (\mathbf{P}_\mathcal{K} \cup \{\mathbf{K}\,\pi(\mathcal{O})\}, \mathrm{KA}(\mathcal{K}) \setminus \mathbf{N}_\mathcal{K})$ *is the well-founded partition of* $\mathcal{K}$.

It can be shown that the well-founded partition is consistent and that the least fixpoint can be computed directly from the greatest one and vice-versa similarly to the alternating fixpoint of normal logic programs [16].

Proposition 4.2. *Let* $\mathcal{K}$ *be a nondisjunctive DL-safe hybrid MKNF knowledge base. Then* $\mathbf{P}_\mathcal{K} = \Gamma_\mathcal{K}(\mathbf{N}_\mathcal{K})$ *and* $\mathbf{N}_\mathcal{K} = \Gamma_\mathcal{K}(\mathbf{P}_\mathcal{K})$.

Example 4.1. Let us consider the following hybrid MKNF knowledge base

$$NaturalDeath \sqsubseteq Pay \qquad Suicide \sqsubseteq \neg Pay$$

$$\mathbf{K}\,Pay(x) \leftarrow \mathbf{K}\,murdered(x), \mathbf{K}\,benefits(y, x), \mathbf{not}\,responsible(y, x)$$
$$\mathbf{K}\,Suicide(x) \leftarrow \mathbf{not}\,NaturalDeath(x), \mathbf{not}\,murdered(x)$$
$$\mathbf{K}\,murdered(x) \leftarrow \mathbf{not}\,NaturalDeath(x), \mathbf{not}\,Suicide(x)$$

based on which a life insurance company decides whether to pay or not the insurance. Additionally, we know that Mr. Jones who owned a life insurance was found death in his living room, a revolver on the ground. Then ¬*NaturalDeath*(*jones*) and the last two rules offer us a choice between commitment of suicide or murder. While there are two MKNF models in such a scenario, one concluding for payment and the other one not, the three-valued framework allows to assign **u** to both so that we delay this decision until the evidence is evaluated. Assume that the police investigation reveals that the known criminal Max is responsible for the murder, though not being detectable, so we cannot conclude *Suicide*(*jones*) while $\mathbf{K}\,responsible(max, jones)$ and $\mathbf{K}\,murdered(jones)$ hold. Unfortunately (for the insurance company), the person benefitting from the insurance is the nephew Thomas who many years ago left the country, i.e. $\mathbf{K}\,benefits(thomas, jones)$. Computing the well-founded partition yields thus $\mathbf{K}\,Pay(jones)$, so the company has to contact the nephew. However, being not satisfied with the payment, they also hire a private detective who finds out that Max is Thomas, having altered his personality long ago, i.e. we have *thomas* ≈ *max* in the hybrid KB. Due to $D_\mathcal{K}$ and grounding we now obtain a well-founded partition which contains $\mathbf{K}\,responsible(thomas, jones)$ and $\mathbf{K}\,benefits(max, jones)$ being true and the insurance is not paid any longer.

Further examples, and an involved discussion of the importance of hybrid MKNF knowledge bases for modeling knowledge in the semantic web can be found in [10]. Apart from that, [7] and [8] provide arguments for the usefullness of epistemic reasoning the way it is done in MKNF logics.

We can also show that the well-founded partition yields a three-valued model.

Theorem 4.1. *Let* $\mathcal{K}$ *be a consistent nondisjunctive DL-safe hybrid MKNF KB and* $(\mathbf{P}_\mathcal{K} \cup \{\mathbf{K}\,\pi(\mathcal{O})\}, \mathrm{KA}(\mathcal{K}) \setminus \mathbf{N}_\mathcal{K})$ *be the well-founded partition of* $\mathcal{K}$. *Then* $(I_P, I_N) \models \pi(\mathcal{K})$ *where* $I_P = \{I \mid I \models \mathrm{ob}_{\mathcal{K}, \mathbf{P}_\mathcal{K}}\}$ *and* $I_N = \{I \mid I \models \mathrm{ob}_{\mathbf{N}_\mathcal{K}}\}$.

Note that the DL-part of the knowledge base is not used for forming I_N. Otherwise I_N could be inconsistent since our approach might contain an undefined modal atom $\mathbf{K}\,\varphi$ even though φ is first-order false in the DL part. We are aware that this will need to be improved in further investigations. However, the deficiency is not severe in terms of the contribution of this work, since I_N is not used to evaluate the DL-part.

This result can be combined with our previously proven proposition to obtain that the well-founded partition gives us in fact a three-valued MKNF model.

Theorem 4.2. *Let $\mathcal{K}$ be a consistent nondisjunctive DL-safe hybrid MKNF KB and $(\mathbf{P}_\mathcal{K} \cup \{\mathbf{K}\,\pi(\mathcal{O})\}, \mathsf{KA}(\mathcal{K}) \setminus \mathbf{N}_\mathcal{K})$ be the well-founded partition of $\mathcal{K}$. Then (I_P, I_N) where $I_P = \{I \mid I \models \mathsf{ob}_{\mathcal{K},\mathbf{P}_\mathcal{K}}\}$ and $I_N = \{I \mid I \models \mathsf{ob}_{\mathbf{N}_\mathcal{K}}\}$ is an MKNF model – the well-founded MKNF model.*

In fact, it is not just any three-valued MKNF model but the least one with respect to derivable knowledge. For that, we show that the partition induced by a three-valued MKNF model is a fixpoint of $\Gamma_\mathcal{K}^2$.

Lemma 4.1. *Let $\mathcal{K}$ be a consistent nondisjunctive DL-safe hybrid MKNF KB and let (T, F) be the partition induced by an MKNF model (M, N) of $\mathcal{K}$. Then T and F are fixpoints of $\Gamma_\mathcal{K}^2$.*

Proof. (sketch) We show the argument for T; F follows dually. The set T contains all modal $\mathbf{K}$-atoms from $\mathsf{KA}(\mathcal{K})$ which are true in the MKNF model. We know that $\Gamma_\mathcal{K}^2$ is monotonic, i.e. $T \subseteq \Gamma_\mathcal{K}^2(T)$. Assume that $T \subset \Gamma_\mathcal{K}^2(T)$ so there are new consequences when applying $\Gamma_\mathcal{K}^2$ to T. Basically these modal atoms are justified either by $R_\mathcal{K}$ or by $D_\mathcal{K}$ or by some other new consequence of the application of $\Gamma_\mathcal{K}^2$. It is well known for this kind of rule languages (from logic programming theory) that consequences which depend on each other are established in a well-defined manner meaning that there are no cyclic dependencies (otherwise these modal atoms would remain undefined). We thus directly consider only basic new consequences which depend on no other new modal atom and restrict to the first two cases starting with $R_\mathcal{K}$.

If $\mathbf{K}\,H$ is a new consequence by means of $R_\mathcal{K}$ then it is because some rule of the form $\mathbf{K}\,H \leftarrow \mathbf{K}\,A_1, \ldots, \mathbf{K}\,A_n$ occurs in $\mathcal{K}_G/T$. Since $\mathbf{K}\,H$ does not depend on any other new consequence we already know that all $\mathbf{K}\,A_i$ are present in T. But $\mathbf{K}\,H$ does not occur in T, so there must be a modal atom $\mathbf{not}\,B_j$ in the corresponding rule such that $\mathbf{K}\,B_j \notin T$; otherwise (M, N) would be no model. This $\mathbf{K}\,B_j$ cannot occur in F either, otherwise (M, N) would again be no model of $\mathcal{K}$. Then $\mathbf{K}\,B_j$ is undefined but it is not possible to derive the falsity of $\mathbf{K}\,B_j$ and from that further conclusions in one step of $\Gamma_\mathcal{K}^2$.

Alternatively, $\mathbf{K}\,H$ is a consequence of $D_\mathcal{K}$. But then, according to the definition of that operator, any new consequence depends on some prior consequences or on the DL knowledge base itself. It is easy to see that thus new consequences derived from $D_\mathcal{K}$ depend on a new consequence introduced from the rules part; otherwise (M, N) would not be a model of $\mathcal{K}$.

From that we immediately obtain that the well-founded MKNF model is the least MKNF model wrt. derivable knowledge.

Theorem 4.3. *Let $\mathcal{K}$ be a consistent nondisjunctive DL-safe hybrid MKNF KB. Among all three-valued MKNF models, the well-founded MKNF model is the least wrt. derivable knowledge from $\mathcal{K}$.*

Proof. (sketch) We have shown that any three-valued MKNF model induces a partition which yields the MKNF model again (via the objective knowledge). Since this partition (T, F) consists of two fixpoints of $\Gamma_{\mathcal{K}}^2$ and we know that the well-founded partition (P_W, N_W) contains the least fixpoint (minimally necessary true knowledge) and the greatest fixpoint (minimally necessary false knowledge) we conclude that $P_W \subseteq T$ and $N_W \subseteq F$. It is straightforward to see that an MKNF model containing more true (and false) modal atoms allows for a greater set of logical consequences.

Thus, the well-founded partition can also be used in the algorithms presented in [14] for computing a subset of that knowledge which holds in all partitions corresponding to a two-valued MKNF model.

One of the open questions in [14] was that MKNF models are not compatible with the well-founded model for logic programs. Our approach, regarding knowledge bases just consisting of rules, does coincide with the well-founded model for the corresponding (normal) logic program.

Finally the following theorem is obtained straightforwardly from the data complexity results for positive nondisjunctive MKNF knowledge bases in [14] where data complexity is measured in terms of A-Box assertions and rule facts.

Theorem 4.4. *Let $\mathcal{K}$ be a nondisjunctive DL-safe hybrid MKNF KB. Assuming that entailment of ground DL-atoms in $\mathcal{DL}$ is decidable with data complexity C the data complexity of computing the well-founded partition is in P^C.*

For comparison, the data complexity for reasoning with MKNF models in nondisjunctive programs is shown to be $\mathcal{E}^{\mathrm{P}^C}$ where $\mathcal{E} = \mathrm{NP}$ if $C \subseteq \mathrm{NP}$, and $\mathcal{E} = C$ otherwise. Thus, computing the well-founded partition ends up in a strictly smaller complexity class than deriving the MKNF models. In fact, in case the description logic fragment is tractable,[7] we end up with a formalism whose model is computed with a data complexity of P.

5   Conclusions and Future Work

We have continued the work on hybrid MKNF knowledge bases providing an alternating fixpoint restricted to nondisjunctive rules within a three-valued extension of the MKNF logic. We basically achieve better complexity results by having only one model which is semantically weaker than any MKNF model defined in [14], but bottom-up computable. The well-founded model is not only a sound approximation of any three-valued MKNF model but a partition of modal atoms which can seemlessly be integrated in the reasoning algorithms presented for MKNF models in [14] thus reducing the difficulty of guessing the 'right' model. Future work shall include the extension to disjunctive rules, handling of paraconsistency, and a study on top-down querying procedures.

---

[7] See e.g. the W3C member submission on tractable fragments of OWL 1.1 at
http://www.w3.org/Submission/owl11-tractable/

References

1. Baader, F., Calvanese, D., McGuinness, D.L., Nardi, D., Patel-Schneider, P.F. (eds.): The Description Logic Handbook: Theory, Implementation, and Applications. Cambridge University Press, Cambridge (2003)
2. de Bruijn, J., Eiter, T., Polleres, A., Tompits, H.: Embedding non-ground logic programs into autoepistemic logic for knowledge-base combination. In: IJCAI 2007. Proceedings of the Twentieth International Joint Conference on Artificial Intelligence, Hyderabad, India, January 6–12, 2007, AAAI Press, Stanford, California, USA (2007)
3. Eiter, T., Lukasiewicz, T., Schindlauer, R., Tompits, H.: Combining answer set programming with description logics for the semantic web. In: Dubois, D., Welty, C., Williams, M.-A. (eds.) KR 2004, pp. 141–151. AAAI Press, Stanford, California, USA (2004)
4. Eiter, T., Lukasiewicz, T., Schindlauer, R., Tompits, H.: Well-founded semantics for description logic programs in the semantic web. In: Antoniou, G., Boley, H. (eds.) RuleML 2004. LNCS, vol. 3323, pp. 81–97. Springer, Heidelberg (2004)
5. Fitting, M.: The family of stable models. Journal of Logic Programming 17(2/3&4), 197–225 (1993)
6. Gelfond, M., Lifschitz, V.: The stable model semantics for logic programming. In: Kowalski, R.A., Bowen, K.A. (eds.) ICLP, MIT Press, Cambridge (1988)
7. Grimm, S., Hitzler, P.: Semantic matchmaking of web resources with local closed-world reasoning. International Journal of e-Commerce (to appear)
8. Grimm, S., Motik, B., Preist, C.: Matching semantic service descriptions with local closed-world reasoning. In: Sure, Y., Domingue, J. (eds.) ESWC 2006. LNCS, vol. 4011, pp. 575–589. Springer, Heidelberg (2006)
9. Heymans, S., Nieuwenborgh, D.V., Vermeir, D.: Guarded open answer set programming. In: Baral, C., Greco, G., Leone, N., Terracina, G. (eds.) LPNMR 2005. LNCS (LNAI), vol. 3662, pp. 92–104. Springer, Heidelberg (2005)
10. Horrocks, I., Motik, B., Rosati, R., Sattler, U.: Can OWL and logic programming live together happily ever after? In: Cruz, I., Decker, S., Allemang, D., Preist, C., Schwabe, D., Mika, P., Uschold, M., Aroyo, L. (eds.) ISWC 2006. LNCS, vol. 4273, pp. 501–514. Springer, Heidelberg (2006)
11. Knorr, M., Alferes, J.J., Hitzler, P.: A well-founded semantics for MKNF knowledge bases. In: Calvanese, D., Franconi, E., Haarslev, V., Lembo, D., Motik, B., Turhan, A.-Y., Tessaris, S. (eds.) Description Logics 2007, CEUR–WS, pp. 417–425 (2007)
12. Lifschitz, V.: Nonmonotonic databases and epistemic queries. In: IJCAI 1991, pp. 381–386 (1991)
13. Motik, B., Rosati, R.: Closing semantic web ontologies. Technical report, University of Manchester, UK (2006)
14. Motik, B., Rosati, R.: A faithful integration of description logics with logic programming. In: IJCAI 2007. Proceedings of the Twentieth International Joint Conference on Artificial Intelligence, Hyderabad, India, January 6–12, 2007, pp. 477–482. AAAI Press, Stanford, California, USA (2007)
15. Rosati, R.: Dl+Log: A tight integration of description logics and disjunctive datalog. In: Doherty, P., Mylopoulos, J., Welty, C. (eds.) KR 2006, AAAI Press, Stanford, California, USA (2006)
16. van Gelder, A.: The alternating fixpoint of logic programs with negation. In: Principles of Database Systems, pp. 1–10. ACM Press, New York (1989)
17. van Gelder, A., Ross, K.A., Schlipf, J.S.: The well-founded semantics for general logic programs. Journal of the ACM 38(3), 620–650 (1991)

Optimal Brain Surgeon for General Dynamic Neural Networks

Christian Endisch, Christoph Hackl, and Dierk Schröder

Institute for Electrical Drive Systems, Technical University of Munich,
Arcisstraße 21, 80333 München, Germany
christian.endisch@tum.de

Abstract. This paper presents a pruning algorithm based on optimal brain surgeon (OBS) for general dynamic neural networks (GDNN). The pruning algorithm uses Hessian information and considers the order of time delay for saliency calculation. In GDNNs all layers have feedback connections with time delays to the same and to all other layers. The parameters are trained with the Levenberg-Marquardt (LM) algorithm. Therefore the Jacobian matrix is required. The Jacobian is calculated by real time recurrent learning (RTRL). As both LM and OBS need Hessian information, a rational implementation is suggested.

Keywords: System identification, dynamic neural network, optimization, Levenberg Marquardt, real time recurrent learning, pruning, optimal brain surgeon, quasi-online learning.

1 Introduction

In static neural networks the output is directly calculated from the input through forward connections. Dynamic neural networks include delay lines between the layers. So the output depends also on previous inputs or previous states of the network. In this paper general dynamic neural networks (GDNN) are considered for system identification. In GDNN all layers have feedback connections with many time delays, see Fig. 1. As we do not know the structure of the plant we use a larger network architecture than necessary. During the identification process the network architecture is reduced to find a model for the plant as simple as possible. For this architecture reduction several pruning algorithms are known [1]-[4]. Two well known methods in feedforward neural networks are optimal brain damage (OBD) [1] and optimal brain surgeon (OBS) [2]. Both methods are based on weight ranking due to the saliency, which is defined as the change in the output error using Hessian information. The OBD method calculates the saliency only with the pivot elements of the Hessian without retraining after the pruning step. The OBS uses the complete Hessian information to calculate the saliency, which is regarded as a continuation of the OBD method. After an OBS pruning step the remaining weights are retrained. In OBS pruning the calculation of the inverse Hessian causes a great amount of computation. To overcome this disadvantage of OBS this paper suggests to use the OBS pruning

J. Neves, M. Santos, and J. Machado (Eds.): EPIA 2007, LNAI 4874, pp. 15–28, 2007.

algorithm together with the LM training algorithm. So OBS gets the inverse Hessian for free. In addition to that the OBS approach in this paper considers the order of time delay for the saliency calculation. Reducing the model size has a lot of advantages:

- the generalization is improved
- the training speed is getting faster
- larger networks have better local and global minima, the goal is to keep the low cost function during pruning
- since the optimization problem after a pruning step is changed it is possible to overcome a local minimum

The next section presents the recurrent neural network used in this paper. Administration matrices are introduced to manage the pruning process. Section 3 deals with parameter optimization and the quasi-online approach used throughout this paper. In Section 4 the OBS algorithm is discussed and a modification for dynamic networks is suggested. Identification examples are shown in section 5. Finally, in Section 6 we summarize the results.

2   General Dynamic Neural Network (GDNN)

De Jesus described in his doctoral theses [7] a broad class of dynamic networks, he called the framework layered digital dynamic network (LDDN). The sophisticated formulations and notations of the LDDN allow an efficient computation of the Jacobian matrix using real-time recurrent learning (RTRL). Therefore we follow these conventions suggested by De Jesus. In [5]-[8] the optimal network topology is assumed to be known. In this paper the network topology is unknown and so we choose an oversized network for identification. In GDNN all feedback connections exist with a complete tapped delay line (from first-order time delay element z^{-1} up to the maximum order time delay element $z^{-d_{max}}$). The output of a tapped delay line (TDL) is a vector containing delayed values of the TDL input. Also the network inputs have a TDL. During the identification process the optimal network architecture should be found. Administration matrices show us, which weights are valid. Fig. 1 shows a three-layer GDNN. The simulation equation for layer m is

$$
\underline{n}^m(t) = \sum_{l \in L_m^f} \sum_{d \in DL^{m,l}} \underline{LW}^{m,l}(d) \cdot \underline{a}^l(t-d) + \\
\sum_{l \in I_m} \sum_{d \in DI^{m,l}} \underline{IW}^{m,l}(d) \cdot \underline{p}^l(t-d) + \underline{b}^m
\tag{1}
$$

$\underline{n}^m(t)$ is the summation output of layer m, $\underline{p}^l(t)$ is the l-th input to the network, $\underline{IW}^{m,l}$ is the input weight matrix between input l and layer m, $\underline{LW}^{m,l}$ is the layer weight matrix between layer l and layer m, $\underline{b}^m$ is the bias vector of layer m, $DL^{m,l}$ is the set of all delays in the tapped delay line between layer l and layer m, $DI^{m,l}$ is the set of all input delays in the tapped delay line between

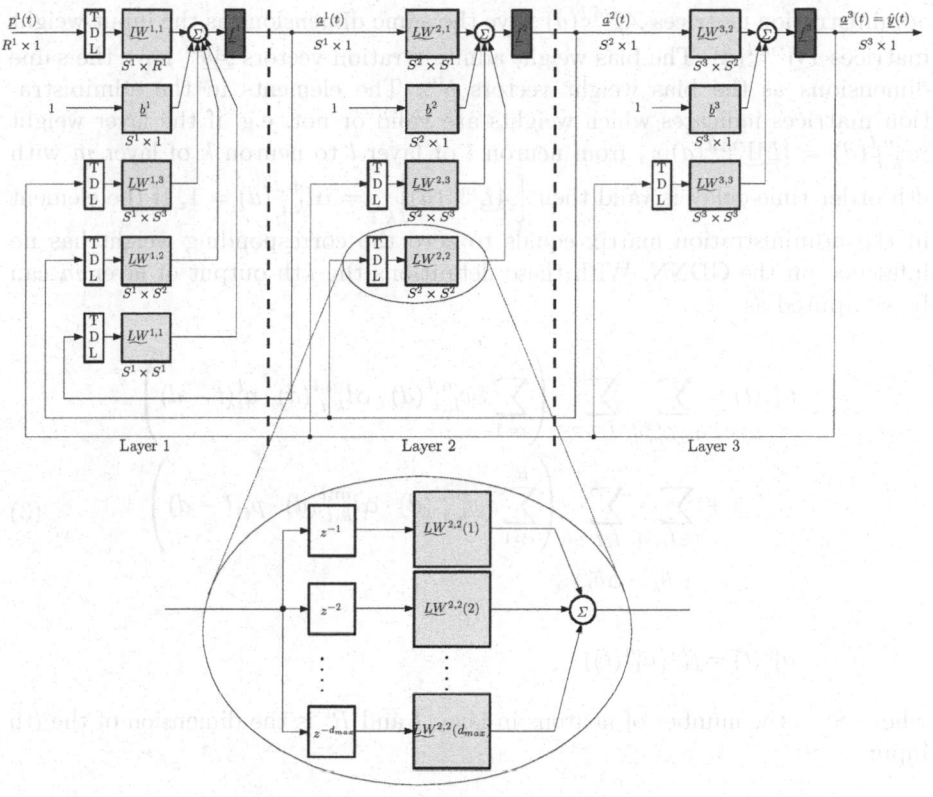

Fig. 1. Three-layer GDNN (two hidden layers)

input l and layer m, I_m is the set of indices of input vectors that connect to layer m, L_m^f is the set of indices of layers that directly connect forward to layer m. The output of layer m is

$$\underline{a}^m(t) = \underline{f}^m(\underline{n}^m(t)) \qquad (2)$$

where $\underline{f}^m(\cdot)$ are nonlinear transfer functions. In this paper we use $tanh$-functions in the hidden layers and linear transfer functions in the output layer. At each point of time the equations (1) and (2) are iterated forward through the layers. Time is incremented from $t = 1$ to $t = Q$. (See [7] for a full description of the notation used here). In Fig. 1 the information under the matrix-boxes and the information under the arrows denote the dimensions. R^m and S^m respectively indicate the dimension of the input and the number of neurons in layer m. $\hat{\underline{y}}$ is the output of the GDNN.

2.1 Administration Matrices

The layer weight administration matrices $\underline{AL}^{m,l}(d)$ have the same dimensions as the layer weight matrices $\underline{LW}^{m,l}(d)$ of the GDNN. The input weight administration matrices $\underline{AI}^{m,l}(d)$ have the same dimensions as the input weight matrices $\underline{IW}^{m,l}(d)$. The bias weight administration vectors $\underline{Ab}^m$ have the same dimensions as the bias weight vectors $\underline{b}^m$. The elements in the administration matrices indicates which weights are valid or not. e.g. if the layer weight $lw_{k,i}^{m,l}(d) = [\underline{LW}^{m,l}(d)]_{k,i}$ from neuron i of layer l to neuron k of layer m with dth-order time-delay is valid then $\left[\underline{AL}^{m,l}(d)\right]_{k,i} = al_{k,i}^{m,l}(d) = 1$. If the element in the administration matrix equals to zero the corresponding weight has no influence on the GDNN. With these definitions the kth output of layer m can be computed as

$$n_k^m(t) = \sum_{l \in L_m^f} \sum_{d \in DL^{m,l}} \left(\sum_{i=1}^{S^l} lw_{k,i}^{m,l}(d) \cdot al_{k,i}^{m,l}(d) \cdot a_i^l(t-d) \right)$$

$$+ \sum_{l \in I_m} \sum_{d \in DI^{m,l}} \left(\sum_{i=1}^{R^l} iw_{k,i}^{m,l}(d) \cdot ai_{k,i}^{m,l}(d) \cdot p_i^l(t-d) \right) \qquad (3)$$

$$+ b_k^m \cdot ab_k^m$$

$$a_k^m(t) = f_k^m(n_k^m(t))$$

where S^l is the number of neurons in layer l and R^l is the dimension of the lth input.

2.2   Implementation

For the simulations throughout this paper the graphical programming language *Simulink (Matlab)* was used. GDNN, Jacobian calculation, optimization algorithm and pruning were implemented as S-function in *C*.

3   Parameter Optimization

First of all a quantitative measure of the network performance has to be defined. In the following we use the squared error

$$E(\underline{w}_k) = \frac{1}{2} \cdot \sum_{q=1}^{Q} (\underline{y}_q - \underline{\hat{y}}_q(\underline{w}_k))^T \cdot (\underline{y}_q - \underline{\hat{y}}_q(\underline{w}_k))$$

$$= \frac{1}{2} \cdot \sum_{q=1}^{Q} \underline{e}_q^T(\underline{w}_k) \cdot \underline{e}_q(\underline{w}_k) \qquad (4)$$

where q denotes the pattern in the training set, $\underline{y}_q$ and $\underline{\hat{y}}_q(\underline{w}_k)$ are, respectively, the desired target and actual model output on the qth pattern. The vector $\underline{w}_k$ is composed of all weights in the GDNN. The cost function $E(\underline{w}_k)$ is small if the training (and pruning) process performs well and large if it performs poorly. The cost function forms an error surface in a $(n+1)$-dimensional space, where n is equal to the number of weights in the GDNN. In the next step this space has to be searched in order to reduce the cost function.

3.1 Levenberg-Marquardt Algorithm

All Newton methods are based on the second-order Taylor series expansion about the old weight vector $\underline{w}_k$:

$$E(\underline{w}_{k+1}) = E(\underline{w}_k + \Delta \underline{w}_k)$$
$$= E(\underline{w}_k) + \underline{g}_k^T \cdot \Delta \underline{w}_k + \frac{1}{2} \cdot \Delta \underline{w}_k^T \cdot \underline{H}_k \cdot \Delta \underline{w}_k \tag{5}$$

If a minimum at the error surface is found, the gradient of the expansion (5) with respect to $\Delta \underline{w}_k$ is zero:

$$\nabla E(\underline{w}_{k+1}) = \underline{g}_k + \underline{H}_k \cdot \Delta \underline{w}_k = 0 \tag{6}$$

Solving (6) for $\Delta \underline{w}_k$ gives the Newton method

$$\Delta \underline{w}_k = -\underline{H}_k^{-1} \cdot \underline{g}_k^T$$
$$\underline{w}_{k+1} = \underline{w}_k - \underline{H}_k^{-1} \cdot \underline{g}_k \tag{7}$$

The vector $-\underline{H}_k^{-1} \cdot \underline{g}_k^T$ is known as the Newton direction, which is a descent direction, if the Hessian matrix $\underline{H}_k$ is positive definite. There are several difficulties with the direct application of Newton's method. One problem is that the optimization step may move to a maximum or saddle point if the Hessian is not positive definite and the algorithm could become unstable. There are two possibilities to solve this problem. Either the algorithm uses a line search routine (e.g. Quasi-Newton) or the algorithm uses a scaling factor (e.g. Levenberg Marquardt).

The direct evaluation of the Hessian matrix is computationally demanding. Hence the Quasi-Newton approach (e.g. BFGS formula) builds up an increasingly accurate term for the inverse Hessian matrix iteratively, using first derivatives of the cost function only. The Gauss-Newton and Levenberg-Marquardt approach approximate the Hessian matrix by [9]

$$\underline{H}_k \approx \underline{J}^T(\underline{w}_k) \cdot \underline{J}(\underline{w}_k) \tag{8}$$

and it can be shown that

$$\underline{g}_k = \underline{J}^T(\underline{w}_k) \cdot \underline{e}(\underline{w}_k) \tag{9}$$

where $\underline{J}(\underline{w}_k)$ is the Jacobian matrix

$$
\underline{J}(\underline{w}_k) = \begin{bmatrix} \dfrac{\partial e_1(\underline{w}_k)}{\partial w_1} & \dfrac{\partial e_1(\underline{w}_k)}{\partial w_2} & \cdots & \dfrac{\partial e_1(\underline{w}_k)}{\partial w_n} \\ \dfrac{\partial e_2(\underline{w}_k)}{\partial w_1} & \dfrac{\partial e_2(\underline{w}_k)}{\partial w_2} & \cdots & \dfrac{\partial e_2(\underline{w}_k)}{\partial w_n} \\ \vdots & \vdots & \ddots & \vdots \\ \dfrac{\partial e_Q(\underline{w}_k)}{\partial w_1} & \dfrac{\partial e_Q(\underline{w}_k)}{\partial w_2} & \cdots & \dfrac{\partial e_Q(\underline{w}_k)}{\partial w_n} \end{bmatrix} \tag{10}
$$

which includes first derivatives only. n is the number of all weights in the neural network and Q is the number of time steps evaluated, see subsection 3.2.

With (7), (8) and (9) the Gauss-Newton method can be written as

$$
\underline{w}_{k+1} = \underline{w}_k - \left[\underline{J}^T(\underline{w}_k) \cdot \underline{J}(\underline{w}_k) \right]^{-1} \cdot \underline{J}^T(\underline{w}_k) \cdot \underline{e}(\underline{w}_k) \tag{11}
$$

Now the Levenberg-Marquardt (LM) method can be expressed with the scaling factor μ_k

$$
\underline{w}_{k+1} = \underline{w}_k - \left[\underline{J}^T(\underline{w}_k) \cdot \underline{J}(\underline{w}_k) + \mu_k \cdot \underline{I} \right]^{-1} \cdot \underline{J}^T(\underline{w}_k) \cdot \underline{e}(\underline{w}_k) \tag{12}
$$

where $\underline{I}$ is the identity matrix. As the LM algorithm is the best optimization method for small and moderate networks (up to a few hundred weights), this algorithm is used for all simulations in this paper. In the following subsections the calculation of the Jacobian matrix $\underline{J}(\underline{w}_k)$ and the creation of the error vector $\underline{e}(\underline{w}_k)$ are considered.

3.2 Quasi-online Learning

The quasi-online learning approach is shown in Fig. 2. In every optimization step the last Q errors are used for the Jacobian calculation. It is like a time window that slides along the time axis. In every time step the eldest training pattern drops out of the time window. It is assumed, that the training data of the time window describes the optimization problem definitely and so the error surface is constant. The data content of the time window is used for parameter optimization. With this simple method we are able to train the LM algorithm online. As this quasi-online learning method works surprisingly well, it is not necessary to use a recurrent approach like [10]. In the simulations in this paper the window size is set to 250 steps (using a sampling time of $0.01sec$), but it can be varied from 10 to 1000 or more. A change of the window size does not have too much influence, as the matrix $\underline{J}^T(\underline{w}_k) \cdot \underline{J}(\underline{w}_k)$ (which have to be inverted) has $n \times n$ elements.

3.3 Jacobian Calculations

To create the Jacobian matrix, the derivatives of the errors have to be computed, see Eq. (10). The GDNN has feedback elements and internal delays, so that the Jacobian cannot be calculated by the standard backpropagation algorithm. There are two general approaches to calculate the Jacobian matrix in dynamic systems: By backpropagation-through-time (BPTT) [11] or by real-time recurrent learning (RTRL) [12]. For Jacobian calculations the RTRL algorithm

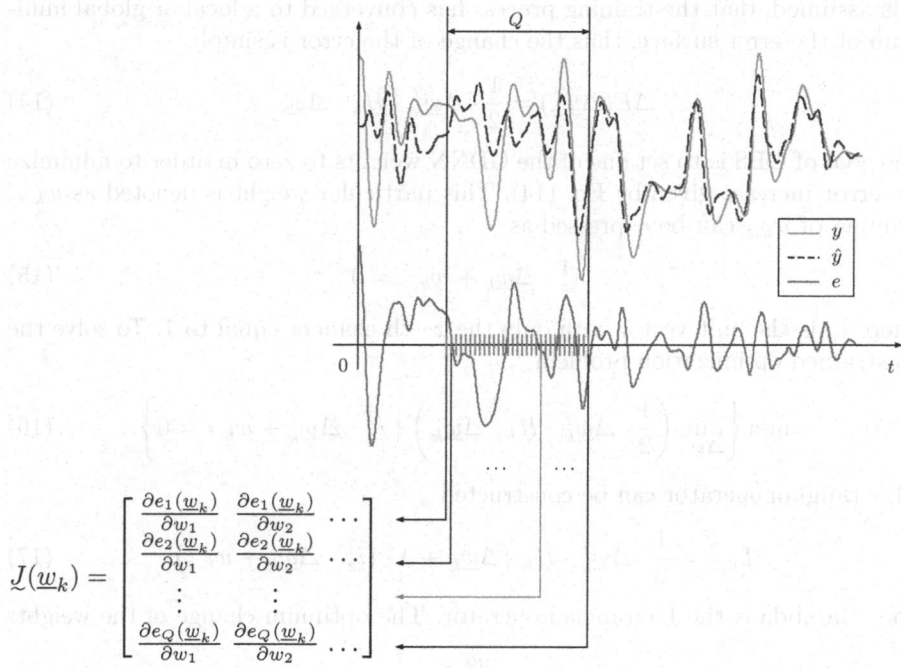

Fig. 2. Quasi-online learning

is more efficient than the BPTT algorithm [8]. According to this, the RTRL algorithm is used in this paper. Therefore we make use of the developed formulas of the layered digital dynamic network. The interested reader is referred to [5]-[8] for further details.

4 Architecture Selection

Pruning algorithms [1]–[4] have initially been designed for static neural networks to obtain good generalization. Due to the quasi-online approach 3.2 we do not have overtraining problems. This paper is focused on the question: Is it possible to find the optimal structure of a nonlinear dynamic system? In the following subsection we review the mathematical formulation of saliency analysis suggested by Hassibi [2] for OBS. In Subsection 4.2 we present a simple but necessary modification for dynamic systems.

4.1 Optimal Brain Surgeon

In Eq. (5) a second order Taylor series expansion is used. When the weight vector is changed by $\Delta \underline{w}_k$ the change of the error is approximately

$$\Delta E(\Delta \underline{w}_k) = \underline{g}_k^T \cdot \Delta \underline{w}_k + \frac{1}{2} \cdot \Delta \underline{w}_k^T \cdot \underline{H}_k \cdot \Delta \underline{w}_k \qquad (13)$$

It is assumed, that the training process has converged to a local or global minimum of the error surface, thus the change of the error is simply

$$\Delta E(\Delta \underline{w}_k) = \frac{1}{2} \cdot \Delta \underline{w}_k^T \cdot \underline{H}_k \cdot \Delta \underline{w}_k \qquad (14)$$

The goal of OBS is to set one of the GDNN weights to zero in order to minimize the error increase given by Eq. (14). This particular weight is denoted as $w_{k,z}$. Pruning of $w_{k,z}$ can be expressed as

$$\underline{i}_z^T \cdot \Delta \underline{w}_k + w_{k,z} = 0 \qquad (15)$$

where $\underline{i}_z$ is the unit vector with only the z-th element equal to 1. To solve the constrained optimization problem

$$\min_z \left\{ \min_{\Delta \underline{w}_k} \left(\frac{1}{2} \cdot \Delta \underline{w}_k^T \cdot \underline{H}_k \cdot \Delta \underline{w}_k \right) \mid \underline{i}_z^T \cdot \Delta \underline{w}_k + w_{k,z} = 0 \right\} \qquad (16)$$

a Lagrangian operator can be constructed

$$L_{k,z} = \frac{1}{2} \cdot \Delta \underline{w}_k^T \cdot \underline{H}_k \cdot \Delta \underline{w}_k + \lambda \cdot \left(\underline{i}_z^T \cdot \Delta \underline{w}_k + w_{k,z} \right) \qquad (17)$$

where lambda is the Lagrangian operator. The optimum change of the weights is

$$\Delta \underline{w}_k = - \frac{w_{k,z}}{\left[\underline{H}_k^{-1} \right]_{z,z}} \cdot \underline{H}_k^{-1} \cdot \underline{i}_z \qquad (18)$$

and the saliency of the weight $w_{k,z}$ is calculated by

$$S_{k,z} = \frac{1}{2} \cdot \frac{w_{k,z}^2}{\left[\underline{H}_k^{-1} \right]_{z,z}} \qquad (19)$$

4.2   Weight Saliencies in GDNN

The OBS algorithm in subsection 4.1 requires the complex calculation of the inverse Hessian. The OBS approach suggested in this subsection makes use of the approximation of the Hessian, see Eq. (8). This calculation is already used in the LM optimization method (see subsection 3.1), so the OBS approach can be received with low computational cost.

When training a GDNN, the final model should be as simple as possible. It can be observed, that after the training process with the standard OBS algorithm (see subsection 4.1) the GDNN found reduces the error very well, but the final GDNN-model includes time delays of high order (there are many models possible, which generate the same input-output behavior). To avoid this problem, two weights of similar saliency – but different delay orders – should have a different influence on the pruning process. As we want to have a model as simple as possible, networks weights with higher order of delay d should produce a smaller saliency:

$$Sn_{k,z} = \frac{w_{k,z}^2}{(d+1) \cdot \left[\left(\underline{J}^T(\underline{w}_k) \cdot \underline{J}(\underline{w}_k) + \mu_k \cdot \underline{I} \right)^{-1} \right]_{z,z}} \qquad (20)$$

Bias-weights and forward-weights have no TDL, hence $d = 0$. All other weights have at least one time delay and are more attractive to pruning. With (8), (12) and (18) the optimum change in weights can be approximated by

$$\Delta \underline{w}_k = -\frac{w_{k,z} \cdot (\underline{J}^T(\underline{w}_k) \cdot \underline{J}(\underline{w}_k) + \mu_k \cdot \underline{I})^{-1} \cdot \underline{i}_z}{\left[(\underline{J}^T(\underline{w}_k) \cdot \underline{J}(\underline{w}_k) + \mu_k \cdot \underline{I})^{-1} \right]_{z,z}} \tag{21}$$

5 Identification

A crucial task in identification with recurrent neural networks is the selection of an appropriate network architecture. How many layers and how many nodes in each layer should the neural network have? Should the layer have feedback connections and what about the taped delay lines between the layers? Should every weight in a layer weight matrix or an input matrix exist? It is very difficult to decide a priori what architecture and what size are adequate for an identification task. There are no general rules for choosing the structure before identification. A common approach is to try different configurations until one is found and works well. This approach is very time consuming if many topologies have to be tested. In this paper we start with a "larger than necessary" network and remove unnecessary parts, so we only require an estimate of what size is "larger than necessary". The excess degrees of freedom of the large initial GDNN is leading to a fast reduction of the cost function (with high computational effort for one optimization step).

In this section two simulation results for plant identification are presented, using the modified OBS algorithm suggested in (21) and (22), Jacobian calculation with real time recurrent learning and the Levenberg Marquardt algorithm (12). The first example to be identified is a simple linear PT_2 system. This example is chosen to test the architecture selection ability of the OBS approach suggested in subsection 4.2. The second example is the identification of a nonlinear dynamic system.

5.1 Excitation Signal

In system identification it is a very important task to use an appropriate excitation signal. A nonlinear dynamic system requires, that all combinations of frequencies and amplitudes (in the system's operating range) are represented in the signal. In this paper an APRBS-signal (Amplitude Modulated Pseudo Random Binary Sequence) was used, uniformly distributed from -1 to +1. For more information see [13].

5.2 Identification of PT_2 Plant

To identify the simple PT_2-plant $\frac{10}{0.1 \cdot s^2 + s + 100}$ a three-layer GDNN (with three neurons in the hidden layers and $d_{max} = 2 \Rightarrow n = 93$) is used. This network architecture is obviously larger than necessary. The identification and pruning

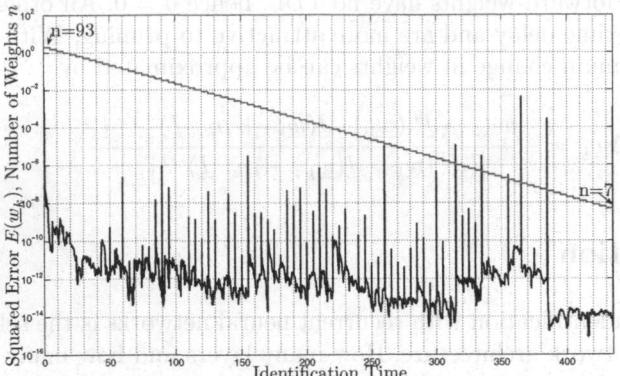

Fig. 3. Squared Error $E(\underline{w}_k)$ and number of weights n for $PT2$-identification example (at the beginning: $n = 93$, at the end: $n = 7$)

process should find a model for the plant as simple as possible. Fig. 3 shows the discrete weight reduction (every 500 optimization steps one pruning step is performed) from $n = 93$ to $n = 7$ and the identification error, which is also reduced although the model is becoming smaller.

Most of the weights are deleted. The change in the cost function is evaluated in every pruning step. The last pruning step is cancelled after an abrupt rise (over two decades) in the cost function $E(\underline{w}_k)$. Therefore the GDNN weights and the layer-outputs (these are the states of the dynamic system) have to be stored! Fig. 4 shows the outputs of the PT2-plant y (solid gray line) and the GDNN-model $\hat{y}$ (dashed black line) during the first 2.5 seconds. The output of the GDNN-model follows the output of the plant almost immediately. Fig. 5 shows the identification result and the successful weight reduction of the suggested OBS algorithm. Each circle in the signal flow chart includes a summing junction and

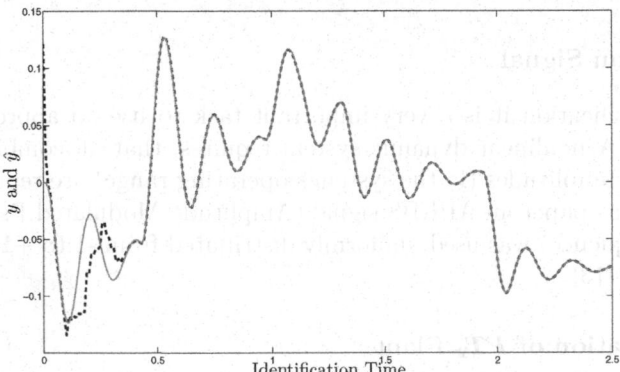

Fig. 4. Outputs of the PT2-plant y (solid gray line) and the GDNN-model $\hat{y}$ (dashed black line) for the first 2.5 seconds

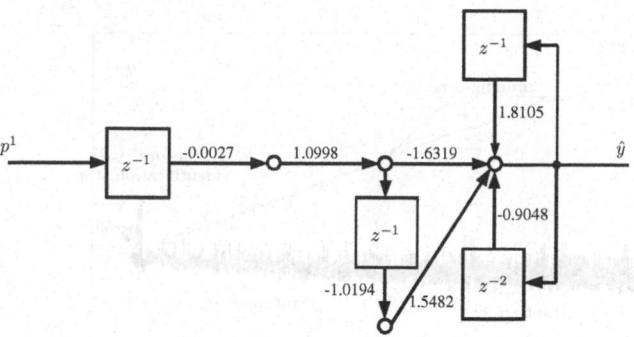

Fig. 5. Identification result: Model of PT2 example

a transfer function. The model found uses the first and the second time delay from the input and from the output. If the continuous-time model $\frac{10}{0.1 \cdot s^2 + s + 100}$ is discretized using zero-order hold [14] with a sample time of $10ms$ the discrete model can be expressed as

$$y(k) = 0.004798 \cdot u(k-1) + 0.00464 \cdot u(k-2) + 1.81 \cdot y(k-1) - 0.9048 \cdot y(k-2) \quad (22)$$

The factors 1.81 and -0.9048 of the difference equation (22) correspond to the feedback weights of the GDNN model (see Fig. 5). Because of the small input weight -0.0027 in the GDNN only the linear region (with gradient 1) of the $tanh$-functions is crucial. Note: All weights in the GDNN-model correspond very closely to the coefficients of the difference equation (e.g. $(-0.0027) \cdot 1.0998 \cdot (-1.0194) \cdot 1.5482 = 0.0046865 \approx 0.00464$).

5.3   Identification of a Nonlinear Plant

The example in the last subsection is chosen to emphasize the weight reduction ability of the suggested pruning algorithm. In this subsection a nonlinear dynamic system presented by Narendra [15] is considered to show the identification ability:

$$y(k) = \frac{y(k-1) \cdot y(k-2) \cdot y(k-3) \cdot u(k-2) \cdot [y(k-3) - 1] + u(k-1)}{1 + y^2(k-2) + y^2(k-3)} \quad (23)$$

For this identification example a three-layer GDNN (with four neurons in the hidden layers and $d_{max} = 3 \Rightarrow n = 212$) is used. The pruning interval is 500 (every 500 optimization steps one pruning step is performed). In every pruning step the previous weight reduction is evaluated. If the error increase is to much (greater than 5), then the last pruning step is cancelled. Fig. 6 shows the weight reduction from $n = 212$ to $n = 53$ and the identification error, which increases if the GDNN size is smaller than about $n = 100$ network weights. The identification and pruning process is stopped after 915 seconds because of the abrupt rise (over four decades) in the cost function $E(\underline{w}_k)$. The last pruning step is cancelled. Fig. 7 shows the outputs of the nonlinear plant y (solid gray line) and the GDNN-model $\hat{y}$ (dashed black line) during the first 2.5 seconds.

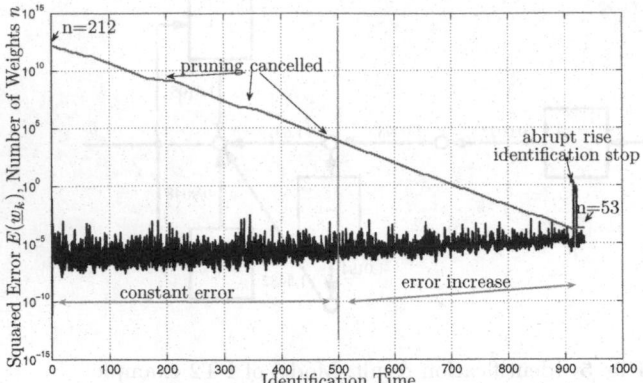

Fig. 6. Squared Error $E(\underline{w}_k)$ and number of weights n for nonlinear identification example (at the beginning: $n = 212$, at the end: $n = 53$)

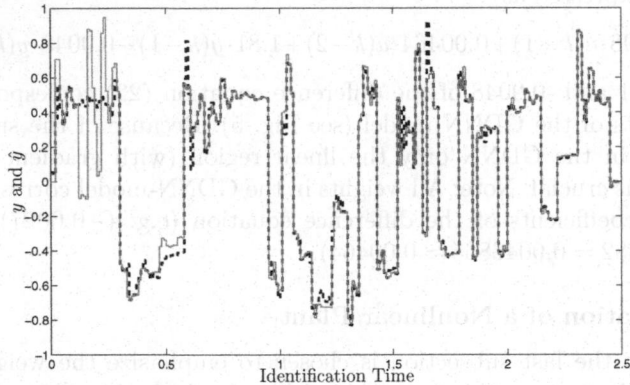

Fig. 7. Outputs of the nonlinear plant y (solid gray line) and the GDNN-model $\hat{y}$ (dashed black line) for the first 2.5 seconds

6   Conclusion

In this paper it is shown, that network pruning not only works in static neural networks but can also be applied to dynamic neural networks. For this a rational pruning algorithm based on OBS is suggested. The pruning algorithm uses Hessian information and considers the order of time delay for saliency calculation. As the LM training algorithm already approximates the Hessian matrix by the Jacobian, it is suggested to use this information for a rational OBS implementation. After a pruning step the remaining weights are retrained. In every pruning step the success of the last pruning step is checked. If the increase in error is too much, the last pruning step is cancelled. For this revision, in every pruning step the weights and the layer-outputs (states of the

GDNN) have to be stored. The Jacobian matrix is calculated quasi-online with RTRL. Quasi-online learning uses a time window for parameter optimization. With this approach the LM algorithm can be trained online. To manage the pruning process with GDNNs, administration matrices are introduced. These matrices show us the actual GDNN and indicates, which weights are already deleted. The suggested algorithm find accurate models of low order. The simple linear system identification example is chosen to show the weight reduction ability. The input weight of the final GDNN is very small to get an almost linear behavior of the model in the system's operating range. In addition to that the weights of the GDNN model correspond very closely to the coefficients of the according difference equation. If the GDNN is too small, the error increases with the pruning process. This is shown in the nonlinear plant identification example. In that case the user has to know the quality demand of the final GDNN model.

References

1. Le Cun, Y., Denker, J.S., Solla, S.A.: Optimal Brain Damage. In: Touretzky, D.S. (ed.) Advances in Neural Information Processing Systems, pp. 598–605. Morgan Kaufmann, San Francisco (1990)
2. Hassibi, B., Stork, D.G., Wolff, G.J.: Optimal Brain Surgeon and General Network Pruning. In: IEEE International Conference on Neural Networks, vol. 1, pp. 293–299 (April 1993)
3. Reed, R.: Pruning Algorithms – A Survey. IEEE Transactions on Neural Networks 4(5), 740–747 (1993)
4. Attik, M., Bougrain, L., Alexandre, F.: Optimal Brain Surgeon Variants For Feature Selection. In: Proceedings of the International Joint Conference on Neural Networks, pp. 1371–1374. IEEE, Los Alamitos (2004)
5. De Jesús, O., Hagan, M.: Backpropagation Algorithms Through Time for a General Class of Recurrent Network. In: IEEE Int. Joint Conf. Neural Network, pp. 2638–2643. IEEE Computer Society Press, Washington (2001)
6. De Jesús, O., Hagan, M.: Forward Perturbation Algorithm For a General Class of Recurrent Network. In: IEEE Int. Joint Conf. Neural Network, pp. 2626–2631. IEEE Computer Society Press, Washington (2001)
7. De Jesús, O.: Training General Dynamic Neural Networks. Ph.D. dissertation, Oklahoma State University, Stillwater, OK (2002)
8. De Jesús, O., Hagan, M.: Backpropagation Algorithms for a Broad Class of Dynamic Networks. IEEE Transactions on Neural Networks 18(1), 14–27 (2007)
9. Hagan, M., Mohammed, B.M.: Training Feedforward Networks with the Marquardt Algorithm. IEEE Transactions on Neural Networks 5(6), 989–993 (1994)
10. Ngia, L.S.H., Sjöberg, J.: Efficient Training of Neural Nets for Nonlinear Adaptive Filtering Using a Recursive Levenberg-Marquardt Algorithm. IEEE Transactions on Signal Processing 48(7), 1915–1927 (2000)
11. Werbos, P.J.: Backpropagation Through Time: What it is and how to it. Proc. IEEE 78(10), 1550–1560 (1990)
12. Williams, R.J., Zipser, D.: A Learning Algorithm for Continually Running Fully Recurrent Neural Networks. Neural Computing 1, 270–280 (1989)

13. Nelles, O.: Nonlinear System Identification. Springer, New York (2001)
14. Schröder, D.: Elektrische Antriebe - Regelung von Antriebssystemen, 2nd edn. Springer, Heidelberg (2001)
15. Narendra, K.S., Parthasarathy, K.: Identification and Control of Dynamical Systems Using Neural Networks. IEEE Transactions on Neural Networks 1(1), 4–27 (1990)

Answer-Set Programming Based Dynamic User Modeling for Recommender Systems

João Leite and Manoela Ilić

CENTRIA, Universidade Nova de Lisboa, Portugal

Abstract. In this paper we propose the introduction of dynamic logic programming – an extension of answer set programming – in recommender systems, as a means for users to specify and update their models, with the purpose of enhancing recommendations.

1 Introduction

Recommender systems have become an essential tool for finding the needle in the haystack of the World Wide Web - the information or item one is searching for [22]. Finding the desired item is a daunting task considering the amount of information that is present in the WWW and its databases, and almost every E-commerce application will provide the user with the help of a recommender system to suggest products or information that the user might want or need [23].

Recommender systems are employed to recommend products in online stores, news articles in news subscription sites or financial services. Common techniques for selecting the right item for recommendation are: collaborative filtering (e.g. [21,16]) where user ratings for objects are used to perform an inter-user comparison and then propose the best rated items; content-based recommendation (e.g.[3,6]) where descriptions of the content of items is matched against user profiles, employing techniques from the information retrieval field; knowledge-based recommendation (e.g. [9,14]) where knowledge about the user, the objects, and some distance measures between them, i.e. relationships between users and objects, are used to infer the right selections; and, as always, hybrid versions of these (e.g. [8,25]) where two or more of these techniques (collaborative filtering being usually one of them) are used to overcome their individual limitations. For further details on this subject the reader is referred to [10].

Recommender systems can also be categorised, according to the way they interact with the user [7], as being *single shot systems* where for each request the system make its interpretation of information and proposes recommendations to the user without taking in account any previous interaction, and *conversational systems* where recommendations are made on the basis of the current request of the user and some feedback provided as a response to previously proposed items.

The extent to which users find the recommendations satisfactory is, ultimately, the key feature of a recommendation system, and the accuracy of the user models that are employed is of significant importance to this goal. Such user models represent the user's taste and can be implicit (e.g. constructed from information

J. Neves, M. Santos, and J. Machado (Eds.): EPIA 2007, LNAI 4874, pp. 29–42, 2007.

about the user behavior), or explicit (e.g. constructed from direct feedback or input by the user, like ratings). The accuracy of a user model greatly depends on how well short-term and long-term interests are represented [5], making it a challenging task to include both sensibility to changes of taste and maintenance of permanent preferences. While implicit user modeling disburdens the user of providing direct feedback, explicit user modeling may be more confidence inspiring to the user since recommendations are based on a conscious assignment of preferences.

Though most of the recommender systems are very efficient from a large-scale perspective, the effort in user involvement and interaction is calling for more attention. Moreover, problems concerning trust and security in recommender systems could be approached with a better integration of the user and more control over the user model. For more details on the subject about security and manipulation in recommender systems the reader is referred to [17].

In this paper we will concentrate on explicit user modeling for recommender systems, guided by the following three claims:

1. Recommender systems should provide users with a way (language) to specify their models and preferences. This language should be expressive enough to allow for specifications that exceed the mere assignment of ratings to products. It should allow for the usage of existing concepts (e.g. product characteristics) as well as for the definition of new concepts (e.g. own qualitative classifications based on product characteristics). The language should allow for the specification of rules that use these concepts to define the policies regarding the recommendations made by the system. The language should also include some form of negation to allow for the specification of both positive as well as negative information.
2. Users should be able to update their models by specifying new rules. The system must consider that users are not consistent along time i.e., some newer rules may directly contradict previously specified ones, possibly representing an evolution of the user's tastes and needs, and these "contradictions" should be dealt with by the system, relieving the user from any consistency requirements, always difficult to impose, and often a discouraging factor.
3. Recommender systems should not depend solely on the model specified by the user. Other approaches and existing systems that do not require the user specification should be used as complement. Their outputs should be taken into consideration since they may already encode large amounts of data that should not be disregarded, and would be particularly useful in the absence of user specified knowledge. At the same time the output of the recommender system should take into strict consideration the user specifications which, if violated, would turn the user away from the recommendation system.

In this paper, we borrow concepts from the area of knowledge representation and non-monotonic reasoning, to set forth a proposal that aims at extending systems with the possibility of allowing users to specify their individual models and preferences, while taking into account the three claims above. Concretely, we adopt the paradigm of *Dynamic Logic Programming* [2,18,1] to suit our needs.

Dynamic Logic Programming (DLP) is an extension of Answer-set Programming (ASP) [15] introduced to deal with knowledge updates. ASP is a form of declarative programming that is similar in syntax to traditional logic programming and close in semantics to non-monotonic logic, that is particularly suited for knowledge representation[1]. In contrast to Prolog, where proofs and substitutions are at its heart, the fundamental idea underlying ASP is to describe a problem declaratively in such a way that models of the description provide solutions to problems. Enormous progress concerning both the theoretical foundations of the approach and implementation issues have been made in recent years. The existence of very efficient ASP solvers (e.g. DLV [19] and SMODELS[2] [20]) make it finally possible to investigate some serious applications in the area of e.g. data integration [11], data source selection [12] and also the Semantic Web [24].

According to *DLP*, as extension of ASP, knowledge is given by a series of theories, encoded as generalized logic programs[3] (or answer-set programs), each representing distinct states of the world. Different states, sequentially ordered, can represent different time periods, thus allowing DLP to represent knowledge undergoing successive updates. As individual theories may comprise mutually contradictory as well as overlapping information, the role of *DLP* is to employ the mutual relationships among different states to determine the declarative semantics for the combined theory comprised of all individual theories at each state. Intuitively, one can add, at the end of the sequence, newer rules leaving to *DLP* the task of ensuring that these rules are in force, and that previous ones are valid (by inertia) only so far as possible, i.e. that they are kept for as long as they are not in conflict with newly added ones, these always prevailing.

DLP can provide an expressive framework for users to specify rules encoding their model, preferences and their updates, while enjoying several properties, some discussed below, such as: a simple extendable language; a well defined semantics; the possibility to use default negation to encode non-deterministic choice, thus generating more than one set of recommendations, facilitating diversity each time the system is invoked; the combination of both strong and default negation to reason with the closed and open world assumptions; easy connection with relational databases (ASP can also be seen as a query language, more expressive than SQL); support for explanations; amongst others.

Since we are providing users with a way to express themselves by means of rules, we can also provide the same rule based language to the owners of the recommendation system, enabling them to specify some policies that may not be captured by the existing recommendation system (e.g. preference for

---

[1] The main difference between traditional logic programming (e.g. Prolog) and ASP is how negation as failure is interpreted. In traditional logic programming, negation-as-failure indicates the failure of a derivation; in ASP, it indicates the consistency of a literal. In contrast to Prolog, the semantics of ASP do not depend on a specific order of evaluation of the rules and of the atoms within each rule. For more on ASP, namely its semantics and applications, the reader is referred to [4] and [26].

[2] Available at www.dlvsystem.com and www.tcs.hut.fi/Software/smodels/ resp.

[3] LPs with default and strong negation both in the body and head of rules.

recommending certain products). Even though this was not at the heart of our
initial goal, it is an extra feature provided by our proposal, with no added cost.

In a nutshell, we want to propose a system, with a precise formal specification
and semantics, composed of three modules, namely the output of an existing
recommendation system, a set of rules specified by the owner of the recommendation system and a sequence of rules specified by the user, for which we provide
an expressive language. The modules are combined in a way such that they produce a set of recommendations that obeys certain properties such as obedience
to the rules specified by the user, removal of contradictions specified by the user
along time, keep the result of the initial recommendation module as much as
possible in the final output, among others.

The remainder of this paper is organised as follows. In Sect. 1, for self-
containment, we briefly recap the notion of *DLP*, establishing the language used
in the paper. In Sect. 3 we define our framework and its semantics. In Sect. 4
we preset a brief illustrative example. In Sect. 5 we discuss some properties and
conclude in Sect. 6.

2   Dynamic Logic Programming

For self containment, in this Appendix, we briefly recap the notion and semantics
of Dynamic Logic Programming needed throughout. More motivation regarding
all these notions can be found in [2,18].

Let $\mathcal{A}$ be a set of propositional atoms. An **objective literal** is either an
atom A or a strongly negated atom $\neg A$. A **default literal** is an objective
literal preceded by *not*. A **literal** is either an objective literal or a default
literal. A **rule** r is an ordered pair $H(r) \leftarrow B(r)$ where $H(r)$ (dubbed the
head of the rule) is a literal and $B(r)$ (dubbed the body of the rule) is a finite set of literals. A rule with $H(r) = L_0$ and $B(r) = \{L_1, \ldots, L_n\}$ will
simply be written as $L_0 \leftarrow L_1, \ldots, L_n$. A **tautology** is a rule of the form
$L \leftarrow Body$ with $L \in Body$. A **generalized logic program** (GLP) P, in $\mathcal{A}$,
is a finite or infinite set of rules. If $H(r) = A$ (resp. $H(r) = not\, A$) then
$not\, H(r) = not\, A$ (resp. $not\, H(r) = A$). If $H(r) = \neg A$, then $\neg H(r) = A$.
By the **expanded generalized logic program** corresponding to the GLP P,
denoted by $\mathbf{P}$, we mean the GLP obtained by augmenting P with a rule of
the form $not\, \neg H(r) \leftarrow B(r)$ for every rule, in P, of the form $H(r) \leftarrow B(r)$,
where $H(r)$ is an objective literal. An **interpretation** M of $\mathcal{A}$ is a set of
objective literals that is consistent i.e., M does not contain both A and $\neg A$.
An objective literal L is true in M, denoted by $M \models L$, iff $L \in M$, and
false otherwise. A default literal $not\, L$ is true in M, denoted by $M \models not\, L$,
iff $L \notin M$, and false otherwise. A set of literals B is true in M, denoted
by $M \models B$, iff each literal in B is true in M. An interpretation M of $\mathcal{A}$
is an **answer set** of a GLP P iff $M' = least(\mathbf{P} \cup \{not\, A \mid A \notin M\})$, where
$M' = M \cup \{not\_A \mid A \notin M\}$, A is an objective literal, and $least(.)$ denotes
the least model of the definite program obtained from the argument program,

replacing every default literal $not\,A$ by a new atom not_A. Let $AS(P)$ denote the set of answer-sets of P.

A **dynamic logic program** (DLP) is a sequence of generalized logic programs. Let $\mathcal{P} = (P_1, ..., P_s)$ and $\mathcal{P}' = (P_1', ..., P_n')$ be DLPs. We use $\rho(\mathcal{P})$ to denote the multiset of all rules appearing in the programs $\mathbf{P}_1, ..., \mathbf{P}_s$, and $\mathcal{P} \cup \mathcal{P}'$ to denote $(P_1 \cup P_1', ..., P_s \cup P_s')$ and $(P_i', P_j', \mathcal{P})$ to denote $(P_i', P_j', P_1, ..., P_n)$. We can now set forth the definition of a semantics, *based on causal rejection of rules*, for DLPs. We start by defining the notion of conflicting rules as follows: two rules r and r' are **conflicting**, denoted by $r \bowtie r'$, iff $H(r) = not\,H(r')$, used to accomplish the desired rejection of rules:

Definition 1 (Rejected Rules). *Let* $\mathcal{P} = (P_1, ..., P_s)$ *be a* DLP *and* M *an interpretation. We define:*

$$Rej(\mathcal{P}, M) = \{r \mid r \in \mathbf{P}_i, \exists r' \in \mathbf{P}_j, i \le j, r \bowtie r', M \vDash B(r')\}$$

We also need the following notion of default assumptions.

Definition 2 (Default Assumptions). *Let* $\mathcal{P} = (P_1, ..., P_s)$ *be a* DLP *and* M *an interpretation. We define (where* A *is an objective literal):*

$$Def(\mathcal{P}, M) = \{not\,A \mid \nexists r \in \rho(\mathcal{P}), H(r) = A, M \vDash B(r)\}$$

We are now ready to define the semantics for DLPs based on the intuition that some interpretation is a model according iff it obeys an equation based on the least model of the multiset of all the rules in the (expanded) DLP, without those rejected rules, together with a set of default assumptions. The semantics is dubbed *(refined) dynamic stable model semantics* $(RDSM)$.

Definition 3 (Semantics of DLP). *Let* $\mathcal{P} = (P_1, ..., P_s)$ *be a* DLP *and* M *an interpretation.* M *is a* (refined) dynamic stable model *of* $\mathcal{P}$ *iff*

$$M' = least\,(\rho(\mathcal{P}) - Rej(\mathcal{P}, M) \cup Def(\mathcal{P}, M))$$

where $M', \rho(.)$ *and* $least(.)$ *are as before. Let* $RDSM(\mathcal{P})$ *denote the set of all refined dynamic stable models of* $\mathcal{P}$.

3   Framework and Semantics

Our goal is to take the strengths of DLP as a knowledge representation framework with the capabilities of allowing for the representation of updates, and put it at the service of the user and the company, while at the same time ensuring some degree of integration with the output of other recommendation modules, possibly based on distinct paradigms (e.g. statistical, etc).

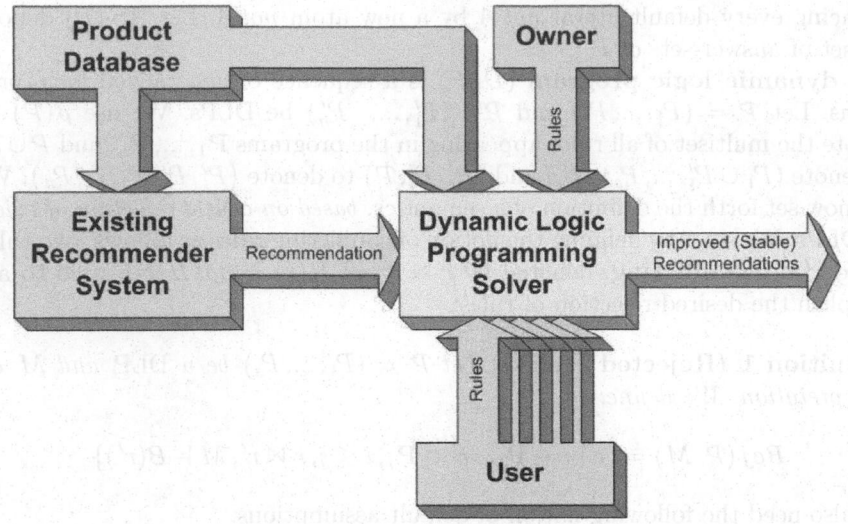

Fig. 1. Outline of the system

To make the presentation of our ideas simpler, we will make some assumptions and simplifications that, in our opinion, do not compromise our proposal and can be subsequently lifted (not in this paper though).

We start by assuming a layered system where the output of the existing recommendation module is simply used as the input to our system, and where no feedback to the initial module exists. We are aware that allowing for feedback from our system to the existing module could benefit its output, but such process would greatly depend on such existing module and we want to make things as general as possible, and concentrate on other aspects of the system. This takes us to the next assumption, that of the output of such existing module. We consider it to be an interpretation, i.e. a consistent set of objective literals representing the recommendations. For simplicity, we can assume that the language contains a reserved predicate of the form $rec/1$ where the items are the terms of the predicate, and the interpretation will contain those predicates corresponding to the recommended items. For example, the interpretation $M = \{rec("Takk"), rec("Underground"), rec("The Köln Concert"), rec("Paris, Texas"), rec("Godfather")\}$ encodes the recommendations for the films "*Godfather*", "*Underground*" and "*Paris, Texas*", as well as for the music albums "*Takk*" and "*The Köln Concert*". It would be straightforward to extend this case to one where some value would be associated with each recommendation, e.g. using a predicate of the form $rec(item, value)$. However, to get our point across, we keep to the simplified version where the output of the existing module is simply a set of recommendations. An outline is given in figure 1.

What we have, then, is an initial interpretation, representing the output of the initial module, which we dub the *initial model*, a generalised logic program representing the owner's policy, and a dynamic logic program, representing the rules (and their evolution) specified by the user. The Product Database is, typically, a relational database[4] easily represented by a set of facts in a logic program. For simplicity, without loss of generality, we assume such database to be part of the generalised logic program representing the owner's policy. To formalise this notion, we introduce the concept of Dynamic Recommender Frame:

Definition 4 (Dynamic Recommender Frame). *Let $\mathcal{A}$ be a set of propositional atoms. A Dynamic Recommender Frame (DRF), over $\mathcal{A}$, is a triple $\langle M, P_0, \mathcal{P} \rangle$ where M is an interpretation of $\mathcal{A}$, P_0 a generalised logic program over $\mathcal{A}$, and $\mathcal{P}$ a dynamic logic program over $\mathcal{A}$.*

The semantics of a Dynamic Recommender Frame is given by the set of stable models of its transformation into a Dynamic Logic Program. This transformation is based on a few underlying assumptions concerning the way these three modules should interact and be combined. In particular, we want the rules specified by the user to be the most relevant and be able to supersede both those issued by the owner and the recommendation issued by the existing module. This is a natural principle as users would not accept a recommendation system that would explicitly violate their rules (e.g. recommend a horror movie when the user said that no horror movies should be recommended, just because the owner wants to push horror movies). This limits the impact of recommendations made by the initial module and the company to those not directly constrained by the user, or to those whose rules specified by the user allow for more than one alternative. The other principle concerns the relationship between the initial model and the policy specified by the owner. Here, we will opt for giving a prevailing role to the rules specified by the owner. The rational for this is rather natural: the owner must be given the option to supersede the initial recommendation module (e.g. preference to specify products of a given brand over those of another because of higher profit margins), and it may be impossible to have such control inside the initial module (e.g. a sub-symbolic system such as a neural network).

With these principles in mind, we first define a transformation from a Dynamic Recommender Frame into a Dynamic Logic Program:

Definition 5 (Υ). *Let $\mathcal{R} = \langle M, P_0, \mathcal{P} \rangle$ be a Dynamic Recommender Frame. Let $\Upsilon(\mathcal{R})$ be the DLP $(P_M, P_0, \mathcal{P})$ where $P_M = \{A \leftarrow: A \in M\}$.*

Intuitively, we construct a DLP where the initial knowledge is the program obtained from the initial model. Such initial program is followed (updated with) the owner's policy specification (P_0), in turn followed by the sequence of user specifications ($\mathcal{P}$). We define the semantics of a DRF as follows:

Definition 6 (Stable Recommendation Semantics). *Let $\mathcal{R} = \langle M, P_0, \mathcal{P} \rangle$ be a Dynamic Recommender Frame and M_R an interpretation. M_R is a stable*

---

[4] Recent developments have formalised an extension of Answer-Set Programming that allows for the interface with ontologies specified in Description Logics [13], making our work easily extensible to the case where product information is available in the Semantic Web, instead of a local relational database.

recommendation iff M_R is a dynamic stable model of $\Upsilon(\mathcal{R})$. Let $SR(\mathcal{R})$ denote the set of all stable recommendations.

4   Short Illustrative Example

In this Section we present a small and very simplified example, with the purpose of illustrating some (very few) features of our proposal.

Let's consider an on-line dish and wine recommender, with some existing recommendation system based on statistical analysis performed over the years. We consider the output of such module to be a set of recommended items, that, for our purposes, we will consider constant throughout this example. Let the interpretation M represent such output, i.e. the initial model, and be: $M = \{rec(d_1), rec(d_2), rec(w_1), rec(w_2), rec(d_7), rec(w_6)\}$ where d_i are dishes offered in a restaurant and w_i are wines. Then, the owner can define some further recommendation policies, e.g. that the system should:

− recommend at most one dish of a given restaurant, encoded by the rule (1):

$$not\, rec(X) \leftarrow restaurant(X, E), restaurant(Y, E), X \neq Y, rec(Y). \quad (1)$$

− non-deterministically, recommend at least one of dishes d_3 and d_4, and one of the wines w_6 and w_7 encoded by the rules $(2 - 5)^5$:

$$
\begin{array}{ll}
rec(d_3) \leftarrow not\, rec(d_4). & rec(w_6) \leftarrow not\, rec(w_7). \\
rec(d_4) \leftarrow not\, rec(d_3). & rec(w_7) \leftarrow not\, rec(w_6).
\end{array} \quad (2\text{-}5)
$$

− always recommend the wine that goes with a recommended dish, and vice versa, in case such a relation exists (defined by the predicate $rel/2$), encoded by the rules (6,7):

$$
\begin{array}{l}
rec(Y) \leftarrow type(Y, wine), type(X, dish), rel(X, Y), rec(X). \\
rec(X) \leftarrow type(Y, wine), type(X, dish), rel(X, Y), rec(Y).
\end{array} \quad (6\text{-}7)
$$

The owner program P_0 contains the previous rules, together with the site's relational database where the relations $type/2$, $rel/2$, etc, are defined. Here, we consider $restaurant(d_2, rest_1), restaurant(d_1, rest_1), restaurant(d_5, rest_1), rel(d_5, w_4)$, $taste(w_7, \text{"buttery"})$, $grapesort(w_4, \text{"Chardonnay"})$, $type(w_i, wine)$ and $type(d_i, dish)$. Without rules specified by the user, the frame has four stable recommendations resulting from the effect of the owner's rules on the initial model[6]:

$$
\begin{array}{l}
M_{R1} = \{rec(d_1), rec(d_3), rec(w_1), rec(w_2), rec(d_7), rec(w_6)\} \\
M_{R2} = \{rec(d_2), rec(d_3), rec(w_1), rec(w_2), rec(d_7), rec(w_6)\} \\
M_{R3} = \{rec(d_1), rec(d_4), rec(w_1), rec(w_2), rec(d_7), rec(w_6)\} \\
M_{R4} = \{rec(d_2), rec(d_4), rec(w_1), rec(w_2), rec(d_7), rec(w_6)\}
\end{array}
$$

---

[5] This encoding uses the classic even loop through negation which, in ASP, produces two models each with one of the propositions belonging to it.

[6] Here, we restrict the models to the propositions of the form $rec/1$.

When taking a closer look at the four stable recommendations, we observe that the first rule specified by the owner removed the recommendation for either d_1 or d_2, and the second and third rules introduced either $rec(d_3)$ or $rec(d_4)$. The combination of these generated four stable recommendations. Note that the fourth and fifth rules (for products w_6 and w_7) did not cause any change because $rec(w_6)$ belonged to the initial recommendation and the semantics tends to keep recommendations by inertia. The system would, e.g. non-deterministically, choose one of these stable recommendations to present to the user thus add diversity to the recommendation system which, otherwise, would only have one set of recommendations to present in case of consecutive equal requests.

Let's now consider the user. Initially, being only looking for dish recommendations, the user states that she doesn't want any recommendations for wines, encoded by the rule (8): $not\,rec(X) \leftarrow type(X, wine)$. This rule alone will override all the initial recommendations for wines and also all the rules of the owner specifying wine recommendations such as rules 4 and 5. If $\mathcal{P}_1 = (P_1)$, where P_1 contains rule 8, the four stable recommendations of the frame $\langle M, P_0, \mathcal{P}_1 \rangle$ are:

$$M_{R5} = \{rec(d_1), rec(d_3), rec(d_7)\} \quad M_{R7} = \{rec(d_1), rec(d_4), rec(d_7)\}$$
$$M_{R6} = \{rec(d_2), rec(d_3), rec(d_7)\} \quad M_{R8} = \{rec(d_2), rec(d_4), rec(d_7)\}$$

Later on, the user decides to get some recommendations for wines. For this, she decides to define the concept of what good wines are. Initially, she considers a good wine to be one with the grape "'Chardonnay"' or one with a buttery taste. She writes the following two rules (9 and 10):

$$good(X) \leftarrow type(X, wine), grapesort(X, \text{“Chardonnay”}).$$
$$good(X) \leftarrow type(X, wine), taste(X, \text{“buttery”}). \tag{9-10}$$

Furthermore, she decides that she wants to get at least one recommendation for a good item. She writes the rules (11-14)[7]:

$$rec(X) \leftarrow good(X), not\,n\_rec(X). \qquad rec\_at\_least\_one \leftarrow good(X), rec(X).$$
$$n\_rec(X) \leftarrow good(X), not\,rec(X). \qquad \leftarrow not\,rec\_at\_least\_one$$

If $\mathcal{P}_2 = (P_1, P_2)$, where P_2 contains rule 9-14, the stable recommendations of the frame $\langle M, P_0, \mathcal{P}_2 \rangle$ are:

$$M_{R9} = \{rec(w_4), rec(d_5), rec(d_3), rec(d_7)\}$$
$$M_{R10} = \{rec(w_7), rec(d_1), rec(d_4), rec(d_7)\}$$
$$M_{R11} = \{rec(w_7), rec(d_2), rec(d_4), rec(d_7)\}$$
$$M_{R12} = \{rec(w_4), rec(w_7), rec(d_5), rec(d_3), rec(d_7)\}$$

---

[7] These rules are a classic construct of ASP. The first two rules state that each good item X is either recommended or n_rec. Then, the third rule makes the proposition $rec\_at\_least\_one$ true if at least one good item is recommended. The fourth rule, an integrity constraint, eliminates all models where $rec\_at\_least\_one$ is not true. The recommendation system would have special syntactical shortcuts for these kind of specifications, since we cannot expect the user to write this kind of rules.

Note that all four sets have at least one recommendation for a good wine. Furthermore, M_{R9} and M_{R12} have, besides the recommendation for w_4, also the recommendation for d_5, as specified by rule 7 since they are related. Also note that rules 11 and 12 cause a direct contradiction with rule 8 for good wines. The semantics based on the causal rejection of rules deals with this issue.

Also worth noting is that, for each recommendation, there is an explanation based on user rules, on owner rules, or on the initial recommendation (as is the case of product d_7). This and other properties are explored in the next Section.

5   Properties

The use of formal semantics enables a rigorous account of its behaviour. In this Section we discuss some properties of the Stable Recommendation Semantics.

We start with conservation, stating that if the initial recommendation is a dynamic stable model of the DLP consisting of the owner rules followed by the user DLP, then it is a stable recommendation. This ensures that the semantics will keep the results of the initial module if they are agreed by owner and user.

Proposition 1 (Conservation). *Let* $\mathcal{R} = \langle M, P_0, \mathcal{P} \rangle$ *be a DRF[8]. Then,*

$$M \in RDSM((P_0, \mathcal{P})) \Rightarrow SR(\mathcal{R}) \supseteq \{M\}$$

We now establish the relationship between the stable recommendation semantics and DLP for the case of no existing recommendation module, ensuring the transfer of all properties of DLP, and ASP, namely expressiveness results.

Proposition 2 (Generalisation of DLP). *Let* P_0 *be a generalised logic program and* $\mathcal{P}$ *a dynamic logic program. Then,*

$$SR(\langle \emptyset, \emptyset, \mathcal{P} \rangle) = RDSM(\mathcal{P}) \quad SR(\langle \emptyset, P_0, \mathcal{P} \rangle) = RDSM((P_0, \mathcal{P}))$$

An important issue in recommendation systems is that of explaining the output to the user (and the owner). The fact that the stable recommendation semantics is well defined already provides a formal basis to support its results. However, we can state stronger results concerning the justification for the existence and absence of recommendations. If a recommendation belongs to a stable recommendation then, either there is a user rule supporting it, or there is an owner rule supporting it and no user rule overriding it, or it is in the output of the initial module and no rules are overriding it, i.e. there is always an explanation.

Proposition 3 (Positive Supportiveness). *Let* $\mathcal{R} = \langle M, P_0, \mathcal{P} \rangle$ *be a DRF and* $A \in M_R \in SR(\mathcal{R})$. *Then:*

$\exists r \in \rho(\mathcal{P}) : H(r) = A, M_R \vDash B(r)$ *or*
$\exists r' \in P_0 : H(r') = A, M_R \vDash B(r') \wedge \not\exists r \in \rho(\mathcal{P}) : H(r) = not\ A, M_R \vDash B(r)$ *or*
$A \in M \wedge \not\exists r \in \rho((P_0, \mathcal{P})) : H(r) = not\ A, M_R \vDash B(r)$

---

[8] Lack of space prevents us from presenting proofs for the propositions.

Likewise for absence of recommendations. If a recommendation belongs to the output of the initial system and is not part of a stable recommendation, then there must be a rule overriding it (either from the user or from the owner).

Proposition 4 (Negative Supportiveness). *Let $\mathcal{R} = \langle M, P_0, \mathcal{P} \rangle$ be a DRF and $A \notin M_R \in SR(\mathcal{R})$ and $A \in M$. Then,*

$$\exists r \in \rho((P_0, \mathcal{P})) : H(r) = not\, A, M_R \vDash B(r)$$

Going back to the property of conservation, stating that when the output of the existing module is a dynamic stable model of the DLP $(P_0, \mathcal{P})$, then it is a stable recommendation, it could be desirable to have a stronger result, namely that the semantics obey a notion of strong conservation according to which it would be *the only* stable recommendation. It turns out that the stable recommendation semantics does not obey such property:

Proposition 5 (Strong Conservation). *Let $\mathcal{R} = \langle M, P_0, \mathcal{P} \rangle$ be a DRF. Then,*

$$M \in RDSM((P_0, \mathcal{P})) \not\Rightarrow SR(\mathcal{R}) = \{M\}$$

Likewise, if we could establish a distance measure between sets of recommendations, we could argue for a semantics that would only keep those stable recommendations that are closer to the output of the existing module, i.e., impose minimal change. If such distance measure is given by:

Definition 7 (Distance between interpretations). *Let $\mathcal{A}$ be a set of propositional atoms and M, M_1, and M_2 interpretations of $\mathcal{A}$. We say that M_1 is closer to M than M_2, denoted by $M_1 \sqsubseteq_M M_2$ iff*

$$(M_1 \backslash M \cup M \backslash M_1) \subset (M_2 \backslash M \cup M \backslash M_2)$$

The stable recommendation semantics does not obey minimal change:

Proposition 6 (Minimal Change). *Let $\mathcal{R} = \langle M, P_0, \mathcal{P} \rangle$ be a DRF. Then,*

$$M_1, M_2 \in RDSM((P_0, \mathcal{P})) \wedge M_1 \sqsubseteq_M M_2 \not\Rightarrow M_2 \notin SR(\mathcal{R})$$

This is not necessarily a bad property as there are cases where one may argue for stable recommendations that involve non-minimal change i.e., the owner/user rules introduce other alternatives, even though the initial model is one of them. Consider a very simple example with four products, w_1, w_2, w_3 and w_4 and where the initial set of recommendations is $M = \{rec(w_1), rec(w_3)\}$. The product database contains $property(a, w_2)$ and $property(b, w_1)$ encoding the fact that product w_2 (resp. w_1) has property a (resp b). Let us now consider that the owner of the system specifies the following policy:

$$rec(w_2) \leftarrow rec(w_1). \qquad rec(w_4) \leftarrow rec(w_3), not\, recom\_prop(a).$$
$$recom\_prop(P) \leftarrow rec(W), property(P, W).$$

encoding that if product w_1 is recommended then so should product w_2 and that if product w_3 is being recommended and no product with property a is being recommended, then product w_4 should be recommended. Let us now assume that the user specifies that she doesn't want to receive recommendations for products with property b if no product with property a is being recommended. This can be encoded as: $not\,rec\,(W) \leftarrow not\,recom\_prop\,(a)\,,property(b,W)$. With this scenario, there are two stable recommendations, namely

$$M_1 = \{rec\,(w_1)\,,rec\,(w_2)\,,rec\,(w_3)\} \qquad M_2 = \{rec\,(w_3)\,,rec\,(w_4)\}$$

where $M_1 \sqsubseteq_M M_2$, but where it may be argued that M_2 should also be accepted since it obeys the specified rules, provides both positive as well as negative explanations, and adds to the diversity of the possible recommendations. However, if desired, such stable recommendations can be eliminated and we can define a refined semantics that obeys such properties. Formally:

Definition 8 (Minimal Change SR Semantics). *Let $\mathcal{R} = \langle M, P_0, \mathcal{P} \rangle$ be a DRF and M_R an interpretation. M_R is a minimal change stable recommendation iff $M_R \in SR\,(\mathcal{R})$ and $\nexists M'_R \in RDSM\,(\Upsilon\,(\mathcal{R})) : M'_R \sqsubseteq_M M_R$. $SR^m\,(\mathcal{R})$ denotes the set of all minimal change stable recommendations.*

This new semantics obeys both strong conservation and minimal change. As for the remaining properties, the minimal change stable recommendation semantics obeys conservation and positive and negative supportiveness. As expected, it no longer generalises Dynamic Logic Programming as it is well known that DLP accepts non-minimal dynamic stable models that would be eliminated.

6   Conclusions and Future Work

In most large-scale recommender systems, performance and effectiveness are of high importance and, therefore, implicit user model creation is essential. That might lead to imperfect results which can be refined by means of explicit user modeling. In this paper we proposed a system that can be seen as a complemental module for existing recommender systems, allowing for an improvement of results by means of explicit user modeling. The proposed system can thus be seen as part of a knowledge based (or hybrid) recommender system, with several characteristics, namely:

- allowing user personalisation with complex and expressive rules, improving the quality of recommendations;
- allowing for interaction with the user by means of updates to those rules, automatically removing inconsistencies;
- taking into account the output of other recommendation modules;
- allowing for customisation by the owner of the system;
- providing a semantics with multiple recommendation sets, facilitating diversity and non-determinism in recommendations;

- enjoying a formal, well defined semantics which supports justifications;
- enjoying all the other formal properties mentioned in the previous section, and many more inherited from DLP and ASP such as the expressiveness of the language and the efficient implementations.

Our future work will concentrate on integrating the module in a trial application in order to investigate its functionality and effectiveness. Such a system should contain a user interface for an easier specification and handling of the rules, as well as a justification option for recommender results. Moreover, a suitable way of connecting to external recommender systems should be included.

Subsequently, inspired by the idea of a meta-recommender system that integrates diverse information, we will consider a more developed architecture that receives input from more than one source. Data integration and information source selection has been investigated in the context of ASP [12,11], with promising results. We will use such results to move away from the idea of a recommender system being limited to one product domain and to embrace the integration of interconnected information.

We believe, however, that our proposal already encodes several important concepts that bring an added value to existing recommender systems.

References

1. Alferes, J.J., Banti, F., Brogi, A., Leite, J.A.: The refined extension principle for semantics of dynamic logic programming. Studia Logica 79(1) (2005)
2. Alferes, J.J., Leite, J.A., Pereira, L.M., Przymusinska, H., Przymusinski, T.: Dynamic updates of non-monotonic knowledge bases. Journal of Logic Programming 45(1-3) (2000)
3. Balabanović, M., Shoham, Y.: Fab: content-based, collaborative recommendation. Communications of the ACM 40(3), 66–72 (1997)
4. Baral, C.: Knowledge Representation, Reasoning and Declarative Problem Solving. Cambridge University Press, Cambridge (2003)
5. Billsus, D., Pazzani, M.J.: User modeling for adaptive news access. User Model. User-Adapt. Interact 10(2-3), 147–180 (2000)
6. Billsus, D., Pazzani, M.J.: Content-based recommendation systems. In: The Adaptive Web. LNCS, vol. 4321, pp. 325–341. Springer, Heidelberg (2007)
7. Bridge, D., Kelly, J.P.: Diversity-enhanced conversational collaborative recommendations. In: Creaney, N. (ed.) Procs. of the Sixteenth Irish Conference on Artificial Intelligence and Cognitive Science, University of Ulster, pp. 29–38 (2005)
8. Burke, R.D.: Integrating knowledge-based and collaborative-filtering recommender systems. In: AAAI Workshop on AI in Electronic Commerce, pp. 69–72. AAAI, Stanford, California, USA (1999)
9. Burke, R.D.: Knowledge-based recommender systems. In: Kent, A., Dekker, M. (eds.) Encyclopedia of Library and Information Systems, vol. 69, ch. suppl. 32 (2000)
10. Burke, R.D.: Hybrid recommender systems: Survey and experiments. User Model. User-Adapt. Interact 12(4), 331–370 (2002)
11. Eiter, T.: Data integration and answer set programming. In: Baral, C., Greco, G., Leone, N., Terracina, G. (eds.) LPNMR 2005. LNCS (LNAI), vol. 3662, pp. 13–25. Springer, Heidelberg (2005)

12. Eiter, T., Fink, M., Sabbatini, G., Tompits, H.: A generic approach for knowledge-based information-site selection. In: KR, pp. 459–469 (2002)
13. Eiter, T., Lukasiewicz, T., Schindlauer, R., Tompits, H.: Combining answer set programming with description logics for the semantic web. In: KR 2004. Procs. of 9th Int. Conference on Principles of Knowledge Representation and Reasoning, pp. 141–151. AAAI Press, Stanford, California, USA (2004)
14. Felfernig, A., Kiener, A.: Knowledge-based interactive selling of financial services with FSAdvisor, pp. 1475–1482. AAAI, Stanford, California, USA (2005)
15. Gelfond, M., Lifschitz, V.: Logic programs with classical negation. In: Procs. of ICLP 1990, MIT Press, Cambridge (1990)
16. Goldberg, D., Nichols, D., Oki, B.M., Terry, D.: Using collaborative filtering to weave an information tapestry. Communications of the ACM 35(12), 61–70 (1992) Special Issue on Information Filtering
17. Lam, S.K., Frankowski, D., Riedl, J.: Do you trust your recommendations? An exploration of security and privacy issues in recommender systems. In: Procs. of the 2006 Int. Conference on Emerging Trends in Information and Communication Security, Freiburg, Germany, ETRICS (2006)
18. Leite, J.A.: Evolving Knowledge Bases. IOS Press, Amsterdam (2003)
19. Leone, N., Pfeifer, G., Faber, W., Calimeri, F., Dell'Armi, T., Eiter, T., Gottlob, G., Ianni, G., Ielpa, G., Koch, S.P.C., Polleres, A.: The dlv system. In: Flesca, S., Greco, S., Leone, N., Ianni, G. (eds.) JELIA 2002. LNCS (LNAI), vol. 2424, Springer, Heidelberg (2002)
20. Niemelä, I., Simons, P.: Smodels: An implementation of the stable model and well-founded semantics for normal LP. In: Fuhrbach, U., Dix, J., Nerode, A. (eds.) LPNMR 1997. LNCS, vol. 1265, Springer, Heidelberg (1997)
21. Resnick, P., Iacovou, N., Suchak, M., Bergstorm, P., Riedl, J.: GroupLens: An Open Architecture for Collaborative Filtering of Netnews. In: Proceedings of ACM 1994 Conference on Computer Supported Cooperative Work, Chapel Hill, North Carolina, pp. 175–186. ACM Press, New York (1994)
22. Resnick, P., Varian, H.R.: Recommender systems. Communications of the ACM 40(3), 56–58 (1997)
23. Schafer, J.B., Konstan, J.A., Riedl, J.: E-commerce recommendation applications. Data Min. Knowl. Discov. 5(1/2), 115–153 (2001)
24. Schindlauer, R.: Answer-Set Programming for the Semantic Web. Phd thesis, Vienna University of Technology, Austria (December 2006)
25. Smyth, B., Cotter, P.: Personalized electronic program guides for digital TV. AI Magazine 22(2), 89–98 (2001)
26. Working group on Answer-Set Programming. http://wasp.unime.it

Application of Logic Wrappers to Hierarchical Data Extraction from HTML⋆

Amelia Bădică[1], Costin Bădică[2], and Elvira Popescu[2]

[1] University of Craiova, Business Information Systems Department
A.I.Cuza 13, Craiova, RO-200585, Romania
ameliabd@yahoo.com
[2] University of Craiova, Software Engineering Department
Bvd.Decebal 107, Craiova, RO-200440, Romania
{badica_costin,popescu_elvira}@software.ucv.ro

Abstract. Logic wrappers combine logic programming paradigm with efficient XML processing for data extraction from HTML. In this note we show how logic wrappers technology can be adapted to cope with hierarchical data extraction. For this purpose we introduce *hierarchical logic wrappers* and illustrate their application by means of an intuitive example.

1 Introduction

The Web is extensively used for information dissemination to humans and businesses. For this purpose Web technologies are used to convert data from internal formats, usually specific to data base management systems, to suitable presentations for attracting human users. However, the interest has rapidly shifted to make that information available for machine consumption by realizing that Web data can be reused for various problem solving purposes including common tasks like searching and filtering, and also more complex tasks like analysis, decision making, reasoning and integration.

Two emergent technologies that have been put forward to enable automated processing of information available on the Web are semantic markup [14] and Web services [15]. Note however that most of the current practices in Web publishing are still based on the combination of traditional HTML – *lingua franca* for Web publishing [10], with server-side dynamic content generation from databases. Moreover, many Web pages are using HTML elements that were originally intended for use to structure content (e.g. those elements related to tables), for layout and presentation effects, even if this practice is not encouraged in theory.

Therefore, techniques developed in areas like information extraction, machine learning and wrapper induction are expected to play a significant role in tackling the problem of Web data extraction. Research in this area resulted in a large number of Web data extraction approaches that differ at least according to the task domain, the degree of automation and the technique used [6].

⋆ This is an extended version of the paper: Amelia Bădică, Costin Bădică, Elvira Popescu: Using Logic Wrappers to Extract Hierarchical Data from HTML. In: *Advances in Intelligent Web Mastering. Proc.AWIC'2007*, Fontainebleu, France. *Advances in Soft Computing* 43, 25–40, Springer, 2007.

J. Neves, M. Santos, and J. Machado (Eds.): EPIA 2007, LNAI 4874, pp. 43–52, 2007.

Our recent work in the area of Web data extraction was focused on combining logic programming with efficient XML processing [8]. The results were: i) definition of *logic wrappers* or L-wrappers for data extraction from the Web ([2]); ii) the development of a methodology for the application of L-wrappers on real problems ([2]); iii) design of efficient algorithms for semi-automated construction of L-wrappers ([1]); iv) efficient implementation of L-wrappers using XSLT technology ([3,7]).

So far we have only applied L-wrappers to extract relational data – i.e. sets of tuples or flat records. However, many Web pages contain hierarchically structured presentations of data for usability and readability reasons. Moreover, it is generally appreciated that hierarchies are very helpful for focusing human attention and management of complexity. Therefore, as many Web pages are developed by knowledgeable specialists in human-computer interaction design, we expect to find this approach in many designs of Web interfaces to data-intensive applications.

Let us consider for example a real-world document describing printer information from Hewlett Packard's Web site[1] for the 'HP Deskjet D4230 Printer'. The printer information is represented in multi-section two column HTML tables (see figure 1). Each row contains a pair (feature name, feature value). Consecutive rows represent related features that are grouped into feature classes. For example, there is a row with the feature name 'Print technology' and the feature value 'HP Thermal Inkjet'. This row has the feature class 'Print quality/technology'. So actually this table contains hierarchically structured triples (feature class, feature name, feature value) with some triples having identical feature classes.

In this note we propose an approach for utilizing L-wrappers to extract hierarchical data. The advantage would be that extracted data is suitably annotated to preserve its visual hierarchical structure, as found in the Web page. Basically we show how L-wrappers can be used for converting input HTML documents that were formatted for humans into well-structured output XML documents, formatted according to a well-defined schema. Further processing of this data would benefit from additional metadata to allow for more complex tasks, rather than simple searching and filtering by populating a relational database. For example, one can imagine the application of this technique to the task of ontology extraction, as ontologies are assumed to be natively equipped with the facility of capturing taxonomically structured knowledge. For example, as shown in figure 1, hierarchical data extraction can be used to learn both the attributes of a certain type of printer and also how these attributes can be semantically categorized into attributes related to the printing speed, attributes related to the printing technology and attributes related to the supported media.

In our opinion, the main achievement of this work is to show that standard logic programming (i.e. Prolog) when combined with XSLT technology and supported by appropriated tools (i.e. XSLT transformation engines) can be effectively used for hierarchical data extraction from the Web.

The paper is structured as follows. Section 2 briefly reviews L-wrappers. Section 3 presents our approach for extending L-wrappers to be able to extract hierarchically structured data. Section 4 introduces hierarchical logic wrappers (*HL-wrappers*) by means of an intuitive example. Formal definitions are also included. Section 5 briefly

---

[1] http://www.hp.com

HP Deskjet D2430 Printer (CB614A) - Product Specifications

Glossary of terms	
Glossary	Definitions of specifications and terms to assist in your buying process.
Speed/monthly volume	
Print speed, black (draft quality mode)	Up to 12.5 ppm
Print speed, black (normal quality mode)	Up to 2.1 ppm
Print speed, black (best quality mode)	Up to 1 ppm
Print speed, color (draft color mode)	Up to 12.5 ppm
Print speed, color (normal quality mode)	Up to 2.1 ppm
Monthly duty cycle	Up to 1000 pages
Print quality/technology	
Print technology	HP Thermal Inkjet
Print quality, black	Up to 1200 rendered dpi black
Print quality, color	Up to 4800 x 1200 optimized dpi color and 1200 input dpi
Cartridge colors	cyan, magenta, yellow
Ink types	Pigment-based, dye-based
Paper handling/media	
Paper tray (s), standard	1

Fig. 1. Hierarchically structured data describing HP printers

compares our approach with related works found in the literature. The last section concludes and points to future works.

2   L-Wrappers Background

L-wrappers are the basis of a new technology for constructing wrappers for data extraction from the Web, inspired by the logic programming paradigm. L-wrappers can be semi-automatically generated using inductive logic programming and efficiently implemented using XSLT transformation language. So far we have adopted a standard relational model by associating to each Web data source a set of distinct attributes. An L-wrapper for extracting relational data operates on a target Web document represented as a labeled ordered tree and returns a set of relational tuples of nodes of this document.

The use of L-wrappers assumes that Web documents are converted through a preprocessing stage to well-formed XML (i.e. to a tree) before they are processed for extraction. The document tree is then represented in logic programming as a set of facts.

An *L-wrapper* can be formally defined as a set of patterns represented as logic rules. Assuming that N is the set of document tree nodes and Σ is the set of HTML tags, an *L-wrapper* is a logic program defining a relation $extract(N_1, \ldots, N_k) \subseteq N \times \ldots \times N$ (see section 3 for an example). For each clause, the head is $extract(N_1, \ldots, N_k)$ and the body is a conjunction of literals in a set $(child, next, first, last, (tag_\sigma)_{\sigma \in \Sigma})$ of relations. Relations $child, next, first, last$ and tag_σ are defined as follows:

i) $child \subseteq N \times N$, $(child(P, C) = true) \Leftrightarrow (P$ is the parent of $C)$.
ii) $next \subseteq N \times N$, $(next(L, N) = true) \Leftrightarrow (L$ is the left sibling of $N)$.

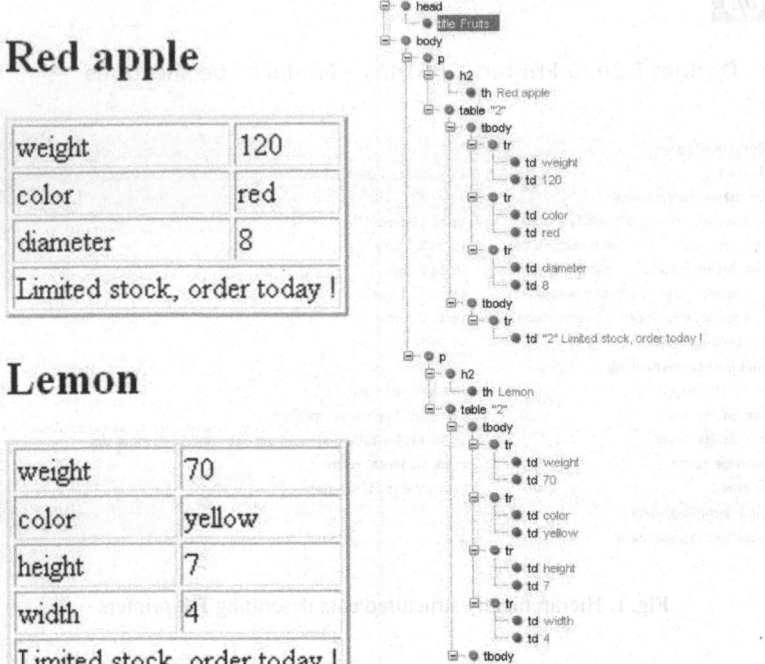

Fig. 2. A sample HTML document containing hierarchically structured data

iii) $first \subseteq N$, $(first(X) = true) \Leftrightarrow (X$ is the first child of its parent node).
iv) $last \subseteq N$, $(last(X) = true) \Leftrightarrow (X$ is the last child of its parent node).
v) $tag_\sigma \subseteq N$, $(tag_\sigma(N) = true) \Leftrightarrow (\sigma$ is the tag of node N).

Note that L-wrapper patterns can also be represented using graphs. Within this framework, a pattern is a directed graph with labeled arcs and vertices that corresponds to a rule in the logic representation. Arc labels denote conditions that specify the tree delimiters of the extracted data, according to the parent-child and next-sibling relationships (eg. is there a parent node ?, is there a left sibling ?, a.o). Vertex labels specify conditions on nodes (eg. is the tag label td ?, is it the first child ?, a.o). A subset of graph vertices is used for selecting the items for extraction.

An arc labeled 'n' denotes the "next-sibling" relation (*next*) while an arc labeled 'c' denotes the "parent-child" relation (*child*). As concerning vertex labels, label 'f' denotes "first child" condition (*first*), label 'l' denotes "last child" condition (*last*) and label denotes "equality with tag σ" condition (*tag$_\sigma$*). Small arrows attached to leaf nodes associate attributes with vertices used extracting items (see figure 3 for examples).

More details about L-wrappers including properties, construction and evaluation can be found in [2].

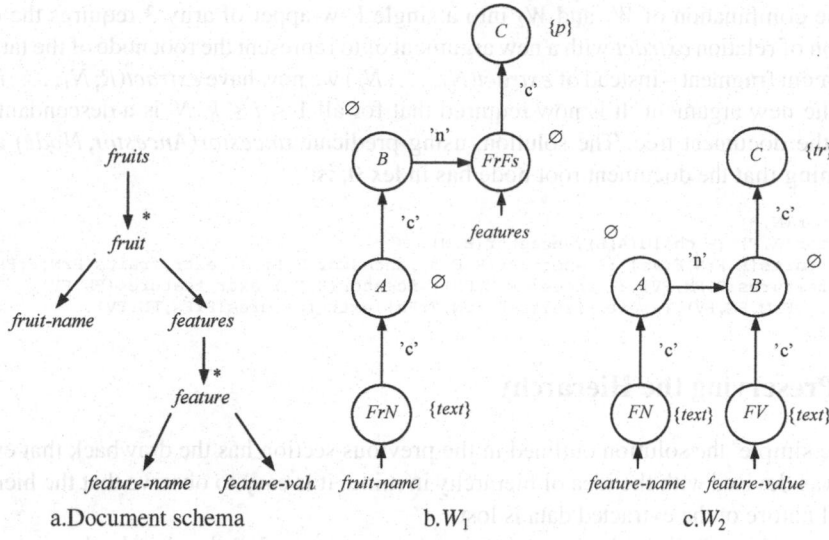

a.Document schema   b.W_1 c.W_2

Fig. 3. Wrappers for the document shown in figure 2

3   Extending L-Wrappers

L-wrappers assume a relational model by associating to each Web data source a set of distinct attributes. An L-wrapper operates on a target Web document represented as a labeled ordered tree and extracts a set of relational tuples of document nodes. See [2,3] for details about definition, implementation and evaluation of L-wrappers.

Let us now consider a very simple HTML document that contains hierarchical data about fruits (see figure 2). A fruit has a name and a sequence of features. Additionally, a feature has a name and a value. This is captured by the schema shown in figure 3a. Note that features can be fruit-dependent; for example, while an apple has an average *diameter*, a lemon has both an average *width* and an average *height*.

Abstracting the hierarchical structure of data, we can assume that the document shown in figure 2 contains triples: *fruit-name*, *feature-name* and *feature-value*. So, an L-wrapper of arity 3 would suffice to wrap this document.

Following the hierarchical structure of this data, the design of an L-wrapper of arity 3 for this example can be done in two stages: i) derive a wrapper W_1 for binary tuples (*fruit-name,list-of-features*); ii) derive a wrapper W_2 for binary tuples (*feature-name,feature-value*). Note that wrapper W_1 is assumed to work on documents containing a list of tuples of the first type (i.e. the original target document), while the wrapper W_2 is assumed to work on document fragments containing the list of features of a given fruit (i.e. a single table from the original target document).

For example, wrappers W_1 and W_2 can be designed as in figure 3 (following the graph notation introduced in [2,3]). Their expression in logic programming is:

```
extr_fruits(FrN,FrFs) :-
    tag(FrN,text),child(A,FrN),child(B,A),next(B,FrFs),child(C,FrFs),tag(C,p).
extr_features(FN,FV) :-
    tag(FN,text),tag(FV,text),child(A,FN),child(B,FV),next(A,B),child(C,B),tag(C,tr).
```

The combination of W_1 and W_2 into a single L-wrapper of arity 3 requires the extension of relation *extract* with a new argument of to represent the root node of the target document fragment – instead of *extract*$(N_1, \ldots, N_k)$ we now have *extract*$(R, N_1, \ldots, N_k)$, R is the new argument. It is now required that for all $1 \le i \le k$, N_i is a descendant of R in the document tree. The solution, using predicate *ancestor*(*Ancestor, Node*) and assuming that the document root node has index 0, is:

```
ancestor(N,N).
ancestor(A,N) :- child(A,B),ancestor(B,N).
extr_fruits(R,FrN,FrFs) :- ancestor(R,FrN),ancestor(R,FrFs),extr_fruits(FrN,FrFs).
extr_features(R,FN,FV) :- ancestor(R,FN),ancestor(R,FV),extr_features(FN,FV).
extract(FrN,FN,FV) :- extr_fruits(0,FrN,FrFs),extr_features(FrFs,FN,FV).
```

4 Preserving the Hierarchy

While simple, the solution outlined in the previous section has the drawback that even if it was devised with the idea of hierarchy in mind, it is easy to observe that the hierarchical nature of the extracted data is lost.

Assuming a Prolog-like execution engine, we can solve the drawback using the *findall* predicate. *findall*(X, G, Xs) returns the list Xs of all terms X such that goal G is true (it is assumed that X occurs in G). The solution and the result are shown below. We assume that i) the root node of the document has index 0 and ii) predicate *text*(*TextNode, Content*) is used to determine the content of a text node.

```
extract_all(Res) :-                        ?-extract_all(Res).
  extr_fruits_all(0,Res).                   Res = fruits(
extr_fruits_all(Doc,fruits(Res)) :-  [fruit(name('Red apple'),
  findall(                                     features(
    fruits(name(FrN),FrFs),                      [feature(name('weight'),value('120')),
    (extr_fruits(Doc,NFrN,NFrFs),                 feature(name('color'),value('red')),
     text(NFrN,FrN),                              feature(name('diameter'),value('8'))
     extr_features_all(NFrFs,FrFs)),              ])),
    Res).                                   fruit(name('Lemon'),
extr_features_all(Doc,features(Res)) :-features(
  findall(                                     [feature(name('weight'),value('70')),
    feature(name(FN),value(FV)),              feature(name('color'),value('yellow')),
    (extr_features(Doc,NFN,NFV),              feature(name('height'),value('7')),
     text(NFN,FN),text(NFV,FV)),              feature(name('width'),value('4'))
    Res).                                     ])])
```

Let us now formally introduce the concept of *hierarchical logic wrapper* or *HL-wrapper*. We generalize the data source schema from flat relational to hierarchical and we attach to this schema a set of L-wrappers.

Definition 1. *(Schema tree) Let* $\mathcal{W}$ *be a set denoting all vertices. A schema tree* S *is a directed graph defined as a quadruple* $\langle A, V, L, \lambda_a \rangle$ *s.t.* $V \subseteq \mathcal{W}$, $A \subseteq V \times V$, $L \subseteq V$ *and* $\lambda_a : A \to \{'*', '1'\}$. *The set of schema trees is defined inductively as follows:*

i) *For all* $n \ge 1$, *if* $u, v, w_i \in \mathcal{W}$ *for all* $1 \le i \le n$ *then* $S = \langle A, V, L, \lambda_a \rangle$ *such that* $V = \{u, v, w_1, \ldots, w_n\}$, $A = \{(u, v), (v, w_1), \ldots, (v, w_n)\}$, $L = \{w_1, \ldots, w_n\}$, $\lambda_a((u, v)) =' *'$, *and* $\lambda_a((v, w_1)) = \ldots = \lambda_a((v, w_n)) =' 1'$ *is a schema tree.*

ii) *If* $S = \langle A, V, L, \lambda_a \rangle$ *is a schema tree,* $n \ge 1$, $u \in L$ *and* $v, w_i \in \mathcal{W} \setminus V$ *for all* $1 \le i \le n$ *then* $S' = \langle A', V', L', \lambda_a' \rangle$ *defined as* $V' = V \cup \{v, w_1, \ldots, w_n\}$, $A' = A \cup \{(u, v), (v, w_1), \ldots, (v, w_n)\}$, $L' = (L \setminus \{u\}) \cup \{w_1, \ldots, w_n\}$, $\lambda_a'((u, v)) =' *'$,

$\lambda'_a((v, w_1)) = \ldots = \lambda'_a((v, w_n)) =' 1'$ and $\lambda'_a(a) = \lambda_a(a)$ for all $a \in A \setminus A'$ then S' is also a schema tree.

If Σ is a set of tag symbols denoting schema concepts and S is a schema tree then a pair consisting of a schema tree and a mapping of schema tree vertices to Σ is called a *schema*. For example, for the schema shown in figure 3a, $\Sigma = \{fruits, fruit, features, feature, feature-name, feature-value\}$ (note that labels $'1'$ are not explicitly shown on that figure). For an L-wrapper corresponding to the relational case ([2]) if D is the set of attribute names then $\Sigma = D \cup \{result, tuple\}$. Also it is not difficult to see that in an XML setting a schema would nicely correspond to the document type definition of the output document that contains the extracted data.

Definition 2. *(HL-wrapper) An HL-wrapper consists of a schema and an assignment of L-wrappers to split vertices of the schema tree. A vertex v of the schema tree is called split vertex if it has exactly one incoming arc labeled $'*'$ and $n \geq 1$ outgoing arcs labeled $'1'$. An L-wrapper assigned to v must have arity n to be able to extract tuples with n attributes corresponding to outgoing neighbors of v.*

An HL-wrapper for the example document considered in this paper consists of: i) schema shown in figure 3a, ii) L-wrapper W_1 assigned to the vertex labeled with symbol *fruit* and iii) L-wrapper W_2 assigned to vertex labeled with symbol *feature*.

Let us now outline a solution for HL-wrapper implementation using XSLT technology. For this purpose we combine the idea of the hierarchical Prolog implementation with the translation of L-wrappers to XSLT introduced in [3]. The resulted XSLT code is presented in the appendix.

Following [3], a single-pattern L-wrapper for which the pattern graph has n leaves, can be mapped to an $XSLT_0$ stylesheet consisting on $n + 1$ constructing rules ([5]). In our example, applying this technique to each of the wrappers W_1 and W_2 we get three rules for W_1 (start rule, rule for selecting *fruit name* and rule for selecting *features*) and three rules for W_2 (start rule, rule for selecting *feature name* and rule for selecting *feature value*). Note that additionally to this separate translation of W_1 and W_2 we need to assure that W_2 selects feature names and feature values from the document fragment corresponding to a given fruit – i.e. the document fragment corresponding to the *features* attribute of wrapper W_1). This effect can be achieved by including the body of the start rule corresponding to wrapper W_2 into the body of the rule for selecting features, in-between tags `<features>` and `</features>` (see appendix). Actually this operation corresponds to a join of wrappers W_1 and W_2 on the attribute *features* (assuming L-wrappers are extended with an argument representing the root of the document fragment to which they are applied – see section 3).

5   Related Work

Two earlier works on hierarchical data extraction from the Web are hierarchical wrapper induction algorithm Stalker [12] and visual wrapper generator Lixto [4].

Stalker ([12]) uses a hierarchical schema of the extracted data called *embedded catalog formalism* that is similar to our approach (see section 3). However, the main difference is that Stalker abstracts the document as a string rather than a tree and therefore their approach is not able to benefit from existing XML processing technologies. Extraction rules of Stalker are based on a special type of finite automata called *landmark automata*, rather than logic programming – as our L-wrappers.

Lixto [4] is a software tool that uses an internal logic based extraction language – Elog. While in Elog a document is abstracted as a tree, the differences between Elog and L-wrappers are at least two fold:

i) L-wrappers are only devised for the extraction task and they use a classic logic programming approach – an L-wrapper can be executed without any modification by a Prolog engine. Elog was devised for both crawling and extraction and has a customized logic programming-like semantics, that is more difficult to understand. This basically means that an L-wrapper can be implemented using any available XSLT transformation engine (many are open-source), while ELog is useless in practice, unless the customized Elog engine is either purchased from Lixto Software GmbH [11] or re-implemented from scratch.

ii) L-wrappers are efficiently implemented by translation to XSLT, and for Elog the implementation approach is different – a custom interpreter has been devised from scratch. Moreover, while the main motivation behind the specialized Elog engine were to increase performance, we were not able to find implementation details and results of Elog; only theoretical results are available [9].

6   Concluding Remarks

In this note we introduced HL-wrappers – an extension of L-wrappers to extract hierarchical data from Web pages. The main benefit is that all results and techniques already derived for L-wrappers can be applied to the hierarchical case. As future work we plan: i) to do an experimental evaluation of HL-wrappers (but, as they are based on L-wrappers, we have reasons to believe that the results will be good); ii) to incorporate HL-wrappers into an information extraction tool.

References

1. Bădică, C., Bădică, A., Popescu, E.: A New Path Generalization Algorithm for HTML Wrapper Induction. In: Last, M., Szczepaniak, P. (eds.) AWIC 2006. Proc. of the 4th Atlantic Web Intelligence Conference, Israel. Studies in Computational Intelligence Series, vol. 23, pp. 10–19. Springer, Heidelberg (2006)
2. Bădică, C., Bădică, A., Popescu, E., Abraham, A.: L-wrappers: concepts, properties and construction. A declarative approach to data extraction from web sources. Soft Computing - A Fusion of Foundations, Methodologies and Applications 11(8), 753–772 (2007)
3. Bădică, C., Bădică, A., Popescu, E.: Implementing Logic Wrappers Using XSLT Stylesheets. In: ICCGI 2006. International Multi-Conference on Computing in the Global Information Technology, Romania, IEEE Computer Society, Los Alamitos (2006), http://doi.ieeecomputersociety.org/10.1109/ICCGI.2006.127
4. Baumgartner, R., Flesca, S., Gottlob, G.: The Elog Web Extraction Language. In: Nieuwenhuis, R., Voronkov, A. (eds.) LPAR 2001. LNCS (LNAI), vol. 2250, pp. 548–560. Springer, Heidelberg (2001)

5. Bex, G.J., Maneth, S., Neven, F.: A formal model for an expressive fragment of XSLT. Information Systems 27(1), 21–39 (2002)
6. Chang, C.-H., Kayed, M., Girgis, M.R., Shaalan, K.: A Survey of Web Information Extraction Systems. IEEE Transactions on Knowledge and Data Engineering 18(10), 1411–1428 (2006)
7. Clark, J.: XSLT Transformation (XSLT) Version 1.0, W3C Recommendation (November 16, 1999), http://www.w3.org/TR/xslt
8. Extensible Markup Language. http://www.w3.org/XML/
9. Gottlob, G., Koch, C.: Logic-based web information extraction. ACM SIGMOD Record 33(2), 87–94 (2004)
10. HyperText Markup Language. http://www.w3.org/html/
11. Lixto Software GmbH: http://www.lixto.com/
12. Muslea, I., Minton, S., Knoblock, C.: Hierarchical Wrapper Induction for Semistructured Information Sources. Journal of Autonomous Agents and Multi-Agent Systems 4(1-2), 93–114 (2001)
13. Oxygen XML Editor. http://www.oxygenxml.com/
14. Semantic Web. http://www.w3.org/2001/sw/
15. Web Services. http://www.w3.org/2002/ws/

Appendix: XSLT Code of the Sample Wrapper

The XSLT code of the sample HL-wrapper for the "fruits" example is shown below:

```
<?xml version="1.0" encoding="UTF-8"?>
<xsl:stylesheet xmlns:xsl="http://www.w3.org/1999/XSL/Transform" version="1.0">
    <xsl:template match="html">
        <fruits><xsl:apply-templates select=
            "//p/*/preceding-sibling::*[1]/*/text()" mode="select-fruit-name"/></fruits>
    </xsl:template>
    <xsl:template match="node()" mode="select-fruit-name">
        <xsl:variable name="var-fruit-name" select="."/>
        <xsl:apply-templates mode="select-features" select=
            "parent::*/parent::*/following-sibling::*[position()=1]">
            <xsl:with-param name="var-fruit-name" select="$var-fruit-name"/>
        </xsl:apply-templates>
    </xsl:template>
    <xsl:template match="node()" mode="select-features">
        <xsl:param name="var-fruit-name"/>
        <xsl:variable name="var-features" select="."/>
        <fruit><name><xsl:value-of select="normalize-space($var-fruit-name)"/></name>
        <features><xsl:apply-templates
            select="$var-features//tr/*/preceding-sibling::*[1]/text()"
            mode="select-feature-name"></xsl:apply-templates></features></fruit>
    </xsl:template>
    <xsl:template match="node()" mode="select-feature-name">
        <xsl:variable name="var-feature-name" select="."/>
        <xsl:apply-templates mode="select-feature-value" select=
            "parent::*/following-sibling::*[position()=1]/text()">
            <xsl:with-param name="var-feature-name"
                select="$var-feature-name"/></xsl:apply-templates>
    </xsl:template>
    <xsl:template match="node()" mode="select-feature-value">
        <xsl:param name="var-feature-name"/>
        <xsl:variable name="var-feature-value" select="."/>
        <feature><name><xsl:value-of select="normalize-space($var-feature-name)"/>
            </name><value><xsl:value-of select=
                "normalize-space($var-feature-value)"/></value></feature>
    </xsl:template>
</xsl:stylesheet>
```

For wrapper execution we can use any of the available XSLT transformation engines. In our experiments we have used Oxygen XML editor, a tool that incorporates some of these engines [13]. The result of applying the wrapper to the "fruits" example document is the XML document shown below:

```
<?xml version="1.0" encoding="utf-8"?>
<fruits>
    <fruit>
        <name>Red apple</name>
        <features>
            <feature>
                <name>weight</name>
                <value>120</value>
            </feature>
            <feature>
                <name>color</name>
                <value>red</value>
            </feature>
            <feature>
                <name>diameter</name>
                <value>8</value>
            </feature>
        </features>
    </fruit>
    <fruit>
        <name>Lemon</name>
        <features>
            <feature>
                <name>weight</name>
                <value>70</value>
            </feature>
            <feature>
                <name>color</name>
                <value>yellow</value>
            </feature>
            <feature>
                <name>height</name>
                <value>7</value>
            </feature>
            <feature>
                <name>width</name>
                <value>4</value>
            </feature>
        </features>
    </fruit>
</fruits>
```

Note that the DTD of this document closely corresponds to the extraction schema presented in figure 3a.

Relaxing Feature Selection in Spam Filtering by Using Case-Based Reasoning Systems

J.R. Méndez[1], F. Fdez-Riverola[1], D. Glez-Peña[1], F. Díaz[2], and J.M. Corchado[3]

[1] Dept. Informática, University of Vigo, Escuela Superior de Ingeniería Informática,
Edificio Politécnico, Campus Universitario As Lagoas s/n, 32004, Ourense, Spain
{moncho.mendez,riverola,dgpena}@uvigo.es
[2] Dept. Informática, University of Valladolid, Escuela Universitaria de Informática,
Plaza Santa Eulalia, 9-11, 40005, Segovia, Spain
fdiaz@infor.uva.es
[3] Dept. Informática y Automática, University of Salamanca,
Plaza de la Merced s/n, 37008, Salamanca, Spain
corchado@usal.es

Abstract. This paper presents a comparison between two alternative strategies for addressing feature selection on a well known case-based reasoning spam filtering system called SPAMHUNTING. We present the usage of the k more predictive features and a percentage-based strategy for the exploitation of our amount of information measure. Finally, we confirm the idea that the percentage feature selection method is more adequate for spam filtering domain.

1 Introduction and Motivation

With the boom of Internet, some advanced communication utilities have been introduced in our society in order to improve our quality of life. Nowadays, there is no doubt of the large utility of some services like the World Wide Web (WWW), the Instant Messaging (IM) and Internet Relay Chat (IRC) tools as well as the e-mail possibilities. However, due to the great audience of these recent technologies and the lack of international legislation for their regulation, Internet has also been used as the basis for some illegal activities.

In this context, spam can be viewed as a generic concept referred to the usage of Internet technologies in order to promote illegal/fraudulent products and/or disturb Internet users. The low cost associated to these technologies is the main attraction for the spammers (the senders of spam messages). The most common forms of spam are (*i*) the insertion of advertisements in blog comments, (*ii*) the delivery of announce- ment mobile messages (SMS), (*iii*) the usage of advertising bots in IM or IRC channels, (*iv*) the delivery of advertisement messages on newsgroups and (*v*) the dis- tribution of spam e-mail.

Since the most common and oldest form of spam is the usage of e-mail for disturb- ing Internet users, most of the efforts on spam filtering have been focused in this direction. In the same way, as the most extensive form of spam is the delivery of spam messages, this work has also been bounded to this area.

Previous research work on spam filtering have shown the advantages of disjoint feature selection in order to affectively capture the information carried by e-mails

J. Neves, M. Santos, and J. Machado (Eds.): EPIA 2007, LNAI 4874, pp. 53–62, 2007.
© Springer-Verlag Berlin Heidelberg 2007

[1, 2, 3]. These methods are based on representing each message using only the words that better summarize its content. This kind of message representation has been successfully used in combination with some latest technologies in order to build a Case Based Reasoning (CBR) spam filtering system called SPAMHUNTING [3]. The usage of a poor feature selection method can significantly reduce the performance of any good classifier [1].

Traditionally, the feature selection stage has been carried out from the training dataset by using measures of global performance for each possible feature (term). The most well-known metrics used are Information Gain (IG), Mutual Information (MI), Document Frequency (DF) and Chi square test (χ^2) [1]. Features can be selected either using a threshold over the metric or a best k number of features with k established before training the model [4]. The latest one is the most common in spam filtering domain [1, 4, 5, 6, 7].

As showed in [1, 2], when a new message is received SPAMHUNTING extracts its tokens and computes a measure of the Achieved Information for each term included into the message, $AI(t)$. Then, terms are sorted descending by its $AI(t)$ score. Finally, starting from the n terms extracted from the message, only the first k terms having an amount of information greater than a percentage p of the total information achieved by using all terms. During our past experiments, we had found 60% as a good value for the percentage p.

Taking into consideration the SPAMHUNTING feature selection technique, a static number k of features could be used instead of a percentage selection of attributes. In this paper we test the benefits of using a percentage feature selection approach instead of selecting a pre-established number of features for every message. As these alternatives have not been compared yet, we are interested in a deep analysis of performance in order to select the most appropriate way for guaranteeing the most accurate results.

The rest of the paper is organized as follows: Section 2 introduces previous work on feature selection used in conjunction with CBR systems. Section 3 shows the available corpus for empirical evaluation as well as some miscellaneous configuration details used during our experimental stage. Section 4 presents the experimental protocol and the results of the experiments carried out, discussing the major findings. Finally, Section 5 exposes the main conclusions reached by the analysis of the experiments carried out.

2   Feature Selection on Spam Filtering CBR Systems

Recent research works have shown that CBR systems are able to outperform some classical techniques in spam filtering domain [1, 2, 3, 6, 7]. Moreover, some previous works state that CBR systems work well for disjoint concepts as spam (spam about *porn* has little in common with spam offering *rolex*) whereas older techniques try to learn a unified concept description [6]. Another important advantage of this approach is the ease with which it can be updated to tackle the *concept drift* problem and the changing environment in the anti-spam domain [7]. Due to the relevance of these works, subsection 2.1 presents different feature selections strategies followed by some well-known CBR systems used for spam filtering. Subsection 2.2 explains the approaches for addressing feature selection that will be compared in this work.

2.1 Previous Work on Feature Selection

Motivated for the relevance of some previous work on spam filtering CBR systems [1, 2, 3, 6, 7], we had analyzed the weaknesses and strengths of several feature selection methods implemented by these successful classifiers.

In the work of [7] a new CBR spam filter was introduced. One of the most relevant features included in ECUE, (*E-mail Classification Using Examples*), was the ability for dynamically updating the knowledge. The feature selection approach used in this system has been designed as a classical spam filtering technique, by using the 700 word-attributes from training corpus having the highest IG score. The IG measure for a term, t, is defined by Expression (1).

$$IG(t) = \sum_{c \in \{spam, legitimate\}} P(t \wedge c) \cdot \log \frac{P(t \wedge c)}{P(t) \cdot P(c)}$$
(1)

where $P(t \wedge c)$ is the frequency of the documents belonging to the category c (legitimate or spam) that contains the term t, $P(t)$ represents the amount of documents having the term t and, finally, $P(c)$ is the frequency of documents belonging to the class c. Moreover, ECUE uses a similarity retrieval algorithm based on Case Retrieval Nets (CRN) [8], an efficient implementation of a k-nearest neighbourhood strategy. Finally, the system uses a unanimous voting strategy to determine whether a new e-mail is spam or not. It also defines a combination of two Case-Based Editing techniques known as Blame Based Noise Reduction (BBNR) and Conservative Redundancy Reduction (CRR) [9].

The ECUE system represents the evolution from a previous Odds Ratio (OR) based filter [6]. The main difference between both systems is the feature selection method used. This fact supports the idea of the relevance about feature selection methods in spam filtering domain. The feature selection method used in the oldest version of ECUE is based on selecting the 30 terms having the highest OR measure for each category. The OR measure for a term, t, in the class c (spam or legitimate) is computed as showed in Expression (2).

$$OR(t,c) = \frac{P(t \wedge c) \cdot \left[1 - P(t \wedge \overline{c})\right]}{P(t \wedge \overline{c}) \cdot \left[1 - P(t \wedge c)\right]}$$
(2)

where $P(t \wedge c)$ represents the frequency of the documents belonging to category c (legitimate or spam) that contains the term t, and $P(t \wedge \overline{c})$ stands for the frequency of documents containing the term t that are not included in category c.

Another relevant system can be found in [2], where the authors present a lazy learning hybrid system for accurately solving the problem of spam filtering. The model, known as SPAMHUNTING, follows an Instance-Based Reasoning (IBR) approach. The main difference between SPAMHUNTING and other approaches can be found during the feature selection stage. Instead of a global feature selection, SPAMHUNTING addresses this phase as an independent process for each email. Therefore, when a new message e arrives, SPAMHUNTING extracts its terms $\{t_i \in e\}$ and computes for each of them, the Amount of Information (AI) achieved when used for the representation of the target e-mail e [1]. This estimation is made by means of using Expression (3).

$$AI(t,e) = P(t \wedge e) \cdot \left[\frac{|P(spam) \cdot P(t \wedge spam) - P(legitimate) \cdot P(t \wedge legitimate)|}{P(t)} \right] \quad (3)$$

where $P(t,e)$ represents the frequency of the term t in the message e, $P(spam)$ and $P(legitimate)$ are in that order, the frequency of spam and legitimate messages stored in the system memory, $P(t,spam)$ and $P(t,legitimate)$ stand for the frequency of spam and legitimate messages stored the system memory having the term t and, finally, $P(t)$ represents the frequency of instances from CBR knowledge containing the term t. The AI estimation summarizes the information of the relevance of the term t in the message e (by using $P(t,e)$) as well as the global relevance of the term in the whole corpus (the remaining expression).

Each message is represented using a set of attributes that better describe its content as a set of keywords describe a research paper. For this purpose we had defined a measure for the total amount of information of a message, $AI(e)$, computed as the sum of the AI measure for each of the terms extracted from e. This measure can be computed as Expression (4) shows.

$$AI(e) = \sum_{t_i \in e} AI(t_i, e) \quad (4)$$

where $AI(t_i, e)$ stands for the amount of information of the term t_i in the message e defined in Expression (3). The selection of features for the message e has been defined in [2] as the list of terms t_i with the highest $AI(t_i, e)$ rate having an amount of achieved information greater than a certain percentage p of the total amount of information $AI(e)$. Expression (5) demonstrates how to carry out the feature selection process for a given message e, $FS(e)$.

$$FS(e) = \left\{ t_i \in e \middle| \begin{array}{c} \left\{ AI(t_j,e) \geq AI(t_k,e) \wedge t_k \in FS(e) \right\} \rightarrow t_j \in FS(e) \\ \\ \nexists FS' \subset FS(e) \middle| \sum_{t_j \in FS'} AI(t_j,e) > \frac{p}{100} \cdot \sum_{t_k \in e} AI(t_k,e) \end{array} \right\} \quad (5)$$

where $AI(t_k, e)$ stands for the achieved information score of the term t_k in the message e (showed in Expression (3)). Previous work using this approach has shown good performance results when $p=60$ [1, 2].

According to the disjoint feature selection method used by SPAMHUNTING, it comprises an Enhanced Instance Retrieval Network known as EIRN as a primary way of managing knowledge [10]. This indexing structure guarantees the representation of disjoint information and implements an efficient similarity metric defined as the number of relevant features found in a set of given messages. Similarly with ECUE, the reuse of retrieved messages has been defined as a simple unanimous voting strategy

[7]. Finally, the revision stage comprises the usage of a measure of the quality of the available information for classifying the target message [11].

Once the ground works has been exposed, next subsection presents a different approach for carrying out the feature selection using the background ideas of our successful SPAMHUNTING system process.

2.2 Best k Feature Selection on SPAMHUNTING

This subsection presents a new approach for feature selection in our previous successful SPAMHUNTING system. This proposal for feature selection will be compared with the original (showed in Expression (5)) during the experimentation stage.

The SPAMHUNTING feature selection stage has been addressed as an independent process executed every time a message is received. The main instrument for this computation is a percentage p of the global AI from the target message e, $AI(e)$. Despite this approach is reasonable and has led to obtain good results, authors do not justify the requirement of the percentages.

In this work we plain to use a fixed number of attributes (k) for representing each email. In order to do this, we will select these attributes holding the greatest AI rate, $AI(t_i, e)$, for each given message e. In our new proposal, the selected terms for a message e will be computed as showed in Expression (6)

$$FS(e) = \left\{ t_i \in e \middle| \left[\begin{array}{c} numTerms(e) \geq k \rightarrow size(FS(e)) = k \\ numTerms(e) \geq k \rightarrow size(FS(e)) = numTerms(e) \\ AI(t_i, e) \geq AI(t_j, e) \wedge t_j \in FS(e) \rightarrow t_i \in FS(e), \; \forall t_j, t_i \in e \end{array} \right] \right\} \quad (6)$$

where $FS(e)$ represents the list of selected features for the message e, $size(x)$ is the length of x and $numTerms(e)$ stands for the amount of terms extracted from the target message e.

This paper contains a brief empirical analysis of the analyzed approaches shown in Expressions (5) and (6). The final goal is to determine the best approach for spam filtering CBR systems or identify the circumstances that improve their performance. Next section presents the available corpora for spam filtering, the main criterions for the dataset selection and some miscellaneous configuration issues used during the experiments.

3 Available Corpus and Experimental Setup

Due to privacy issues (mainly from the usage of legitimate messages), the spam filtering domain presents some difficulties in order to guarantee the relevance of the results achieved. For addressing this difficulty, only publicly available datasets should be used. This section contains a summary of the available dataset repositories for spam filtering. Moreover, it includes a discussion about the decisions adopted for the experimental datasets and several SPAMHUNTING configuration details used during our experimentation.

There are several publicly available corpora of e-mails including LingSpam, Junk E-mail, PU corpuses, DivMod or SpamAssassin[1]. Most of them, according with the RFC (*Request for Comments*) 822, are distributed as they were sent through Internet. Although some researchers and students from our University had built a corpus called SING, most of the messages have been qualified as private by its owners and we are not authorized to publish it. This corpus can only be used for parameter optimization purposes while the results should be generated using publicly available datasets. Moreover, there are some corpuses shared after some preprocessing steps that can cause the loosing of some information. Table 1 shows a brief description of all of them focusing on preprocessing and distribution issues.

Table 1. Comparison of some publicly available datasets

Name	Public	Message amount	Contains dates	Legitimate percentage	Spam percentage	Distribution - preprocessing
Ling-Spam	YES	2893	NO	83.3	16.6	tokens
PU1	YES	1099	NO	56.2	43.8	token ids
PU2	YES	721	NO	80	20	token ids
PU3	YES	4139	NO	51	49	token ids
PUA	YES	1142	NO	50	50	token ids
SpamAssassin	YES	9332	YES	84.9	15.1	RFC 822
Spambase	YES	4601	NO	39.4	60.6	feature vectors
Junk-Email	YES	673	YES	0	100	XML
Bruce Guenter	YES	171000	YES	0	100	RFC 822
Judge	YES	782	YES	0	100	RFC 822
Divmod	YES	1247	YES	0	100	RFC 822
Grant Taylor	YES	2400	YES	0	100	RFC 822
SING	NO	20130	YES	69.7	39.3	XML y RFC 822

As we can see from Table 1, most of the available corpus do not contain legitimate messages and should be discarded. Moreover, in order to use some preprocessing issues introduced in [12], an RFC 822 distribution of the corpus need to be selected. Due to the above mentioned reasons, we have selected the SpamAssassin corpus containing 9332 different messages from January 2002 up to and including December 2003.

Following the findings of [12], hyphens and punctuation marks are not removed during the tokenizing stage. We have used "\\S+" as the regular expression for determining a token on the message text. This expression means that we define tokens as character lists containing the greatest amount of non white space consecutive characters. Moreover, attending to [12], we have executed a stopword removal process over the identified tokens [13].

Finally, continuous updating strategies have been used during the experiments. Every time a message is classified, a new instance is generated using the e-mail content and the generated solution. Next section presents the experimental design and the benchmark results.

---

[1] Available at http://www.spamassassin.org/publiccorpus/

4 Experimental Protocol and Results

The final goal of our experiments is to measure the differences between the feature selection strategies showed in Expressions (5) and (6). This section presents the experimental methodology and the results achieved from the tests carried out. All the experiments have been carried out using a 10-fold stratified cross-validation [14] in order to increase the confidence level of the results obtained.

We have executed our SPAMHUNTING system using Expression (5) applying the values 30, 40, 50, 60 and 70 for p and Expression (6) using values from 4 to 14 stepping 2 for k. Then, we have selected the best of them using the area under the Receiver Operating Characteristic (ROC) curve [15]. ROC curves have been computed using the amount of spam votes as target measure for the discrimination between spam and legitimate messages. In order to provide information about the theoretical maximum performance reachable by using each analyzed configuration, we have also computed the sensitivity and specificity for the best classification criterion. Table 2 summarizes the area under ROC curves for the different model configurations and the theoretical maximum performances achieved.

As we can see from Table 2, the difference between the analyzed configurations of each feature selection strategy is very small. Although there are some configurations able to achieve some important values on singular measures, we have selected those achieving the highest global performance level (evaluated through the area under the ROC curve). In this sense, $p=40$ and $k=12$ have shown to be the best configurations for the approach.

After the preliminary analysis, we have compared the results achieved by using the best configuration for the two selected proposals. This task has been carried out using a complete ROC curve analysis [11, 15]. Then, we have carried out a comparison of the areas under ROC curves and executed a statistic test in order to determine the significance of the differences found. Figure 1 shows the ROC curves plot for the best configuration of the analyzed approaches.

As we can see from Figure 1, there is a difference between the analyzed strategies. As we have previously mentioned, we have executed a statistic test considering the

Table 2. Preliminary comparison of ROC curves achieved by the SPAMHUNTING system

	Configuration	Area under ROC	Sensitivity	Specificity	Best criterion
PERCENTAGE SELECTION	30%	0.985	96.9	98.5	>0,0714
	40%	0.987	97.4	98.9	>0.0417
	50%	0.984	96.9	99.3	>0.0426
	60%	0.983	96.8	99.4	>0.0417
	70%	0.980	96.1	99.7	>0.0741
BEST k SELECTION	4	0.990	97.4	97.6	>0.2
	6	0.989	97.7	98.3	>0.1644
	8	0.992	97.8	98.9	>0.1648
	10	0.992	97.9	99.2	>0.1603
	12	0.994	98.7	98.7	>0.0545
	14	0.993	98.4	99.1	>0.0972

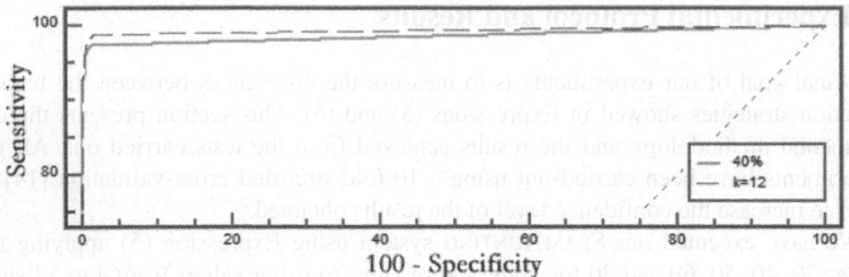

Fig. 1. ROC curves plot for the best configuration of the two feature selection approaches

equality of the areas under ROC curves showed as the null hypothesis. As the computed *p*-value is lower than 0.01 (*p*=0.001), the null hypothesis must be discarded and we can state that, from a statistically point of view, there is a very significant difference between both methods for feature selection.

In order to provide a more detailed analysis, we have computed positive Likelihood Ratio (+LR), Negative Likelihood ratio (-LR) and δ index [16] for the best cut values of the analyzed proposals. These measures have been included in Table 3.

Table 3. ROC analysis for the selected feature selection strategies

	Percentage 40%	Best k=12
+LR	86.76	73.07
-LR	0.03	0.01
δ index	2.76	2.91

As we can realize from Table 3, the detection of a legitimate message has a great confidence level when best *k* feature selection is used, whereas the percentage method presents a great confidence level when detecting spam messages (reducing the false positive error rate). Moreover, analyzing the δ index we can see that the best *k* feature selection alternative can theoretically achieve a better amount of correct classifications.

Finally, six well-known metrics [17] have been used in order to evaluate the performance (efficacy) of both models: percentage of correct classifications (%OK), percentage of False Positives (%FP), percentage of False Negatives (%FN), spam *recall*, spam *precision* and *total cost ratio* (TCR) with three different cost values. In order to achieve a down-to-earth approximation of the performance level, these measures has been computed considering the results of the unanimous voting strategy instead of the best criterion identified by using the ROC curve analysis. Table 4 shows a comparison of both feature selection techniques using percentages of correct/fail classifications, precision, recall and TCR measures.

As we can see from Table 4, despite the differences between best *k* and percentage feature selection are small, they support the conclusions achieved during the ROC analysis. Percentage feature selection is suitable to address FP error reduction while best *k* feature selection can increase the percentage of correct classifications.

Finally, keeping in mind the relevance of false positive errors on the target domain, TCR measure achieved for the highest cost factor (λ=999) supports the convenience of the percentage feature selection.

Table 4. Performance measures for the analyzed feature selection strategies in SPAMHUNTING

	Classification percentages					TCR		
	%OK	%FP	%FN	Recall	Precision	λ=1	λ=9	λ=999
Percentage 40%	97.51	0.09	2.39	0.90	0.99	10.54	8.70	4.76
Best k=12	97.82	0.10	2.09	0.92	0.99	12.22	8.84	2.25

5   Conclusions and Further Work

This work has presented, discussed and analyzed two different proposals for addressing the feature selection stage on a previous successful spam filtering CBR system. Moreover, we have discussed the main differences between several feature selection schemes applied in the context of the target domain. We had also analyzed the current SPAMHUNTING feature selection heuristic in order to find new improvements. Finally, we have executed a benchmark finding the percentage approach as the most suitable alternative for spam filtering domain.

The performance of compared feature selection strategies are based on an analysis of the discrimination capabilities of each term. When the most relevant terms have lower discerning capabilities, the percentage selection strategy is able to select a greater amount of terms in order to guarantee the existence of a minimum amount of information for accomplishing the classification. Moreover, the best k feature selection approach can not guarantee a minimum of information.

Spammers often include terms that can be understood as reliable signs for classifying their messages as legitimate. These messages present a large amount of information because they combine a greater number of legitimate and spam signs. Percentage selection is able to include both spam and legitimate signs for an adequately classification avoiding the possibilities of finding disagreement between the terms. Moreover, best k feature selection will include only the legitimate evidences storing these inconsistencies in the system memory. Although the global performance can increase, those inconsistencies could generate a greater amount of FP errors. These findings are supported by the +LR and –LR tests carried out and the TCR comparisons for λ=9.

We believe in the relevance of a disjoint and percentage feature selection strategy for the spam filtering domain. In this sense future works should be focused in improving our feature selection methods [1, 2], using semantic information and addressing the noise reduction during the feature selection stage.

References

1. Méndez, J.R., Fdez-Riverola, F., Iglesias, E.L., Díaz, F., Corchado, J.M.: A Comparative Performance Study of Feature Selection Methods for the Anti-Spam Filtering Domain. In: Proc. of the 6th Industrial Conference on Data Mining, pp. 106–120 (2006)
2. Méndez, J.R., Fdez-Riverola, F., Díaz, F., Iglesias, E.L., Corchado, J.M.: Tracking Concept Drift at Feature Selection Stage in SpamHunting: an Anti-Spam Instance-Based Reasoning System. In: Proc. of the 8th European Conference on Case-Based Reasoning, pp. 504–518 (2006)
3. Fdez-Riverola, F., Iglesias, E.L., Díaz, F., Méndez, J.R., Corchado, J.M.: SpamHunting: An Instance-Based Reasoning System for Spam Labeling and Filtering. Decision Support Systems (in press, 2007) http://dx.doi.org/10.1016/j.dss.2006.11.012

4. Méndez, J.R., Corzo, B., Glez-Peña, D., Fdez-Riverola, F., Díaz, F.: Analyzing the Performance of Spam Filtering Methods when Dimensionality of Input Vector Changes. In: Proc. of the 5th International Conference on Data Mining and Machine Learning (to appear, 2007)
5. Metsis, V., Androutsopoulos, I., Paliouras, G.: Spam Filtering with Naive Bayes – Which Naive Bayes? In: Proc. of the 3rd Conference on Email and Anti-Spam, pp. 125–134 (2006), http://www.ceas.cc
6. Cunningham, P., Nowlan, N., Delany, S.J., Haahr, M.: A Case-Based Approach to Spam Filtering that Can Track Concept Drift. In: Ashley, K.D., Bridge, D.G. (eds.) ICCBR 2003. LNCS, vol. 2689, Springer, Heidelberg (2003)
7. Delany, S.J., Cunningham, P., Coyle, L.: An Assessment of Case-base Reasoning for Spam Filtering. In: AICS 2004. Proc. of Fifteenth Irish Conference on Artificial Intelligence and Cognitive Science, pp. 9–18 (2004)
8. Lenz, M., Burkhard, H.D.: Case Retrieval Nets: Foundations, properties, implementation and results. Technical Report: Humboldt University, Berlin (1996)
9. Delany, S.J., Cunningham, P.: An Analysis of Case-Based Editing in a Spam Filtering System. In: Proceedings of the 7th European Conference on Case-Based Reasoning, pp. 128–141 (2004)
10. Fdez-Riverola, F., Iglesias, E.L., Díaz, F., Méndez, J.R., Corchado, J.M.: Applying Lazy Learning Algorithms to Tackle Concept Drift in Spam Filtering. ESWA: Expert Systems With Applications 33(1), 36–48 (2007)
11. Méndez, J.R., González, C., Glez-Peña, G., Fdez-Riverola, F., Díaz, F., Corchado, J.M.: Assessing Classification Accuracy in the Revision Stage of a CBR Spam Filtering System. Lecture Notes on Artificial Intelligence (to appear, 2007)
12. Méndez, J.R., Iglesias, E.L., Fdez-Riverola, F., Díaz, F., Corchado, J.M.: Tokenising, Stemming and Stopword Removal on the Spam Filtering Domain. In: Proc. of the 11th Conference of the Spanish Association for Artificial Intelligence, pp. 449–458 (2005)
13. Baeza-Yates, R., Ribeiro-Neto, B.: Modern Information Retrieval. Addison-Wesley, Reading (1999)
14. Kohavi, R.: A study of cross-validation and bootstrap for accuracy estimation and model selection. In: IJCAI 1995. Proceedings of the 14th International Joint Conference on Artificial Intelligence, pp. 1137–1143 (1995)
15. Egan, J.P.: Signal Detection Theory and ROC Analysis. Academic Press, New York (1975)
16. Hasselband, V., Hedges, L.: Meta-analysis of diagnostics test. Psychological Bulletin 117, 167–178 (1995)
17. Androutsopoulos, I., Paliouras, G., Michelakis, E.: Learning to Filter Unsolicited Commercial E-Mail. Technical Report 2004/2, NCSR "Demokritos" (2004)

Gödel and Computability*

Luís Moniz Pereira

Centro de Inteligência Artificial – CENTRIA
Universidade Nova de Lisboa, Portugal

Abstract. We discuss the influence of Gödel and his results on the surfacing of the rigorous notion of computability afforded by Turing. We also address the debate about the limits of Artificial Intelligence spurned by Roger Penrose, on the basis of Gödel's theorems, and the views of Gödel himself. We conclude by touching upon the use of logic as a tool with which to approach the description of mind.

Hilbert's Questions

The research programme that David Hilbert presented to the international congress of mathematics in Bologna in 1928 was essentially an extension of the work he had initiated in the 1890s. He was not intent on answering the question that Frege and Russell had posed, about what mathematics *really was*. In that regard, he showed himself to be less philosophical and less ambitious. On the other hand he was much more penetrating, as he raised profound and hard questions *about* the systems of the kind that Russell had produced. In fact, Hilbert begot the question about what were, in principle, the limitations of a scheme like *Principia Mathematica*. Was there any way to find out what could, and could not, be demonstrated from within such theory?

Hilbert's approach was deemed *formalist*, as he treated mathematics as a game, a matter of form. The allowed steps in a demonstration should be looked at in the same way as the possible moves in the game of chess, with the axioms being the initial state of the game. In this analogy, 'playing chess' stood for 'executing mathematics', but the *statements* about chess (such as 'two knights cannot achieve checkmate') stood for statements *about* the scope of mathematics. It was with the latter statements that Hilbert's programme was concerned.

In that congress of 1928, Hilbert made his questions very precise. First, was mathematics *complete*? In the sense of whether all mathematical statements (such as "all integers are the sum of four square numbers") could be demonstrated, or falsified. Second, he asked whether mathematics was *consistent*, which amounts to asking whether any contradictory statements, like '2+2=5' could be demonstrated by a correct application of valid steps of the rules for derivation. And third he asked was mathematics *decidable*? By this he meant to ascertain if there was a well-defined

---

* Translation of an invited article for a special issue of the "Bulletin of the Portuguese Mathematical Society" dedicated to Kurt Gödel, to appear end of 2006 on the occasion of the 100th anniversary of his birth.

J. Neves, M. Santos, and J. Machado (Eds.): EPIA 2007, LNAI 4874, pp. 63–72, 2007.
© Springer-Verlag Berlin Heidelberg 2007

method which, in principle, if applied to any assertion, could be guaranteed to produce a correct decision about the truth of that assertion.

In 1928, none of these questions had an answer. Hilbert hypothesized that the answer was 'yes' to every one of them. In his opinion, "there were no unsolvable problems". But it didn't take long before the young Austrian mathematician, Kurt Gödel, presented results which struck a severe blow to Hilbert's programme.

Gödel's Answers

Kurt Gödel was able to show that every formalization of arithmetic must be *incomplete*: that there are assertions that cannot be demonstrated nor rebutted. Starting from Peano's axioms for the integer numbers, he extended them to a simple type theory, in such a way that the obtained system could represent sets of integers, sets of sets of integers, and so forth. However, his argument could be applied to any formal system sufficiently rich to include number theory, and the details of the axioms were not crucial.

Next he showed that all operations in a 'proof', the 'chess-like' deduction rules, could be represented in arithmetic. This means that such rules would be constituted only by operations like counting and comparing, in order to verify if an expression (itself represented arithmetically) had been correctly replaced by another – just like the validity check of a chess move is only a matter of counting and comparing. In truth, Gödel showed that the formulas of his system could be codified as integers, to represent assertions *about* integers. This is his main idea. Subsequently, Gödel showed how to codify demonstrations as integers, in such a way that he had a whole theory of arithmetic, codified *from within* arithmetic. He was using the fact that, if mathematics was to be regarded as a pure symbol game, then one could employ numerical symbols instead of any others. He managed to show that the property of 'being a demonstration' or being 'provable' was as much arithmetical as that of 'being square' or 'being prime'.

The result of this codification process was that it became possible to write arithmetical assertions that referred to *themselves*, just like when a person says "I'm lying". In fact, Gödel constructed a specific assertion that had exactly that property, as it effectively said "This assertion is not provable." It followed that it could not be shown *true*, as that would lead to a contradiction. Neither could it be shown *false*, by the same reason. It was thus an assertion that could neither be proved nor refuted by logical deduction from the axioms. Gödel had just showed that arithmetic was *incomplete*, in Hilbert's technical sense.

But there is more to be said about the matter, because there was something special about that particular assertion of Gödel, more precisely the fact that, considering that it is not provable, it is in some sense *true*. But to assert that it is *true* requires an observer that can look at the system from the outside. It cannot be shown *from within* the axiomatic system.

Another important issue is that this argument assumes that arithmetic is *consistent*. In fact, if arithmetic were inconsistent then *any* assertion would be provable, since in elementary logic everything follows from contradiction. More precisely, Gödel had just shown that arithmetic, once formalized, had to be either *inconsistent* or *incomplete*[1]. He could also show that arithmetic could not be proved consistent from within its own axiomatic system. To prove consistency, it would suffice to show that there was some proposition (say, '2+2=5') that could not be proven true.

But Gödel managed to show that such an existential statement had the same characteristic of the assertion that claimed its own non-provability. In this way, he «discharged» both the first and second questions posed by Hilbert. Arithmetic *could not* be proved consistent, and for sure it *wasn't* consistent *and* complete. This was a spectacular turn of events in research on the matter, and quite disturbing for those that wished for an absolutely perfect and unassailable mathematics.

Gödel, Computability, and Turing

The third of Hilbert's questions, that of *decidability*, still remained open, although now it had to be formulated in terms of *provability*, instead of *truth*. Gödel's results did not eliminate the possibility that there could be some way of distinguishing the provable from the non-provable assertions. This meant that the peculiar self-referencing assertions of Gödel might in some way be separated from the rest. Could there be a well-defined *method*, i.e. a *mechanizable* procedure, that could be applied to any mathematical statement, and that could decide if that statement was, or was not, derivable in a given formal system?

From 1929 to 1930, Gödel had already solved most of the fundamental problems raised by Hilbert's school. One of the issues that remained was that of finding a precise concept that would characterize the intuitive notion of computability. But it was not clear at the time that this problem would admit a definitive answer. Gödel was probably surprised by Turing's solution, which was more elegant and conclusive than what he had expected. Gödel fully understands, at the beginning of the '30s, that the concept of formal system is intimately tied up with that of mechanizable procedure, and he considers Alan Turing's 1936 work (on computable numbers) as an important complement of his own work on the limits of formalization.

In 1934, Gödel gave lectures at the Institute of Advanced Studies in Princeton where he recommended the Turing analysis of mechanizable procedures (published in 1936) as an essential advance that could raise his incompleteness theorems to a more finished form. In consequence of Turing's work, those theorems can be seen to "apply to *any* consistent formal system that contains part of the finitary number theory".

It is well known that Gödel and Alonzo Church, probably in the beginning of 1934, discussed the problem of finding a precise definition to the intuitive concept of computability, a matter undoubtedly of great interest to Gödel. In accordance with Church, Gödel specifically raised the issue of the relation between that concept and Church's definition of *recursivity*, but he was not convinced that both ideas could be satisfactorily made equivalent, "except heuristically". Over the years, Gödel regularly credited Turing's 1936 article as the definitive work that captures the intuitive concept of computability, he being the only author to present persuasive arguments about the adequacy of the precise concept he himself defined.

---

[1] In truth, the fact that a Gödel assertion cannot be refuted formally depends on the omega-consistency of the system, not just on its consistency. Rosser, later, by means of a slightly different kind of assertion, showed that one can do without the omega-consistency requirement.

Regarding the concept of mechanizable procedure, Gödel's incompleteness theorems also naturally begged for an exact definition (as Turing would come to produce) by which one could say that they applied to every formal system, i.e. every system on which proofs could be verified by means of an automatic procedure. In reality, Hilbert's programme included the *Entscheidungsproblem* (literally, 'decision problem'), which aimed to determine if there was a procedure to decide if, in elementary logic, any proposition was derivable or not by Frege's rules (for elementary logic, also known as first-order logic). This requires a precise concept of automatic procedure, in case the answer is negative (as is the case).

To this end, in 1934, Gödel introduces the concept of general recursive functions, which was later shown to capture the intuitive concept of mechanizable computability. Gödel suggested the concept and Kleene worked on it. The genesis of the concept of general recursive functions was implicit in Gödel's proof about the incompleteness of arithmetic. When Gödel showed that the concept of proof using "chess-like" rules was an 'arithmetical' concept, he was in fact saying that a proof could be realized by a 'well-defined' method. This idea, once formalized and somewhat extended, gave rise to the definition of 'recursive function'. It was later verified that these were exactly equivalent to the computable functions.

After some time, Gödel came to recognize that the conception of 'Turing machine' offers the most satisfactory definition of 'mechanic procedure'. According to Gödel himself, Turing's 1936 work on computable numbers is the first to present a convincing analysis of such a procedure, showing the correct perspective by which one can clearly understand the intuitive concept.

On the rigorous definition of the concept of computability performed by Turing, Gödel says: "This definition was by no means superfluous. Although before the definition the term 'mechanically calculable' did not have a clear meaning, though not analytic [i.e. synthetic], so the issue about the adequacy of Turing's definition would not make perfect sense, instead, with Turing's definition the term's clarity undoubtedly receives a positive answer".

Once the rigorous concept, as defined by Turing, is accepted as the correct one, a simple step suffices to see that, not only Gödel's incompleteness theorems apply to formal systems in general, but also to show that the *Entscheidungsproblem* is insoluble. The proof of this insolubility by Turing himself showed that a class of problems that cannot be solved by his "A-machines" (from "**A**utomatic machines", today known as Turing machines) can be expressed by propositions in elementary logic.

In his Ph.D. thesis, Turing is interested in finding a way out from the power of Gödel's incompleteness theorem. The fundamental idea was that of adding to the initial system successive axioms. Each 'true but not demonstrable' assertion is added as a new axiom. However, in this way, arithmetic acquires the nature of a Hydra, because, once the new axiom is added, a new assertion of that type will be produced that takes it now into consideration. It is then not enough to add a *finite* number of axioms, but it is necessary to add an infinite number, which was clearly against Hilbert's finitary dream. If it were possible to produce a finite generator of such axioms, then the initial theory would also be finite and, as such, subject to Gödel's theorem.

Another issue is that there exist an infinite number of possible sequences by which one can add such axioms, leading to different and more complete theories. Turing described his different extensions to the axioms of arithmetic as 'ordinal logics'. In these, the rule to generate the new axioms is given by a 'mechanical process' which can be applied to 'ordinal formulas', but the determination whether a formula is 'ordinal' was *not* mechanizable. Turing compared the identification of an 'ordinal' formula to the work of intuition, and considered his results disappointingly negative, for although there existed 'complete logics', they suffered from the defect that one could not possibly count how many 'intuitive' steps were needed to prove a theorem[2].

This work, however, had a pleasantly persistent side effect, namely the introduction of the concept of 'oracle Turing machine', precisely so it could be allowed to ask and obtain from the exterior the answer to an insoluble problem from within it (as that of the identification of an 'ordinal formula'). It introduced the notion of relative computability, or relative insolvability, which opened a new domain in mathematical logic, and in computer science.

The connection, made by S.A. Cook, in 1971, between Turing machines and the propositional calculus would give rise to the study of central questions about computational complexity. In 1936, Gödel also noted that theorems with long proofs in a given system could very well obtain quite shorter proofs if one considered extensions to the system. This idea has been applied in Computer Science to demonstrate the inefficiency of some well known decision procedures, to show that they require an exponential (or worse) amount of time to be executed. This is a particular instance of one of Gödel's favourite general ideas: problems raised in a particular system can be handled with better results in a richer system.

Mens ex-machina

It is worth mentioning that, contrary to Turing, Gödel was not interested in the development of computers. Their mechanics is so connected with the operations and

---

[2] I would like to make a reference here to my own ongoing work. The basic idea is to use an implemented logic programming language, EVOLP (Alferes et al. 2002), which has the capacity of *self-updating*, in order to obtain this self-evolution predicted by Turing.

The model theory can be defined with the revised stable models semantics (Pereira e Pinto, 2005), with the introduction in each program of the following causal scheme, involving the usual *not* construct of logic programming, representing non-provability:

$$GA \leftarrow \text{gödel\_assertion}(GA), not\ GA$$

This scheme, representing its the *ground* instances, states that if GA is a Gödel assertion, and it is not demonstrable, then GA is true, where we take as a given the codification of the predicate gödel_assertion(GA), which tests, once codified, if the literal GA can be constructed using the Gödelization method. The conclusion is obtained by *reductio ad absurdum*, i.e. if *not* GA was true, then GA would be true; since this is a contradiction and we use a two-valued logic, then the premise *not* GA is false and the conclusion GA is true.

The proof system is derived from this intuition, and uses the following EVOLP rule, where the literal GA can itself be a full-blown rule:

$$assert(GA) \leftarrow \text{gödel\_assertion}(GA), not\ GA$$

This means that each GA that complies with the premises will be part of the next state of the self-evolving program, and so forth. Such a role takes the place of the intuitive step necessary for evolution, conceptually realizing it by *reductio ad absurdum* followed by an *update*, which is allowed expression within the EVOLP language itself.

concepts of logic that, nowadays, it is quite commonplace for logicians to be involved, in some way or other, in the study of computers and computer science. However, Gödel's famous incompleteness theorem demonstrates and establishes the rigidness of mathematics and the limitations of formal systems and, according to some, of computer programs. It relates to the known issue of whether mind surpasses machine. Thus, the growing interest given to computers and Artificial Intelligence (AI) has lead to a general increase in interest about Gödel's own work. But, so Gödel himself recognizes, his theorem does not settle the issue of knowing if the mind surpasses the machine. Actually, Gödel's work in this direction seems to favour (instead of countering) the *mechanist* position (and even *finitism*) as an approach to the automation of formal systems.

Gödel contrasts *insight* with *proof*. A proof can be explicit and conclusive for it has the support of axioms and of rules of inference. In contrast, insights can be communicated only via "pointing" at things. Any philosophy expressed by an exact theory can be seen a special case of the application of Gödel's conceptual realism. According to him, its objective should be to give us a clear perspective of all the basic metaphysical concepts. More precisely, Gödel claims that this task consists in determining, through intuition, the primitive metaphysical concepts C and in making correspond to them a set of axioms A (so that only C satisfies A, and the elements in A are implied by the original intuition of C). He further admits that, from time to time, it would be possible to add new axioms.

Gödel also advocates an 'optimistic rationalism'. His justification appeals to (1) "The fact that those parts of mathematics which have been systematically and completely developed … show an astonishing degree of beauty and perfection." As such (2) It is not the case "that human reason is irredeemably irrational by asking itself questions to which it cannot answer, and at the same time emphatically asserting that only reason can answer them." It follows that (3) There are no "undecidable questions of number theory for the human mind." So (4) "The human mind surpasses all machines."

However, the inference from (1) to (2) seems to be obtained from accidental successes in very limited fields to justify an anticipation of universal success. Besides, both (2) and (3) concern only a specific and delimited part of mind and reason which refer just to mathematical issues.

Gödel understands that his incompleteness theorem by itself does not imply that the human mind surpasses all machines. An additional premise is necessary. Gödel presents three suggestions to that end: (a) It is sufficient to accept his 'optimistic realism' (b) By appealing "to the fact that *the mind, and the way it's used, is not static, but finds itself in constant development,*" Gödel suggests that "there is no reason to claim that" the number of mental states "cannot converge to infinity in the course of its development." (c) He believes that there is a mind separate from matter, and that such will be demonstrated "scientifically (maybe by the fact that there are insufficient nerve cells to account for all the observable mental operations)."

There is a known ambiguity between the notion of *mechanism* confined to the mechanizable (in the precise sense of computable or recursive) and the notion of materialist *mechanism*. Gödel enounces two propositions: (i) The brain operates basically like a digital computer. (ii) The laws of physics, in their observable consequences, have a finite limit of precision. He is of the opinion that (i) is very likely, and that (ii) is practically certain. Perhaps the interpretation Gödel assigns to (ii) is what makes it compatible with the existence of non-mechanical physical laws, and in the same breath he links it to (i) in the sense that, as much as we can observe of the brain's behaviour, it functions like a digital computer.

Is Mathematical Insight Algorithmic?

Roger Penrose (1994) claims that it is *not*, and supports much of his argument, as J. R. Lucas before him (revisited in 1996), on Gödel's incompleteness theorem: It is *insight* that allows us to *see* that a Gödel assertion, undecidable in a given formal system, is accordingly true. How could this intuition be the result of an algorithm? Penrose insists that his argument would have been "certainly considered by Gödel himself in the 1930s and was never properly refuted since then ..."

However, in his Gibbs lecture delivered to the *American Mathematical Society* in 1951, Gödel openly contradicts Penrose:

"On the other hand, on the basis of what has been proven so far, it remains possible that a theorem proving machine, indeed equivalent to mathematical insight, can exist (and even be empirically discovered), although that cannot be *proven*, nor even proven that it only obtains correct theorems of the finitary number theory."

In reality, during the 1930s, Gödel was especially careful in avoiding controversial statements, limiting himself to what could be proven. However, his Gibbs lecture was a veritable surprise. Gödel insistently argued that his theorem had important philosophical implications. In spite of that, and as the above citation makes it clear, he never stated that mathematical insight could be shown to be non-algorithmic.

It is likely that Gödel would agree with Penrose's judgment that mathematical insight could not be the product of an algorithm. In fact, Gödel apparently believed that the human mind could not even be the product of natural evolution. However, Gödel never claimed that such conclusions were consequence of his famous theorem.

Due to the current prevailing trend to restrict discussion about the limits of rationality, in contraposition to insight, to the well-defined and surprising advances in computer science technology and programming, the perspective of considering reason in terms of mechanical capabilities has received much attention in the last decades. Such is recognised as being core to the study of AI, which is clearly relevant to Gödel's wish of separating mind and machine.

In this stance, AI would be primarily interested in what is feasible from the viewpoint of computability, whose formal concern involves only a very limited part of mathematics and logic. However, the study of the limitations of AI cannot be reduced to this restriction of its scope. In this regard, it is essential to distinguish between *algorithms for problem-solving*, and *algorithms simpliciter*, as sets of rules to follow in a systematic and automatic way, which are eventually self-modifiable, and *without* necessarily having a specific and well-defined problem to solve.

Logic Consciousness

If one asks "How can we introduce the unconscious in computers?" some colleagues will answer that computers are totally unconscious. In truth, what we don't know is how to introduce consciousness in algorithms, because we use computers as unconscious appendices to our own consciousness. The question is as such premature, seeing that we can only refer to the human conception of unconscious after we introduce consciousness into the machine. Indeed, we understand much better the computational unconscious than our own unconscious.

These fertile questions point to the complexity of our thought processes, including those of creativity and intuition, that in great measure we do not understand, and pose a much richer challenge to AI, which can help us by providing an indispensable symbiotic mirror.

The translation into a computational construction, of some functional model, as alluded above, of an introspective and thus self-referent consciousness, would be permitted using whatever methodologies and programming paradigms currently at our disposal. In the light of this realization, one might be inclined to ask why the use a logic paradigm, via logic programming, which is precisely the one we prefer (e.g. Lopes and Pereira (2006), plus ongoing work). There are several arguments which can be raised against its use. Hence we will try to reproduce in this section the most relevant, rebut them, and present our own in its favour.

The first argument to be raised in these discussions is that regular human reasoning does not use logic, there existing complex, non-symbolic processes in the brain that supposedly cannot be emulated by symbolic processing. Following this line of thought, many models have been produced based on artificial neural networks, on emergent properties of purely reactive systems, and many others, in an attempt to escape the tyranny of GOFAI ('Good Old Fashioned AI'). There is a catch, however, to these models: Their implementation by its proponents ends up, with no particular qualms, being on a computer, which cannot help but use symbolic processing to simulate these other paradigms.

The relationship of this argument to logic is ensured by the philosophical perspective of functionalism: logic itself can be implemented on top of a symbol processing system, independently of the particular physical substrate supporting it. Once a process is described in logic, we can use its description to synthesize an artefact with those very same properties. So long it is a computational model, any attempt to escape logic will not prove itself to be inherently more powerful.

On the other hand, there is an obvious human capacity for understanding logical reasoning, a capacity developed during the course of brain evolution. Its most powerful expression today is science itself, and the knowledge amassed from numerous disciplines, each of which with their own logic nuances dedicated to reasoning in their domain. From nation state laws to quantum physics, logic, in its general sense, has become the pillar on which human knowledge is built and improved, the ultimate reward for our mastery of language.

Humans can use language without learning grammar. However, if we are to understand linguistics, knowing the logic of grammar, syntax and semantics is vital. Humans do use grammar without any explicit knowledge of it, but that does not mean it cannot be described. Similarly, when talking about the movement of electrons we surely do not mean that a particular electron knows the laws it follows, but we are certainly using symbolic language to describe the process, and it is even possible to use the description to implement a model and a simulation which exhibits precisely the same behaviour. Similarly, even if human consciousness does not operate directly on logic, that does not mean we won't be forced to use logic to provide a rigorous *description* of that process. And, if we employ logic programming for the purpose, such a description can function as an executable specification.

Once obtained a sufficiently rigorous description of the system of consciousness, we are supposedly in possession of all *our* current (temporary) knowledge of that system, reduced to connections between minimal black boxes, inside which we know not yet how to find other essential mechanisms. Presently, no one has managed to adequately divide the black box of our consciousness about consciousness in the brain, but perhaps we can provide for it a sufficiently rigorous description so that we can model a functional system its equivalent. If a division of that epistemic cerebral black box into a diversity of others is achieved later on, we are sure to be able, in this perspective, to describe new computational models equivalent to the inherent functional model.

In its struggle for that rigorous description, the field of Artificial Intelligence has made viable the proposition of turning logic into a programming language (Pereira 2002). Logic can presently be used as a specification language which is not only executable, but on top of which we can demonstrate properties and make proofs of correction that validate the very descriptions we produce with it. Facing up to the challenge, AI developed logic beyond the confines of monotonic cumulativity, far removed from the artificial paradises of the well-defined, and well into the real world purview of incomplete, contradictory, arguable, reviseable, distributed and updatable, demanding, among other, the study and development of non-monotonic logics, and their computer implementation.

Over the years, enormous amount of work has been carried out on individual topics, such as logic programming language semantics, belief revision, preferences, evolving programs with updates, and many other issues that are crucial for a computational architecture of the mind. We are in the presence of a state-of-the-art from whence we can begin addressing the more general issues with the tools already at hand, unifying such efforts into powerful implementations exhibiting promising new computational properties. Computational logic has shown itself capable to evolve to meet the demands of the difficult descriptions it is trying to address.

The use of the logic paradigm also allows us to present the discussion of our system at a sufficiently high level of abstraction and generality to allow for productive interdisciplinary discussions both about its specification and its derived properties.

As previously mentioned, the language of logic is universally used both by the natural sciences and the humanities, and more generally is at the core of any source of human derived common knowledge, so that it provides us with a common ground on which to reason about our theories. Since the field of cognitive science is essentially a joint effort on the part of several distinct knowledge fields, we believe such language and vocabulary unification efforts are not just useful, but mandatory.

Consulted Oeuvres and References

Alferes, J.J., Brogi, A., Leite, J.A., Pereira, L.M.: Evolving Logic Programs. In: Flesca, S., Greco, S., Leone, N., Ianni, G. (eds.) JELIA 2002. LNCS (LNAI), vol. 2424, pp. 50–61. Springer, Heidelberg (2002)

Casti, J.L., DePauli, W.: Gödel – A life of Logic. Basic Books, New York (2000)

Davis, M.: Is mathematical insight algorithmic? Behavioral and Brain Sciences 13(4), 659–660 (1990)

Davis, M.: How subtle is Gödel's theorem. More on Roger Penrose. Behavioral and Brain Sciences 16, 611–612 (1993)

Davis, M.: The Universal Computer: The Road from Leibniz to Turing. W.W. Norton & Co. (2000)

Dennett, D.C.: Sweet Dreams – philosophical obstacles to a science of consciousness. MIT Press, Cambridge (2005)

Glymour, C.: Thinking things through. MIT Press, Cambridge (1992)

Hodges, A.: Alan Turing – the enigma. Simon and Schuster (1983)

LaForte, G., Hayes, P.J., Ford, K M.: Why Gödel's theorem cannot refute computationalism. Artificial Intelligence 104, 265–286 (1998)

Lopes, G., Pereira, L.M.: Prospective Programming with ACORDA. In: Empirically Successful Computerized Reasoning (ESCoR 2006) workshop, at The 3rd International Joint Conference on Automated Reasoning (IJCAR 2006), Seattle, USA (2006)

Lucas, J.R.: Minds, Machines, and Gödel: A Retrospect. In: Millican, P., Clark, A. (eds.) Machines and Thought, vol. 1, pp. 103–124. Oxford University Press, Oxford (1996)

McDermott, D.: Mind and Mechanism. MIT Press, Cambridge (2001)

Minsky, M.L.: Computation: Finite and Infinite Machines. Prentice-Hall, Englewood Cliffs (1967)

Nagel, E., Newman, J.R.: Gödel's Proof. New York University Press (2001)

Nagel, E., Newman, J.R., Gödel, K., Girard, J.-Y.: Le Théorème de Gödel.Éditions du Seuil (1989)

Penrose, R.: Shadows of the Mind: a search for the missing science of consciousness. Oxford University Press, Oxford (1994)

Pereira, L.M.: Philosophical Incidence of Logical Programming. In: Gabbay, D., et al. (eds.) Handbook of the Logic of Argument and Inference. Studies in Logic and Practical Reasoning series, vol. 1, pp. 425–448. Elsevier Science, Amsterdam (2002)

Pereira, L.M., Pinto, A.M.: Revised Stable Models – a semantics for logic programs. In: Bento, C., Cardoso, A., Dias, G. (eds.) EPIA 2005. LNCS (LNAI), vol. 3808, pp. 29–42. Springer, Heidelberg (2005)

Turing, A., Girard, J.-Y.: La Machine de Turing. Éditions du Seuil (1995)

Wang, H.: From Mathematics to Philosophy. Routledge (1973)

Wang, H.: Reflections on Kurt Gödel. MIT Press, Cambridge (1987)

Webb, J.C.: Mechanism, Mentalism and Meta-mathematics. Reidel, Dordrechtz (1980)

Prospective Logic Agents

Luís Moniz Pereira and Gonçalo Lopes

Centro de Inteligência Artificial - CENTRIA
Universidade Nova de Lisboa, 2829-516 Caparica, Portugal
goncaloclopes@gmail.com,
lmp@di.fct.unl.pt

Abstract. As we face the real possibility of modelling agent systems capable of non-deterministic self-evolution, we are confronted with the problem of having several different possible futures for any single agent. This issue brings the challenge of how to allow such evolving agents to be able to *look ahead*, prospectively, into such hypothetical futures, in order to determine the best courses of evolution from their own present, and thence to prefer amongst them. The concept of prospective logic programs is presented as a way to address such issues. We start by building on previous theoretical background, on evolving programs and on abduction, to construe a framework for prospection and describe an abstract procedure for its materialization. We take on several examples of modelling prospective logic programs that illustrate the proposed concepts and briefly discuss the ACORDA system, a working implementation of the previously presented procedure. We conclude by elaborating about current limitations of the system and examining future work scenaria.

1 Introduction

Continuous developments in logic programming (LP) language semantics which can account for evolving programs with updates [2, 3] have opened the door to new perspectives and problems amidst the LP and agents community. As it is now possible for a program to talk about its own evolution, changing and adapting itself through non-monotonic self-updates, one of the new looming challenges is how to use such semantics to specify and model logic based agents which are capable of anticipating their own possible future states and of preferring among them in order to further their goals, prospectively maintaining truth and consistency in so doing. Such predictions need to account not only for changes in the perceived external environment, but need also to incorporate available actions originating from the agent itself, and perhaps even consider possible actions and hypothetical goals emerging in the activity of other agents.

While being immersed in a world (virtual or real), every proactive agent should be capable, to some degree, of conjuring up hypothetical *what-if* scenaria while attending to a given set of integrity constraints, goals, and partial observations of the environment. These scenaria can be about hypothetical observations (what-if this observation were true?), about hypothetical actions (what-if this action were performed?) or hypothetical goals (what-if this goal was pursued?). As we are dealing with non-monotonic logics, where knowledge about the world is incomplete and revisable, a way to represent predictions about the future is to consider possible scenaria as tentative evolving

J. Neves, M. Santos, and J. Machado (Eds.): EPIA 2007, LNAI 4874, pp. 73–86, 2007.

hypotheses which *may* become true, pending subsequent confirmation or disconfirmation on further observations, the latter based on the expected consequences of assuming each of the scenaria.

We intend to show how rules and methodologies for the synthesis and maintenance of abductive hypotheses, extensively studied by several authors in the field of Abductive Logic Programming [9, 11, 17, 16], can be used for effective, yet defeasible, prediction of an agent's future. Note that we are considering in this work a very broad notion of abduction, which can account for any of the types of scenaria mentioned above. Abductive reasoning by such prospective agents also benefits greatly from employing a notion of simulation allowing them to derive the consequences for each available scenario, as the agents imagine the possible evolution of their future states prior to actually taking action towards selecting one of them.

It is to be expected that a multitude of possible scenaria become available to choose from at any given time, and thus we need efficient means to prune irrelevant possibilities, as well as to enact preferences and relevancy preorders over the considered ones. Such preference specifications can be enforced either a priori or a posteriori w.r.t hypotheses making. A priori preferences are embedded in the knowledge representation theory itself and can be used to produce the most interesting or relevant conjectures about possible future states. Active research on the topic of preferences among abducibles is available to help us fulfill this purpose [6, 7] and results from those works have been incorporated in the presently proposed framework.

A posteriori preferences represent meta-reasoning over the resulting scenaria themselves, allowing the agent to actually make a choice based on the imagined consequences in each scenario, possibly by attempting to confirm or disconfirm some of the predicted consequences, by attributing a measure of interest to each possible model, or simply by delaying the choice over some models and pursuing further prospection on the most interesting possibilities which remain open. At times, several hypotheses may be kept open simultaneously, constantly updated by information from the environment, until a choice is somehow forced during execution (e.g. by using escape conditions), or until a single scenario is preferred, or until none are possible.

In prospective reasoning agents, exploration of the future is essentially an open-ended, non-deterministic and continuously iterated process, distinct from the one-step, best-path-takes-all planning procedures. First, the use of abduction can dynamically extend the theory of the agent during the reasoning process itself in a context-dependent way so that no definite set of possible actions is implicitly defined. Second, the choice process itself typically involves acting upon the environment to narrow down the number of available options, which means that the very process of selecting futures can drive an agent to autonomous action. Unlike Rodin's thinker, a prospective logic agent is thus proactive in its look ahead of the future, acting upon its environment in order to anticipate, pre-adapt and enact informed choices efficiently. These two features imply that the horizon of search is likely to change at every iteration and the state of the agent itself can be altered during this search.

The study of this new LP outlook is essentially an innovative combination of fruitful research in the area, providing a testbed for experimentation in new theories of program evolution, simulation and self-updating, while launching the foundational seeds for modeling rational self-evolving prospective agents. Preliminary research results have proved themselves useful for a variety of applications and have led to the development of the ACORDA[1] system, successfully used in modelling diagnostic situations [13]. This paper presents a more formal abstract description of the procedure involved in the design and implementation of prospective logic agents.Some examples are also presented as an illustration of the proposed system capabilities, and some broad sketches are laid out concerning future research directions.

2 Logic Programming Framework

2.1 Language

Let $\mathcal{L}$ be any first order language. A domain literal in $\mathcal{L}$ is a domain atom A or its default negation *not A*, the latter expressing that the atom is false by default (CWA). A domain rule in $\mathcal{L}$ is a rule of the form:

$$A \leftarrow L_1, \ldots, L_t \ (t \geq 0)$$

where A is a domain atom and $L_1, \ldots, L_t$ are domain literals. An integrity constraint in $\mathcal{L}$ is a rule of the form:

$$\perp \leftarrow L_1, \ldots, L_t \ (t > 0)$$

where $\perp$ is a domain atom denoting falsity, and $L_1, \ldots, L_t$ are domain literals.

A (logic) program P over $\mathcal{L}$ is a set of domain rules and integrity constraints, standing for all their ground instances. Every program P is associated with a set of *abducibles* $\mathcal{A} \subseteq \mathcal{L}$, consisting of literals which (without loss of generality) do not appear in any rule head of P. Abducibles may be thought of as hypotheses that can be used to extend the current theory, in order to provide hypothetical solutions or possible explanations for given queries.

2.2 Preferring Abducibles

An abducible can be assumed only if it is a considered one, i.e. it is expected in the given situation, and moreover there is no expectation to the contrary [6, 7].

$$consider(A) \leftarrow expect(A), not \ expect\_not(A).$$

The rules about expectations are domain-specific knowledge contained in the theory of the agent, and effectively constrain the hypotheses (and hence scenaria) which are available.

To express preference criteria among abducibles, we consider an extended first order language $\mathcal{L}^*$. A preference atom in $\mathcal{L}^*$ is one of the form $a \lhd b$, where a and b are abducibles. $a \lhd b$ means that the abducible a is preferred to the abducible b. A preference rule in $\mathcal{L}^*$ is one of the form:

$$a \lhd b \leftarrow L_1, \ldots, L_t \ (t \geq 0)$$

---

[1] ACORDA literally means "wake-up" in Portuguese. The *ACORDA* system project page is temporarily set up at: http://articaserv.ath.cx/

where $a \lhd b$ is a preference atom and every $L_i(1 \leq i \leq t)$ is a domain or preference literal over $\mathcal{L}^*$.

Although the program transformation in [6, 7] accounted only for mutually exclusive abducibles, we have extended the definition to allow for sets of abducibles, so we can generate *abductive stable models* [6, 7] having more than a single abducible. For a more detailed explanation of the adapted transformation, please consult the ACORDA project page, mentioned in the previous footnote.

3 Prospective Logic Agents

We now present the abstract procedure driving evolution of a prospective logic agent. Although it is still too early to present a complete formal LP semantics to this combination of techniques and methodologies, as the implemented system is undergoing constant evolution and revision, it is to be expected that such a formalization will arise in the future, since the proposed architecture is built on top of logically grounded and semantically well-defined LP components. The procedure is illustrated in Figure 1, and is the basis for the implemented ACORDA system, which we will detail in Section 5.

Each prospective logic agent has a knowledge base containing some initial program over $\mathcal{L}^*$. The problem of prospection is then one of finding abductive extensions to this initial theory which are both:

- *relevant* under the agent's current desires and goals
- *preferred* extensions w.r.t. the preference rules in the knowledge base

We adopt the following definition for the relevant part of a program P under a literal L:

Definition 1. *Let L, B, C be literals in $\mathcal{L}^*$. We say L directly depends on B iff B occurs in the body of some rule in P with head L. We say L depends on B iff L directly depends on B or there is some C such that L directly depends on C and C depends on B. We say that $Rel_L(P)$, the relevant part of P, is the logic program constituted by the set of all rules of P with head L or some B on which L depends on.*

Given the above definition, we say that an abductive extension Δ of P (i.e. $\Delta \subseteq \mathcal{A}_P$) is *relevant* under some query G iff all the literals in Δ belong to $Rel_G(P \cup \Delta)$. The first step thus becomes to select the desires and goals that the agent will possibly attend to during the prospective cycle.

3.1 Goals and Observations

Definition 2. *An observation is a quaternary relation amongst the observer; the reporter; the observation name; and the truth value associated with it.*

$$observe(Observer, Reporter, Observation, Value)$$

Observations can stand for actions, goals or perceptions. The *observe*/4 literals are meant to represent observations reported by the environment into the agent or from one agent to another, which can also be itself (self-triggered goals). We also introduce the

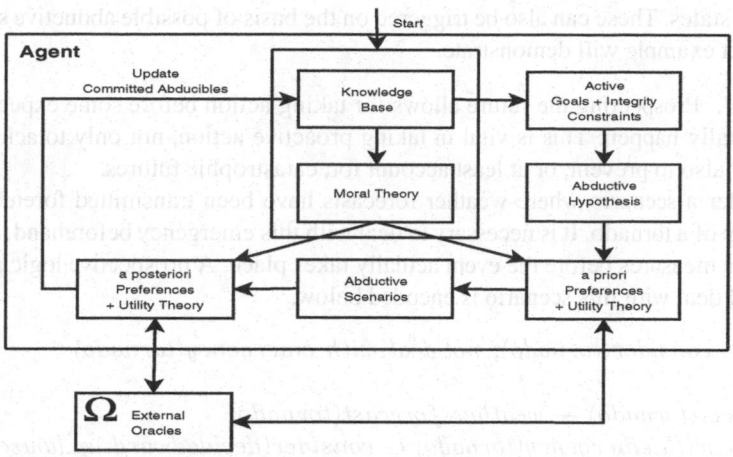

Fig. 1. Prospective agent cycle

corresponding *on_observe*/4 literal, which we consider as representing active goals or desires that, once triggered, cause the agent to attempt their satisfaction by launching the queries standing for the observations contained inside.

The prospecting mechanism thus polls for the *on_observe*/4 literals which are satisfied under the initial theory of the agent. In an abstract representation, we are interested in those *on_observe*/4 literals which belong to the Well-Founded Model of the evolving logic program at the current knowledge state.

Definition 3. *The set of active goals of initial program P is:*

$$Goals(P) = \{G : on\_observe(agent, agent, G, true) \in WFM(P)\}$$

By adopting the more skeptic Well-Founded Semantics at this stage, we guarantee a unique model for the activation of *on_observe*/4 literals. It should be noted that there can be many situations where more than one active goal is derived under the current knowledge theory of the agent. Since we are dealing with the combinatorial explosion of all possible abductive extensions, it is possible that, even if no combination of abducibles satisfies the entire conjunction of active goals, that at least a subset of those goals will be satisfied in some models. In order to allow for the generation of all these possible scenaria, we actually transform active goals into *tentative* queries, encoded in the following form:

$$try(G) \leftarrow G \qquad try(G) \leftarrow not\ try\_not(G)$$
$$try\_not(G) \leftarrow not\ try(G)$$

In this way, we guarantee that computed scenaria will provide all possible ways to satisfy the conjunction of desires, *or* possible subsets of desires, allowing us then to apply selection rules to qualitatively determine which abductive extensions to adopt based on the relative importance or urgency of activated goals. Integrity constraints are also considered, so as to ensure the agent always performs transitions into valid evolution states. These can also be triggered on the basis of possible abductive scenaria, as the next example will demonstrate.

Example 1. Prospecting the future allows for taking action before some expected scenaria actually happen. This is vital in taking proactive action, not only to achieve our goals, but also to prevent, or at least account for, catastrophic futures.

Consider a scenario where weather forecasts have been transmitted foretelling the possibility of a tornado. It is necessary to deal with this emergency beforehand, and take preventive measures before the event actually takes place. A prospective logic program that could deal with this scenario is encoded below.

$$\perp \leftarrow consider(tornado), \; not \; deal\_with\_emergency(tornado)$$

$$expect(tornado) \leftarrow weather\_forecast(tornado)$$
$$deal\_with\_emergency(tornado) \leftarrow consider(decide\_board\_up\_house)$$

$$expect(decide\_board\_up\_house) \leftarrow consider(tornado)$$
$$\perp \leftarrow decide\_board\_up\_house, \; not \; boards\_at\_home, \; not \; go\_buy\_boards$$

The first sentence expresses that, in case a tornado scenario is considered, the program should deal with the emergency. A possible way to deal with this emergency is deciding to board up the house. This hypothesis is only made available in the event of a tornado, since we do not want in this case to account for this decision in any other situation (we could change the corresponding $expect/1$ rule to state otherwise). The weather forecast brings about that a tornado is expected, and there being no contrary expectation to this scenario, the above program presents two possible predictions about the future. In one of the scenaria, the tornado is absent, but in the scenario where it is actually confirmed, the decision to board up the house follows as a necessity.

If we commit to the decision of boarding up the house, by assuming the tornado scenario is more relevant, and we do not have boards at home, it is necessary that we go and buy the boards. This is reflected by the second integrity constraint, which in fact would launch a subgoal for buying boards. As such, even if no goals were active, the possibility of considering certain scenaria can trigger integrity constraints, and also contextual abducibles which may in turn be used, once they are confirmed, to support activation of other goals.

3.2 Generating Scenaria

Once the set of active goals for the current state is known, the next step is to find out which are the relevant abductive extensions which are considered in the situation. They can be found by reasoning backwards from the goals into abducibles which come up under $consider/1$ literals. Each abducible represents a choice: the agent can either assume it true, or assume it false, meaning that it may potentially face a number of interpretations equal to all possible combinations of relevant abducibles. In practice, the combinatorial explosion of possible interpretations is contained and made tractable by a number of factors.

To begin with, the simple fact that all abducibles are constrained to the relevant part of the program under the active goals already leaves all the irrelevant abducibles out of the generation of scenaria. Secondly, the context-dependent rules presented in Section 2.2 for considering abducibles further excludes those abducibles which are not relevant to the actual situation of the agent. Furthermore, it is often the case that available abducibles are contradictory, i.e. considering an abducible actually precludes considering another one, for instance, when choosing between drinking coffee or drinking tea [6, 7]. Finally, this step includes the application of a priori preferences in the form of contextual preference rules among the available abducibles.

In each possible interpretation, or scenario, thus generated, we also reason forwards from abducibles to obtain the relevant consequences of actually committing to each of them. Each abductive stable model is characterized by the abducible choices contained in it, but is in fact a whole model of the program sent to it. Information about each of the models will then be used to enact preferences over the scenarios *a posteriori*, taking into account the consequences in each scenario

3.3 Preferring a Posteriori

Once each possible scenario is actually obtained, there are a number of different strategies which can be used to choose which of the scenaria leads to more favorable consequences. A possible way to achieve this was first presented in [16], using numeric functions to generate a quantitative measure of utility for each possible action. We allow for the application of a similar strategy, by making a priori assignments of probability values to uncertain literals and utilities to relevant consequences of abducibles. We can then obtain a posteriori the overall utility of a model by weighing the utility of its consequences by the probability of its uncertain literals. It is then possible to use this numerical assessment to establish a preorder among remaining models.

Although such numerical treatment of a posteriori preferences can be effective in some situations, there are occasions where we do not want to rely on probability and utility alone, especially if we are to attribute tasks of responsibility to such autonomous agents. In particular, it may become necessary to endow such agents with a set of behaviour precepts which are to be obeyed at all times, no matter what the quantitative assessments may say. This is the role of the moral theory presented in the figure. Although being clearly outside the scope of the presented work, we regard it as a growing concern which must be weighed as more intelligent and autonomous agents are built and put to use. A more detailed analysis of this moral perspective can be found in [15].

Both qualitative and quantitative evaluations of the scenarios can be greatly improved by merely acquiring additional information to make a final decision.We next consider the mechanism that our agents use to question external systems, be they other agents, actuators, sensors or other procedures. Each of these serves the purpose of an *oracle*, which the agent can probe through observations of its own, of the form

$$observe(agent, oracle\_name, query, Value) \leftarrow oracle, L_1, \ldots, L_t \ (t \geq 0)$$

representing that the agent is performing the observation *query* on the oracle identified by *oracle_name*, whenever oracle observations are allowed (governed by the reserved toggle literal *oracle*) and given that domain literals $L_1, \ldots, L_t$ hold in the current knowledge state. Following the principle of parsimony, it is not desirable that the oracles be consulted ahead of time in any situation. Hence, the procedure starts by using its available local knowledge to generate the preferred abductive scenaria (i.e. the toggle is turned off), and then extends the search to include available oracles, by toggling *oracle* on. Each oracle mechanism may in turn have certain conditions specifying whether it is available for questioning. At the next iteration, this toggle is turned off, as more consequences will be computed using the additional information.

Whenever the agent acquires additional information to deal with a problem at hand, it is possible, and even likely, that ensuing side-effects may affect its original search. Some considered abducibles may now be disconfirmed, but it is also possible that some new abducibles which were previously unavailable are now triggered by the information obtained by the oracle observations. To ensure all possible side-effects are accounted for, a second round of prospection takes place, by relaunching the whole conjunctive query. Information returned from the oracle may change the preferred scenaria previously computed, which can in turn trigger new questions to oracles, and so on, in an iterated process of refinement, which stops if no changes to the models have been enacted, and there are no new oracle questions to perform, or user updates to execute.

Even after extending the search to allow for experiments, it may still be the case that some abducibles are tied in competition to explain the active goals, e.g. if some available oracle was unable to provide a crucial deciding experiment. In this case, the only remaining possible action is to branch the simulation into two or more possible update sequences, each one representing an hypothetical world where the agent simulates commitment to the respective abducible. This means delaying the choice, and keeping in mind the evolution of the remaining scenaria until they are gradually defeated by future updates, or somehow a choice is enforced. Exactly how these branches are kept updated and eventually eliminated is not trivial, and this is why we purposefully leave undefined the procedure controlling the evolution of these branching prospective sequences. Another interesting possibility would be to consider those abductions common to all the models and commit to them, in order to prune some irrelevant models while waiting for future updates to settle the matter.

3.4   Prospective Procedure

We conclude this section by presenting the full abstract procedure defining the cycle of a prospective logic agent.

Definition 4. *Let P be an evolving logic program, representing the knowledge theory of an agent at state S. Let oracle be the propositional atom used as a toggle to restrict access to additional external observations. A prospective evolution of P is a set of updates onto P computed by the following procedure:*

1. *Let O be the (possibly empty) set of all on_observe/4 atoms which hold at S.*
2. *Obtain the set of stable models of the residual program derived by evaluating the conjunction $Q = \{G_1, \ldots, G_n, not\bot\}, n \geq 0$, where each G_i represents the goal contained in a distinct observe/4 literal obtained from the corresponding on_observe/4 in O.*

3. *If the set contains a single model, update the abductive choices characterizing the model onto P as facts, toggle the oracle off and stop.*
4. *Otherwise, if oracle currently holds and no new information from the oracles or from the scenaria is derived, for each abductive stable model M_i create a new branching evolution sequence P_i and update the abductive choices in M_i onto P_i. Execute the procedure starting from step 1 on each branching sequence P_i.*
5. *Otherwise, toggle the oracle on and return to 2.*

4 Modelling Prospective Logic Agents

4.1 Accounting for Emergencies

Example 2. Consider the emergency scenario in the London underground [11], where smoke is observed, and we want to be able to provide an explanation for this observation. Smoke can be caused by fire, in which case we should also consider the presence of flames, but smoke could also be caused by tear gas, in case of police intervention. The tu literal in observation values stands for true or undefined.

$$smoke \leftarrow consider(fire) \qquad smoke \leftarrow consider(tear\_gas)$$
$$flames \leftarrow consider(fire) \qquad eyes\_cringing \leftarrow consider(tear\_gas)$$

$$expect(fire)$$
$$expect(tear\_gas) \qquad \bot \leftarrow observation(smoke), not\ smoke$$
$$fire \lhd tear\_gas \qquad observation(smoke)$$

$$\bot \leftarrow flames, not\ observe(program, user, flames, tu)$$
$$\bot \leftarrow eyes\_cringing, not\ observe(program, user, eyes\_cringing, tu)$$

This example illustrates how an experiment can be derived in lieu of the consequences of an abduction. In order for fire to be abduced, we need to be able to confirm the presence of flames, which is a necessary consequence, and hence we trigger the observation to confirm flames, expressed in the second integrity constraint. Only in case this observation does not disconfirm flames are we allowed to abduce fire.

4.2 Automated Diagnosis

Prospective logic programming has direct application in automated diagnosis scenaria, as previously shown in [13]. Another illustration is that of a use case in ongoing research on diagnosis of self-organizing industrial manufacturing systems [4].

Example 3. Consider a robotic gripper immersed in a collaborative assembly-line environment. Commands issued to the gripper from its controller are updated to its evolving knowledge base, as well as regular readings from the sensor. After expected execution of its commands, diagnosis requests by the system are issued to the gripper's prospecting controller, in order to check for abnormal behaviour. When the system is confronted with multiple possible diagnosis, requests for experiments can be asked of the controller. The gripper can have three possible logical states: open, closed or something intermediate. The available gripper commands are simply *open* and *close*. This scenario can be encoded as the initial prospective program below.

$open \leftarrow request\_open,\ not\ consider(abnormal(gripper))$
$open \leftarrow sensor(open),\ not\ consider(abnormal(sensor))$

$intermediate \leftarrow request\_close, manipulating\_part,$
$\qquad\qquad\qquad not\ consider(abnormal(gripper)),\ not\ consider(lost\_part)$
$intermediate \leftarrow sensor(intermediate),\ not\ consider(abnormal(sensor))$

$closed \leftarrow request\_close,\ not\ manipulating\_part,$
$\qquad\qquad not\ consider(abnormal(gripper))$
$closed \leftarrow sensor(closed),\ not\ consider(abnormal(sensor))$

$\bot \leftarrow open, intermediate \qquad \bot \leftarrow open, closed$
$\bot \leftarrow closed, intermediate$

$expect(abnormal(gripper)) \qquad expect(lost\_part) \leftarrow manipulating\_part$
$expect(abnormal(sensor))$
$expect\_not(abnormal(sensor)) \leftarrow$
$\qquad manipulating\_part, observe(system, gripper, ok(sensor), true)$

$observe(system, gripper, Experiment, Result) \leftarrow$
$\qquad oracle, test\_sensor(Experiment, Result)$

$abnormal(gripper) \lhd abnormal(sensor) \leftarrow$
$\qquad request\_open,\ not\ sensor(open),\ not\ sensor(closed)$
$lost\_part \lhd abnormal(gripper) \leftarrow$
$\qquad observe(system, gripper, ok(sensor), true), sensor(closed)$
$abnormal(gripper) \lhd lost\_part \leftarrow not\ (lost\_part \lhd abnormal(gripper))$

For each possible logical state, we encode rules predicting that state from requested actions and from provided sensor readings. We consider that execution of actions may fail, or that the sensor readings may be abnormal. There are also situations where mechanical failure did not occur and sensor readings are also correct, but there was some other failure, like losing the part the robot was manipulating, by dropping it.

In this case, there is an available experiment to test whether the sensor is malfunctioning, but resorting to it should be avoided as much as possible, as it will imply occupying additional resources from the assembly-line coalition. As expected, evaluation is context-dependent on the situation. Consider this illustrative update set:

$$U = \{manipulating\_part, request\_close, sensor(closed)\}.$$

It represents the robot in the process of manipulating some part, receiving an order to close the gripper in order to grab it, but the sensor reporting the gripper is completely closed. This violates an integrity constraint, as the gripper should be in an intermediate state, taking hold of the part. At the start of a diagnosis, three abductive hypotheses are expected and considered,

$$\mathcal{A}_P = \{lost\_part, abnormal(gripper), abnormal(sensor)\}.$$

Without further information, abducible *abnormal(gripper)* is preferred to *lost_part*, but still no single scenario has been determined. Activating oracle queries, the system finds the experiment to test the sensor. If it corroborates closed, not only the abducible *abnormal(sensor)* is defeated, but also *abnormal(gripper)*, since *lost_part* is preferred. However, failure to confirm the sensor reading would result in no single scenario being abduced for this situation, and other measures would have to be taken.

4.3 Encoding Actions

Another interesting possibility in future prospection is to consider the dynamics of actions. To perform an action, a prospective agent needs not just to consider the necessary preconditions for executing it in the present, but also to look ahead at the consequences it will entail in a future state. These two verifications take place on different reasoning moments. While the preconditions of an action can be evaluated immediately when collecting the relevant abducibles for a given knowledge state, its postconditions can only be taken into consideration after the model generation, when the consequences of hypothetically executing an action are known.

The execution of an action can be encoded in EVOLP by means of *assert/1* rules, of the form:

$$assert(A) \leftarrow L_1, \ldots, L_t \quad (t \geq 0)$$

where A is a domain atom representing the name of the action and $L_1, \ldots, L_t$ are domain literals representing the preconditions for the action. The preconditions can themselves contain other $assert/1$ literals in their bodies, allowing lookahead into future updates. The postconditions of a given action can be encoded as integrity constraints on the name of the action and will be triggered during generation of the stable models.

Example 4. Consider an agent choosing an activity in the afternoon. It can either go to the beach, or to the movies, but not both, and it can only go see a movie after buying tickets to it. The abducibles in this case are $\mathcal{A}_P = \{go\_to\_beach, go\_to\_movies\}$. There is a single integrity constraint stating that tickets cannot be bought without money. In ACORDA syntax:

> $afternoon\_activity \leftarrow assert(beach)$
> $afternoon\_activity \leftarrow assert(movies)$

> $assert(beach) \leftarrow consider(go\_to\_beach)$ $expect(go\_to\_beach)$
> $assert(movies) \leftarrow tickets$ $expect(go\_to\_movies)$
> $assert(tickets) \leftarrow consider(go\_to\_movies)$ $\perp \leftarrow tickets, not\ money$

The abduction of either *go_to_beach* or *go_to_movies* fulfills, respectively, the preconditions for the action *beach* and the action *tickets*. The consequence of buying the tickets is that the precondition for going to the movies is fulfilled. However, that consequence may also trigger the integrity constraint if the agent does not have money. Fortunately, by simulating the consequences of actions in the next state, the agent can effectively anticipate that the constraint will be violated, and proceed to choose the only viable course of action, that is going to the beach.

5   Implementing the ACORDA System

The basis for the developed ACORDA system is an EVOLP meta-interpreter on which we can evaluate literals for truth according to three- and two-valued semantics. Both this meta-interpreter and the remaining components were developed on top of XSB Prolog, an extensively used and stable LP inference engine implementation, following the Well-Founded Semantics (WFS) for normal logic programs.

The tabling mechanism [18] used by XSB not only provides significant decrease in time complexity of logic program evaluation, but also allows for extending WFS to other non-monotonic semantics. An example of this is the XASP interface (standing for XSB Answer Set Programming), which extends computation of the WFM, using Smodels [14] to compute two-valued models from the *residual program* resulting from querying the knowledge base [5]. This residual program is represented by delay lists, that is, the set of undefined literals for which the program could not find a complete proof, due to mutual dependencies or loops over default negation for that set of literals, detected by the XSB tabling mechanism. It is also possible to access Smodels by building up a clause store in which a normal logic program is composed, parsed and evaluated, with the computed stable models sent back to the XSB system.

This integration allows one to maintain the relevance [8] property for queries over our programs, something that the Stable Models semantics does not originally enjoy. In Stable Models, by the very definition of the semantics, it is necessary to compute all the models for the whole program. Furthermore, since computation of all the models is NP-complete, it would be unwise to attempt it in practice for the whole knowledge base in a logic program, which can contain literally thousands of rules and facts and unlimited abducibles. In our system, we sidestep this issue, using XASP to compute the relevant residual program on demand, usually after some degree of transformation. Only the resulting program is then sent to Smodels for computation of possible futures. The XSB side of the computation also plays the role of an efficient grounder for rules sent to Smodels, that otherwise resorts to Herbrand base expansion, which can be considerably hastened if we can provide a priori the grounding of domain literals. Also, the stable models semantics is not cumulative [8], which is a prohibitive restriction when considering self-evolving logic programs, in which it is extremely useful to store previously deduced conclusions as lemmas to be reused.

6   Conclusions and Future Work

As far as we know, the only other authors taking a similar LP approach to the derivation of the consequences of candidate abductive hypotheses are [11, 10], and [16, 17]. Both represent candidate actions by abducibles and use logic programs to derive their possible consequences, to help in deciding between them. However, they do not derive consequences of abducibles that are not actions, such as observations for example. Nor do they consider the possibility of determining the value of unknown conditions by consulting an oracle or by some other process.

Poole uses abduction, restricted to acyclic programs, to provide explanations for positive and negative goals. An explanation represents a set of independent choices, each of which is assigned a probability value. The probability of a goal can be found by considering the set of abductively generated possible worlds containing an abductive explanation for the goal. His main concern is to compute goal uncertainty, with a view to decision making, taking into account both the probabilities of the abductive assumptions and the utilities of their outcomes.

Kowalski argues that an agent can be more intelligent if it is able to reason preactively - that is to say, to reason forward from candidate actions to derive their possible consequences. These consequences, he recognizes, may also depend upon other conditions over which the agent has no control, such as the actions of other agents or unknown states of the environment. He considers the use of Decision Theory, like Poole, to choose actions that maximise expected utility. But he has not explored ways of obtaining information about conditions over which the agent does not have control, nor the use of preferences to make choices [12].

Compared with Poole and Kowalski, one of the most interesting features of our approach is the use of Smodels to perform a kind of forward reasoning to derive the consequences of candidate hypotheses, which may then lead to a further cycle of abductive exploration, intertwined with preferences for pruning and for directing search.

With branching update sequences we have begun to address the problem of how to arbitrarily extend the future lookahead within simulations. Independent threads can evolve on their own by commiting to surviving assumptions and possibly triggering new side-effects which will only take place after such commitment.Nevertheless, some issues in the management of these branching sequences must still be tackled, namely envolving coordination and articulation of information shared among threads belonging to a common trunk, as well as the control of the lifetime of each individual thread.

Preferences over observations are also desirable, since not every observation costs the same for the agent. For example, in the industrial manufacture example, the experiment for testing the sensor was costly, but additional and cheaper experiments could eventually be developed, and they should be preferred to the more expensive one whenever possible. Furthermore, abductive reasoning can be used to generate hypotheses of observations of events possibly occurring in the future along the lines of [1].

Prospective LP accounts for abducing the possible means to reach an end, but the converse problem is also of great interest, that is, given the observations of a set of actions, abduce the goal that led to the selection of those actions. This would be invaluable in abducing the intentions of other agents from the sequence of actions they exhibit.

Although we are currently dealing only with prospection of the future, prospective simulations of the past can also be of interest to account for some learning capabilities based on counterfactual thought experiments. This means that we can go back to a choice point faced in the past and relaunch the question in the form "'What would happen if I knew then what I know now?'", incorporating new elements on reevaluating past dilemmas. This could allow for debugging of prospective strategies, identifying experiments that could have been done as well as alternative scenarios that could have been pursued so that in the future the same errors are not repeated.

References

[1] Alberti, M., Gavanelli, M., Lamma, E., Mello, P., Torroni, P.: Abduction with hypotheses confirmation. In: IJCAI 2005. Proc. of the 19th Intl. Joint Conf. on Artificial Intelligence, pp. 1545–1546 (2005)

[2] Alferes, J.J., Brogi, A., Leite, J.A., Pereira, L.M.: Evolving logic programs. In: Flesca, S., Greco, S., Leone, N., Ianni, G. (eds.) JELIA 2002. LNCS (LNAI), vol. 2424, pp. 50–61. Springer, Heidelberg (2002)

[3] Alferes, J.J., Leite, J.A., Pereira, L.M., Przymusinska, H., Przymusinski, T.C.: Dynamic updates of non-monotonic knowledge bases. J. Logic Programming 45(1-3), 43–70 (2000)

[4] Barata, J., Ribeiro, L., Onori, M.: Diagnosis on evolvable production systems. In: ISIE 2007. Procs. of the IEEE Intl. Symp. on Industrial Electronics, Vigo, Spain (forthcoming, 2007)

[5] Castro, L., Swift, T., Warren, D.S.: XASP: Answer Set Programming with XSB and Smodels. http://xsb.sourceforge.net/packages/xasp.pdf

[6] Dell'Acqua, P., Pereira, L.M.: Preferential theory revision. In: Pereira, L.M., Wheeler, G. (eds.) Procs. Computational Models of Scientific Reasoning and Applications, pp. 69–84 (2005)

[7] Dell'Acqua, P., Pereira, L.M.: Preferential theory revision (ext.). J. Applied Logic (2007)

[8] Dix, J.: A classification theory of semantics of normal logic programs: i. strong properties, ii. weak properties. Fundamenta Informaticae 22(3), 227–255, 257–288 (1995)

[9] Kakas, A., Kowalski, R., Toni, F.: The role of abduction in logic programming. In: Gabbay, D., Hogger, C., Robinson, J. (eds.) Handbook of logic in Artificial Intelligence and Logic Programming, vol. 5, pp. 235–324. Oxford University Press, Oxford (1998)

[10] Kowalski, R.: How to be artificially intelligent (2002-2006), http://www.doc.ic.ac.uk/~rak/

[11] Kowalski, R.: The logical way to be artificially intelligent. In: Toni, F., Torroni, P. (eds.) Proceedings of CLIMA VI. LNCS (LNAI), pp. 1–22. Springer, Heidelberg (2006)

[12] Kowalski, R.: Private communication (2007)

[13] Lopes, G., Pereira, L.M.: Prospective logic programming with ACORDA. In: Sutcliffe, G., Schmidt, R., Schulz, S. (eds.) Procs. of the FLoC 2006 Ws. on Empirically Successful Computerized Reasoning, 3rd Intl. J. Conf. on Automated Reasoning, CEUR Workshop Procs., vol. 192 (2006)

[14] Niemelä, I., Simons, P.: Smodels: An implementation of the stable model and well-founded semantics for normal logic programs. In: Fuhrbach, U., Dix, J., Nerode, A. (eds.) LPNMR 1997. LNCS, vol. 1265, pp. 420–429. Springer, Heidelberg (1997)

[15] Pereira, L.M., Saptawijaya, A.: Modelling morality with prospective logic. In: Neves, J.M., Santos, M.F., Machado, J.M. (eds.) EPIA 2007. Procs. 13th Portuguese Intl. Conf. on Artificial Intelligence. LNCS (LNAI), Springer, Heidelberg (2007)

[16] Poole, D.: The independent choice logic for modelling multiple agents under uncertainty. Artificial Intelligence 94(1-2), 7–56 (1997)

[17] Poole, D.: Abducing through negation as failure: Stable models within the independent choice logic. Journal of Logic Programming 44, 5–35 (2000)

[18] Swift, T.: Tabling for non-monotonic programming. Annals of Mathematics and Artificial Intelligence 25(3-4), 201–240 (1999)

An Iterative Process for Building Learning Curves and Predicting Relative Performance of Classifiers

Rui Leite and Pavel Brazdil

LIAAD-INESC Porto L.A./Faculty of Economics,
University of Porto, Rua de Ceuta, 118-6,
4050-190 Porto, Portugal
rleite@liacc.up.pt, pbrazdil@liacc.u.pt

Abstract. This paper concerns the problem of predicting the relative performance of classification algorithms. Our approach requires that experiments are conducted on small samples. The information gathered is used to identify the nearest learning curve for which the sampling procedure was fully carried out. This allows the generation of a prediction regarding the relative performance of the algorithms. The method automatically establishes how many samples are needed and their sizes. This is done iteratively by taking into account the results of all previous experiments - both on other datasets and on the new dataset obtained so far. Experimental evaluation has shown that the method achieves better performance than previous approaches.

1 Introduction

The problem of predicting the relative performance of classification algorithms continues to be an issue of both theoretical and practical interest. There are many algorithms that can be used on any given problem. Although the user can make a direct comparison between the considered algorithms for any given problem using a cross-validation evaluation, it is desirable to avoid this, as the computational costs are significant.

It is thus useful to have a main method that helps to determine which algorithms are more likely to lead to the best results on a new problem (dataset). The common approach of many methods is to store previous experimental results on different datasets. The datasets, including the one in question, are characterized using a set of measures. A (meta-)learning method is used to generate a prediction, for instance, in the form of a relative ordering of the algorithms.

Some methods rely on dataset characteristics such as statistical and information-theory measures [5,3]. However, these measures need to be identified beforehand, which is a non-trivial task.

These difficulties have lead some researchers to explore alternative ways to achieve the same goal. Some authors have used simplified versions of the algorithms referred to as *landmarks* [1,13]. Other researchers have proposed to use

J. Neves, M. Santos, and J. Machado (Eds.): EPIA 2007, LNAI 4874, pp. 87–98, 2007.

the algorithms performance on simplified versions of the data, which are sometimes referred to as *sampling landmarks*. The performance of the algorithms on samples can be used again to estimate their relative performance. The use of previous information about learning curves on representative datasets is essential. Without this, sampling may lead to poor results [12].

Our previous approach [10] used the performance information given by partial learning curves to predict the relative performance on the whole dataset. This method performed better than the one that uses *dataset characteristics*. However, this had one shortcoming. It required the user to choose the sizes of the partial learning curves, *i.e.* the number and sizes of the samples used by the base algorithms.

This choice is not trivial since there is a *tradeoff* between using more information (more samples and larger samples) leading to better predictions and the cost of obtaining this information.

In this paper we propose a new method that comprises two functions AMDS and SetGen: the function AMDS uses the performance of the base algorithms on some given samples to determine which algorithm to use on the whole data, and the function SetGen establishes the sample sizes that should be used by AMDS in order to obtain better performance.

The method uses performance information concerning learning curves on other datasets and also involves conducting some experiments on a new dataset

The planning of these experiments is build up gradually, by taking into account the results of all previous experiments – both on other datasets and on the new dataset obtained so far. This is one important novelty of our approach. Experimental evaluation has shown that the new method can achieve higher performance levels when compared to other approaches.

2   Using Sampling to Predict the Outcome of Learning

The aim is to decide which of the two algorithms (Ap and Aq) is better on a new dataset d. The performance of the algorithms on a set of samples sheds some light to the final results of the learning curves (see Fig. 1). Intuitively this approach should work better if several samples of different sizes are used for each algorithm, letting us perceive the shape of the two learning curves. We assume the existence of available information about how both algorithms perform on different datasets $d_1 \cdots d_n$ for several samples with sizes $s_1 \cdots s_m$. This information defines the relation between the performance of the algorithms on partial learning curves and the relative performance on the whole dataset. The method previously developed [10] and the enhanced version presented here are both based on this intuitive idea. The main difference between the methods is that the new method establishes the sizes of the samples. More samples are introduced if it improves the confidence of the prediction. This method also tries to minimize the costs associated with the training of the algorithms on the chosen samples.

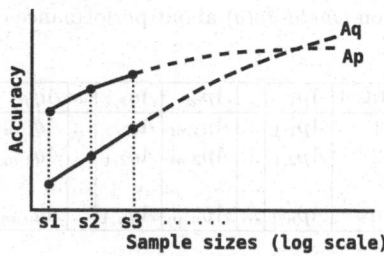

Fig. 1. Partially known learning curves

The improved method comprises two functions: function AMDS which uses the performance of the algorithms on given samples; and function SetGen which establishes the sample sizes that should be used by AMDS. Both are described in more detail in the following section.

2.1 Using Performance on Samples of Specified Sizes for Decision Making (ADMS)

The function AMDS accepts a sequence of sample sizes[1] as an input. For example, consider the case shown in Figure 1. The input can be represented as $\{Ap_1, Ap_2, Ap_3, Aq_1, Aq_2, Aq_3\}$ corresponding to samples of sizes s_1, s_2, s_3 for both algorithms. Bellow we give an overview of the method. The reader can find additional details about this method in our previous work [10].

1. Describe the new dataset d. This step involves the estimation of the algorithms performance when trained on several samples of given sizes.
2. Compute the distances between d and all the other datasets $d_1 \cdots d_n$. Identify the subset of k nearest datasets.
3. For each of the k nearest datasets identified in step 2, adapt each pair of learning curves to the new partial learning curves build for dataset d.
4. For each pair of adapted curves decide which algorithm is better. This is decided by looking at the relative performance at the end of the curves.
5. Identify the algorithm to use on the new dataset d, by considering the results on all k neighbour datasets.

It is clear from the overview that AMDS is a k-NN classifier. The classification problem is predict the algorithm to use (from a set of 2 - Ap and Aq) on new dataset. Our training cases are associated with the datasets and are described by the performances of the base algorithms on several samples (see Table 1).

In the following we give more details concerning this function in particular how we compute distances between the cases (datasets) and how we adapt the retrieved learning curves.

---

[1] The sequences of possible sample sizes are finite (e. g.$\{s_1, s_2, \cdots, s_m\}$) and the sizes grow as a geometric sequence $a_0 \times r^n$. Here we have used sizes 91, 128 etc.

Table 1. Stored information (*meta-data*) about performance of algorithms Ap and Aq on different datasets

Dataset	Ap_1	...	Ap_m	Aq_1	...	Aq_m
1	$Ap_{1,1}$	...	$Ap_{1,m}$	$Aq_{1,1}$	...	$Aq_{1,m}$
2	$Ap_{2,1}$	...	$Ap_{2,m}$	$Aq_{2,1}$	...	$Aq_{2,m}$
...	...	...	...	...	...	...
n	$Ap_{n,1}$	...	$Ap_{n,m}$	$Aq_{n,1}$	...	$Aq_{n,m}$

Compute the Distances Between Datasets

In function AMDS we use a Manhattan distance function. Assuming that the indices of the features computed to characterize the new case are I_p for algorithm Ap and I_q for Aq the following formula computes the distance between datasets d_i and d_j:

$$d(d_i, d_j) = \underbrace{\sum_{s \in I_p} |Ap_{i,s} - Ap_{j,s}|}_{Ap} + \underbrace{\sum_{s \in I_q} |Aq_{i,s} - Aq_{j,s}|}_{Aq} \tag{1}$$

Adapting the Retrieved Learning Curves

Once the nearest cases are identified we can take the decision based on what algorithm was better in most of the times (in these neighborhood).

However the quality of the final decision will be poor if the most similar cases are not similar enough to the given one. The decision will be inaccurate even if we use more information to identify the cases. This fact explains the somewhat surprising finding that if we use more information (i.e. more samples and larger samples), this may not always lead to improvements. This could be one of the reasons for the somewhat discouraging results reported in literature [6,14]. The strategy to overcome this drawback exploits the notion of *adaptation* as in Case-based Reasoning [7,9].

The main idea behind this is not only to retrieve a pair of curves, but to *adapt* it to new circumstances. One straightforward way of doing this is by *moving* each curve of the retrieved pair to the corresponding partial curve available for the new dataset (see Figure 2). This adaptation can be seen as a way of combining the information of the retrieved curve with the information available for the new dataset.

We have designed a simple adaptation procedure which modified a retrieved curve using a *scale coefficient* (named f).

Suppose that the we are considering the curves related to algorithm Ap and the new dataset is d_i. The corresponding partial learning curve is defined by the vector $< Ap_{i,s} >$ where $s \in I_{p,q}$. Suppose also that d_r is one retrieved case and the corresponding partial learning curve (for Ap) is defined by $< Ap_{r,s} >$ where $s \in I_{p,q}$. We adapt the retrieved curve by multiplying each point of $Ap_{r,i}$ by f ($i \in \{1, \cdots, m\}$).

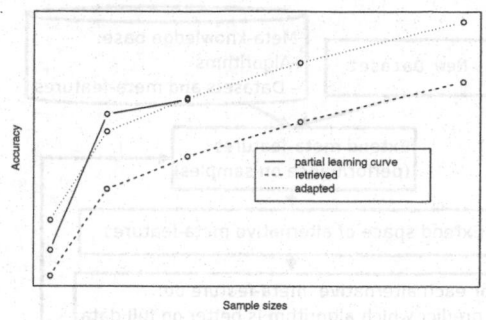

Fig. 2. Adaptation of a retrieved learning curve

The scale coefficient f is the one that minimises the Euclidean distance between the two partial learning curves. Besides we consider that each point has a different weight. The idea is to give more importance to points occurring later on the learning curve. The weights increases as the sample sizes increases. An obvious way to express this idea is to define the weight as the sample size associated to the considered point ($w_j =| S_j |$).

The following equation determines f:

$$ f = \frac{\sum_{s \in I_{p,q}} \left(Ap_{i,s} \times Ap_{r,s} \times w_s^2 \right)}{\sum_{s \in I_{p,q}} \left(Ap_{r,s}^2 \times w_s^2 \right)} \tag{2} $$

The Aq's learning curves are adapted in similar way as it was for Ap.

Identify the Algorithm to Use

After adapting all the retrieved curves the final decision is straighforward. For each pair of learning curves (there is one for each identified case - dataset) we identify which algorithm is the better. The final decision is to use the algorithm that was was better in most of the times (on the adapted curves pairs).

2.2   Generator of Attribute Sequences (SetGen)

We have a method (AMDS) that predicts the relative performance of classification algorithms using their on different samples. The samples are randomly drawn from the given datasets and can only assume some specific sizes $(s_1, s_2, \cdots, s_m)$. The sample sizes are passed to AMDS as an input (e.g. <s1,s2,s3> related to Ap and <s2,s5> related to Aq).

The aim of the algorithm SetGen is to determine how many samples should be used and what their sizes should be. This is not solved using a regular feature selection method (a forward selection) that only tries to improve the method accuracy. In our case it also deals with feature costs.

The samples are chosen taking into account the improvement of the prediction confidence. At the same time the system tries to identify those candidate

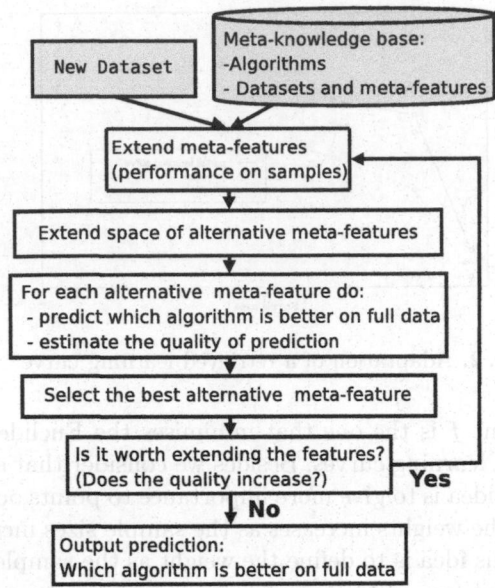

Fig. 3. Iterative process of characterizing the new dataset and determining which algorithm is better

solutions that increase the computational costs by the least amount. The desired sequence of samples is identified using a hill climbing approach.

An overview of the method is presented in Fig.3. It starts with a given sequence of meta-features for the two algorithms (Ap, Aq). Each sequence consists of performance values on a given sequence of samples. The algorithm works in an iterative fashion. In each step it tries to determine how the sequence should be extended. Various alternatives are considered and evaluated. As we have pointed out before, the aim is to improve the confidence of the prediction (which algorithm to use). The best extension is selected and the process is repeated while the confidence of the prediction increases.

Fig. 4 shows more details about how the search is conducted. The algorithm starts with an initial set of attributes $(a0)$. At each step it expands the current state into its successors. The successor states are the sets of attributes obtained by adding a new attribute to the current set. In the example the successors of $(a0)$ are $(a0, a1)$, $(a0, a2)$ etc. For simplicity we use a_j to represent a *generic attribute*. For each state (represented by the particular set of attributes) the (meta-)accuracy of the method is estimated, as well as the cost of obtaining the attributes.

Once a state is defined the next one is chosen among the successors that improve the meta–accuracy by at least Δ (for instance, the value 0.01). Finally, the one with the minimum cost is chosen. In our example (Fig. 4) the algorithm moves from state $(a0)$ to state $(a0, a2)$. In the next step it chooses the state $(a0, a2, a4)$. This continues until none of the successors improves the accuracy by at least Δ.

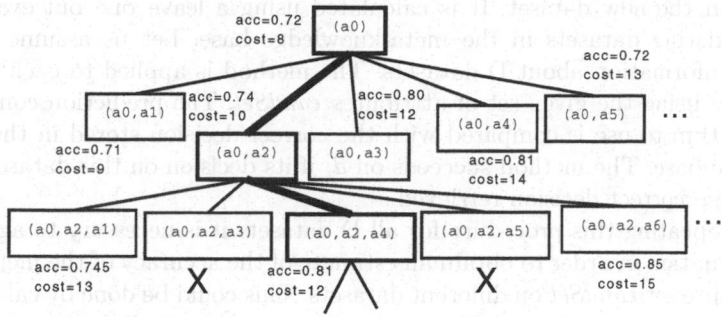

Fig. 4. An example of search in the attribute space

Let us clarify the issue of what the generic attributes used here represent. Each particular a_j represents either a particular attribute of algorithm Ap_j (representing the performance of Ap on sample of size s_j), or some attribute of algorithm Aq_k or a combination of both attributes (for instance (Ap_4 and Aq_6)). This means that on each iteration the SetGen method can either insert a new attribute related to one particular algorithm (say Ap) or insert two attributes one related to Ap and another related to Aq. This allows the method to extend either one of the partial learning curves or both at the same time. In the example presented in Fig. 4 the generic attributes are defined using the following matrix:

		Ap			
		0	1	2	3 ⋯
	0	a0	a1	a4	a9 ⋯
	1	a2	a3	a5	a10 ⋯
Aq	2	a6	a7	a8	a11 ⋯
	3	a12	a13	a14	a15 ⋯
	⋯	⋯	⋯	⋯	⋯

For instance the generic attribute $a2$ is defined using just the attribute Aq_1 while $a5$ is defined using attributes Ap_2 and Aq_1.

Another aspect needs still to be clarified. When the method attempts to extend a particular state, it verifies whether the extension is *valid*. Valid extensions introduce attributes that are in general *more informative* than those used so far. For instance, the attribute Ap_3 is considered more informative than Ap_2, as the corresponding sample is larger. Therefore, the system can, add Ap_3 to the set containing Ap_2, but the opposite is not possible. In Fig. 4 some extensions of states that were considered, but are in fact invalid, are marked by "X".

Evaluation of Attribute Sequences

As described earlier, each state is characterized by two different values, accuracy and computational cost. The algorithm needs to anticipate what would be the accuracy of the method if it used a specific set of attributes (*candSet*). This estimate, evaluated using *EAcc* function, expresses the chances of success of the method on the new dataset. It is calculated using a leave–one–out evaluation on the existing datasets in the meta-knowledge base. Let us assume that it contains information about D datasets. The method is applied to each dataset d_i of D by using the given set of attributes *candSet*. The prediction concerning the algorithm to use is compared with the correct decision stored in the meta-knowledge base. The method succeeds on d_i if its decision on this dataset is the same as the correct decision retrieved.

After repeating this procedure for all D datasets it is necessary to aggregate this information in order to obtain an estimate of the accuracy of the method for the attribute set *candSet* on different datasets. This could be done by calculating an average. However, this would give equal importance to all datasets, without regarding whether they are similar to the dataset in question or not. For this reason the method uses a *weighted average* that corrects this shortcoming. The following equations define the *EAcc* function.

$$EAcc(c) = (1 + \sum_{d_i \in D} [success(d_i, candSet) \times w_i])/2 \qquad (3)$$

where D is the set of known datasets, w_i represents a weight and $success(...)$ is +1 (-1) if the method gives the correct (wrong) answer for d_i using the attributes indicated in candSet. The weight is calculated as follows: $w_i = w_i'/\sum_k w_k'$ where $w_i' = 1/(dist(d, d_i) + 1)$. The computational cost estimate is calculated by using a similar leave–one–out mode. For each dataset d_i the meta–database is used to retrieve the training time spent on evaluating all the attributes in *candSet*. The final value is the weighted average of times for the specified attributes.

$$ECost(candSet) = \sum_{d_i \in D} Cost(d_i, candSet) \times w_i \qquad (4)$$

This provides an estimate for the training time needed to compute the same attributes on a new dataset.

3 Empirical Evaluation

The decision problem concerning whether algorithm A_p is better than A_q can be seen as a classification problem which can have three different outcomes: 1 (or -1) if algorithm Ap gives significantly better (or worse) results than A_q, or 0 if they are not significantly different.

In the experimental study our first aim was to determine the accuracy and the computational cost of our approach (AMDS–SetGen). Additionally we also aimed at comparing these results to a previous method [3] which relies on dataset characteristics instead. Since the method was designed to rank algorithms we made some changes to fit our problem. This method can be breefly described by:

1. Compute the characterization measures for all datasets (including the new one).
2. Compute the distance between the new dataset and the stored ones.

3. Choose the k stored datasets (neighbours) that are "nearest" to the new dataset (according to the distance).
4. Use the algorithm that was most often the best one on the identified nearest datasets.

The predicted class was compared with the true classification determined by a usual cross–validation evaluation procedure on each dataset for the two given algorithms. A statistical test (t–test) was used to compute the statistical significance.

Instead of using the usual accuracy measure, we have used a different measure that is more suited for our classification task with 3 possible outcomes. The errors are called *penalties* and are calculated as follows. If a particular method (e.g. AMDS–SetGen) classifies some case as +1 (or -1), while the true class is 0 (the given algorithms are not significantly different) then, from a practical point of view the method did not fail, because any decision is a good one. Therefore the penalty is 0. However if the method classifies the dataset as 0, while the true class is +1 (or -1) then we consider that the method partially failed. The penalty is 0.5. If the classification is +1 (or -1), while the true class is -1 (or +1), this counts as a complete failure, and the penalty is set to 1. The corresponding accuracy, referred to as *meta–accuracy*, is computed using this formula $1 - \sum_{i \in D} \frac{penalty(i)}{|D|}$.

In this empirical study we have used the following 5 classification algorithms, all implemented within Weka [15] machine learning tools: J48 (C4.5 implemented in weka), JRip - rule set learner (RIPPER [4]), logistic [8], MLP - multi-layer perceptron, and Naive Bayes. Using this setting we get 10 classification problems, one for each pair of algorithms. We have used 40 datasets in the evaluation. Some come from UCI [2], others were used in the project METAL [11]. The evaluation procedure used to estimate the accuracy of the methods is the leave–one–out[2] to on the datasets.

In Figure 5 we present the performance reached by the methods in terms of meta–accuracies. The AMDS–SetGen results have been obtained with the parameter Δ=0.07. The figure also shows the performance of MDC and the default accuracy on each meta-classification problem. The default accuracy measures the performance of the decision based on the most often best algorithm on the known datasets. For instance if Ap is on 60% of the times better than Aq, and worst on 40% of the times then the accuracy is 60%.

On average with AMDS-SetGen the accuracy improves by 12.2% (from 79.6% to 91.8%) when compared to the default accuracy. Moreover, our method improves the performance by 11% when compared to MDC (that reaches 80.8%), while MDC only improves the performance by 1.2% when compared to the default accuracy. Our method gives better results than MDC on all problems. The largest improvements are observed on the hardest problems where the default accuracy is close to 50%. In such situations our method can attain nearly 25% of improvement.

---

[2] We plan to use a Bootstrap-based error estimator to reduce the variance usually associated with leave-one-out error estimators.

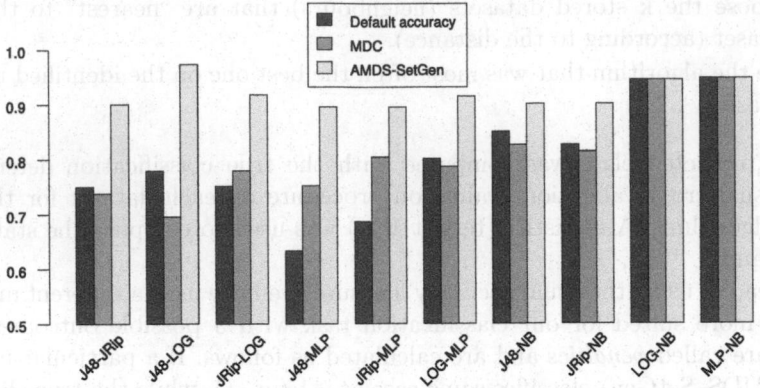

Fig. 5. Result concerning meta–accuracies

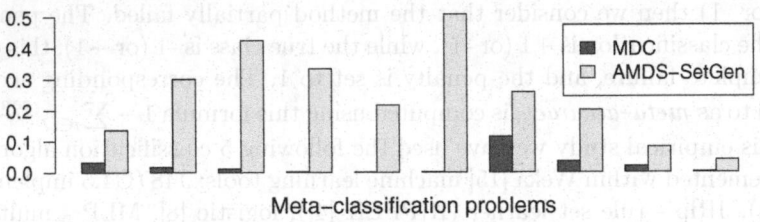

Fig. 6. Results concerning costs

Regarding computational costs the results of ADMS-SetGen and MDC are compared to the cost of deciding which algorithm is better using cross–validation (CV). To express the run–time savings, Figure 6 shows the ratio between of the time spent by AMDS–SetGen and the time spent by CV. In most cases AMDS–SetGen is much faster than CV (the ratio is on average about 0.2).

3.1 Further Experimental Results and Discussion

Variation of Δ and its effects: We briefly discuss why we have chosen Δ=0.07. We have conducted further experiments by using the following values for the Δ parameter: 0.2, 0.1, 0.07, 0.05, 0.02, 0.01, 0.005, 0.002, 0.001. If we use small Δ values the meta–accuracy increases, at the same time as the increasing of computational time. The value of $\Delta = 0.07$ seems to be the best compromise, leading, on average, to a meta–accuracy of 0.92 and to a computational time (expressed as a ratio) of 0.2.[3]

---

[3] If the Δ is set to 0.05 the meta–accuracy remains almost the same (0.92) but the time increases (ratio 0.3). If on the other hand Δ is adjusted to 0.1, the meta–accuracy decreases to 0.88 and the method becomes faster (ratio 0.16).

AMDS using fixed samples: The user can choose some simple samples like s_j and also combinations like $(s1,s2)$, $(s1,s2,s3)$ etc. Although the method performs reasonably well the SetGen method is generally more accurate.

A typical case of selected samples: Typically the samples selected by Set-Gen are not the same for both algorithms and the sample sequences can have gaps. The following example shows the selected samples obtained for J48 *vs* JRip while making the prediction for dataset *musk*: J48 samples=$(s1, s5)$ JRip samples=$(s1, s4)$. This is a typical case. Having gaps which happens frequently represents a useful feature as it allows time savings.

Extension to N classifiers: Although this paper focuses on the problem of determining which of the 2 given classification algorithms should be used on a new dataset, it does not represent a serious limitation. If we had N classification algorithms and wanted to find out which one(s) should be used, one could conduct pairwise tests and use the one that wins more often. If the number of alternatives is larger, it is advisable to divide the algorithms into subgroups and establish a winner of each subgroup and then repeat the process. However, this issue exceeds the aims of this paper.

4   Conclusions

In this paper we have described a meta-learning scheme that exploits sampling to determine which of the two given methods is better. The method automatically establishes how many samples are needed and their sizes.

The method uses stored performance information concerning learning curves on other datasets to guide the construction of the sequences of samples to be used on a new dataset. The method involves conducting some experiments on the new dataset. The plan of these experiments is built up gradually, by taking into account the results of all previous experiments - both on other datasets and on the new dataset obtained so far.

The method extend the sequence of samples to improve the results. It tries to determine whether this would improve the confidence of the prediction, while controlling the process and looking for extensions with minimal costs. This is one important novelty of this approach.

Experimental evaluation has shown that the new method achieves good performance. The (meta-)accuracy is on average 91.8% , which is higher than any of the approaches reported before. Besides, the user does not have to be concerned with the problem of determining the sizes of the samples. This represents a significant improvement over previous methods in dealing with the problem of predicting the relative performance of learning algorithms in a systematic manner.

References

1. Bensussan, H., Giraud-Carrier, C.: Discovering task neighbourhoods through landmark learning performances. In: Zighed, A.D.A., Komorowski, J., Żytkow, J.M. (eds.) PKDD 2000. LNCS (LNAI), vol. 1910, pp. 325–330. Springer, Heidelberg (2000)

2. Blake, C., Merz, C.: UCI repository of machine learning databases (1998), http://www.ics.uci.edu/~mlearn/mlrepository.html
3. Brazdil, P., Soares, C., Costa, J.: Ranking learning algorithms: Using ibl and meta-learning on accuracy and time results. Machine Learning 50, 251–277 (2003)
4. Cohen, W.W.: Fast effective rule induction. In: Prieditis, A., Russell, S. (eds.) Proc. of the 12th International Conference on Machine Learning, Tahoe City, CA, July 9–12, pp. 115–123. Morgan Kaufmann, San Francisco (1995)
5. Michie, D., Spiegelhalter, D.J., Taylor, C.C.: Machine Learning, Neural and Statistical Classification. Ellis Horwood (1994)
6. Fürnkranz, J., Petrak, J.: An evaluation of landmarking variants. In: IDDM 2001. Proceedings of the ECML/PKDD Workshop on Integrating Aspects of Data Mining, Decision Support and Meta-Learning, pp. 57–68. Springer, Heidelberg (2001)
7. Kolodner, J.: Case-Based Reasoning. Morgan Kaufmann, San Francisco (1993)
8. le Cessie, S., van Houwelingen, J.: Ridge estimators in logistic regression. Applied Statistics 41(1), 191–201 (1992)
9. Leake, D.B.: Case-Based Reasoning: Experiences, Lessons & Future Directions. AAAI Press, Stanford, California, USA (1996)
10. Leite, R., Brazdil, P.: Predicting relative performance of classifiers from samples. In: ICML 2005. Proceedings of the 22nd international conference on Machine learning, pp. 497–503. ACM Press, New York (2005)
11. Metal project site (1999), http://www.metal-kdd.org/
12. Perlich, C., Provost, F., Simonoff, J.S.: Tree induction vs. logistic regression: a learning-curve analysis. J. Mach. Learn. Res. 4, 211–255 (2003)
13. Pfahringer, B., Bensusan, H., Giraud-Carrier, C.: Meta-learning by landmarking various learning algorithms. In: ICML 2000. Proceedings of the 17th International Conference on Machine Learning, Stanford, CA, pp. 743–750 (2000)
14. Soares, C., Petrak, J., Brazdil, P.: Sampling-based relative landmarks: Systematically test-driving algorithms before choosing. In: Brazdil, P.B., Jorge, A.M. (eds.) EPIA 2001. LNCS (LNAI), vol. 2258, pp. 88–94. Springer, Heidelberg (2001)
15. Witten, I., Frank, E., Trigg, L., Hall, M., Holmes, G., Cunningham, S.: Weka: Practical machine learning tools and techniques with java implementations (1999)

Modelling Morality with Prospective Logic

Luís Moniz Pereira[1] and Ari Saptawijaya[2]

[1] CENTRIA, Universidade Nova de Lisboa, 2829-516 Caparica, Portugal
lmp@di.fct.unl.pt
[2] Fakultas Ilmu Komputer, Universitas Indonesia, 16424 Depok, Jawa Barat, Indonesia
saptawijaya@cs.ui.ac.id

Abstract. This paper shows how moral decisions can be drawn computationally by using prospective logic programs. These are employed to model moral dilemmas, as they are able to prospectively look ahead at the consequences of hypothetical moral judgments. With this knowledge of consequences, moral rules are then used to decide the appropriate moral judgments. The whole moral reasoning is achieved via a priori constraints and a posteriori preferences on abductive stable models, two features available in prospective logic programming. In this work we model various moral dilemmas taken from the classic trolley problem and employ the principle of double effect as the moral rule. Our experiments show that preferred moral decisions, i.e. those following the principle of double effect, are successfully delivered.

1 Introduction

Morality no longer belongs only to the realm of philosophers. Recently, there has been a growing interest in understanding morality from the scientific point of view. This interest comes from various fields, e.g. primatology [4], cognitive sciences [11, 18], neuroscience [23], and other various interdisciplinary perspectives [12, 14]. The study of morality also attracts the artificial intelligence community from the computational perspective, and has been known by several names, including machine ethics, machine morality, artificial morality, and computational morality. Research on modelling moral reasoning computationally has been conducted and reported on, e.g. AAAI 2005 Fall Symposium on Machine Ethics [10, 22].

There are at least two reasons to mention the importance of studying morality from the computational point of view. First, with the current growing interest to understand morality as a science, modelling moral reasoning computationally will assist in better understanding morality. Cognitive scientists, for instance, can greatly benefit in understanding complex interaction of cognitive aspects that build human morality or even to extract moral principles people normally apply when facing moral dilemmas. Modelling moral reasoning computationally can also be useful for intelligent tutoring systems, for instance to aid in teaching morality to children. Second, as artificial agents are more and more expected to be fully autonomous and work on our behalf, equipping agents with the capability to compute moral decisions is an indispensable requirement. This is particularly true when the agents are operating in domains where moral dilemmas occur, e.g. in health care or medical fields.

J. Neves, M. Santos, and J. Machado (Eds.): EPIA 2007, LNAI 4874, pp. 99–111, 2007.

Our ultimate goal within this topic is to provide a general framework to model morality computationally. This framework should serve as a toolkit to codify arbitrarily chosen moral rules as declaratively as possible. We envisage that logic programming is an appropriate paradigm to achieve our purpose. Continuous and active research in logic programming has provided us with necessary ingredients that look promising enough to model morality. For instance, default negation is suitable for expressing exception in moral rules, abductive logic programming [13, 15] and stable model semantics [8] can be used to generate possible decisions along with their moral consequences, and preferences are appropriate for preferring among moral decisions or moral rules [5, 6].

In this paper, we present our preliminary attempt to exploit these enticing features of logic programming to model moral reasoning. In particular, we employ prospective logic programming [16, 19], an on-going research project that incorporates these features. For the moral domain, we take the classic trolley problem of Philippa Foot [7]. This problem is challenging to model since it contains a family of complex moral dilemmas. To make moral judgments on these dilemmas, we model the principle of double effect as the basis of moral reasoning. This principle is chosen by considering empirical research results in cognitive science [11] and law [18], that show the consistency of this principle to justify similarities of judgments by diverse demographically populations when given this set of dilemmas.

Our attempt to model moral reasoning on this domain shows encouraging results. Using features of prospective logic programming, we can conveniently model both the moral domain, i.e. various moral dilemmas of the trolley problem, and the principle of double effect declaratively. Our experiments on running the model also successfully deliver moral judgments that conform to the human empirical research results.

We organize the paper as follows. First, we discuss briefly and informally prospective logic programming, in Section 2. Then, in Section 3 we explain the trolley problem and the principle of double effect. We detail how we model them in prospective logic programming together with the results of our experiments regarding that model, in Section 4. Finally, we conclude and discuss possible future work, in Section 5.

2   Prospective Logic Programming

Prospective logic programming enables an evolving program to look ahead prospectively its possible future states and to prefer among them to satisfy goals [16, 19]. This paradigm is particularly beneficial to the agents community, since it can be used to predict an agent's future by employing the methodologies from abductive logic programming [13, 15] in order to synthesize and maintain abductive hypotheses.

Figure 1 shows the architecture of agents that are based on prospective logic [19]. Each prospective logic agent is equipped with a knowledge base and a moral theory as its initial theory. The problem of prospection is then of finding abductive extensions to this initial theory which are both relevant (under the agent's current goals) and preferred (w.r.t. preference rules in its initial theory). The first step is to select the goals that the agent will possibly attend to during the prospective cycle. Integrity constraints are also considered here to ensure the agent always performs transitions into valid evolution states. Once the set of active goals for the current state is known, the next step

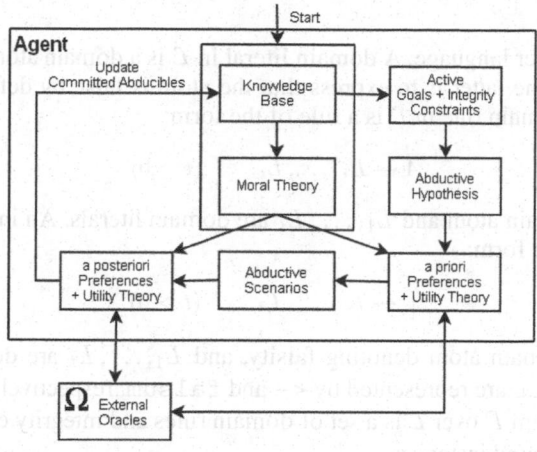

Fig. 1. Prospective logic agent architecture

is to find out which are the relevant abductive hypotheses. This step may include the application of a priori preferences, in the form of contextual preference rules, among available hypotheses to generate possible abductive scenarios. Forward reasoning can then be applied to abducibles in those scenarios to obtain relevant consequences, which can then be used to enact a posteriori preferences. These preferences can be enforced by employing utility theory and, in a moral situation, also moral theory. In case additional information is needed to enact preferences, the agent may consult external oracles. This greatly benefits agents in giving them the ability to probe the outside environment, thus providing better informed choices, including the making of experiments. The mechanism to consult oracles is realized by posing questions to external systems, be they other agents, actuators, sensors or other procedures. Each oracle mechanism may have certain conditions specifying whether it is available for questioning. Whenever the agent acquires additional information, it is possible that ensuing side-effects affect its original search, e.g. some already considered abducibles may now be disconfirmed and some new abducibles are triggered. To account for all possible side-effects, a second round of prospection takes place.

ACORDA is a system that implements prospective logic programming and is based on the above architecture. ACORDA is implemented based on the implementation of EVOLP [1] and is further developed on top of XSB Prolog[1]. In order to compute abductive stable models [5, 6], ACORDA also benefits from the XSB-XASP interface to Smodels[2].

In this section, we discuss briefly and informally prospective logic programming and some constructs from ACORDA that are relevant to our work. For a more detailed discussion on prospective logic programming and ACORDA, interested readers are referred to the original paper [16, 19].

---

[1] XSB Prolog is available at http://xsb.sourceforge.net

[2] Smodels is available at http://www.tcs.hut.fi/Software/smodels

2.1 Language

Let $\mathcal{L}$ be a first order language. A domain literal in $\mathcal{L}$ is a domain atom A or its default negation $not\ A$. The latter is to express that the atom is false by default (close world assumption). A domain rule in $\mathcal{L}$ is a rule of the form:

$$A \leftarrow L_1, \ldots, L_t. \quad (t \geq 0)$$

where A is a domain atom and $L_1, \ldots, L_t$ are domain literals. An integrity constraint in $\mathcal{L}$ is a rule of the form:

$$\perp \leftarrow L_1, \ldots, L_t. \quad (t > 0)$$

where $\perp$ is a domain atom denoting falsity, and $L_1, \ldots, L_t$ are domain literals. In ACORDA, $\leftarrow$ and $\perp$ are represented by `<-` and `falsum`, respectively.

A (logic) program P over $\mathcal{L}$ is a set of domain rules and integrity constraints, standing for all their ground instances.

2.2 Abducibles

Every program P is associated with a set of abducibles $A \subseteq \mathcal{L}$. Abducibles can be seen as hypotheses that provide hypothetical solutions or possible explanations of given queries.

An abducible A can be assumed only if it is a considered one, i.e. it is expected in the given situation, and moreover there is no expectation to the contrary [5, 6].

$$consider(A) \leftarrow expect(A), not\ expect\_not(A).$$

The rules about expectations are domain-specific knowledge contained in the theory of the program, and effectively constrain the hypotheses which are available.

In addition to mutually exclusive abducibles, ACORDA also allows sets of abducibles. Hence, an abductive stable model may contain more than a single abducible. To enforce mutually exclusive abducibles, ACORDA provides predicate `exclusive/2`. The use of this predicate will be illustrated later, when we model morality in a subsequent section.

2.3 A Posteriori Preferences

Having computed possible scenarios, represented by abductive stable models, more favourable scenarios can be preferred among them a posteriori. Typically, a posteriori preferences are performed by evaluating consequences of abducibles in abductive stable models. The evaluation can be done quantitatively (for instance by utility functions) or qualitatively (for instance by enforcing some rules to hold). When currently available knowledge is insufficient to prefer among abductive stable models, additional information can be gathered, e.g. by performing experiments or consulting an oracle.

To realize a posteriori preferences, ACORDA provides predicate `select/2` that can be defined by users following some domain-specific mechanism for selecting favoured abductive stable models. The use of this predicate to perform a posteriori preferences in a moral domain will be discussed in a subsequent section.

3   The Trolley Problem and the Principle of Double Effect

Several interesting results have emerged from recent interdisciplinary studies on morality. One common result from these studies shows that morality has evolved over time. In particular, Hauser, in his recent work, argues that a moral instinct, playing the role of generating rapid judgments about what is morally right or wrong, has evolved in our species [11].

Hauser [11] and Mikhail [18] propose a framework of human moral cognition, known as universal moral grammar, analogously to Chomsky's universal grammar in language. Universal moral grammar, which can be culturally adjusted, provides universal moral principles that enable an individual to unconsciously evaluate what actions are permissible, obligatory, or forbidden. To support this idea, Hauser and Mikhail independently created a test to assess moral judgments of subjects from demographically diverse populations, using the classic trolley problem. Despite their diversity, the result shows that most subjects widely share moral judgments when given a moral dilemma from the trolley problem. Although subjects are unable to explain the moral rules in their attempts at justification, their moral judgments are consistent with a moral rule known as the principle of double effect.

The trolley problem presents several moral dilemmas that inquire whether it is permissible to harm one or more individuals for the purpose of saving others. In all cases, the initial circumstances are the same [11]:

> There is a trolley and its conductor has fainted. The trolley is headed toward five people walking on the track. The banks of the track are so steep that they will not be able to get off the track in time.

Given the above initial circumstance, in this work we consider six classical cases of moral dilemmas, employed for research on morality in people [18].

1. **Bystander.** Hank is standing next to a switch, which he can throw, that will turn the trolley onto a parallel side track, thereby preventing it from killing the five people. However, there is a man standing on the side track with his back turned. Hank can throw the switch, killing him; or he can refrain from doing this, letting the five die. Is it morally permissible for Hank to throw the switch?
2. **Footbridge.** Ian is on the footbridge over the trolley track. He is next to a heavy object, which he can shove onto the track in the path of the trolley to stop it, thereby preventing it from killing the five people. The heavy object is a man, standing next to Ian with his back turned. Ian can shove the man onto the track, resulting in death; or he can refrain from doing this, letting the five die. Is it morally permissible for Ian to shove the man?
3. **Loop Track.** Ned is standing next to a switch, which he can throw, that will temporarily turn the trolley onto a loop side track. There is a heavy object on the side track. If the trolley hits the object, the object will slow the train down, giving the five people time to escape. The heavy object is a man, standing on the side track with his back turned. Ned can throw the switch, preventing the trolley from killing the five people, but killing the man. Or he can refrain from doing this, letting the five die. Is it morally permissible for Ned to throw the switch?

4. **Man-in-front.** Oscar is standing next to a switch, which he can throw, that will temporarily turn the trolley onto a side track. There is a heavy object on the side track. If the trolley hits the object, the object will slow the train down, giving the five people time to escape. There is a man standing on the side track in front of the heavy object with his back turned. Oscar can throw the switch, preventing the trolley from killing the five people, but killing the man. Or he can refrain from doing this, letting the five die. Is it morally permissible for Oscar to throw the switch?

5. **Drop Man.** Victor is standing next to a switch, which he can throw, that will drop a heavy object into the path of the trolley, thereby stopping the trolley and preventing it from killing the five people. The heavy object is a man, who is standing on a footbridge overlooking the track. Victor can throw the switch, killing him; or he can refrain from doing this, letting the five die. Is it morally permissible for Victor to throw the switch?

6. **Collapse Bridge.** Walter is standing next to a switch, which he can throw, that will collapse a footbridge overlooking the tracks into the path of the trolley, thereby stopping the train and preventing it from killing the five people. There is a man standing on the footbridge. Walter can throw the switch, killing him; or he can refrain from doing this, letting the five die. Is it morally permissible for Walter to throw the switch?

Interestingly, although all cases have the same goal, i.e. to save five albeit killing one, subjects come to different judgments on whether the action to reach the goal is permissible or impermissible. As reported by Mikhail [18], the judgments appear to be widely shared among diverse demographically populations, the summary being given in Table 1.

Table 1. Summary of moral judgments for the trolley problem

Case	Judgment
1. Bystander	Permissible
2. Footbridge	Impermissible
3. Loop Track	Impermissible
4. Man-in-front	Permissible
5. Drop Man	Impermissible
6. Collapse Bridge	Permissible

Although subjects have difficulties to uncover which moral rules they apply for reasoning in these cases, their judgments appear to be consistent with the so-called the principle of double effect. The principle can be expressed as follows [11]:

Harming another individual is permissible if it is the foreseen consequence of an act that will lead to a greater good; in contrast, it is impermissible to harm someone else as an intended means to a greater good.

The key expression here is "intended means". We shall refer in the subsequent sections to the action of harming someone as an intended means, as an intentional killing.

4   Modelling Morality in ACORDA

It is interesting to model the trolley problem in ACORDA due to the intricacy that arises from the dilemma itself. Moreover, there are similarities and also differences between cases. Some cases even exhibit subtle differences. Consequently, this adds complexity to the process of modelling them in order to deliver appropriate moral decisions through reasoning. By appropriate moral decisions we mean the ones that conform with those the majority of people make, in adhering to the principle of double effect.

We model each case of the trolley problem in ACORDA separately. The principle of double effect is modelled via a priori constraints and a posteriori preferences. To assess how flexible is our model of the moral rule, we additionally model another variant for the cases of Footbridge and Loop Track. Even for these variants, our model of the moral rule allows the reasoning to deliver moral decisions as expected.

In each case of the trolley problem, there are always two possible decisions to make. One of these is the same for all cases, i.e. letting the five people die by merely watching the train go straight. The other decision depends on the cases, i.e. between throwing the switch or shoving a heavy man with the purpose to save the five people, but also harming a person in the process.

In this work, these two possible decisions are modelled in ACORDA as abducibles. Moral decisions are made by computing abductive stable models and then preferring among them those models with the abducibles and consequences that conform to the principle of double effect.

Due to space constraints, in subsequent sections we only detail the model for the cases of Bystander and Footbridge in ACORDA. We also show how to model the principle of double effect. Then we present some results of running our models in the ACORDA system.

4.1   Modelling the Bystander Case

Facts to describe that there is a man (here, named John) standing on the side track can be modelled simply as the following:

```
side_track(john).
human(john).
```

The clauses `expect(watching)` and `expect(throwing_switch)` in the following model indicate that watching and throwing the switch, respectively, are two available abducibles, that represent possible decisions Hank has. The other clauses represent the chain of actions and consequences for every abducible.

The predicate `end(die(5))` represents the final consequence if `watching` is abduced, i.e. it will result in five people dying, whereas `end(save_men,ni_kill(N))` represents the final consequence if `throwing_switch` is abduced, i.e. it will save the five people without intentionally killing someone. The way of representing these two consequences is chosen differently, because the different nature of these two abducibles. Merely watching the trolley go straight is an omission of action that just has negative consequence, whereas throwing the switch is an action that is performed to achieve a goal and additionally has negative consequence. Since abducibles in other cases of the trolley problem also share this property, this way of representation will be used throughout them. The predicate `observed_end` is used to encapsulate these two different means of representation, useful later when we model the principle of double effect, to avoid floundering.

```
expect(watching).
train_straight <- consider(watching).
end(die(5)) <- train_straight.
observed_end <- end(X).

expect(throwing_switch).
redirect_train <- consider(throwing_switch).
kill(1) <- human(X), side_track(X), redirect_train.
end(save_men,ni_kill(N)) <- redirect_train, kill(N).
observed_end <- end(X,Y).
```

We can model the exclusiveness of the two possible decisions, i.e. Hank has to decide either to throw the switch or merely watch, by using the `exclusive/2` predicate of ACORDA:

```
exclusive(throwing_switch,decide).
exclusive(watching,decide).
```

Note that the exclusiveness between two possible decisions also holds in other cases.

4.2 Modelling the Footbridge Case
We represent the fact of a heavy man (here, also named John) on the footbridge standing near to Ian similarly to the Bystander case:

```
stand_near(john).
human(john).
heavy(john).
```

We can make this case more interesting by additionally having another (inanimate) heavy object, e.g. rock, on the footbridge near to Ian and see whether our model of the moral rule still allows the reasoning to deliver moral decisions as expected:

```
stand_near(rock).
inanimate_object(rock).
heavy(rock).
```

Alternatively, if we want only to have either a man or an inanimate object on the footbridge next to Ian, we can model it by using an even loop over default negation:

```
stand_near(john) <- not stand_near(rock).
stand_near(rock) <- not stand_near(john).
```

In the following we show how to model the action of shoving an object as an abducible, together with the chain of actions and consequences for this abducible. The model for the decision of merely watching is the same as in the case of Bystander. Indeed, since the decision of watching is always available for other cases, we use the same modelling in every case.

```
expect(shove(X)) <- stand_near(X).
on_track(X) <- consider(shove(X)).
stop_train(X) <- on_track(X), heavy(X).
kill(1) <- human(X), on_track(X).
kill(0) <- inanimate_object(X), on_track(X).
end(save_men,ni_kill(N)) <- inanimate_object(X), stop_train(X),
                                kill(N).
end(save_men,i_kill(N)) <- human(X), stop_train(X), kill(N).
observed_end <- end(X,Y).
```

Note that the action of shoving an object is only possible if there is an object near Ian to shove, hence the clause `expect(shove(X)) <- stand_near(X)`. We also have two clauses that describe two possible final consequences. The clause with the head `end(save_men,ni_kill(N))` deals with the consequence of reaching the goal, i.e. saving five, but not intentionally killing someone (in particular, without killing anyone in this case). To the contrary, the clause with the head `end(save_men,i_kill(N))` expresses the consequence of reaching the goal but involving an intentional killing.

4.3 Modelling the Principle of Double Effect

The principle of double effect can be modelled by using a combination of integrity constraints and a posteriori preferences.

Integrity constraints are used for two purposes. First, we need to observe the final consequences or endings of each possible decision to enable us later to morally prefer decisions by considering the greater good between possible decisions. To achieve this, we can use the integrity constraint `falsum <- not observed_end`. This integrity constraint enforces all available decisions to be abduced together with their consequences, by computing all possible observable hypothetical endings using all possible abductions. Indeed, to be able to reach a moral decision, all hypothetical scenarios afforded by the abducibles must lead to an observable ending. Second, we also need to rule out impermissible actions, i.e. actions that involve intentional killing in the process of reaching the goal. This can be enforced by specifying the integrity constraint `falsum <- intentional_killing`. Intentional killing can be easily defined as follows:

```
intentional_killing <- end(save_men,i_kill(Y)).
```

The above integrity constraints serve as the first filtering function of our abductive stable models, by ruling out impermissible actions (the latter being coded by abducibles). In other words, integrity constraints already afford us with just those abductive stable models that contain only permissible actions.

Additionally, one can prefer among permissible actions those resulting in greater good. This can be realized by a posteriori preferences that evaluate the consequences of permissible actions and then prefer the one with greater good. The following definition of `select/2` achieves this purpose. The first argument of this predicate refers to the set of initial abductive stable models to prefer, whereas the second argument refers to the preferred ones. The auxiliary predicate `select/3` only keeps abductive stable models that contain decisions with greater good of consequences. In the trolley problem, the greater good is evaluated by a utility function concerning the number of people that die as a result of possible decisions. This is realized in the definition of predicate `select/3` by comparing final consequences that appear in the initial abductive stable models. The first clause of `select/3` is the base case. The second clause and the third clause together eliminate abductive stable models containing decisions with worse consequences, whereas the fourth clause will keep those models that contain decisions with greater good of consequences.

```
select(Xs,Ys) :- select(Xs,Xs,Ys).
```

```
select([],_,[]).
select([X|Xs],Zs,Ys) :-
    member(end(die(N)),X),
    member(Z,Zs),
    member(end(save_men,ni_kill(K)),Z), N > K,
    select(Xs,Zs,Ys).
select([X|Xs],Zs,Ys) :-
    member(end(save_men,ni_kill(K)),X),
    member(Z,Zs),
    member(end(die(N)),Z), N =< K,
    select(Xs,Zs,Ys).
select([X|Xs],Zs,[X|Ys]) :- select(Xs,Zs,Ys).
```

Recall the variant of the case Footbridge, where either a man or an inanimate object is on the footbridge next to Ian. This exclusive alternative is specified by an even loop over default negation and we have an abductive stable model that contains the consequence of letting die the five people when a rock next to Ian. This model is certainly *not* the one we would like our moral reasoner to prefer. The following replacement definition of `select/2` accomplishes this case.

```
select([],[]).
select([X|Xs],Ys) :-
    member(end(die(N)),X),
    member(stand_near(rock),X),
    select(Xs,Ys).
select([X|Xs],[X|Ys]) :- select(Xs,Ys).
```

It is important to note that in this case, since either a man or a rock is near to Ian, and the model with shoving a man is already ruled out by our integrity constraint, there is no need to consider greater good in terms of the number of people that die. This means, as shown subsequently, that only two abductive stable models are preferred, i.e. the model with watching as the abducible whenever a man is standing near to Ian, the other being the model with shoving the rock as the abducible.

4.4   Running the Models in ACORDA

We report now on the experiments of running our models in ACORDA. Table 2 gives a summary of all cases of the trolley problem. Column Initial Models contains info about the abductive stable models obtained before a posteriori preferences are applied,

Table 2. Summary of experiments in ACORDA

Case	Initial Models	Final Models
Bystander	`[throwing_switch],[watching]`	`[throwing_switch]`
Footbridge(a)	`[watching],[shove(rock)]`	`[shove(rock)]`
Footbridge(b)	`[watching,stand_near(john)],` `[watching,stand_near(rock)],` `[shove(rock)]`	`[watching,stand_near(john)],` `[shove(rock)]`
Loop Track(a)	`[throwing_switch(right,rock)]` `[watching]`	`[throwing_switch(right,rock)]`
Loop Track(b)	`[watching,side_track(john)],` `[watching,side_track(rock)],` `[throwing_switch(rock)]`	`[watching,side_track(john)],` `[throwing_switch(rock)]`
Man-in-front	`[watching],` `[throwing_switch(rock)]`	`[throwing_switch(rock)]`
Drop Man	`[watching]`	`[watching]`
Collapse Bridge	`[watching]` `[throwing_switch(bridge)]`	`[throwing_switch(bridge)]`

whereas column Final Models those after a posteriori preferences are applied. Here, only relevant literals are shown.

Note that entry Footbridge(a) refers to the variant of Footbridge where both a man and a rock are near to Ian, and Footbridge(b) where either a man or a rock is near to Ian. Loop Track(a) refers to the variant of Loop Track where there are two loop tracks, with a man on the left loop track and a rock on the right loop track. Loop Track(b) only considers one loop track where either a man or a rock is on the single loop track.

These results comply with the results found for most people in morality laboratory experiments.

5 Conclusions and Future Work

We have shown how to model moral reasoning using prospective logic programming. We use various dilemmas of the trolley problem and the principle of double effect as the moral rule. Possible decisions in a dilemma are modelled as abducibles. Abductive stable models are then computed which capture abduced decisions and their consequences. Models violating integrity constraints, i.e. models that contain actions involving intentional killing, are ruled out. Finally, a posteriori preferences are used to prefer models that characterize more preferred moral decisions, including the use of utility functions. These experiments show that preferred moral decisions, i.e. the ones that follow the principle of double effect, are successfully delivered. They conform to the results of empirical experiments conducted in cognitive science and law.

Much research has emphasized using machine learning techniques, e.g. statistical analysis [22], neural networks [10], case-based reasoning [17] and inductive logic programming [2] to model moral reasoning from examples of particular moral dilemmas. Our approach differs from them as we do not employ machine learning techniques to deliver moral decisions.

Powers proposes to use nonmonotonic logic to specifically model Kant's categorical imperatives [21], but it is unclear whether his approach has ever been realized in a working implementation. On the other hand, Bringsjord et. al. propose the use of deontic logic to formalize moral codes [3]. The objective of their research is to arrive at a methodology that allows an agent to behave ethically as much as possible in an environment that demands such behaviour. We share our objective with them to some extent as we also would like to come up with a general framework to model morality computationally. Different from our work, they use an axiomatized deontic logic to decide which moral code is operative to arrive at an expected moral outcome. This is achieved by seeking a proof for the expected moral outcome to follow from candidates of operative moral codes.

To arrive at our ultimate research goal, we envision several possible future directions. We would like to make a more declarative specification of a posteriori preferences, i.e. a specification that may encapsulate the details of predicate select/2 from the viewpoint of users (cf. [20] for preliminary results). We also want to explore how to express metarule and metamoral injunctions. By metarule we mean a rule to resolve two existing conflicting moral rules in deriving moral decisions. Metamorality, on the other hand, is used to provide protocols for moral rules, to regulate how moral rules interact with one another. Another possible direction is to have a framework for generating precompiled moral rules. This will benefit fast and frugal moral decision making which is sometimes needed, cf. heuristics for decision making in law [9], rather than to have full deliberative moral reasoning every time.

We envision a final system that can be employed to test moral theories, and also can be used for training moral reasoning, including the automated generation of example tests and their explanation. Finally, we hope our research will help in imparting moral behaviour to autonomous agents.

References

[1] Alferes, J.J., Brogi, A., Leite, J.A., Pereira, L.M.: Evolving logic programs. In: Flesca, S., Greco, S., Leone, N., Ianni, G. (eds.) JELIA 2002. LNCS (LNAI), vol. 2424, pp. 50–61. Springer, Heidelberg (2002)

[2] Anderson, M., Anderson, S., Armen, C.: MedEthEx: A prototype medical ethics advisor. In: IAAI 2006. Procs. 18th Conf. on Innovative Applications of Artificial Intelligence (2006)

[3] Bringsjord, S., Arkoudas, K., Bello, P.: Toward a general logicist methodology for engineering ethically correct robots. IEEE Intelligent Systems 21(4), 38–44 (2006)

[4] de Waal, F.: Primates and Philosophers, How Morality Evolved. Princeton U. P. (2006)

[5] Dell'Acqua, P., Pereira, L.M.: Preferential theory revision. In: Pereira, L.M., Wheeler, G. (eds.) Procs. Computational Models of Scientific Reasoning and Applications, pp. 69–84 (2005)

[6] Dell'Acqua, P., Pereira, L.M.: Preferential theory revision (extended version). Journal of Applied Logic (to appear, 2007)

[7] Foot, P.: The problem of abortion and the doctrine of double effect. Oxford Review 5, 5–15 (1967)

[8] Gelfond, M., Lifschitz, V.: The stable model semantics for logic programming. In: Kowalski, R., Bowen, K.A. (eds.) 5th Intl. Logic Programming Conf., MIT Press, Cambridge (1988)

[9] Gigerenzer, G., Engel, C. (eds.): Heuristics and the Law. MIT Press, Cambridge (2006)

[10] Guarini, M.: Particularism and generalism: how AI can help us to better understand moral cognition. In: Anderson, M., Anderson, S., Armen, C. (eds.) Machine ethics: Papers from the AAAI Fall Symposium, AAAI Press, Stanford, California, USA (2005)

[11] Hauser, M.D.: Moral Minds, How Nature Designed Our Universal Sense of Right and Wrong. Little Brown, Boston – USA (2007)

[12] Joyce, R.: The Evolution of Morality. MIT Press, Cambridge (2006)

[13] Kakas, A., Kowalski, R., Toni, F.: The role of abduction in logic programming. In: Gabbay, D., Hogger, C., Robinson, J. (eds.) Handbook of Logic in Artificial Intelligence and Logic Programming, vol. 5, pp. 235–324. Oxford University Press, Oxford (1998)

[14] Katz, L.D. (ed.): Evolutionary Origins of Morality, Cross-Disciplinary Perspectives. Imprint Academic (2002)

[15] Kowalski, R.: The logical way to be artificially intelligent. In: Toni, F., Torroni, P. (eds.) Procs. of CLIMA VI. LNCS (LNAI), p. 122. Springer, Heidelberg (2006)

[16] Lopes, G., Pereira, L.M.: Prospective logic programming with ACORDA. In: Procs. of the FLoC 2006, Workshop on Empirically Successful Computerized Reasoning, 3rd Intl. J. Conf. on Automated Reasoning (2006)

[17] McLaren, B.M.: Computational models of ethical reasoning: Challenges, initial steps, and future directions. IEEE Intelligent Systems 21(4), 29–37 (2006)

[18] Mikhail, J.: Universal moral grammar: Theory, evidence, and the future. Trends in Cognitive Sciences 11(4), 143–152 (2007)

[19] Pereira, L.M., Lopes, G.: Prospective logic agents. In: Neves, J.M., Santos, M.F., Machado, J.M. (eds.) EPIA 2007. Procs. 13th Portuguese Intl.Conf. on Artificial Intelligence. LNCS (LNAI), Springer, Heidelberg (2007)

[20] Pereira, L.M., Saptawijaya, A.: Moral decision making with ACORDA. In: Dershowitz, N., Voronkov, A. (eds.) LPAR 2007. Short papers call, Local Procs. 14th Intl. Conf. on Logic for Programming Artificial Intelligence and Reasoning (2007)

[21] Powers, T.M.: Prospects for a Kantian machine. IEEE Intelligent Systems 21(4), 46–51 (2006)

[22] Rzepka, R., Araki, K.: What could statistics do for ethics? The idea of a commonsense-processing-based safety valve. In: Anderson, M., Anderson, S., Armen, C. (eds.) Machine ethics: Papers from the AAAI Fall Symposium, AAAI Press, Stanford, California, USA (2005)

[23] Tancredi, L.: Hardwired Behavior, What Neuroscience Reveals about Morality. Cambridge University Press, Cambridge (2005)

Change Detection in Learning Histograms from Data Streams

Raquel Sebastião[1] and João Gama[1,2]

[1] LIAAD-INESC Porto L.A., University of Porto
[2] Faculty of Economics, University of Porto
Rua de Ceuta, 118 - 6
4050-190 Porto, Portugal
{raquel,jgama}@liacc.up.pt

Abstract. In this paper we study the problem of constructing histograms from high-speed time-changing data streams. Learning in this context requires the ability to process examples once at the rate they arrive, maintaining a histogram consistent with the most recent data, and forgetting out-date data whenever a change in the distribution is detected. To construct histogram from high-speed data streams we use the two layer structure used in the Partition Incremental Discretization (PiD) algorithm. Our contribution is a new method to detect whenever a change in the distribution generating examples occurs. The base idea consists of monitoring distributions from two different time windows: the reference time window, that reflects the distribution observed in the past; and the current time window reflecting the distribution observed in the most recent data. We compare both distributions and signal a change whenever they are greater than a threshold value, using three different methods: the Entropy Absolute Difference, the Kullback-Leibler divergence and the Cosine Distance. The experimental results suggest that Kullback-Leibler divergence exhibit high probability in change detection, faster detection rates, with few false positives alarms.

1 Introduction

Histograms are one of the most used tools for exploratory data analysis. Data analysis is complex, interactive, and exploratory over very large volumes of historic data. Traditional pattern discovery process requires online ad-hoc queries, not previously defined, that are successively refined. Due to the exploratory nature of these queries, an exact answer may not be required. A user may prefer a fast approximate answer. Histograms are one of the techniques used in data stream management systems to solve range queries and selectivity estimation (the proportion of tuples that satisfy a query), two illustrative examples where fast but approximate answers are more useful than slow and exact ones.

Another aspect is that data arrives continuously in data streams and it's necessary to evaluate it, detecting if there is a change in the distribution. Therefore it's not reasonable to allow processing algorithms enough memory capacity to

J. Neves, M. Santos, and J. Machado (Eds.): EPIA 2007, LNAI 4874, pp. 112–123, 2007.

store the full history of the stream. So based on a previous work [1], we perform incremental discretization and use it to detect changes. Change detection in data streams was the main motivation for this work. The paper is organized as follows. The next section presents an algorithm to continuously maintain histograms over a data stream. In Section 3 we extend the algorithm for predictive data mining. Section 4 presents preliminary evaluation of the algorithm in benchmark datasets and one real-world problem. Last section concludes the paper and presents some future research lines.

2   Histograms

Histograms are one of the most used tools in exploratory data analysis. They present a graphical representation of data, providing useful information about the distribution of a random variable. A histogram is visualized as a bar graph that shows frequency data. The basic algorithm to construct a histogram consists of sorting the values of the random variable and places them into *bins*. Next we count the number of data samples in each bin. The height of the bar drawn on the top of each bin is proportional to the number of observed values in that bin.

A histogram is defined by a set of non-overlapping intervals. Each interval is defined by its boundaries and a frequency count. In the context of open-ended data streams, we never observe all values of the random variable. For that reason, and allowing to consider extreme values and outliers, we define an histogram as a set of break points $b_1, ..., b_{k-1}$ and a set of frequency counts $f_1, ..., f_{k-1}, f_k$ that define k intervals in the range of the random variable:
$$] - \infty, b_1],]b_1, b_2], ...,]b_{k-2}, b_{k-1}],]b_{k-1}, \infty[.$$

The most used histograms are either *equal width*, where the range of observed values is divided into k intervals of equal length ($\forall i, j : (b_i - b_{i-1}) = (b_j - b_{j-1})$), or *equal frequency*, where the range of observed values is divided into k bins such that the counts in all bins are equal ($\forall i, j : (f_i = f_j)$).

When all the data is available, there are exact algorithms to construct histograms [2]. All these algorithms require a user defined parameter k, the number of bins. Suppose we know the range of the random variable (domain information) and the desired number of intervals k. The algorithm to construct *equal width* histograms traverses the data once; whereas in the case of *equal frequency* histograms a sort operation is required.

One of the main problems of using histograms is the definition of the number of intervals. A rule that has been used is the Sturges' rule: $k = 1 + log_2 n$, where k is the number of intervals and n is the number of observed data points. This rule has been criticized because it is implicitly using a binomial distribution to approximate an underlying normal distribution[1]. Sturges rule has probably survived because, for moderate values of n (less than 200) produces reasonable

---

[1] Alternative rules for constructing histograms include Scott's (1979) rule for the class width: $k = 3.5sn^{-1/3}$ and Freedman and Diaconis's (1981) rule for the class width: $k = 2(IQ)n^{-1/3}$ where s is the sample standard deviation and IQ is the sample interquartile range.

histograms. However, it does not work for large n. In exploratory data analysis, histograms are used iteratively. The user tries several histograms using different values of k (the number of intervals), and choose the one that better fits his purposes.

2.1 The Partition Incremental Discretization

The *Partition Incremental Discretization* algorithm (*PiD* for short) is composed by two layers. The first layer simplifies and summarizes the data; the second layer constructs the final histogram.

The first layer is initialized without seeing any data. As described in [1], the input for the initialization phase is the number of intervals (that should be much larger than the desired final number of intervals) and the range of the variable.

Instead of initialize the algorithm with the number of bins, we manage a way such that the numbers of bins depends on the upper bound on relative error and on the desirable confidence level:

$$nBins1 = 50INT(ln(\tfrac{1}{\delta}) * ln(\tfrac{1}{\epsilon^2})).$$

Figure 1 shows that the number of bins increase when the relative error (ϵ) decreases and the confidence $(1 - \delta)$ increases.

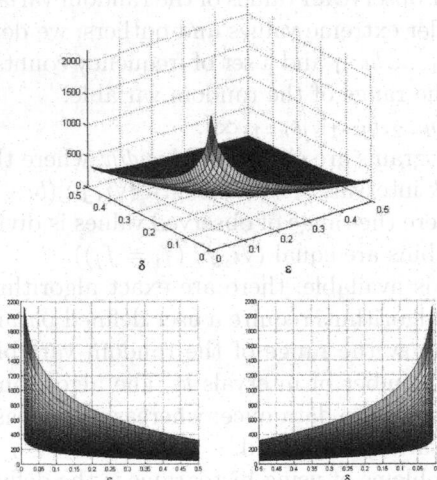

Fig. 1. Representation of the number of bins of $layer_1$. The top figure shows the dependency from ϵ and δ and bottom figures shows it according to only one variable.

Figure 1 (top) represents the number of bins of $layer_1$ in function of ϵ and δ. The bottom figures give a projection of the number of bins according with the variables ϵ and δ (respectively).

So, the input for the initialization phase is a pair of parameters (that will be used to express accuracy guarantees) and the range of the variable:

- The upper bound on relative error ϵ.
- The desirable confidence level δ.
- The range of the variable.

The range of the variable is only indicative. It is used to initialize the set of breaks using an equal-width strategy. Each time we observe a value of the random variable, we update $layer_1$. The update process determines the interval corresponding to the observed value, and increments the counter of this interval. Whenever the counter of an interval is above a user defined threshold (a percentage of the total number of points seen so far), a split operator triggers. The split operator generates new intervals in $layer_1$.

If the interval that triggers the split operator is the first or the last, a new interval with the same step is inserted. In all the other cases, the interval is split into two, generating a new interval. Figure 2 describes the $layer_1$ update process. The process of updating $layer_1$ works online, performing a single scan over the data stream. It can process infinite sequences of data, processing each example in constant time and space.

The second layer merges the set of intervals defined by the first layer. It triggers whenever it is necessary (e.g. by user action). The input for the second

```
Update-Layer1(x, breaks, counts, NrB, alfa, Nr)
x - Observed value of the random variable
breaks - Vector of actual set of break points
counts - Vector of actual set of frequency counts
NrB    - Actual number of breaks
alfa   - Threshold for Split an interval
Nr     - Number of observed values

 If (x < breaks[1]) k = 1; Min.x = x
 Else If (x > breaks[NrB]) k = NrB; Max.x = x
 Else k = 2 + integer((x - breaks[1]) / step)

 while(x < breaks[k-1]) k <- k - 1
 while(x > breaks[k]) k <- k + 1

 counts[k] = 1 + counts[k]
 Nr = 1 + Nr
 If ((1+counts[k])/(Nr+2) > alfa)  {
     val = counts[k] / 2
     counts[k] = val
     if (k == 1) {
        breaks = append(breaks[1]-step, breaks)
        counts <- append(val,counts)
     }
     else {
        if(k == NrB) {
           breaks <- append(breaks, breaks[NrB]+step)
           counts <- append(counts,val)
        }
        else {
           breaks <- Insert((breaks[k]+ breaks[k+1])/2, breaks, k)
           counts <- Insert(val, counts, k)
        }
     }
   NrB = NrB + 1
}
```

Fig. 2. The PiD algorithm for updating $layer_1$

layer is the breaks and counters of $layer_1$, the type of histogram (equal-width or equal-frequency) and the number of intervals. The algorithm for the $layer_2$ is very simple. For equal-width histograms, it first computes the breaks of the final histogram, from the actual range of the variable (estimated in $layer_1$). The algorithm traverses the vector of breaks once, adding the counters corresponding to two consecutive breaks. For equal-frequency histograms, we first compute the exact number F of points that should be in each final interval (from the total number of points and the number of desired intervals). The algorithm traverses the vector of counters of $layer_1$ adding the counts of consecutive intervals till F.

The two-layer architecture divides the histogram problem into two phases. In the first phase, the algorithm traverses the data stream and incrementally maintains an equal-width discretization. The second phase constructs the final histogram using only the discretization of the first phase. The computational costs of this phase can be ignored: it traverses once the discretization obtained in the first phase. We can construct several histograms using different number of intervals and different strategies: equal-width or equal-frequency. This is the main advantage of PiD in exploratory data analysis.

2.2 Analysis of the Algorithm

The histograms generated by PiD are not exact. There are two sources of error:

1. The set of boundaries. The breaks of the histogram generated in the second layer are restricted to the set of breaks defined in the first layer.
2. The frequency counters. The counters of the second layer are aggregations of the frequency counters in the first layer. If the splitting operator does not trigger, counters in first layer are exact, and also counters in second layer. The splitting operator can produce inexact counters. If the merge operation of the second layer aggregate those intervals, final counters are correct.

A comparative analysis of the histograms produced by PiD and histograms produced by exact algorithms using all the data reveals some properties of the PiD algorithm. Assuming a equal-width discretization (that is the split operator did not trigger) for the first layer and any method for the second layer, the error of PiD boundaries (that is the sum of absolute differences between boundaries between PiD and batch discretization) is bound, in the worst case, by: $R * N_2/(2 * N_1)$, where N_1 denotes the number of intervals of $layer_1$, N_2 the number of intervals of $layer_2$, and R is the range of the random variable. This indicates that when N_1 increases, the error decreases. The algorithm guarantees that frequencies at second layer are exact (for the second layer' boundaries). We should note that the splitting operator will always decrease the error.

The time complexity of PiD depends on the discretization methods used in each layer. The time complexity of the second layer is constant because its input is the first layer that has a (almost) fixed number of intervals. The time complexity for the first layer is linear in the number of examples.

3   Change Detection

The algorithm described in the previous section assumes that the observations came from a stationary distribution. When data flows over time, and at least for large periods of time, it is unlikely the assumption that the observations are generated at random according to a stationary probability distribution. At least in complex systems and for large time periods, we should expect changes in the distribution of the data.

3.1   Related Work

A fundamental question when monitoring a stream of values is: *Are the observations we are receiving now from the same distribution we have observed in the past?*.

There are several methods in machine learning to deal with changing concepts [3,4,5,6]. In machine learning drifting concepts are often handled by time windows or weighted examples according to their age or utility. In general, approaches to cope with concept drift can be classified into two categories: *i*) approaches that adapt a learner at regular intervals without considering whether changes have really occurred; *ii*) approaches that first detect concept changes, and next, the learner is adapted to these changes. Examples of the former approaches are *weighted examples* and *time windows* of fixed size. Weighted examples are based on the simple idea that the importance of an example should decrease with time (references about this approach can be found in [4,6,7]). When a time window is used, at each time step the learner is induced only from the examples that are included in the window. Here, the key difficulty is how to select the appropriate window size: a small window can assure a fast adaptability in phases with concept changes but in more stable phases it can affect the learner performance, while a large window would produce good and stable learning results in stable phases but can not react quickly to concept changes. In the latter approaches, with the aim of detecting concept changes, some indicators (e.g. performance measures, properties of the data, etc.) are monitored over time (see [5] for a good classification of these indicators). If during the monitoring process a concept drift is detected, some actions to adapt the learner to these changes can be taken. When a time window of adaptive size is used these actions usually lead to adjusting the window size according to the extent of concept drift [5]. As a general rule, if a concept drift is detected the window size decreases, otherwise the window size increases.

3.2   Monitoring Distributions on Two Different Time Windows

Most of the methods in this approach monitor the evolution of a distance function between two distributions: from past data in a *reference window* and in a current window of the most recent data points. An example of this approach, in the context of learning from Data Streams, has been present by [8]. The author proposes algorithms (statistical tests based on Chernoff bound) that examine samples drawn from two probability distributions and decide whether these distributions are different.

3.3 Entropy Based Change Detection

As a measurement of information (from Information Theory [9]), we conveniently adapted the entropy as a measurement of change detection. Entropy between the probabilities absolute difference measures the dispersion of the Distributions Differences and is defined by the following equation:

$$H(p||q) = -\sum_i |q(i) - p(i)| * log_2(|q(i) - p(i)|)$$

were q_i and p_i represents the probability of a point belongs to the bin i of the current window and the probability of belongs to the correspondent reference bin. From the properties of the above equation, we can derive the bounds: $0 \leq H(q||p) \leq 2$. Smaller values of $H(p||q)$, corresponds to smaller dispersion between the distributions of the two variables.

3.4 Kullback-Leibler Based Change Detection

From information theory [9], the Relative Entropy is one of the most general ways of representing the distance between two distributions [10]. Also known as the Kullback-Leibler's distance or divergence, it measures the distance between two probability distributions and so it can be used to test for change. Given a reference window with empirical probabilities p_i, and a sliding window with probabilities q_i, the KL distance is:

$$KL(p||q) = \sum_i p(i)log_2 p(i)/q(i).$$

The KL divergence is non negative and asymmetric and as higher is his value, the more distinct the distribution of the two variables. A higher value of the distance represents distributions that are further apart.

3.5 Cosine Distance Based Change Detection

Instead of comparing the distance between two probability distributions we compare the angle between them. The cosine distance [11] is derived from the dot product of two vectors and is given by the following equation:

$$C(p||q) = 1 - \frac{\sum_i p(i)q(i)}{||p||||q||}$$

We considered this definition in order to have non-negative values and guarantee that $0 \leq C(q||p) \leq 2$. This measure is symmetric and a lower value means that two distributions are closer.

4 Experimental Evaluation

The advantage of the two-layer architecture of PiD is that after generating the $layer_1$, the computational costs, in terms of memory and time, to generate the final histogram (the $layer_2$) is low: only depends on the number of intervals of the $layer_1$. From $layer_1$, we can generate histograms with different number of intervals and using different strategies (equal-width or equal-frequency). We should note that the standard algorithm to generate equal-frequency histograms requires a sort operation, which could be costly for large n. This is not the case of PiD. Generation of equal-frequency histograms from the $layer_1$ is straightforward.

4.1   Methodology and Design of Experiments

In order to analyze the capacity to detect changes using different kinds of distributions we design three different datasets:

- 100000 points of a random variable from a Normal distribution. This dataset is composed by two samples from a normal distribution (with different parameters) of 50000 values each one. The initial 50000 points are generated from the standard normal distribution and then we force an abrupt change using parameters distinct enough to guarantee that in spite of the similar shape of distributions, the algorithm should detect changes.
- 100000 points of a random variable from a Normal distribution. This dataset is composed by two samples from a normal distribution (with different parameters) of 50000 values each one. The initial 50000 points are generated from the standard normal distribution and then we force a smooth change, using similar parameters expecting that the algorithm should not detect changes.
- 100000 points of a random variable from LogNormal and Normal distributions of 50000 points each one.

The data is received at any time producing an equal-with histogram. The number of bins were defined according to the equation on section 2.1, setting both the variables ϵ and δ as 5%. We considered that the initial data points should be used as a representation of data and that the number of initial points should be chosen according to the number of intervals of the $layer_1$. So we decided that the first $30 * nBins1$ data points are part of a stabilization process and that no change occurs in this range. The windows size was also defined as dependent on the number of intervals of the $layer_1$, being half of these ones: $\frac{nBins1}{2}$.

Assuming that sample in the stabilization set has distribution P and that the current windows has distribution Q, we use as a measure to detect whether has occurred a change in the distribution the measures described above. For all the 3 measures the more distinct the distributions P and Q the higher is the distance between them. According to this, we define that had occurred a change in the distribution of the current window relatively to the reference distribution if the distance computed based on those distributions is greater that $\mu + z_{1-\frac{\alpha}{2}}\sigma$, where μ and σ represents the sample mean and the estimate standard deviation, and $z_{1-\frac{\alpha}{2}}$ represents the point on the standard normal density curve such that the probability of observing a value lower than $z_{1-\frac{\alpha}{2}}$ is equal to $1 - \frac{\alpha}{2}$. For instance, we established a confidence level of 95% ($\alpha = 0.05$), so $z_{1-\frac{\alpha}{2}} = 1.96$. If no change occurs we maintain the reference distribution and consider more $\frac{nBins1}{2}$ data points in the current window, and start a new comparison. If we detect a change we clean the reference data set and initialize the process of search for changes.

4.2 Controlled Experiments with Artificial Data

In this Section we evaluate the 3 change detection methods in controlled experiments using artificial data. The goals of these experiments are:

1. Capacity to Detect and React to drift.
2. Resilience to False Alarms when there is no drift, that is not detect drift when there is no change in the target concept.
3. The number of examples required to detect a change after the occurrence of a change.

Table 1. Results of the 3 change detection methods using artificial data

Datasets	TP			FP			Ne (mean)			Ne (std)		
	Cos	Ent	KLD	Cos	Ent	KLD	Cos	Ent	KLD	Cos	Ent	KLD
D1	1	1	1	0.1	0	0	1964	707	258	189.3	0	0
D2	1	0.9	1	0	0	0	7038	12400	5691	1427	5228	933.5
D3	0.95	0.95	0.95	0.65	0	0	9285	801	258	8906	188	0

Table 1 shows the results of the 3 change detection methods using the described artificial data. It can be observed that the use of the Kullback-Leibler Distance as a measure to decide if there is a drift achieve better results and reaches those requiring a smaller number of examples. It also shows that all the measures have a good resilience to false alarms when there are no drifts. Notice that even in the second dataset, where the change was very smooth, the cosine and the Kullback-Leibler distances detected it in all the experiments. The entropy of the absolute differences also detected it almost the times. Although the cosine distance detects changes when there not existing, the performance of other measures were very consistent and precise. The results obtained with those data sets clearly show that the cosine distance presents the worse results.

To evaluate the performance of the 3 algorithms we also use quality metrics such as *Precision* and *Recall*. The *Precision* gives a ratio between the correct detected changes and all the detected changes and *Recall* is defined as a ratio between the correct detected changes and all the occurred changes:

$$Precision = \frac{TP}{TP+FP}$$

$$Recall = \frac{TP}{TP+FN}$$

Table 2 shows the precision and recall achieved by the 3 change detection methods using artificial data. For both quality metrics, the closest to one, the better are the results. Those results sustain the above observations, showing that Kullback-Leibler Distance is the algorithm that reaches better results and that the algorithm based on the cosine distance is the worst one.

Table 2. The *Precision* and the *Recall* obtained using the 3 algorithms to detect changes

Datasets	Precision			Recall		
	Cos	Ent	KLD	Cos	Ent	KLD
D1	0.9091	1	1	1	1	1
D2	1	1	1	1	0.9000	1
D3	0.5938	1	1	1	1	1

4.3   Data and Experimental Setup

To obtain data, tests were carried out in a Kondia HS1000 machining centre equipped with a Siemens 840D open-architecture CNC. The blank material used for the tests was a 170 mm profile of Aluminum with different hardness. The maximum radial depth of cut was 4.5 mm using Sandvik end-mill tools with two flutes and diameter 8, 12 and 20 mm. Tests were done with different cutting parameters, using sensors for registry vibration and cutting forces.

A multi-component dynamometer with an upper plate was used to measure the in-process cutting forces and piezoelectric accelerometers in the X and Y axis for vibrations measure. A Karl Zeiss model Surfcom 130 digital surface roughness instrument was used to measure surface roughness.

Each record includes information on the following seven main variables used in a cutting process:

- Fz - feed per tooth
- $Diam$ - tool diameter
- ae - radial depth of cut
- HB - Hardness on the type of material
- $Geom$ - tools geometry
- rpm - spindle speed
- Ra - average surface roughness

This factors was proceeding the Design of Experiment explained in [12] was used for validation a bayesian model for prediction of surface roughness. A multi-component dynamometer with an upper plate was used to measure the in-process cutting forces (X and Y axis). The cutting speed on X and Y axis was also measured.

We use these sensors measures to detect when a change had occurred in the experiments. We must point out that a change in the activity does not mean a change in the data coming from sensors.

Our goal is to study the three detection methods in a real data problem. For each sensor we record the points where each method signals a change. We study concordances in change points for the different methods. As the sensors response to different incentives, the agreement in change points for the sensors was not carried out.

For the second sensor, neither the entropy of the absolute differences nor cosine distances detected any change.

The obtained results shows that the changes detected by Kullback-Leibler distance for the data points from the second sensor agree with the results obtained with the first sensor, supporting that this measure is consistent.

Comparing the results obtained with the three detected methods on the data points from the first sensor, we found out 21 conformities in changes points between the Kullback-Leibler and the cosine distances. The entropy of the absolute differences reached 23 and 21 accordance's with the Kullback-Leibler and the cosine distances, respectively.

5   Conclusion and Future Work

The main conclusion we found is that the Kullback-Leibler Distance reach better performances than the other algorithms for all the kinds of artificial datasets we had study and that the cosine distance were the worst algorithm. From the artificial data results we can also conclude that both entropy of the absolute differences algorithm and Kullback-Leibler Distance has a good resilience to false alarms when there are no drifts on the datasets.

From the experiments with real data, we must take into consideration that we don't have information if there are changes and when they occur; and so on we can't extract strong or well based conclusions from the results obtained with that data. Although this lack, we can conclude that the concordances in changes points between the detection methods supports their capacity to detect changes. As a final conclusion, one can say that the results achieved so far are quite encouraging and motivate the continuation of the work. As an improvement of this initial work we intend to apply the described algorithms into dataset collect in a medical and in an industrial context. We also intend to improve the algorithms in order to reduce the number of points that are needed to detect a change.

Acknowledgments

Thanks to the financial support given by the FEDER, the Plurianual support attributed to LIAAD, project ALES II (POSI/EIA/55340/2004), and project RETINAE.

References

1. Gama, J., Pinto, C.: Discretization from Data Streams: applications to Histograms and Data Mining. In: ACM Symposium on Applied Computing, pp. 662–667. ACM Press, New York (2006)
2. Pestana, D.D., Velosa, S.F.: Introdução á Probabilidade e á Estatística. Fundação Calouste Gulbenkian (2002)
3. Klinkenberg, R.: Learning drifting concepts: Example selection vs. example weighting. Intelligent Data Analysis 8(3), 281–300 (2004)

4. Klinkenberg, R., Joachims, T.: Detecting concept drift with support vector machines. In: Langley, P. (ed.) Proceedings of ICML 2000. 17th International Conference on Machine Learning, Stanford, US, pp. 487–494. Morgan Kaufmann Publishers, San Francisco (2000)
5. Klinkenberg, R., Renz, I.: Adaptive information filtering: Learning in the presence of concept drifts. In: Learning for Text Categorization, pp. 33–40. AAAI Press, Stanford, California, USA (1998)
6. Widmer, G., Kubat, M.: Learning in the presence of concept drift and hidden contexts. Machine Learning 23, 69–101 (1996)
7. Maloof, M., Michalski, R.: Selecting examples for partial memory learning. Machine Learning 41, 27–52 (2000)
8. Kifer, D., Ben-David, S., Gehrke, J.: Detecting change in data streams. In: VLDB 2004: Proceedings of the 30th International Conference on Very Large Data Bases, pp. 180–191. Morgan Kaufmann Publishers Inc., San Francisco (2004)
9. Berthold, M., Hand, D.: Intelligent Data Analysis - An Introduction. Springer, Heidelberg (1999)
10. Dasu, T., Krishnan, S., Venkatasubramanian, S., Yi, K.: An Information-Theoretic Approach to Detecting Changes in Multi-Dimensional Data Streams. In: Interface 2006 (Pasadena, CA) Report (2006)
11. Tan, P.-N., Steinbach, M., Kumar, V.: Introduction to Data Mining. Addison-Wesley, Reading (2006)
12. Correa, M., de Ramírez, M.J., Bielza, C., Pamies, J., Alique, J.R.: Prediction of surface quality using probabilistic models. In: 7th Congress of the Colombian Association of Automatic, Cali, Colombia, March 21 24, 2007 (2007) (in Spanish)
Domingos, P., Hulten, G.: Learning from infinite data in finite time. In: Advances in Neural Information Processing Systems 14. MIT Press, Cambridge, MA (2002)

Real-Time Intelligent Decision Support System for Bridges Structures Behavior Prediction

Hélder Quintela, Manuel Filipe Santos, and Paulo Cortez

Department of Information System, University of Minho
Campus de Azurém
4800-058 Guimarães, Portugal
{hquintela,mfs,pcortez}@dsi.uminho.pt

Abstract. There is an increasing need of deploying automatic real-time decision support systems for civil engineering structures like bridges, making use of prediction models based in Artificial Intelligence techniques (e.g., Artificial Neural Networks) to support the monitoring and prediction activities. Past experiments with Data Mining (DM) techniques and tools opened room for the development of such a real-time Decision Support Systems. However, it is necessary to test this approach in a real environment, using real-time sensors monitoring. This study presents the development of prediction models for structures behavior and a novel architecture for operating in a real-time system.

Keywords: Knowledge Discovery in Databases, Data Mining, Decision Trees, Neural Networks, Intelligent Decision Support Systems, Civil Engineering Structures.

1 Introduction

In the field of Civil Engineering the increasement on the number of bridges in service, and the appearance of widespread failures has highlighted the importance of condition assessment using a new generation of tools that incorporate technology into real-time monitoring and decision support using expert systems – a new generation of Bridge Management Systems (BMS) that can help in maintenance, safety and management.

A bridge maintenance policy needs to incorporate multidisciplinary subjects from structural engineering, economics and computer science [1]. The future in inspection and condition assessment shows the need of incorporating into BMS models based in Artificial Intelligence, such as Artificial Neural Networks (ANN), for prediction of condition deterioration, developed using data collected: (i) in real time environment by sensors in the structure, (ii) in visual inspections by experts, (iii) in destructive and non-destructive tests. However the development of such models may not be an easy task using statistical techniques for historical data analysis, due to the high complexity of data and relations among several variables [2].

Regarding that, the application of Data Mining (DM) techniques to analyze civil engineering data has gained an increasing interest, due to the intrinsic characteristics such as ability to deal with non-linear relationships. The results attained by previous studies (some of them presented in the Previous Work section) are encouraging and

J. Neves, M. Santos, and J. Machado (Eds.): EPIA 2007, LNAI 4874, pp. 124–132, 2007.

open room for the development of Intelligent Decision Support Systems for Bridge Management Systems.

According to Supratim and Vankayala [3], Real-Time Business Intelligence systems are gaining an increasing attention from the research community. Indeed, developing and maintaining these systems is a challenging task, with a critical issue that lies in assuring that these systems can really work with an almost zero latency.

Tools for support decision making processes and problem-solving activities have proliferated and evolved over the last decades from applications based in spreadsheet software, to decision-support systems incorporating optimization models and, finally, to systems with components from artificial intelligence and statistics [4].

This paper presents the second stage of a bigger project, called SIISEC - an Intelligent System for Civil Engineering Structures, with the goal of deploying forecasting models using Knowledge Discovery from Databases (KDD)/Data Mining technologies for structure damage detection. In particular, the DM techniques considered include: Artificial Neural Networks (ANNs) and Decision Trees (DTs).

This paper is organized as follows: first, is presented an overview of some experiments with simple structural elements using DM techniques; then the experiments with a structure prototype are described, being the results analyzed in terms of several criteria; next, is presented a novel architecture for a new Intelligent Decision Support System for structures monitoring; finally, closing conclusions are drawn and future research activities are depicted.

2 Previous Work

The first experiments in the context of the SIISEC project were centered on the development of predicting models for: (i) ultimate resistance of steel beams subjected to concentrated loads, and (ii) critical shear stress in tapered plate girders.

In the conception and design of civil engineering structures several factors should be considered: aesthectics, functionality, deformability, durability, resistance and cost. In general, this exercise is conditioned to the search of the safest solution with the lowest cost. This concern, associated to the evolution of the materials' properties and of the computational tools, led to the use of more refined design methods. However, the error presented by the current design formulas is significant due to: the influence of several independent parameters in the behavior; the insufficient number of experimental data that allows a parametric analysis; and the calibration of simplified models.

While parametric analysis is an intensive and hard work, the construction of models using DM techniques in a KDD process, could be used to induce predictive models in a more flexible and efficient fashion. Hence, the following studies, opened room for the development of the study presented in this paper, since the accuracy of the proposed DM models are better than the actual analytical formulations, enabling the development of models not for isolated elements of a civil structure (e.g., beams), but for the structure (e.g., bridge) as a whole.

2.1 Prediction of Ultimate Resistance of Steel Beams

The models developed to predict the ultimate resistance of steel beams subjected to concentrated loads based in ANN presented a better accuracy than the design formula

Table 1. Models' performance for ultimate resistance prediction

	Model A		Model B	
	ANN	Roberts	ANN	Roberts
Mean Error (%)	9,9	19,3	6,6	15,3
Standard Deviation (% mean error)	5,11		3,32	

proposed by Roberts [5] as presented in Table 1. Model A is adjusted to beams with a web thickness lower or equal to 2.12 and Model B for beams with a web thickness higher.

In previous work [6] we have introduced a clustering approach [7], in order to search and explore some kind of homogeneity in the data set, producing in this way predicted models with a better degree of accuracy - what is a main issue -, to assure a high level of safety and usability. The proposed approach includes a macro-analysis in the DM step of the KDD process, using a Kohonen ANN [8], to segment the available examples.

2.2 Prediction of Ultimate Shear Resistance of Non-prismatic Tapered Plate Girders

Tapered plate girders with slender webs are used in steel and composite structures, when there is necessity of large span lengths, and/or when the structure is required to bear heavy loads. The design of this kind of structures requires in-depth study of their buckling to ensure their correct behavior, both in situations close to collapse and in service situations.

The models proposed were developed using the clustering approach presented before and based in ANNs.

In [9], it was used a dataset with 176 girders with geometric parameters and the results of buckling coefficient and of the critical shear stress, using the Zárate Model

Table 2. Models' performance for ultimate shear prediction

Model A			Model B		
Error	ANN	Zárate	Error	ANN	Zárate
MAD	0,658846	0,87	MAD	0,223529	0,566471
SSE	38,0461	22,7292	SSE	1,9988	13,8585
MSE	1,415938	0,8742	MSE	0,11757647	0,815206
RMSE	1,189932	0,934987	RMSE	0,34289426	0,902888

and the Finite Element Method. One of the lessons learned stressed that even with a small number of samples, the ANN are an applicable machine learning paradigm in prediction tasks, since the available data is representative of the universe in study.

The models (Table 2) were compared with the study of Zárate [10], using some of the widespread measures used in regression tasks. The accuracy of the predicting models generated making use of machine learning techniques in a KDD process proven to be better for Model B than the previous model proposed by Zárate, while for Model A the results are almost the same.

3 Data Mining Models for Structures Behavior Prediction

The main questions addressed in the present work are:

- What is the weight that the structure was subjected?;
- What is the point on the structure where the load was placed?;
- What are the values for dislocation and deformation of the structure, taking in account the weight of the load and the point where the load was placed in the structure?

3.1 Materials and Methods

The dataset used in this study considers 1050 experiments in a prototype of a civil structure with 7 sensors for real-time collection of data, with 9 variables (Table 3). The sensors were used for deformation and dislocation measuring when the structure was subjected to a load.

According to the Figure 1, deformation sensors were placed in positions T_1 (point 2250 mm), T_2 (point 3250 mm) and T_4 (point 1250 mm); and the extensometers for dislocation measuring were placed in positions EXT_1 (point 2850 mm), EXT_2 (point 3250 mm), EXT_3 (point 3700 mm) and EXT_4 (point 1250 mm).

During the experiments were placed loads of 45 and 77 kg in the following points: 250 mm, 1250, 2250 mm, 3250 mm, 4250 mm, and when the structure was stabilized the data of deformation and dislocation values were collected to the database.

The analysis of the domain of values of the T and EXT sensors indicates that the higher scale of the last ones could influence the results in the modeling phase. However, in the pre-processing tasks it was decided to keep the original values, and evaluate the results of the deployed models in the DM phase before any re-scaling activity.

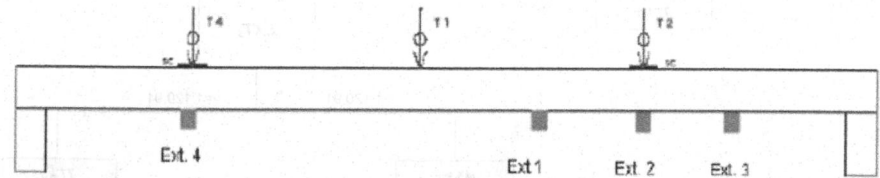

Fig. 1. Prototype of a civil structure with a length of 4.5 meters, with sensors location - T_i are the deformation sensors and EXT_j are the extensometers

Table 3. Sensors variables and domain of values

Name	Min	Max	Mean
T_1	2.82	3.45	3.01
T_2	2.05	3.06	2.69
T_4	2.47	3.02	2.66
EXT_1	1120.4	1146.9	1127.4
EXT_2	1445.4	1476.2	1453.5
EXT_3	1291.1	1309.3	1295.9
EXT_4	799.8	960.5	928.8

3.2 What Is the Weight That the Structure Was Subjected?

The first DM experiments intend to develop a classification model that can predict the value of the weight that the structure was subjected; using the values of the sensors and the point where the load was applied. It was decided to use two techniques: DTs [11] and ANNs [12], as available in the SAS Enterprise Miner environment [13].

When using the DTs technique the split criteria used was the *Gini Reduction*. The obtained model (Figure 2) presents 7 leaves and a misclassification rate of 0. The most important variable with an importance degree of 0.569 is the EXT_1 sensor, located almost in the middle of the structure, measuring the deformation of the structure when a load is applied.

In the ANN modelling, the popular Multi-Layer Perceptron (MLP) architecture [14] was adopted, where the number of nodes in the hidden layer was automatically

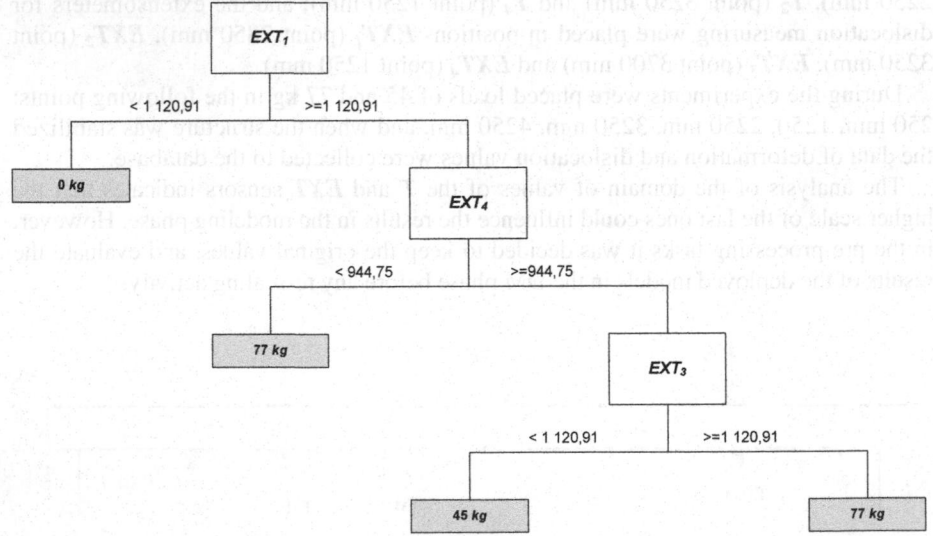

Fig. 2. Example of the Decision Tree based model for classification of the weight

Table 4. Confusion Matrix for Decision Tree and Artificial Neural Network based models

Desired/Predicted	0	45	77
0	102	0	0
45	0	102	0
77	0	0	96

set by the SAS tool, and the accuracy was estimated using the holdout method by randomly dividing the available data into three exclusive partitions: training, validation and testing.

The final ANN based model for weight prediction presents architecture of 6 nodes in the hidden layer, with a Maximum Absolute Error of 0.13. For evaluation of the two proposed models was used the Confusion Matrix [15] (Table 4) and calculated the Precision measure.

Since the two models have an accuracy of 100%, they can be used in this task, but the DT presents the advantage of easy understanding by civil engineering field specialists, which can evaluate and validate the captured knowledge, while the ANN based model is like a black box.

3.3 What Is the Point on the Structure Where the Load Was Placed?

In order to plan the interventions on a structure could be important to know where the impact was applied. Similar to the previous task, both DTs and ANNs were used. However the results after several experiments, with different parameters and architectures, were not enough to validate this approach, since the error is very high.

3.4 What Are the Values for Dislocation and Deformation of the Structure?

For the last question addressed in the DM phase, it was decided to use the ANN technique, to generate a model that predicts the value for dislocation (T_i) and deformation (EXT_j). It was adopted the following framework: (i) a **model A** with 2 inputs (weight and point) and 7 outputs $(T_i,$ and $EXT_j)$; (ii) a **model B** with 2 inputs (weight and point) and 4 outputs (EXT_j), rejecting the T variables; and (iii) a **model C** **with** 2 inputs (weight and point) and 3outputs (T_i), rejecting the EXT variables.

The best predictive performance was attained by the model A. Table 5 presents the predictive performance in test set for all the output variables of model A, using the Root Mean Squared Error (RMSE) measure (Equation 1). The RMSE is one of the most commonly used measures on regression tasks. This metric, which will be adopted in this work, has the property of being more sensitive to high error values.

$$\sqrt{\frac{(a_1-c_1)^2+(a_2-c_2)^2+...+(a_n-c_n)^2}{(a_1-\overline{c})^2+(a_2-\overline{c})^2+...+(a_n-\overline{c})^2}} \qquad (1)$$

The results shown that the EXT_4 variable presents the higher error and that can be due to a noise problem of the sensor in data collection during the experiments, what

Table 5. The predictive test performance

Output	RMSE
T_1	0.11
T_2	0.20
T_4	0.10
EXT_1	5.47
EXT_2	5.84
EXT_3	3.79
EXT_4	35.63
Overall	12.79

can influence the value of the overall RMSE error. To test that hypothesis the variable was rejected, and the new model presents a better overall RMSE Error of 3.47, while the errors for the other variables are almost the same.

4   The Future: SIISEC System

The results attained by the prediction models for structures behavior opened room for the deployment of a system that can monitor in real-time a civil engineering structure (e.g., bridges), based in a Services Oriented Architecture [16], with the following modules, presented in Figure 3:

Data entry: capture all the data using remote sensors, and mobile systems for inspections variables introduced by the specialists, and store them in a transactional database.

Intelligent Monitoring: this module uses the DM module, since the predicting modules should be invoked in a continuous way for online monitoring. In case of abnormal values a set of alerts should be displayed, and the alert(s) should be delivered for the system managers by mobile systems.

Intelligent Decision Support: this module implements the necessary functions to help managers to take decisions using the information available in the system, concerning the bridge security and interventions plans. It implements the tools for reporting and uses the DM module for predicting models invocation and to overview their historic performance.

Data Mining: this module implements mechanism for models invocation, and functions to adapt the existing DM models with the new data collected. The results of the invocation and evolution should be stored in the knowledge database, enabling the performance analysis.

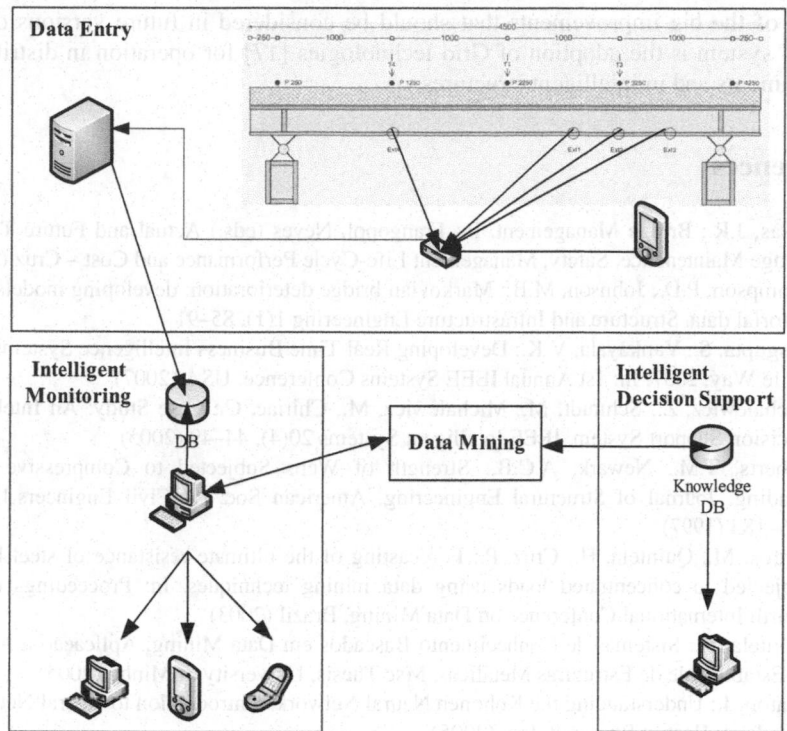

Fig. 3. SIISEC system architecture

5 Conclusions

This paper presented an overview of some experiments that were done in order to develop a real-time Intelligent Decision Support System for civil structures monitoring and behavior prediction. The results attained in the past with simple elements (e.g., beams) opened room for the development of prediction models of deformation and dislocation for a bridge structure prototype with encouraging results, as presented in this study.

Bridge Management is in the starting phase, but there are a lot of works in this area worldwide on inspection and health monitoring, capacity assessment, life-cycle cost analysis and maintenance repair. As proved in the experiments in this study, there are space for the use of techniques from Artificial Intelligence in this kind of systems, to improve the accuracy of deterioration models to predict the future condition sate and safety, in an accurate, fashion and economic way.

In the future we intend to use: data from multiple sources (e.g., traffic data, costs, predicted future conditions) and smart structures that enables a non costly and real-time monitoring of more parameters that can be used to build more reliable systems, and improve the accuracy of the proposed SIISEC system, that will be developed in the next phase of this project.

One of the big improvements that should be considered in future versions of the SIISEC system is the adoption of Grid technologies [17] for operation in distributed environments and in intelligent structures.

References

1. Casas, J.R.: Bridge Management. In: Frangopol, Neves (eds.) Actual and Future Trends, Bridge Maintenance, Safety, Management Life-Cycle Performance and Cost – Cruz (2006)
2. Thompson, P.D., Johnson, M.B.: Markovian bridge deterioration: developing models from historial data. Structure and Infrastructure Engineering 1(1), 85–91
3. Dasgupta, S., Vankayala, V.K.: Developing Real Time Business Intelligence Systems: The Agile Way, 2007. In: 1st Annual IEEE Systems Conference, USA (2007)
4. Michalewicz, Z., Schmidt, M., Michalewicz, M., Chiriac, C.: Case Study: An Intelligent Decision Support System. IEEE Intelligent Systems 20(4), 44–49 (2005)
5. Roberts, T.M., Newark, A.C.B.: Strength of Webs Subjected to Compressive Edge Loading. Journal of Structural Engineering, American Soc. Of Civil Engineers 123(2), 176–183 (1997)
6. Santos, M., Quintela, H., Cruz, P.: Forecasting of the ultimate resistance of steel beams subjected to concentrated loads using data mining techniques. In: Proceedings of the Fourth International Conference on Data Mining, Brazil (2003)
7. Quintela, H.: Sistemas de Conhecimento Baseados em Data Mining; Aplicação à Análise da Estabilidade de Estruturas Metálicas, Msc Thesis, University of Minho (2005)
8. Heaton, J.: Understanding the Kohonen Neural Networks, Introduction to Neural Networks with Java, Heaton Research, Inc. (2005)
9. Cruz, P.J.S., Lourenço, L., Santos, M., Quintela, H., Cortez, P.: Strength of corroded tapered plate girders under pure shear. In: Frangopol, Neves (eds.) Bridge Maintenance, Safety, Management Life-Cycle Performance and Cost – Cruz (2006)
10. Zárate, V.A.: Un modelo para el dimensionamento de vigas armadas de inércia variable de alma esbelta, Ph.D. Thesis, Departamento de Ingeniería de la Construcción, Universitat Politécnica de Catalunya, Barcelona, Spain (2002)
11. Quinlan, J.R.: Bagging Boosting and C4.5. In: Proceedings of Fourteenth National Conference on Artificial Intelligence (1996)
12. Pompe, P., Feelders, A.: Using Machine Learning, Neural Networks, and Statistics to Predict Corporate Bankruptcy. Microcomputers in Civil Engineering 12, 267–276 (1997)
13. Cerrito, P.B.: Introdution to Data Mining Using SAS Enterprise Miner, SAS Publishing (2007)
14. Matignon, R.: Neural Network Modeling using SAS Enterprise Miner, AuthorHouse (2005)
15. Souza, J., Matwin, S., Japkowicz, N.: Evaluating Data Mining Models: A Pattern Language. In: Proceedings of the 9th Conference on Pattern Language of Programs, USA (2002)
16. Newcomer, E., Lomow, G.: Understanding SOA with Web Services, Independent Technology Guides. Addison-Wesley Professional, Reading (2004)
17. Foster, I., Kesselman, C.: The anatomy of the grid: Enabling scalable virtual organizations. Tuecke, S.: Int. J. Supercomput. Appl. 15(3) (2001)

Semi-fuzzy Splitting in Online Divisive-Agglomerative Clustering

Pedro Pereira Rodrigues[1,2] and João Gama[1,3]

[1] LIAAD - INESC Porto L.A.
Rua de Ceuta, 118 - 6 andar, 4050-190 Porto, Portugal
[2] Faculty of Sciences of the University of Porto
[3] Faculty of Economics of the University of Porto
pprodrigues@fc.up.pt, jgama@fep.up.pt

Abstract. The Online Divisive-Agglomerative Clustering (ODAC) is an incremental approach for clustering streaming time series using a hierarchical procedure over time. It constructs a tree-like hierarchy of clusters of streams, using a top-down strategy based on the correlation between streams. The system also possesses an agglomerative phase to enhance a dynamic behavior capable of structural change detection. However, the split decision used in the algorithm focus on the crisp boundary between two groups, which implies a high risk since it has to decide based on only a small subset of the entire data. In this work we propose a semi-fuzzy approach to the assignment of variables to newly created clusters, for a better trade-off between validity and performance. Experimental work supports the benefits of our approach.

Keywords: fuzzy clustering, streaming time series, hierarchical models.

1 Introduction

The task of clustering variables over data streams, or streaming time series, is not widely studied. Data streams usually consist of variables producing examples continuously over time. The basic idea behind it is to find groups of variables that behave similarly through time, which is usually measured in terms of time series similarities. Clustering time series has been already studied in various fields of real world applications. Many of them, however, could benefit from a data stream approach. For example:

- in electrical supply systems, clustering *demand profiles* (ex: industrial or urban) decreases the computational cost of predicting each individual sub-network load [5,6];
- in medical systems, clustering *medical sensor data* (such as ECG, EEG, etc.) is useful to determine correlation between signals [14];
- in financial markets, clustering *stock prices* evolution helps preventing bank-ruptcy [10];

J. Neves, M. Santos, and J. Machado (Eds.): EPIA 2007, LNAI 4874, pp. 133–144, 2007.

All of these problems address data coming from a stream at high rate. Hence, data stream approaches should be considered to solve them.

In this work we address the problem of clustering streaming series assuming data is gathered by a centralized process while it is becoming available for online analysis, as it was already targeted by recent research. Wang and Wang introduced an efficient method for monitoring composite correlations, i.e., conjunctions of highly correlated pairs of streams among multiple time series [15]. They use a simple mechanism to predict the correlation values of relevant stream pairs at the next time position, using an incremental computation of the correlation, and rank the stream pairs carefully so that the pairs that are likely to have low correlation values are evaluated first. Beringer and Hüllermeier proposed an online version of *k-means* for clustering parallel data streams, using a Discrete Fourier Transform approximation of the original data [1]. The basic idea is that the cluster centers computed at a given time are the initial cluster centers for the next iteration of *k-means*, applying a procedure to dynamically update the optimal number of clusters at each iteration. Clustering On Demand *(COD)* is another framework for clustering streaming series which performs one data scan for online statistics collection and has compact multi-resolution approximations, designed to address the time and the space constraints in a data stream environment [3]. It is divided in two phases: a first online maintenance phase providing an efficient algorithm to maintain summary hierarchies of the data streams and retrieve approximations of the sub-streams; and an offline clustering phase to define clustering structures of multiple streams with adaptive window sizes. In this paper, however, we focus on the Online Divisive-Agglomerative Clustering *(ODAC)* system, a hierarchical procedure which dynamically expands and contracts clusters based on their diameters [13].

In the next section we present an overview on ODAC and its main characteristics, while Section 3 proposes the new semi-fuzzy assignment criterion. Section 4 enunciates the validity indices used in Section 5 to validate our proposal, while Section 6 presents some concluding remarks.

2   ODAC Overview

The Online Divisive-Agglomerative Clustering *(ODAC)* is an incremental approach for clustering streaming time series using a hierarchical procedure [13]. It constructs a tree-like hierarchy of clusters of streams, using a top-down strategy based on the correlation between streams. The system also possesses an agglomerative phase to enhance a dynamic behavior capable of structural change detection. The splitting and agglomerative operators are based on the diameters of existing clusters and supported by a significance level given by the Hoeffding bound [8]. Accordingly, we observe that:

- the update time and memory consumption does not depend on the number of examples, as it gathers sufficient statistics to compute the correlations within each cluster; moreover, anytime a split is reported, the system becomes faster as less correlations must be computed;

- the system possesses an anytime compact representation, since a binary hierarchy of clusters is available at each time stamp, and does not need to store anything more than the sufficient statistics and the last example to compute the first-order differences;
- an agglomerative phase is included to react to structural changes; these changes are detected by monitoring the diameters of existing clusters;
- this online system was not designed to include new streams along the execution; however, it could be easily extended to cope with this feature;
- given its hierarchical core, the system possesses a inherently adaptable configuration of clusters;

As reported by the authors, this is one of the systems clearly proposed to address clustering of multiple streams. It copes with high-speed production of examples and reduced memory requirements, with constant time update. It also presents adaptability to new data, detecting and reacting to structural drift.

2.1 Dissimilarity Measure

The system must analyze distances between incomplete vectors, possibly without having any of the previous values available. Thus, these distances must be incrementally computed. The system uses Pearson's correlation coefficient [12] between time series as *similarity* measure. This way, the *sufficient statistics* needed to compute the correlation are easily updated at each time step.

2.2 Splitting Criterion

One problem that usually arises with approximate models is the definition of a minimum number of observations necessary to assure convergence. One approach is to apply techniques based on the Hoeffding bound [8] to solve this problem. The Hoeffding bound has the advantage of being independent of the probability distribution generating the observations [4], stating that after n independent observations of a real-valued random variable r with range R, and with confidence $1 - \delta$, the true mean of r is at least $\bar{r} - \epsilon$, where $\bar{r}$ is the observed mean of the samples and

$$\epsilon = \sqrt{\frac{R^2 ln(1/\delta)}{2n}} . \tag{1}$$

As each leaf is fed with a different number of examples, each cluster c_k will possess a different value for ϵ, designated ϵ_k.

Let $d(a, b)$ be the distance measure between pairs of time series, and $D_k = \{(x_i, x_j) \mid x_i, x_j \in c_k, i < j\}$ be the set of pairs of variables included in a specific leaf c_k. After seeing n samples at the leaf, let

$$(x_1, y_1) = \underset{(x,y) \in D_k}{\mathrm{argmax}}\ d(x, y)$$

be the pair of variables with maximum dissimilarity within the cluster c_k, and in the same way considering $D'_k = D_k \backslash \{(x_1, y_1)\}$, let

$$(x_2, y_2) = \operatorname*{argmax}_{(x,y) \in D'_k} d(x, y)$$

be the second top-most dissimilar pair of variables. Consider $d_1 = d(x_1, y_1)$ and $d_2 = d(x_2, y_2)$ in $\Delta d = d_1 - d_2$, a new random variable consisting on the difference between the observed values through time. Applying the Hoeffding bound to Δd, if $\Delta d > \epsilon_k$, one can confidently say that, with probability $1 - \delta$, the difference between d_1 and d_2 is larger than zero, and select (x_1, y_1) as the pair of variables representing the diameter of the cluster. With this rule, the ODAC system will only split the cluster when the true diameter of the cluster is known with statistical confidence given by the Hoeffding bound. However, to prevent the hierarchy from growing unnecessarily, another criterion is defined in ODAC which has to be fulfilled in order to perform the splitting, which falls out of the scope of this work.

2.3   Assigning Criterion

When a split point is reported, the *pivots* are variables x_1 and y_1 where $d_1 = d(x_1, y_1)$, which are separated into each of the newly created clusters. The system then assigns each of the remaining variables of the old cluster to the cluster which has the closest *pivot*. This crisp assignment is the key object of our proposal in this work. When considering the expansion of the structure, the strict splitting of variables appears as a possible drawback, in the sense that a previous decision of moving a variable to a leaf, when there was no statistical confidence on the decision of assignment, may split variables that should be together. Left plot of figure 1 presents an example of a possible configuration where the problem could arise. An approach based on fuzzy sets [17] would let forthcoming examples decide what to do with those variables. Section 3 introduces our proposal to deal with this uncertainty.

2.4   Aggregation Criterion

The main setting of the system is the monitoring of existing clusters' diameters. On stationary data streams, the diameter of a cluster decreases every time a split occurs. Nevertheless, usual real-world problems deal with non-stationary data streams, where time series that were correlated in the past are no longer correlated to each other, in the current time period, and might be approaching time series of other clusters. The strategy that is adopted in ODAC to detect changes in the structure is based only on the analysis of the diameters. In fact, the diameter of each two new clusters should be less or equal than their parent's diameter. In this way, no computation is needed between the variables of the two siblings.

3   ODAC with a Semi-fuzzy Assignment Criterion

When a split point is reported, ODAC determines two variables as *pivots* and assigns each of the remaining variables to the cluster which has the closest *pivot*.

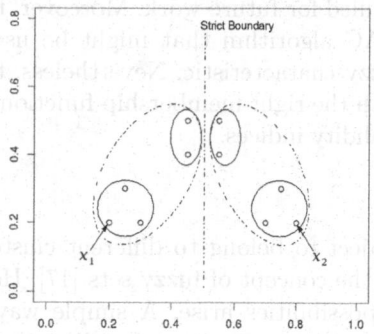 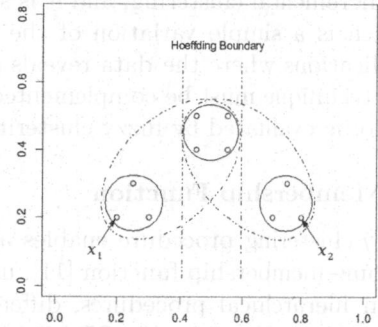

Fig. 1. Example of a dissimilarity structure between ten variables, produced by three clusters, and the comparison on the assignment method: *strict* (left) and *semi-fuzzy* (right). Variables x_1 and x_2 are the chosen *pivots* for splitting at first level this set of variables. Dot-dashed lines represent the first-level splitting while continuous lines present a second-level splitting.

This is usually a good heuristic, as it often finds an optimal border hyperplane. It is a lot faster than the heuristic performed by DIANA [9], since it is not needed to compute the average distances to decide which leaf will receive each variable. However, this may lead to erroneous situations if the moving variable is equally distant from the two pivots, there is no way of determining to which cluster it should be assigned.

This issue has a possible solution. The Hoeffding bound can be used to control the expansion of a cluster. We could include this notion of the Hoeffding bound as a decision support tool to the decision of moving a variable considering the two pivots. Let x and y be the pivots of the clusters a and b, respectively, and m be the moving variable. The expansion is decided as follows:

– if $d(y, m) - d(x, m) > \epsilon_k$ move variable to cluster a;
– else if $d(x, m) - d(y, m) > \epsilon_k$ move variable to cluster b;
– else move m into **both** clusters a and b, with a given degree of membership;

An example of application of this *semi-fuzzy* assignment is explained in the right plot of figure 1. This option may in fact allow the system to try different combinations of objects. However, this expansion eliminates the characteristic of speeding up the process with structure growth. Another example of this behavior follows when, given a *crisp* data set, the final specification is presented in figure 2 illustrating the difference between the two approaches. Top plot shows the result for *strict* clustering, where nodes 4, 5, 6 and 7 represent the real clusters defined for the *crisp* data set.

The *fuzzy* approach enables a wider observation on the relations between clusters, as they appear in different configurations. However, if crisp sets are *fuzzified*, a later pruning action could be considered. The diameters of the first ancestors of the leaves or the leaves themselves could act as a post-prune criterion. Preliminary results suggest it may have a very important role in time series incremental clustering, and it is scheduled for future work. Moreover, this approach is a simple variation of the ODAC algorithm that might be useful in applications where the data reveals a fuzzy characteristic. Nevertheless, this simple technique must be complemented with the right membership function, in order to be evaluated by fuzzy clustering validity indices.

3.1   Membership Function

A *fuzzy* clustering procedure enables an object to belong to different clusters, with some membership function [11], using the concept of fuzzy sets [17]. However, in hierarchical procedures, different possibilities arise. A simple way of applying *fuzzy* clustering in ODAC is to consider that, at each split where a variable is not clearly closer to one *pivot* than the other, the degree of membership of the variable to the new clusters should be equal, assigning the variable to *both* clusters with same probability. This will enable the definition of accurate validity indices for fuzzy clustering structures.

The result of a fuzzy clustering procedure is usually defined as a matrix $U = [u_{ic}]$ where each u_{ic} is the degree of membership of a vector x_i to cluster c. In our case, the vectors are the variables. At the root level, all variables belong to one cluster, so if cluster r is the root node, $u_{ir} = 1$ for all variables i. Every time a split occurs on a cluster p, the membership of variable i to each offspring cluster c (it was assigned to) should depend on the degree of membership u_{ip}. So, our approach is to compute it as

$$u_{ic} = u_{ip} * \beta_{ip} \tag{2}$$

where p is the parent cluster of c and β_{ip} is the distribution of membership due to possible fuzzy assignment of variable i when splitting cluster p. The β function can be computed based on several parameters, including the diameters of p, c and c's siblings. However, a first approach will consider $\beta = (n_{ip})^{-1}$, where n_{ip} is the number of new clusters to which variable i was assigned when splitting cluster p. For a clear strict assignment, we consider $\beta = 1^{-1} = 1$. For a fuzzy assignment after a binary split, we would have $\beta = 2^{-1} = 0.5$.

For the example presented in figure 2, the values for U are presented in table 1. It is easy to observe that the sum of membership values for each variable to all final clusters is 1. Values of U for non-leaf nodes are also presented to enable a clear insight on the splitting procedure.

4   Fuzzy Cluster Validity

Simple insights on the fuzziness of the clustering structure can be extracted using only the memberships values u_{ij}. Simple indices have been proposed such as the partition coefficient (PC) and the partition entropy (PE), defined next.

The partition coefficient index [2] is defined as

$$PC = \frac{1}{N} \sum_{i=1}^{N} \sum_{c=1}^{nc} u_{ic}^2 \tag{3}$$

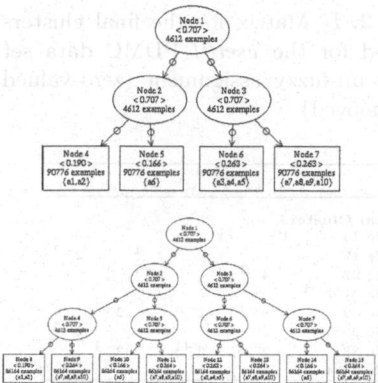

Fig. 2. ODAC structure comparison: *strict* (top) vs *fuzzy* (bottom) clustering (*crisp* data set)

Table 1. U Matrix for the fuzzy clustering structure gathered for *crisp* data set (zero-valued cells removed)

	a1	a2	a3	a4	a5	a6	a7	a8	a9	a10
High-level Nodes, u_{ic}										
node 1	1	1	1	1	1	1	1	1	1	1
node 2	1	1				2^{-1}	2^{-1}	2^{-1}	2^{-1}	2^{-1}
node 3			1	1	2^{-1}	2^{-1}	2^{-1}	2^{-1}	2^{-1}	
node 4	1	1				4^{-1}	4^{-1}	4^{-1}	4^{-1}	
node 5					2^{-1}	4^{-1}	4^{-1}	4^{-1}	4^{-1}	
node 6			1	1	1	4^{-1}	4^{-1}	4^{-1}	4^{-1}	
node 7					2^{-1}	4^{-1}	4^{-1}	4^{-1}	4^{-1}	
Final Clusters, u_{ic}										
node 8	1	1								
node 9						4^{-1}	4^{-1}	4^{-1}	4^{-1}	
node 10					2^{-1}					
node 11						4^{-1}	4^{-1}	4^{-1}	4^{-1}	
node 12			1	1	1					
node 13						4^{-1}	4^{-1}	4^{-1}	4^{-1}	
node 14					2^{-1}					
node 15						4^{-1}	4^{-1}	4^{-1}	4^{-1}	

with range in $[1/nc, 1]$, where nc is the number of clusters. The closer the index is to 1 the crisper the clustering is. In case that all membership values to a fuzzy partition are equal, the closer the value of PC is to $1/nc$, and the fuzzier the clustering is.

The partition entropy coefficient [7] is a slight variation defined as

$$PE_a = -\frac{1}{N} \sum_{i=1}^{N} \sum_{c=1}^{nc} u_{ic} \cdot log_a(u_{ic}) \qquad (4)$$

for values of U greater than zero, where a is the base of the logarithm, thus the index values range in $[0, log_a(nc)]$. The closer the value of PE is to 0, the crisp the clustering is.

Regarding the correspondence between the fuzzy clustering and the real data, a more robust criteria is the *Xie-Beni* index [16], also called the *compactness and separation* validity function. It is based on several measures of the clustering structure, with respect to the real data. The *fuzzy deviation* of variable i from cluster c, f_{ic}, is the distance between i and the center of cluster c, v_c, weighted by the fuzzy membership degree of data point i to cluster c, i. e.:

$$f_{ic} = u_{ic} \cdot d(i, v_c). \qquad (5)$$

We can compute the variation of cluster c as $\sigma_c = \sum_{i=1}^{n_c} f_{ic}^2$. The sum of all the variations of final clusters, $\sigma = \sum_{c=1}^{nc} \sigma_c$, is called *total variation* of the data set. The *compactness* of cluster c is calculated as the average variation in cluster c, $\pi_c = \sigma_c/n_c$, where n_c is the number of variables belonging to cluster c. Hence the *compactness* of the whole partition is $\pi = \sigma/n$. The *separation* of the fuzzy partition, d_{min}, is defined as the minimum centroid linkage between any two clusters. The index is defined as

$$XB = \pi/d_{min}. \qquad (6)$$

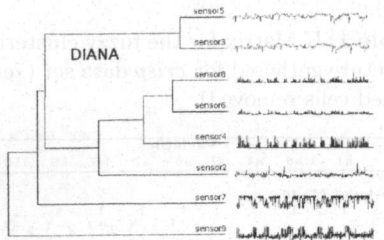

Fig. 3. Batch DIvisive ANAlysis clustering structure for the *user01* in PDMC data set, using the same correlation-based dissimilarity measure

Table 2. U Matrix for the final clusters obtained for the *user01* PDMC data set using semi-fuzzy assignment (zero-valued cells removed)

	Sensors							
	s2	s3	s4	s5	s6	s7	s8	s9
Final Clusters, u_{ic}								
node 14	1		1					
node 15					2^{-1}		2^{-1}	
node 27			4^{-1}		2^{-1}			
node 31	2^{-1}							
node 32			4^{-1}	2^{-1}		2^{-1}		
node 29	4^{-1}							2^{-1}
node 33			2^{-1}	2^{-1}		2^{-1}		
node 34	4^{-1}							

Small values of XB are expected for compact and well-separated clusters. However, attention should be paid as it monotonically decreases with the number of clusters nc.

5   Experimental Evaluation

In order to compare the *fuzziness* of the structures gathered by the system, we have applied the algorithm to a real data set published in the 2004 ICML Physiological Data Mining Competition, which has no clear crisp structure. Visually, we can only stress that there is high correlation between *sensor3* and *sensor5*. For the data belonging to *user01* (93344 observations), presented in figure 3, it is easy to note that only *sensor3* and *sensor5* are clearly correlated. The resulting ODAC structure is the right plot of figure 5, and the membership values are presented in table 2. The indices for both the *crisp* and *user01* data sets present concordant directions. While the *crisp* data set was partitioned with $PC = 0.650$ and $PE_2 = 0.900$, the *user01* set produced a partitioning with $PC = 0.594$ and $PE_2 = 0.875$. One can note that the PC index reveals a fuzzier structure in the outcome of *user01*, as expected. Accordingly, the PE_2 shows that the entropy of *crisp* data set result is higher, revealing that less fuzzy structure was found. Evidence appears that the *user01* data set has some inherit fuzziness, which was absent in the *crisp* data set. The proposed method relies on a confidence test based on the Hoeffding bounds. This way, sensitivity analysis must be performed to assess the level of dependence of the method to this parameter. Figure 4 presents the analysis on the *user01* data set. We can observe that, for usual values of δ, the system reveals low sensitivity, being the best results observed for δ parameter values between 0.02 and 0.06. From this point, we chose to fix $\delta = 0.05$.

However, in order to assess the sensitivity of the method to different levels of fuzziness, we have applied the algorithm to two other users' data from the PDMC data set (80182 and 141251 observations). Figures 5 to 7 present the resulting structures for the three users, using strict (left) and semi-fuzzy (right)

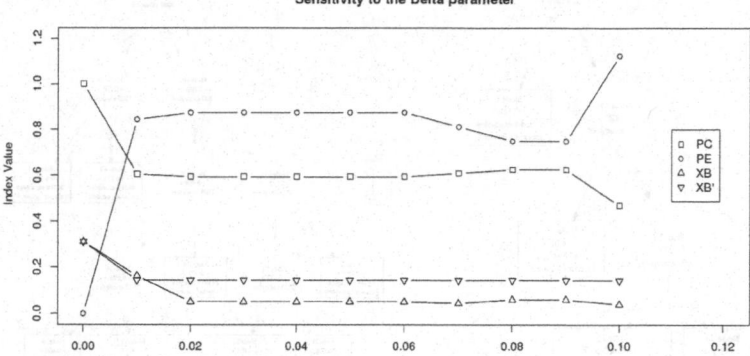

Fig. 4. ODAC quality sensitivity to the δ parameter, for the *user01* data, for PC, PE_2 and the *Xie-Beni* index for the *semi-fuzzy* (XB) and *strict* (XB') partitioning

assignment method. The resulting crisp clusters are the same in the three sets, although the hierarchy may be different. When using semi-fuzzy assignment, *user25* revealed much less fuzziness in the final structure than the remaining two sets, with almost the same clusters as the strict assignment method. Table 3 presents the values for the two partition indices and the *Xie-Beni* index for the three users, with the XB index also computed for the *strict* assignment method. While for the *user01* and *user06* data sets the semi-fuzzy approach resulted in better values of XB, the *user25* appears to contradict our approach. The strict assignment resulted in a much better value for the index, indicating that a fuzzy clustering would probably not be a good approach. Accordingly, when looking to the resulting hierarchy we can state that the semi-fuzzy assignment resulted in an almost strict partition of the variables. In fact, the loss beyond the crisp definition of final clusters is only an extra cluster with one single variable, supporting the notion that, although semi-fuzzy, this approach will nonetheless find crisp partitions when data is inherently crisp.

Table 3. Fuzzy validity indices for three users from the PDMC data set ($\delta = 0.05$). Table presents the values for each index and the number of final clusters for each user. Last line presents the Xie-Beni index for the resulting structure using strict assignment.

	user01	user06	user25
Validity Index, value (nc)			
PC	0.594 (8)	0.672 (9)	0.938 (5)
PE_2	0.875 (8)	0.688 (9)	0.125 (5)
XB	0.051 (8)	0.123 (9)	0.134 (5)
$XB(strict)$	0.059 (4)	0.199 (4)	0.059 (4)

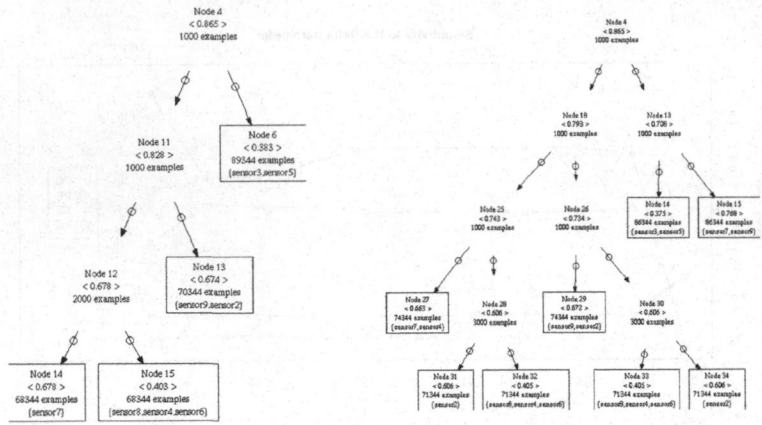

Fig. 5. ODAC strict (left) and semi-fuzzy (right) structure for the PDMC *user01* data

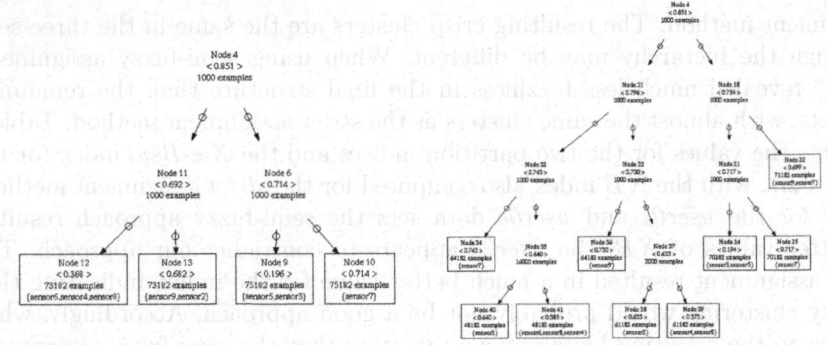

Fig. 6. ODAC strict (left) and semi-fuzzy (right) structure for the PDMC *user06* data

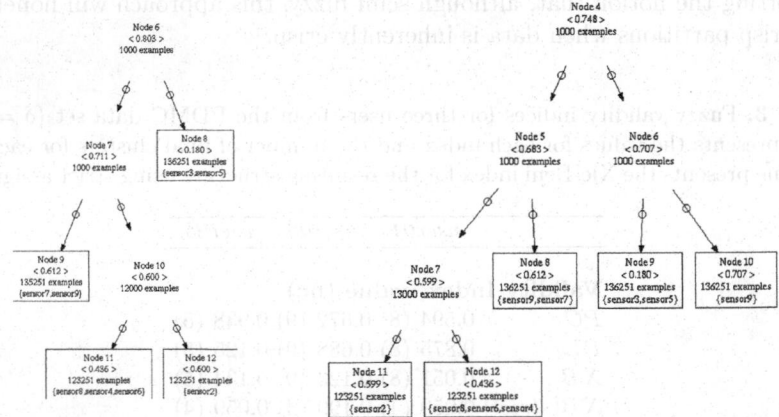

Fig. 7. ODAC strict (left) and semi-fuzzy (right) structure for the PDMC *user25* data

6 Concluding Remarks

In this paper we have presented a semi-fuzzy variation to the assignment criterion of ODAC, a clustering system for streaming time series. ODAC uses a top-down strategy to construct a binary tree hierarchy of clusters with the goal of finding highly correlated sets of variables. The main underlying concept in ODAC is the clusters' diameter. The split decision used in the algorithm focus on the crisp boundary between two groups, generating uncertainty in the assignment since it has to decide based on only a small subset of the entire data. In this work we propose a semi-fuzzy approach to the assignment of variables to newly created clusters, for a better trade-off between validity and performance. Experimental results show that this new assignment criterion will find better hierarchies when data is inherently fuzzy, without much loss in the quality of structures when the data is inherently crisp. Current and future work is concentrated on several areas, such as: the study of more complex membership functions; the inclusion of the membership function in the decision parameters of the entire system, such as the distance between variables and diameters computation; a post-prune criterion to reduce the size of the structure focusing on repeated instances of the same cluster; and a complete *fuzzy* assignment criterion for a complete *fuzzy* system.

Acknowledgment

Pedro P. Rodrigues is supported by a PhD grant by the Portuguese Foundation for Science and Technology (SFRH/BD/29219/2006). The authors also wish to acknowledge the participation of projects ALES II (POSC/EIA/55340/2004) and RETINAE (PRIME/IDEIA/70/00078).

References

1. Beringer, J., Hüllermeier, E.: Online clustering of parallel data streams. Data and Knowledge Engineering 58(2), 180–204 (2006)
2. Bezdek, J.C., Ehrlich, R., Full, W.: Fcm: The fuzzy c-means clustering algorithm. Computers and Geoscience 10(2), 191–203 (1984)
3. Dai, B.-R., Huang, J.-W., Yeh, M.-Y., Chen, M.-S.: Adaptive clustering for multiple evolving streams. IEEE Transactions on Knowledge and Data Engineering 18(9), 1166–1180 (2006)
4. Domingos, P., Hulten, G.: Mining high-speed data streams. In: Proceedings of the Sixth ACM-SIGKDD International Conference on Knowledge Discovery and Data Mining, Boston, MA, pp. 71–80. ACM Press, New York (2000)
5. Gama, J., Rodrigues, P.P.: Stream-based electricity load forecast. In: PKDD 2007. Proceedings of the 11th European Conference on Principles and Practice of Knowledge Discovery in Databases. LNCS (LNAI), vol. 4702, pp. 446–453. Springer, Heidelberg (2007)
6. Gerbec, D., Gasperic, S., Smon, I., Gubina, F.: An approach to customers daily load profile determination. In: Power Engineering Society Summer Meeting, pp. 587–591. IEEE Computer Society Press, Los Alamitos (2002)

7. Halkidi, M., Batistakis, Y., Varzirgiannis, M.: On clustering validation techniques. Journal of Intelligent Information Systems 17(2-3), 107–145 (2001)
8. Hoeffding, W.: Probability inequalities for sums of bounded random variables. Journal of the American Statistical Association 58(301), 13–30 (1963)
9. Kaufman, L., Rousseeuw, P.J.: Finding Groups in Data: An Introduction to Cluster Analysis. John Wiley and Sons, Inc., New York (1990)
10. Mantegna, R.N.: Hierarchical structure in financial markets. The European Physical Journal B 11(1), 193–197 (1999)
11. Nascimento, S.: Fuzzy Clustering Via Proportional Membership Model. In: Frontiers in Artificial Intelligence and Applications, vol. 119, IOS Press, Amsterdam (2005)
12. Pearson, K.: Regression, heredity and panmixia. Philosophical Transactions of the Royal Society 187, 253–318 (1896)
13. Rodrigues, P.P., Gama, J., Pedroso, J.P.: ODAC: Hierarchical clustering of time series data streams. In: SDM 2006. Proceedings of the Sixth SIAM International Conference on Data Mining, pp. 499–503. SIAM (April 2006)
14. Sherrill, D.M., Moy, M.L., Reilly, J.J., Bonato, P.: Using hierarchical clustering methods to classify motor activities of copd patients from wearable sensor data. Journal of Neuroengineering and Rehabilitation 2(16) (2005)
15. Wang, M., Wang, X.S.: Efficient evaluation of composite correlations for streaming time series. In: Dong, G., Tang, C.-j., Wang, W. (eds.) WAIM 2003. LNCS, vol. 2762, pp. 369–380. Springer, Heidelberg (2003)
16. Xie, X.L., Beni, G.: A validity measure for fuzzy clustering. IEEE Transactions Pattern Analysis and Machine Intelligence 13, 841–847 (1991)
17. Zadeh, L.A.: Fuzzy sets. Information and Control 8(3), 338–353 (1965)

On the Use of Rough Sets for User Authentication Via Keystroke Dynamics

Kenneth Revett[1], Sérgio Tenreiro de Magalhães[2], and Henrique M.D. Santos[2]

[1] University of Westminster
Harrow School of Computer Science
London, UK HA1 3TP
revettk@westminster.ac.uk
[2] Universidade do Minho
Department of Information Systems
Campus de Azurem
4800-058 Guimaraes, Portugal
{psmagalhaes,hsantos}@dsi.uminho.pt

Abstract. Keystroke dynamics is a behavioral biometric that is based on *how* a user enters their login details. In this study, a set of eight attributes were extracted during the course of entering login details. This collection of attributes was used to form a reference signature (a biometrics identification record) for subsequent authentication requests. The algorithm for the authentication step entails transforming the attributes into a discretised form based on the amino acid alphabet. A set of bioinformatics based algorithms are then used to perform the actual authentication test. In addition, the use of rough sets was employed in this study to determine if subsets of attributes were more important in the classification (authentication) than others. Lastly, the results of this study indicate that the error rate is less than 1% in the majority of the cases.

Keywords: behavioral biometrics, keystroke dynamics, multiple sequence alignment, reducts, rough sets.

1 Introduction

The concept of keystroke dynamics emerged from studies of the typing patterns exhibited by users when entering text into a computer using a keypad entry device. Researchers in the field focused on the keystroke pattern, in terms of keystroke duration and keystroke latencies. Evidence from preliminary studies indicated that typing patterns were sufficiently unique as to be easily distinguishable from one another, much like a person's written signature [1]. The basic idea is to extract a characteristic signature from a particular user's entry of a login ID – and use this information along with the login ID and password in deciding whether a login attempt is legitimate. There are two critical issues that must be addressed in the development of viable biometric: the selection of discriminatory attributes and the classification algorithm. In this paper, the attributes are selected based on latency (both digraphs and trigraphs are employed), dwell times (the amount of time a key is depressed, typing speed for both the login ID and password, length of time for login ID,

J. Neves, M. Santos, and J. Machado (Eds.): EPIA 2007, LNAI 4874, pp. 145–159, 2007.

password, and total time for both ID and password. These attributes are then discretised into an alphabet containing 20 discrete elements (the amino acid alphabet of molecular biology) for subsequent classification. The classification algorithm employed in this study is based on the rough set paradigm first presented to the literature by Z. Pawlak in the early 1980s [2],[3]. The result of this combined approach will be explored in this paper and compared with our previous work and that of author researchers in the filed. In the next section, we provide an overview of keystroke dynamics, followed by a brief description of the relevant bioinformatics material, and then a description of the rough sets approach to classification.

2 Keystroke Dynamics

Keystroke dynamics is a particular instance of a behavioural biometric that captures the typing style of a user. The dynamics of a user's interaction with a keyboard input device yields quantitative information with respect to dwell time (how long a key is pressed) and time-of-flight (the time taken to enter successive keys). By collecting the dynamic aspects acquired during the login process, one can develop a model that captures potentially unique characteristics that can be used for the identification of an individual. To facilitate the development of the model of how the user enters their details, an enrollment phase is required, when the user is asked to enter his/her login id/password until a steady value is obtained (usually limited to 10-15 trials – but this is implementation dependent). Once this data has been collected, a reference 'signature' is generated for this user. The reference signature is then used to authenticate the user account on subsequent login attempts. The user with that particular login id/password combination has their keystroke dynamics extracted and then compared with the stored reference signature. If they are within a prescribed tolerance limit – the user is authenticated. If not - then the system can decide whether to lock up the workstation - or take some other suitable action.

When devising such a biometric solution - there is always a trade off between being overly stringent - rejecting every attempt to login in and being overly lenient - allowing imposters to access the computer. The former is usually reported as a measure of false rejection - a type I error and the later a false acceptance or type II error. Another measure - called the cross over error rate (CER) – sometimes referred to as the equal error rate (EER) is also reported [4],[5]. The EER provides a quantitative measure of how sensitive the biometric is at balancing ease of use for the authentic user while at the same time reducing the imposter access rate. All extant biometric systems yield a trade-off between these two measures - those that reject imposters effectively (low FAR) are usually accompanied by a high FRR and vice versa. The critical research issue is 'how can the FAR be reduced independently of the FAR?' The next section presents a brief overview of the historical context in which keystroke dynamics has developed highlighting advances made in the methodological approaches to this critical question.

2.1 Background

Gaines was the first to report the results of a properly controlled study in the field of keystroke dynamics [1]. His study examined the typing patterns of seven professional typists – with the goal of determining if there were unique typing styles that could be used to distinguish between the typists. Although the results were not on par with current techniques, the deployment of digraphs – the time taken to enter two successive characters was a breakthrough. Joyce & Gupta presented in 1990 [6] an algorithm based on digraphs – but with a larger cohort and the results were significantly improved with respect to the Gaines study [1]. In 1997 Monrose and Rubin use the Euclidean Distance and probabilistic calculations based on the assumption that the latency times for one-digraph exhibits a Normal Distribution [7]. Later, in 2000, they also present an algorithm for identification, based on the similarity models of Bayes, and in 2001 they present an algorithm that uses polynomials and vector spaces to generate complex passwords from a simple one, using the keystroke pattern [8]. Various fuzzy logic algorithms have been applied – mapping the variability in typing patterns to a fuzzy concept. For instance, Hussein et al [9] used a combination of fuzzy clustering algorithms - obtaining an error rate (EER) of approximately 5-10%-depending on the number of samples they acquired per login id/password combination. Another study [10] employed a fuzzy rule set in order to classify login id/password combination with somewhat better success than Hussein - although they report only their preliminary results. Techniques based on neural networks have been explored - focusing on ART-2 and multi-layer perceptrons trained with the backpropagation algorithm. For instance, Obadiat provides data that suggests that the error rate can be reduced to approximately 2.4-4.2%, depending on the exact pre-processing performed using a non-standard neural network [11]. Other researchers have also applied neural networks (using standard backpropagation) to keystroke dynamics, generating error rates on the order of 2-4%[12], [13]. Other machine learning approaches, based on support vector machines (SVM) have been used to address the classification problem presented by keystroke dynamics.   De Oliveira et al [12] applied SVM to a small keystroke dataset and compare their results to standard neural network technology. The authors claim that the SVM classifier is more efficient and at least as accurate as neural network technologies. Sung et al. have also applied SVM to this domain, reporting an error rate of approximately 8-10% [13]. Bergadano et al. have employed an edit distance approach to user authentication. The edit distance is a measure of the entropy between two characters (in this case tri-graphs) contained within tow or more strings [14]. Revett et al. have used the rough sets algorithm to extract rules that form models for predicting the validity of a login ID/password attempt [15]. The results indicate that the error rate can be as low as 2% in many cases. In addition, the use of various bioinformatics based approaches such as motifs and multiple sequence alignments have yielded success with respect to user authentication and identification [16]. The next section describes the fundamentals of the bioinformatics approach that was undertaken in this study.

2.2 Sequence Alignment Algorithms

The central dogma in molecular biology is fairly straight forward: DNA begets RNA and RNA begets protein. How this dogma is manifested is as amazing as it is complex. Proteins are complex molecules that are involved in all life supporting activities - from respiration to thinking. The structure of a protein is usually represented schematically as a sequence of units that are joined together through chemical bonding. The units are termed amino acids and in humans - there are approximately 20 naturally occurring amino acids. It is the specific combination of amino acids - which is specified through the sequence of nucleotides found in DNA, that impart specificity in terms of function to proteins. All organisms need to respire and perform similar functions - which implies that there are certain genes in common across most organisms. Similarity between the genome of organisms is a major theme within the bioinformatics community. From a computer science perspective - this issue is one of pattern-matching and search - how closely related are two strings which represent the amino acid sequence of a particular gene between two species? What is the difference between a person with a disease that has a genetic basis from normal controls? These are interesting questions in their own right - and form the basis for the bioinformatics approach used in this study.

To place these ideas into the current context - imagine that a login attempt can be converted into a string of characters. To do this, the attributes extracted from the login attempt must be mapped onto a string within a given alphabet. The principle attributes extracted from keystroke dynamics include digraphs, trigraphs, and keypress durations (dwell time). These attributes are then mapped onto an alphabet – in this work, the amino acid alphabet, which contains 20 elements. This mapped version of the input attributes becomes a query sequence and the task then is to determine the closest match to a database of sequences derived in exactly the same manner from each user of the system. We then have a similar situation to that of a bioinformaticist - can we determine which sequence in the database is a closest match to the query sequence? This type of question has been addressed repeatedly within the bioinformatics community for over three decades now – with quite satisfactory results achieved in most cases. The principle tool of the bioinformaticist is the sequence alignment problem, which is illustrated in Figure 1.

Fly: GAKKVIISAPSAD-APM--F
Human: GAKRVIISAPSAD-APM--F
Yeast: GAKKVVSTAPSS-TPM--F

Fig. 1. Multiple sequence alignment of a portion of the glyceraldehyde3-phosphate dehydrogenase (GADPH) protein from three different animal species. Note the dashes indicate gaps - insertions/deletions in the alignment. Also note that there is a considerable amount of sequence identity - that is symbols that are the same across all three sequences.

The strings of characters in figure 1 represent a short hand notation for amino acids - which can be viewed as a string of these symbols. Of course there is biological meaning to them - but for the most part - bioinformaticists can treat them symbolically simply as a string of [17], [18]. In this work, the digraph times, dwell time, and trigraph times are discretised into an amino acid alphabet - which yields a string of characters similar to that found in a protein - but considerably shorter (42 residues in length). We can then apply the huge amount practical and theoretical research that has been successfully developed in bioinformatics for authentication.

In order to determine the relative alignment between a given query sequence and a database of 1,000's of sequences, one must have a metric - called the alignment score. For instance, if two elements match at a given position (in vertical register), then a score of +1 is recorded. If the characters at a particular position do not match a score of -1 is recorded. Gaps can be introduced, in global alignments, that serve to increase the overall score. If for instance two sequences were similar at some point but one had an insertion/deletion mutation, then these mutations would cause a misalignment between the sequences from this point forward - reducing the overall alignment score. Gaps can be placed in either the query or the target sequence and as many as required can be added. They also serve to ensure that the two sequences are the same length. When gaps are added they are penalised and reduce the score, but if they are effective at re-aligning the sequence further downstream, then the penalty may be cancelled out by a higher match score. Otherwise, the effect is to reduce the overall score, which is required because any two sequences can be aligned given enough gaps placed within them.

In this work, the sequence alignment algorithm employed is against a set of sequences stored as motifs (obtained via a position specific scoring matrix (PSSM)). This process entails aligning each column in the series of sequences and calculating the frequency of each amino acid within each column. A matrix is generated that scores the relative frequency of each of the 20 amino acids at each column position. This data can be used to generate a motif – where high scoring positions (corresponding to frequently occurring residues) are important features of the data. There are two dimensions that can be employed in PSSM: the magnitude of the frequency within a column and the number of columns (and whether they are consecutive or not). The process of generating the motif from the training data was extremely fast - on the order of a few milliseconds - which was a constraint placed on this system – it must be able to operate in a real-time environment. One can vary the mapping alphabet -in order to vary the resolution of the resulting mapped string. For instance, a mapping onto the digits 0-9 would produce a much coarser map that would generate longer motifs. The actual extraction of these motifs is described in the methods section, where the experimental methods are described, including a brief description of the dataset.

2.3 Rough Sets

Rough set theory is a relatively new data-mining technique used in the discovery of patterns within data first formally introduced by Pawlak in 1982 [2], [3]. Since its inception, the rough sets approach has been successfully applied to deal with vague or imprecise concepts, extract knowledge from data, and to reason about knowledge derived from the data [19]. We demonstrate that rough sets has the capacity to evaluate the importance (information content) of attributes, discovers patterns within data, eliminates redundant attributes, and yields the minimum subset of attributes for the purpose of knowledge extraction.

The first step in the process of mining any dataset using rough sets is to transform the data into a decision table. In a decision table (DT), each row consists of an observation (also called an object) and each column is an attribute, one of which is the decision attribute for the observation {d}. Formally, a DT is a pair A = $(U,\ A \cup \{d\})$ where d ϖ A is the *decision attribute*, U is a finite non-empty set of objects called the *universe* and A is a finite non-empty set of attributes such that a:U->V_a is called the value set of A. Once the DT has been produced, the next stage entails cleansing the data.

There are several issues involved in small datasets – such as missing values, various types of data (categorical, nominal and interval) and multiple decision classes. Each of these potential problems must be addressed in order to maximise the information gain from a DT. Missing values is very often a problem in biomedical datasets and can arise in two different ways. It may be that an omission of a value for one or more subject was intentional – there was no reason to collect that measurement for this particular subject (i.e. 'not applicable' as opposed to 'not recorded'). In the second case, data was not available for a particular subject and therefore was omitted from the table. We have 2 options available to us: remove the incomplete records from the DT or try to estimate what the missing value(s) should be. The first method is obviously the simplest, but we may not be able to afford removing records if the DT is small to begin with. So we must derive some method for filling in missing data without biasing the DT. In many cases, an expert with the appropriate domain knowledge may provide assistance in determining what the missing value should be – or else is able to provide feedback on the estimation generated by the data collector. In this study, we employ a conditioned mean/mode fill method for data imputation. In each case, the mean or mode is used (in the event of a tie in the mode version, a random selection is used) to fill in the missing values, based on the particular attribute in question, conditioned on the particular decision class the attribute belongs to. There are many variations on this theme, and the interested reader is directed to [3,4] for an extended discussion on this critical issue. Once missing values are handled, the next step is to discretise the dataset. All values that lie within a given range are mapped onto the same value, transforming interval into categorical data. As an example of a discretisation technique, one can apply equal frequency binning, where a number of bins n is selected and after examining the histogram of each attribute, n-1 cuts are generated so that there is approximately the same number of items in each bin. See the discussion in [19] for details on this and other methods of discretisation that have been successfully applied in rough sets. Now that the DT has been pre-processed, the rough sets algorithm can be applied to the DT for the purposes of supervised classification.

The basic philosophy of rough sets is to reduce the elements (attributes) in a DT based on the information content of each attribute or collection of attributes (objects) such that the there is a mapping between similar objects and a corresponding decision class. In general, not all of the information contained in a DT is required: many of the attributes may be redundant in the sense that they do not directly influence which decision class a particular object belongs to. One of the primary goals of rough sets is to eliminate attributes that are redundant. Rough sets use the notion of the lower and upper approximation of sets in order to generate decision boundaries that are employed to classify objects. Consider a decision table $A = (U, A \cup \{d\})$ and let $B \subseteq A$ and $X \subseteq U$. What we wish to do is to approximate X by the information contained in B by constructing the B-lower (B_L) and B-upper (B^U) approximation of X. The objects in B_L (B_LX) can be classified with certainty as members of X, while objects in B^U are not guaranteed to be members of X. The difference between the 2 approximations: B^U - B_L, determines whether the set is rough or not: if it is empty, the set is crisp otherwise it is a *rough set*. What we wish to do then is to partition the objects in the DT such that objects that are similar to one another (by virtue of their attribute values) are treated as a single entity. One potential difficulty arises in this regard is if the DT contains inconsistent data. In this case, antecedents with the same values map to different decision outcomes (or the same decision class maps to two or more sets of antecedents). This is unfortunately the norm in the case of small biomedical datasets, such as the one used in this study. There are means of handling this and the interested reader should consult [19] for a detailed discussion of this interesting topic. The next step is to reduce the DT to a collection of attributes/values that maximises the information content of the decision table. This step is accomplished through the use of the indiscernibility relation *IND(B)* and is defined for any subset $B \subseteq A$ ($B \subseteq A \cup \{d\}$) as follows:

$$IND(B) = \{(x, y) \in U \times U : \text{for every } a \in B \ a(x) = a(y)\} \qquad (1)$$

The elements of *IND(B)* correspond to the notion of an equivalence class. The advantage of this process is that any member of the equivalence class can be used to represent the entire class – thereby reducing the dimensionality of the objects in the DT. This leads directly into the concept of a *reduct*, which is the minimal set of attributes from a DT that preserves the equivalence relation between conditioned attributes and decision values. It is the minimal amount of information required to distinguish objects with in U. The collection of all reducts that together provide classification of all objects in the DT is called the *CORE*(A). The CORE specifies the minimal set of elements/values in the DT which are required to correctly classify objects in the DT. Removing any element from this set reduces the classification accuracy. It should be noted that searching for minimal reducts is an NP-hard problem, but fortunately there are good heuristics that can compute a sufficient amount of reducts in reasonable time to be usable. In the software system that we employ an order based genetic algorithm (o-GA) which is used to search through the decision table for approximate reducts [20]. The reducts are approximate because we do not perform an exhaustive search via the o-GA which may miss one or more attributes that should be included as a reduct. Once we have our set of reducts, we are ready to produce a set of rules that will form the basis for object classification.

Rough sets generate a collection of 'if..then' decision rules that are used to classify the objects in the DT. These rules are generated from the application of reducts to the decision table, looking for instances where the conditionals match those contained in the set of reducts and reading off the values from the DT. If the rules are too detailed (i.e. they incorporate reducts that are maximal in length), they will tend to overfit the training set and classify weakly on test cases. What are generally sought in this regard are rules that possess low cardinality, as this makes the rules more generally applicable. This idea is analogous to the building block hypothesis used in genetics algorithms, where we wish to select for highly accurate and low defining length gene segments. There are many variations on rule generation, which are implemented through the formation of alternative types of reducts such as *dynamic* and *approximate* reducts. Discussion of these ideas is beyond the scope of this paper and the interested reader is directed towards [19] for a detailed discussion of these alternatives. The rules that are generated are in the traditional conjunctive normal form and are easily applied to the objects in the DT. What we are interested in is the accuracy of the classification process – how well has the training rule set classified new objects? In the next section, we describe the dataset that was used in this study.

3 Methods

The dataset we examined consisted of a group of 30 subjects (all university students in a computer science department) – all acting as both authentic users and as imposters. The users were asked to enter a login id/password that was assigned to them (both 8 characters generated randomly by the software system) with an enrollment of 10 trials. We utilised the following attributes: digraph times, trigraph times, dwell times, total login ID typing time, total password typing time, total login ID + password time, typing speed for login ID, and typing speed for password time. The data samples were collected over a 14-day period, throughout various periods of the day. These attributes were used to form a reference signature for each user and were stored in the database for use during the authentication process (can also be sued for identification as well). In addition, in order to capture changes in typing patterns with familiarity with the login ID/passwords, a rolling average of the last ten successful authentication steps are maintained – with the most recent replacing the oldest (total of ten at all times are stored in the database). In order to calculate the EER, both the false acceptance rate (FAR) and the false rejection rate (FRR) must be calculated and plotted as a function of an appropriate decision parameter. To calculate the FAR, each user is requested to log into their own accounts 100 times over the 2-week period. For calculating FRR, each account is logged into by all other participants an equal number of times such that each account. The values of FAR and FRR are reported in the results section – and these values allow us to compare our results to other similar studies.

The basis of the experiments are as follows: the attribute set collected from the enrolment trials are quite substantial compared to most other studies – with a total of eight attributes (see Table 1 for a listing). The question to address is whether all of these attributes are required for the classification (i.e. authentication) process. To test this – the raw data after being discretised into the amino acid alphabet is used for the authentication process. Then, the same discretised data is processed using an implementation of rough sets prior to the authentication step. The purpose of applying rough sets is to reduce the dimensionality of the data. The reduced set of attributes was then subjected to the same authentication process and the results between the two tests are reported. Next, a brief description of the discretisation process is described. For details, please see [15].

Table 1. Attribute classes that were employed for the generation of the AA sequences. The order of the attribute classes was fixed throughout the experiments presented in this paper. The second row indicates the number of elements within this attribute class. The total number of elements was 47.

Di-graphs (ID, pass-word)	Tri-graphs (ID, pass-word)	Dwell time	Total ID time	Total password time	Sum of ID and password time	ID speed	Password speed
7,7	6,6	16	1	1	1	1	1

After the enrolment process, the data was discretised using the 20-letter amino acid (AA) alphabet. To do this, the largest digraph time was determined (with a resolution of 1 mS) and used to normalise all of the attributes recorded during the enrollment process. Next the normalised attributes were mapped onto one of the 20 elements from the AA alphabet (which were arranged in ascending alphabetical order). This yielded a sequence of 10 strings that contained as many characters as were available in the attributes for the username and password (47 in this study), which were maintained in a fixed order throughout the experiments.

The next stage is the development of a motif for each of the enrollment entries – which will be stored in the query database for verification and/or identification. A position specific scoring matrix (PSSM) was generated from the enrollment trials (10 for each user). Briefly, the frequency of each amino acid in each column was calculated and stored in a separate table (the PSSM). The data was not converted into a log-likelihood value as the expected frequencies were not available from a limited set of data. Instead, the values can be interpreted as probabilities of a given residue appearing at each position. There are two parameters that were examined with respect to motif formation: the frequency of each amino acid residue at each position – and the number of elements within the motif (whether they are consecutive or not). These are tunable parameters that can automatically be set based on the data from the enrollment process. The stringency of the motif based signature is based on these parameters: for high level security application, positions with a very high frequency – i.e. greater than 80% and for a minimum of 50% of the residues can be deployed. Likewise, reduced stringency is accomplished by relaxing these values. If during the enrollment process, the frequency within a column and the number of consistent columns was below some minimal threshold (50%), then the user would either be requested to re-enroll, or the normalisation time could be increased iteratively until the minimal threshold was achieved.

The normalisation time (and hence the bin times) was stored with each database entry (all 3 stored separately) in order to allow for a re-mapping of the attributes to sequence values if it became necessary. The mapped entries from the enrolment process and the resulting motifs for all 20 users formed the query database which was used for both verification and identification. The run-time for the motif extraction phase was fairly constant at approximately 10-20 milliseconds. The efficiency in this particular case is related to the short length of the sequences (many proteins contain 100s of residues) and the large degree of similarity between the sequences - these were generated from the enrolment sequence. The generation of the motif is performed immediately after enrolment and for all intensive purposes is so short that the user can then login to the system straight away and be authenticated by the resulting model.

The authentication algorithm works as follows: a user enters their login credentials. It is discretised into the amino acid alphabet based on the normalisation time associated with the login details associated with a given login ID/password combination. The authentication sequence is then compared with the stored motif for the login ID and given a score based on the algorithm specified in (1) using a simple global alignment algorithm. Generally, a match is scored +1 and a mismatch is scored -1, and '-' in either the probe of the query sequence has a score of 0. The score is computed and if it above a specified threshold q then the entry is authentication, else it is rejected. Note that the score 'q' can be based on both the frequency threshold at a specific column and the number of columns (as well as whether the columns are contiguous or not). The user has three attempts before system lockout. The value of q was set equal to the motif length (number of non-blank '-' entries) in this particular study - although in principle it could be varied in order to control the level of security required. The identification algorithm works as follows: the normalisation factor was extracted from the user during the authentication attempt. Then we proceed exactly as in the authentication phase, except we compare the resulting motif against all motifs stored in the database. Duplicate motifs are expected to be a rare event considering the number of expected entries in the motifs and the range of values at each. In this pilot study - the average cardinality of the motifs was 47. Each element of the motif could contain up to 20 different values - yielding an average of $47^{20}/l$ where 'l' is the motif length, possible motifs. In actuality, the full range of the sequence alphabet may not be fully covered. If there was a tie between 2 or more entries from the database then this login attempt is rejected and the user is asked to re-enter their login details.

The next stage involves the application of rough sets. The required decision table consisted of the attributes (discretised) and the decision was a label associated with each user (numeric values from 0-29), indicating the legitimate owner of the attribute. For a detailed discussion of the use of RSES, please consult Szczuka et al. [21]. The next section describes the results obtained from this study, and fill in experimental details as required.

4 Results

The first experiment is to determine the FAR/FRR/EER of the discretised data. All 30 users were requested to enroll and authenticate during a 2-week period of this study. Table 2 presents a sample of a typical enrollment dataset that consists solely of the first 14 enrollment elements (in this case the digraph times) for the username/password extracted during the enrollment period. The first stage in our algorithm is to discretise the enrollment data - since the values obtained are essentially continuous (to a resolution of 1 mS). This stage requires obtaining the largest time for each category of attributes and normalising all elements of each authentication attribute. Then binning was performed, where each digraph was assigned its ordinal position within the discretisation alphabet: A= "ACDEFGHIKLMNPQRSTVWY" - the single letter code for amino acids in ascending lexical order. This resulted in the following dataset displayed in Table3. In our previous work, the maximal frequency of each element within each column - this corresponds to the range of values that were obtained for each attribute. This same approach was taken in this study - as it proved to be quite effective with the previous data collected from a new participant cohort. The consensus sequence for the example in Table 3 is 'EGEEF---HIEI—', where the '-' symbol indicates that there is no unique dominant symbol within the column(s) – although the exact frequency of each amino acid is maintained across all columns. The sequence in (1) above is the consensus sequence for this enrollment instance. It represents the average behavior - in that it captures the most frequently entered values across all attributes. This amounts to an unweighted voting scheme, with a threshold for inclusion (set to 0.5 in this study).

Table 2. Discretisation of an enrollment entry (corresponding to the raw data presented in Table 1) using the amino acid alphabet for a given normalisation value. The motif (consensus sequence) for this data, using a threshold of 0.5 is: 'E G E E F - - - H I E I - -.'

F	G	E	E	G	H	N	M	H	N	E	I	Q	E
E	F	F	E	F	M	S	P	I	H	D	I	Q	F
Y	G	E	E	F	H	Q	Q	H	I	E	H	R	G
F	G	F	F	H	I	L	Q	I	I	D	H	P	F
E	F	F	E	G	I	H	P	K	I	E	I	N	E
E	H	E	E	F	Q	Q	P	I	H	D	I	P	G
E	G	E	F	F	I	H	L	H	H	C	H	Q	G
E	F	F	F	F	Q	E	K	H	H	D	H	R	F
E	E	F	E	F	I	F	G	H	I	E	I	R	H
E	E	E	E	F	N	N	L	I	I	E	I	N	G

The entries within the consensus sequence (motif) also possess information regarding the typing speed and the consistency possessed by the person entering it. This is the value of the normalisation factor and influences the spread of the attributes (and hence amino acid symbols) within the generated sequence. If an imposter attempted to input this particular username/password, the maximal attribute value would more than likely differ from this particular one, and hence the normalisation process may yield a different set of elements within the consensus sequence. Please note that the maximal attribute class value for each enrollment is stored with the user's data and it is this value that is used to discretise all subsequent authentic successful login attempts. Also note that for each subsequent successful login attempt, the data for this account is updated such that the oldest entry is removed and the consensus sequence is updated, along with the longest attribute value (the normalisation factor). This allows the system to evolve with the user as his/her typing of their login details changes over time - which invariably happens as the user becomes more familiar with their login details.

For the authentication processes (either verification and identification), the username and password entered is discretised using the stored maximal value for each class attribute time. Equation (1) provides the values used when performing the motif matching algorithm. The results from our preliminary study indicate that it took on average 21 mS to compare the motifs over the 30 entries - yielding a value of 48 matches/second. It should be noted that if there were less then a threshold level of unique entries in the consensus sequence (a system-defined parameter set at a default value of 0.5), then the system automatically re-calculates the normalisation factor. This re-scaling occurs iteratively until there is at least a threshold level of entries in the consensus sequence. Also note that the amino acid alphabet is just one of many choices – the greater the cardinality the more refined the bin size for a constant normalisation value. The system as just described resulted in an overall FAR of 0.15% and an FRR of 0.2% - a total error rate of 0.35% (for a consistency threshold of 0.5). This result is comparable to that found in other keystroke dynamic based systems ([12], [14]). Please note that this result is measured for individual login attempts - that is 8 attempts were unsuccessful on an individual login attempt (FRR) - but no user was locked out because they failed to login within 3 attempts.

Table 3. Summary of FAR/FRR results as a function of the consistency threshold for elements within the consensus sequence (motif). These results are randomly selected values from a series of one-hundred experiments. Note that these results are from the un-processed (no application of rough sets) dataset.

Threshold	FAR	FRR
1.0	0.0%	0.6%
0.80	0.0%	0.5%
0.60	0.1%	0.1%
0.40	0.25%	0.0%

When the consistency threshold was raised to 0.80 (rounded upwards when necessary), the FAR was 0.0% but the FRR increased to 0.5% - yielding a total error rate of 0.5%. In Table 4 the results for FAR/FRR are summarised with respect to the threshold for dominant consensus elements. Note that in this experiment, the same 30 users were used and each logged into their own accounts 100 times (FRR measurement) and each participant logged into the other 29 accounts 100 times each (a total of 2,900 FRR attempts/account and 100 FAR attempts/account).

In addition to the frequency threshold data that were presented in Table 3, another parameter, based on the total number of elements in the consensus sequence was also investigated in this study. The two parameters are somewhat related – in that for a given position to be significant, it must have a specific frequency value. In this part of the study, a frequency value of 0.5 was used as a first approximation, and the focus was on the total number of matching entries within the motifs stored with a particular use ID/login sequence. The data in Table 4 present a summary of the data as a function of the fractional percentage of the motif cardinality.

Lastly, the data was processed using the RSES implementation of rough sets [20]. The columns in the decision table were the 47 attribute values (discretised into the AA alphabet) and the decision class (which was a coded value for each of the login ID/password sequences). There were no missing values – and the data was already discretised – so the decision table did not require any further processing. or each login ID/password, the FAR and FRR values were used as the negative and positive examples in a two-class decision system (labelled '0' for legitimate owner – for FRR measurements and '1' for imposter – FAR measurement). This process was repeated across all subjects employed in the study. Reducts were generated using the genetic algorithm approach and were used to generate the decision rules for subsequent classification. After processing the data with rough sets, a subset of the attributes were selected and these were used for subsequent analysis. Then the alignment algorithm as stated above was re-applied to the new motifs and the FAR/FRR were recomputed. The results from the FAR/FRR on the reduced motifs are summarised in Table 5.

Table 4. Sample results from an experiment in which the length (total number) of consensus cites matched the stored record for a set of randomly selected login attempts. Note that these results are from the un-processed (no application of rough sets) dataset.

Threshold	FAR	FRR
40	0.0%	1.0%
30	0.0%	0.4%
20	0.1%	0.15%
10	0.3%	0.0%

Table 5. Summary of FAR/FRR results as a function of the consistency threshold for elements within the consensus sequence (motif). These results are randomly selected values from a series of one-hundred experiments. Note that these results are from the *processed* (application of rough sets) dataset.

Threshold	FAR	FRR
1.0	0.0%	0.5%
0.80	0.0%	0.3%
0.60	0.0%	0.0%
0.40	0.1%	0.0%

Of the 47 attributes included in the decision table, on average, 14-18 were most strongly associated with the correct decision class. This correlation was based on the support of the attributes in the decision mapping process. Support indicates the number of times an attributes is contained within the decision rules. In the interest of conserving space, the attributes, in decreasing order of support were: login ID di-graphs (4-6), total login ID time, password di-graphs (3-5), password typing speed, login ID dwell time (5). The numbers in parentheses refer to the range of the respective attribute – as these values were calculated for all 30 users separately.

5 Conclusion

The results from this study indicate that keystroke dynamics can be a very effective means for user authentication. The FAR/FRR results are approximately less then 1% in most cases. These results are superior to our previous work employing keystroke dynamics employing purely statistical measures [4] as well as other reports [9]-[13]. This superiority in performance could be due in part to the relatively large number of attributes that were employed in this study. A total of 47 attributes were used to encode each login ID/password sequence, falling within 8 separate categories (see Table 1). This results in a set of sequences that are 47 characters long – by typical bioinformatics database standards, this yields a significant computational load when performing searches. The rationale for the deployment of rough sets was to reduce the number of attributes to their minimal level – which rough sets have been shown to do very well [15], [19]. The reduced dataset (set of attributes) was then processed as in the control situation and the resulting classification accuracy (FAR/FRR) actually improved slightly. In addition, the execution time was reduced by as much as 40%, yielding a total authentication time of approximately 100 mS to form the motif and 50 mS to authenticate (data not shown). In addition, in a linearly weighted voting scheme, having values for the relative importance of each of the attributes (even within attribute classes) is a significant advantage. This feature will be explored in subsequent work.

It should be noted that the motif-matching algorithm does not require a two-class dataset – as do most neural network based classification schemes. This is a significant advantage when the system is initially started – as there are at best only enrollment datasets. The computational speed of this system is sufficient to allow it to be used in an on-line fashion – not withstanding the roughs sets pre-processing step. The rough sets aspect of this processing pipeline is potentially an issue. In the current work, rough sets was applied in a two-class approach to a fairly large number of samples to select from. In this work, there were 100 legitimate samples from authentic owners and 2,900 imposter logins. The rough sets was applied in such a way that each of the 100 authentic owners were in effect tested by all the imposters, selected 100 at a time randomly and cross-validated. The purpose of applying rough sets in this work was to investigate the variety of attributes that were significant with respect to authentication accuracy. The data from this experiment indicated that first of all, not all attributes were required. The reduction in attribute number was on the order of 60-70% - quite significant. In future work, the thresholds for the determination of significant entries within the motif will be examined using rough sets in greater detail to see if this parameter itself can be determined from the data.

References

[1] Gaines, R., et al.: Authentication by keystroke timing: Some preliminary results. Rand Report R-256-NSF, Rand Corp. (1980)

[2] Pawlak, Z.: Rough Sets. International Journal of Computer and Information Sciences 11, 341–356 (1982)

[3] Pawlak, Z.: Rough sets – Theoretical aspects of reasoning about data. Kluwer, Dordrecht (1991)

[4] Magalhães, S.T., Santos, H.D.: An improved statistical keystroke dynamics algorithm. In: Proceedings of the IADIS MCCSIS 2005 (2005)

[5] Magalhães, S.T., Revett, K.: Password Secured Sites – Stepping Forward with Keystroke Dynamics. In: International Conference on Next Generation Web Service Practises, Seoul, Korea (2005)

[6] Joyce, R., Gupta, G.: Identity authorization based on keystroke latencies. Communications of the ACM 33(2), 168–176 (1990)

[7] Monrose, F., Rubin, A.D.: Authentication via Keystroke Dynamics. In: Proceedings of the Fourth ACM Conference on Computer and Communication Security, Zurich, Switzerland (1997)

[8] Monrose, F., Rubin, A.D.: Keystroke Dynamics as a Biometric for Authentication. Future Generation Computing Systems (FGCS) Journal: Security on the Web (2000)

[9] Hussien, B., Bleha, S., McLaren, R.: An application of fuzzy algorithms in a computer access security system. Pattern Recognition Letters 9, 39–43 (1989)

[10] de Ru, W.G., Eloff, J.: Enhanced Password Authentication through Fuzzy Logic. IEEE Expert 12(6), 38–45 (1997)

[11] Obaidat, M.S., Sadoun, B.: A Simulation Evaluation Study of neural network techniques to Computer User Identification. Information Sciences 102, 239–258 (1997)

[12] de Oliveira, M.V.S., Kinto, E., Hernandez, E.D.M., de Carvalho, T.C.: User Authentication Based on Human Typing Patterns with Artificial Neural Networks and Support Vector Machines. In: SBC 2005 (2005)

[13] Sung, K.S., Cho, S.: GA SVM Wrapper Ensemble for Keystroke Dynamics Authentication. In: International Conference on Biometrics, Hong Kong, pp. 654–660 (2006)

[14] Bergadano, F., Gunetti, D., Picardi, C.: User authentication through keystroke dynamics. ACM Transactions on Information and System Security 5(4), 367–397 (2002)

[15] Revett, K.: On the Use of Multiple Sequence Alignment for User Authentication via Keystroke Dynamics. In: International Conference on Global eSecurity (ICGeS), April 16-18, 2007, University of East London, pp. 112-120 (2007)

[16] Revett, K., Magalhaes, S., Santos, H.: Developing a Keystroke Dynamics Based Agent Using Rough Sets. In: The 2005 IEEE/WIC/ACM International Joint Conference on Web Intelligence and Intelligent Agent Technology Workshop on Rough Sets and Soft Computing in Intelligent Agents and Web Technology, Compiegne, France, September 19-22, 2005, pp. 56–61 (2005)

[17] Needleman, S.B., Wunsch, C.D.: A general method applicable to the search for similarities in the amino acid sequence of two proteins. Mol. Biol. 48, 443–453 (1970)

[18] Smith, T.F., Waterman, M.S.: Identification of common molecular subsequences. J. Mol. Biol. 147, 95–197 (1981)

[19] Slezak, D.: Approximate Entropy Reducts. Fundamenta Informaticae (2002)

[20] Bazan, J., Szczuka, M.: The Rough Set Exploration System. In: Peters, J.F., Skowron, A. (eds.) Transactions on Rough Sets III. LNCS, vol. 3400, pp. 37–56. Springer, Heidelberg (2005) http://logic.mimuw.edu.pl/ rses

[21] Bazan, J., Nguyen, H.S., Nguyen, S.H., Synak, P., Wróblewski, J.: Rough set algorithms in classification problems. In: Studies in Fuzziness and Soft Computing, vol. 56, pp. 49–88. Physica-Verlag, Heidelberg (2000)

The Halt Condition in Genetic Programming

José Neves, José Machado, Cesar Analide, António Abelha, and Luis Brito

Department of Informatics, University of Minho, Braga, Portugal
{jneves,jmac,analide,abelha,lbrito}@di.uminho.pt

Abstract. In this paper we address the role of divergence and convergence in creative processes, and argue about the need to consider them in Computational Creativity research in the Genetic or Evolutionary Programming paradigm, being one´s goal the problem of the Halt Condition in Genetic Programming. Here the candidate solutions are seen as evolutionary logic programs or theories, being the test whether a solution is optimal based on a measure of the quality-of-information carried out by those logical theories or programs. Furthermore, we present Conceptual Blending Theory as being a promising framework for implementing convergence methods within creativity programs, in terms of the logic programming framework.

Keywords: Computational Creativity, Genetic or Evolutionary Programming, Extended Logic Programming, Quality-of-Information, Conceptual Blending Theory.

1 Introduction

While the discussion around the phenomenon of creativity runs about fundamental issues like clarification of concepts, evaluation, psychological factors or philosophical questions, the quest for creativity in Artificial Intelligence (AI) has begun, raising its unavoidable subjects such as knowledge representation, search methods, domain modelling, just to name a few. In this paper, we propose a theme around the subject of modelling creativity, from the point of view of the process. We start by considering the divergence/convergence characteristics of the creative process as an argument for the need of divergent methods that, at some point, are able to detect a convergent solution as a way of goal accomplishment. Although this may seem the description of search methods in general, it is clear that we may deal with wider amplitudes of divergence in tasks that demand higher creativity. These tasks do not necessarily have to present a particular form or be of a specific kind. However, the quest for a previously unseen and correct solution is surely expected. It is a solution that traditional methods do not seem to find. We think that some qualitative jump must be made in AI, such that classical methods become more able to diverge or at least to combine with other processes, in order to enter the realms of creativity.

On the other hand, *Genetic Programming (GP)* may be seen as one of the most useful, general-purpose problem solving techniques available nowadays. It has been used to solve a wide range of problems, such as symbolic regression, data mining, optimization, and emergent behavior in biological communities. *GP* is one instance of

J. Neves, M. Santos, and J. Machado (Eds.): EPIA 2007, LNAI 4874, pp. 160–169, 2007.

the class of techniques called evolutionary algorithms, which are based on insights from the study of natural selection and evolution. Evolutionary algorithms solve problems not by explicit design and analysis, but by a process akin to natural selection. An evolutionary algorithm solves a problem by first generating a large number of random problem solvers (programs). Each problem solver is executed and rated according to a fitness metric defined by the developer. In the same way that evolution in nature results from natural selection, an evolutionary algorithm selects the best problem solvers in each generation and breeds them. Genetic programming and genetic algorithms are two different evolutionary algorithms. Genetic algorithms involve encoded strings that represent particular problem solutions. These encoded strings are run through a simulator and the best strings are mixed to form a new generation. Genetic programming, the subject of this article, follows a different approach. Instead of encoding a representation of a solution, *GP* breeds executable computer programs, here defined in terms of logic programs or logical theories.

2   The Problem

A *genetic* or *evolutionary algorithm* applies the principles of evolution found in nature to the problem of finding an optimal solution to a solver problem. In a *genetic algorithm* the problem is encoded in a series of bit strings that are manipulated by the algorithm; in an *evolutionary algorithm* the decision variables and problem functions are used directly. A drawback of any evolutionary algorithm is that a solution is *better* only in comparison to other, presently known solutions; such an algorithm actually has no concept of an *optimal solution*, or any way to test whether a solution is optimal. This also means that an evolutionary algorithm never knows for certain when to stop, aside from the length of time, or the number of iterations or candidate solutions, that one may wish to explore. In this paper it is addressed the problem of the *halt condition in genetic programming*, where the candidate solutions are seen as evolutionary logic programs or theories, being the test whether a solution is optimal based on a measure of the quality-of-information carried out by those *logical theories* or programs.

2.1   The Past

For example, let us consider the case where we had a series of data that was produced from a set of water's reservoir results taken over time. In this case we have the values that define the output of the function, and we can guess at some parameters which might inter-operate to produce these values - such as a measure of the acidity or alkalinity of the water's reservoir, the dissolved oxygen, the *nitrates*, the *phosphates*, the *chlorophyll*, and so on. We might like to predict how the water's reservoir results will fare in the future, or we may want to fill in some missing data in the series. To do so, we need to find the relationship between the parameters which will generate values as close to the observed values as possible. Therefore, our *optimum* will be fit to the observed values.

To give a very different example, consider the problem a glazier might have in deciding how best to cut up a large sheet of glass in order to achieve the minimum wastage if a known number of different sized panes are to be produced from the sheet. The function is therefore the area of glass needed, and the parameters are the sizes of the panes required. The *optimum* is the minimum area of glass left over after all the panes have been cut out. Whereas traditional search algorithms seek to search such problem spaces in a linear fashion, one search route at a time, i.e., *Genetic Algorithms* maintain a population of candidate solutions, all competing against one another. At each iteration of the GA search engine, each candidate solution is evaluated against the optimality criteria specified, and assigned a measure of goodness. All the candidate solutions are then replaced with new candidates by a reproduction process which seeks to combine parts of one solution with parts of another. By a stochastic process, only the better candidates are selected to perform in this process. As a result, some of the good candidate solutions will be replicated, and some new candidates will be produced based on the existing ones. This forms a kind of directed search with controlled stochastic effects to provide exploration where needed.

The GA operates upon a *population* of candidate solutions to the problem. These solutions can be held in the type of their parameter representation. For example, if the candidate solutions were for a function optimisation problem in which the function took a fixed number of floating point parameters, then each candidate could be represented as an array of such floating point numbers.

Clearly, we want to search only the most promising search paths into the population, although we must remain aware that sometimes non-promising search paths can be the best route to the result we are looking for. In order to work out which are the most promising candidates, we evaluate each candidate solution using a user supplied evaluation function. In general, this assigns a single numeric *goodness measure* to each candidate, so that their relative merit is readily ascertained during the application of the genetic operators. Undoubtedly, the amount of meaning and the interpretation that can be gleaned from this single value, is crucial to a successful search [3].

2.2 The Future

With respect to the computational paradigm it were considered extended logic programs with two kinds of negation, classical negation, $\neg$, and default negation, *not*. Intuitively, *not p* is true whenever there is no reason to believe p *(close world assumption)*, whereas $\neg p$ requires a proof of the negated literal. An extended logic program (program, for short) is a finite collection of rules and integrity constraints, standing for all their ground instances, and is given in the form:

$$p \leftarrow p_1 \wedge \ldots \wedge p_n \wedge not\ q_1 \wedge \ldots \wedge not\ q_m;\ and$$
$$?\ p_1 \wedge \ldots \wedge p_n \wedge not\ q_1 \wedge \ldots \wedge not\ q_m,\ (n,m \geq 0)$$

where *?* is a domain atom denoting falsity, the p_i, q_j, and p are classical ground literals, i.e. either positive atoms or atoms preceded by the classical negation sign $\neg$ [7]. Every program is associated with a set of abducibles. Abducibles can be seen as hypotheses that provide possible solutions or explanations of given queries, being given here in the form of exceptions to the extensions of the predicates that make the program.

These extended logic programs or theories stand for the population of candidate solutions to model the universe of discourse. Indeed, in our approach to *GP*, we will not get a solution to a particular problem, but rather a logic representation (or program) of the universe of discourse to be optimized. On the other hand, logic programming enables an evolving program to predict in advance its possible future states and to make a preference. This computational paradigm is particularly advantageous since it can be used to predict a program evolution employing the methodologies for problem solving that benefit from abducibles [8,9], in order to make and preserve abductive hypotheses. It is on the preservation of the abductive hypotheses that our approach will be based to present a solution to the problem of *The Halt Condition in GP*.

Designing such a selection regime presents, still, unique challenges. Most evolutionary computation problems are well defined, and quantitative comparisons of performance among the competing individuals are straightforward. By contrast, in selecting an abstract and general logical representation or program, performance metrics are clearly more difficult to devise. Individuals (i.e., programs) must be tested on their ability to adapt to changing environments, to make deductions and draw inferences, and to choose the most appropriate course of action from a wide range of alternatives. Above all they must learn how to do these things on their own, not by implementing specific instructions given to them by a programmer, but by continuously responding to positive and negative environmental feedback.

In order to accomplish such goal, i.e., to model the universe of discourse in a changing environment, the breeding and executable computer programs will be ordered in terms of the quality-of-information that stems out of them, when subject to a process of conceptual blending [10]. In blending, the structure or extension of two or more predicates is projected to a separate blended space, which inherits a partial structure from the inputs, and has an emergent structure of its own. Meaning is not compositional in the usual sense, and blending operates to produce understandings of composite functions or predicates, the conceptual domain, i.e., a conceptual domain has a basic structure of entities and relations at a high level of generality (e.g., the conceptual domain for journey has roles for traveler, path, origin, destination). In our work we will follow the normal view of conceptual metaphor, i.e., metaphor will carry structure from one conceptual domain (the source) to another (the target) directly.

Let i (i $\in$ {1,...,m}) denote the predicates whose extensions make an extended logic program or theory that model the universe of discourse, and j (j $\in$ {1,...,n}) the attributes for those predicates. Let $x_j \in [min_j, max_j]$ be a value for attribute j. To each predicate it is also associated a scoring function $V^i_j [min_j, max_j] \rightarrow 0...1$, that gives the score predicate i assigns to a value of attribute j in the range of its acceptable values, i.e., its domain (for the sake of simplicity, scores are kept in the interval 0...1), here given in the form *all(attribute exception list, sub expression, invariants)*. This states that *sub expression* should hold for each combination of the exceptions to the extension of the predicate of the attributes in the *attribute exception list* and the *invariants*. This is further translated by introducing three new predicates. The first predicate creates a list of all possible value combinations (e.g., pairs, triples) as a list of sets determined by the domain size (and the invariants). The second predicate recurses through this list, and makes a call to the third predicate for each exception combination. The third predicate denotes *sub expression* and is constructed accordingly. The quality of the information with respect to a generic predicate K is therefore given by $Q_K=1/Card$, where *Card* denotes the cardinality of the exception set for K, if the exception set is not disjoint. If the exception set is disjoint, the quality of information is given by:

$$Q_K = \frac{1}{C_1^{Card} + ... + C_{card}^{card}}$$

where C_{card}^{card} is a card-combination subset, with *card* elements.

The next element of the model to be considered, it is the relative importance that a predicate assigns to each of its attributes under observation; w_j^i stands for the relevance of attribute j for predicate i (it is also assumed that the weights of all predicates are normalized, i.e. [1],

$$\sum_{1 \le j \le n} w_j^i = 1, \text{ for all i.}$$

It is now possible to define a predicate's scoring function, i.e., for a value $x = (x_1,...,x_n)$ in the multi-dimensional space defined by the attributes domains, which is given in the form:

$$V^i(x) = \sum_{1 \le j \le n} w_j^i V_j^i(x_j).$$

It is now possible to measure the quality of the information that stems from a logic program or theory, by posting the $V^i(x)$ values into a multi-dimensional space, whose axes denote the logic program or theory, with a numbering ranging from 0 (at the center) to 1. For example, one may have what is illustrated by Figure 1, where the dashed area stands for the quality of information that springs from an extended logic program or theory P, built on the extension of 5 (five) predicates, here named as $p_1...p_5$. It not only works out which are the most promising extended logic programs or theories to model the universe of discourse, making the *Halt Condition in Genetic Programming*.

As an example, let us now consider the case referred to above, where we had a series of data that was produced from a set of water's reservoir results taken over time. In this case we will not have the values that define the output of the function, but a measure of the quality of the water's reservoir (Program 1).

pH(january,0.32).

$\neg pH(X,Y) \leftarrow$
    not pH(X,Y) $\wedge$
    not exception$_{pH}$(X,Y).

$\neg(pH(X,Y) \wedge Y \ge 0 \wedge Y \le 1).$ */ This invariant states that pH takes values on the interval 0...1/*

$\neg((exception_{pH}(X,Y) \vee exception_{pH}(X,Z)) \wedge \neg (exception_{pH}(X,Y) \wedge exception_{pH}(X,Z))).$

/This invariant states that the exceptions to the predicate pH follow an exclusive or/

Program 1. The extended logic program for pH with respect to January

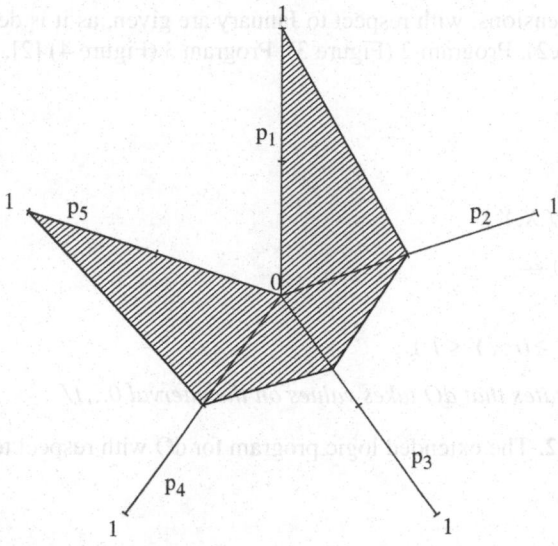

Fig. 1. A measure of the quality-of-information for logic program or theory P

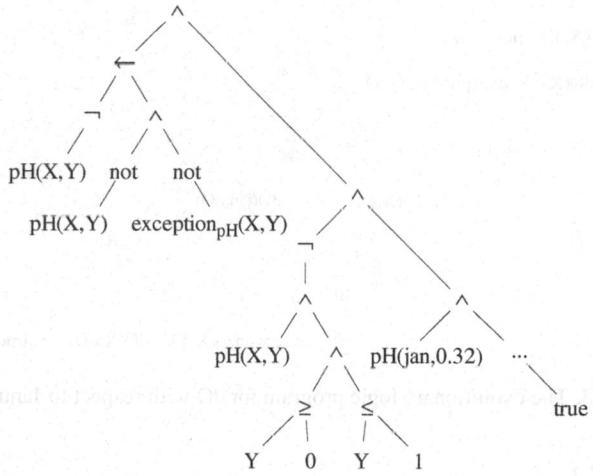

Fig. 2. The evolutionary logic program for pH with respect to January

Therefore we will not guess at some parameters which might inter-operate to produce these values - such as a measure of the acidity or alkalinity of the water's reservoir, the *dissolved oxygen*, the *nitrates*, the *phosphates*, the *chlorophyll*, and so on - but to have such parameters aggregated and giving rise to a set of predicates, which will be given in the form (for the sake of simplicity it will be considered only three predicates, namely those denoting the acidity or alkalinity of the water's reservoir (i.e., the *pH*), the dissolved oxygen (i.e., the *dO*), and the phosphates (i.e., the *tP*) whose extensions, with respect to January are given, as it is depicted below, in Program 1 (Figure2), Program 2 (Figure 3), Program 3 (Figure 4) [2].

dO(january,dO).

¬ dO(X,Y) ←
   not dO(X,Y) ∧
   not exception$_{dO}$(X,Y).

exception$_{dO}$(X,Y) ←
   dO(X,dO).

¬(dO(X,Y) ∧ Y ≥ 0 ∧ Y ≤ 1).

/This invariant states that dO takes values on the interval 0...1/

Program 2. The extended logic program for dO with respect to January

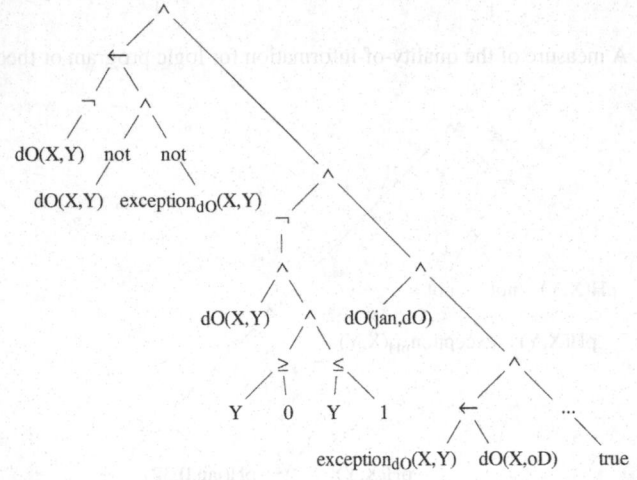

Fig. 3. The evolutionary logic program for dO with respect to January

tP(january,0.21).

¬ tP(X,Y) ← /The closed word assumption is being softened/
   not tP(X,Y) ∧
   not exception$_{tP}$(X,Y).

¬(tP(X,Y) ∧ Y ≥ 0 ∧ Y ≤ 1). */This invariant states that pH takes values on the interval 0...1/*

Program 3. The extended logic program for tP with respect to January

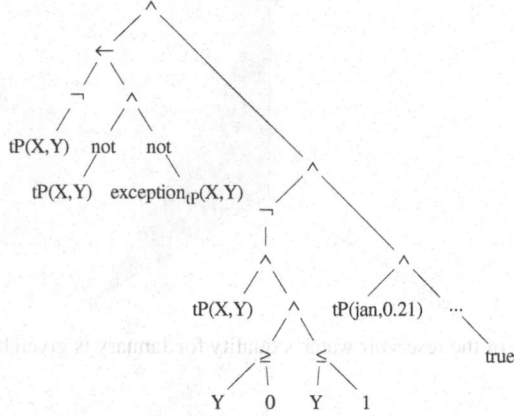

Fig. 4. The evolutionary logic program for tP with respect to January

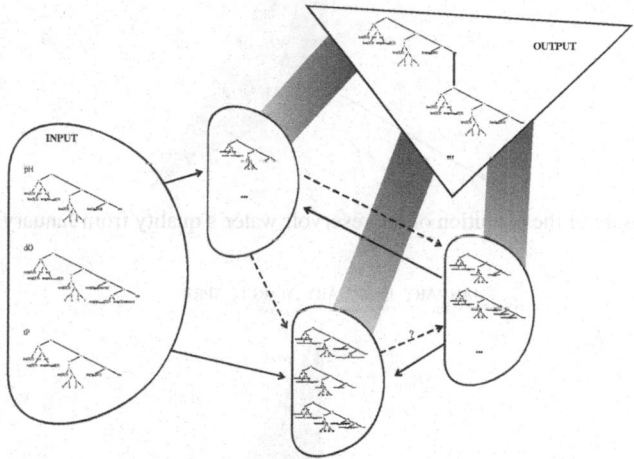

Fig. 5. A blended of the extensions of the predicates pH, dO and tP

Considering what it is illustrated by Figure 5, we might now want to predict how the water's reservoir results will fare in the future, or we may want to fill in some missing data into the series. To do so, we need to evolve the logic theories or logic programs, evolving the correspondent evolutionary logic programs, according to the rules of GP, resulting in what it is depicted by Figures 6,7 [4].

What we are doing, is to put evolution to work for us as well. Indeed, in the present case there is nones change on the quality-of-information when one moves from March to April, i.e., the computational process must stop, once the halt condition is accomplished (Figure 7).

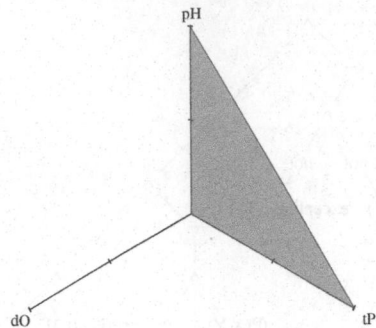

Fig. 6. A measure of the reservoir water's quality for January is given by the paint area

Fig. 7. A measure of the evolution of the reservoir water's quality from January to February

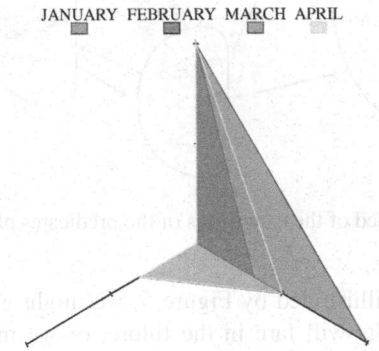

Fig. 8. A measure of the evolution of the reservoir water's quality from January to April

3 Conclusions

This paper shows how to construct a dynamic virtual world of complex and interacting populations, entities that are built as evolutionary logic programs that compete against one another in a rigorous selection regime, where the halt condition of the search process is formalized. In order to produce the optimal solution to a particular problem, one must evolve the logic program or theory that models the universe of discourse, in which fitness is judged by one criterion alone, the quality-of-information.

Clearly, we work out:

What is the model?

A model in this context is to be understood as the composition of the predicates that denote the objects and the relations that may be established between them, that model the universe of discourse.

What parameters are we seeking to discover?

The extensions of predicates of the kind just referred to above [5].

What do we mean by optimal?

By optimal we mean the logic program or theory that models the universe of discourse and maximizes its quality-of-information factor [6].

How can we measure and assign values to possible solutions?

Via mechanical theorem proving, and program blending [10].

References

[1] Jennings, N.R., Faratin, P., Johnson, M.J., Norman, T.J., Brien, O., Wiegand, M.E.: Journal of Cooperative Information Systems 5(2-3), 105–130 (1996)

[2] Angeline, P.J.: Parse Trees. In: Bäck, T., et al. (eds.) Evolutionary Computation 1: Basic Algorithms And Operators, Institute of Physics Publishing, Bristol (2000)

[3] Rudolph, G.: Convergence Analysis of Canonical Genetic Algorithms. IEEE Transactions on Neural Networks, Special Issue on Evolutionary Computation 5(1), 96–101 (1994)

[4] Teller, A.: Evolving programmers: The co-evolution of intelligent recombination operators. In: Kinnear, K., Angeline, P. (eds.) Advances in Genetic Programming 2, MIT Press, Cambridge (1996)

[5] Analide, C., Novais, P., Machado, J., Neves, J.: Quality of Knowledge in Virtual Entities. In: Encyclopedia of Communities of Practice in Information and Knowledge Management, pp. 436–442. Idea Group Inc., USA (2006)

[6] Mendes, R., Kennedy, J., Neves, J.: Avoiding the Pitfalls of Local Optima: How topologies can Save the Day. In: ISAP 2003. Proceedings of the 12th Conference on Intelligent Systems Application to Power Systems, IEEE Computer Society, Lemnos, Greece (2003)

[7] Neves, J.C.: A Logic Interpreter to Handle Time and Negation in Logic Data Bases. In: Proceedings of ACM 1984 Annual Conference, San Francisco, USA (October 24-27, 1984)

[8] Kakas, A., Kowalski, R., Toni, F.: The role of abduction in logic programming. In: Gabbay, D., Hogger, C., Robinson, J. (eds.) Handbook of logic in Artificial Intelligence and Logic Programming, vol. 5, pp. 235–324. Oxford University Press, Oxford (1998)

[9] Kowalski, R.: The logical way to be artificially intelligent. In: Toni, F., Torroni, P. (eds.) Proceedings of CLIMA VI. LNCS (LNAI), Springer, Heidelberg (2006)

[10] Turner, M., Fauconnier, G.: Conceptual Integration and Formal Expression. Johnson, M.: Journal of Metaphor and Symbolic Activity 10(3) (1995)

Two Puzzles Concerning Measures of Uncertainty and the Positive Boolean Connectives

Gregory Wheeler

Artificial Intelligence Center - CENTRIA
Department of Computer Science, Universidade Nova de Lisboa
2829-516 Caparica, Portugal
grw@fct.unl.pt

Abstract. The two puzzles are the *Lottery Paradox* and the *Amalgamation Paradox*, which both point out difficulties for aggregating uncertain information. A generalization of the lottery paradox is presented and a new form of an amalgamation reversal is introduced. Together these puzzles highlight a difficulty for introducing measures of uncertainty to a variety of logical knowledge representation frameworks. The point is illustrated by contrasting the constraints on solutions to each puzzle with the structural properties of the preferential semantics for non-monotonic logics (System P), and also with systems of normal modal logics. The difficulties illustrate several points of tensions between the aggregation of uncertain information and aggregation according to the monotonically positive Boolean connectives, $\wedge$ and $\vee$.

1 Introduction

Uncertainty is a fundamental and unavoidable feature of our relationship to the world, so it is important to know how to represent uncertainty and how to reason about it. Within AI we see various attempts to combine measures of uncertainty with logical calculi in multi-agent systems [11], robotics [34], logic programming [19], security and verification [14], causal and non-monotonic reasoning [30,17] among other fields. But measure functions behave very differently than logical truth functions, and combining both into one framework is a surprisingly subtle undertaking. This paper presents a pair of puzzles designed to highlight some rudimentary difficulties for reconciling aggregation according to combinations of probabilities with aggregation according to the monotonically positive Boolean connectives, $\wedge$ and $\vee$. The two puzzles are Henry Kyburg's *lottery paradox* [21] and the *amalgamation paradox* [13].

This essay presents a generalized version of the lottery paradox, and introduces a new type of amalgamation reversal. The generalized version of the lottery paradox is important because this form of the problem subverts several recent strategies for avoiding it. This point is discussed briefly here in connection with the treatment of *conjunction* and the rule of adjunction within probabilistic logic [35], within cumulative non-monotonic logics [26,17], and within normal modal logics [4,24]. This generalized form of the lottery paradox is discussed in more detail in [36].

J. Neves, M. Santos, and J. Machado (Eds.): EPIA 2007, LNAI 4874, pp. 170–180, 2007.

The importance of the new amalgamation reversal is that this subverts the classic experimental design strategies introduce by I. J. Good and Y. Mittal [13] for avoiding reversal effects. The reason is that the classification categories of the new version are not (necessarily) exclusive, which reflects the behavior of Boolean disjunction in cumulative non-monotonic logics, and also reflects the behavior of Boolean disjunction within *classical* modal logics [27,32,9]. This point is discussed briefly in [24]. The aim of this essay is to discuss this issue in more detail. By highlighting recent work on the lottery paradox, together with this new type of amalgamation reversal, the essay offers a note of caution to applied logicians concerning attempts to combine measure theory and logic for the purpose of representing cogent reasoning under conditions of uncertainty.

The essay is organized as follows. Section 2 presents the original lottery paradox and a generalized variant. Section 3 presents an example of a standard amalgamation reversal. Following Good and Mittal, a generalized form of the puzzle is presented in terms of two-by-two contingency tables, and the example is represented in these terms. Representing examples in these terms helps to illustrate the scope of the amalgamation paradox, since a variety of measures can induce a reversal. We also mention Good and Mittal's analytical results that they use to specify necessary and sufficient conditions for avoiding reversal effects. In section 4 sub-structural conditions for cumulative non-monotonic logics are presented, and the properties for conjunction [**And**] and disjunction [**Or**] are compared with the lottery paradox, and with the amalgamation paradox, respectively.

2   The Lottery Paradox

Henry Kyburg's *lottery paradox* [21, p. 197] arises from considering a fair 1000 ticket lottery that has exactly one winning ticket. If this much is known about the execution of the lottery it is therefore rational to accept that one ticket will win. Suppose that an event is very likely if the probability of its occurring is greater than 0.99. On these grounds it is presumed rational to accept the proposition that ticket 1 of the lottery will not win. Since the lottery is fair, it is rational to accept that ticket 2 won't win either—indeed, it is rational to accept for any individual ticket i of the lottery that ticket i will not win. However, accepting that ticket 1 won't win, accepting that ticket 2 won't win, ..., and accepting that ticket 1000 won't win entails that it is rational to accept that no ticket will win, which entails that it is rational to accept the contradictory proposition that one ticket wins and no ticket wins.

The lottery paradox was designed to demonstrate that three attractive principles governing *rational acceptance* lead to contradiction, namely that

1. It is rational to accept a proposition that is very likely true,
2. It is not rational to accept a proposition that you are aware is inconsistent, and
3. If it is rational to accept a proposition A and it is rational to accept another proposition A', then it is rational to accept $A \wedge A'$,

are jointly inconsistent.

The lottery paradox has sparked an enormous literature [36, see bibiography]. Kyburg's own view was to accept the first two principles and reject the third aggregation principle, and his innovative theory of probability is based upon these commitments [21,22]. Orthodox Bayesians tend to resolve the paradox by accepting the second and third principles and rejecting the first, and with it the notion of rational acceptance underpinning the Kyburgian approach to probability [18,8]. Thus, in the 1960s and 1970s, discussion of the 'lottery paradox' was associated with debates about the merits of different interpretations of probability and various frameworks for representing uncertainty. Since then epistemologists have tended to follow the orthodox line, which is why the paradox is linked to discussions of skepticism, and conditions for (reasonably) asserting knowledge claims. And AI researchers, economists, and mathematical psychologists have followed this track, too, often drawn by obvious mathematical advantages. But it is far less clear that these formal advantages translate to accurate representations of uncertainty.

Philosophical logicians, on the other hand, have tended to look for ways to weaken one or more of the governing principles for rational acceptance, and there are several approaches one may consider, including Jim Hawthorne and Luc Bovens's logic of belief [16], Bryson Brown's application of preservationist paraconsistent logic [7], Horacio Arló-Costa's appeal to classical modalities [4], Joe Halpern's use of first-order probability [15], Igor Douven and Timothy Williamson's appeal to cumulative consequence relations [10], and this author's use of 1-monotone capacities [35].

The generalization of the lottery paradox follows from pointing out some inessential features of the original thought experiment. For instance, the fact that one and only one ticket will be drawn is inessential to the paradox.[1] Furthermore, the puzzle is not necessarily avoided by turning to decision theory. The force of the lottery paradox is the relationship between uncertain evidence and reasonable belief, rather than the relationship between uncertain belief and rational action. We can illustrate this point with the following generalization of the puzzle.

When I book a flight from Philadelphia to Denver I judge it rational to accept that I will arrive without serious incident since I stand rough odds of $1 : 350,000$ of dying on a plane trip in the U.S. any given year,[2] and stand considerably better odds than this of arriving without incident if I book passage on a regularly scheduled U.S. commercial flight. I judge these odds to be those that characterize my flight from Philadelphia to Denver. Nevertheless, I believe that there will be a fatality from an airline accident on a U.S. carrier in the coming year even though I don't believe that there must be at least one fatal accident each year. Indeed, 2002 marked such an exception.[3]

---

[1] Compare to [16].

[2] According to the *National Safety Council*.

[3] According to the 2002 *National Transportation Safety Board* there were 34 accidents on U.S. commercial airlines during 2002 but zero fatalities, a first in twenty years.

Notice that I am not speaking of my decision to board the plane, but my belief that I will get off the flight alive. The reason why I believe I will land safely in Denver is that I stand odds better than $1 : 350,000$ of surviving and, as far as I know, it is reasonable to view my trip as a random member of the class of domestic trips on US domestic carriers. My belief that I will land safely is distinct from what actions I may be willing to take given this stance, such as booking a car or hotel, or even the act of getting on that flight. In short, I believe that I will survive this flight to Denver. My belief that I will survive the trip is part of my background knowledge I have to base my decisions about actions on options I will, or should, take while in Denver. My belief that there will be an accidental fatality on US carriers in this coming year is unlikely to serve as a basis for actions I might take, but it is a belief that I nonetheless have. This belief is revealed to me by my surprise, which perhaps the reader shares, when I learned that there was a recent year in which there were no fatalities on US airline carriers.

We think that the notion of accepting a claim on the basis of uncertain but compelling evidence is fundamental [24]. But if one holds that a knowledge representation framework should accommodate a representation of full but uncertain belief, this leads to structural difficulties. We shall return to the consequences from adopting this view in section 4. But first let's introduce our other puzzle, the amalgamation paradox.

3 Amalgamation Paradox

Imagine that your university wishes to increase the number of women on faculty. To achieve this goal, all hiring departments are instructed to discriminate in favor of women. Suppose that the university advertises positions in the Department of Mathematics and in the Department of Sociology, and only those departments. Five men apply for the openings in Mathematics and one is hired, while eight women apply and two are hired. Therefore the success rate for men is twenty percent, and the success rate for women is twenty-five percent. Thus, the Mathematics Department is in compliance with the University policy. In the Sociology Department eight men apply and six are hired, and five women apply and four are hired. Therefore the success rate for men is seventy-five percent and for women it is eighty percent. Thus, the Sociology Department is also in compliance with University policy. An equal number of female and male candidates applied for jobs, 13, but overall 7 men and 6 women are hired. Thus the success rate for male applicants is approximately 54% but the success rate for female applicants is approximately 46%. Thus, the university is *not* in compliance with its new hiring policy.[4]

The amalgamation paradox is I.J. Good and Y. Mittal's generalization [13] of a puzzle first noticed by Pearson [31, 277-8] and also by Yule [38], which is sometimes referred to as 'Simpson's paradox' or the 'reversal paradox'. Yule

---

[4] This example is based upon a sex discrimination lawsuit that was brought against the University of California, Berkeley, and discussed in [6].

Table 1. University Recruitment Example

	Men	Women
Mathematics	$\frac{1}{5}$ $<$	$\frac{2}{8}$
Sociology	$\frac{6}{8}$ $<$	$\frac{4}{5}$
University	$\frac{7}{13}$ $>$	$\frac{6}{13}$

pointed out that a pair of attributes may not necessarily exhibit a relationship within a population at large even when it is observed in every subpopulation, which is precisely what is illustrated by the University example, while Pearson stressed an analogous point about correlation measures for continuous (i.e., non-categorical) data.

Following Good and Mittal, the paradox can be expressed in terms of a two-by-two contingency table, such as $\mathbf{t} = [a, b; c, d]$, where the entries a, b, c, d sum to N, the sample size, $abcd \neq 0$, and N is assumed large enough to ignore sampling variation.

A measure of association m of $\mathbf{t}$, $m(\mathbf{t})$, is a function of a, b, c, d to $\{1, 0\}$. So, $m(\mathbf{t})$ is either 1 or 0, either 'true' or 'false'. Amalgamation is then defined for mutually exclusive sets of two-by-two contingency tables.

Definition 1 (Amalgamation). *Suppose* $\mathbf{t_i} = [a_i, b_i; c_i, d_i]$, *for* $i = 1, \ldots, n$, *is a set of size* n *of mutually exclusive two-by-two contingency tables. Amalgamation is a single table* $\mathbf{T}$ *composed of the sums of each coordinate in the tables, elements of* $\mathbf{T} = [A, B; C, D] = [\sum_i^n a_i, \sum_i^n b_i; \sum_i^n c_i, \sum_i^n d_i]$.

Note that $A + B + C + D = N$. Let $a_i + b_i + c_i + d_i = N_i$; for simplicity, assume that N_i is proportional to the fraction p_i of the population that makes up the ith subpopulation.[5]

Then the Amalgamation paradox occurs if the maximum measure of association among a set of contingency tables is less than the measure of association of the amalgamation of those tables, or the measure of association for the amalgamation of a set of contingency tables is less than the minimal measure of association among a set of contingency tables, that is if

$$\max_i m(\mathbf{t_i}) < m(\mathbf{A}) \quad \text{or} \quad m(\mathbf{A}) < \min_i m(\mathbf{t_i}).$$

In words, but of slightly stronger form, it is possible for *every* subpopulation to have an association measure pointing in one direction (i.e., either all equal to 0

---

[5] This is not always a reasonable assumption. We might want to sample people from Portugal and people from India, but it would be unreasonable to insist upon taking sample sizes that were proportional to each population. If the sample sizes are sufficiently large, however, we can scale the tables to force N_i to be proportional to p_i. And this scaling is reasonable if sample sizes are large enough to ignore sampling variation, which we assume.

or all equal to 1), but for the population itself to record a 'reverse' association measure (i.e., a measure of 1 or of 0, respectively).

Example 1 (University Recruitment Example). Let $\mathbf{t_i} = [a_i, b_i; c_i, d_i]$ for $i = 1, 2$ where the '1' is the table for the subpopluation of Mathematics Department recruitment data, '2' the subpopulation of Sociology Department recruitment data, a_i represents the number of males hired by department i, b_i the number of males not hired by i, c_i the number of females hired by i, and c_i the number of females not hired by i. $\mathbf{T} = [A, B; C, D]$ is the amalgamation of Mathematics and Sociology recruitment data, representing the University recruitment statistics. Thus,

(i) $\mathbf{t_1} = [1, 4; 2, 6]$; $\mathbf{t_2} = [6, 2; 4, 1]$; $\mathbf{T} = [7, 6; 6, 7]$;
(ii) $N_1 = N_2 = 13$; $N = 26$;
(iii) $m(\mathbf{t}) = \begin{cases} 1 & \text{if } \dfrac{a}{a+b} < \dfrac{c}{c+d}; \\ 0 & \text{otherwise.} \end{cases}$ and
(iv) $m(\mathbf{t_1}) = m(\mathbf{t_2}) = 1$, but $m(\mathbf{T}) = 0$.
(v) Therefore, $m(\mathbf{T}) < \min_i m(\mathbf{t_i})$, for all $i = 1, 2$.   □

The advantage to representing the amalgamation paradox in terms of contingency tables is that one may then consider a variety of measures m that induce reversals. Good and Mittal do just this in [13], where they consider a (non-exhaustive) list of measures. They also define two conditions, *row-uniformity*, for some λ and $i = 1, \ldots, n$:
$$\frac{a_i + b_i}{c_i + d_i} = \lambda,$$
and *column-uniformity*, for some μ and $i = 1, \ldots$, n:
$$\frac{a_i + c_i}{b_i + d_i} = \mu.$$

Whether or not these conditions can be satisfied depends upon the geometry of the tables *and* the sampling procedures used in their construction, and Good and Mittal discuss some situations under which these conditions can be built into the design of an experiment and, hence, avoid reversal effects for the amalgamated table $\mathbf{T}$.

But Good and Mittal's results also depend upon the mutual exclusivity of each of the n contingency tables, $\mathbf{t_i} = [a_i, b_i; c_i, d_i]$. The form of reversal that we presented in [24] and which is discussed in more detail in the next section trades on the mutual-exclusivity of categories. Unlike Pearson's version, our example concerns categorical data. Unlike Yule's version, which is the form generalized by Good and Mittal, the categories fail to be mutually exclusive. Determining whether or not contingency tables satisfy the mutual exclusivity condition may not appear to be problematic from a experimental point of view, since (typically) enough is known about classification categories and experimental design to warrant assumptions that cells are mutually exclusive. However, from a logical point of view, we typically *do not* have this information available when considering the aggregation of information. Indeed, that is the point behind a logical connective. Since the truth conditions for Boolean disjunction are open with respect to items satisfying one or both categories, this means that there is no guarantee within the logical representation that the categories are mutually exclusive. We will return to this point in the next section, when we consider the 'Rotten Apple' example.

4   System P and Systems of Modal Logic

There is a long history behind providing probabilistic semantics for logical calculi [1,33], which was taken up in AI [28,25] in terms of providing a probabilistic semantics for satisfying the axioms of System P for non-monotonic conditionals, an axiom system first discussed in [20]. System P consists of a number of axioms and rules of inference that are taken by many[6] to be a conservative core any nonmonotonic system should contain.

Let $\vdash\!\sim$ be a nonmonotonic consequence relation. Let $\vee, \wedge, \rightarrow$ and $\leftrightarrow$ be standard Boolean connectives in a classical propositional logic, $\rightarrow$ being the truth functional conditional. Let $\models \alpha$ denote α is valid. Finally, $\models \ulcorner \alpha \rightarrow \beta \urcorner$ can be equivalently expressed as $\alpha \models \beta$. The axioms of System P are:

$$\alpha \vdash\!\sim \alpha \qquad \textbf{[Reflexivity]}$$

$$\frac{\models \alpha \leftrightarrow \beta; \ \alpha \vdash\!\sim \gamma}{\beta \vdash\!\sim \gamma} \qquad \textbf{[Left Logical Equivalence]}$$

$$\frac{\models \alpha \rightarrow \beta; \ \gamma \vdash\!\sim \alpha}{\gamma \vdash\!\sim \beta} \qquad \textbf{[Right Weakening]}$$

$$\frac{\alpha \vdash\!\sim \beta; \ \alpha \vdash\!\sim \gamma}{\alpha \vdash\!\sim \beta \wedge \gamma} \qquad \textbf{[And]}$$

$$\frac{\alpha \vdash\!\sim \gamma; \ \beta \vdash\!\sim \gamma}{\alpha \vee \beta \vdash\!\sim \gamma} \qquad \textbf{[Or]}$$

$$\frac{\alpha \vdash\!\sim \beta; \ \alpha \vdash\!\sim \gamma}{\alpha \wedge \beta \vdash\!\sim \gamma} \qquad \textbf{[Cautious Monotonicity]}$$

Both Ernest Adams [1,2] and Judea Pearl [28,29] have sought to make a connection between high probability and logic, but both have taken probabilities *arbitrarily* close to one as corresponding to knowledge. Although Pearl bases his approach on infinitesimal probabilities and Lehmann and Magidor base theirs on a non-standard probability calculus, each shares the view that 'acceptance' or 'full belief' is to be identified with maximal probability, a view that has been developed by [12] [3], defended in [15] and studied by [5], where the latter includes an important limitative result.

But this conception of 'full-belief' conflicts with the notion of rational acceptance that drives Kyburg's lottery paradox. Suppose that $\alpha \vdash\!\sim \beta$ denotes that

---

[6] For example, see [26,15,10].

together with the suppressed background knowledge, α is good, but not necessarily certain, evidence for β. The idea then is that rational acceptance of β is determined relative to a fixed threshold for acceptance. But the lower bound of the probability of the conjunction $\beta \wedge \gamma$ is not higher than and can be lower than the smaller of the two lower bounds of the probabilities of β and of γ. Indeed, [**And**] is simply the rule of adjunction, the third purported principle for governing rational acceptance, and this principle is in direct conflict with full but uncertain belief. Some studies of sub-P, weakly aggregative systems include [35,24,17,4].

Turning to modal logic, note that the schema $(\Box\phi\wedge\Box\psi) \leftrightarrow \Box(\phi\wedge\psi)$ is valid in *all* classes of normal modal logics. So, if we interpret $\Box$ to be a threshold-valued belief operator, we then see that the left-to-right direction,

(**C**) $(\Box\phi \wedge \Box\psi) \rightarrow \Box(\phi \wedge \psi)$,

is the adjunction rule. It is worth mentioning that (**C**) is *not* valid in all classes of *classical* modal logics [27,32,9], and the Scott-Montague semantics have been explored to characterize acceptance principles that restrict adjunction [4,23].

Up to this point we have reviewed how the aggregation of probabilities clashes with the treatment of aggregation in systems of cumulative non-monotonic logic and with the distribution properties of Kripke structures. We have also remarked on the general reaction among applied logicians to the incompatibility between rational acceptance and the rule of adjunction [37]. There is a tendency to view attractive formal properties of logical systems to also be normative constraints of whatever the logic is imagined to represent. In this case, one finds the view that the clash between rational acceptance and the rule of adjunction is so much the worse for the principle of rational acceptance.

But the structural dissimilarities between probability and logical calculi are not restricted to a conflict over [**And**]—and, by extension, over [**Cautious Monotonicity**]. The rule [**Or**] is problematic as well in that it is susceptible to a form of amalgamation reversal that exploits the fact that Boolean disjunction does not guarantee that categories are mutually exclusive. Furthermore, the problematic behavior appears not just in normal modal logics but also in the more general class of classical modal logics.

Consider first a counter-example to [**Or**], the Rotten Apple example.

Example 2 (Rotten Apple Example). Suppose the probability that a Jonathan apple (α) is not rotten (γ) is at least 90%, and the probability that an Ohio apple (β) is not rotten is at least 90%. Actually all non-rotten apples are Jonathan apples *and* from Ohio. In this case the lower bound of the probability that an apple is not rotten given that it is a Jonathan *or* from Ohio $(\alpha \vee \beta)$ is 81.8%.

A joint probability distribution for this counter-example is as follows.

$\alpha\ \beta\ \gamma$	Probability	$\alpha\ \beta\ \gamma$	Probability
0 0 0	0	1 0 0	1/11
0 0 1	0	1 0 1	0
0 1 0	1/11	1 1 0	0
0 1 1	0	1 1 1	9/11

Given a lower bound requirement of 90%, we have $\alpha \mathrel{\vert\!\sim} \gamma$ ($\frac{9/11}{10/11}$) and $\beta \mathrel{\vert\!\sim} \gamma$ ($\frac{9/11}{10/11}$) but not $\alpha \vee \beta \mathrel{\vert\!\sim} \gamma$ ($\frac{9/11}{11/11}$). □

Turning to modal logic, observe that the range of modal systems captured by the Scott-Montague neighborhood semantics—i.e., classical, monotone, regular, normal—is not sufficient to mitigate against the type of reversal effect demonstrated in Example 2. The reason is that α and β here are non-tautological sentences, and facts about deducibility of classical modal logics ($\vdash_C$) entail that if $\alpha \vdash_C \gamma$ and $\beta \vdash_C \gamma$, then $\alpha \vee \beta \vdash_C \gamma$: deducibility is *monotonic*, and since both $\alpha \vdash_C \gamma$ and $\beta \vdash_C \gamma$ and $\alpha \subset \{\alpha \cup \beta\} \supset \beta$, then $\{\alpha \cup \beta\} \vdash_C \gamma$.

Finally, a word about 'exclusive or'. One might consider resolving the conflict with System P by replacing $\vee$ with a connective $\circ$, where $p \circ q$ is true if and only if $(p \vee q) \wedge \neg(p \wedge q)$. Notice, however, that $\circ$ is *not* a suitable logical connective for ensuring, compositionally, that at most one of its arguments is true: $p \circ (q \circ r)$ is true when p, q and r are all true.

5 Conclusion

Most of the warnings about combining probability and logic have concentrated upon the non-compositional character of calculating compound events, on the one hand, and upon the difference between how probabilities and extensional truth assignments are aggregated. The lottery paradox has long been viewed to encapsulate the crux of the problem. However, the introduction of this amalgamation reversal presents a new difficulty to efforts to combine logic and probability, and one that appears to have consequences of wider reach. The class of sub-P non-monotonic systems charted by [17] are predicated upon the [Or] axiom holding, for instance, and we see that the counter-example applies beyond the class of normal logics to include *all* systems of classical modal logics.[7]

References

1. Adams, E.: Probability and the logic of conditionals. In: Hintikka, J., Suppes, P. (eds.) Aspects of Inductive Logic, pp. 265–316. North Holland, Amsterdam (1966)
2. Adams, E.: The Logic of Conditionals. Reidel, Dordrecht (1975)
3. Arló-Costa, H.: Bayesian epistemology and epistemic conditionals. Journal of Philosophy 98(11), 555–598 (2001)
4. Arló-Costa, H.: First order extensions of classical systems of modal logic; the role of the Barcan schemas. Studia Logica 71(1), 87–118 (2002)
5. Arló-Costa, H., Parikh, R.: Conditional probability and defeasible inference. Journal of Philosophical Logic 34, 97–119 (2005)
6. Bickel, P.J., Hjammel, E.A., Connell, J.W.: Sex bias in graduate admissions: Data from berkeley. Science 187, 398–404 (1975)
7. Brown, B.: Adjunction and aggregation. Nous 33(2), 273–283 (1999)

---

[7] This research was supported by grant SFRH/BPD/34885/2007 from *Fundação para Ciência e a Tecnologia*, and by an award from *The Leverhulme Trust*.

8. Carnap, R.: The Logical Foundations of Probability, 2nd edn. University of Chicago Press, Chicago (1962)
9. Chellas, B.: Modal Logic. Cambridge University Press, Cambridge (1980)
10. Douven, I., Williamson, T.: Generalizing the lottery paradox. The British Journal for the Philosophy of Science 57(4), 755–779 (2006)
11. Fagin, R., Halpern, J.Y., Moses, Y., Vardi, M.Y.: Reasoning About Knowledge. MIT Press, Cambridge (2003)
12. Van Fraassen, B.C.: Fine-grained opinion, probability, and the logic of belief. Journal of Philosophical Logic 95, 349–377 (1995)
13. Good, I.J., Mittal, Y.: The amalgamation and geometry of two-by-two contingency tables. The Annals of Statistics 15(2), 694–711 (1987)
14. Haenni, R.: Web of trust: Applying probabilistic argumentation to public-key cryptography. In: Symbolic and Quantitative Approaches to Reasoning with Uncertainty, pp. 243–254. Springer, Heidelberg (2004)
15. Halpern, J.Y.: Reasoning about Uncertainty. MIT Press, Cambridge (2003)
16. Hawthorne, J., Bovens, L.: The preface, the lottery, and the logic of belief. Mind 108, 241–264 (1999)
17. Hawthorne, J., Makinson, D.C.: The quantitative/qualitative watershed for rules of uncertain inference. Studia Logica (2007)
18. Jeffrey, R.: Valuation and acceptance of scientific hypotheses. Philosophy of Science 23(3), 237–246 (1956)
19. Kersting, K., Raedt, L.D.: Bayesian logic programming: Theory and tool. In: Getoor, L., Taskar, B. (eds.) Introduction to Statistical Relational Learning, MIT Press, Cambridge (2007)
20. Kraus, S., Lehman, D., Magidor, M.: Nonmonotonic reasoning, preferential models and cumulative logics. Artificial Intelligence 44, 167–207 (1990)
21. Kyburg Jr., H.E.: Probability and the Logic of Rational Belief. Wesleyan University Press, Middletown, CT (1961)
22. Kyburg Jr., H.E., Teng, C.M.: Uncertain Inference. Cambridge University Press, Cambridge (2001)
23. Kyburg Jr., H.E., Teng, C.M.: The logic of risky knowledge. In: Electronic Notes in Theoretical Computer Science, vol. 67, Elsevier Science, Amsterdam (2002)
24. Kyburg Jr., H.E., Teng, C.M., Wheeler, G.: Conditionals and consequences. Journal of Applied Logic (2007)
25. Lehman, D., Magidor, M.: What does a conditional knowledge base entail? Artificial Intelligence 55, 1–60 (1990)
26. Makinson, D.C.: General patterns in nonmonotonic reasoning. In: Gabbay, D., Hogger, C., Robinson, J.A. (eds.) Handbook of Logic in Artificial Intelligence and Uncertain Reasoning, vol. 3, Clarendon Press, Oxford (1994)
27. Montague, R.: Universial grammer. Theoria 36, 373–398 (1970)
28. Pearl, J.: Probabilistic Reasoning in Intelligent Systems. Morgan Kaufmann, San Francisco (1988)
29. Pearl, J.: System Z: A natural ordering of defaults with tractable applications to default reasoning. Theoretical Aspects of Reasoning about Knowedge, 121–135 (1990)
30. Pearl, J.: Causality. Cambridge University Press, Cambridge (2000)
31. Pearson, K.: Theory of genetic (reproductive) selection. Philosophical Transactions of The Royal Society of London, Ser. A 192, 260–278 (1899)
32. Scott, D.: Advice in modal logic. In: Lambert, K. (ed.) Philosophical Problems in Logic, pp. 143–173. Reidel, Dordrecht (1970)

33. Suppes, P.: A Probabilistic Theory of Causality. North-Holland Publishing Co., Amsterdam (1970)
34. Thrun, S., Burgard, W., Fox, D.: Probabilistic Robotics. MIT Press, Cambridge (2005)
35. Wheeler, G.: Rational acceptance and conjunctive/disjunctive absorption. Journal of Logic, Language and Information 15(1-2), 49–63 (2006)
36. Wheeler, G.: A review of the lottery paradox. In: Harper, W.L., Wheeler, G. (eds.) Probability and Inference: Essays in Honour of Henry E. Kyburg, Jr., King's College Publications (2007)
37. Wheeler, G.: Applied logic without psychologism. Studia Logica (forthcoming)
38. Yule, G.U.: Notes on the theory of association of attributes in statistics. Biometrika 2, 121–134 (1903)

Chapter 2 - First Workshop on AI Applications for Sustainable Transportation Systems (AIASTS 2007)

Nonlinear Models for Determining Mode Choice
(Accuracy Is Not Always the Optimal Goal)

Elke Moons[1], Geert Wets[1], and Marc Aerts[2]

[1] Transportation Research Institute, Hasselt University, Science Park 5/6,
3590 Diepenbeek, Belgium
elke.moons@uhasselt.be, geert.wets@uhasselt.be
[2] Center for Statistics, Hasselt University, University Campus - Building D,
3590 Diepenbeek, Belgium
marc.aerts@uhasselt.be

Abstract. Due to the increasing complexity in transportation systems, one needs to search for different ways to model the separate components of these systems. A general transportation system comprises components/models concerning mode choice, travel duration, trip distance, departure time, accompanying individuals, etc. This paper tries to discover whether semi- and nonlinear models bring an added value to transportation analysis in general and mode choice modelling in particular. Linear (logistic regression), semi-linear (multiple fractional polynomials) and nonlinear (support vector machines and classification and regression trees) models are applied to several binary settings and compared to each other based on sensitivity (i.e. the proportion of positive cases that are predicted correctly). In general, one can state that on skewed data sets, linear and semi-linear models tend to perform better, whereas on more balanced data sets both nonlinear models yield better results. Future research will take a closer look at other extensions of the well-established linear regression model.

1 Introduction

At the beginning of the 21st century, transportation planners are facing an increasing complexity in trying to predict travel demand. On the one hand, transportation systems are inherently very complex systems, including a large number of components that needs to be dealt with, such as efficient, safe and reliable transportation, but also the impact on the environment and on the surrounding communities. Above all this, the economic growth gives rise to an ever-increasing travel demand, creating even more problems (Sadek, 2007). On the other hand, at the same time, different algorithms and models have been developed in different research fields, such as artificial intelligence, statistics, machine learning, etc. The transport analysts are thus faced with a wide variety of types of models to choose from to model the separate components of the transportation systems. It is within this context that artificial intelligence (AI) paradigms can and need to be used to address some of the aforementioned problems that are quite challenging to solve by means of traditional and classical solution methods. This paper

J. Neves, M. Santos, and J. Machado (Eds.): EPIA 2007, LNAI 4874, pp. 183–194, 2007.

focusses on one of the most important application areas of AI within transportation, i.e. decision making, which happens multiple times every day when a trip is planned. With whom do I undertake this trip, when do I carry it out, where do I go to, what will the duration be, which transport mode do I take, etc.? In this paper, the focus lies on the latter decision, i.e. the mode choice. However, the results of the study can be applied to any other component in a general transportation system that concerns with decision making.

In general, AI can be split up into two broad categories. The *symbolic AI* that focuses on the development of knowledge-based systems, and the *computational AI*, which includes neural networks, fuzzy systems and genetic algorithms. In this paper, we will focus on the latter category of methods, more in specific on the use of non- and semilinear models for prediction. It is often the case that transportation problems show a non- or a semilinear relation between input and output. We know, e.g. that income is inversely related to the probability of using slow transport, but whether this inverse relationship is cubic, quadratic, linear, or determined by some square root is not sure.

Linear (logistic) regression models, a legacy of pre-computer times, are well established in the field of mode choice modelling. Ben Akiva and Lerman (1985) introduced discrete choice analysis in general in travel demand analysis (and more in particular: the logit (logistic regression) model) and it has been used as reference model ever since. Recently, some other alternatives to the well-known logit model have been proposed, such as the mixed logit (Hensher and Greene, 2003), the paired combinatory logit (Koppelman and Wen, 2000) or a spatially correlated logit model (Bhat and Guo, 2003). However, in this paper, the comparison will be confined to the traditional reference model, the binary logit model. There are two different ways to extend these models, parametrically, by means of statistical models, and non-parametrically, by means of artificial intelligence/machine learning algorithms. A first parametric extension to linear models that provides more flexibility can be found in fractional polynomials (Royston and Altman, 1994; Royston et al., 1999). The advantage of this extension is that it offers a better prediction, because of its flexibility and it is still interpretable, in the same way as linear models. Fractional polynomials are therefore often called a semi-linear modelling approach.

Apart from this parametrical extension, there are also other models, developed in the AI/machine learning community, that have proven to be very useful in other research fields. Some examples are: classification and regression trees (CART; Breiman et al., 1984), support vector machines (SVM's; Vapnik, 1996), neural networks (Zurada, 1992), multivariate adaptive regression splines (MARS; Friedman, 1991), ... Here, we will confine to fractional polynomials on the one hand and to support vector machines and CART on the other hand, since these techniques have not been explored extensively in the field of mode choice modelling. The use of neural networks, another AI technique, has been explored some more in mode choice analysis. E.g. Nijkamp et al. (1996) discuss the use of feedforward neural networks in comparison to the most widely used discrete choice model, i.e. a logit model. They conclude that for mode choice purposes the neural network performs slightly better than the logit model, but at the same time they admit that it is probably because of the structure of the data set. Hensher and Ton (2000) also compared the predictive potential of neural networks in comparison to nested logit models for commuter mode choice.

'Do the semi- and nonlinear models perform worse than the widely used logit model or do they yield better results?' - is the key question throughout this paper. The next Section describes the different models used, Section 3 gives a short introduction to the data and discusses how the performance of the different methods can be compared to one another. In Section 4, the results are presented, while the final Section provides the conclusions and some avenues for future research.

2 Models

2.1 Logistic Regression

At first, a short introduction will be given to the logistic regression model. Suppose the binary response variable is denoted by Y and the two possible outcomes for that variable by a '0' and a '1'. Such a variable is called a Bernouilli variable and its distribution is defined by the probability of 'success' ($\pi = P(Y = 1)$) and the probability of 'failure' ($1 - \pi = P(Y = 0)$). The relationship between the probability of success ($\pi(\mathbf{x})$ and the given input variables ($\mathbf{x} = (\mathbf{x_1}, \mathbf{x_2}, \ldots, \mathbf{x_p})$) is more often non-linear, so therefore one typically models this relationship by means of an S-shaped curve in function of the explanatory variables:

$$\text{logit}[\pi(\mathbf{x})] = \log\left(\frac{\pi(\mathbf{x})}{1 - \pi(\mathbf{x})}\right) = \beta_0 + \beta_1 \mathbf{x_1} + \beta_2 \mathbf{x_2} + \ldots + \beta_p \mathbf{x_p}.$$

This function is called the *logistic regression function*. The link function is the logit transform $\log[\frac{\pi}{1-\pi}]$ of π, so therefore these logistic regression models are often also called *logit models*. π is in range limited to $(0, 1)$, but the logit of π can take any real value.

The interpretation of these logit models in terms of odds ratio's etc. is not so straightforward, but this would lead us too far, so we refer to Neter *et al.* (1996) and Agresti (1996) or to Ben Akiva and Lerman (1985) for the discrete choice view of the model under discussion.

2.2 Fractional Polynomials

Linear models have been used extensively, almost routinely, by applied statisticians and researchers (Royston and Altman, 1994). However, often is the relationship between a dependent variable and one or more continuous covariates curved. Usually, one attempts to represent curvature in regression models by means of polynomials of the covariates, typically quadratics. However, in general, low order polynomials offer a limited family of shapes, while high order polynomials may fit poorly at the extreme values of the covariates. Various attempts have been made to devise more acceptable models. Box and Tidwell (1962) developed an appropriate linearisation of each variable in a multiple regression model, though for models with more than one covariate, there are considerable difficulties in estimating the powers reliably. Nonparametric (Hastie and Tibshirani, 1990) and spline smoothers (Reinsch, 1967) are other flexible and powerful tools that impose few limitations on the functional form, though the fitting process may be computationally intensive. This paper searches for models that are flexible, easy to understand and parsimonious.

Fractional polynomials (Royston and Altman, 1994; Royston et al., 1999), an extended family of curves whose power terms are restricted to a small predefined set of values, may provide a solution for this. They provide much more flexible shaped curves than conventional polynomials, but in cases where the extension is not necessary, this family essentially reduces to conventional polynomials. A particular feature of the fractional polynomials is that they provide a wide class of functional forms, with only a small number of terms (Royston and Altman, 1994; Sauerbrei and Royston, 1999).

Let x be a continuous covariate, a *fractional polynomial of degree m* is then defined to be

$$\phi_m(x, \zeta, \mathbf{p}) = \zeta_0 + \sum_{i=1}^{m} \zeta_i H_i(x),$$

where m is a positive integer, $\mathbf{p} = (p_1, \ldots, p_m)$ a real-valued vector of powers with $p_1 \leq \ldots \leq p_m$ and $\zeta = (\zeta_1, \ldots, \zeta_m)$ coefficients. We set $H_0(x) = 1$, $p_0 = 0$ and then, for $i = 1, \ldots, m$

$$H_i(x) = \begin{cases} x^{(p_i)} & \text{if } p_i \neq p_{i-1} \\ H_{i-1}(x) \ln(x) & \text{if } p_i = p_{i-1} \end{cases}$$

The round bracket notation indicates the Box-Tidwell transformation:

$$x^{(p_i)} = \begin{cases} x^{p_i} & \text{if } p_i \neq 0 \\ \ln(x) & \text{else} \end{cases}$$

This full definition includes possible 'repeated powers' which involve powers of ln(x). E.g. a fractional polynomial of the third degree ($m = 3$) with powers (1,1,2) is of the form $\zeta_0 + \zeta_1 x + \zeta_2 x \ln(x) + \zeta_3 x^2$. Experience suggests (Royston and Altman, 1994) that $p_i \in \{-2, -1, -\frac{1}{2}, 0, \frac{1}{2}, 1, 2, \max(3, m)\}$ is sufficiently rich to cover many practical cases adequately.

2.3   Support Vector Machines

Support vector machines (SVM's) are learning systems that use a hypothesis space of linear functions in a high dimensional feature space, trained with a learning algorithm from optimisation theory that implements a learning bias derived from statistical learning theory (Hastie et al., 2001). This learning strategy, that has been introduced by Vapnik and co-workers (1996), is a principled and very powerful method that in the few years since its introduction has already outperformed most other systems in a wide variety of applications.

In order to clearly understand the procedure of support vector machines, one first has to discuss the technique for constructing an *optimal separating hyperplane* between two classes that are perfectly separable by a linear boundary. Now, consider N pairs or the training data $(x_1, y_1), (x_2, y_2), \ldots, (x_N, y_N)$, with $x_i \in \mathbb{R}^p$ and $y_i \in \{-1, 1\}$ (for logit models, usually $y_i \in \{0, 1\}$). Define a hyperplane by

$$\{x : f(x) = x^T \beta + \beta_0 = 0\},$$

where β is a unit vector: $\|\beta\| = 1$. The *optimal separating hyperplane* separates the two classes and maximises the distance to the closest point from either class (Vapnik, 1996). The solution vector β, defining the hyperplane, is determined by a linear combination of the *support points (vectors)* x_i, these are the points that are defined to be on the boundary of the margin. This *optimal separating hyperplane* produces a function $\hat{f}(x) = x^T \hat{\beta} + \hat{\beta}_0$, for classifying new observations (from the test sample), and it defines the *support vector classifier*:

$$\hat{G}(x) = \text{sign}\hat{f}(x).$$

Intuition learns that a large margin on the training data will lead to good separation of the test data.

When the data are not separable, an alternative formulation is necessary. This will be provided by the *support vector machine* that allows for an overlap, still maximising the margin, but allowing for some points to be on the wrong side of the margin (Hastie *et al.*, 2001; Cristianini and Shawe-Taylor, 2000).

The *support vector classifier* that has been described so far, finds linear boundaries in the input feature space. This procedure can be made more flexible by enlarging the feature/covariate space using basis expansions such as polynomials or splines. This leads to better training-class separation and translates to nonlinear boundaries in the original space. Once the basis functions $h_m(x), m = 1, \ldots, M$ are selected, the procedure is the same as before. The hyperplane now defines a nonlinear function $\hat{f}(x) = h(x)^T \hat{\beta} + \hat{\beta}_0$.

The *support vector machine* is an extension of this idea, where the dimension of the enlarged space is allowed to get very large, infinite in some cases. The solution can be represented in a special way that only involves the input features via inner products. Thus, only the knowledge of the kernel function $K(x, x') = \langle h(x), h(x') \rangle$ that computes the inner product in the transformed space is required. Three popular choices for K in the SVM literature are:

$$d\text{-th degree polynomial: } K(x, x') = (\kappa + \gamma\langle x, x' \rangle)^d,$$
$$\text{Radial basis: } K(x, x') = \exp(-\frac{\|x - x'\|^2}{\gamma}),$$
$$\text{Neural network: } K(x, x') = \tanh(\kappa_1 \langle x, x' \rangle + \kappa_2).$$

2.4 Classification And Regression Trees (CART)

In general, classification and regression trees (Breiman *et al.*, 1984) describe a way of partitioning the parameter space. Classification trees are used for categorical dependent variables, while regression trees are applied to continuous y-variables. In summary, a tree consists of different layers of nodes. It starts from the *root node* in the first layer, the first parent node. In a binary tree, a parent node is split into 2 *daughter nodes* on the next layer. Each of these 2 daughter nodes become in turn parent nodes. This recursive partitioning algorithm continues until a node is *terminal* and has no offspring (determined by a stopping criterion). Nodes in deeper layers are getting more and more homogeneous, less 'impure', with respect to the response. An internal node is split by considering all allowable splits for all variables and the best split is the one with the most homogeneous daughter nodes. The 'goodness' of a split can be defined as the reduction in impurity

$$\Delta i(\tau) = i(\tau) - P(\tau_L)i(\tau_L) - P(\tau_R)i(\tau_R)$$

with $i(\tau)$ denoting the impurity of the node τ and $P(\tau_L)$ (and $P(\tau_R)$) the probability that a subject falls into the left (resp. right) daughter node τ_L (resp. τ_R) of node τ. A popular example of such an impurity measure is the entropy measure $i(\tau) = -p_\tau \log(p_\tau) - (1 - p_\tau) \log(1 - p_\tau)$, with $p_\tau = P(Y = 1|\tau)$. In the pruning process, the initial tree is then pruned recursively, leading to a sequence of pruned and nested subtrees. Several methods exist for controlling the tree: the minimum number of observations that must exist in a node in order for a split to be attempted (this is set to 20 in the performed analyses), the minimum number of observations in any terminal leaf node (usually about a third of the minimum split size), the maximum depth of the tree, etc. For more details on classification trees and the recursive partitioning and pruning process, we refer to Breiman *et al.* (1984) and Zhang and Singer (1999).

3   Data and Model Comparison

3.1   The Data

The data set used for the analyses describes the mode choice for work purposes from the original Albatross data set (Arentze and Timmermans, 2000). These data stem from an activity diary survey, carried out in February 1997 among 1649 randomly chosen respondents in two municipalities in the South Rotterdam region in the Netherlands. The respondents are asked to write down in their activity diary, for each successive activity, the nature of the activity, the day, start and end time, the location where it took place, the transport mode and, if relevant, accompanying individuals. Further information was gathered via a household questionnaire. The data set contains 1025 observations. It has been split into three different data sets: one for the prediction of using 'slow' transport in order to go to work (18.93%), one for the prediction of using public transport (12.29%) and the third data set contains cases when the car is used to drive to work (68.78%). Both being a car passenger and all different kinds of public transport are considered as public transport in this case.

All data sets are split in a training and a test set. The training set contains a random sample comprising 70% of the total data set, while the remaining 30%

makes up the test set, as commonly used in practice. The training set will be used to build the model, the test set is used for validation.

The explanatory variables in the data sets are person and household related variables, such as the age of the oldest person in the household, the presence of children in the household, gender, the number of cars in the household. Apart from these variables, also characteristics of the work pattern play a role. Variables that are included are: day of the week, possible duration of the work pattern, the number of non-work, out-of-home activities in the work pattern, whether there is a shopping or service, social/leisure or a bring/get activity in the work pattern, accompanying individuals and some variables describing travel time ratio's between the different choice alternatives.

3.2   How to Compare Non-parametric AI Algorithms and Statistical Parametric Models?

In order to make an 'honest' comparison between the results of the AI algorithms (that only provide accuracies as a result of the classification) and the semi-linear models, we have to come up with some kind measure that can be used for both AI and for statistical models.

Two classical diagnostics that are often used in logistic regression analyses are sensitivity and specificity. The *sensitivity* is defined as the probability that a positive case ($y = 1$) is predicted, given that it is observed, hence

$$\text{Sensitivity} = P(y_{predicted} = 1 | y_{observed} = 1).$$

In the same way, the *specificity* is the conditional probability on a negative ($y = 0$) predicted case, given that the observed case is also negative. The *prevalence* is determined by the number of positive observed cases respective to the total number of cases, such that the accuracy can be written as:

$$\text{Accuracy} = \text{Prevalence} \times \text{Sensitivity} + (1 - \text{Prevalence}) \times \text{Specificity}.$$

Thus, if the parametric model predictions can be turned into some sort of classification, we can use the above equation as well. So define

$$y_{i,predicted} = 1 \text{ if } \pi_i \geq \text{cut-off}$$
$$= 0 \text{ if } \pi_i < \text{cut-off}.$$

As stated in Neter *et al.* (1996) using 0.5 as cut-off is not always best. Using 0.5 is only appropriate when it is equally likely in the population of interest that outcomes zero and one will occur. This certainly is not the case here. Neter *et al.* (1996) suggest to take the following proportion as a cut-off value

$$\text{cut-off} = \frac{\sum_{i=1}^{N_{\text{train}}} y_i}{N_{\text{train}}},$$

i.e. the proportion of 'successes' at the training set. This can be seen as a Bayesian estimate (the use of prior information) for the number of 'successes'.

For the application to mode choice models, the sensitivity can be regarded as the most important diagnostic of the three. The fact is that the main purpose of the models in the next section is the prediction of the positive cases. That the negative cases are predicted well comes in handy, though the aim is on the prediction of the positive cases. Suppose that the data set is very skewed and that you have a rather low prevalence (as is the case in the following data sets). If most negative cases are predicted well, but none of the positive cases is correctly predicted, then you have a very high specificity and accuracy, but this was not the intention of the model.

4 Results

In this paper, the focus is on the comparison of semi- and nonlinear models in the context of mode choice models. The use of fractional polynomials, as a parametrical extension to linear models will be investigated on the one hand, and on the other hand, the use of two non-parametrical techniques, support vector machines and classification and regression trees, will be explored as well. The performance of these models will be compared to that of the widely used logit model (see Moons *et al.*, 2004a, 2004b).

4.1 Public Transport

The logistic regression model and the multiple fractional polynomial (mfp) models are both interpretable in terms of their parameters and they show that, when all remaining variables are kept constant, the probability of choosing public transport to go to work is higher if the traveller is a woman and if the activity is conducted with other people. On the other hand, the probability of choosing public transport to go to work decreases when 3 or 4 non-work out-home activities are conducted during the work activity pattern, when there are 1 or more cars available per adult or if the pattern has a bring/get activity.

The linear optimal separating hyperplane needs 380 support vectors to make up its classifier, but it automatically classifies each case to outcome zero, making this model useless since we want to predict the probability of using public transport. The polynomial kernel of degree 3 requires 199 support vectors, the radial kernel 320 and the neural net kernel only 185. The classification tree algorithm results in a final tree of depth eight with nine final nodes.

All results are summarised in Table 1. Note that here the logistic regression model provides the best result. The polynomial SVM comes in second best, despite the obvious over-fitting on the training data. If a model follows the training data too closely and the performance is worse on the test set, one may judge that the model is over-fitting on the training data. The other non-linear models do not perform well. This is because they are trained on accuracy, and not on sensitivity. This bad performance is also due to the skewness of the data set, only 12.29 % of the total sample uses public transport to go to work.

Table 1. Performance values for the models on Public Transport

	accuracy		sensitivity		specificity	
	training set	test set	training set	test set	training set	test set
Linear	0.708	0.668	0.739	0.824	0.703	0.648
Mfp	0.699	0.847	0.652	0.029	0.706	0.949
SVM - Linear	0.872	0.889	0.000	0.000	1.000	1.000
SVM - Polynomial	0.999	0.801	1.000	0.294	0.998	0.864
SVM - Radial basis	0.876	0.889	0.033	0.000	1.000	1.000
SVM - Neural net	0.845	0.857	0.011	0.000	0.968	0.963
CART	0.900	0.863	0.293	0.118	0.989	0.956

4.2 Slow Transport

When interpreting the (semi-)linear models, it turns out that the probability of choosing a slow transport mode (walk/bike) to go to work decreases if there is one or more cars per adult in the household and if there is at least one shopping or service activity in the work activity pattern. The same is true for social/leisure out-home activities, but a bring/get activity in the activity pattern increases the probability of slow transport, except when this activity is the first in the concerned tour. The probability decreases also when the activity is pursued with other members of the household.

The linear SVM is made up based on 209 support vectors. This time it performs better then only classifying each case to the majority class. The polynomial kernel of degree three needs 207 support vectors, the radial kernel 297, and the neural net kernel 213. The second nonparametric model, the classification tree algorithm results in a tree with eleven final nodes and of depth five. To get a clear overview of all results on this data set, a summary is provided in Table 2.

Table 2. Performance values for the models on Slow Transport

	accuracy		sensitivity		specificity	
	training set	test set	training set	test set	training set	test set
Linear	0.843	0.847	0.853	0.882	0.840	0.840
Mfp	0.834	0.847	0.839	0.882	0.833	0.840
SVM - Linear	0.876	0.863	0.608	0.549	0.943	0.926
SVM - Polynomial	0.999	0.824	1.000	0.588	0.998	0.871
SVM - Radial basis	0.918	0.880	0.692	0.510	0.974	0.953
SVM - Neural net	0.805	0.863	0.476	0.569	0.887	0.922
CART	0.897	0.860	0.678	0.549	0.951	0.922

4.3 Car Driver

According to the parametric models, when all other variables are kept constant, the probability of using the car as transport mode to go to work is lower for females, when compared to men. The probability increases with the number of cars per adult, if there is at least one shopping or service activity in the activity pattern, at least one social or leisure out-home activity in the concerned tour, if the first activity of the tour is a bring/get activity and if the travel time by bike increases. The mfp model performs slightly better than the linear model.

The linear SVM is determined by 336 support vectors, leading to a sensitivity of 0.936 on the training set and a respective measure of 0.928 on the test set. The polynomial kernel SVM is based on 335 support vectors. The resulting accuracy, specificity and the sensitivity on the training set equal all three 1.000, indicating something similar as in the previous two data sets, probably an over-classification on the training set. The radial kernel SVM generally needs the highest number of support vectors. 450 support vectors are required here. A sensitivity of 0.973 on training and of 0.946 on test set depict that this SVM is the best model so far. The neural net kernel SVM produces the following results based on 343 support vectors: a training-sensitivity of 0.865 and a test-sensitivity of 0.901.

The resulting classification tree has nine final nodes and a depth of six. Its performance can be situated in between the parametric models and the SVM.

All results of the different classifications are summarised in Table 3.

Table 3. Performance values for the models on Car Driver

	accuracy		sensitivity		specificity	
	training set	test set	training set	test set	training set	test set
Linear	0.731	0.733	0.747	0.730	0.698	0.741
Mfp	0.769	0.788	0.795	0.811	0.715	0.729
SVM - Linear	0.815	0.831	0.936	0.928	0.566	0.576
SVM - Polynomial	1.000	0.723	1.000	0.793	1.000	0.541
SVM - Radial basis	0.859	0.831	0.973	0.946	0.626	0.529
SVM - Neural net	0.730	0.792	0.865	0.901	0.451	0.506
CART	0.813	0.769	0.915	0.864	0.604	0.518

5   Conclusion

In this paper, we tried to discover whether semi- and nonlinear models could add something to transportation analysis in general and to mode choice analysis in the Netherlands in particular. Linear, semi-linear and nonlinear models were fitted and compared to each other by means of three diagnostics (sensitivity, accuracy and specificity) in decreasing order of importance.

In general, the results showed that on very skewed data sets, the performance of (semi-)linear models are usually better than the results of the support vector machines and CART. Since (semi-)linear models are fitted based on a maximum likelihood principle and AI models are derived in order to obtain a high accuracy, it seems only logical that these different models show different performances with respect to different measures. However, the main idea of any model applied to a setting with a binary response variable is to predict the positive cases well. Since the SVM models are especially derived to achieve an accuracy (instead of sensitivity) as high as possible, this may conflict with the purpose of the modeler. Therefore, one always needs to bear in mind what the aim of the study is. On better balanced data sets (as the Car Driver data), the performance of the SVM and the CART models are comparable and usually somewhat better than the results of the (semi-)linear models. These latter models have the advantage of being better interpretable, while the SVM's are simply a black box approach. CART is also interpretable, but not in terms of the parameters as in (semi-)linear models.

Further research will take a closer look at extensions of semi-linear and nonlinear models developed in the AI community: Neural (Zurada, 1992) and Bayesian (Pearl, 1988) networks, to name a few. We also need to acknowledge that (mode choice) decision making is often not about a choice between two choices, but between n independent alternatives. A comparison of the multinomial logit model to parametric and nonparametric extensions is also an opportunity for further research. Another possible avenue for future research is to investigate whether these models are transferable to other countries. If the variables that appeared important in order to model e.g. the probability of using public transport, are the same in all countries, then perhaps policy measures can be undertaken to advise certain groups of people (e.g. commuters with flexible working hours) to use the public transport system.

References

Agresti, A.: An Introduction to Categorical Data Analysis. Wiley Series in Probability and Statistics. Wiley, Chichester (1996)

Arentze, T.A., Timmermans, H.J.P.: Albatross: A Learning-Based Transportation Oriented Simulation System. Eindhoven University of Technology (2000)

Ben Akiva, M., Lerman, S.R.: Discrete Choice Analysis Theory and Application to Travel Demand. MIT Press, Cambridge (1985)

Bhat, C.R., Guo, J.: A mixed spatially correlated Logit model: formulation and application to residential choice modeling. Paper presented at the 82nd Annual Meeting of the Transportation Research Board, Washington, D.C (2003)

Breiman, L., Friedman, J.H., Olshen, R.A., Stone, C.J.: Classification and Regression Trees. Wadsworth Statistics/Probability Series (1984)

Box, G.E.P., Tidwell, P.W.: Transformations of the independent variables. Technom. 4, 531–550 (1962)

Cristianini, N., Shawe-Taylor, J.: An Introduction to Support Vector Machines and Other Kernel-Based Learning Methods. Cambridge University Press, Cambridge (2000)

Friedman, J.H.: Multivariate adaptive regression splines. An. Stat. 19(1), 1–67 (1991)

Hastie, T.J., Tibshirani, R.J.: Generalized Additive Models. Chapman and Hall, London (1990)

Hastie, T., Tibshirani, R., Friedman, J.: The Elements of Statistical Learning; Data Mining, Inference, and Prediction. Springer Series in Statistics. Springer, Heidelberg (2001)

Hensher, D.A., Ton, T.T.: A comparison of the predictive potential of artificial neural networks and nested logit models for commuter mode choice. Transp. Res. 36E, 155–172 (2000)

Hensher, D., Greene, W.H.: The mixed logit model: The state of practice. Transportation 30(2), 133–176 (2003)

Koppelman, F., Wen, C.-H.: The Paired Combinatorial Logit model: Properties, estimation and application. Transp. Res. B. 34(2), 75–89 (2000)

Moons, E., Wets, G., Aerts, M.: Nonlinear models in transportation. In: Proc. of Conf. on Progress in Activity-Based Analysis, Maastricht, The Netherlands (2004a)

Moons, E., Aerts, M., Wets, G.: The application of fractional polynomials and support vector machines in transportation analysis. Paper accepted for presentation at the Joint Stat. Meetings, Toronto, Canada (2004b)

Neter, J., Kutner, M.H., Nachtsheim, C.J., Wasserman, W.: Applied Linear Statistical Models. Irwin (1996)

Nijkamp, P., Reggiani, A., Tritapepe, T.: Modelling inter-urban transport flows in Italy: A Comparison between neural network analysis and logit analysis. Transpn. Res.-C 4(6), 323–338 (1996)

Pearl, J.: Probabilistic Reasoning in Intelligent Systems: Networks of Plausible Inference. Morgan Kaufmann, San Mateo, California (1988)

Reinsch, C.H.: Smoothing by spline functions. Num. Math. 10, 177–183 (1967)

Royston, P., Altman, D.G.: Regression using fractional polynomials of continuous covariates: parsimonious parametric modeling. Appl. Stat. 43, 429–467 (1994)

Royston, P., Ambler, G., Sauerbrei, W.: The use of fractional polynomials to model continuous risk variables in epidemiology. Int. J. of Epi. 28, 964–974 (1999)

Sadek, A.W.: Artificial Intelligence Applications in Transportation. In: Transportation Research Circular E-C113, TRB (2007)

Sauerbrei, W., Royston, P.: Building multivariable prognostic and diagnostic models: transformation of the predictors using fractional polynomials. J. R. Stat. Soc., Series A 162, 71–94 (1999)

Vapnik, V.: The Nature of Statistical Learning Theory. Springer, New York (1996)

Zhang, H.P., Singer, B.: Recursive Partitioning in the Health Sciences. Springer, New York (1999)

Zurada, J.M.: Introduction to Artificial Neural Systems. W. Publishing Company (1992)

Adaptation in Games with Many Co-evolving Agents

Ana L.C. Bazzan[1], Franziska Klügl[2], and Kai Nagel[3]

[1] Instituto de Informática, UFRGS
Caixa Postal 15064, 91.501-970 Porto Alegre, RS, Brazil
bazzan@inf.ufrgs.br
[2] Dep. of Artificial Intelligence, University of Würzburg
Am Hubland, 97074 Würzburg, Germany
kluegl@informatik.uni-wuerzburg.de
[3] Inst. for Land and Sea Transport Systems, TU Berlin Sek SG12
Salzufer 17–19, 10587 Berlin, Germany
nagel@vsp.tu-berlin.de

Abstract. Despite the recent results on formalizing multiagent reinforcement learning using stochastic games, the exponential increase of the space of joint actions prevents the use of this formalism in systems of many agents. In fact, most of the literature concentrates on repeated games with single state and few joint actions. However, many real-world systems are comprised of a much higher number of agents. Also, these are normally not homogeneous and interact in environments which are highly dynamic. This paper discusses the implications of co-evolution between two classes of agents in stochastic games using learning automata. These agents interact in a urban traffic scenario where approaches based on the standard stochastic games are prohibitive. The approach was tested in a network with different traffic conditions.

1 Motivation and Introduction

Learning in systems with two or more agents has a long history in game-theory. Thus, it seems natural to the reinforcement learning community to explore the existing formalisms behind stochastic (Markov) games (SG) as an extension for Markov Decision Processes (MDP's). Despite the inspiring results achieved so far, it is not clear what kind of question the multiagent system community is addressing [12]. It seems that, at this stage, the focus is on what Shoham et al call "the equilibrium" agenda, although SG is not the only approach possible here [13]. In any case, everybody agrees that the problems posed by many agents in multi-agent reinforcement learning (MARL) are inherently more complex than those regarding single agent reinforcement learning (SARL). This complexity has many consequences (note that by SARL we mean an environment with only one agent). First, the approaches proposed for the case of general sum SG require that several assumptions be made regarding the game structure (agents' knowledge, self-play etc.). These assumptions constraint the convergence results

J. Neves, M. Santos, and J. Machado (Eds.): EPIA 2007, LNAI 4874, pp. 195–206, 2007.

to common payoff games and other special cases such as zero-sum games, besides focussing on two-agent stage games. Otherwise, an oracle is needed if one wants to deal with the problem of equilibrium selection when two or more equilibria exist. Second, despite the recent results on formalizing multiagent reinforcement learning using stochastic games, these cannot be used for systems of more than a few agents agents, *if any flavor of joint-action is explicitly considered*, unless the exigence of visiting all pairs of state-action is relaxed, which has impacts on the convergence. The problem with using a high number of agents happens mainly due to the exponential increase in the space of *joint* actions. In fact, most of the literature concentrates on repeated games with two-players and a single state. Third, while the agents themselves must not be cooperative, we may be interested in improving the system's performance. This is a well-known issue. Tumer and Wolpert [14] for instance have shown that there is no general approach to deal with the complex question of collectives.

Up to now, these issues have prevented the use of MARL in real-world problems, unless simplifications are made, such as letting each agent learn *individually* using single-agent based approaches (thus, SARL). As known, this approach is not effective, since agents converge to sub-optimal states.

The aim of this paper is threefold. First, we tackle a many-agent system. Second, we discuss the implications of co-evolution between two classes of agents. Third, we want to pursue the adaptation road for MARL, as a trade-off between the complexity of learning with convergence guarantees, and effectiveness. This road is not the most efficient for systems with small number of agents which interact in well-behaved environments, that means those where the non-determinism does not arise only from the non coordinated actions of the agents. For these, the MARL community has proposed nice and efficient solutions. Rather, our approach targets real-world systems problems with the following characteristics: they are comprised of a high number of agents; agents are normally not homogeneous, i.e., several types of agents having different learning or adaptation algorithms co-exist (thus it is not the case of self-play); agents act and interact in environments which are highly dynamic. In particular, this paper uses an urban traffic scenario to illustrate the use and results of the approach proposed.

This is a relevant scenario because urban mobility is one of the key topics in modern society. Our long term agenda is to propose a methodology to integrate behavioral models of human travelers reacting to traffic patterns and control measures of these traffic patterns, focusing on distributed and decentralized methods. Classically, this is done via network analysis.

To this aim, it is assumed that individual road users seek to optimize their individual costs regarding the trips they make by selecting the "best" route. This is the basis of the well known traffic network analysis based on Wardrop's equilibrium principle [18]. There are many variants of the Wardrop equilibrium, such as the dynamic user equilibrium, or the so-called stochastic user equilibrium (which is, in effect, a deterministic distribution of traffic streams across alternatives). It is even possible to apply the dynamic user equilibrium in a truly stochastic situation, where the traffic situation changes from day to day. In that situation, however, the definition of the game is such that the players can at best play strategies that optimize average reward. Although it is clearly possible to simulate situations where either drivers or traffic lights or both are within-day adaptive to vairable traffic, few if any investigations exist that attempt to clarify the overall system effects of such adaptiveness.

In summary, as equilibrium-based concepts generally overlook the within-day variability regarding demand and capacity, it seems obvious that they are not adequate to be used in microscopic, decentralized approaches. However, the price to be payed when one moves from the former to the latter is an increase in complexity which prevents the use of the approaches based on stochastic games as currently proposed, and demands simplifications and a change in the paradigm, from long-term learning to fast adaptation. This shift is further justified by the fact that convergence to an equilibrium is not the main issue. Rather, we are interested in the design of efficient or at least effective agents for this kind of environments. Here, as more than one class of agent co-exist and co-evolve, general questions are whether co-evolution pays off, and, if so, what kind of evolutionary approach should be used, thus sheding light in two issues recently raised by Shoham and colleagues, namely the "AI agenda" for multiagent reinforcement learning and the learning–teaching aspects of this problem.

In the next two sections we review approaches to SG, and briefly introduce some concepts about traffic assignment, simulation, and control. Section 4 discusses the approach, while the scenario and the results appear in sections 5 and 6 respectively. The last section presents the concluding remarks.

2   Learning in Multiagent Systems

Most of the research on MARL so far is based on a static, single state stage game (i.e. a repeated game) with common payoff (payoff is the same for agent and opponent) as in [6]. The zero-sum case is based on [8] and attempts at generalizations to general-sum SG appeared in [7], among many others (as a comprehensive description is not possible here, we refer the reader to [12] and references therein).

Some works have similar motivation to ours: In [16] the authors tackle a particular kind of game (coordination game) by means of an exploration technique based on learning automata and reduction of the action space. The approach in [17] deals with multiple opponents but they assume that the full game structure and payoffs are known to all agents. Besides, the algorithm is based on joint strategy for all the self-play agents (those who learn using the same algorithm) so that the action space is exponential in the number of self-play agents. Specifically for traffic, a simple stage game is discussed in [2]. In that setting, since the goal is to coordinate neighbor traffic lights so that they synchronize their green phases, it makes sense to model the interaction as a coordination game. For the general case (no a priori coordination), there is no formulation for scenarios with more than a few agents. Camponogara and Kraus [5] have studied a simple scenario with only two intersections, using stochastic game-theory and reinforcement learning.

Shoham and colleagues single out some problems due to focussing on what they call the "Bellman heritage". Two issues are important from our perspective: The first is the focus on convergence to equilibrium regarding the stage game: "If the process (of playing a game) does not converge to equilibrium play, should we be disturbed?" Also, most of the research so far has been focussing on the play to which agents converge, not on the payoff agents obtain. The second issue is that "In a multi-agent setting, one cannot separate learning from teaching" because agent i's action selections both arise from information about agent j's past behavior and impacts j's future actions' selections. Unless i and j are completely unaware of the presence of each other, both can teach and learn how to play in mutual benefit. Therefore it is suggested that a more neutral term would be *multi-agent adaptation* (rather than learning). This is an important point because it agrees with a view that some issues in traffic (mainly related to short time control) are more a quest of adaptation than of optimization. Since the latter is hard to achieve, it is often the case that this cannot be done in real-time. A further point in favor of adaptation is that most of the work on MARL has been assuming static environments. In this kind of environment it may make sense to evaluate the MARL algorithms by the criteria proposed in [4]: convergence to a stationary policy, and convergence to a best response if the opponent converges to a stationary policy. Although other criteria are being proposed (in fact the discussion is just starting; see [17] for other criteria), it certainly makes little sense to evaluate a learning or adaptation algorithm by such criteria when the environment is itself dynamic, as it is the case of the traffic scenario discussed here.

3   Towards Agent-Based Traffic Assignment, Simulation, and Control

Transportation engineering has seen a boom regarding methodologies for microscopic, agent-based modeling. On the side of *demand* forecasting, the arguably most used computational method is the so-called 4-step-process consisting of the four steps: trip generation, destination choice, mode choice, and route assignment. The 4-step-process has several drawbacks. For a discussion of these issues see [1]. Agent-based approaches promise to fill this gap as they allow to simulate individual decision-making. However, until now agent-based simulations with high-level agents on the scale required for traffic simulation of real-world networks have not been developed. Some steps towards that goal is to use concepts of microeconomics to approach decision-making and how drivers adapt to the previous experiences. Basically, simple binary scenarios have been used, based on approaches with minority-game flavors. However, when the coordination emerges out of individual self-interest, sometimes a user equilibrium is achieved, but in general no system optimum.

From the side of *control*, a popular method is to use traffic lights. Several signal plans are normally required for an intersection to deal with changes in traffic volume. Thus, there must be a mechanism to select one of these plans. Readers can find a review in [2].

Besides the works already mentioned in the previous section, the following also tackle optimization of traffic lights via reinforcement learning: In [10] a set of techniques were tried in order to improve the learning ability of the agents in a simple scenario with few agents. [19] describes the use of reinforcement learning by the traffic light controllers in order to minimize the overall waiting time of vehicles in a small grid. The ideas and some of the results presented in that paper are important. However, strong assumptions hinder its use in the real world. First, the kind of communication and knowledge (or, more appropriate, communication *for* knowledge formation) has a high cost; traffic light controllers are suppose to know vehicles destination in order to compute expected waiting times for each. Besides, there is no account of the experience collected by the drivers based on their local perceptions only. Finally, drivers being autonomous, it is not reasonable to expect that all will use the best policy computed, given the value function, which for this sake, was computed by the traffic light and not by the driver itself.

Regarding *integration* of traffic assignment and control, there are a number of works which represent different views of this issue. In [15], a two-level, three-player game is discussed. The control part involves two players, namely two road authorities, while the population of drivers is seen as the third player. Complete information is assumed, which means that all players (including the population of drivers) have to be aware of the movements of others. Moreover, it is questionable whether the same mechanism can be used in more complex scenarios, as claimed. The reason for this is the fact that when the network is composed of tens of links, the number of routes increases and so the complexity of the route choice, given that now it is not trivial to compute the network and user equilibria.

Liu and colleagues [9] describe a modeling approach which integrates microsimulation of individual trip-makers' decisions and individual vehicle movements across the network. Their focus is on the description of the methodology which incorporates both demand and supply dynamics, so that the applications are only briefly described and not many options for the operation and control of traffic lights are reported. One scenario described deals with a simple network with four possible routes and two control policies.

Ben-Akiva and co-workers have investigated in some detail the issue of so-called self-consistent anticipatory route guidance [3]. In this, a loop "traffic control – driver reaction – network loading" is defined. The loop is closed by the traffic control being reactive to the result of the network loading. The resulting problem is defined as a fixed point problem: A solution is found if the traffic control, via driver reaction and network loading, generates the same traffic pattern that was the basis for the traffic control. The approach, however, focuses on information as control input, not traffic signals.

Papageorgiou and co-workers look into the problem with a control-theoretic approach [11]. In that language, human behavior and network loading are combined into the dynamical update of the system, and the goal is to search for a control input that optimizes some aspect of the output from the system. However, human behavior is by necessity of the mathematical formulation very much reduced, and no results about the emergent properties from system-wide signal control seem to be known.

4   Multiagent Adaptation in Stochastic Games

The generalization of a MDP for n agents is a SG, represented by the tuple (N, S, A, R, T) where:

$N = 1..., i..., n$ is the set of agents
S is the discrete state space (set of n-agent stage games)
$A = \times A^i$ is the discrete action space (set of joint actions)
R is the reward function (R determines the payoff for agent i as $r^i : S \times A^1 \times \ldots \times A^n \to \Re$)
T is the transition probability map (set of probability distributions over the state space S).

As said, many attempts to use SG for MARL are grounded on all or some of these assumptions: players know the stochastic game they are playing (or at least its structure); players have information about others' actions and/or rewards; joint actions are observable. Especially the latter is a strong assumption which not only has consequences on the communication load, but also implies that the size of Q-learning tables is exponential in the number of agents.

Instead of assuming that joint actions are observable and that rewards are known by all agents, we propose a learning automata (LA) based approach to stochastic games. A similar approach appears in [16] but the authors deal with multi-stage, common payoff games defined in normal form. In common payoff games, the rewards received by two agents are correlated. Thus, it is possible to verify whether and when there is a convergence to the social optimum. The authors use exploration techniques associated with the learning automata.

A learning automata formalizes stochastic systems and aims at guiding the action selection at any given time t in terms of the last action selected and the environment response (the reward r^t). This response is used to update the actions probabilities. A well known update scheme is the linear reward-inaction scheme (L_{R-I}), which increases the probability of an action if it results in a success (otherwise the probability remains the same). A learning automata consists of a vector of probabilities $p^{i,t} = (p_1^{i,t}, p_2^{i,t}, \ldots, p_m^{i,t})$ over the set of m actions $a_1^{i,t}, \ldots, a_m^{i,t}$. At each time t, $p^{i,t}$ is used by agent i to select an action $a^{i,t}$.

The L_{R-I} scheme is defined as following:

$$\forall_{j \neq i} : \quad \begin{aligned} p^{i,t+1} &= p^{i,t} + \alpha(1 - p^{i,t}) \\ p^{j,t+1} &= p^{j,t}(1 - \alpha) \end{aligned} \tag{1}$$

where $\alpha \in [0, 1]$.

In dynamic and/or unknown environments, as it is the case of the domain here, one drawback of the learning automata update scheme L_{R-I} is that it may discard actions, i.e. one action may never be used again. Thus we use a responsive LA which has the property that all probabilities associated with the actions are positive because the responsive update scheme never discards actions ($\forall_j a_j^i > 0$). This is important if the environment changes. The responsive LA modifies the L_{R-I} scheme so that no action has probability less than α_{min}.

In the next section, we discuss some changes in the basic SG/LA framework in order to deal with the particularities of our scenario.

5 Learning Automata Based Stochastic Game: Application in an Urban Traffic Scenario

The traffic scenario targets a game with two classes of agents: drivers and traffic lights. Notice that, due to the number of learning agents, scenarios of this size are seldom tackled by the RL community. The goal of all agents is to select actions which maximizes individual rewards. Although each one knows the set of available actions and are able to perceive their rewards, there is no communication among them so that non-local rewards or no joint actions are explicitly observed. However, actions selected by the agents do have an effect on each other. Moreover, the two classes of agents have two different types of actions, different learning paces, and the adaptation algorithms tailored for the specific purposes of each class of agents.

A driver's action is to select a route to minimize travel time. Each driver d has a choice of up to m_r routes, that means, this is the maximum number but some drivers may be aware of a smaller number m_d. The choice of action is probabilistic. We use two different schemes to set these probabilities:

- **"random drivers"**: The probabilities of selecting the m_d routes are constant over time and identical between options: $p_d^j = 1/m_d$.
- **"LA drivers"**: The update of these probabilities is done each time a route is completed (we call this a trip), using the responsive L_{R-I} scheme (Eq. 1) substituting α for α_d.

The traffic lights have a "north-south/south-north" phase and an "east-west/west-east" phase, with fractions of time $f_{tl}^{\updownarrow}$ and $f_{tl}^{\leftrightarrow} = 1 - f_{tl}^{\updownarrow}$. At the end of each phase, the following is done:

- If the phase was "successful" (i.e. traffic volume improves (locally) in this direction), then that phase is expanded according to a scheme based on Eq. 1 substituting α for α_t:

$$f_{tl}^{i,t+1} = f_{tl}^{i,t} + \alpha_t \left(1 - f_{tl}^{i,t}\right)$$

Each time one phase is expanded, the other phases are implicitly shortened. Thus implicitly, the actions of the traffic lights are to priorize one of the two traffic directions.

– If the phase was not "successful", then nothing changes.

It is important to notice that, roughly, while the traffic lights adapt in a time frame of minutes, the drivers update once a trip (day). Therefore, in Eq. 1, α must have different values (thus α_t and α_d).

Formally, the SG defined in the previous section has the following particular setting:

– $N = \mathcal{D} \cup \mathcal{T}$ (set of agents is the union of the set of drivers and set of traffic light agents)
– each agent i has only a local, individual perception of the whole environment so that the state space is actually the cartesian product over the individual state sets $(\times S^i)$
– the action space is the cartesian product over all actions of the drivers and the traffic lights

With these figures, it is obvious that, if we use a dynamic programming based approach which considers all states and actions, each agent needs to maintain tables which are exponential in the number of agents: $|S^1| \times \ldots \times |S^k| \times |A^1| \times \ldots \times |A^k|$. Let us assume a very simple mapping of states, namely that all traffic light agents can map the local states to either jammed or not jammed, i.e. $|S^i| = 2$ for $i = 1, \ldots, |\mathcal{T}|$, and that drivers cannot perceive more than one state. Thus, the cartesian product over the states has a size $2^{|\mathcal{T}|} \times 1^{|\mathcal{D}|}$. As the traffic lights have two actions (two signal plans) and the drivers have at most five actions (five routes to choose from), the size of Q tables is $2^{|\mathcal{T}|} \times 2^{|\mathcal{T}|} \times 5^{|\mathcal{D}|}$. Already the last term makes this approach computationally intractable as the number of drivers tends to grow to the hundreds at least, not to speak about the communication demand. Therefore, the learning automata approach proposed here is able to deal with these figures as it does not consider the joint actions and states. Instead, the L_{R-I} scheme defined in the previous section is used, which considers only the individual set of actions for each agent.

We have implemented the simulation in the agent-based simulation environment SeSAm. The movement of vehicles is queue-based.

To exemplify the approach, we use a typical commuting scenario where drivers repeatedly select a route to go from an origin to a destination in a grid-like network. We use a grid to avoid a simple scenario such as a two-route (binary decision) as in [15]. The grid is reasonably more complex and captures desirable properties real scenarios have regarding the aim of this study, namely the co-evolution among drivers and traffic lights. Next we detail the particular scenario used.

We use a grid where the 36 nodes are tagged from A1 to F6, as in Figure 1. All links are one-way and drivers can turn in each crossing. This kind of scenario is a realistic one and, in fact, from the point of view of route choice and equilibrium computation, it is very complex as the number of possible routes between two nodes is high.

Moreover, contrarily to simple two-route scenarios, in the grid one it is possible to set arbitrary origins and destinations. Each driver has one particular

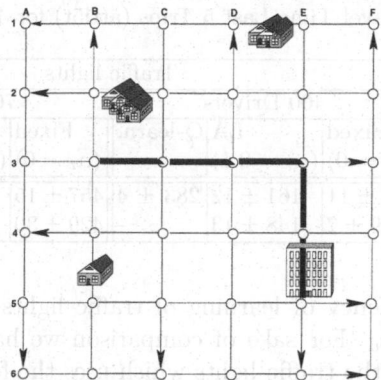

Fig. 1. Grid 6x6 showing the main destination (E4E5), the three main origins (B5B4, E1D1, C2B2), and the "main street"

origin and destination. To render the scenario more realistic, there is one main destination: on average, 60% of the road users have the link labelled as E4E5, associated with node E4, as destination (see Figure 1). Other links have, each, 1.7% probability of being a destination. Origins are nearly equally distributed in the grid, with three exceptions: links B5B4, E1D1, and C2B2 have approximately 5% probability of being an origin. The remaining links have each a probability of 1.5%. This was done to model residential neighborhoods. Regarding capacity, all links can hold up to 15 vehicles, except those located in the so called "main street" which can hold up to 45. This main street is formed by the links B3 to E3, E4, and E5 (thicker links in Figure 1).

6   Results and Discussion

6.1   Metrics and Parameters

In order to evaluate the experiments, four quantities were measured: the number of drivers who have arrived at their destinations up to the time out t_{out} for each particular trip; the mean travel time over all drivers for a given trip, as well as the mean of the average of the travel time over all possible routes, over all drivers. Plots for these two are not shown here due to lack of space. Rather, we show the mean travel time over only the last $T_t = 5$ trips to give a reference of the travel time at the end of the experiments. All experiments were repeated 50 times.

The other parameters used were: $|\mathcal{T}| = 36$; $m_r = 5$ (maximum number of known routes, generated via an algorithm that computes the shortest path (one route) and the shortest path via arbitrary detours (four others)). We have run simulations with $|\mathcal{D}| = 400$ and $|\mathcal{D}| = 700$ drivers. In these cases $t_{out} = 300$ and $t_{out} = 500$ respectively.

Regarding the learning automata, we experimented several values of α_d and α_t; here we show the results with the best values. Due to lack of space we only discuss the case where $\alpha_d = 0.4$ (drivers) and $\alpha_t = 0.05$ (traffic lights).

Table 1. Average Travel Time Last 5 Trips (attl5t) for 400 and 700 Drivers

Drivers	Traffic lights					
	400 Drivers			700 Drivers		
	Fixed ($\alpha_t = 0$)	LA ($\alpha_t = 0.1$)	Q-learn.	Fixed ($\alpha_t = 0$)	LA ($\alpha_t = 0.1$)	Q-learn.
random ($\alpha_d = 0$)	157 ± 11	161 ± 12	283 ± 4	457 ± 15	375 ± 65	480 ± 6
LA ($\alpha_d = 0.4$)	139 ± 7	148 ± 13	–	429 ± 30	423 ± 31	–

The fact that the frequency of learning of traffic lights is lower than that of drivers requires $\alpha_t < \alpha_d$. For sake of comparison we have implemented a Q-learning mechanism for the traffic lights which uses the following values for the parameters: the learning rate is $\beta = 0.1$ and the discount rate is $\gamma = 0.9$. Available actions are to open the phase serving either one direction or the other. The states are the combination of states in both approaching links, i.e. $\{D_1\text{-}jammed, D_1\text{-}not\text{-}jammed\} \times \{D_2\text{-}jammed, D_2\text{-}not\text{-}jammed\}$. The reward is one minus the average occupancy in the incoming links of a given node. Contrarily to the traffic lights, the drivers cannot assess the state of the network (not even locally) from their individual travel times. Thus Q-learning was not implemented for the drivers.

6.2   Overall Discussion

In Table 1 we summarize the average travel time over the last $T_t = 5$ trips (henceforward *attl5t*) for different conditions, for 400 and 700 drivers. These figures correspond to an overall occupancy of 38% and 78% of the network.

TLs Fixed / Random drivers. When $\alpha_d = 0$, drivers select a route with equal probability; $\alpha_t = 0$ means that the traffic lights run a signal plan which priorizes no direction. This is used here to benchmark the next variants.

TLs with LA / Random drivers. As expected, adaptive traffic lights have no effect in under saturated networks (400 drivers); the *attl5t* is 161. The effect of this adaptation is clear when there are 700 drivers (*attl5t* = 375).

Drivers with LA / Fixed TLs. While the traffic lights remain fix, drivers can improve their performance by using the learning automata because they are able to choose other routes, possibly with less drivers. Travel times (*attl5t*) drop to 139 (400 drivers) and 429 (700 drivers).

LA both. This is a typical commuting scenario where the control tries to adapt to the drivers and these to the control. Once a better control is achieved, too many drivers try to exploit this fact and end up flocking to given links, with a negative impact in the performance. The control however has not so much room to act in oversaturated situations (remember that signal plan has to serve all directions for at least a minimum green time), which occur in parts of the network (e.g. links close to the main destination).

Q-learning traffic lights. The low performance of Q-learning in traffic scenarios is due basically to the non-stationary environment and too many agents learning simultaneously.

7 Conclusion

Many tools for management of traffic flow exist (e.g. control of traffic lights). It is possible to combine these approaches with intelligent traffic assignment, e.g. via information to the drivers. Important issues then are how drivers process this information in order to make decision, and how they proceed in order to adapt to their environments.

However, there are few attempts and no conclusive results concerning what happens when *both* the driver and the traffic light use some adaptive mechanism in the same scenario or environment, especially if *no central control exist*, i.e. the co-evolution happens in a decentralized fashion, in which case some form of auto-organization may arise. This is an important issue because, although intelligent transportation systems have reached a high technical standard, the reaction of drivers to these systems is fairly unknown. In general, the optimization measures carried out in the network both affect and are affected by drivers' reactions to them. This leads to a feedback loop which has received little attention to date.

When one tries to approach this problem using traditional SG based MARL, one gets stuck on the computational complexity. Therefore, in the present paper we have investigated that loop by means of a multiagent adaptation using learning automata. The results show an improvement regarding travel time when agents adapt. This improvement is not very significant when all agents co-evolve, especially in saturated networks, as expected, for the reasons already explained. This was compared with situations in which either only drivers or only traffic lights evolve, in different scenarios.

This work can be extended in two main directions. First, we plan to integrate the tools developed by the authors independly for control and traffic assignment. The second extension relates to the use of heuristics about the the network in order to improve its performance.

Acknowledgments

The authors would like to thank CAPES (Brazil) and DAAD (Germany) for their support to the joint, bilateral project "Large Scale Agent-based Traffic Simulation for Predicting Traffic Conditions". Ana Bazzan is partially supported by CNPq and Alexander von Humboldt Stiftung.

References

1. Balmer, M., Nagel, K., Raney, B.: Large–scale multi-agent simulations for transportation applications. Journal of Intelligent Transportation Systems: Technology, Planning, and Operations 8(4), 205–221 (2004)
2. Bazzan, A.L.C.: A distributed approach for coordination of traffic signal agents. Autonomous Agents and Multiagent Systems 10(1), 131–164 (2005)

3. Bottom, J., Ben-Akiva, M., Bierlaire, M., Chabini, I.: Generation of consistent anticipatory route guidance. In: Proceedings of TRISTAN III, San Juan, Puerto Rico, vol. 2 (June 1998)
4. Bowling, M.H., Veloso, M.M.: Rational and convergent learning in stochastic games. In: Nebel, B. (ed.) Proceedings of the Seventeenth International Joint Conference on Artificial Intelligence, Seattle, pp. 1021–1026. Morgan Kaufmann, San Francisco (2001)
5. Camponogara, E., Kraus Jr., W.: Distributed learning agents in urban traffic control. In: Pires, F.M., Abreu, S.P. (eds.) EPIA 2003. LNCS (LNAI), vol. 2902, pp. 324–335. Springer, Heidelberg (2003)
6. Claus, C., Boutilier, C.: The dynamics of reinforcement learning in cooperative multiagent systems. In: Proceedings of the Fifteenth National Conference on Artificial Intelligence, pp. 746–752 (1998)
7. Hu, J., Wellman, M.P.: Multiagent reinforcement learning: Theoretical framework and an algorithm. In: Proc. 15th International Conf. on Machine Learning, pp. 242–250. Morgan Kaufmann, San Francisco (1998)
8. Littman, M.L.: Markov games as a framework for multi-agent reinforcement learning. In: Proceedings of the 11th International Conference on Machine Learning ML, New Brunswick, NJ, pp. 157–163. Morgan Kaufmann, San Francisco (1994)
9. Liu, R., Van Vliet, D., Watling, D.: Microsimulation models incorporating both demand and supply dynamics. Transportation Research Part A: Policy and Practice 40(2), 125–150 (2006)
10. Nunes, L., Oliveira, E.C.: Learning from multiple sources. In: Jennings, N., Sierra, C., Sonenberg, L., Tambe, M. (eds.) Proceedings of the 3rd International Joint Conference on Autonomous Agents and Multi Agent Systems, AAMAS, vol. 3, pp. 1106–1113. IEEE Computer Society, New York (2004)
11. Papageorgiou, M.: Traffic control. In: Hall, R.W. (ed.) Handbook of Transportation Science, vol. 8, pp. 243–277. Kluwer Academic Pub., Dordrecht (2003)
12. Shoham, Y., Powers, R., Grenager, T.: If multi-agent learning is the answer, what is the question? Artificial Intelligence 171(7), 365–377 (2007)
13. Stone, P.: Multiagent learning is not the answer. It is the question. Artificial Intelligence 171(7), 402–405 (2007)
14. Tumer, K., Wolpert, D.: A survey of collectives. In: Tumer, K., Wolpert, D. (eds.) Collectives and the Design of Complex Systems, pp. 1–42. Springer, Heidelberg (2004)
15. van Zuylen, H.J., Taale, H.: Urban networks with ring roads: a two-level, three player game. In: Proc. of the 83rd Annual Meeting of the Transportation Research Board. TRB (January 2004)
16. Verbeeck, K., Nowé, A., Peeters, M., Tuyls, K.: Multi-agent reinforcement learning in stochastic games and multi-stage games. In: Kudenko, D., Kazakov, D., Alonso, E. (eds.) Adaptive Agents and Multi-Agent Systems II. LNCS (LNAI), vol. 3394, pp. 275–294. Springer, Heidelberg (2005)
17. Vu, T., Powers, R., Shoham, Y.: Learning against multiple opponents. In: Proceedings of the Fifth International Joint Conference on Autonomous Agents and Multiagent Systems, pp. 752–760 (2006)
18. Wardrop, J.G.: Some theoretical aspects of road traffic research. Proceedings of the Institute of Civil Engineers 2, 325–378 (1952)
19. Wiering, M.: Multi-agent reinforcement learning for traffic light control. In: ICML 2000. Proceedings of the Seventeenth International Conference on Machine Learning, pp. 1151–1158 (2000)

Chapter 3 - Third Workshop on Artificial Life and Evolutionary Algorithms (ALEA 2007)

Symmetry at the Genotypic Level and the Simple Inversion Operator

Cristian Munteanu and Agostinho Rosa

Instituto de Sistemas e Robótica, Instituto Superior Técnico
Av. Rovisco Pais, 1 - Torre Norte, 6.21
1049-001 Lisboa, Portugal
{cmunteanu,acorsa}@isr.ist.utl.pt

Abstract. Classical Genetic Algorithm theory was built on four operators: proportional selection, one-point crossover, mutation and inversion. While the role of inversion was questioned, the use of the other remaining operators has thrived, some of these newly designed operators being motivated by good empirical results, some having a solid theory to support their use. In this paper we present a Simple Inversion Operator, and we investigate its potential mixing capabilities for problems where the optimum consists of juxtaposed Symmetric Building Blocks. Both theoretical investigation and experimental results obtained, indicate that our operator is quite powerful in finding the right building blocks that compose the optimum, whenever symmetrical building blocks play an important role in the discovery of the global solution.

1 Introduction

Holland proposed inversion in his seminal work [1], as a method to increase the linkage of highly fitted schemata that initially exhibit a long defining length. This gives a better chance for a building block to survive the disruptive effect of one-point crossover, and to eventually combine with other potentially useful building blocks, as the building block hypothesis predicts [2]. Several authors before Holland, reported different results on the application of inversion in Genetic Algorithms (GA), such as Bagley (1967), Frantz (1972) and Cavicchio [2], while later, several researchers, such as Whitley in [3], and Wienholt, in [4] included inversion in their algorithm's structure. Syswerda in [5] advocated the use of inversion as a proper method to change the representation of the chromosome, before applying one point crossover, as a modified representation may be beneficial for the GA, especially when tight linkage is required. Inversion (reordering genes) has also been applied to ordering problems like Traveling Salesman Problem, in [6]. However, inversion is still not a commonly used operator, and this is due partly to the fact that genes' proper ordering and linkage was solved by other approaches like messy GA [7], learning and exploiting the genes' linkage [8], [9], [10], co-evolving the proper representation and consequently the most suitable linkage, for a specific problem and a specific representation [11].

In this paper we take a different approach to inversion, considering it a simple reordering operation capable of attaining very good performance on problems compliant with the Symmetrical Building Block Paradigm, that we introduced in [12]

J. Neves, M. Santos, and J. Machado (Eds.): EPIA 2007, LNAI 4874, pp. 209–222, 2007.

and [13]. Our approach is different from the classical view of inversion benefits, in that it is not concerned with increasing linkage between different genes, but rather aims at the direct discovery of building blocks with a specific symmetry. Inversion together with crossover works to properly mix and align the building blocks, to form the optimal solution. The main advantages of our approach are the application with good results of a fairly simple and unpretentious operator, and the clear identification of the class of problems where this operator has the highest potential. This class of problems requires the optimum to consist of several juxtaposed Symmetrical Building Blocks (SBBs). Apart from this requirement which loosely speaking means that the optimum presents some kind of symmetry at the genotypic level, the class is rather general, the representation of the chromosome could be of any kind, such as natural binary coding, Gray coding, integer or real coding. More recently symmetry at the genotypic level has been studied in [14] where the authors recognize a certain difficulty for the unspecialized Evolutionary Algorithm (EA) for problems that present symmetry at the alphabet level, while these findings were later confirmed in studies such as [15]. The practical importance of coping with symmetry in the fitness functions, comes from the fact that several combinatorial optimization problems such as the max-cut problem, graph partitioning, graph coloring, bin-packing, workshop layouting, and some Ising problems induce symmetric fitness functions.

2   How Does Simple Inversion Exploit Building Block Symmetry

Inversion has been devised in order to increase the linkage of highly fitted schemata that exhibit a loose linkage between alleles, or a long defining length equivalently. Since an above average schema gets selected with a high probability, and since it has a long defining length, the probability of disrupting the respective schema, by applying crossover[1], is high too. To keep the linkage high, as to make the schema less vulnerable to the disruptive effect of crossover, inversion may be applied [1]. Whitley acknowledges another role that inversion might have in a real featured search space, that "it provides a type of nondestructive noise that helps crossover to escape local maxima" [3]. Wienholt in [4] included inversion into a refined GA that resembles in more detail the biological model, using the argument that inversion works at the DNA level, and he creates a genotype, as well as a phenotype inversion, that produce different kind of new information at two different levels.

Inversion, as in the classical GA theory [1] works by randomly selecting two points within a chromosome and inverting the order of genes between selected points, but remembering the gene's meaning or functionality. To identify the genes in a chromosome, information about their original position is added to each gene. A GA with inversion searches for the best arrangement of genes to correctly transmit the building blocks of the problem.

Our Simple Inversion Operator (SIO) works much the same as the usual inversion operator but without keeping track of the initial position or functionality of genes in

---

[1] Note that inversion has been proposed when one-point crossover, or another highly positional biased crossover is applied, as positional bias requires tight linkage between genes, to safely transmit the building blocks.

the chromosome structure. Without adding the information regarding the initial "meaning" of genes, our operator simply reverses the order of genes in the chromosomes between two randomly chosen loci. For a binary encoding, SIO performs a simple bit reordering. We found that a GA with SIO is better suited than a Standard Genetic Algorithm (SGA) without inversion, for problems where the optimum consists of juxtaposed Symmetrical Building Blocks (SBB), where we first introduced SBBs in [12]. In what follows we review the formal definition of a SBB: Let l be the length of the chromosomes in the population and $\#_k$ a "don't care" symbol at locus k (that means the gene at locus k may take any value in $\{0,1\}$).

Definition 1. A SBB of length δ, at position q (denoted as SBB_q^δ) is the string:

$$\mathrm{SBB}_q^\delta = \#_1\#_2...\#_{q-1}\, a^1 a^2 ...a^i ...a^{\delta-1} a^\delta\, \#_{q+\delta-1}...\#_l \tag{1}$$

$$\begin{cases} l \text{ is divisible by } \delta \\ q = j\cdot\delta+1, \quad j\in\left\{0,1,...,\dfrac{l-\delta}{\delta}\right\} \\ a^i \in\{0,1,\#\}, \quad i\in\{1,...,\delta\} \\ a^i = a^{\delta-i+1}, \quad \text{for} \quad i\in\left\{1,...,\dfrac{\delta}{2}\right\} \text{if } \delta=2r \text{ or } i\in\left\{1,...,\dfrac{\delta-1}{2}\right\} \text{if } \delta=2r+1, r\in\mathbf{N}^* \end{cases} \tag{2}$$

where $\#$ is a "don't care" value in $\{0,1\}$. We call the sequence of all a^i with $i\in\{1,...,\delta\}$ the kernel of the SBB. Eq. (2) defines in a formal fashion the symmetry of the kernel while Fig. 1 gives a graphical depiction of a SBB and its kernel. In Fig. 1, the kernel of the SBB has $\delta = 5$ symbols: hollow circle, hollow and filled square, symbols that can take any value in the set $\{0, 1, \#\}$. The string of symbols forming the kernel corresponds to the string $a^1... a^\delta$ in (1) and (2). The remaining 3 cells of the chromosome (the white cells in Fig. 1) have each δ symbols "#" and the length of the chromosome is $l = 4, \delta = 20$.

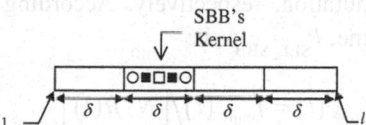

Fig. 1. An example of a SBB with its kernel

Definition 2. The optimum string which is SBB-compliant is a string x_{opt} that is formed by juxtaposing SBB_q^δ with δ fixed:

$$x_{\mathrm{opt}}=A_1 A_2 ...A_{l/\delta} \text{ where } A_1\equiv A_2\equiv...\equiv A_{l/\delta}=a^1 a^2...a^{\delta-1}a^\delta \tag{3}$$

If $\exists a^i$, $i\in\{1,...,\delta\}$, so that $a^i = \#$, then we refer to a multimodal problem.

2.1 Probability to Generate a New SBB

A new SBB is a SBB that contains a kernel properly aligned on the chromosome according to (2), on free positions that previously did not contain another kernel (before applying SIO, there was no chromosome in the population that did contain a SBB's kernel on the respective positions). SIO has the role of moving an existent SBB to a new location on the same chromosome, while crossover may combine a copy of the non-inverted string with the newly inverted string, such as the two SBBs would appear aligned on the same chromosome. The following considerations apply only to problems that are SBB compliant, that is, their optimum is of the form given in eq. (3).

Proposition 1. Consider that in the current population $P(t)$, at generation t, we have *one single* SBB, (i.e., we have a chromosome that contains this SBB). The length δ is fixed and let o stand for the order of the SBB (i.e. the number of defined positions) and Δ the actual defining length of the SBB (note that $\Delta \leq \delta$ and $\Delta = \delta$ iff (a^1 and $a^\delta \neq$ #)). The existing SBB will be denoted as $SBB^\delta{}_i$ with fitness $f_{SBB_i^\delta}(t)$. N stands for the total number of chromosomes in the population, $\hat{\mu}(t)$ is the mean fitness of the individuals in $P(t)$. We consider applying the genetic operators, as follows: a single string selection, followed by SIO with probability P_{SIO}, crossover with probability P_c and mutation with probability P_m. Hence, we may calculate the probability of generating a *new* SBB, call it $SBB^\delta{}_j$, from the existing one $SBB^\delta{}_i$, as follows:

$$P(t)_{SBB_j^\delta} = \left\{ 2 \cdot f_{SBB_i^\delta} \big/ \left[(1+l) \cdot l \cdot N \cdot \hat{\mu}(t) \right] \right\} P_{SIO} \left(1 - P_c \, \Delta/(l-1) \right) \left(1 - P_m \right)^o \tag{4}$$

Proof: The probability of generating a new SBB ($SBB^\delta{}_j$) from an existing one ($SBB^\delta{}_i$) in the current t^{th} generation, is the product of the probability $P_{SEL\_SBB_i^\delta}$ to select the chromosome that contains $SBB^\delta{}_i$, the probability $P_{GENERATE}$ to generate $SBB^\delta{}_j$ from $SBB^\delta{}_i$ by applying SIO, and the probabilities $P_{SURVIVE\_C}$, $P_{SURVIVE\_M}$ for the $SBB^\delta{}_j$ surviving crossover and mutation, respectively. According to [1], in a one string proportional selection scheme, $P_{SEL\_SBB_i^\delta}$ is:

$$P_{SEL\_SBB_i^\delta}(t) = f_{SBB_i^\delta}(t) \big/ \left[N \cdot \hat{\mu}(t) \right] \tag{5}$$

To derive $P_{GENERATE}$ we should look at Fig. 2 that shows how SIO may generate SBB_j^δ from the existing SBB_i^δ by inverting genes between the unique specific loci i and n. We have:

$$P_{GENERATE} = P(\{\text{invert genes between loci } i \text{ and } n\} | \text{SIO}) P_{SIO} \tag{6}$$

where P_{SIO} is the fixed rate or probability of applying SIO. Probability $P(\{\text{invert genes between loci } i \text{ and } n\} | \text{SIO})$ is a conditional probability that given the fact that SIO actually applies on the selected string, the inversion is done between the two specific positions i and n. The total number of possible ways of combining two inverting limit positions in a string of length l (including the cases where the inverting limits are the same) is: $\binom{l}{2} + \binom{l}{1} = 2\big/\left[l(l+1)\right]$ and there is only one[2] favorable case: inverting between i and n. Therefore, we have:

$$P\big(\{\text{invert genes between loci } i \text{ and } n\} \mid \text{SIO}\big) = 2\big/\left[l(l+1)\right] \tag{7}$$

The new SBB discovered after applying an effective simple inversion should survive crossover and mutation. By the well known disruption terms in the Schema Theorem [2], we have that the probability of *not* disrupting SBB_j^δ is:

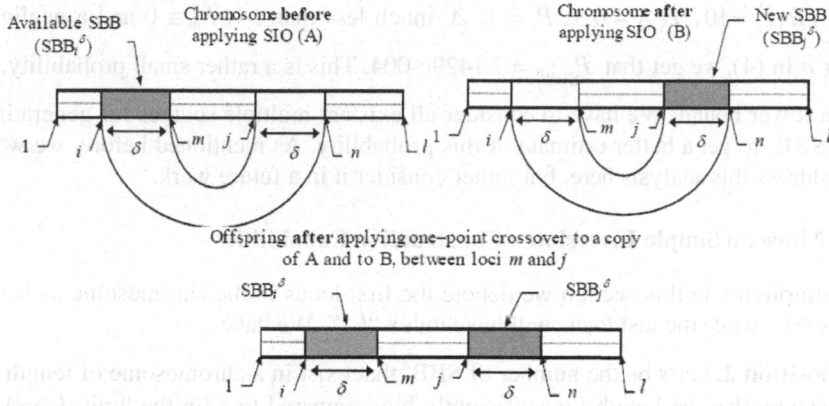

Fig. 2. The construction of a new SBB from an existing one, by applying SIO, and the resulting offspring obtained by applying one-point crossover

$$P_{\text{SURVIVE\_C}} = 1 - P_c\, \Delta/(l-1) \tag{8}$$

Similarly, the probability for SBB_j^δ surviving mutation is:

$$P_{\text{SURVIVE\_M}} = \left(1 - P_m\right)^o \tag{9}$$

By tacking the product for all right hand terms of eq. (5) – (9) we get eq. (4) which gives a lower bound for the probability $P(t)_{\text{SBB}_j^\delta}$ because we have taken into account only the constructive potential of SIO, alone. Crossover and mutation work more as disruptive operators, in the classical GA theory sense. The new SBB, discovered after SIO, has to "survive" the disruptive effect of one-point crossover and mutation. Eq. (4) may be extended by considering more than one instance of SBB_i^δ in $P(t)$ and

---

[2] In general, there is *at least* one favorable case: inverting between loci i and n. The analysis gets too complicated if we include analogous inversion events, like inversions between loci $i - r$ and $n + r$ for some integer r, which don't hold in general, for example when we consider creating $\text{SBB}_{l-\delta}$ from SBB_1. Therefore, P_{GENERATE} should be regarded as a lower bound on the respective probability.

moreover, considering a pool of different SBB that may lead to a new SBB in the next generation. In this case, however, the analysis gets more complicated as the different multiple sources of generating a new building block are not necessarily independent. For example, two SBBs placed on the same chromosome means that we have two *dependent* sources of generating a new SBB. This problem arises when convergence of the population happens, as many chromosomes have many SBB aligned. We shall address this more complicated analysis in a future work, but for now it suffices to note that according to (4), there is a nonzero probability of generating a new SBB from an existing one, SIO together with the other genetic operators increasing the good genetic material in the population. Thus, eq. (4) gives a *nonzero* lower bound on the discovery of a new SBB. To get a feeling on the order of magnitude of the probability in (4), we may consider a typical example: If we let $N = l = 20$, $f_{SBB_i^\delta}/\hat{\mu}(t) = 10$, $P_{SIO} = 0.3$, $P_c = 1$, Δ much less than l-1, $P_m \cong 0$ and a negligible order o in (4), we get that $P_{SBB_j^\delta} = 7.1429\text{e-}004$. This is a rather small probability, but it is a lower bound. We have to consider all existent multiple sources for generating a new SBB, to get a better estimate of this probability. As mentioned before, we would not address this analysis here, but rather consider it in a future work.

2.2   More on Simple Inversion's Constructive Capabilities

For simplicity, in this section we denote the first locus in the chromosome as having index "0", while the last locus will have index "l-1". We have:

Proposition 2. Let s be the number of SBB$^\delta$ that exist in a chromosome of length l. If we assume that the length l is sufficiently big compared to s (in the limit: $l \rightarrow \infty$), we have always the possibility to produce a maximum of s new SBBs through SIO.

Proof: To proceed with the proof, let us first look at Fig. 3, where a chromosome with the s SBBs, scattered along its length, is depicted. These s SBBs exist in the chromosome (hence, in the population), *before* applying SIO. The cells in the chromosome will be denoted as C. Each cell has the length of the SBB's kernel, that is δ. The index attached to each cell indicates the position in the chromosome where the cell starts.

The s SBB's kernels

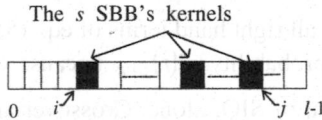

Fig. 3. The chromosome containing s SBBs. The white cells are empty cells, while the kernel of a SBB occupies the black cells

If we take k cells of length δ at the left of position i, or at the right of position j, we can generate s new SBB$^\delta$ by applying SIO between positions i-k and j, or between i and j+k, respectively, if conditions 1) and 2), respectively, are obeyed:

$$\begin{array}{c} 1)\ \text{If}\ C_{i-k+r} \in \text{SBB}^\delta \Rightarrow C_{j-r-1} \notin \text{SBB}^\delta\ \text{and reciprocal}\ 1^*), \\ 2)\ \text{If}\ C_{i+r} \in \text{SBB}^\delta \Rightarrow C_{j+k-r-1} \notin \text{SBB}^\delta\ \text{and reciprocal}\ 2^*), \\ r = 0, \left\lceil \dfrac{j-i+k+1}{2} \right\rceil \begin{array}{l} 1^*)\ \text{If}\ C_{j-r-1} \in \text{SBB}^\delta \Rightarrow C_{i-k+r} \notin \text{SBB}^\delta \\ 2^*)\ \text{If}\ C_{j+k-r-1} \in \text{SBB}^\delta \Rightarrow C_{i+r} \notin \text{SBB}^\delta \end{array} \end{array} \qquad (10)$$

In (10), non-appartenence or appartenance of a cell to a SBB, means that the respective cell is free, or occupied by a SBB's kernel, respectively. Basically, the formal definition of (10), states that we can double the number of SBBs by applying SIO, if any of the occupied cells will fall, after inversion, on an unoccupied cell. We can find the minimum k for which conditions in (10) hold, and in the limit we have $k = j-i+1$, that guarantees the number of SBBs is doubled. As $k = 0, j-i+1$, it is sufficient to take the smallest value of k for which condition 1) or 2), in (10), holds. It is obvious from the previous considerations that the most SIO can do is doubling the number of SBBs that exist in a chromosome. It is also obvious that (10) will hold much more probably in the beginning of the run, than in the end, when most of the cells are occupied by SBB's kernels.

To further investigate the cases in which SIO can double the number of available SBBs, we can proceed by denoting the *starting* position of the *first* occupied cell in a chromosome by S_1 and the *ending* position of the *last* occupied cell by S_2. The string between positions S_1 and S_2 will be called a *sequence*. The sequence contains all SBBs in the chromosome. We have the following proposition:

Proposition 3. The sequence will double[3] itself whenever one of the following events occur:

a)   Inversion between positions i and j with:

$$i = S_1 - r, r = \overline{0, S_1}, j = 2S_2 - S_1 + 1 + k\delta + r, k = 0, \left\lceil \frac{l - 2(S_2 + 1) + S_1 - r}{\delta} \right\rceil \qquad (11)$$

b)   Inversion performed when the pivot, that is in-between the two inverting points, at the middle, is chosen such that each cell to the right / left of the pivot should have its symmetrical cell free (the cell at the left / right of the pivot). The pivot, denoted as *piv*, should lie also inside[4] the sequence, between S_1 and S_2. Once we have found the pivot, one of the inverting points i_1 should be chosen such that

$$i_1 = \begin{cases} S_1, & \text{if } piv - S_1 \geq S_2 - piv \\ S_2, & \text{otherwise} \end{cases}$$ while the other inverting point i_2 should be

chosen at the symmetric position around the pivot.

---

[3] Here, the word "double" means rather that the new sequence will appeared moved in another place of the chromosome, without overlapping the old sequence. Hopefully, crossover will take two chromosomes containing these sequences and will do the proper mixing of the information, so that the sequence will appear doubled in the same offspring.

[4] If the pivot lies outside the sequence than we get a subset of the inversion events defined in a).

Condition a) states that we can choose the first inversion point i as a reference point, and move the point j in increments of δ to the right, such as the sequence would appear translated to the right. Also, i can move to the left by r positions, but then we have to translate j with r position to the right so that inversion would move the sequence at the same spot as when $r = 0$. An analogous condition holds when letting j be the reference point while moving the first inversion point i to the left.

This analysis attempted to shed some light into the constructive potential of the simple inversion operator. The constructive potential of SIO will be supported also by the experimental results obtained on several test functions, given in the next section.

3 Test Set and Results

3.1 Problems That Are SBB-Compliant

We have tested the SIO with a simple GA on several test functions that are SBB-compliant (e.g. their optimum is of the form given in (3)).

The first function R1 is the Royal Road Function [16] and it consists of 8 building blocks, each block comprising 8 "1" bits. The blocks ought to be juxtaposed onto a string of length 64. The basic building blocks are the 8 bit, all-ones, building blocks, and the problem rewards each discovery of a new basic building block with a fixed amount. Progressively bigger building blocks receive bigger rewards as they are found from the smaller basic building blocks. The optimum is a string of 64 bits, all ones, having all building blocks aligned and the optimum fitness of 256 (see [16] for details). This problem is SBB-compliant as we have, according to (3), that:

$$\delta = 8 \text{ and } A_1 \equiv A_2 \equiv .. \equiv A_8 = 11111111 \tag{12}$$

The SBB's kernels $A_1 ... A_8$ are the basic building blocks of the function.

The next test functions are implemented directly according to the definition of SBB-compliance in (3). The SBBF (Symmetrical Building Block Functions), are:

$$\text{SBBF1: } \delta = 5; l = 25; A_1 \equiv A_2 \equiv .. \equiv A_5 = 01\#10 \tag{13}$$

$$\text{SBBF2: } \delta = 5; l = 25; A_1 \equiv A_2 \equiv .. \equiv A_5 = 1\#\#\#1 \tag{14}$$

Fitness is computed by summing the number of bits that are correctly matched by the respective optimal pattern x_{opt} in (3) and (13), for SBBF1, and in (3) and (14), for SBBF2, respectively. The optimum fitness for SBBF1 is 20, while for SBBF2 is 10. We chose these two functions to differ in the order o of the SBBs, as it would be instructive to compare how SIO works on similar problems having different levels of specification for their building blocks.

The fourth problem is not apparent to be SBB-compliant. SBB-compliance becomes apparent only when a certain coding of the chromosome is used. The function is defined as follows:

$$f(\mathbf{x}) = 1 \bigg/ \sum_{i=1}^{6} (x_i - 27)^2 + 1 \tag{15}$$

with $\mathbf{x} = [x_1 \ x_2 \ x_3 \ x_4 \ x_5 \ x_6]$ an array of 6 variables x_i each coded with 5 bits, with $0 \le x_i \le 31$ for $i \in \{1,...,6\}$. The function in (15) will be denoted as FUNC, for brevity, and the optimum is the string $x_{opt} = [27\ 27\ 27\ 27\ 27\ 27]$. Taking into account this particular coding we see that this string is $x_{opt} = A_1 A_2 A_3 A_4 A_5 A_6$ with $A_1 = A_2 = A_3 = A_4 = A_5 = A_6 = 1\ 1\ 0\ 1\ 1$. From (3) FUNC is SBB-compliant.

The fifth and sixth functions (called Royal Road function Level 3 and 4, respectively), are the *revised* versions of the Royal Road functions defined by Holland [17]. The difference from the original version of R1 in [16] lies in adding some non-intervening bits between the basic building blocks. Again, the basic building blocks are groups of 8 bits "1", but these groups are alternating with groups of 7 non-intervening bits (or "don't care" bits equivalently). The number of basic blocks is 16, but adding the 16 groups of 7 "don't care" bits, the total length of the chromosome will be $l = 240$. The fitness is calculated similar to R1, in two steps called PART calculation, which rewards only the basic, 8 bit building blocks, and BONUS, respectively, which rewards combinations of building blocks into higher order blocks. Level 3 has an optimum fitness of 11.8, while Level 4 has an optimum fitness of 12.8, the difference being that in the Level 4 the complete 16 basic building block found, is rewarded additionally. Details regarding the rewarding mechanism are given in [17]. These two functions are quite interesting for our purpose, as they are only *partly* SBB-compliant, in that not all bits in the optimum chromosome can be put in the form x_{opt} in (3). We can imagine having different types of SBBs for this problem, such as: $1_8 \#_7 1_8$, $\#_7 1_8 \#_7$ or 1_8, where the symbol a_b denotes a string of b values equal to a. Any of these SBBs, cannot cover completely the string of length 240, so there are "don't care" bits that are left over from the symmetry pattern defined in (3). We can even imagine having two different types of SBBs, in the same optimal chromosome, like: $1_8 \#_7 1_8$ alternating with $\#_7 1_8 \#_7$. Apart from being a more complicated variant of the original Royal Road functions, Level 3 and 4 pose an interesting challenge for our experimental setup: SIO may deal with problems that are not perfectly SBB-compliant? According to experimental results obtained the answer is affirmative.

The last two functions tested are two versions of the 4-order fully deceptive problem defined in [18]. The 4-order fully deceptive problem is defined as follows:

$$f(\mathbf{x}) = \sum_{i=1}^{M} f_{4_i}(x_{I_i}) \text{ with } f_4(u) = \begin{cases} -u+3, \text{for } 0 \le u \le 3 \\ 4u-12, \text{for } 3 < u \le 4 \end{cases} \text{ and index set:} \tag{16}$$

$$I_1 = \{1,2,3,4\}, I_{i+1} = I_i + 4$$

We employ two versions, namely DF1 (Deceptive Function 1) with $M = 10$ subfunctions of unitation, and DF2 (Deceptive Function 2) with $M = 100$. As $f_4(u)$ is based on unitation, with the optimal solution - a string having all the bits equal to 1, this problem is also SBB-compliant with: $\delta = 4$, $l = M \cdot \delta$, and $x_{opt} = A_1 A_2 \ldots A_M$ with $A_1 = A_2 = \ldots = A_M = 1\ 1\ 1\ 1\ 1$. Each SBB kernel $A_1 - A_M$ corresponds to the subgroup of 4 bits x_{l_i} in (16). We have chosen the original "natural" representation of the problem, which means that the linkage between genes is maximum: the optimal blocks are the groups of 4 consecutive bits "1". This representation is opposed to the "ugly" deceptive representation given in [19], where genes belonging to the same subgroup of 4 bits are scattered apart, along the chromosome.

3.2 Experimental Results

In this subsection we present and discuss the results obtained by testing SIO with a GA (SIOGA), in comparison to a simple GA (SGA). The parameters of the strategies were tuned to achieve a high convergence rate and a low average of Number of Function Evaluations (NFE) for SIOGA. The goal was to show that applying SIO yields a good convergence rate with a quite low average of NFE, and that SGA (that is SIOGA with SIO switched off) is incapable of finding the optimum with the same rate and for the same, low, threshold of NFE.

3.2.1 Results for the Royal-Road Function

For R1, we compared SIOGA to SGA, both strategies having the following common parameters: one-point crossover with rate $P_c = 0.7$, bit-flip mutation with rate $P_m = 0.005$, population size set according to [16] as $N = 128$, proportional selection with sigma-truncation. For SIOGA, the rate of applying SIO is $P_{SIO} = 0.3$. We did 50 independent runs for each strategy the results (mean number of generations to optimum $\overline{N}_{opt}$ and its respective Standard Error of the Mean: SEM(N_{opt})) being tabulated in Table 1. For the R1 function, results in Table 1 indicate that applying SIO gives an average speedup more than one order of magnitude bigger when compared to the same algorithm but without SIO. Our result (mean NFE = 3752.9) is again, nearly more than one order of magnitude better than the best result obtained by Mitchell in [16], that is: mean NFE = 37453.

Table 1. Results for R1 function over 50 independent runs; in parentheses we give the NFE

$\overline{N}_{opt}$ SGA	SEM(N_{opt}) SGA	$\overline{N}_{opt}$ SIOGA	SEM(N_{opt}) SIOGA
351 (44928)	7.97 (1020)	29.32 (3752.9)	147.6 (18892)

3.2.2 Results for SBBF1 and SBBF2

For the SBBF1 and SBBF2 we did 30 independent runs for each strategy. Each strategy was allowed to run for 100 generations. Proportional selection (no scaling) was employed. The common parameters for both strategies are: $P_c = 0.7$, $P_m = 0$, and for SIOGA alone: $P_{SIO} = 0.3$. The size of the population is $N = 20$ for SBBF1 and $N = 10$, for SBBF2. In Table 2 we gave the Number of Convergent Runs (NCR) to optimum with average $\overline{N}_{opt}$, as well as the number of runs that Do Not Converge to Optimum (DNCO) in 100 generations. We also gave the number of runs that get Stuck in a SubOptimal point (SSO). This case has been frequently encountered (note that $P_m = 0$). From Table 2 SIOGA outperformed SGA, for both functions, the convergent rate being 100%, while for SGA on SBBF1 the convergence rate is 0%,

Table 2. Results for SBBF1 and SBBF2 over 30 independent runs

Strategy	NCR/ $\overline{N}_{opt}$	No. DNCO	No. SSO
	SBBF1		
SGA	0 / –	3	27
SIOGA	30 / 54.2	0	0
	SBBF2		
SGA	1 / 13	0	29
SIOGA	30 / 29.56	0	0

Table 3. Results for FUNC over 30 independent runs

	SGA	SIOGA
CR [%]	0	73.33

and for SBBF2 convergence rate is only 3.3%. The results confirm the intuition that SBBF2 should be simpler to solve than SBBF1, as it has more "don't care" values in its definition (see eq. (14) compared to (13)). Indeed, on the simpler function SBBF2, SGA starts to find the optimum in the given time, while SIOGA finds the optimum harder on SBBF1 (mean NFE=54.2) than on SBBF1 (mean NFE=29.56).

3.2.3 Results for FUNC

For the fourth problem, that is function FUNC in (15) the following parameters are common to both strategies: population size N = 50, proportional selection with sigma-truncation, P_c = 0.75, P_m = 0.005, and for SIOGA alone: P_{SIO} = 0.35. Maximum number of generations was 100. Table 3 gives the percentage of Convergent Runs: CR [%], from a total of 30 independent runs for each strategy. On FUNC, SIOGA still beats the algorithm without SIO which has a convergence rate of 0%, but the convergence rate for SIOGA is less than 100%, namely 73.3%.

3.2.4 Results for Royal Road Level 3 and Level 4

For these functions, we have the following common parameters: binary tournament selection plus k-elitism with k = 5, (see [20] for explanation on k-elitism), one-point crossover with rate P_c = 0.9, bit flip mutation with rate P_m = 0.005, and SIO with rate P_{SIO} = 0.4. The maximum number of generation was set to 250, and we compared our results with results found in literature and obtained with other variants of GA. For Level 3 the population comprises N = 50 individuals. We performed 60 independent runs with SIOGA. For Level 4 the population is N = 80 individuals. Results are given in Table 4, in terms of minimum, average, maximum, standard error of the mean for the Number of Function Evaluations (NFE) to optimum, as well as the percentage of Convergent Runs: CR [%]. For the Royal Road functions SIOGA outperforms other variants of GAs that obtained good results on this problem. For the most difficult variant (Level 4) the mean NFE to optimum (9479.2) is below Holland's challenging threshold, namely 10000 NFE [21], which means a very good result achieved by SIOGA. Also, taking into account the minimum NFE (i.e. 3440), and the relatively small standard error of the mean (i.e. 3508), the superiority of SIOGA becomes apparent. We compared the results with other reported good results on Royal Road, such as Culbertson's Genetic Invariance GA (GIGA), described in [22], and the Selfish Gene (SG) algorithm [23], based on Richard Dawkin's "selfish gene" concept. As it may be seen from Fig. 4, SIOGA outperforms by far the other strategies. Note that we used a logarithmic scale to display the Average NFE in Fig. 4.

Table 4. Results for Royal Road Level 3 and 4 over 60 independent runs

Royal Road	Min NFE	Avg NFE	Max NFE	SEM NFE	CR [%]
Level 3	2550	6912.8	12400	2496.9	81.67
Level 4	3440	9479.2	18400	3508.7	85

3.2.5 Results for the 4-Order Fully Deceptive Problems

For the last two deceptive functions results are given in Table 5 and Table 6. Both strategies have the following common parameters: types of selection, crossover and mutation are the same as for the Royal Road functions (Level 3 and 4), $P_c = 0.8$, $P_m = 0.01$, $P_{SIO} = 0.6$. For $M = 10$ the maximum number of generations was 150, population size $N = 50$ and the number of independent runs was 20. For $M = 100$ the maximum number of generation was 15000, the population size $N = 100$, and the, number of independent runs: 10. In Tables 5 and 6 results in terms of Number of Function Evaluations (NFE) were given. If the strategies failed to reach the global optimum in the allowed running time, the fitness of best solution, in the respective run, was given in parentheses. For the 4-order fully deceptive problem, SIOGA achieved again, much

Table 5. Results for 4-order fully deceptive function ($M = 10$); in parentheses, fitness of the best solution per run was given, whenever optimum was not found

SIOGA		SGA		SIOGA		SGA	
#	NFE	#	NFE	#	NFE	#	NFE
1	1000	1	>7500(38)	11	1100	11	>7500(36)
2	2550	2	>7500(38) (38)	12	1600	12	>7500(36)
3	1850	3	>7500(35)	13	2950	13	>7500(32)
4	3250	4	>7500(37)	14	1700	14	>7500(34)
5	2800	5	>7500(35)	15	4100	15	>7500(37)
6	1650	6	>7500(36)	16	2100	16	>7500(35)
7	2100	7	>7500(36)	17	1200	17	>7500(35)
8	800	8	>7500(36)	18	1600	18	>7500(37)
9	1650	9	>7500(35)	19	2050	19	>7500(35)
10	1750	10	>7500(38)	20	850	20	>7500(36)

Table 6. Results for 4-order fully deceptive function ($M = 100$); in parentheses, fitness of the best solution per run was given, whenever optimum was not found

SIOGA		SGA	
#	NFE	#	NFE
1	7123	1	> 15000 (346)
2	6423	2	> 15000 (340)
3	6366	3	> 15000 (339)
4	11215	4	> 15000 (341)
5	12342	5	> 15000 (329)
6	7239	6	> 15000 (334)
7	5624	7	> 15000 (335)
8	> 15000 (396)	8	> 15000 (332)
9	12458	9	> 15000 (341)
10	> 15000 (399)	10	> 15000 (339)

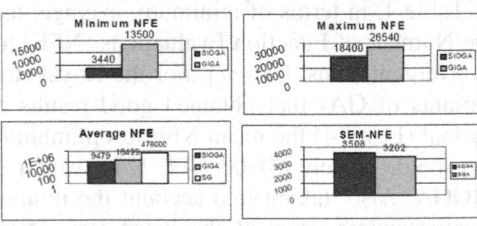

Fig. 4. Comparison for Royal Road Level 4 between SIOGA, GIGA and SG

better results than SGA. We performed only 10 runs for the case when $M = 100$, due to excessive computational cost of each run, and we haven't given any statistics for such a small sample. However, from Table 5, and 6 it is clear that SIOGA is able to find the optimum for the small NFE threshold imposed, while SGA achieves 0% rate convergence at the same threshold.

4 Conclusions

We have shown both theoretically and experimentally, that a GA with SIO performs better than a SGA on problems consistent with definition (3). One cannot know in advance if a problem is consistent or not with (3), but we argue that it is worthwhile applying SIO, because x_{opt} in (3) may be viewed not just as the optimal *string*, but as the optimal *schema* of the respective problem. Therefore, a large number of optimization problems may have their optimum "matched" by a specific x_{opt} and so, applying SIO on those problems, would make sense. The sheer efficiency of SIO stems from the way it operates on SBBs. We experimentally shown that SIOGA performs very good, even on problems where the optimum consists of a partial sequence of identical SBB or the optimum does not have a perfect symmetry, and this was the case of the Royal Roads Level 3 and 4. We have also shown how coding can turn a problem that was apparently not consistent with the SBB paradigm (i.e., the FUNC function), into a problem that was perfectly SBB-compliant. For future work we will concentrate on extending our static analysis of SIO, as well as on analyzing SIOGA's dynamics with Markov chains. On the experimental level we will check the capabilities of SIOGA on real-world problems.

Acknowledgments. This work was supported in part by the Portuguese Ministry of Science & Technology – "Fundação para a Ciência e a Tecnologia" through Grant POCTI - SFRH/BPD/30347/2006.

References

1. Holland, J.: Adaptation in Natural and Artificial Systems. MIT Press, Cambridge (1992)
2. Goldberg, D.E.: Genetic Algorithms in Search, Optimization and Machine Learning. Addison-Wesley, Reading (1989)
3. Whitley, D.: Using Reproductive Evaluations to Improve Genetic Search and Heuristic Discovery. In: Genetic Algorithms and Their Applications - Proceedings of the Second ICGA, pp. 108–115. Lawrence Erlbaum Assoc., Mahwah (1987)
4. Wienholt, W.: A Refined Genetic Algorithm for Parameter Optimization Problems. In: Proceedings of the Fifth ICGA, pp. 589–596. Morgan Kaufmann, San Francisco (1993)
5. Syswerda, G.: Uniform Crossover in genetic Algorithms. In: Proceedings of the Third ICGA, pp. 2–9. Morgan Kaufmann, San Francisco (1989)
6. Sirag, D.J., Weisser, P.T.: Toward a Unified Thermodynamic Genetic Operator. In: Genetic Algorithms and Their Applications - Proceedings of the Second ICGA, pp. 116–122. Lawrence Erlbaum Associates Publ., Mahwah (1987)
7. Goldberg, D.E.: Don't Worry, Be Messy. In: Proceedings of the Fourth International Conference on Genetic Algorithms, pp. 24–30. Morgan Kaufmann, San Francisco (1991)

8. Harik, G., Goldberg, D.E.: Learning linkage. In: Foundations of Genetic Algorithms, vol. 4, pp. 247–262. Morgan Kaufmann, San Francisco (1996)
9. Lobo, F.G., Deb, K., Goldberg, D.E., Harik, G., Wang, L.: Compressed introns in linkage learning genetic algorithm. In: Proceedings of the Third Annual Conference on Genetic Programming, pp. 551–558. Morgan Kaufmann, San Francisco (1998)
10. Salman, A., Mehrotra, K., Mohan, C.K.: Linkage Crossover for Genetic Algorithms. In: Proceedings of GECOO 1999, pp. 564–571. Morgan Kaufmann, San Francisco (1999)
11. Paredis, J.: The Symbiotic Evolution of Solutions and their Representations. In: Proceedings of the Sixth ICGA, pp. 359–365. Morgan Kaufmann, San Francisco (1995)
12. Munteanu, C., Lazarescu, V.: Simple Inversion Genetic Algorithm an Effective Genetic Strategy. In: Proceedings of ICSC NC 1998, pp. 151–156 (1998)
13. Munteanu, C., Rosa, A.: Symmetrical Building Blocks and the Simple Inversion Operator. In: Proceedings of GECCO 2000, p. 365. Morgan Kaufmann Publ., San Francisco (2000)
14. Van Hoyweghen, C., Naudts, B., Goldberg, D.E.: Spin-flip symmetry and synchronization. Evol. Comput. 10, 317–344 (2002)
15. Choi, S.S., Kwon, Y.-K., Moon, B.-R.: Properties of symmetric fitness functions. In: Proceedings of GECCO 2006, pp. 1117–1124. ACM Press, New York (2006)
16. Mitchell, M., Forrest, S., Holland, J.: The Royal Road for genetic Algorithms: Fitness Landscapes and GA Performance. In: Proceedings of the First European Conference on Artificial Life, pp. 245–254. MIT Press, Cambridge (1991)
17. Jones, T.: Evolutionary Algorithms, Fitness Landscapes and Search. PhD thesis, University of New Mexico, Albuquerque NM (1995)
18. Goldberg, D.E., Deb, K., Clark, J.H.: Genetic algorithms, noise, and the sizing of populations. Complex Systems 6, 333–362 (1992)
19. Deb, K., Goldberg, D.E.: Sufficient conditions for deceptive and easy binary functions. Annals of mathematics and Artificial Intelligence 10, 385–408 (1994)
20. Bäck, T., Hoffmeister, F.: Extended Selection Mechanisms in Genetic Algorithms. In: Proceedings of the Fourth ICGA, pp. 92–99. Morgan Kaufmann, San Francisco (1991)
21. Holland, J.: Royal Road Functions. Genetic Algorithms Digest 7(22) (1993)
22. Culberson, J.: Holland's Royal Road and GIGA. GA Digest 7(23) (1993)
23. Corno, F., Sonza Reorda, M., Squillero, G.: Optimizing Deceptive Functions with the SG-Clans Algorithms. In: Proceedings of IEEE CEC 1999, IEEE Press, Los Alamitos (1999)

A Genetic Programming Approach to the Generation of Hyper-Heuristics for the Uncapacitated Examination Timetabling Problem

Nelishia Pillay[1] and Wolfgang Banzhaf[2]

[1] School of Computer Science, Univesity of KwaZulu-Natal, Pietermaritzburg Campus,
Pietermaritzburg, KwaZulu-Natal, South Africa
pillayn32@ukzn.ac.za
[2] Department of Computer Science, Memorial University of Newfoundland, St. John's, NL
A1B 3X5, Canada
banzhaf@cs.mun.ca

Abstract. Research in the field of examination timetabling has developed in two directions. The first looks at applying various methodologies to induce examination timetables. The second takes an indirect approach to the problem and examines the generation of heuristics or combinations of heuristics, i.e. hyper-heuristics, to be used in the construction of examination timetables. The study presented in this paper focuses on the latter area. This paper presents a first attempt at using genetic programming for the evolution of hyper-heuristics for the uncapacitated examination timetabling problem. The system has been tested on 9 benchmark examination timetabling problems. Clash-free timetables were found for all 9 nine problems. Furthermore, the performance of the genetic programming system is comparable to, and in a number of cases has produced better quality timetables, than other search algorithms used to evolve hyper-heuristics for this set of problems.

Keywords: hyper-heuristics, genetic programming, examination timetabling.

1 Introduction

Research applying evolutionary algorithms to the domain of examination timetabling has generally focused on using an evolutionary algorithm to evolve a timetable that meets the hard constraints and minimizes the soft constraints of a specific examination timetabling problem. Ross et al. [15] and Burke et al. [5] suggest that evolutionary algorithms would be more effective at inducing hyper-heuristics that can be used to construct the timetable rather than evolving the actual timetable. The main contribution of this paper is an evaluation of genetic programming as a means of evolving hyper-heuristics for the uncapacitated examination timetabling problem (ETP). This study takes a similar approach to that employed by Cowling et al. [10], in that the hyper-heuristics evolved by the GP system are in the form of a sequence of low-level heuristics. In the study presented in this paper each heuristic sequence is composed of one or more of the following low-level heuristics: largest degree, largest

J. Neves, M. Santos, and J. Machado (Eds.): EPIA 2007, LNAI 4874, pp. 223–234, 2007.

weighted degree, largest enrollment, saturation degree and highest cost. During the timetable construction process, each heuristic is used to schedule n examinations, where n is the total number of examinations divided by the number of heuristics in the sequence. The GP system was used to generate hyper-heuristics for 9 benchmark problems. The system produced feasible timetables for all 9 problems. Furthermore, the quality of the timetables produced by the system was comparative to, and in a number cases better than, those induced by other hyper-heuristic systems.

The following section provides an overview of the examination timetabling problem and previous work investigating the generation of hyper-heuristics for the uncapacitated examination timetabling problem. Section **3** proposes a GP system for the induction of hyper-heuristics for the uncapacitated ETP. The overall methodology employed in the study is presented in section **4** and the performance of the GP-based hyper-heuristic system on the 9 benchmarks is discussed in section **5**. Section **6** summarizes the findings of the study and describes future extensions of the project.

2 The Uncapacitated Examination Timetabling Problem (ETP) and Hyper-Heuristics

This section provides an overview of the uncapacitated examination timetabling problem and previous work investigating the generation of hyper-heuristics for this domain.

2.1 The Uncapacitated Examination Timetabling Problem (ETP)

The ETP requires n examinations to be scheduled in m timeslots so as to satisfy the hard constraints of the problem and minimize the soft constraints violated [14]. The hard constraints of a problem must be met in order for the timetable to be feasible. The following are examples of hard constraints:

• There must be no clashes, i.e. two students cannot be scheduled to write two or more examinations during the same period.
• Room capacities for each sitting must not be exceeded.

Soft constraints are often contradictory and we aim to minimize the number of soft constraints violated. For example:

• Examinations are well-spaced for each group of students.
• Scheduling examinations with large enrollments earlier in the examination period.

The hard and soft constraints usually differ for each institution. The uncapacitated version of the problem does not require room capacities to be catered for. The following section describes previous studies researching the induction of hyper-heuristics for the uncapacitated ETP.

2.2 Hyper-Heuristics and the ETP

Hyper-heuristics are heuristics that are used to choose one or more low-level heuristics for a particular problem ([5], [15] and [16]). These selections do not rely on domain knowledge and hence provide a more general solution to the ETP. The first attempt at using evolutionary algorithms to evolve instructions for constructing an examination timetable rather than inducing the actual timetable was the study conducted by Terashima-Marin et al. [17] to generate solutions to the capacitated ETP. Each element of the population is a combination of one of three timetable construction strategies, condition/s indicating when to change strategy as well as heuristics for deciding which examination to allocate next and which timeslot to assign an examination to. This section provides an overview of previous work investigating the generation of hyper-heuristics for the uncapacitated examination timetabling problem. The studies described are those that are most relevant to that presented in the paper, i.e. the methodologies employed generate combinations of low-level heuristics and have been tested on the same set of benchmarks.

Asmuni et al. [2] have used a fuzzy expert system to induce hyper-heuristics for the uncapacitated ETP. The hyper-heuristic is in the form of a fuzzy weight combining two of the following low-level heuristics: largest degree, largest enrollment and largest saturation degree. An exhaustive search is used to fine tune the fuzzy terms. The examinations are sorted in descending order according to their fuzzy weight and scheduled in this order. Each examination is allocated to the minimum penalty slot. In the case of clashes examinations are de-allocated and re-scheduled so as to remove the clash.

Qu et al. [13] use a variable neighborhood search (VNS) to search the space of hyper-heuristics for the uncapacitated ETP. The hyper-heuristic output by the VNS is used to construct the timetable. Each hyper-heuristic is a sequence of two or more low-level heuristics (color degree, largest degree, largest enrollment, largest weighted degree, saturation degree and random ordering). Each low-level heuristic in the sequence is used to decide which examination should be scheduled next. A heuristic maybe used to schedule one or more examinations. In each case the examination is allocated to the minimum penalty slot. During the search process the VNS uses one of two neighborhood sets, namely, VNS1 and VNS2. VNS1 randomly changes 2 to 5 heuristics in a sequence, whereas VNS2 randomly changes 2 to 5 heuristics in a subsequence.

Kendall et al. [11] and Burke et al. [7] have implemented Tabu searches to generate hyper-heuristics for the uncapacitated ETP. In the study conducted by Kendall et al. the process begins by creating an initial solution by allocating examinations according to the largest degree or saturation degree heuristics. This initial solution may not be complete in that all examinations may not be scheduled. The neighborhood of the initial solution is than examined so as to improve the timetable. The Tabu search is used to determine which low-level heuristic to apply next. One of four types of low-level heuristics can be applied, namely, heuristics to select and schedule an exam; heuristics to move an exam; a heuristic to swap exams and a heuristic to remove an exam. The Tabu inactive heuristic that produces the best improvement is applied next. The refinement process continues until either a time limit has been exceeded or there are no more improvements.

The Tabu search employed by Burke et al. [7] is used to search the hyper-heuristic search space in order to identify a heuristic sequence that produces the best quality timetable. Each list is comprised of two or more of the following low level heuristics: least saturation degree, largest colour degree, largest degree, largest weighted degree, largest enrollment and random ordering. Each heuristic in the list is used to schedule two examinations. The initial heuristic list for all experiments is composed of only the saturation degree heuristic.

The following section presents a GP system for the generation of hyper-heuristics. Section 5 compares the performance of the GP-based hyper-heuristic system and the hyper-heuristic systems described in this section on a set of 9 benchmark problems.

3   Evolving Hyper-Heuristics

This section describes the GP system that has been implemented to evolve hyper-heuristics for the uncapacitated ETP. Genetic programming systems generally produce a program which when executed provides a solution to the problem at hand ([3] and [12]). In this study the program is a sequence of low-level heuristics which specify the order in which examinations should be scheduled when constructing an examination timetable. The GP system uses the generational control model and a run is terminated once the maximum number of generations has been reached. The hyper-heuristic that has produced the timetable with lowest hard constraint and soft constraint costs during the run is returned as the solution.

3.1   Representation and Initial Population Generation

Each element of the population is a string of variable length composed of characters representing one of the following low-level heuristics:

- Largest degree (l) – The examination with the largest number of conflicts is scheduled first.
- Largest enrollment (e) – The examination with the largest student enrollment is scheduled first.
- Largest weighted degree (w) – The examination with the largest number of students involved in clashes is scheduled first.
- Saturation degree (s) – The examination with the least number of feasible (i.e. will not result in a clash) timeslot options is scheduled first.
- Highest cost (h) - The examination with the highest proximity cost is scheduled first. The pseudo-code of the function for calculating the proximity cost is listed in **Figure 1** and the weight function used in calculating the proximity cost is defined in **Figure 2**.

An example of a hyper-heuristic is *hsseel*. The hyper-heuristic strings are randomly created during initial population generation. A limit is set on the maximum length of each hyper-heuristic.

```
function calc_cost( exam e, period p, total number of students n)
begin
   cost = 0
   for each exam e_j other than e
   begin
      if(e_j has students in common with e and e_j has already been scheduled)
      begin
         dist = the absolute value of the distance between p and the period e_j
                has been allocated to
         ecost = weight(dist) * the number of students common to both exams
         cost = cost + ecost
      endif
   endfor
   return cost/n
end
```

Fig. 1. Pseudo-code for the proximity cost

```
function weight (dist d)
begin
   case of d
      1: return 16
      2: return 8
      3: return 4
      4: return 2
      5: return 1
      default : return 0
   endcase
end
```

Fig. 2. Weight function

3.2 Evaluation and Selection

Each element of the population is evaluated by using the hyper-heuristic to construct a timetable. Each low-level heuristic in the hyper-heuristic is used to schedule n examinations, where n is the total number of examinations divided by the length of the hyper-heuristic. Each examination is allocated to the minimum penalty slot, i.e. the slot that does not cause a clash and results in the lowest proximity cost. The raw fitness of a hyper-heuristic is calculated by applying equation (1) to the timetable constructed using the hyper-heuristic:

$$(number\_of\_clashes+1)*proximity\_cost \tag{1}$$

The proximity cost is a measure of how well the examinations are spaced and is calculated using equation (2) in section **4**. Tournament selection is used to choose the parents of the next generation.

3.3 Genetic Operators

The mutation and crossover operators are used to create the next generation. A limit is not set on the length of the offspring produced by the genetic operators. The mutation operator changes a randomly chosen heuristic. This process is illustrated in **Figure 3**.

---

1. Randomly choose a heuristic to change in the parent *hssess*

2. Replace the chosen heuristic with a randomly selected heuristic (from *h, s, l, e, w*)

   wssess

Fig. 3. The mutation process

The crossover operator is depicted in **Figure 4**.

---

1.   Randomly choose crossover points in both the parents

      P$_1$: *lwesh*   P$_2$: *sshhle*

2.   Swap the fragments at the crossover points to create offspring

      O$_1$: *lwshhle*   O$_2$: *sesh*

3.   Return the fitter of the two offspring

Fig. 4. The crossover process

The crossover operator randomly selects crossover points in each of the chosen parents. Two offspring are created by swapping the fragments at the crossover points. The fitter of the two offspring is returned as the result of the operation.

4 Experimental Setup

The GP-based hyper-heuristic system was tested on 9 of the Carter benchmarks [9]. These benchmarks are data sets for real-world exam timetabling problems from various universities and high schools. The characteristics of the data sets used are listed in **Table 1**. The density of the conflict matrix is to some extent an indication of the problem difficulty and is calculated to be the ratio of the number of examinations involved in clashes and the total number of exams.

The hard constraint for this problem is that no student must be scheduled to write two or more examinations at the same time, i.e. there must be no clashes. The soft

Table 1. Carter benchmarks

Data Set	Institution	Periods	No. of Exams	No. of Students	Density of Conflict Matrix
ear-f-83 I	Earl Haig Collegiate Institute, Toronto	24	190	1125	0.27
hec-s-92 I	Ecole des Hautes Etudes Commerciales, Montreal	18	81	2823	0.42
kfu-s-93	King Fahd University of Petroleum and Minerals, Dharan	20	461	5349	0.06
lse-f-91	London School of Economics	18	381	2726	0.06
rye-s-93	Ryerson University, Toronto	23	486	11483	0.08
sta-f-83 I	St Andrew's Junior High School, Toronto	13	139	611	0.14
tre-s-92	Trent University, Peterborough, Ontario	23	261	4360	0.18
ute-s-92	Faculty of Engineering, University of Toronto	10	184	2749	0.08
yor-f-83 I	York Mills Collegiate Institute, Toronto	21	181	941	0.29

constraint requires the examinations to be well-spaced. The soft constraint cost is referred to as the proximity cost and is calculated using the following equation:

$$\frac{\sum w(|e_i - e_j|)N_{ij}}{S} \tag{2}$$

where:

1) $|e_i - e_j|$ is the distance between the periods of each pair of examinations (e_i, e_j) with common students.
2) N_{ij} is the number of students common to both examinations.
3) S is the total number of students
4) $w(1) = 16$, $w(2) = 8$, $w(3) = 4$, $w(4) = 2$ and $w(5) = 1$, i.e. the smaller the distance between periods the higher the weight allocated.

Note that this equation calculates the proximity cost of a complete timetable whereas the pseudo-code in **Figure 1** calculates the cost of scheduling a particular examination in a partially constructed timetable, given the examinations that have already been allocated to a timeslot at that point. For problems with different hard and soft constraints the function for calculating the raw fitness will be different from that defined in section **3.2**.

The GP parameter values are listed in **Table 2**. These values have been obtained empirically by performing test runs for each of the 9 data sets.

Table 2. GP Parameters

Population size	500
No. of generations	50
Maximum initial length	5
Tournament size	10
Mutation rate	40%
Crossover rate	60%

The system was implemented in Java using JDK1.4.2 and simulations were run on a Windows XP machine with an Intel Pentium M with 512 MB of RAM.

5 Results and Discussion

Due to the randomness associated with GP and hence the possibility of selection noise[1] ten runs were performed for each data set. The duration of a run ranges from about 15 minutes for the smaller data sets to about 4 hours for the larger data sets. The system evolved feasible timetables for all 9 problems. **Table 3** lists the average proximity costs and the best individual and its proximity cost for each of the data sets. The timetables constructed for each data set using the best hyper-heuristic obtained can be found at http://saturn.cs.unp.ac.za/~nelishiap/et/hyper_heuristics.htm.

The saturation degree (s) and the highest cost (h) heuristics occur most frequently in the best hyper-heuristics found for each of the data sets. **Figure 5** illustrates the distribution of calls to low-level heuristics in the best hyper-heuristic found for all data sets. Note that the saturation degree, highest cost and largest degree heuristics are invoked in the best hyper-heuristics for almost all the data sets with the highest cost and saturation degree occurring most frequently. The saturation degree heuristic appears to speed up the construction of clash-free timetables while the highest cost heuristic reduces the soft constraint cost.

We compare the performance of the GP system to other methodologies applied to generating hyper-heuristics for the uncapacitated ETP. **Table 4** lists the performance of the GP system and other methods employed to generate hyper-heuristics (details of

Table 3. Performance of the GP system on the Carter benchmarks

Data Set	Average Proximity Cost	Best Proximity Cost	Best Hyper-Heuristic
ear-f-83 I	36.94	36.74	hsshsssshslssh
hec-s-92 I	11.64	11.55	hhlssll
kfu-s-93	14.25	14.22	hhslsssssll
lse-f-91	10.97	10.90	hsshshshshhhshhshhshhshhhllhhhww
rye-s-93	9.39	9.35	hssshshshees
sta-f-83 I	158.35	158.22	hsehshehshhhhshhshwhswl
tre-s-92	8.51	8.48	hssshshshees
ute-s-92	27.61	26.65	hhssshsslessshssslshshsssseslllsswssh
yor-f-83 I	41.82	41.57	ssshh

---

[1] Please note that the reason for performing more than one run is not to show statistical significance.

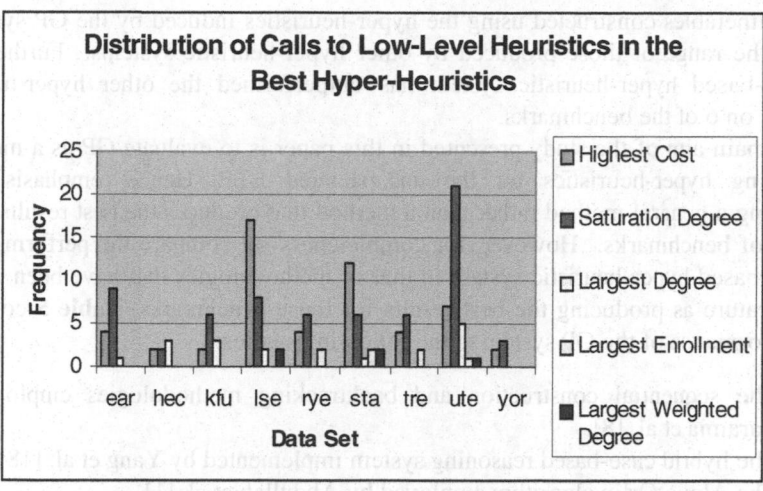

Fig. 5. Distribution of calls to low-level heuristics

these systems are provided in section **2.2**) for the same set of benchmarks. The best soft constraint cost obtained for each of the data sets is highlighted. These methodologies are:

- FES – The fuzzy expert system implemented by Asmuni et al. [2].
- VNS – The variable neighborhood search implemented by Qu et al. [13].
- TS1 – The Tabu search applied by Kendall et al. [11]. In this study each simulation was run for 4 hours.
- TS2 – The Tabu search implemented by Burke et al. [7].

Note that the methodologies that the GP system is being compared to employ very different search mechanisms from that used by the system and a direct comparison of the parameters used is therefore not feasible.

It is evident from **Table 4** that the performance of genetic programming is comparable to the other search methods used to generate hyper-heuristics. The quality

Table 4. The performance of the GP system and other methodologies used to induce hyper-heuristics for the uncapacitated ETP

Data Set	GP	FES	VNS	TS1	TS2
ear-f-83 I	**36.74**	37.02	37.29	40.18	38.19
hec-s-92 I	**11.55**	11.78	12.23	11.86	12.72
kfu-s-93	**14.22**	15.81	15.11	15.84	15.76
lse-f-91	**10.90**	12.09	12.71	-	13.15
rye-s-93	**9.35**	10.35	-	-	-
sta-f-83 I	158.22	160.42	**139.3**	157.38	141.08
tre-s-92	8.48	8.67	8.67	**8.39**	8.85
ute-s-92	**26.65**	27.78	29.68	27.60	31.65
yor-f-83 I	41.57	40.66	43.0	-	**40.13**

of the timetables constructed using the hyper-heuristics induced by the GP system is within the range of those produced by other hyper-heuristic systems. Furthermore, the GP-based hyper-heuristic system has outperformed the other hyper-heuristic systems on 6 of the benchmarks.

The main aim of the study presented in this paper is to evaluate GP as a means of generating hyper-heuristics for the uncapacitated ETP. Hence, emphasis is on producing a general method rather than a method that produces the best results on the the set of benchmarks. However, for completeness we compare the performance of the GP-based hyper-heuristic system to that of methodologies that have been cited in the literature as producing the best results for these benchmarks. **Table 5** compares the performance of the GP system to the following studies:

- The sequential construction and backtracking methodologies employed by Caramia et al. [8].
- The hybrid case-based reasoning system implemented by Yang et al. [18].
- The Ahuja-Orlin algorithm employed by Abdullah et al. [1].
- The Flex-Deluge algorithm implemented by Burke et al. [6].

Note that the system presented in this paper only performs the construction phase for timetable induction while the other methods listed in **Table 5** also include an improvement phase aimed at reducing the proximity cost of feasible timetables produced during the construction phase. Despite this the results produced by the GP system are still within range of the best results produced for the data sets.

Table 5. A comparison of the results obtained by the GP system and the best results cited for the benchmarks

Problem	GP	Caramia et al., 2001	Yang et al. 2004	Abdullah et al., 2004	Burke et, al. 2006	Difference
ear-f-83 I	36.74	**29.3**	33.71	34.84	32.76	7.44
hec-s-92 I	11.55	**9.2**	10.83	10.28	10.15	2.35
kfu-s-93	14.22	13.8	13.82	13.46	**12.96**	1.26
lse-f-91	10.90	**9.6**	10.35	10.24	9.83	1.30
rye-s-93	9.35	**6.8**	8.53	8.7	-	2.55
sta-f-83 I	158.22	158.2	**151.52**	159.28	157.03	6.7
tre-s-92	8.48	9.4	7.92	8.13	**7.75**	0.73
ute-s-92	26.65	24.4	25.39	**24.21**	24.82	2.44
yor-f-83 I	41.57	36.2	36.53	36.11	**34.84**	6.73

6 Conclusion and Future Work

The study presented in this paper is a first attempt at evaluating genetic programming for the purpose of inducing hyper-heuristics for the uncapacitated ETP. A GP system was implemented to induce hyper-heuristics for this problem and was tested on 9 of the Carter benchmarks. This study has revealed the potential of genetic programming as a means of evolving hyper-heuristics for the uncapacitated ETP. The GP system generated hyper-heuristics that produced feasible examination timetables with soft constraint costs within the range of other search methods employed for this purpose. Furthermore, the GP-based hyper-heuristic system outperformed the other hyper-heuristic systems on 6 of the 9 problems.

One of the drawbacks of this system is the runtime for larger data sets. For example, the runtime for the *kfu-s-93* set is approximately three and a half hours and just over four hours for the *rye-s*-93 data set. Future work will address improving the runtime of the overall system and testing it on additional benchmarks and problems with different hard and soft constraints. In this study a limit was not set on the size of the offspring produced and this did not appear to result in bloating. However, a closer look needs to be taken into the effect of not using such a limit and the overall effect of introns and bloat in this domain. Future extensions of this study will also investigate evolving programs that apply each heuristic more than once, e.g. h4s2, will apply the highest cost heuristic four times and the saturation degree heuristic twice.

Acknowledgments. The authors would like to thank the reviewers for their helpful comments and suggestions. This material is based on work financially supported by the National Research Foundation (NRF) of South Africa.

References

1. Abdullah, S., Ahmadi, S., Burke, E.K., Dror, M.: Investigating Ahuja-Orlin's Large Neighbourhood Search for Examination Timetabling. Technical Report NOTTCS-TR-2004-8, School of CSiT, University of Nottingham, U.K (2004)
2. Asmuni, H., Burke, E.K., Garibaldi, J.M.: Fuzzy Multiple Ordering Criteria for Examination Timetabling. In: Burke, E.K., Trick, M.A. (eds.) PATAT 2004. LNCS, vol. 3616, pp. 147–160. Springer, Heidelberg (2005)
3. Banzhaf, W., Nordin, P., Keller, R.E., Francone, F.D.: Genetic Programming - An Introduction - On the Automatic Evolution of Computer Programs and its Applications. Morgan Kaufmann Publishers, Inc., San Francisco (1998)
4. Burke, E.K., Petrovic, S.: Recent Research Directions in Automated Timetabling. European Journal of Operational Research 140(2), 266–280 (2002)
5. Burke, E., Hart, E., Kendall, G., Newall, J., Ross, P., Schulenburg, S.: Hyper-Heuristics: An Emerging Direction in Modern Research. In: Handbook of Metaheuristics, ch. 16, pp. 457–474. Kluwer Academic Publishers, Dordrecht (2003)
6. Burke, E.K., Bykov, Y.: Solving Exam Timetabling Problems with the Flex-Deluge Algorithm. In: Burke, E.K., Rudova, H. (eds.) Proceedings of PATAT 2006, pp. 370–372 (2006)
7. Burke, E.K., McCollum, B., Meisels, A., Petrovic, S., Qu, R.: A Graph-Based Hyper-Heuristic for Educational Timetabling Problems. European Journal of Operational Research 176, 177–192 (2007)
8. Caramia, M., Dell'Olmo, P., Italiano, G.: New Algorithms for Examination Timetabling. In: Näher, S., Wagner, D. (eds.) WAE 2000. LNCS, vol. 1982, pp. 230–241. Springer, Heidelberg (2001)
9. Carter, M.W., Laporte, G., Lee, S.Y.: Examination Timetabling: Algorithmic Strategies and Applications. Journal of the Operational Research Society 47(3), 373–383 (1996)

10. Cowling, P., Kendall, G., Han, L.: An Investigation of a Hyperheuristic Genetic Algorithm Applied to a Trainer Scheduling Problem. In: Proceedings of Congress on Evolutionary Computation (CEC), Hilton Hawaiian Village Hotel, Honolulu, Hawaii, May 12 -17, 2002, pp. 1185–1190 (2002) ISBN 0-7803-7284-2

11. Kendall, G., Mohd Hussin, N.: An Investigation of a Tabu Search Based on Hyper-Heuristics for Examination Timetabling. In: Proceedings of MISTA (Multidisciplinary International Conference on Scheduling) 2003, Nottingham, UK (2003)

12. Koza, J.R.: Genetic Programming I: On the Programming of Computers by Means of Natural Selection. MIT Press, Cambridge (1992)

13. Qu, R., Burke, E.K.: A Hybrid Neighbourhood Hyper-Heuristic for Exam Timetabling Problems. In: Proceedings of MIC 2005: The 6th Metaheuristics International Conference, Vienna, Austria (2005)

14. Qu, R., Burke, E., McCollum, B., Merlot, L.T.G., Lee, S.Y.: A Survey of Methodologies and Automated Approaches for Examination Timetabling. Technical Report NOTTCS-TR-2006-4 (2006), http://www.cs.nott.ac.uk/TR-cgi/TR.cgi

15. Ross, P., Hart, E., Corne, D.: Some Observations about GA-based Exam Timetabling. In: Burke, E.K., Carter, M. (eds.) PATAT 1997. LNCS, vol. 1408, pp. 115–129. Springer, Heidelberg (1998)

16. Ross, P.: Hyper-heuristics. In: Burke, E.K., Kendall, G. (eds.) Search Methodologies: Introductory Tutorials in Optimization and Decision Support Methodologies, ch. 17, pp. 529–556. Kluwer, Dordrecht (2005)

17. Terashima-Marin, H., Ross, P., Valenzuela-Rendon, M.: Evolution of Constraint Satisfaction Strategies in Examination Timetabling. In: Banzhaf, W., et al. (eds.) Proceedings of GECCO 1999: Genetic Programming and Evolutionary Computation Conference, pp. 635–642. Morgan Kaufmann, San Francisco (1999)

18. Yang, Y., Petrovic, S.: A Novel Similarity Measure for Heuristic Selection in Examination Timetabling. In: Burke, E.K., Trick, M. (eds.) PATAT 2004. LNCS, vol. 3616, pp. 247–269. Springer, Heidelberg (2005)

Asynchronous Stochastic Dynamics and the Spatial Prisoner's Dilemma Game

Carlos Grilo[1] and Luís Correia[2]

[1] Dep. Eng. Informática, Escola Superior de Tecnologia e Gestão
Instituto Politécnico de Leiria
Morro do Lena, 2411-901 Leiria, Apartado 4163, Portugal
grilo@estg.ipleiria.pt
[2] LabMag, Dep. Informática, Faculdade Ciências da Universidade de Lisboa
Edifício C6, Campo Grande, 1749-016 Lisboa
Luis.Correia@di.fc.ul.pt

Abstract. We argue that intermediate levels of asynchronism should be explored when one uses evolutionary games to model biological and sociological systems. Usually, only perfect synchronism and continuous asynchronism are used, assuming that it is enough to test the model under these two opposite update methods. We believe that biological and social systems lie somewhere between these two extremes and that we should inquire how the models used in these situations behave when the update method allows more than one element to be active at the same time but not necessarily all of them. Here, we use an update method called Asynchronous Stochastic Dynamics which allows us to explore intermediate levels of asynchronism and we apply it to the Spatial Prisoner's Dilemma game. We report some results concerning the way the system changes its behaviour as the synchrony rate of the update method varies.

1 Introduction

The explanation of how cooperation could ever emerge on nature and human societies by means of natural evolution has been a difficult problem to solve [1]. Evolutionary game theory [15] has been largely used as a tool to study this problem. In this area interactions between agents are usually modeled as a game and the Prisoner's Dilemma game is one of the most used metaphors to study the evolution of cooperation. In this game there are two possible strategies: Cooperate (C) or Defect (D). Figure 1 shows the payoff matrix of the game, where the following conditions must be met: $T > R > P > S$.

When panmitic populations are used[1] and when the players play the game just once on each encounter without remembering what happened on previous encounters, theory says that the C strategy is completely dominated by the D one, until complete extinction [15]. However, in [13] Nowak and May showed that

---

[1] On panmitic populations each agent can interact with any other agent in the population.

J. Neves, M. Santos, and J. Machado (Eds.): EPIA 2007, LNAI 4874, pp. 235–246, 2007.
© Springer-Verlag Berlin Heidelberg 2007

	C	D
C	R, R	S, T
D	T, S	P, P

Fig. 1. Payoff matrix for the Prisoner's Dilemma game. The row player gets the first value of each matrix element.

cooperation can be maintained when the game is played in a two-dimensional spacial grid, in which agents can only interact with their immediate neighbours. This work was almost immediately contested in [7], the reason being that these results were only possible because a synchronous system was used which, according to the authors, is an artificial feature. In a synchronous system all the agents in the population interact and are updated exactly at the same time. Instead, the authors of this work used a continuous asynchronous updating method called *uniform choice* [14] in which, at each time step, only a randomly chosen individual, with reposition, is able to interact with its neighbours, and the reported output was that cooperation is no longer sustainable even if only a D agent exists in the initial population. Given that these results were obtained with a single combination of payoff values, Nowak and colleagues counter-answered in [12], testing several conditions, namely, different payoff values, synchronous and asynchronous (*uniform choice*) update methods, and different levels of determinism in the transition rule (see parameter m in section 3). The results show that cooperation can be maintained for many different conditions, including asynchronism, but they are presented through system snapshot images, which render difficult to measure the exact way they are affected by the modification from a synchronous to a continuous asynchronous discipline. In spite of that, this and other subsequent works [10][17] allowed spatial structure to be viewed as a feature that, in certain situations, can be beneficial to the evolution of cooperation. Also, since the criticism made in [7], it's common to see papers [16][6][11] where both synchronous and continuous update methods are used.

In this paper we argue that both perfect synchronism and continuous asynchronism are equally artificial ways of simulating the global dynamics of a population of interacting agents. This doesn't mean that the above mentioned practice of presenting results achieved with these two methods is not a positive one. However, it takes for granted that it is enough to test the system under these two opposite methods. We believe that biological and social systems lie somewhere between these two extremes and that we should inquire how the models used in these situations behave when the update method allows more than one element to be "active" at the same time but not necessarily all of them. In order to do that, we used one update method named *asynchronous stochastic dynamics* [4] that allows us to cover all the spectrum between synchronous and continuous updating. We applied the method to a model similar to the one used in [12] and we report some results obtained.

The paper is structured as follows: in Sect. 2 we explain why we think that synchronism and continuous asynchronism are both equally artificial and why methods like *asynchronous stochastic dynamics* should be explored when studying populations of interacting agents. In Sect. 3 we describe the model we used in our experiments and in Sect. 4 we present and discuss some results. Finally, some conclusions are drawn and future work is advanced.

2 Asynchronous Stochastic Dynamics

Beyond the evolution of cooperation, the influence of the update method has been studied in areas like, for example, cellular automata [8][3] and evolutionary algorithms [9][5]. In some of these works, especially the ones about cellular automata and evolution of cooperation, the utilization of synchronous update methods is criticized, the argument being that the real world is not synchronous and, so we can not entirely rely on results achieved that way. As an alternative, continuous asynchronous methods are usually used (see [14] for an analysis of several asynchronous methods). This procedure corresponds to choosing small enough time intervals so that at each one exactly one element of the population can interact with its neighbours and be updated [7].

We think that continuous asynchronism when applied to biological and sociological environments can be considered as artificial as perfect synchronism. There are two reasons for this: the first reason is that, when we have a population of interacting agents, many interactions can be occurring at the same time. If interactions were an instantaneous phenomena we could model the dynamics of the system as if interactions occurred one after another but that is not the case. Interactions can take some time, which means that their output is not available to other ongoing interactions. Even if we consider interactions as being instantaneous, the time that information takes to be transmitted and perceived implies that interactions' consequences are not immediately available. Another reason is the determinism of continuous update methods: on what basis can we say that, at each time step, exactly 1, 2, or n elements are "active"? Even if we have an idea of the level of activity of the system being modeled, it's doubtful that it is always exactly the same. This, of course, is not a problem for evolutionary algorithms, where, given a problem to solve, the goal is to achieve the best solution as fast as possible, independently of the methods we use. But social systems are not as predictable and some sort of nondeterminism should be used.

A feasible alternative to perfect synchronism and continuous asynchronism is a method named *asynchronous stochastic dynamics* (ASD) in which, at each time step, each element of the population has a given probability $0 < \alpha \leq 1$ of being selected to interact with its neighbours, after which it is updated using a given transition rule. The α parameter is called the *synchrony rate* and is the same for all the elements of the population. After this selection procedure both the interactions and the application of the transition rule are done as if they occurred simultaneously, i.e., synchronously. The stochastic nature of the method implies that the number of selected elements may vary from time step to time step. Also, the α parameter allows us to explore intermediate levels of asynchronism. This was done, for example, in [4] for studying the robustness of elementary cellular automata to asynchronism. The authors found that some automata are very sensitive to small changes in the α parameter. Besides, it can happen that a given automata has a similar behaviour to, for example, $\alpha = 1$ and $\alpha = 0.1$, which are values at almost opposite sides of the α domain, but has a very different behaviour say, for example, for $\alpha = 0.5$ [2]. This made us question if this could be the case for spatial evolutionary games, since these models resemble cellular automata in many aspects [13]. Evolutionary games are used as metaphors to model real situations. It's difficult to know exactly the update discipline of the modeled system's elements. Therefore, it's useful to know if the model is robust to changes in the update method. Even if it's robust, in the sense that it doesn't change significantly its behaviour to small changes on the update method, it can gradually change it's behaviour as we change the synchrony rate. This change can be such that the system has a very different behaviour under the two extremes of the update method (synchronous vs. continuous) and the ASD method gives us the possibility of knowing how this change happens.

3   The Model

The model we used in the experiments is very similar to the one used in [12] so that the results could be compared. Agents are placed in a toroidal two dimensional grid so that each agent occupies one cell. Each time step can be divided in three stages: the activation stage, the interaction stage and the update stage.

In the first stage we decide, using the ASD method, which agents will be active at the current time step, i.e., which are the agents that will interact with their neighbours and, therefore, will be updated in the following stages. The utilization of this update method is the main modification we made to the original model. Recall that in the original model the update methods used were synchronous update and *uniform choice*. The ASD method equals synchronous update when $\alpha = 1$ and approaches *uniform choice* as $\alpha \to \frac{1}{n}$ where n is the population size.

In the second stage, the selected agents play a one round Prisoner's Dilemma game with all their 8 surrounding neighbours. This type of neighbourhood is usually called the *Moore* neighbourhood. In some works the neighbourhood is allowed to include the agent itself. This is justified by considering that each cell can represent not a single agent but a set of similar agents that may interact with each other. Here, we do not consider self-interaction since we are interested in modeling cells as individual agents.

Agents can only play C or D and the only way they can change their strategy is by way of the application of the transition rule in the third stage. It is common practice to define the game's payoff values as $R = 1$, $T = b$ ($b > 1$) and $S = P = 0$. The b parameter represents the advantage of D players over C ones when

---

[2] The measure used to compare the behaviour of an automaton under different conditions was the mean number of cells with value 1 during a given sampling period after a transient period.

these play the game with each other. Defining payoff values this way has the advantage that the game can be characterized by just one parameter.

Besides the regular two dimensional grid, we also made experiments with small-world networks (SWNs) [18] in order to verify if the results were dependent on the underlying interaction topology. We build SWNs as in [16]: first, a regular two dimensional grid is built so that each agent is linked to its 8 surrounding neighbours by undirected links; then, with probability p, each link is replaced by another one linking two randomly selected agents. Parameter p is called the rewiring probability. During the rewiring process we do not allow the creation of self links, as we do not allow self-interaction. Repeated links and disconnected graphs are also avoided. The rewiring process may create long range links connecting distant agents. For simplicity, we will call neighbours to all interconnected agents, even if they are not located at adjacent cells. By varying p from 0 to 1 we are able to build from completely regular networks to random ones. SWNs have the property that, even for very small values of the rewiring probability, the mean path length between any two nodes is much smaller than in a regular network, maintaining however a high clustering coefficient observed in many real systems including social ones.

The third stage is used to model the fact that agents tend to imitate the most successful agents they know. It can also be interpreted as the selection step of an evolutionary process in which the least successful strategies tend to be replaced by the most successful ones. This is done by synchronously applying a transition rule to the agents selected in the first stage. The transition rule used here (and also in [12]) is a generalization of the proportional update rule. Let G_i be the average payoff earned by agent i in the interaction stage, N_i be the set of neighbours of agent i, s_i be equal to 1 if i's strategy is C and 0 otherwise, and m a positive number. The probability that in the next time step agent i adopts C as its strategy is then given by

$$p_C = \frac{\sum_{l \in N_i \cup i} s_i G_l^m}{\sum_{l \in N_i \cup i} G_l^m}. \tag{1}$$

The m parameter acts as a weight that favors the most successful neighbour's strategy B in the update process: the bigger m, the larger is the probability that i adopts B. When $m = +\infty$ we have a deterministic best neighbour rule such that i always adopts B as its next strategy. When $m = 1$ we have the proportional update rule. It can be viewed, as well, as the deterministic degree of the transition rule. We use average payoffs instead of total payoffs because the rewiring process used to build small-world topologies may result in agents having a different number of neighbours.

4   Simulations and Results

As in [12], the simulations were done with populations of $80 \times 80 = 6400$ agents. When the system is running synchronously, i.e., when $\alpha = 1$ we let the system run during a transient period of 200 iterations. After this, we let the system run during 100 more iterations, and at the end we take as output the average proportion of cooperators during this period, which is called the sampling period. When $\alpha \neq 1$ the number of selected agents at each time step may not be equal to the size of the population and it may vary between two consecutive time steps. In order to guarantee that the runs with $\alpha \neq 1$ are equivalent to the synchronous ones in what concerns to the total number of individual updates, we let the system first run until $200 * 6400 = 1280000$ individual updates have been done. After this, we sample the proportion of cooperators during $100 * 6400 = 640000$ individual updates and we average by the number of time steps needed to do these updates.

Each point in the charts of Fig. 2 is the result for a combination of the b, m and α values and $p = 0$ (regular grid), averaged over 30 runs. We used 10 different b values varying from 1.02 to 1.7. α values vary from 1 to 0.1 by steps of 0.1. For the m parameter we used values $+\infty$, 100, 10, 8, 6, 4, 2 and 1.

In [4] a given cellular automata is said to be robust if the average density of 1's in the sampling period doesn't change by more than 0.1 when the synchrony rate α parameter is changed by a small value (they also change α by 0.1 steps). If we take these same values, but using the frequency of cooperators instead of the density of 1's as the measure of interest, the first conclusion we can derive from Fig. 2 is that, in general, the system is robust to small changes of the α value. There are, however, some situations of non-robustness: when $m = +\infty$ ($b = 1.15$, $b = 1.35$ and $b = 1.61$) and $m = 100$ ($b = 1.55$ and $b = 1.61$) and several when $m = 2$ and $m = 1$. The big jumps that can be observed for the $m = 2$ and $m = 1$ cases are due to the large difference in the frequency of cooperators for opposite values of α: in order to get from one point to the other, big jumps must be made. This is not the real justification when $b = 1.02$. In this case we can see that the real reason for the non-robustness is that the frequency of cooperators doesn't change uniformly as we change the α value: when we move from $\alpha = 1$ to $\alpha = 0.1$ the frequency of cooperators grows very quickly until it stabilizes in the maximum value. This means that, for this particular combination of the m and b values, cooperation hegemony is the dominant result for a significant fraction of the α domain.

Another result that we can derive from the charts is that, as we change the α parameter, the variation of the frequency of cooperators is not always monotonic. For example, there are some situations in which, when we move from $\alpha = 1$ to $\alpha = 0.1$, the frequency of cooperators first decreases but then, at some point, it starts increasing. The most significant of these situations happens for ($m = 8$, $b = 1.35$) and ($m = 6$, $b = 1.3$). In the first situation, the difference in the frequency of cooperators obtained with $\alpha = 1$ and $\alpha = 0.1$ is 0.022, but the difference of the values obtained with $\alpha = 1$ and $\alpha = 0.5$ is 0.231. Excepting these situations, which happen for a relatively low value of m, non-monotonicity happens mainly for large values of m, that is, when the probability that an agent imitates its most successful neighbour is high. Nevertheless, we can say that, in general, the system responds monotonically as we change α from one extremity to the other.

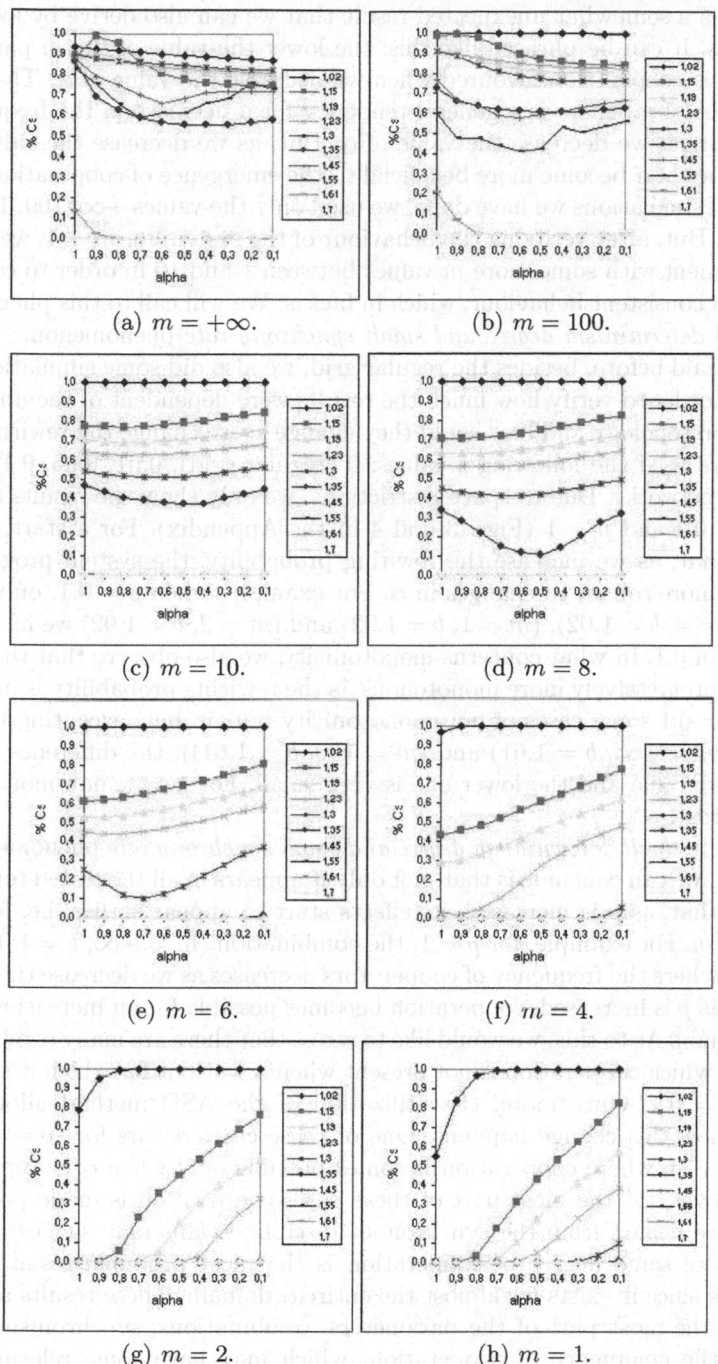

Fig. 2. % of cooperators for $p = 0$ and different combinations of m, b and α

There is a somewhat unexpected result that we can also derive by looking at the charts. It can be phrased like this: the lower the value of the m parameter, the more is cooperation favoured when we decrease the value of α. That is, for high values of m, there is a general tendency for a decrease in the frequency of cooperators as we decrease the value of α. But, as we decrease the value of m, lower values of α become more beneficial to the emergence of cooperation. In the first set of simulations we have done, we used only the values $+\infty$, 100, 10, and 1 as in [12]. But, after verifying the behaviour of the system for $m = 1$, we decided to experiment with some more m values between 1 and 10 in order to confirm if this was a consistent behaviour, which in fact is. We will call to this phenomenon the *small determinism degree and small synchrony rate* phenomenon.

As we said before, besides the regular grid, we also did some simulations with SWNs in order to verify how much the results were dependent of the underlying interaction topology and how could they change as we change the rewiring probability. We used the following p values: 0 (regular grid), 0.01, 0.05, 0.1 and 1.0 (random network). Due to space restrictions, we only show the results achieved with $p = 0.1$ and $p = 1$ (Figs. 3 and 4 in the Appendix). For a start, we may observe that, as we increase the rewiring probability, the system progressively becomes more robust to changes in α. For example, when $p = 0.1$, only for the cases $(m = 4, b = 1.02)$, $(m = 1, b = 1.02)$ and $(m = 2, b = 1.02)$ we have jumps larger than 0.1. In what concerns monotonicity, we also observe that the system becomes progressively more monotonous as the rewiring probability is increased. When $p = 0.1$ some cases of non-monotonicity remain but, excepting two situations $((m = +\infty, b = 1.61)$ and $(m = 100, b = 1.61))$, the difference between the largest value and the lower one is very small. For $p = 1$, non-monotonicity doesn't exist.

As to the *small determinism degree and small synchrony rate* phenomenon, the first thing we can conclude is that, not only it appears in all the tested topologies, but also that, as p is increased, its effects start to appear earlier, i.e, for larger values of m. For example, for $p = 1$, the combination $(m = +\infty, b = 1.02)$ is the only one where the frequency of cooperators decreases as we decrease the value of α. Also, as p is increased, cooperation becomes possible for an increasing part of the b domain. As to this, we would like to stress that there are many combinations (m,b) for which cooperation is not present when $\alpha = 1$ but for which it is present when $\alpha = 0.1$. Once more, the utilization of the ASD method allows us to analyze how this change happens. One of these cases occurs for $(p = 0.1, m = +\infty, b = 1.7)$, where cooperation becomes possible only when α is approaching 0.1. However, for the most part of these cases cooperation is made possible as soon as we depart from the synchronous discipline. This may suggest that the existence of some degree of cooperation is the most probable result in these situations since it exists for almost the entire α domain. These results also show that, for the most part of the parameters' combinations, synchronism renders difficult the emergence of cooperation, which may have some relevant social consequences. For example, this may suggest that two negotiators submitting their proposals by the way of a mediator will have more difficulty at arriving to an agreement than if they do it asynchronously.

5 Conclusion and Future Work

In this work we argued that intermediate levels of asynchronism should be explored in the study of models such as spacial evolutionary games, since we believe that real systems, and specially sociological ones, lie somewhere between perfect synchronism and continuous updating. We used an update method called *asynchronous stochastic dynamics* which allows the exploration of all the space between these two extremes and applied it to spacial versions of the Prisoner's Dilemma game. This method allows us derive some results concerning, for example, robustness and monotonicity of the system that are not possible to derive if only synchronous and continuous updating are used. We found that, in general, the spacial Prisoner's Dilemma game responds robustly and monotonically to changes in the synchrony rate. This behaviour is usually taken for granted but results obtained in recent works on elementary cellular automata systems, with which spatial evolutionary games have many resemblances, show that the behaviour of these systems may be very different on intermediate levels of the update discipline. This method also allows us to analyze how the synchrony rate affects the system behaviour as other system parameters change. We found, for example, that lower values of the synchrony rate become more beneficial to the evolution of cooperation as the level of determinism of the generalized proportional update transition rule diminishes. Finally, in situations in which some degree of cooperation exists under one of the extremes of the update rule but not under the other, this method allows us to understand how the change from one type of result to the other happens and what is the most common output throughout the synchrony rate spectrum.

One of our first future extensions to this work will be to explore the ASD method with other very used games in order to verify if some of the results achieved with the Prisoner's Dilemma game as, for example, the *small determinism degree and small synchrony rate* phenomenon, can be generalized. The results achieved in [16] with the Snowdrift game, where the best-neighbour (equivalent to $m = +\infty$) and the simple proportional update ($m = 1$) transition rules, as well as synchronous and continuous updating were used, seem to indicate that this is the case. However, only by exploring intermediate levels of asynchronism and intermediate levels of determinism of the transition rule we can confirm this. We also plan to use scale free networks [2] as the underlying interaction topology. This will allow us to compare, for example, the robustness of other types of topologies to changes in the synchrony rate.

References

1. Axelrod, R.: The Evolution of Cooperation. Penguin Books (1984)
2. Barabasi, A.-L., Albert, R.: Emergence of scaling in random networks. Science 286, 509 (1999)

3. Bersini, H., Detours, V.: Asynchrony induces stability in cellular automata based models. In: Maes, P., Brooks, R. (eds.) Proceedings of the Artificial Life IV Conference, pp. 382–387. MIT Press, Cambridge (1994)
4. Fatés, N.A., Morvan, M.: An experimental study of robustness to asynchronism for elementary cellular automata. Complex Systems 16(1), 1–27 (2005)
5. Giacobini, M.: Artificial Evolution on Network Structures: How Time and Space Influence Dynamics. PhD thesis, Université de Lausanne, Switzerland (2005)
6. Hauert, C., Doebeli, M.: Spatial structure often inhibits the evolution of cooperation in the snowdrift game. Nature 428, 643–646 (2004)
7. Huberman, B., Glance, N.: Evolutionary games and computer simulations. Proceedings of the National Academy of Sciences USA 90, 7716–7718 (1993)
8. Ingerson, T.E., Buvel, R.L.: Structure in asynchronous cellular automata. Physica D Nonlinear Phenomena 10, 59–68 (1984)
9. De Jong, K.A., Sarma, J.: Generation gaps revisited. In: Whitley, L.D. (ed.) Foundations of Genetic Algorithms 2, pp. 19–28. Morgan Kaufmann, San Mateo, CA (1993)
10. Killingback, T., Doebeli, M.: Spatial Evolutionary Game Theory: Hawks and Doves Revisited. Royal Society of London Proceedings Series B 263, 1135–1144 (1996)
11. Luthi, L., Giacobini, M., Tomassini, M.: Synchronous and asynchronous network evolution in a population of stubborn prisoners. In: Kendall, G., Lucas, S. (eds.) IEEE Symposium on Computational Intelligence and Games, pp. 225–232 (2005)
12. Nowak, M., Bonhoeffer, S., May, R.M.: More spatial games. International Journal of Bifurcation and Chaos 4(1), 33–56 (1994)
13. Nowak, M.A., May, R.M.: Evolutionary games and spatial chaos. Nature 359, 826–829 (1992)
14. Schönfich, B., de Roos, A.: Synchronous and asynchronous updating in cellular automata. BioSystems 51(3), 123–143 (1999)
15. Smith, J.M.: Evolution and the Theory of Games. Cambdridge University Press, Cambdridge (1982)
16. Tomassini, M., Luthi, L., Giacobini, M.: Hawks and doves on small-world networks. Physical Review E 73(1), 016132 (2006)
17. Vainstein, M.H., Arenzon, J.J.: Disordered environments in spatial games. Physical Review E 64, 051905 (2001)
18. Watts, D., Strogatz, S.H.: Collective dynamics of small-world networks. Nature 393, 440–442 (1998)

A   Results for $p = 0.1$ and $p = 1$

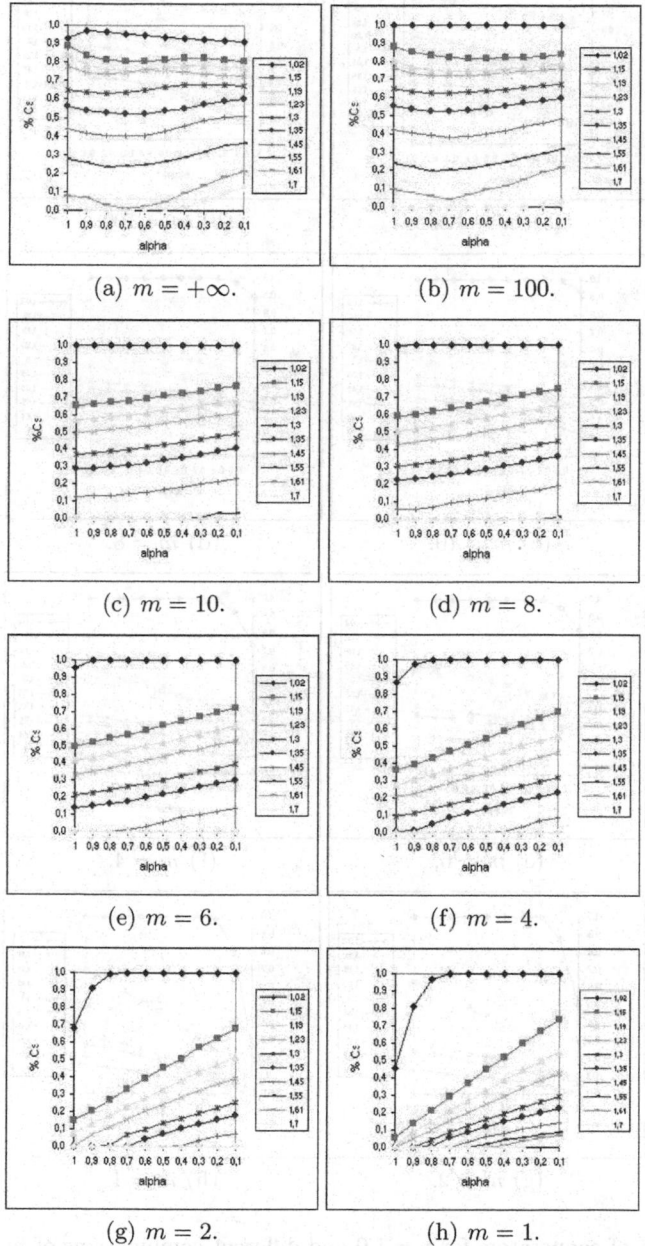

(a) $m = +\infty$.   (b) $m = 100$.

(c) $m = 10$.   (d) $m = 8$.

(e) $m = 6$.   (f) $m = 4$.

(g) $m = 2$.   (h) $m = 1$.

Fig. 3. % of cooperators for $p = 0.1$ and different combinations of m, b and α

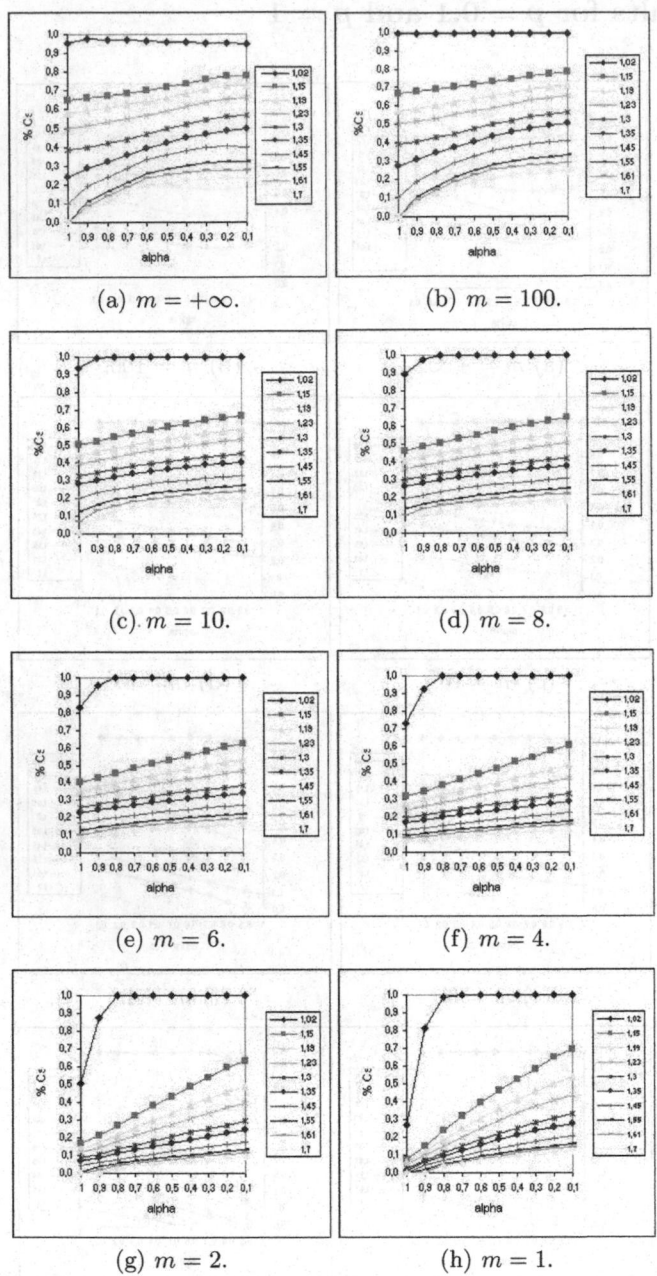

(a) $m = +\infty$.

(b) $m = 100$.

(c) $m = 10$.

(d) $m = 8$.

(e) $m = 6$.

(f) $m = 4$.

(g) $m = 2$.

(h) $m = 1$.

Fig. 4. % of cooperators for $p = 1.0$ and different combinations of m, b and α

Improving Evolutionary Algorithms with Scouting

Konstantinos Bousmalis[1], Gillian M. Hayes[2], and Jeffrey O. Pfaffmann[3]

[1] Department of Informatics, The University of Edinburgh, Edinburgh, UK
K.Bousmalis@sms.ed.ac.uk
[2] Institute of Perception, Action and Behaviour(IPAB), Department of Informatics,
The University of Edinburgh, Edinburgh, UK
gmh@inf.ed.ac.uk
[3] Department of Computer Science, Lafayette College, Easton, PA 18042, USA
pfaffmaj@cs.lafayette.edu

Abstract. The goal of an Evolutionary Algorithm(EA) is to find the
optimal solution to a given problem by evolving a set of initial potential
solutions. When the problem is multi-modal, an EA will often become
trapped in a suboptimal solution(premature convergence). The Scouting-
Inspired Evolutionary Algorithm(SEA) is a relatively new technique that
avoids premature convergence by determining whether a subspace has
been explored sufficiently, and, if so, directing the search towards other
parts of the system. Previous work has only focused on EAs with point
mutation operators and standard selection techniques. This paper examines the effect of scouting on EA configurations that, among others, use
crossovers and the Fitness-Uniform Selection Scheme(FUSS), a selection
method that was specifically designed as means to avoid premature convergence. We will experiment with a variety of problems and show that
scouting significantly improves the performance of all EA configurations
presented.

1 Introduction

1.1 Evolutionary Algorithms

Evolutionary Algorithms are a family of optimization techniques that attempt to
solve a given problem by evolving a set of solutions. A typical EA randomly ini-
tializes a population of potential solutions-individuals that are subsequently *(a)*
assigned a measure of merit(fitness value), *(b)* selected for reproduction based
on that merit, and *(c)* varied via crossover(exchange of genes), mutation, and
deletion, to produce a new generation of individuals. The cycle of fitness assignment, selection, reproduction and deletion usually continues for a preset number
of generations, or until the global optimum of a certain objective function is
reached. The goal is to find this global optimum, but often, a typical EA gets
trapped in local optima, a problem this paper suggests a solution for.

Generation-based EAs replace the entire population in each generation,
whereas steady-state EAs replace only one or two individuals. The Fitness-
Uniform Selection Scheme, which will be one of the focal points of the work

J. Neves, M. Santos, and J. Machado (Eds.): EPIA 2007, LNAI 4874, pp. 247–258, 2007.

presented here, has been tested only on steady-state EAs.[5] This work focuses on the use of generation-based EAs, with future work potentially expanding into the area of steady-state EAs.

1.2   Scouting-Inspired Evolutionary Algorithm: Previous Work

Scouting was originally introduced as a mechanism for automated exploration of complex phenomena, using a conservative number of samples. It uses an evolutionary technique, which, instead of focusing on finding an optimum, searches for regions of the search space that exhibit "surprising" behavior.[2] "Surprise" is defined as the difference between an estimated sampling result and the actual returned value. It was apparent from the introduction of the technique, that an experience database, namely a database of observations, could create a simple model of a system.

Figure 1 provides an illustration of a Scouting Algorithm(SA), which consists of an Evolution Strategy(ES) that evaluates individuals, creates new generations solely through mutation, and selects only the best individual for reproduction.[2] As observations are made on a given system, the results are saved in an experience database. A characteristic of an ES is self-adaptation of the mutation strength.[8] The way the SA achieves that is by adapting the Gaussian distribution of the mutation range based on the surprise value of each individual. As already mentioned, surprise is defined as the absolute difference of the estimated and the actual fitness values. The estimate is calculated via a weighted k-nearest neighbor algorithm, which uses the experiences stored in the database. When the surprise is high, the search continues with a small mutation range and vice versa. Hence, the technique will explore a subspace until the results gained are no longer "surprising," according to the experience the search has already had in the system at hand. Scouting has also been the focus of two more recent papers, which presented an SA as means to automated experimentation in biological systems. [3][4]

Pfaffmann et al.[1] defined the Scouting-Inspired Evolutionary Algorithm as an evolutionary technique that uses scouting to model a given search space, and

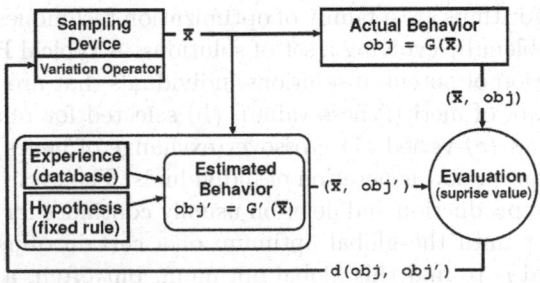

Fig. 1. A Scouting Algorithm varies individuals based on their surprise value. $\bar{x}$ is a new individual, $G(\bar{x})$ the actual, and $G'(\bar{x})$ the estimated objective values. d is their difference.[1]

provide a simple way to avoid premature convergence in deceptive and multi-modal problems. It was defined as an EA that uses the Roulette Wheel selection scheme, scouting-driven mutation as the only genetic operator, and deletion of the entire population in each generation (generation-based EA). For each mutation operation, the mutation strength is randomly chosen from a Gaussian distribution, the standard deviation(σ) of which changes based on the surprise value of the parent. Given minimum and maximum standard deviation, σ_{min} and σ_{max}, the active σ for a given parent with surprise s_{ind}, scaled to [0,1], is chosen by the modulator function

$$\sigma(s_{ind}) = \sigma_{max} - s_{ind} \times (\sigma_{max} - \sigma_{min}) \ . \tag{1}$$

As one can easily determine in (1), when the surprise is minimal, σ approaches σ_{max}, whereas when the surprise is close to its maximum value of one, it approaches σ_{min}.

1.3   Fitness-Uniform Selection Scheme (FUSS)

The SEA was designed to promote genetic diversity in the overall pool of individuals via regulating mutation. It was hypothesized that this would help find fitter solutions faster, in multi-modal and deceptive problems. The Fitness-Uniform Selection Scheme(FUSS), a relatively new selection method, has also been proved successful in maintaining genetic diversity, and helping an EA perform better in such problems.[5]

The way FUSS achieves high genetic diversity, is by allowing only a small number of fitness-similar individuals in a population. Similarity between two individual fitness values is simply defined as their absolute difference. This particular selection technique works in two stages. Firstly, a random fitness value f is selected uniformly from the interval $[f_{min}, f_{max}]$, where f_{min} and f_{max} are the minimum and maximum fitness values of the given population. Subsequently, the individual with fitness nearest to f is selected. FUSS, as opposed to traditional selection techniques, does not have an inherent goal to achieve populations with the highest average fitness possible. Hutter et al.[5] show that by focusing selection pressure towards less represented fitness values, premature convergence can be avoided and the path towards fitter solutions can remain open.

2   A Closer Look to the SEA

2.1   A New Standard Deviation(σ) Modulator

Reproduction and analysis of the results presented in [1] showed that the vast majority of all surprise values are in the lowest 10% of the surprise range. Table 1 shows the number of individuals in each surprise level, from the $29,950,012$ individuals created throughout a set of experiments that used scouting-driven mutation.

As one can see in Fig.2, the linear mapping from surprise to standard deviation in (1) makes the SEA behave similarly to a traditional EA with a mutation

Table 1. The number of individuals in each surprise level throughout a set of experiments. (10 individuals per generation, 5000 generations per experiment).

Surprise Level	Number of Individuals	Percentage(%)
[0, 0.1]	29,064,234	97.04
(0.1, 0.2]	479,374	1.6
(0.2, 0.3]	211,509	0.71
(0.3, 0.4]	146,481	0.49
(0.4, 0.5]	41,365	0.14
(0.5, 0.6]	5,953	0.02
(0.6, 0.7]	911	0.003
(0.7, 0.8]	128	0.0004
(0.8, 0.9]	48	0.0001
(0.9, 1]	8	0.00002

standard deviation of σ_{max}. In order to achieve a fairer comparison with an EA that uses σ_{min}, we ought to regulate the effect of scouting such that when $s_{ind} = 0$, $\sigma = \sigma_{max}$, when $s_{ind} = 0.10$, σ approaches σ_{min}, and slowly decreases, to equal σ_{min} when $s_{ind} = 1$. We will therefore introduce a new standard deviation modulator, which, as seen in Fig.2, meets the requirements set above.

$$\sigma(s_{ind}) = \sigma_{max} - (s_{ind})^{\gamma} \times (\sigma_{max} - \sigma_{min}) , \quad 0 < \gamma \leq 1 . \qquad (2)$$

Based on Table 1, $\gamma = 0.01305$ is the optimal value for the new parameter. The reason is that when $s_{ind} = 0.1$, $s_{ind}^{0.01305} = 0.9704$, namely the technique will use 97.04% of the σ range on 97.04% of the individuals. Note that (1) is a case of (2) for $\gamma = 1$.

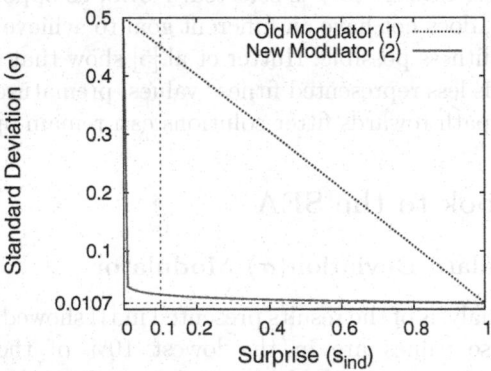

Fig. 2. Standard deviation vs. surprise using the old (1) and new (2) modulators with $\gamma = 0.01305$, $\sigma_{min} = 0.0107$ and $\sigma_{max} = 0.5$. The majority of the individuals created throughout an experiment has surprise $s_{ind} \leq 0.1$.

2.2   Crossovers

A traditional EA usually uses both crossovers and mutation as its genetic operators. As mentioned earlier, previous work has focused on EAs that only use point mutation. We want to show that scouting can improve an EA, and any evidence for this claim would not be complete without the examination of its effect on a configuration that uses crossovers.

In this paper, we examine the effect scouting has on EA configurations with and without single-point crossovers. Crossover and mutation are applied with certain crossover and mutation rates. When an EA configuration uses the crossover operator, and crossover is not chosen for a given pair of parents, mutation is always applied, in order to ensure that the parents are not cloned.

The introduction of crossover does not significantly affect the way scouting controls mutation standard deviation. When scouting-driven mutation is used together with the crossover operator, there arises the issue of determining a s_{ind} to use in (2), without calculating fitness and surprise for intermediate individuals. In the techniques presented here, we use the average surprise values of the two parents used for the crossover.

We will therefore examine the effect of scouting on EAs with crossover and show that the improvement incurred is equivalent to the one observed in a mutation-only EA.

2.3   FUSS and the SEA

Another focal point of this paper is FUSS, mainly because of its similarities to the SEA. Firstly, both techniques aim to avoid premature convergence, by maintaining genetic diversity. Secondly, they have both been designed for deceptive and multi-modal domains, where traditional evolutionary techniques tend to become trapped in suboptimal solutions. Finally, the goal of both FUSS and the SEA is not to explicitly find higher fitness individuals, but to avoid premature convergence allowing for greater fitness levels to be achieved. The first technique succeeds by favoring the least-represented individuals in a population, whereas the latter succeeds by directing the search towards surprising parts of the search space of interest via mutation strength regulation.

One of the goals of this paper is to examine the effect a combination of the two techniques could have on a traditional EA when attempting to solve deceptive and multi-modal problems. We will show that scouting can improve an EA that also uses FUSS, and that even higher fitness levels could potentially be reached when these two techniques work together.

3   Implementation

The EA framework was written in C/C++. The random number generators used are the "Mersenne Twister" [9], and the two versions of the "luxury random number generator" algorithm[10], which are all included in the GNU Scientific Library (gsl)[7]. The problems used to test and compare the techniques were generated by Schmidt and Michalewicz's TCG-2 test-case generator.[6]

3.1   Test Cases and the TCG-2 Package

The problem domain and the reasons for using TCG-2 are fully outlined in [1]. Briefly, TCG-2 is a very configurable C++ software package that can generate a vast variety of nonlinear constrained parameter optimization problems with different levels of complexity. The fitness function used here follows the suggestion by Schmidt and Michalewicz[6] for a static penalty approach:

$$Fit(\bar{x}) = G(\bar{x}) - W \times CV(\bar{x}) \; , \tag{3}$$

where $G(\bar{x})$ is the objective function, W the static penalty, and $CV(\bar{x})$ is the constraint violation function for the given test case.

The test cases generated were varied in the number and width of peaks, in an attempt to get a better idea of the kind of problems the SEA is particularly effective at. We will refer to the width of peaks as σ_{peak} to differentiate it from the standard deviation σ used in mutation. We used the TCG-2 parameters shown in Table 2, and created the test landscapes by setting the parameter σ_{peak} to 0.02, 0.1 and 0.2, and the number of peaks p to 10, 50, 100 and 150. Consequently, twelve different two-dimensional test-case problems were created, as illustrated in Fig. 3.

The number of dimensions was mainly kept to two, for visualization purposes and easier understanding of the behavior of the new techniques. After analysis of the results on the two-dimensional problems, a set of experiments was also run in three dimensions with the parameters outlined in Table 2, for the case of 10 peaks and $\sigma_{peak} = 0.02$, as an example of a higher-dimension, multi-modal and deceptive problem.

3.2   The Evolutionary Algorithm

The implemented Evolutionary Algorithm follows the guidelines provided in the introductory SEA paper, with the necessary changes for the goals of this paper. Selection for parenthood uses either Roulette Wheel or FUSS, and the parents are varied via crossover and/or mutation to create the next generation of individuals. The process loops for a set number of generations—5000 for all experiments presented here.

Mutation varies individual genes with a Gaussian distribution centered around mean $\mu = 0$. The number of genes changed for each individual during mutation is configurable, but all genes of an individual are varied during mutation in the experiments presented. New individuals that are created out of range are rejected, and mutation recreates individuals, until one is within bounds.

Both mutation and crossover rates are set to 0.5 for all experiments, due to the same configuration in the FUSS paper [5]. This is also in accordance with the basis of the original SA, the traditional ES, which uses crossover and mutation with equal importance.[8]

Minimum standard deviation for scouting-driven mutation is set equal to the standard deviation for Random Mutation, $\sigma_{min} = 0.0107$, whereas the maximum one is $\sigma_{max} = 0.5$. The constant parameter γ is set, for reasons explained in the previous section, to $\gamma = 0.01305$ for all experiments.

Table 2. TCG-2 parameters for the two-dimensional experiments

Number of dimensions (n)	: 2
Number of feasible components (m)	: 10
Search space feasibility (ρ)	: 0.5
Search space complexity (c)	: 0
Active constraints at global optimum (a)	: 0
Number of peaks (p)	: 10, 50, 100, 150
Peak width (σ)	: 0.02, 0.1, 0.2
Peak decay (α)	: 0.1
Component minimum distance (d)	: 0.01
Penalty (W)	: 10

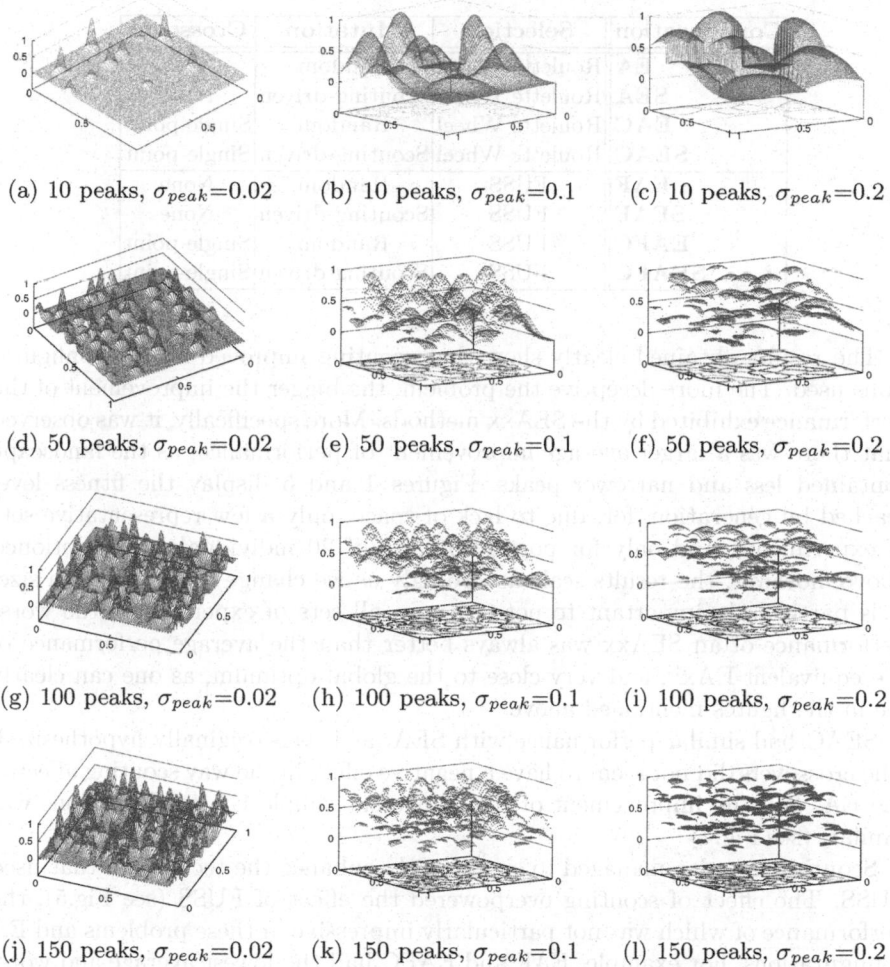

(a) 10 peaks, σ_{peak}=0.02   (b) 10 peaks, σ_{peak}=0.1   (c) 10 peaks, σ_{peak}=0.2

(d) 50 peaks, σ_{peak}=0.02   (e) 50 peaks, σ_{peak}=0.1   (f) 50 peaks, σ_{peak}=0.2

(g) 100 peaks, σ_{peak}=0.02   (h) 100 peaks, σ_{peak}=0.1   (i) 100 peaks, σ_{peak}=0.2

(j) 150 peaks, σ_{peak}=0.02   (k) 150 peaks, σ_{peak}=0.1   (l) 150 peaks, σ_{peak}=0.2

Fig. 3. The objective functions for each two-dimensional test case

4 Results and Analysis

4.1 Two-Dimensional Test Cases

We ran all experiments with eight different EA configurations, as shown in Table 3, for each of the twelve different two-dimensional test cases generated by TCG-2. The experiments were repeated 150 times per set (50 seeds with 3 random-number generator techniques, as suggested in [1]). Different experiment sets were run for a population size of 10, 20, 30 and 100 individuals per generation and it was found, similarly to the results presented in [1], that the results scaled accordingly, as the population size increased.

Table 3. The eight different EA configurations

Configuration	Selection	Mutation	Crossover
EA	Roulette Wheel	Random	None
SEA	Roulette Wheel	Scouting-driven	None
EAC	Roulette Wheel	Random	Single-point
SEAC	Roulette Wheel	Scouting-driven	Single-point
EAF	FUSS	Random	None
SEAF	FUSS	Scouting-driven	None
EAFC	FUSS	Random	Single-point
SEAFC	FUSS	Scouting-driven	Single-point

The results obtained clearly show that scouting improved all EA configurations used. The more deceptive the problem, the bigger the improvement of the performance exhibited by the SEAxx methods. More specifically, it was observed that there was a larger average improvement of performance, as the landscape contained less and narrower peaks. Figures 4 and 5 display the fitness level reached by generation, for, due to lack of space, only a few representative sets of experiments, and only for population size of 20 individuals. As mentioned above, however, the results scale accordingly as we change the population size. It is particularly important to note, that in all sets of experiments, the worst performance of an SEAxx was always better than the average performance of the equivalent EAxx, and very close to the global optimum, as one can clearly see in the figures mentioned above.

SEAC had similar performance with SEA, as it was originally hypothesized. The crossovers did not seem to have a negative effect in the way scouting affected the EA, and the improvement of the equivalent simple EA configurations was similar. (see Fig.4)

Scouting has also managed to significantly enhance the techniques that used FUSS. The effect of scouting overpowered the effect of FUSS (see Fig.5), the performance of which was not particularly impressive in these problems and EA configurations. For example, EAF and EAFC had the lowest average and worst fitness level reached for the test case of 10 peaks and $\sigma_{peak} = 0.02$. While it did enhance the performance of the traditional EA, in certain test cases, the

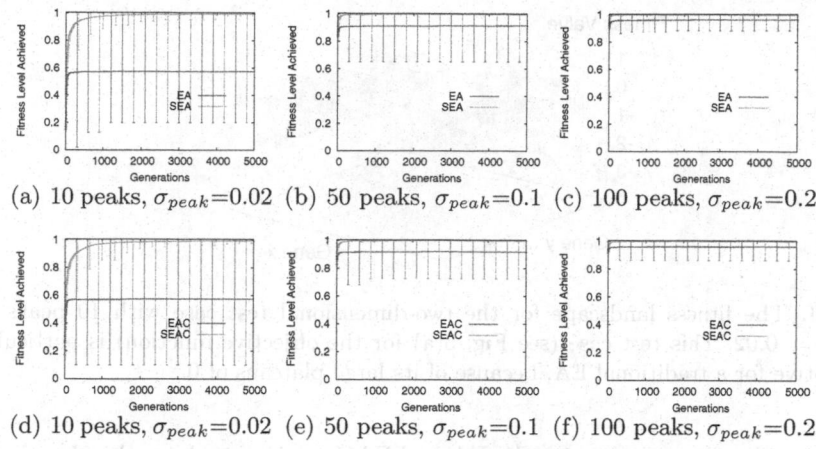

Fig. 4. Fitness level achieved per generation by EA vs SEA(first row), and by EAC vs SEAC(second row), for few of the experiment sets. It is clear that as the peaks and their width decrease, the improvement scouting achieves is more impressive.

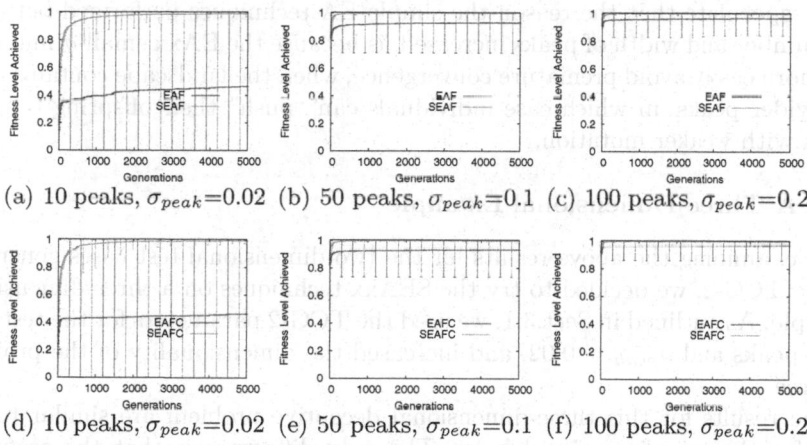

Fig. 5. Same as Fig. 4 but for EAF vs SEAF(first row), and EAFC vs SEAFC(second row)

improvement was not significant. However, it is important to note that FUSS was originally designed for steady-state EAs and for larger population sizes.

The most impressive achievement of scouting was exhibited in the test case with 10 peaks and $\sigma_{peak} = 0.02$. The landscape can be imagined as an almost completely flat surface of objective values of zero, with 10 narrow cones that rise up to 10 different optima, the global of which is at the objective value of 1. One can clearly see the plateaus of the zero fitness level in Fig. 6, which displays the fitness function for this test case—generated using (3). All EAxx configurations

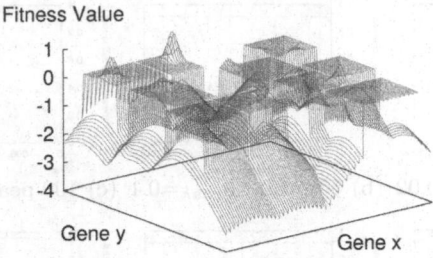

Fitness Value

Fig. 6. The fitness landscape for the two-dimensional test case with 10 peaks and $\sigma_{peak} = 0.02$. This test case (see Fig. 3(a) for the objective function) is particularly deceptive for a traditional EA, because of its large plateaus of 0.

performed rather poorly with the EA and EAC getting stuck in a local optimum, and the EAF and EAFC performing even worse, fitness-wise, but showing signs of slow improvement. (See Figs.4(a), 4(d), 5(a) and 5(d).) All four techniques, once enhanced with scouting, got very close to the global optimum, even from the early stages of the evolution process.

We speculate that the reason the simple EA techniques performed better as the number and width of peaks increased, is because the EAxx small-σ mutation can more easily avoid premature convergence, when the landscape contains more and wider peaks, in which case individuals can "push" their offspring to other peaks with weaker mutation.

4.2   A Three-Dimensional Example

After examining the above results for the two-dimensional test cases generated by the TCG-2, we decided to try the SEAxx techniques on a three-dimensional example. As outlined in Sect.3.1, we used the TCG-2 parameters for the test case of 10 peaks and $\sigma_{peak} = 0.02$, and increased the dimensionality of the problem to three.

The results for this three-dimensional deceptive problem are similar to the ones for the two-dimensional cases. The only difference is that the scouting-aided techniques are a little slower in reaching higher levels of fitness. Figure 7 shows the different fitness levels achieved per generation by the EAC and the SEAC. One can clearly see the improvement scouting has on the EAC. The SEAC catches up with the EAC at ca. 500 generations and continues reaching better fitness levels at a steady pace, whereas the EAC on average converges prematurely.

The behaviors of the EA and the SEA are very similar to the EAC and the SEAC. However, the EAFx techniques performed very poorly, whereas the SEAFx ones performed similarly to the SEA techniques that used Roulette Wheel selection.

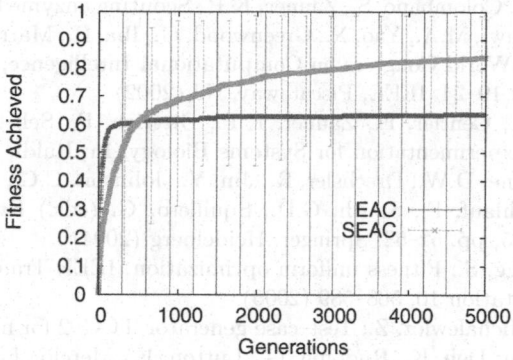

Fig. 7. Fitness level achieved per generation by the EAC and the SEAC with a population size of 20 individuals, for the 3-dimensional case of 10 peaks and $\sigma_{peak} = 0.02$

5   Conclusion-Future Work

The results obtained by all experiments prove that scouting can make a significant positive difference to the performance of an EA in the domain of NLP problems with rugged landscapes and multiple peaks. We have suggested a new mutation strength modulator and shown that scouting improves a variety of different EA configurations for twelve two-dimensional, multi-modal and deceptive problems, and one three-dimensional example. These configurations, among others, include crossover, and the Fitness-Uniform Selection Scheme(FUSS), a scheme specifically designed to improve performance in such problems. We have also examined the kind of problems SEA is particularly good at, and found that it exhibits particularly impressive performance when dealing with landscapes that contain large plateaus.

Future work will extensively examine the performance of the SEAxx techniques on multi-dimensional problems. Another research direction could be the effect of different crossover and mutation rates on SEAxx techniques. Scouting-driven adaptation of these rates is also an interesting path. Finally, another avenue of exploration could include steady-state EAs and higher population sizes, namely configurations that would favor FUSS. Additional testing could also include some of the problems FUSS was originally tested on.[5]

References

1. Pfaffmann, J.O., Bousmalis, K., Colombano, S.: A scouting–inspired evolutionary algorithm. In: CEC 2004. Congress on Evolutionary Computation, 2004. June 19–23, 2004, vol. 2, pp. 1706–1712 (2004)
2. Pfaffmann, J.O., Zauner, K.P.: Scouting context–sensitive components. In: Keymeulen, D., Stoica, A., Lohn, J., Zebulum, R.S. (eds.) The Third NASA/DoD Workshop on Evolvable Hardware–EH-2001, Long Beach, 12–14 July 2001, pp. 14–20. IEEE Computer Society, Los Alamitos (2001)

3. Matsumaru, N., Colombano, S., Zauner, K.P.: Scouting enzyme behavior. In: Fogel, D.B., El–Sharkawi, M.A., Yao, X., Greenwood, G., Iba, H., Marrow, P., Shackleton, M. (eds.) 2002 World Congress on Computational Intelligence, Honolulu, Hawaii, May 12–17, pp. 19–24. IEEE, Piscataway, NJ (2002)

4. Matsumaru, N., Centler, F., Zauner, K.P., Dittrich, P.: Self–Adaptive–Scouting Autonomous Experimentation for Systems Biology. In: Raidl, G.R., Cagnoni, S., Branke, J., Corne, D.W., Drechsler, R., Jin, Y., Johnson, C.G., Machado, P., Marchiori, E., Rothlauf, F., Smith, G.D., Squillero, G. (eds.) EvoWorkshops 2004. LNCS, vol. 3005, pp. 52–62. Springer, Heidelberg (2004)

5. Hutter, M., Legg, S.: Fitness uniform optimization. IEEE Transactions on Evolutionary Computation 10, 568–589 (2006)

6. Schmidt, M., Michalewicz, Z.: Test–case generator TCG–2 for nonlinear parameter optimization. In: Deb, K., Rudolph, G., Lutton, E., Merelo, J.J., Schoenauer, M., Schwefel, H.-P., Yao, X. (eds.) Parallel Problem Solving from Nature-PPSN VI. LNCS, vol. 1917, pp. 539–548. Springer, Heidelberg (2000)

7. Galassi, M., Davies, J., Theiler, J., Gough, B., Jungman, G., Booth, M., Rossi, F.: GNU Scientific Library Reference Manual. Network Theory Ltd., Bristol, UK (2003)

8. Beyer, H.-G.: The Theory of Evolution Strategies. Springer, Berlin (2001)

9. Matsumoto, M., Nishimura, T.: Mersenne twister: A 623-dimensionally equidistributed uniform pseudo-random number generator. ACM Transactions on Modeling and Computer Simulation 8(1), 3–30 (1998)

10. Lüscher, M.: A portable high-quality random number generator for lattice field theory calculations. Computer Physics Communications 79, 1000–1110 (1994)

Stochastic Barycenters and Beta Distribution for Gaussian Particle Swarms

Rui Mendes[1] and James Kennedy[2]

[1] CCTC
Universidade do Minho
Portugal
[2] Bureau of Labor Statistics
USA

Abstract. Recent research has explored methods for modifying the particle swarm algorithm so that it samples from a probability distribution rather than generating the particle's trajectory from averaged differences. The present paper explores a model where a Gaussian sample is taken, with the mean of the distribution varying randomly within some bounds. In particular, these experiments use either a stochastic barycenter or a beta distribution to define the mean of the Gaussian sample.

1 Introduction

Particle swarms were introduced in 95 by Kennedy and Eberhart [8] and use a simple formula that describes how particles explore the solution space. Each individual uses an equation that describes how their position and velocity are updated. The velocity update uses two sources of influence: the previous best position found by the individual and the best position found by the group.

Recently, researchers have been experimenting with variants of particle swarms with new ways of generating solutions [4,6,5,7]. This paper proposes two forms of specifying the social central tendency for the Gaussian particle swarm. One is the stochastic barycenter, where the contributions of the neighbors are averaged using stochastic weights and the other uses the Beta distribution to generate the value that will be used.

The paper is organized as follows: section 2 presents the variants of particle swarm that will be used in the comparison and that inspired the modifications presented. Section 3 proposed the versions using the stochastic barycenter and the beta distribution. Section 4 presents the experiments performed and section 5 discusses the results. Finally, section 6 presents the conclusions.

2 The Particle Swarm

2.1 The Canonical Particle Swarm

The canonical particle swarm uses a constriction coefficient that was introduced in 2002 by Clerc et al [1]. The equations for determining the velocity and position are the following:

J. Neves, M. Santos, and J. Machado (Eds.): EPIA 2007, LNAI 4874, pp. 259–270, 2007.

$$v_{t+1} = \chi(v_t + \mathrm{U}[0,1]\,\frac{\varphi}{2}(p_i - x_t) + \mathrm{U}[0,1]\,\frac{\varphi}{2}(p_g - x_t)) \tag{1}$$

$$x_{t+1} = x_t + v_{t+1} \tag{2}$$

where x_t is the position, v_t the velocity, p_i is the previous best position, p_g is the best position of the group, φ is the acceleration coefficient and χ is the constriction factor.

2.2   The Fully Informed Particle Swarm

There are other models besides the one presented above where the influence comes from the individual itself and the best of its neighborhood. In 2003, Mendes et al [10] proposed a different model where each individual is influenced by the entire neighborhood. In this case, the number of sources of influence is equal to the number of neighbors. The velocity update is given by:

$$c_k = \mathrm{U}[0,1] \tag{3}$$

$$p_m = \frac{\sum_{k \in \mathcal{N}} c_k\, p_k}{\sum_{k \in \mathcal{N}} c_k} \tag{4}$$

$$\phi = \frac{\varphi}{|\mathcal{N}|} \sum_{k \in \mathcal{N}} c_k \tag{5}$$

$$v_{t+1} = \chi(v_t + \phi\,(p_m - x_t)) \tag{6}$$

where $\mathcal{N}$ is the set of neighbors of the current individual and p_k represent their previous best positions. As can be easily understood, p_m is a stochastic barycenter of the previous best positions of the neighbors of the particle.

In fact, the canonical particle swarm may be rewritten using this model by simply setting p_m as the stochastic average of p_i and p_g. Using this general model, the velocity is updated in order to attract the individual to a given target position p_m. In the case of canonical particle swarm, this position is given by:

$$c_1 = \mathrm{U}[0,1] \tag{7}$$

$$c_2 = \mathrm{U}[0,1] \tag{8}$$

$$p_m = \frac{c_1\, p_i + c_2\, p_g}{c_1 + c_2} \tag{9}$$

By substituting in equation 6 we get:

$$v_{t+1} = \chi(v_t + \frac{\varphi}{2}(c_1 + c_2)\,(\frac{c_1\, p_i + c_2\, p_g}{c_1 + c_2} - x_t)) \tag{10}$$

Which, by a process of expansion and simplification, yields equation 1.

2.3   The Gaussian FIPS

In 2003, Kennedy [3] suggested a new way of looking at particle swarms by generating new solutions without using velocity. In this model, the particle's position was updated according to:

$$x_{t+1} = N\left(\frac{p_i + p_g}{2}, |p_i - p_g|\right) \tag{11}$$

In this model, the particle explores the region defined by the social central tendency, defined by its previous best position and the best position found in its group.

By using a normal distribution, the particle incorporates a social dispersion that is scaled by the range between its previous best position and the best position found in its group. This has the advantage of using an adaptive step size. The adaptive step size is a very ingenious process that allows a particle to explore the region around itself and increasing the amount of exploration used if a better solution is found in a larger area.

This characteristic is intrinsic to all forms of particle swarms and is one of the reasons for its success. The adaptive step size allows particles to explore wider areas when diversity is high and to exploit good solutions when clustering together. However, if any particle discovers a new solution in another region of the search space, the particles immediately adopt an exploratory behavior.

However, this model didn't perform as well as the canonical particle swarm. After a certain amount of experiments, it became apparent that the statistical distribution needed larger tails than provided by the normal distribution and the velocity was re-incorporated as a way to provide it.

The algorithm evolved, eventually incorporating ideas from FIPS, becoming the Gaussian FIPS [4,6,5,7]. The rationale behind this system is to explore around the mean of the best positions of the particle's neighbors. The equations of the Gaussian FIPS are:

$$p_m = \frac{\sum_{k \in N} p_k}{|\mathcal{N}|} \tag{12}$$

$$x_{t+1} = \chi (x_t - x_{t-1}) + N\left(p_m, \frac{|p_m - p_i|}{2}\right) \tag{13}$$

3   New Approaches to Central Tendency

3.1   Characteristics of the Stochastic Barycenter

An advantage of the Gaussian FIPS is that it is very simple to understand. It simply generates a new position by adding the velocity to a number following the normal distribution with mean equal to the mean of the previous best positions of the neighborhood and standard deviation equal to the half the range between the mean and the previous best position of the individual.

The characteristics of the probability distribution used by the particles can be modified easily by use of a stochastic barycenter which will increase the range of the search and make it somewhat more flexible. The stochastic barycenter is implemented similarly to FIPS by using equations 3 and 4 instead of equation 12. This algorithm will be called the *Gaussian Stochastic Barycenter FIPS* (GSBF for short).

We can characterize the statistical distribution of the stochastic barycenter, denoted $\text{Bar}(x)$ by the vector $x = x_1, \ldots, x_n$ whose components represent the contributions of each of the particle's neighbors. Random values following the distribution $\text{Bar}(x)$ are generated by using equations 3 and 4.

Without loss of generality, we may state that the points are given in ascending order thus making x_1 the minimum and x_n the maximum. To illustrate the statistical distribution of the stochastic barycenter, we plotted several histograms for different values of the vector x.

Figure 1 shows several histograms of the generation of points from a stochastic barycenter when $x_1 = 0$, $x_4 = 1$ and $0 \leq x_2, x_3 \leq 1$. As can be seen, the distribution is only symmetrical when the mean of the generating points $\bar{x}$ is 0.5, i.e., it coincides with the mean of the range. In fact, other experiments corroborate this observation: the stochastic barycenter distribution is only symmetrical if $\bar{x} = \frac{x_1 + x_n}{2}$ assuming that $x_1 = \min x$ and $x_n = \max x$.

The distribution $\text{Bar}(x)$ may be characterized assuming without loss of generality that $x = x_1, \ldots, x_n$, $x_1 = \min x$ and $x_n = \max x$. Thus, if $X \sim \text{Bar}(x)$ then:

- The expected value of the distribution coincides with the mean of the neighbor's contributions represented in vector x, i.e., $E[X] = \bar{x}$
- The distribution $\text{Bar}(x)$ asymmetry depends on the mean;
- The distribution is supported on a bounded interval, i.e., $X \in [x_1; x_n]$.

3.2   The Beta Distribution

The beta distribution is a continuous distribution, supported on the bounded interval $[0, 1]$ of which the uniform distribution is a special case [2]. This distribution has two parameters α and β and probability density function

$$\frac{x^{\alpha-1}(1-x)^{\beta-1}}{\text{B}(\alpha, \beta)} \tag{14}$$

where B is the Beta function. This distribution has mean and variance

$$E(X) = \frac{\alpha}{\alpha + \beta} \tag{15}$$

$$Var(X) = \frac{\alpha\beta}{(\alpha + \beta)^2(\alpha + \beta + 1)} \tag{16}$$

In particular, given a sample with mean $\bar{x}$ and variance v, it is possible to estimate the parameters of the distribution by using

$$\alpha = \bar{x}\left(\frac{\bar{x}(1 - \bar{x})}{v} - 1\right) \tag{17}$$

$$\beta = (1 - \bar{x})\left(\frac{\bar{x}(1 - \bar{x})}{v} - 1\right) \tag{18}$$

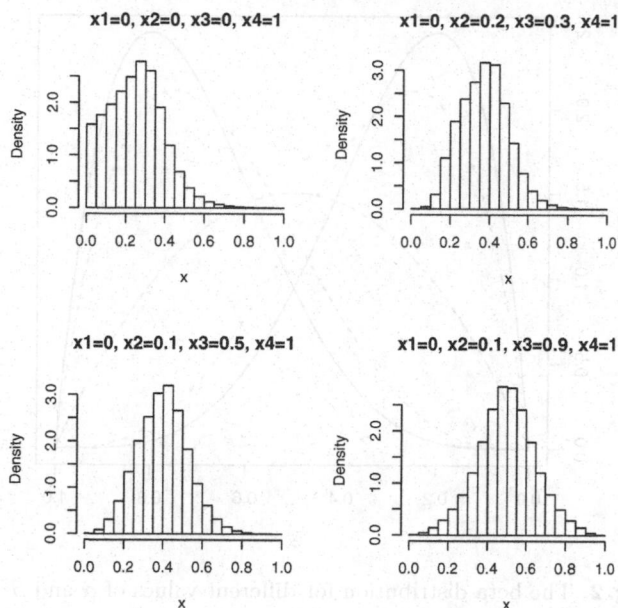

Fig. 1. Histogram of the generation of 1,000,000 values according to the stochastic barycenter. The histograms correspond to different values of x_2 and x_3 while keeping $x_1 = 0$ and $x_4 = 1$

3.3   Approximating the Stochastic Barycenter

The Beta distribution is a good candidate for approximating the stochastic barycenter because it is bounded and, depending on the parameters α and β, it may be asymmetrical or symmetrical.

To approximate a given distribution by the Beta distribution it is necessary to:

- Convert the distribution to a new one whose values lie in the interval $[0; 1]$;
- Estimate the mean of the distribution;
- Estimate the variance of the distribution.

Let us assume, for simplicity sake, that the positions of the neighbors (the parameters of the stochastic barycenter distribution) are given by $x_1, \ldots, x_n$ whose values are in ascending order. Thus, x_1 is the lowest and x_n is the highest. The first thing to do is to convert the values to a new set of values $y_1, \ldots, y_n$ given by

$$y_i = \frac{x_i - x_1}{x_n - x_1} \ \forall i \in 1, \ldots, n \tag{19}$$

and now we are reasoning in the interval $[0, 1]$.

The mean of the stochastic barycenter distribution is very easy to estimate, as it is simply the mean of the points $x_1, \ldots, x_n$. The real difficulty is estimating the variance. For this we decided to experiment with the special case of four

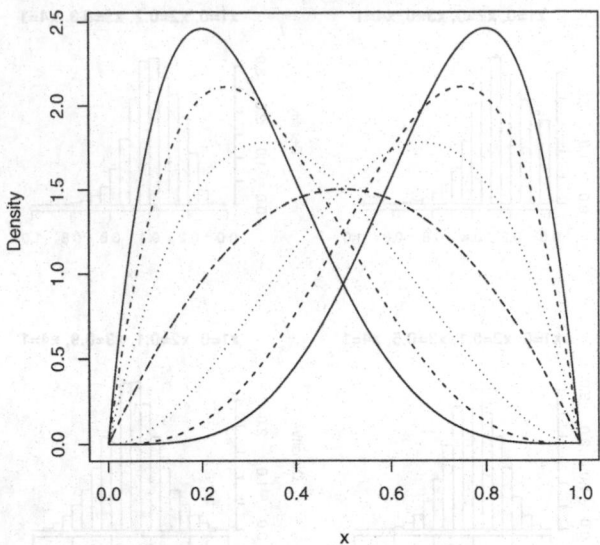

Fig. 2. The beta distribution for different values of α and β

neighbors because the study in [9] suggested that the optimal number of neighbors is four. As it is obvious that $y_1 = 0$ and $y_4 = 1$, we decided to vary y_2 and y_3 in the interval $[0, 1]$ with step 0.01. Using these values, we generated 1,000,000 points using the stochastic barycenter for these values of the vector y and record the variance. The results of the experiment can be seen in figure 3.

As is obvious from the figure, we then took the data and fitted it to an equation of the form

$$var_y = a + b\,(y_2 - 0.5)^2 + c\,(y_3 - 0.5)^2 \tag{20}$$

which gave us $a = 0.01303$ and $b = c = 0.1951$ with an adjusted $R^2 = 0.773$. We then used formulas 17 and 18 in order to estimate the values of α and β from $\bar{y}$ and var_y. This new algorithm exhibited a performance similar to GSBF but still was very complicated. However, it showed us that it was possible to approximate the stochastic barycenter by a beta distribution.

3.4   The Beta FIPS

In order to simplify the algorithm mentioned in the previous section, we decided to use a fixed variance of 0.015. This value was chosen because of the results we got from the experiment described in the previous section, as it was close to the mean variance we found. This is the algorithm that we dubbed Beta FIPS. This algorithm works as follows

$$y_j = \frac{p_j - p_1}{p_n - p_1} \; \forall j \in 1, \dots, n \tag{21}$$

$$\alpha = \bar{y} \left(\frac{\bar{y}\,(1 - \bar{y})}{0.015} - 1 \right) \tag{22}$$

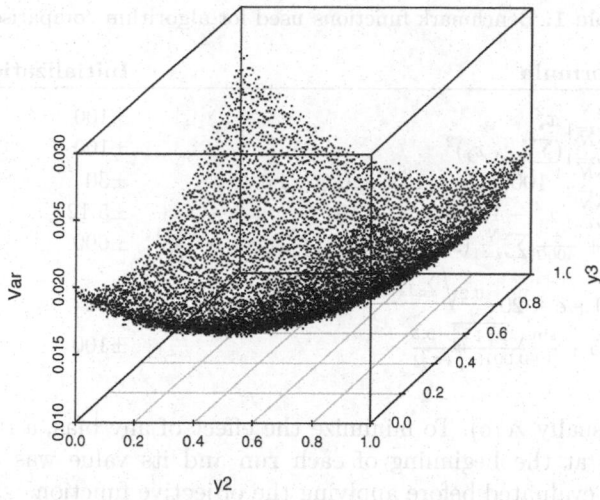

Fig. 3. Variance of the samples where $y_1 = 0$, $y_n = 1$ and y_2 and y_3 are given by the values of the x and y vertex respectively

$$\beta = (1 - \bar{y}) \left(\frac{\bar{y}\,(1 - \bar{y})}{0.015} - 1 \right) \tag{23}$$

$$p_b = p_1 + (p_n - p_1)\,\mathrm{B}(\alpha, \beta) \tag{24}$$

$$x_{t+1} = \chi(x_t - x_{t-1}) + \mathrm{N}\left(p_b, \frac{|\bar{p} - p_i|}{2} \right) \tag{25}$$

where $p_1, \ldots, p_n$ are the best positions of the neighbors given by ascending order (in fact, it is only important that p_1 is the minimum and p_n the maximum). Equations 21 and 24 are necessary to transform the values to and from the interval $[0, 1]$.

4   Experiments

The algorithms were implemented in C++ and the random distributions used were provided by the GNU scientific library. The code is available on `http://omega.di.uminho.pt/pso`.

The parameters were set to the usual values with $\varphi = 4.01$, $\chi = 0.729$ and the population size was 20. The particles were connected using the von Neumann topology where each particle is connected to 4 others following a rectangular torus configuration.

The functions given in table 1 were used to compare the different approaches and the experiments were replicated 100 times. Function f_5 was optimized for both 10 and 30 dimensions because it is actually harder in 10 dimensions because of the product term.

It is well known that some algorithms may present good results on some function benchmarks simply because they have a tendency to converge onto a

Table 1. Benchmark functions used for algorithm comparison

Function	Formula	Initialization Bounds
f_1	$\sum_{i=1}^{N} x_i^2$	± 100
f_2	$\sum_{i=1}^{N} (\sum_{j=1}^{i} x_j)^2$	± 100
f_3	$\sum_{i=1}^{N-1} 100(x_{i+1} - x_i^2)^2 + (x_i - 1)^2$	± 30
f_4	$\sum_{i=1}^{N} x_i^2 - 10\cos 2\pi x_i + 10$	± 5.12
f_5	$1 + \frac{1}{4000} \sum_{i=1}^{N} (x_i - 100)^2 - \prod_{i=1}^{N} \cos \frac{x_i - 100}{\sqrt{i}}$	± 600
f_6	$20 + e - 20 e^{0.2\sqrt{\frac{\sum_{i=1}^{N} x_i^2}{N}}} - e^{\frac{\sum_{i=1}^{N} \cos 2\pi x_i}{N}}$	± 32
f_7	$0.5 + \frac{\sin \sqrt{x_1^2 + x_2^2} - 0.5}{1 + 0.001(x_1^2 + x_2^2)}$	± 100

given value (usually zero). To minimize the effect of any bias, a random vector was generated at the beginning of each run and its value was added to the solution being evaluated before applying the objective function.

5   Results

The notation followed for the algorithms in the presentation of the results was the following: *beta* stands for the Beta FIPS, *can* for the canonical PSO, *fips* for FIPS, *gbary* for the GSBF and *gfips* stands for Gaussian FIPS.

Table 2 presents the mean best function result found after 3,000 iterations, averaged over 100 trials. The results are presented after 3,000 iterations have elapsed and display the 95% confidence intervals of the mean fitness. We believe that convergence plots convey a lot of information in this study and present them in figures 4 and 5.

Function f_1 is very simple as can be seen from the results. In fact, except for *can* and *gfips*, all the algorithms display similar convergence. It is interesting to

Table 2. 95% confidence intervals of the mean value of the objective function after 3,000 iterations over 100 runs

	f_1	f_2	f_3	f_4
beta	9.6e-27±1.5e-27	3.3±0.63	33±5	18±3.1
can	9.1e-24±6.8e-24	8.2±1.5	55±14	60±3.1
fips	8.1e-26±9.2e-27	1.7±0.38	35±6	17±0.87
gbary	7.1e-26±8.3e-27	2.5±0.4	39±9.3	38±7.3
gfips	2.5e-26±3.8e-27	22±11	51±16	17±0.95

	$f_5^{(10)}$	$f_5^{(30)}$	f_6	f_7
beta	0.0087±0.0021	0.00085±0.0006	1.7e-14±1.7e-15	0.00024±0.00028
can	0.049±0.0058	0.0086±0.0023	0.11±0.074	0.0012±0.00063
fips	0.0076±0.0016	0.0028±0.0024	0.019±0.026	0.0017±0.00073
gbary	0.016±0.0048	0.00062±0.00051	6.6e-14±5.8e-15	0.00022±0.00028
gfips	0.014±0.0035	0.05±0.023	0.32±0.1	0.0013±0.00058

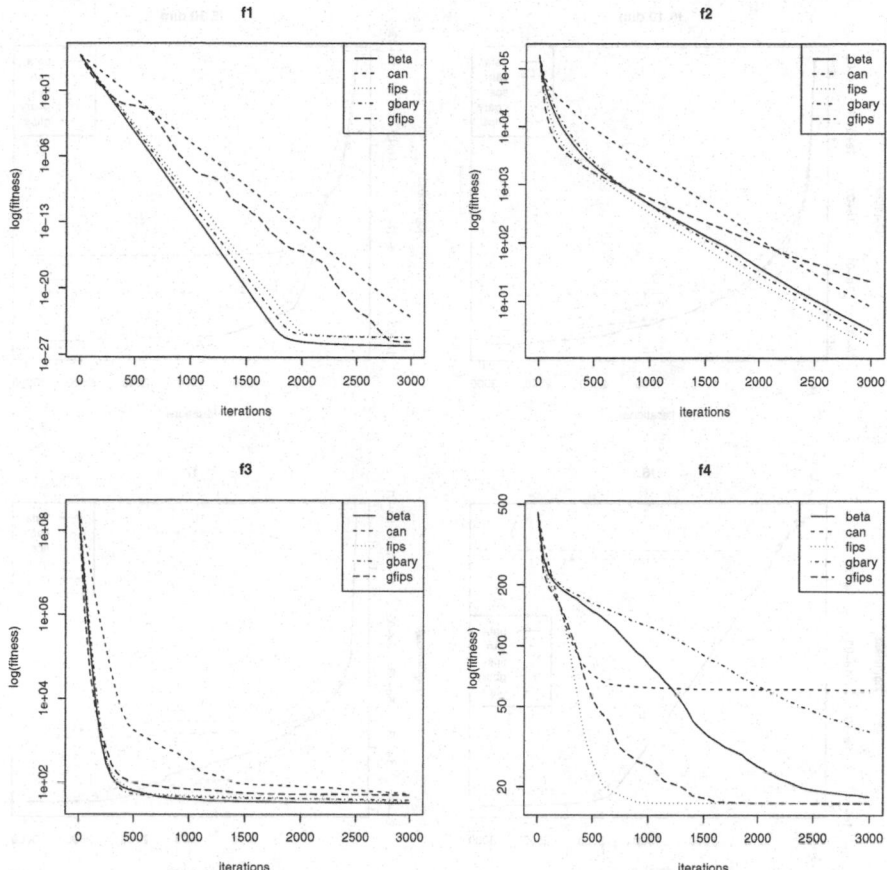

Fig. 4. Convergence plots for functions f_1 through f_4 averaged over 100 runs

notice that except for *gfips*, all algorithms display an exponential behavior on this function. This is because the function is very easy to optimize. The faster algorithms in this problem are *beta* and *gbary*.

The fact that three of the algorithms seem to stop improving after a certain number of iterations have elapsed is due to precision problems related to the the double precision floating point representation. This happens because of the random value that is added to each variable prior to the evaluation that decreases the precision available for the mantissa in some of the variables. The same behavior may be observed on *can* if the algorithm is run for 5,000 iterations and obviously disappears if the random value is not added.

fips, *gbary* and *beta* have very similar convergence behaviors in problem f_2. It is interesting to notice that *gfips* typically displays a very good performance in the beginning of the run and then starts to lag behind all the other algorithms, including *can*. The best algorithm in this problem is *fips*.

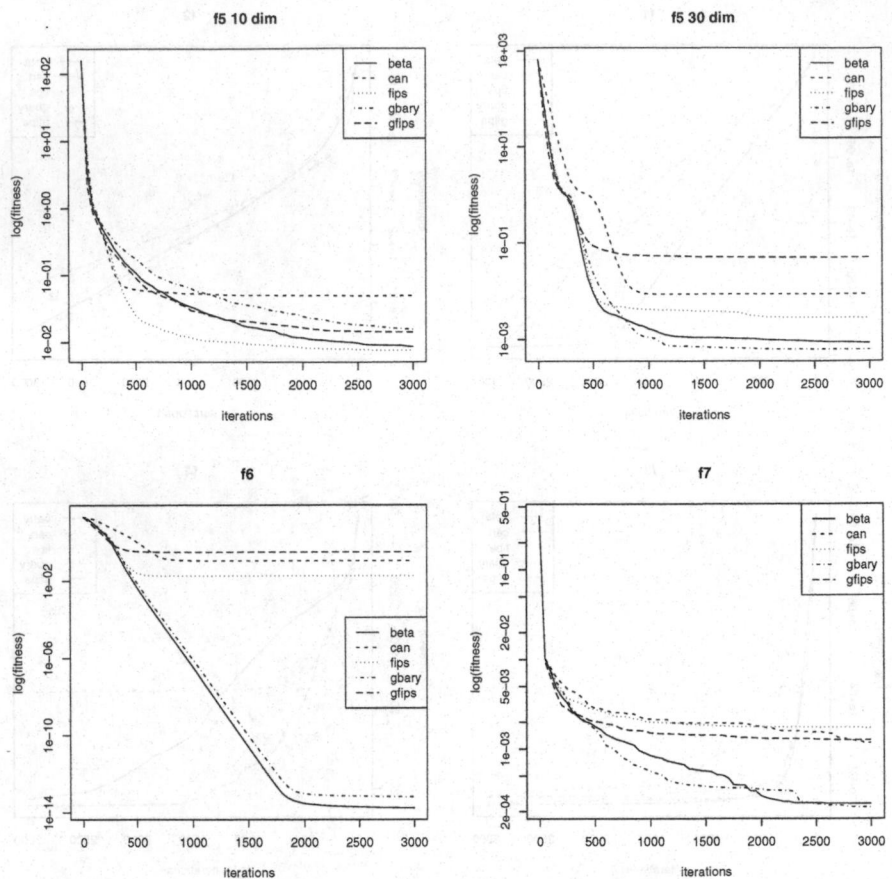

Fig. 5. Convergence plots for functions f_5 through f_7 averaged over 100 runs

All the approaches present similar convergence curves on problem f_3 except for *can*) that is slower. However, if given enough time, it catches up.

fips is the fastest algorithm in problem f_4 but is later joined by *gfips*. *beta* is faster than *gbary* in this problem and would probably find results similar to the other two algorithms with more function evaluations. *can* seems to consistently get stuck in local optima.

fips is the fastest algorithm in problem f_5 in 10 dimensions and *beta* is the second fastest. *gfips* and *gbary* lag a little behind while *can* gets stuck in local optima. The problem seems to be quite different in 30 dimensions where *beta* and *gbary* are clearly superior to the other algorithms. *gfips* is strangely the algorithm that has the most problems to optimize f_5 in 30 dimensions.

beta and *gbary* show a clear superiority in problem f_6. The other algorithms get stuck in local optima far above the better values found by these two approaches. This is only case where some of the algorithms exhibit such a clear superiority over the other ones.

Problem f_7 is a strange artifact and is very hard to optimize because of the large number of deceptive local optima close to the global optimum. Given enough iterations, *beta* and *gbary* seem to be harder to fool than the other algorithms.

6 Conclusions

In the traditional particle swarm, a particle's movement through the search space is defined in relation to its position on the last time-step, the direction it was moving on the previous iteration, and a stochastic mean of the differences between its current position and the previous best positions found by its sources of influence. The effect of the particle's trajectory was to sample around a region centered at the mean of the previous best positions, and with variance scaled to the differences between the influence sources' previous bests.

Thus recent investigations have modified the algorithm by sampling directly from a symmetric distribution, often Gaussian. The "bare-bones" particle swarm did not implement a velocity at all, but simply selected a point in the search space determined by the previous best points of the sources of influence. Performance was improved however when the particle's previous position and direction ("velocity") were included in the process. The observed distribution of sampled points in the canonical algorithm though is not Gaussian, in fact the kurtosis of the distribution increases with the number of iterations, as "bursts of outliers" are generated.

The present paper modifies the Gaussian distribution by centering it on a point that varies randomly, either according to a stochastic barycenter or a beta distribution. This is expected to add flexibility and increase the variance of the samples, with the both distribution's skewness perhaps focusing attention on most promising regions. In fact the beta-distributed central tendency performed well on all test functions; it converged more slowly than others on some functions, but it performed respectably, and better than the canonical particle swarm, on all the problems.

References

1. Clerc, M., Kennedy, J.: The particle swarm - explosion, stability, and convergence in a multidimensional complex space. IEEE Transactions on Evolutionary Computation 6(1), 58–73 (2002)
2. Evans, M., Hastings, N., Peacock, B.: Beta Distribution. In: Statistical Distributions, 3rd edn., pp. 34–42. Wiley, New York (2000)
3. Kennedy, J.: Bare bones particle swarms. In: SIS 2003. Proceedings of the Swarm Intelligence Symposium, Indianapolis, IN, Purdue School of engineering and technology, IUPUI, IEEE Computer Society, Los Alamitos (2003)
4. Kennedy, J.: Probability and dynamics in the particle swarm. In: CEC 2004. Proceedings of the IEEE Congress on Evolutionary Computation, pp. 340–347. IEEE Computer Society, Los Alamitos (2004)

5. Kennedy, J.: Dynamic-probabilistic particle swarms. In: GECCO 2005. Proceedings of The Genetic and Evolutionary Computation Conference, Washington, DC, pp. 201–207 (2005)
6. Kennedy, J.: Why does it need velocity? In: Proceedings of the 2005 IEEE Swarm Intelligence Symposium, Pasadena, CA, IEEE Computer Society, Los Alamitos (2005)
7. Kennedy, J.: Search of the essential particle swarm. In: CEC 2006. Proceedings of the IEEE Congress on Evolutionary Computation, IEEE Computer Society, Los Alamitos (2006)
8. Kennedy, J., Eberhart, R.: Particle swarm optimization. In: Proceedings of IEEE International Conference on Neural Networks, pp. 1942–1948. IEEE Press, Los Alamitos (1995)
9. Mendes, R.: Population Topologies and Their Influence in Particle Swarm Performance. PhD thesis, Escola de Engenharia, Universidade do Minho, May, Phd Thesis (May 2004)
10. Mendes, R., Kennedy, J., Neves, J.: Watch thy neighbor or how the swarm can learn from its environment. In: SIS 2003. Proceedings of the Swarm Intelligence Symposium, pp. 88–94. IEEE Press, Los Alamitos (2003)

Exploiting Second Order Information in Computational Multi-objective Evolutionary Optimization

Pradyumn Kumar Shukla

Institute for Numerical Mathematics
Dresden University of Technology
Dresden, D-01062, Germany
pradyumn.shukla@mailbox.tu-dresden.de

Abstract. Evolutionary algorithms are efficient population based algorithms for solving multi-objective optimization problems. Recently various authors have discussed the efficacy of combining gradient based classical methods with evolutionary algorithms. This is done since gradient information leads to convergence to Pareto-optimal solutions with a linear convergence rate. However none of existing studies have explored how to exploit second order or Hessian information in evolutionary multi-objective algorithms. Second order information though costly, leads to a quadratic convergence to Pareto-optimal solutions. In this paper, we take Levenberg-Marquardt methods from classical optimization and show two possible ways of hybrid algorithms. These algorithms require gradient and Hessian information which is obtained using finite difference techniques. Computational studies on a number of test problems of varying complexity demonstrate the efficiency of resulting hybrid algorithms in solving a large class of complex multi-objective optimization problems.

1 Introduction

Multi-objective optimization is a rapidly growing area of research and application in modern-day optimization. There exist a plethora of non-classical methods which follow some natural or physical principles for solving multi-objective optimization problems, see for example the book by [2]. On the other hand a large amount of studies have been devoted to develop classical methods for solving multi-objective optimization problems ([4]).

Evolutionary algorithms use stochastic transition rules using crossover and mutation search operators to move from one solution to another. In this way global structure of search space is exploited. Classical methods, on the other hand, usually use deterministic (usually gradient/ Hessian based) transition rules to move from one solution to another. Classical methods effectively use local information thus ensuring fast convergence. This however comes up at the cost of requiring gradient or Hessian information which requires a large number of function evaluations. Hence one sees that there is a trade-off between fast

J. Neves, M. Santos, and J. Machado (Eds.): EPIA 2007, LNAI 4874, pp. 271–282, 2007.
© Springer-Verlag Berlin Heidelberg 2007

convergence and number of function evaluations. Hybrid implementations thus continue to be developed and tested (see for example [1,7]).

The limitations of the existing studies is that they only use gradient or the first order information. Second order information is usually ignored. The reason being that second order information is costly in terms of number of function evaluations (if analytical expression are not available). However they give an added advantage that now convergence is also of second order. In order to explore this trade-off, we take in this contribution a quadratically convergent robust method for solving nonlinear equations the Levenberg-Marquardt method and apply it on the KKT system (optimality conditions) of unconstrained multi-objective problem. We use this in a state-of-the-art multi-objective evolutionary algorithm to create a powerful hybrid multi-objective metaheuristics algorithm. We demonstrate their efficiency in solving real valued differentiable problems of varying complexity.

This paper is structured as follows. The next section presents the Levenberg-Marquardt method and gradient/ Hessian estimation methods. Section 3 presents the computational setup while simulation results are discussed in Section 4. Conclusions as well as extensions which emanated from this study are presented in the end of this contribution.

2   A Second Order Method

For the present study we take the Levenberg-Marquardt method [8,9] and use this as an additional search operator in the elitist non-dominated sorting GA or NSGA-II developed by [3]. The gradient/ Hessian of the objective functions are numerically computed by standard finite difference method.

2.1   Levenberg-Marquardt Method

The Levenberg-Marquardt method is an efficient Newton type method for solving the general nonlinear equation

$$F(\mathbf{x}) = 0.$$

It computes a trial step as

$$d = -(J(\mathbf{x})^\top J(\mathbf{x}) + \mu I)^{-1} J(\mathbf{x})^\top F(\mathbf{x}),$$

where $J(\mathbf{x}) = F'(\mathbf{x})$ is the Jacobian, and $\mu = \|F(\mathbf{x})\|$ is a parameter. The Levenberg-Marquardt step is a modification of the standard Newton step. It works even when the matrix $J(\mathbf{x})$ is singular as opposed to the standard Newton method. The parameter μ makes the matrix $(J(\mathbf{x})^\top J(\mathbf{x}) + \mu I)$ positive definite and numerically robust [5]. The Levenberg-Marquardt (LM) method has a quadratic rate of convergence under some mild smoothness conditions [5]. However in order to apply one step of the LM method one has to reformulate minimization of multiple objectives (in the Pareto sense) to a system of nonlinear equations. Let $f = (f_1, f_2, \ldots, f_m)$ be the unconstrained multiple objectives (where m equals the number of objectives). Then the necessary KKT optimality conditions for this multi-objective optimization problem read as:

$$\sum_{i=1}^{m} \lambda_i \nabla f_i(\mathbf{x}) = 0,$$

$$\lambda_i \geq 0.$$

We apply the above LM local search on this nonlinear equation i.e. $F(\mathbf{x}) := \sum_{i=1}^{m} \lambda_i \nabla f_i(\mathbf{x})$. The values of λ vector we choose in two different ways:

Randomly direction. As a first method we generate each λ_i from uniform distribution $U(0, 1)$ and use it in F i.e.

$$F(\mathbf{x}) = \sum_{i=1}^{m} \lambda_i \nabla f_i(\mathbf{x}). \qquad (1)$$

Steepest descent direction. In this method we use the multipliers λ in such way that the resulting direction $-\sum_{i=1}^{m} \lambda_i \nabla f_i(\mathbf{x})$ is a steepest descent direction at point $\mathbf{x}$. The exact details of finding such a λ is not explained here. For an interested reader we refer to the original study [10].

2.2 Hessian Computation Method

In almost all classical algorithms (for both single and multi-objective problems) the gradients and the Hessian of a function (say in general y) are required. One approach for estimating the Hessian is the Finite-difference (FD) method. Let $\mathbf{e}_i$ and $\mathbf{e}_j$ denote a unit vector in the ith and j^{th} direction respectively, then for a variable (say $\mathbf{x}$) of dimension n the FD approximation for the Hessian matrix is given as

$$h_{ji}(\mathbf{x}) = h_{ij}(\mathbf{x}) = \frac{y(\mathbf{x} + \varepsilon \mathbf{e}_i + \varepsilon \mathbf{e}_j) - y(\mathbf{x} + \varepsilon \mathbf{e}_i) - y(\mathbf{x} + \varepsilon \mathbf{e}_j) + y(\mathbf{x})}{\varepsilon^2},$$

This method requires evaluating the y at $\mathbf{x} + \varepsilon \mathbf{e}_i + \varepsilon \mathbf{e}_j$ for all possible i and j and at the points $\mathbf{x} + \varepsilon \mathbf{e}_i$ for all $i = 1, 2, \ldots, n$ and at $\mathbf{x}$. This requires a total of $n(n+1)/2 + n + 1$ function evaluations. From the above function known values one can also compute the one-sided FD gradient estimate given by

$$g_i(\mathbf{x}) = \frac{y(\mathbf{x} + \varepsilon \mathbf{e}_i) - y(\mathbf{x})}{\varepsilon},$$

Using LM method on Equation 1 (in NSGA-II as described in next section) the resulting hybrid algorithm we call as LR-NSGA. Using LM method with λ corresponding to steepest descent direction (in NSGA-II as described in next section) the resulting hybrid algorithm we call as LS-NSGA.

3   Computational Setup

In this section, we compare the above two hybrid methods with the elitist non-dominated sorting GA or NSGA-II [3][1] on a number of unconstrained test problems. The test problems are chosen in such a way so as to systematically investigate various aspects of an algorithm. We consider the two-objective ZDT test problems discussed in [2]. In their initial form these test problems (except ZDT4) are box constrained ones as the Pareto-optimal set lies on box constrains. We use the unconstrained versions of these test problems discussed elsewhere [12]. Table 1 present these test problems.

For all problems solved, we use a population size of 100. For NSGA-II every generation requires 100 function evaluation. For the hybrid algorithms first at end of the first generation we find one non-dominated point on which we apply the LM based local search for the hybrid algorithms. For an n dimensional variable the LM based local search requires total of $n(n+1)/2+n+1$ function evaluations. After this costly generation we do not use the LM based local search in the next $\lfloor \frac{n(n+1)/2+n+1}{100} \rfloor$ generations. Then we again find one non-dominated point on which we apply the LM based local search for the hybrid algorithms. Then we again do not use LM search for next $\lfloor \frac{n(n+1)/2+n+1}{100} \rfloor$ generations, then we again use the LM search and so on. In this way the entire population is expected to be pulled towards the efficient front by the point found out by the LM local search (which is quadratically close to efficient front). We set the upper limit on the number of function evaluations as 4000 for ZDT1, ZDT2, ZDT3 since in about 4000 function evaluations the population reaches the Pareto-optimal front by the best algorithm. For ZDT4 and ZDT6 we set the number of function evaluation to be 40000 and 12000 for the same reason. For all the algorithms, we use a standard real-parameter SBX and polynomial mutation operator with $\eta_c = 10$ and $\eta_m = 10$, respectively [2]. Following the guidelines in [11] we take the step length for computing finite difference estimate of Hessian and Jacobian as $\varepsilon = 10^{-6}$. It is to be noted that (except ε) the hybrid algorithms do not need any additional parameters.

Convergence and diversity are two distinct goals in multi-objective optimization. In order to evaluate convergence we use the Inverted Generational Distance (IGD) metric [2]. This measure of convergence indicated how far is the true Pareto-optimal front from the obtained front by each of the algorithms. Diversity of solutions is evaluated using the Spread (denoted by S) metric [2]. Algorithms A is better than Algorithm B in terms of convergence (diversity) if IGD (S) of Algorithm A is less than IGD (S) of Algorithm B. We run each of the three algorithms for 50 times (using same initial population for each algorithms and different initial populations every time in the 50 runs). After every run we use non-dominated solutions in the end for calculating the average, best worst and standard deviation of IGD and S metric values.

These unary metrices for convergence and diversity are used together with two binary metrices which can detect whether an approximation set is better than

---

[1] Revision 1.1 available from http://www.iitk.ac.in/kangal/codes.shtml

Table 1. Test problems used in this study

Name	Objective functions	$g(\cdot)$ function / n	Variable bounds / Type
ZDT1	$f_1(\mathbf{x}) = x_1$ $f_2(\mathbf{x}) = g(\mathbf{x})\left[2 - \sqrt{\frac{x_1}{g(\mathbf{x})}}\right]$	$g(x) = 1 + \frac{9}{n-1}\sum_{i=2}^{n} x_i^2$ $n = 30$	$[0.01, 1] \times [-1, 1]^{29}$ convex
ZDT2	$f_1(\mathbf{x}) = x_1$ $f_2(\mathbf{x}) = g(\mathbf{x})\left[2 - \left(\frac{x_1}{g(\mathbf{x})}\right)^2\right]$	$g(x) = 1 + \frac{9}{n-1}\sum_{i=2}^{n} x_i^2$ $n = 30$	$[0.01, 1] \times [-1, 1]^{29}$ non-convex
ZDT3	$f_1(\mathbf{x}) = x_1$ $f_2(\mathbf{x}) = g(x)\left[2 - \sqrt{\frac{x_1}{g(\mathbf{x})}} - \frac{x_1}{g(\mathbf{x})}\sin(10\pi x_1)\right]$	$g(x) = 1 + \frac{9}{n-1}\sum_{i=2}^{n} x_i^2$ $n = 30$	$[0.01, 1] \times [-1, 1]^{29}$ convex, disconnected
ZDT4	$f_1(\mathbf{x}) = x_1$ $f_2(\mathbf{x}) = g(\mathbf{x})\left[2 - \sqrt{\frac{x_1}{g(\mathbf{x})}}\right]$	$g(x) = 1 + 10(n-1) + \sum_{i=2}^{n}(x_i^2 - 10\cos(4\pi x_i))$ $n = 10$	$[0.01, 1] \times [-5, 5]^{9}$ convex, multimodal
ZDT6	$f_1(\mathbf{x}) = 1 - \exp(-4x_1)\sin^6(4\pi x_1)$ $f_2(\mathbf{x}) = g(\mathbf{x})\left[2 - \left(\frac{f_1(\mathbf{x})}{g(\mathbf{x})}\right)^2\right]$	$g(x) = 1 + 9\left(\frac{\sum_{i=2}^{n} x_i^2}{n-1}\right)^{0.25}$ $n = 10$	$[0.01, 1] \times [-1, 1]^{9}$ non-convex, non-uniform density

Table 2. Description of relations used to compare algorithms

Relation	Conditions	Description
strictly dominates	$I_\epsilon(A, B) < 1$	every $b \in B$ is strictly dominated by at-least one $a \in A$
dominates	$I_C(A, B) = 1$, $I_C(B, A) = 0$	every $b \in B$ is dominated by at-least one $a \in A$
better	$I_\epsilon(A, B) \leq 1$, $I_\epsilon(B, A) > 1$; $I_C(A, B) = 1$, $I_C(B, A) < 1$	every $b \in B$ is weakly dominated by at-least one $a \in A$ and $A \neq B$
weakly dominates	$I_\epsilon(A, B) \leq 1$; $I_C(A, B) = 1$	every $b \in B$ is weakly dominated by at-least one $a \in A$
incomparable	$I_\epsilon(A, B) > 1$, $I_\epsilon(B, A) > 1$; $I_C(A, B) \in (0, 1)$, $I_C(B, A) \in (0, 1)$	neither A weakly dominates B nor B weakly dominates A
relatively better	$I_\epsilon(A, B) < I_\epsilon(B, A)$; $I_C(A, B) > I_C(B, A)$	approximation set A is relatively better than set B
better in terms of convergence	$IGD(A) < IGD(B)$	convergence of approximation set A is better than that of set B
better in terms of diversity	$S(A) < S(B)$	diversity of approximation set A is better than that of set B

another. We use the multiplicative binary ϵ indicator discussed by Zitzler [14] and the two Set Coverage (SC) ([13]) to assess the performance of the algorithms. Given two outcomes A and B, of different algorithms, the binary ϵ indicator $I_\epsilon(A, B)$ gives the factor by which an approximation set is worse than another with respect to all objectives. The Set Coverage (SC) metric I_C calculates the proportion of solutions produced by Algorithm B, which are weakly dominated by solutions produced by Algorithm A. We use these binary metrices to conclude whether an approximation set produced by an algorithm *strictly dominates,*

dominates, better, weakly dominates or *is incomparable* with the approximation set produced by another algorithm (see [14] and Table 2 for definitions). For example, if $I_\epsilon(A, B) \leq 1$ and $I_\epsilon(B, A) > 1$ occurs then we can conclude that Algorithm A is better than Algorithm B. These conditions are quite difficult to satisfy using binary ϵ indicator values. We will use the binary I_ϵ indicator values to conclude *partial* results: we will say that Algorithm A is *relatively better as per* I_ϵ *metric* than Algorithm B if $I_\epsilon(A, B) < I_\epsilon(B, A)$. For the SC metric we will say that Algorithm A is *relatively better as per SC metric* than Algorithm B if $SC(A, B) > SC(B, A)$. The final combined non-dominated solutions (of the 50 different runs of each algorithm) are used for calculating the binary performance metrices and for producing the plots shown.

For statistical evaluation we use attaintment surface based statistical metric ([6]). We run each algorithm for 51 times and the median attaintment surface (26th) plots are shown.

4 Simulation Results

The unconstrained ZDT1 problem has a convex efficient front for which solutions correspond to $0.01 \leq x_1^* \leq 1$ and $x_i^* = 0$ for $i = 2, 3, \ldots, 30$. In this problem the algorithms face difficulty in tackling a large number of variables. Figure 1

Table 3. Inverted generational distance metric values for test problems

ZDT1	NSGA-II	LR-NSGA	LS-NSGA
best	0.0024	3.57e-4	5.3e-4
worst	0.0252	0.0136	0.0145
average	0.0119	0.0040	0.0064
std. dev.	0.0045	0.0020	0.0030
rank.	3	1	2
ZDT2	NSGA-II	LR-NSGA	LS-NSGA
best	0.0051	5.5e-4	2.5e-4
worst	0.0902	0.0374	0.0142
average	0.0303	0.0064	0.0048
std. dev.	0.0107	0.0051	0.0023
rank.	3	2	1
ZDT3	NSGA-II	LR-NSGA	LS-NSGA
best	0.0015	3.2e-4	2.7e-4
worst	0.0385	0.0118	0.0100
average	0.0089	0.0034	0.0042
std. dev.	0.0072	0.0019	0.0020
rank.	3	1	2
ZDT4	NSGA-II	LR-NSGA	LS-NSGA
best	0.0386	0.0198	0.0543
worst	0.1861	0.0436	0.3453
average	0.0574	0.0276	0.1172
std. dev.	0.0365	0.0281	0.1287
rank.	2	1	3
ZDT6	NSGA-II	LR-NSGA	LS-NSGA
best	0.104	7.1e-8	3.5e-7
worst	0.2243	0.0066	0.0070
average	0.1775	7.8e-4	8.5e-4
std. dev.	0.0477	0.0010	0.0010
rank.	3	1	2

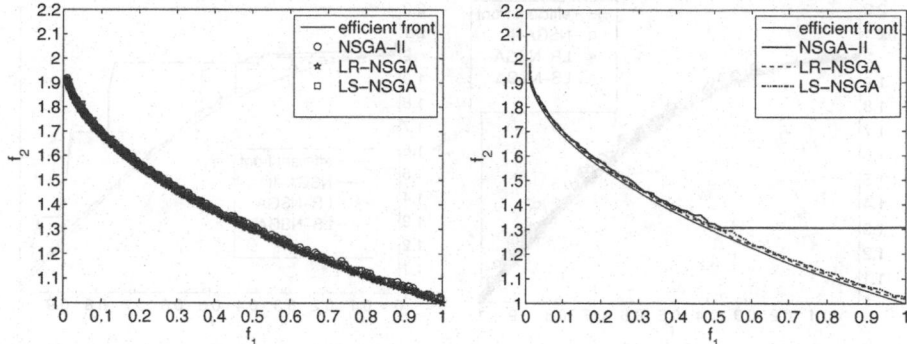

Fig. 1. Performance of the three algorithms on ZDT1

Fig. 2. Median attaintment surface (26th) plots of the three algorithms on ZDT1

shows the performance of all the algorithms after 4000 function evaluations. From the figure it can be visually concluded that all the algorithms perform well on this problem in maintaining a diverse set of solutions close to the efficient front. Table 3 shows the IGD convergence metric values. In terms of average IGD values both LR-NSGA and LS-NSGA perform better than NSGA-II. Table 4 shows the S diversity metric values. In terms of average S values LR-NSGA still performs the best while both LR-NSGA and LS-NSGA perform better than NSGA-II. Hence as far as convergence and diversity separately are considered all the hybrid algorithms perform better than the original NSGA-II.

The binary performance metrices values of all the algorithms are shown in Table 5. For a particular test problem, an element (i, j) in this table (rows and columns corresponding to different algorithms) represents I_ϵ(algorithm j, algorithm i) while the $I_\mathbb{C}$(algorithm j, algorithm i) values are shown in italics. From the table one obtains that the efficient front obtained by both LR-NSGA and LS-NSGA are *better* that of NSGA-II. Figure 2 shows the median attaintment surface plots of all the algorithms. It can be visually seen that with the exception of SM-NSGA all the other two algorithms perform better than NSGA-II stochastically. The median attaintment surface plot also shows that NSGA-II usually does not finds diverse solutions with f_1 values larger than 0.5 while the other algorithm do not have this difficulty.

The unconstrained ZDT2 problem has a non-convex efficient front. Figure 3 shows the performance of all the algorithms after 4000 function evaluations. From the figure it can be visually concluded that all the hybrid algorithms perform better that NSGA-II on this problem in maintaining a diverse set of solutions close to the non-convex efficient front. In terms of average IGD values (Table 3) LS-NSGA performs the best while LR-NSGA also performs better than NSGA-II. From Table 4 one sees that in terms of S spread metric values all hybrid algorithms perform better than NSGA-II. Hence as far as convergence and diversity separately are considered all the hybrid algorithms perform better than original NSGA-II.

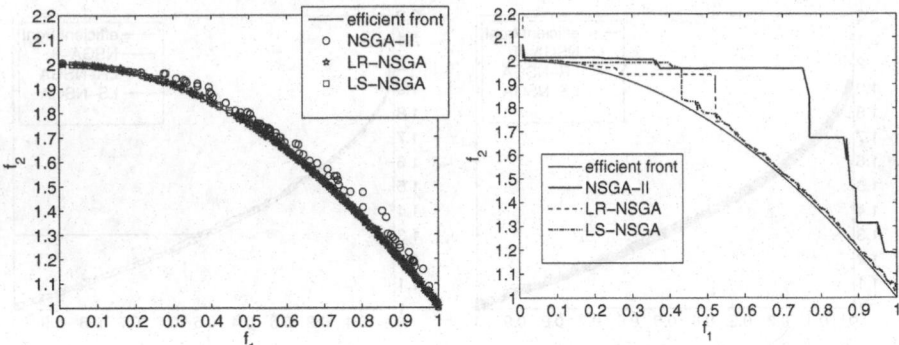

Fig. 3. Performance of the three algorithms on ZDT2

Fig. 4. Median attaintment surface (26th) plots of the three algorithms on ZDT2

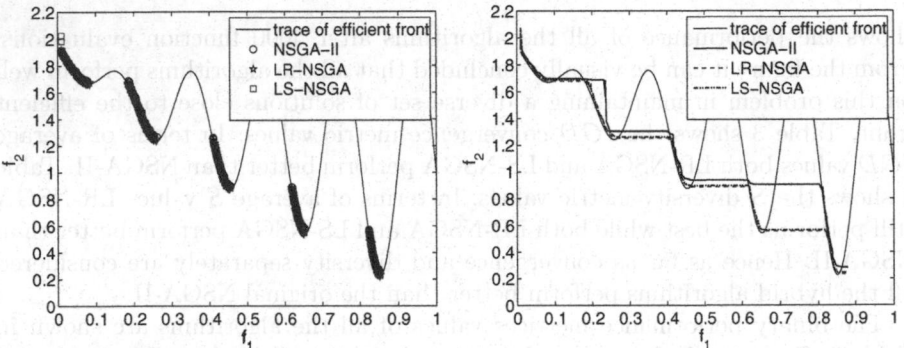

Fig. 5. Performance of the three algorithms on ZDT3

Fig. 6. Median attaintment surface (26th) plots of the three algorithms on ZDT3

Next we consider the binary performance metrices on this problem (Table 5). From the table one obtains that the efficient front obtained by LS-NSGA *dominates* that of NSGA-II. This information is obtained only by the set coverage metric value. For domination relation this is the only binary metric that can evaluate whether one set dominated the other or not justifying the inclusion of it. Moreover the front produced by LR-NSGA is *relatively better* than that of NSGA-II. One may wonder why does some of the algorithms for example LS-NSGA does not strictly dominates NSGA-II (i.e. the condition $I_\epsilon(A, B) < 1$ is never met). This is due to the presence of some solutions by NSGA-II on the $f_1 = 0.01$ weak efficient front, in this case there cannot exist any feasible points which are strictly better in all objective values. Figure 4 shows the median attaintment surface plots of all the algorithms. It can be visually seen that both LS-NSGA and LR-NSGA perform better stochastically than NSGA-II.

Next we consider unconstrained ZDT3, this problem has a convex discontinuous efficient frontier. Figure 5 shows the performance of all the algorithms after

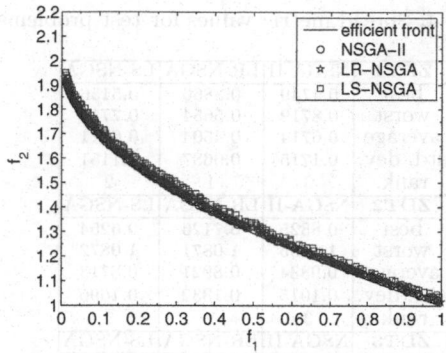

Fig. 7. Performance of the three algorithms on ZDT4

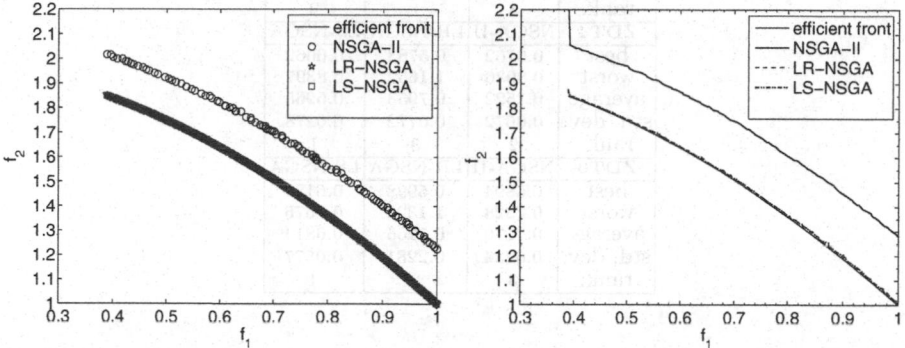

Fig. 8. Performance of the three algo- **Fig. 9.** Median attainment surface plots
rithms on ZDT6 (26th) of the three algorithms on ZDT6

4000 function evaluations. From the figure it can be visually concluded that all
the algorithms perform well on this problem in maintaining a diverse set of solutions close to the efficient front. In terms of average IGD values (Table 3) all the
two hybrid algorithms LR-NSGA and LS-NSGA perform better than NSGA-II
while LR-NSGA performs the best. Similar results are obtained in terms of S
spread metric values (Table 4).

Next we consider the binary performance metrices on this problem (Table 5).
From the table one obtains that the front produced by LS-NSGA and LR-NSGA
is *better* than that of NSGA-II. Figure 6 shows the median attainment surface
plots of all the algorithms. It can be visually seen that the hybrid algorithms
perform better stochastically than NSGA-II over entire efficient front. Moreover
NSGA-II is not able to find the efficient front from $f_1 = 0.6$ to $f_1 = 0.8$ values.

The problem unconstrained ZDT4 has a total of 100 distinct local efficient
fronts in the objective space. The global Pareto-optimal solutions correspond
to $0.01 \leq x_1^* \leq 1$ and $x_i^* = 0$ for $i = 2, 3, \ldots, 10$ (similar to ZDT1). Since
ZDT4 is a complex multi-modal problem in this problem all the algorithms are

Table 4. Spread metric values for test problems

ZDT1	NSGA-II	LR-NSGA	LS-NSGA
best	0.4749	0.3866	0.5156
worst	0.8719	0.5654	0.7716
average	0.6714	0.4504	0.6511
std. dev.	0.1715	0.0687	0.1151
rank.	3	1	2
ZDT2	NSGA-II	LR-NSGA	LS-NSGA
best	0.8325	0.7126	0.6254
worst	1.1166	1.0871	1.0872
average	0.9834	0.8921	0.8713
std. dev.	0.1015	0.1332	0.1690
rank.	3	2	1
ZDT3	NSGA-II	LR-NSGA	LS-NSGA
best	0.7805	0.7368	0.7931
worst	0.9446	0.8715	0.8882
average	0.8721	0.8071	0.8008
std. dev.	0.0684	0.0484	0.0546
rank.	3	1	2
ZDT4	NSGA-II	LR-NSGA	LS-NSGA
best	0.6762	0.5783	0.6062
worst	0.8936	1.1652	0.8397
average	0.7522	0.7953	0.6965
std. dev.	0.0672	0.0772	0.0278
rank.	2	3	1
ZDT6	NSGA-II	LR-NSGA	LS-NSGA
best	0.6894	0.5998	0.6189
worst	0.7328	1.1320	0.7675
average	0.7170	0.7265	0.6814
std. dev.	0.0234	0.2281	0.0577
rank.	3	2	1

run till 40000 function evaluations. Figure 7 shows the performance of all the algorithms. It can be seen that all the algorithms are able to overcome many local efficient fronts. This happens for the hybrid algorithms (which use local search guaranteing local Pareto optimality) due to the crossover and mutation search operators built in the NSGA-II algorithm. For this problem in terms of convergence LR-NSGA performs best while LS-NSGA performs best in terms of spread. In terms of binary metrices NSGA-II performs *relatively better* than LS-NSGA. The attainment surface plots are similar to that in Figure 7 and are thus not presented here.

Next we consider another difficult problem, ZDT6. This problem has a non-convex and non-uniformly spaced Pareto-optimal solutions. The Pareto-optimal solutions correspond to $0.01 \leq x_1^* \leq 1$ and $x_i^* = 0$ for $i = 2, 3, \ldots, 10$. In this problem also we set the number of function evaluations to be 12000. Figure 8 shows the performance of all the algorithms after 12000 function evaluations. From the figure it can be visually concluded that all the three hybrid algorithms perform much better that NSGA-II on this problem in maintaining a diverse set of solutions close to the efficient front. In terms of average IGD values (Table 3) LR-NSGA performs the best. In terms of average S values (Table 4) LS-NSGA performs the best.

Next we consider the binary performance metrices on this problem (Table 5). From the table one obtains that the efficient front obtained by LR-NSGA and

Table 5. Epsilon and coverage metric values for test problems

ZDT1	NSGA-II		LR-NSGA		LS-NSGA	
	I_ϵ	I_C	I_ϵ	I_C	I_ϵ	I_C
NSGA-II	1.00	1.00	1.00	0.94	1.00	0.83
LR-NSGA	1.04	0.01	1.00	1.00	1.01	0.08
LS-NSGA	1.02	0.03	1.00	0.54	1.00	1.00
ZDT2	NSGA-II		LR-NSGA		LS-NSGA	
	I_ϵ	I_C	I_ϵ	I_C	I_ϵ	I_C
NSGA-II	1.00	1.00	1.01	0.98	1.00	1.00
LR-NSGA	1.08	0.00	1.00	1.00	1.00	0.45
LS-NSGA	1.07	0.00	1.02	0.77	1.00	1.00
ZDT3	NSGA-II		LR-NSGA		LS-NSGA	
	I_ϵ	I_C	I_ϵ	I_C	I_ϵ	I_C
NSGA-II	1.00	1.00	1.00	0.83	1.00	0.73
LR-NSGA	1.10	0.01	1.00	1.00	1.01	0.32
LS-NSGA	1.12	0.07	1.00	0.64	1.00	1.00
ZDT4	NSGA-II		LR-NSGA		LS-NSGA	
	I_ϵ	I_C	I_ϵ	I_C	I_ϵ	I_C
NSGA-II	1.00	1.00	1.10	0.16	1.08	0.76
LR-NSGA	1.04	0.37	1.00	1.00	1.03	0.98
LS-NSGA	1.09	0.09	1.01	0.26	1.00	1.00
ZDT6	NSGA-II		LR-NSGA		LS-NSGA	
	I_ϵ	I_C	I_ϵ	I_C	I_ϵ	I_C
NSGA-II	1.00	1.00	1.00	1.00	1.00	1.00
LR-NSGA	1.19	0.00	1.00	1.00	1.00	0.03
LS-NSGA	1.22	0.00	1.01	0.03	1.00	1.00

LS-NSGA *dominates* that of NSGA-II. Figure 9 shows the median attainment surface plots of all the algorithms. It can be visually seen that all the three hybrid algorithms perform much better stochastically than NSGA-II.

5   Conclusions

This study brings into light how the second order Hessian from classical optimization methods can be effectively hybridized in a state-of-the-art evolutionary algorithm. These local search methods require gradient and Hessian information which is numerically evaluated using the Finite-Difference technique. These local search methods are used in addition to the usual global and local search operators used in NSGA-II. The comparison of these methods with NSGA-II on a number of test problems have adequately demonstrated that these methods perform very well when the problem size and search space complexity is large. Multi-modality and non-convexity is also efficiently tackled. Among the two hybrid algorithms, the LR-NSGA use a random directional weighted objective gradient on which a full Levenberg-Marquardt step is performed while in LS-NSGA this step is performed on a suitable steepest descent direction. Finally it is to be mentioned that the LM steps can be also carried out for box-constrained and general nonlinearly constrained problems. However then one can not take the full Levenberg-Marquardt step as that might violate feasibility. Then one could perform some kind of Armijo step length along the direction. This needs controlling more number of parameters. Some such extensions would be an immediate focus for useful research and application in the area of evolutionary multi-objective algorithms.

References

1. Bosman, P.A.N., de Jong, E.D.: Combining gradient techniques for numerical multi-objective evolutionary optimization. In: GECCO 2006. Proceedings of the 8th annual conference on Genetic and evolutionary computation, pp. 627–634. ACM Press, New York (2006)
2. Deb, K.: Multi-objective optimization using evolutionary algorithms. Wiley, Chichester (2001)
3. Deb, K., Agrawal, S., Pratap, A., Meyarivan, T.: A fast and elitist multi-objective genetic algorithm: NSGA-II. IEEE Transactions on Evolutionary Computation 6(2), 182–197 (2002)
4. Ehrgott, M.: Multicriteria Optimization. Springer, Berlin (2000)
5. Fan, J.-y., Yuan, Y.-x.: On the quadratic convergence of the Levenberg-Marquardt method without nonsingularity assumption. Computing 74(1), 23–39 (2005)
6. Fonesca, C.M., Fleming, P.J.: On the performance assessment and comparison of stochastic multiobjective optimizers. In: Ebeling, W., Rechenberg, I., Voigt, H.-M., Schwefel, H.-P. (eds.) PPSN IV. LNCS, vol. 1141, pp. 584–593. Springer, Heidelberg (1996)
7. Harada, K., Sakuma, J., Kobayashi, S.: Local search for multiobjective function optimization: pareto descent method. In: GECCO 2006. Proceedings of the 8th annual conference on Genetic and evolutionary computation, pp. 659–666. ACM Press, New York (2006)
8. Levenberg, K.: A method for the solution of certain non-linear problems in least squares. Quart. Appl. Math. 2, 164–168 (1944)
9. Marquardt, D.W.: An algorithm for least-squares estimation of nonlinear parameters. J. Soc. Indust. Appl. Math. 11, 431–441 (1963)
10. Mukai, H.: Algorithms for multicriterion optimization. IEEE Trans. Automat. Control 25(2), 177–186 (1980)
11. Nocedal, J., Wright, S.J.: Numerical Optimization. Springer, Heidelberg (1999)
12. Shukla, P.K., Deb, K.: On finding multiple pareto-optimal solutions using classical and evolutionary generating methods. European Journal of Operational Research 181(3), 1630–1652 (2007)
13. Zitzler, E., Thiele, L.: Multiobjective optimization using evolutionary algorithms – A comparative case study. In: Eiben, A.E., Bäck, T., Schoenauer, M., Schwefel, H.-P. (eds.) PPSN V. LNCS, vol. 1498, pp. 292–301. Springer, Heidelberg (1998)
14. Zitzler, E., Thiele, L., Laumanns, M., Fonseca, C.M., da Fonseca, V.G.: Performance Assessment of Multiobjective Optimizers: An Analysis and Review. IEEE Transactions on Evolutionary Computation 7(2), 117–132 (2003)

Chapter 4 - First Workshop on Ambient Intelligence Technologies and Applications (AMITA 2007)

Ambient Intelligence – A State of the Art from Artificial Intelligence Perspective

Carlos Ramos

GECAD – Knowledge Engineering and Decision Support Group
Institute of Engineering – Polytechnic of Porto, Portugal
csr@isep.ipp.pt

Abstract. Ambient Intelligence (AmI) deals with a new world where computing devices are spread everywhere (ubiquity), allowing the human being to interact in physical world environments in an intelligent and unobtrusive way. These environments should be aware of the needs of people, customizing requirements and forecasting behaviours. AmI environments may be so diverse, such as homes, offices, meeting rooms, schools, hospitals, control centers, transports, touristic attractions, stores, sport installations, music devices, etc. In the aims of Ambient Intelligence, research envisages to include more intelligence in the AmI environments, allowing a better support to the human being and the access to the essential knowledge to make better decisions when interacting with these environments. This paper can be seen as a State of the Art of Ambient Intelligence, according to an Artificial Intelligence (AI) perspective. We will define Ambient Intelligence; refer some of their prototype and systems; and to analyze how the main Artificial Intelligence areas can be applied.

Keywords: Ambient Intelligence, Artificial Intelligence, Ubiquitous Computing, Context Awareness.

1 Introduction

The European Commission's IST Advisory Group (ISTAG) has introduced the concept of Ambient Intelligence (AmI) [1,2]. ISTAG believes that it is necessary to take a holistic view of Ambient Intelligence, considering not just the technology, but the whole of the innovation supply-chain from science to end-user, and also the various features of the academic, industrial and administrative environment that facilitate or hinder realisation of the AmI vision [3]. Due to the great amount of technologies involved in the Ambient Intelligence concept we may find several works that appeared even before the ISTAG vision pointing in the direction of Ambient Intelligence trends. Other concepts have some overlapping with Ambient Intelligence, namely Ubiquitous Computing, Pervasive Computing, Context Awareness and Embedded Systems.

The concept of Ubiquitous Computing (UbiComp) was introduced by Mark Weiser during his tenure as Chief Technologist of the Palo Alto Research Center (PARC) [4]. Ubiquitous Computing means that we have access to computing devices anywhere in an integrated and coherent way. Ubiquitous Computing was mainly driven by

J. Neves, M. Santos, and J. Machado (Eds.): EPIA 2007, LNAI 4874, pp. 285–295, 2007.

Communications and Computing devices scientific communities but now is involving other research areas. Ambient Intelligence differs from Ubiquitous Computing because sometimes the environment where Ambient Intelligence is considered is simply local. For instance we can imagine a meeting room equipped with Ambient Intelligence but without any ubiquity. However, as ubiquity is an important issue, there is a natural trend to consider it in AmI. In the same example referred before it is quite common that meeting participants fail in attending the meeting because they are travelling to another city or country, contacting partners and doing business. To minimize the negative impacts of these absences, Ambient Intelligence must consider ubiquity, so meeting participants that are abroad may have mobile devices, like PDAs or notebooks, and be aware of the meeting development and interact with the meeting participants, a kind of telepresence in the meeting. Another difference is that Ambient Intelligence makes more emphasis on intelligence than Ubiquitous Computing.

A concept that sometimes is seen as a synonymous of Ubiquitous Computing is Pervasive Computing. According Teresa Dillon, Ubiquitous Computing is best considered as the underlying framework, the embedded systems, networks and displays which are invisible and everywhere, allowing us to 'plug-and-play' devices and tools, On the other hand, Pervasive Computing, is related with all the physical parts of our lives; mobile phone, hand-held computer or smart jacket [5].

Context Awareness means that the system has conscience about the current situation we are dealing with. An example is the automatic detection of the current situation in a Control Centre. Are we in presence of a normal situation or are we dealing with a critical situation , or even an emergency? In this Control Centre the intelligent alarm processor will exhibit different outputs according the identified situation [6]. Automobile Industry is also investing in Context Aware systems, like near-accident detection. Human-Computer Interaction scientific community is paying lots of attention to Context Awareness. Context Awareness is one of the most desired concepts to include in Ambient Intelligence, the identification of the context is important for deciding to act in an intelligent way. However, sometimes the AmI final users do not want this high level of intelligence available in the system.

Embedded Systems mean that electronic and computing devices are embedded in current objects or goods. Today goods like cars are equipped with microprocessors; the same is true for washing machines, refrigerators, toys etc. Embedded Systems community is more driven by electronics and automation scientific communities. Current efforts go in the direction to include electronic and computing devices in the most usual and simple objects we use, like furniture, mirrors etc. Ambient Intelligence differs from Embedded Systems since computing devices may be clearly visible in AmI scenarios. However, there is a clear trend to involve more embedded systems in Ambient Intelligence.

The Encyclopedia of Artificial Intelligence refers the concept of Ambient Intelligence [7] and notice that in the past AI community centred the attention in the hardware (40's and 50's), in the computer (60's), in the network (70's and 80's) and in the Web (90's till now). However, it starts to be clear that Intelligence must be provided to our daily-used environments. We are aware of the push in the direction of Intelligent Homes, Intelligent Vehicles, Intelligent Transportation Systems, Intelligent

Manufacturing Systems, even Intelligent Cities. This is the reason why Ambient Intelligence concept is so important nowadays [7].

Ambient Intelligence is not possible without Artificial Intelligence. On the other hand, AI researchers must be aware of the need to integrate their techniques with other scientific communities techniques (Automation, Computer Graphics, Communications, etc). Ambient Intelligence is a tremendous challenge, needing the better effort of different scientific communities.

In which concerns AI almost all research areas can contribute for the Ambient Intelligence effort. In section 2 we will explain how areas like Machine Learning, Computational Intelligence, Planning, Natural Language, Knowledge Representation, Computer Vision, Intelligent Robotics, Incomplete and Uncertain Reasoning and Multi-Agent Systems can be used in the Ambient Intelligent challenge.

Section 3 will be dedicated to some Ambient Intelligence real-world prototypes and systems.Finally, in section 4 we try to establish some conclusions and further directions.

2 How Artificial Intelligence Can Contribute for the Ambient Intelligence Challenge

Recently Ambient Intelligence is receiving a significant attention from Artificial Intelligence Community. We may refer the Ambient Intelligence Workshops organized by Juan Augusto and Daniel Shapiro at ECAI'2006 (European Conference on Artificial Intelligence) and IJCAI'2007 (International Joint Conference on Artificial Intelligence) and the Special Issue on Ambient Intelligence, coordinated by Carlos Ramos, Juan Augusto and Daniel Shapiro to appear in the March/April'2008 issue of the IEEE Intelligent Systems magazine.

In this section we will analyze the possible contributions of AI community for the Ambient Intelligence effort.

2.1 Knowledge Representation

Knowledge Representation is one of the most important areas in the AI field. After the bad times of AI in the 60's it started to be clear that knowledge is too important for the success of Intelligent Systems. AI reborn at the beginning of the 70's was due to the success of some Knowledge-based Systems, like MYCIN [8] or AUTHORIZER's ASSISTANT [9]. Expert Systems achieved a tremendous success in areas like Medicine, Industry, and Business. During the 90's with the strong development of the Internet and the born of WWW the human being was faced with a critical problem; information achieved huge dimensions and the mapping between information and knowledge was pointed as urgent. AI community started to pay attention to Ontologies and Semantic Web. New areas like Information Retrieval and Text Mining appeared.

The early experience in Intelligent Systems development show us that intelligence is not possible without knowledge. However, little attention has been paid to Knowledge Representation in most of the Ambient Intelligence projects.

2.2 Machine Learning

Machine Learning received attention from the AI community since the beginning. The building of the first artificial neural models and hardware, with the Walter Pitts and Warren McCullock work [10] and Marvin Minsky and Dean Edmonds SNARC system are in the origin of AI. Neural Networks have obtained a great success, namely after the 70's, being applied in many real-world problems, namely in classification. Other techniques have been used with success, using more high-level descriptions, like Inductive Learning, Case-based Reasoning, and Decision Trees based methods.

During the 80's the term Data Mining started to be used. Many people from Databases area preferred to use this term to refer to the Machine Learning techniques (together with some Statistics methods like K-means) in the overall Knowledge Discovery effort. Data Mining is seen just like a phase in Knowledge Discovery (selection, cleaning and pre-processing are phases before Data Mining, while interpretation and evaluation are phases after Data Mining). At the end of the 90's Business Intelligence appeared as a buzzword in Information Systems, covering Data Mining and Knowledge Discovery, but also Warehouses, Enterprise Resource Planning, Client Relationship Management among others.

Nowadays, Machine Learning is widely used, so it is expectable that Ambient Intelligence will need to handle this kind of technology. One aspect very important for AmI is the need to learn from user observation. Several systems understand user commands, but they are not so intelligent to avoid things that the user does not wish to do. The use of basic Machine Learning methods will allow learning from the user observation, making AmI systems more acceptable for users.

2.3 Computational Intelligence

On the last years Computational Intelligence community is very active, claiming for a great success in real world problems. Sometimes this community is involved in the AI field, sometimes appears as an alternative to symbolic AI and more close to Operations Research. Computational Intelligence involves many pattern recognition and optimization oriented methods, like Neural Networks, Genetic Algorithms, Ant Colonies, Particle Swarm Intelligence, Taboo Search, Simulated Annealing, Fuzzy Logic and even Agents. These methods are oriented for specific problems, suffering from tuning, i.e. parameter selection and values choice is crucial for the success of these methods.

Considering that Ambient Intelligence environments will support the possible choices of the human being we may expect that Computational Intelligence will be placed in AmI systems.

2.4 Planning

Planning is the activity by means it is possible to solve a problem in which the solution has the format of a plan. AI Planning studies all the aspects relative to general planning. Allen Newell's General Problem Solver system has defined some important aspects for Planning, however the first system exclusively dedicated to AI Planning was STRIPS. The first Planning systems were dedicated to Blocks World, however, namely after the 80's, real-world problems were treated by AI involving several types of constraints (e.g. resources, time). Plans can be established before the plan execution (off-line) or during the execution (on-line). They can be deliberative (we plan and execute what was planned without considering non-expected events) or reactively (we react to stimulus in a much more basic way), or hybrid (combining the best of deliberative and reactive policies).

Planning is studied in many other areas. In Robotics it is important the Trajectory Plan for robot arms movements or for mobile robots movements, collision avoidance are the main aspect to deal with, and the problem is much more a geometric reasoning problem, different from most AI symbolic reasoning planners. Assembly Planning is another very important area, while having some analogy with Blocks World problems, the geometry is also important here. Manufacturing Systems deal with Planning, in this area the attention is given to plan how products will be manufactured (Process Planning) and produced (Production Planning, that is more a scheduling problem) or even how the layout of the factories will be done (Layout Planning). Planning is studied in many other disciplines, so we listen terms like Treatment Planning in Medicine, Enterprise Resource Planning in Information Systems, and Restoration Planning in Power Systems.

Planning is one of the activities more related with intelligence. It is quite difficult to convince someone that a system is intelligent without the ability to plan how to solve problems. In this way Ambient Intelligence environments will need to support planning in order to give intelligent advices for the users. A clear example is found in Transportation area, inside vehicles where intelligent driving systems will help drivers; and on the road, where route planning will consider many constraints related with traffic, time, and cost.

2.5 Incompleteness and Uncertainty

Real-world problems are affected with incompleteness and uncertainty. Generally we deal with information, some part of this information is correct, some part may be incorrect, and some part is missing. The question is how to proceed with an elaborated reasoning process dealing with these information problems. Many techniques have been used (e.g. Bayesian Networks, Fuzzy Logic, Rough Sets) to handle the problem.

Since AmI environments are real, we are sure that Incompleteness and Uncertainty will be present there, and users expect support from these environments even if these problems exist.

2.6 Speech Recognition and Natural Language

The most common way human beings use to interact is by means of language, by voice or written. So it is clear that this kind of interaction is also expectable in the environments with Ambient Intelligence. Speech Recognition and Natural Language are different and complementary problems, using different techniques.

In Speech Recognition an electric signal is obtained by means of a microphone. The basic problem is the identification of phonemes in this signal, so it is more a signal processing and pattern recognition work. Joining phonemes and identifying words is the next activity. Several Speech Recognition systems are available and can be used with more or less success, depending how the user speaks.

The input of Natural Language is a written sequence, resulting from a speech recognition system or obtained from a keyboard, or even from a written document. The objective of Natural Language is to understand this input. First, it is necessary to do a Syntax analysis, after this Semantics is important. This is a difficult task, namely because some sentences are ambiguous, like the well known "the boy saw the man in the hill with the telescope" (Who was in the hill? Who has the telescope?). If this ambiguity is observed in just one sentence, the problem is much more complex in texts, because the understanding of one sentence depends on the understanding of previous sentences, or posterior sentences, or even from the user knowledge. Knowledge Representation plays an important role in Natural Language. Automatic Translation Systems are one of the most studied areas of Natural Language. Recently there is a trend to use Statistics-based Translation, while being fast and more easy to implement, the results are not good. So, today the combination between statistics approaches and knowledge-oriented approaches is being experimented.

2.7 Computer Vision

Vision is the richest sensorial input of the human being. So, the ability to automate the vision is very important. Basically, Computer Vision is a geometric reasoning problem. Computer Vision comprises many areas, like Image Acquisition, Image Processing, Object Recognition (2D and 3D), Scene Analysis, and Image Flow Analysis.

Computer Vision can be used in different situations in Ambient Intelligence. In Intelligent Transportation Systems it can be used to identify traffic problems on the roads, or in intelligent driving assistance to identify patterns or the approaching to another vehicle. Computer Vision can be used to identify human being gestures used to control the environment equipment or expressions of the human being face to identify the emotional state of somebody.

2.8 Robotics

Robots are widely used in Manufacturing, in this kind of environment Robotics can be viewed according to the automation approach. However, there is a close connection between Robotics and Artificial Intelligence, namely where the attention is to give more focus to all intelligent aspects of the created robots. This resulted in what was called previously Intelligent Robotics and received more recently a new vision, referred as Cognitive Robotics.

Ambient Intelligent environments, like home, can benefit of intelligent robots. This is especially true when persons live alone, are elder people, or have health problems. The creation of intelligent robots, able to perform several tasks or just to act as companion elements is very important. The problem is that it is easy to create robots operating very well in specific tasks, but it is too complex to create robots with the flexibility to do different tasks as the human being. This limitation is more related with physical constraints.

2.9 Multi-agent Systems

In the beginning of the 80's AI community started with a new area, Distributed Artificial Intelligence (DAI), combining AI with Distributed Computing. From DAI emerged the Intelligent Agents and Multi-Agent Systems Paradigms. Agents are expected to support several features, like sensing capabilities, autonomy, reactive and proactive reasoning, social abilities, and learning, among others. Multi-Agent Systems emphasize the social abilities, like communication, cooperation, conflict resolution, negotiation, argumentation, and emotion. Rapidly, Multi-Agent Systems started to be the main paradigm in AI. After the World Wide Web boom in the 90's Agents received even more attention.

Multi-Agent Systems are especially good in modeling real-world and social systems, where problems are solved in a concurrent and cooperative way without the need to obtain optimal solutions (e.g. in traffic or manufacturing).

In Ambient Intelligence environments, Agents are a good way to model meaningful entities, like rooms, cars or even persons.

3 Ambient Intelligence Prototypes and Systems

Here we will analyze some examples of Ambient Intelligence prototypes and systems, divided by the area of application.

3.1 AmI at Home

Domotics is a consolidated area of activity. After the first experiences using Domotics at homes there was a trend to refer the Intelligent Home concept. However, Domotics is too centred in the automation, giving to the user the capability to control the house devices from everywhere. We are still far from the real Ambient Intelligence in homes, at least at the commercial level.

Several organizations are doing extended experiences to achieve the Intelligent Home concept. Some examples are HomeLab from Philips, MIT House_n, Georgia

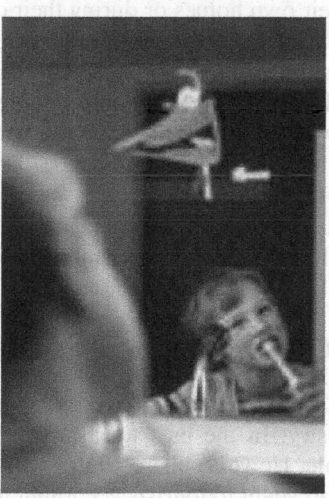

Fig. 1. Entertainment in the bathroom mirror, from Philips HomeLab

Tech Aware Home, Microsoft Concept Home, and e2 Home from Electrolux and Ericsson. **Fig. 1** illustrates a personal health coach detecting the use of a toothbrush and playing cartoon to make brushing enjoyable for children, an example from HomeLab.

3.2 AmI in Vehicles and Transports

Since the first experiences with NAVLAB 1 [11], Carnegie Mellon University has developed several prototypes for Autonomous Vehicle Driving and Assistance. The last one, NAVLAB 11, is an autonomous Jeep. Most of the car industry companies are doing research in the area of Intelligent Vehicles for several tasks like car parking assistance or pre-collision detection.

Another example of AmI application is related with Transports, namely in connection with Intelligent Transportation Systems (ITS). The ITS Joint Program of the US Department of Transportation identified several areas of applications, namely: arterial management; freeway management; transit management; incident management; emergence management; electronic payment; traveler information; information management; crash prevention and safety; roadway operations and management; road weather management; commercial vehicle operations; and intermodal freight. In all these application areas Ambient Intelligence can be used.

3.3 AmI in Elderly and Health Care

Several studies point to the aging of population during the next decades. While being a good result of increasing of life expectation, this also implies some problems. The percentage of population with health problems will increase and it will be very difficult to Hospitals to maintain all patients. Our society is faced with the responsibility to care for these people in the best possible social and economical ways. So, there is a clear interest to create Ambient Intelligence devices and environments allowing the patients to be followed in their own homes or during their day-by-day life.

The medical control support devices may be embedded in clothes, like T-shirts, collecting vital-sign information from sensors (blood pressure, temperature, etc). Patients will be monitored at long distance. The surrounding environment, for example the patient home, may be aware of the results from the clinical data and even perform emergency calls to order an ambulance service.

For instance, we may refer the IST Vivago® system (IST International Security Technology Oy, Helsinki, Finland), an active social alarm system, which combines intelligent social alarms with continuous remote monitoring of the user's activity profile [12].

3.4 AmI in Tourism and Cultural Heritage

Tourism and Cultural Heritage are good application areas for Ambient Intelligence. Tourism is a growing industry. In the past tourists were satisfied with pre-defined tours, equal for all the people. However there is a trend in the customization and the same tour can be conceived to adapt to tourists according their preferences.

Immersive tour post is an example of such experience [13]. MEGA is an user-friend virtual-guide to assist visitors in the Parco Archeologico della Valle del Temple in Agrigento, an archaeological area with ancient Greek temples in Agrigento, located in Sicily, Italy [14]. DALICA has been used for constructing and updating the user profile of visitors of Villa Adriana in Tivoli, near Rome, Italy [15].

3.5 AmI at Work

The human being spends considerable time in working places like offices, meeting rooms, manufacturing plants, control centres, etc.

SPARSE is a project initially created for helping Power Systems Control Centre Operators in the diagnosis and restoration of incidents [6]. It is a good example of context awareness since the developed system is aware of the on-going situation, acting in different ways according the normal or critical situation of the power system. This system is evolving for an Ambient Intelligence framework applied to Control Centres.

Decision Making is one of the most important activities of the human being. Nowadays decisions imply to consider many different points of view, so decisions are commonly taken by formal or informal groups of persons. Groups exchange ideas or engage in a process of argumentation and counter-argumentation, negotiate, cooperate, collaborate or even discuss techniques and/or methodologies for problem solving. Group Decision Making is a social activity in which the discussion and results consider a combination of rational and emotional aspects. ArgEmotionAgents is a project in the area of the application of Ambient Intelligence in the group argumentation and decision support considering emotional aspects and running in the Laboratory of Ambient Intelligence for Decision Support (LAID), seen in **Fig. 2** [16], a kind of Intelligent Decision Room. This work has also a part involving ubiquity support.

Fig. 2. Ambient Intelligence for Decision Support, LAID Laboratory

3.6 AmI in Sports

Sports involve high-level athletes and many more practitioners for hobby of free-time occupancy. Many sports are done without any help of the associated devices, opening here a clear opportunity for Ambient Intelligence to create sports assistance devices and environments.

FlyMaster NAV+ is a free-flight on-board pilot Assistant (e.g. gliding, hangliding, paragliding), using the FlyMaster F1 module with access to GPS and sensorial information. FlyMaster Avionics S.A., a spin-off, was created to commercialize these products (see **Fig. 3**).

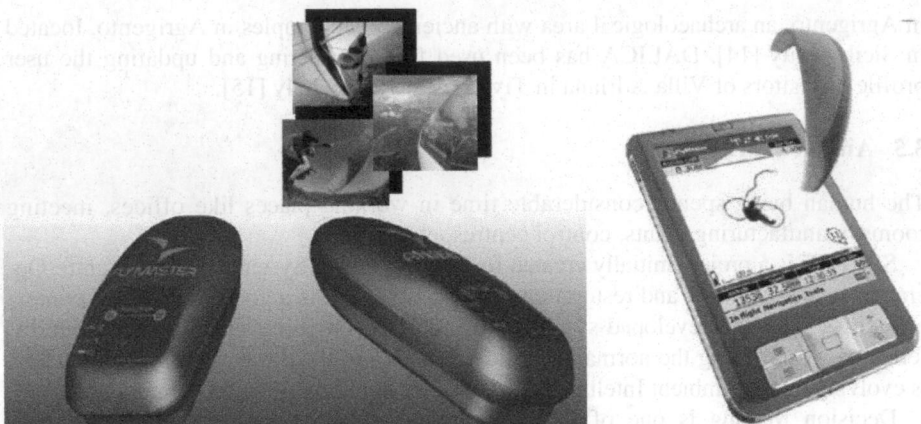

Fig. 3. FlyMaster Pilot Assistant device, from FlyMaster Avionics S.A

4 Conclusions

This article presents the state of the art in which concerns Ambient Intelligence field. After the history of the concept, we established some related concepts definitions, like Ubiquitous and Pervasive Computing, Embedded Systems, Context Awareness. Applications of Ambient Intelligence are presented. We identified several Artificial Intelligence areas important for achieving the Ambient Intelligence concept, namely Knowledge Representation, Machine Learning, Computational Intelligence, Planning, Multi-Agent Systems, Natural Language, Speech Recognition and Computer Vision, Robotics, Incompleteness and Uncertainty.

Ambient Intelligence deals with a futuristic notion for our lives. Most of the practical experiences concerning Ambient Intelligence are still in a very incipient phase, due to the recent existence of this concept. Today, it is not clear the separation between the computer and the environments. Most Ambient Intelligence prototypes involve computers, notebooks, PDAs, interactive displays, keyboards, mouses, pointers. However, for new generations things will be more transparent, and environments with Ambient Intelligence will be more widely accepted. Ambient Intelligence in vehicles and in traffic and travel control, in health and elderly care, in tourism and cultural heritage, at home and at work, will be a reality soon. There is a long way to follow in order to achieve the Ambient Intelligence concept, however, in the future, this concept will be referred as one of the landmarks in the Artificial Intelligence development.

References

1. ISTAG, Scenarios for Ambient Intelligence in 2010, European Commission Report (2001)
2. ISTAG, Strategic Orientations & Priorities for IST in FP6, European Commission Report (2002)
3. ISTAG, Ambient Intelligence: from vision to reality, European Commission Report (2003)

4. Weiser, M.: The Computer for the Twenty-First Century. Scientific American, 94–104 (September 1991)
5. Dillon, T.: Pervasive and Ubiquitous Computing. Futurelab (2006), available at http://www.futurelab.org.uk/viewpoint/art71.htm
6. Vale, Z., Moura, A., Fernandes, M., Marques, A., Rosado, A., Ramos, C.: SPARSE: An Intelligent Alarm Processor and Operator Assistant. IEEE Expert- Special Track on AI Applications in the Electric Power Industry 12(3), 86–93 (1997)
7. Ramos, C.: Ambient Intelligence. In: Rabuñal, J.R., Dorado, J., Pazos, A. (eds.) Encyclopedia of Artificial Intelligence, Idea Group Reference (2007)
8. Shortliffe, E.: Computer-Based Medical Consultations: MYCIN. Elsevier, North Holland (1976)
9. Rothi, J., Yen, D.: Why American Express Gambled on an Expert Data Base. Information Strategy: The Executive´s Journal 6(3), 16–22 (1990)
10. McCulloch, W., Pitts, W.: A Logical Calculus of Ideas Immanent in Nervous Activity. Bulletin of Mathematical Biophysics (5), 115–133 (1943)
11. Thorpe, C., Hebert, M., Kanade, T., Shafer, S.: Vision and navigation for the Carnegie-Mellon Navlab. IEEE Transactions on Pattern Analysis and Machine Intelligence 10(3), 362–373 (1988)
12. Särelä, A., Korhonen, I., Lötjönen, L., Sola, M., Myllymäki, M.: IST Vivago® - an intelligent social and remote wellness monitoring system for the elderly. In: Proceedings of the 4th Annual IEEE EMBS Special Topic Conference on Information Technology Applications in Biomedicine, pp. 362–365 (2003)
13. Park, D., Nam, T., Shi, C., Golub, G., Van Loan, C.: Designing an immersive tour experience system for cultural tour sites, pp. 1193–1198. ACM Press, New York (2006)
14. Pilato, G., Augello, A., Santangelo, A., Gentile, A., Gaglio, S.: An intelligent multimodal site-guide for the Parco Archeologico della Valle del Temple in Agrigento. In: Proc. of the First Workshop in Intelligent Technologies for Cultural HeritageExploitation. European Conference on Artificial Intelligence (2006)
15. Constantini, S., Inverardi, P., Mostarda, L., Tocchio, A., Tsintza, P.: User Profile Agents for Cultural Heritage fruition. Artificial and Ambient Intelligence. In: Proc. of the Artificial Intelligence and Simulation of Behaviour Annual Convention, pp. 30–33 (2007)
16. Marreiros, G., Santos, R., Ramos, C., Neves, J., Novais, P., Machado, J., Bulas-Cruz, J.: Ambient Intelligence in Emotion Based Ubiquitous Decision Making. In: Proc. Artificial Intelligence Techniques for Ambient Intelligence, IJCAI 2007 – Twentieth International Joint Conference on Artificial Intelligence, Hyderabad, India (2007)

Ubiquitous Ambient Intelligence in a Flight Decision Assistance System

Nuno Gomes[1], Carlos Ramos[1], Cristiano Pereira[2], and Francisco Nunes[3]

[1] GECAD – Knowledge Engineering and Decision Support Group
Institute of Engineering – Polytechnic of Porto
[2] IMEDIATA – Multimedia Systems S.A.
3 VISTEON Corporation Portugal
{nbg,csr}@isep.ipp.pt, cristiano.pereira@imediata.pt,
nunes.francisco@netcabo.pt

Abstract. In this paper a new free flight instrument is presented. The instrument named FlyMaster distinguishes from others not only at hardware level, since it is the first one based on a PDA and with an RF interface for wireless sensors, but also at software level once its structure was developed following some guidelines from Ambient Intelligence and ubiquitous and context aware mobile computing. In this sense the software has several features which avoid pilot intervention during flight. Basically, the FlyMaster adequate the displayed information to each flight situation. Furthermore, the FlyMaster has its one way of show information.

Keywords: Personal Flight Assistant, Ambient Intelligence, Mobile Computing, Ubiquitous Computing.

1 Introduction

In the last years flight activities have been growing among the sportive and leisure interests of modern societies. Particularly, free flight related sports like Gliding, Hangliding and especially Paragliding are nowadays present in almost every country's sportive agenda. This growing interest and consequently the increase in the number of pilots turned the free flight related products market desirable for many companies. Mainly due to this factor glider design has been greatly improved in the past years, resulting in exponential growth of pure glider performance. Strangely, in spite of the electronics and software development just marginal changes were verified in terms of free flight instruments. In fact, besides the introduction of the GPS, almost the same basic technology is used for the past 10 years. Considering the most commons equipments, we continue to see monochromatic displays with little resolution, providing the same information, using numbers or simple graphics.

Naturally, there is more than one reason for this fact. The most relevant is that free flight instruments design requirements are very hard to fulfill. Due to the conditions of use, the equipment should be mechanically resistant, support low and high temperatures, support low pressure, be visible under direct sun light, have an autonomy of several hours, be simple to use, have low cost, just to name a few [1].

J. Neves, M. Santos, and J. Machado (Eds.): EPIA 2007, LNAI 4874, pp. 296–308, 2007.

Even at present time and with totally new designs it is not easy to fulfill all these requirements, and almost all the actual flight instruments are evolutions of older versions.

Considering the above we can clearly identify a necessity of new features for free-flight instruments. These features can be in terms of software, like more information for flight optimization particularly in competition, more user interaction capabilities, adequate information providing (the right information at the right time in a usable manner), and also in terms of hardware like high resolution color displays, new sensors for context aware, short and long range communication infrastructure, among others.

These types of features have been quite explored within Mobile Computing area. At this day and age, the computational power of a high end desktop computer of a decade ago, is available on a palm sized computer. Modern Personal Digital Assistants (PDA's), smart phones, or micro portable computers on top of good computational power, high definition color displays, wireless communications, flexible interfaces and good autonomy, and everything at accessible prices. Furthermore, recent advances in context aware mobile computing [2,3] and Ubiquitous Computing [4], have opened new ways for human-computer interaction. As everyone knows, flying is a very demanding task, for which all the pilot attention is needed to control the glider and the exterior parameters. Furthermore, almost all the free flight aircrafts require both hands at the same time to be controlled. Good examples are paraglider and hanglider. This results in an increased difficulty for pilot/instrument interaction. In sum a good flight instrument should provide only the needed information at the right time and manner, in an unobtrusive way. If we consider the definition of Ambient Intelligence (AmI), we can see that it deals with a new world where computing devices are spread everywhere, allowing the human being to interact in physical world environments in an intelligent and unobtrusive way. These environments should be aware of the needs of people, customizing requirements and forecasting behaviours [5].

In this paper a new free flight instrument considering the main features of Ubiquitous Computing and Ambient Intelligence is presented. The instrument named FlyMaster, distinguishes from others not only at hardware level, since is the only one based on a PDA and with an RF interface for wireless sensors, but also at software level once it structure was developed following some guidelines from ubiquitous and context aware mobile computing. In this sense the software has several features which avoid pilot intervention during flight.

Basically, the FlyMaster adequate the displayed information to each flight situation. Furthermore, the FlyMaster has its one way of show information. Taking advantage of the PDA's graphical capabilities it uses less numbers and more colored graphics in order to facilitate reading the information.

2 Flight Information and Flight Instruments

An unmotorized aircraft can not climb on its own. In fact, gravity is unforgiving, constantly pulling it down. A typical descent rate is usually between -0,5 m/s and -3m/s (the – sign indicates "descent" movement, where the + sign indicates "climbing" movement). In order to climb, or at least to maintain altitude, the pilot should look for rising air known as "lift" (were the lift velocity should be superior to the glider descent rate), and stay in the rising air as long as possible. Unfortunately, our senses are not oriented for flying, and we have particular difficulties in 3D spatial orientation [6]. In practical terms, due to the absence of references, the pilot has great difficulties to identify the movement of their glider relative to the air or to the ground, either horizontally or vertically. Eventually, he can detect changes in the glider velocity (acceleration), but even then, just for significative values. In order to overcome this important limitation the pilots use a flight instrument. The most common which is capable of measure vertical velocity is called "Variometer". This instrument belongs to the basic flight instruments class.

2.1 Basic Flight Information and Instruments

A Variometer is a vertical speed indicator which usually measures pressure. Considering that there is a correlation between pressure and altitude it is possible to calculate instantaneous rate of descent or climb. All the commercial basic free flight instruments (BFFI) have Variometer function, since it is the most wanted feature. However, it usually possible to find within the same instruments an Altimeter, a clock a thermometer and some memory for logging basic flight information. Examples of this type of BFFI the 4005 Compact from Flytec [7], or the IQ-One from Brauniger [8]. This instruments use a low resolution monochromatic display to show the information.

Besides simply staying aloft, an experienced free flight pilot usually has higher objectives. With appropriate weather conditions it is possible to travel long distances (Paragliding world record 426 km, rigid Hangliding world record 827 km, open class Gliding world record 2192 km, see www.fai.org), this is called Cross Country flying. A flight of this nature may also require the pilot to fly through several turn points, and cross a finish line (also known as goal line). In a competition the pilot would want to accomplish this in the shortest time possible. All these type of flights can be classified has "Performance Flights", for which the limited information provided by a BFFI is not sufficient.

2.2 Performance Flight Information and Instruments

Performance flights are very demanding flights, were the pilot needs to obtain and analyze enormous amounts of information in order to constantly decide were to go, and how to go. One part of the information, probably the most important, results from direct observation of the environment, and can only be obtained trough the pilot sight sense. Some examples of things to observe are: other pilots, birds, clouds, orography, etc. However, there is another part of the information that can only be obtained by electronics means, since parameters like velocity or position can not be sensed and some calculations must be made. Some of that information is (for a better comprehension of some of the following concepts see [9]): Ground Speed; Position and Bearing; Wind speed and direction (by layer); Average rising air (thermal) vertical velocity; Thermal marking (previous thermals or from a preset thermal map); Glide Ratio relative to air and to ground.

As is common knowledge a normal commercial GPS can provide ground speed and position, which is by itself useful information. By combining the GPS data with data from a BFFI, and performing a simple calculation it is possible to obtain a good estimate of the instantaneous ground Glide Ratio, as also rough estimates of the velocity to the optimal glide and McReady Speed. However this process can be tricky in flight, special when so much has to be done. Considering this, most experienced pilots have at least a free flight instrument with a GPS interface port, such as a IQ-Competition GPS. This type of instrument can be connected to a standard commercial GPS and receive data from it. With that data the instrument calculate and provide the pilot with the information mentioned in 5), 8), 9) and 10).

With the combination of a BFFI and a GPS a "leisure" pilot has almost all the important information for a Cross Country flight. Although most of the free flight pilots practice their sport for recreation, a non negligible number likes to compete.

2.3 Competition Flight Information and Instruments

According to Sporting Code Section 7 from the "Fédération Aéronautique Internationale" (FAI) a competition consists of a set of tasks, each one corresponding to a different flight and score. The same code defines several types of tasks depending on the objectives. Some examples of tasks types can be: Maximum Straight Distance; Race to Goal with turn points; Speed to a goal with turn points, just to name a few. However, in the last years the most commonly used has been by far the Race to Goal with turn point's type or just Race to Goal (RTG). This type of task is also the most demanding in terms of information control and information requirements, reason why we explain it here.

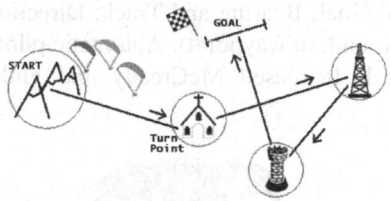

Fig. 1. Race to Goal Task

In a RTG task all pilots start at the same time and the pilot that arrives first at a certain goal wins. Between the start and the goal, the pilots should usually pass through one or more turnpoints like in Figure 1. Turn points are cylinders, projected from the ground up into space, usually with 400 m of radius, and the center given by a map coordinate. As the race usually starts on the air there is also a "Start Cylinder". Depending on the type of "Start", all the pilots should be inside or outside this cylinder after the race starts.

Besides these position constraints the pilot should also respect a time schedule. There are five important time instants during a task. The first two define the "Take Off Window" which correspond to the take off period (e.g. between 13:00 and 14h00). The third one is the "Start time" which is the instant after which the race begins (e.g. 13:00). The last two are the "Race End" and the "Landing Deadline" which correspond respectively to the instant at which the race ends (e.g. 17h30) and at which the pilot should be on the ground (e.g. 19h30)

The control of all of these constraints either by the pilot or by the organization is only possible trough the use of a GPS. In this way the competition organization provides a list of accepted GPS's, or flights instruments with GPS obligatory for track registry.

Considering the above, the combination of a BFFI with a normal commercial GPS is sufficient for participate in competition. However, on the one hand, there is specific information for competition flight optimization that is impossible, to obtain just with a BFFI plus GPS (Height over best Glide, Height above Goal, Glide to Goal, etc), on the other, we can easily imagine how difficult can be to control every parameter related with position, orientation and time, just with a BFFI and a GPS. For example, when the start time is approaching the pilot needs to constantly monitoring the time and calculate the difference to the start instant. At the same time the pilot needs to control the distance to center of the start cylinder and calculate the difference to the cylinder limit. With these two values the pilot should define a strategy (trajectory, speed, etc.) in order to be in the optimal point of the cylinder exactly at the start instant. In order to fulfill to this specific information competition requirements the pilot need to use the so called Integrated Flight Instruments (IFI). These flight instruments are basically an augmented BFFI integrated with a GPS. When compared with a BFFI usually an IFI has a better display, more computation power, and more memory. This allows the CFI to provide all the features of a BFFI and a GPS, as also specific competition features. As an example following are some of the features of one of the bests IFI's, the Brauniger Compeo [10] shown in Figure 2: Wind speed and direction; Time and Flight time; Speed over ground and difference to TAS; Height over Goal and Distance to Goal; Bearing and Track; Direction to best climbing; L/D Ratio (through air, over ground, to waypoint); Automatic pilot notification, when start or a turn point is passed; Increased McCready and glide path features; Three configurable displays

Fig. 2. Brauniger IQ-Compeo

Despite IFI's being a great improvement over BFFI's they still have some drawbacks, mostly due to hardware limitations. In fact, when compared with other available mobile devices (e.g. PDA, measurement equipment, etc) they have a low definition Black and White display, low computation power, low memory and little interface and communications capabilities. Other important drawback is the limited interface capabilities with external sensors like wind probe, or temperature sensors. Whenever this interface exist is by wire connection which can be very impractical. With these limitations is not possible to provide other important features, mainly related with information display and context awareness, but also with different type off information and pilot interface.. Some features examples are: Real time track color display with context aware wind, and thermals adaptive information; Context based specific competition information providing like: Start Optimization, Turn Points optimization, etc; Better user interface capabilities particularly important for instrument configuration, competition route definition, waypoints introduction, etc; Better communications capabilities like Bluetooth, Wireless, GSM, particularly important for security and logistic reasons.

In order to surpass these drawbacks our team has developed a totally new, integrated type, flight instrument. This IFI is based on a new concept and was named FlyMaster.

3 Flymaster – The Concept

As pointed before, the main reason for the other IFI's drawbacks was due to hardware limitations. Better hardware implies higher costs, particularly if we considerer bespoke development and the relatively low market size of IFI's. Considering this, the only chance to have better hardware at lower costs was to use a common PDA. As PDA does not have all the sensors an IFI requires we also needed to develop a hardware module with them. In this way, the FlyMaster concept is based on a hardware module responsible for context information retrieval and preprocessing (and a little more), and a PDA with specific software for information processing and displaying, user interface and communications. The basic FlyMaster diagram can be seen in Figure 3.

In this article we will focus mainly in the software issues, however a short hardware discussion will be made on the next section.

3.1 FlyMaster Hardware Module

As we can see in Fig. 3 the hardware module can be divided in seven modules. Each module has a specific function being the processor responsible for the module control. In fact the Vario Module has its one processor as described below.

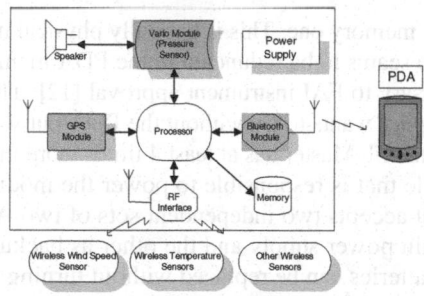

Fig. 3. FlyMaster Diagram

The GPS and the Bluetooth (BT) modules excuse any special commentary. Basically, the processor receives the information in NMEA (National Marine Electronics Association) format [11] from the GPS and sends it to the PDA through the Bluetooth module that works like a wireless rs232 serial port. The inverse can also happen when the PDA wants to configure or make a specific information requirement to the GPS. Due to this type of operation the FlyMaster hardware module can be used like any other Bluetooth GPS module. When the processor wants to send any other information, like vertical speed or temperature, for the PDA, or vice versa, some messages are inserted between the NMEA messages. An important module is the Vario one. This module is a complete Variometer which includes a very sensitive pressure sensor and a dedicated processor which calculates vertical speed (0,1 m/s resolution). The results are sent to the PDA and a proportional acoustic signal is produced in the built in speaker. Whenever the pilots want it can turn off all the modules except the Vario one, turning the FlyMaster in a simple Variometer.

The FlyMaster Hardware has also a RF interface module. This module includes a universal ISM band FSK transmitter/receiver that allows the module to communicate with external sensors. At the moment a temperature sensor and a wind speed probe is being developed. However, other sensors are being studied for a near future. In spite, some of the available IFI's have built in temperature sensors and also the possibility to connect a wired speed probe, the FlyMaster is the only with "wireless" connection possibility. This is a great advantage since it allows reducing measurement errors, and opens new possibilities in terms of were and what we can measure besides being by far more usable and less obstructive.

Fig. 4. FlyMaster Hardware Module Appearance

Another module is the memory one. This is basically physical memory to record every type of data. This module seams to be redundant to the PDA memory, however a built in memory module is necessary to FAI instrument approval [12]. The built in memory it is also important when the pilot wants to fly without the PDA but wants to record its flight. Compared to other IFI's the FlyMaster has at least 4 times more memory.

Finally the last module that is responsible to power the module. This module has a particular feature since it accepts two independent sets of two AA batteries each. One set functions has the main power supply and the other as backup. The switch is made automatically, and the batteries can be replaced without turning off the module.

In spite of not being indicated in Figure 3 the hardware module has also two switches and three informative LEDs. Basically, the switches allow powering the module on and off and some of its blocks (e.g. GPS, Variometer, etc). They also allow controlling the volume and some more little configurations. The LEDs have informative function and allow to verify if the module is on, if it locked onto satellites, if it is communicating trough Bluetooth, or with the sensors. The module appearance can be seen Figure 4.

3.2 FlyMaster Software Module

As referred before the FlyMaster Software (FMS) runs on any PDA with Windows Mobile. In order to offer all its features a FlyMaster hardware module is necessary. However, the FMS can also be used with a normal Bluetooth GPS but with several limitations. In terms of features, the FMS offers almost, not to say all, the features available in the actual IFI's, as also some extra features. The description of the features is beyond the scope of this article, and we will only present the ones which follow guidelines from ubiquitous and context aware mobile computing. Despite the number of features the FMS uses a different approach to information display. As we saw in previews sections during flight a pilot needs a lot of information. Due to the limited size of displays, and also due to practical issues, is not possible to show at the same time all the available information. Considering this, and as the information necessities varies according the flight phase, the other IFI's use several displays with just a few fields each. Each display has a specific usage. The user can configure some of the fields in each display being the others fixed. During flight the user can change from one display to other by pressing a key. Although pressing a key seam a simple act a competition flight can be very rough which difficult it. Specifically, as referred before, the use of both hands is required to control the glider and their use in other function always decreases the security level. Because of this, a lot of pilots prefer to use less information and avoid changing the display. Differently from other IFI's the FMS uses a single display, but it also uses some techniques in order to show all the information that the pilot needs. First, as most of the PDA's has a color display, with higher resolution and contrast, the FMS can have more fields in the display at the same time. As we will see later a bar for example, can be larger or thicker, and have different colors depending the situation. Second, the FMS shows different information according to the flight phase. In Fig. 5 we can see two different flight phases for which different information is displayed. Note that, the passage from one situation to another is completely automatic. Furthermore, in each of the situations we have more than eleven different types of information, were other IFI's usually has less than six. In the next sections we will see some original ways of showing flight information.

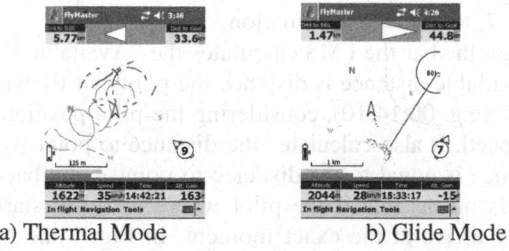

a) Thermal Mode   b) Glide Mode

Fig. 5. Two distant flight phases

4 Tricks and Tips

The FlyMaster has several tricks to show the needed information at the right instants in a useful way. We will introduce the most important ones in the following sections. For lack of space we will not present all the options, since almost all the fields or graphics are optional, besides the user can change the color, text size, etc.

4.1 Start Time Indicator

As referred above usually the race starts already in the air. In order to give equal chances to all the pilots a "Start" is defined. There is more than one type of start. We will explain here the most common being the others very similar in terms of needed information.

The Start is defined by a cylinder and an hour. The pilots can go to the air after the take off hour which precedes the start by a certain period of time. However all the pilots must have a GPS track point inside the cylinder after the start hour.

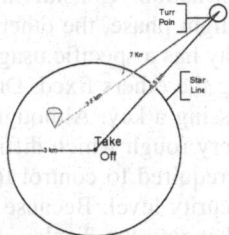

Fig. 6. Start situation

In figure 6 a possible flight situation is represented where the pilot is already flying. Ideally, the pilot should pass through point A exactly at the start hour (e.g. 13:15:00). Considering the pilot position, the existing IFI's shows two information fields in the form of numbers. One number indicates the remaining time to 13h (e.g. 00:14:10), and the other indicates the remaining distance to the turn point (e.g. 7 km). With this information the pilot should make its calculations, in the sense to decide when to depart and at what velocity, in order to arrive to point A at 13h. Naturally, this calculation is not easy without the appropriate means. Differently, the FMS decides everything for the pilot, and instead of numbers it uses the "start bar", represented in Fig. 7, to give the information.

In order to design the bar the FMS calculates the "Available Distance" towards the next waypoint. Available distance is distance the pilot can fly with the time available before "start open" (e.g. 00:14:10), considering the pilot position, wind direction and speed and flight speed. It also calculates the distance to point A (e.g. 3,5 km). When the available distance is equal to the distance to point A the bar will be at 50% of it maximum size, this means that if the pilot where to fly to start at that moment he would theoretically arrive at the exact moment. In Fig. 7 the bar indicates that the pilot should wait, otherwise it will arrive too soon.

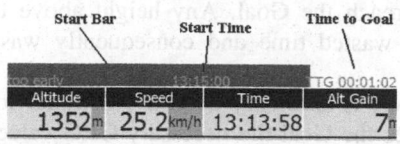

Fig. 7. Start Information

A great advantage of the start bar, besides the direct decision information, is that it represents a relation between distances, therefore, as the pilot get closer to point A the bar becomes more sensitive. Just looking to the bar the pilot knows that, values greater than 50% means the line will be crossed too early, (the pilot should depart and eventually accelerate), values less than 50% means the line will be crossed after the start has opened (the pilot can remains in place or look for a thermal).

4.2 Thermal Mode

When the pilot starts climbing and changing direction the FMS automatically changes to thermal mode. Besides changes in the numeric bottom fields the most important information is obtained through the "Track" (see figure 5 a)). The track shown is a colour representation of where you have passed (it is stored on a second by second interval). The colours indicate whether altitude was lost or gained between each track point. Red gradually going to violet represents climb, stronger towards the violet. Green gradually going to blue represents sink, stronger towards the blue.

A thicker line shows track that was flown at the current altitude to 15m (this value is configurable) below the current altitude. This is particularly useful when trying to get back to a thermal. Optionally, the sink may not be displayed. As the pilot drift in the thermal the alignment becomes apparent. If the pilot looses the thermal or if it passes two times trough the same place it can use the track information to find the thermal again. In order to help the visualization the pilot can make a zoom only by touching in a certain area.

Most other IFI's have a something similar to the FMS thermal mode. Specifically, they usually have a track visualization display. However, as there is no colour and the definition is low, there is no distinction between climbing, sinking and height. The absence of this information, particularly the one related with height, withdraws most of the display utility. As an example, imagine the track from figure 5 a) in Black and White. As the pilot dos not have any information about height he could not know if the indicated thermal is at his height, 500 m below or 500 m above.

a.   Glide Optimization

Most of the times after a thermal climb the pilot parts for a transition, either looking for another thermal, or to in direction to a certain objective (turn point, goal, etc.). The idea in a transition is to maximize the Glide ratio over ground (GRG) (ratio between ground speed and vertical speed) [9], or to go faster, as soon as we have enough height to reach the objective. The last case is particularly important during the final glide to the Goal in a competition. Ideally, in the last thermal the pilot should only climb the necessary to reach the Goal. Any height above the goal at the arriving moment corresponds to wasted time and consequently wasted points in the tasks scoring.

In each task instant there is necessary GRG to reach the Goal (it could be infinite). Once the pilot approaches the Goal the necessary GRG value goes lower. At certain point, usually during a thermal climb, the probable pilot GRG is sufficient to reach the Goal. At this instant the pilot can leave the thermal and go directly to the goal. Naturally, it is very difficult to predict the average GRG, since the air is always moving in all directions affecting directly the GRG. In this way any decision has always a certain risk. In order to evaluate the risk the pilot needs to know the required GRG to goal, its average GRG, the remaining distance to goal and eventually its average vertical and horizontal speed. Most of the IFI's can provide all the needed information. However, at least 6 fields are needed which implies to use of more than one display, and toggle between displays. The FMS solves this problem using graphic information, besides the numeric information.

After the start opens the FMS automatically changes to the "Goal" mode. In this mode the top numeric fields can be configured to show the remaining distance to goal, and the distance to the next waypoint. Further, the Start Bar is replaced by three other bars. These bars represent GRG with configurable scale depending the type of aircraft (eg. in the Figure 8 the "GRG to Goal" is 20:1, the "Current GRG" is approximately 16:1 and the "Long Term GRG" is approx. 7:1).

The top bar is a current glide ratio indicator, which correspond to the effective GRG at the current instant. This pilot can use this value to obtain the ideal speed for best possible glide towards the next waypoint. The colour of the current GRG indicator will become greyed out when ratio is not being calculated, which happens during climbing. The middle bar is a long term GRG indicator. The value corresponds to the average GRG during transitions. Combined with the bottom glide ratio bar, which is GRG to goal, provides estimate to whether enough altitude has been gained for final glide to goal. In other words, when the middle graph is longer that the bottom graph, you have enough altitude and may begin final glide to goal.

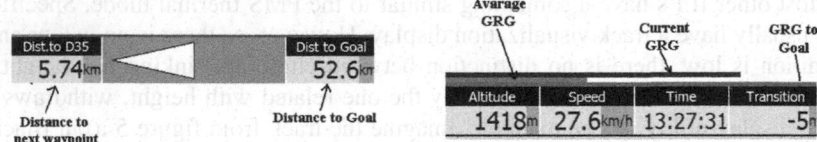

Fig. 8. Flight Information

4.3 General Behavior

Besides the previews features the FMS has the possibility to automate its behavior in several ways. One of the most useful is the automatic Zoom.

There are some situations were a zoom is needed. For example when in Glide mode and when approaching a turn point, the FMS can make an automatic zoom in order the pilot see the line cylinder. In this way the pilot can anticipate the turn and earn time. Another situation is in thermal mode. Sometimes, during climbing the pilot can loose the thermal core. In this case a quick look to the track can be very useful. Considering this the FMS can be configured to make a automatic zoom when enter in to thermal mode. Naturally, the zoom scale can also be changed just by touching an appropriate display area.

Another useful general feature is the automatic information field change. As referred before depending the flight situation different information is needed. In order to avoid the pilot intervention, the FMS has different functioning modes. Some of this modes were already partially described (e.g. Thermal Mode, Start Mode, Gliding Mode). According to the gathered information (e.g. vertical speed, horizontal speed, position, etc.) the FMS tries to identify the situation and automatically change to the adequate mode. For example, if in glide the vertical speed goes to positive values and the pilot starts turning (possible thermal), the FMS automatically changes to thermal mode. After, if the pilot stops turning and follows the same direction for more than 20 seconds (configurable), then the FMS returns to the Glide mode. Naturally, each of the modes has a different information field configuration.

5 Conclusions and Future Work

In this paper we presented a new Flight Instrument for free flight assistance that follows the main characteristics of Ubiquitous Computing and Ambient Intelligence. The instrument, named FlyMaster, distinguish from other free flight instruments not only in terms of hardware concept but also in the software concept. Regarding the hardware the FlyMaster is constituted by a module responsible for data retrieval and preprocessing, which includes several sensors, a GPS and a Bluetooth module. This module communicates through Bluetooth with the FlyMaster Software that runs on a normal PDA. Note, that the FlyMaster module can be used as a normal Bluetooth GPS. In this way after landing the pilot can have a orientation equipment just by changing the software (e.g. TomTom).

Due to the PDA use, when compared with other Free Flight Instruments the FlyMaster has more computation power, a color display and with more resolution, more interface capabilities and also more communications capabilities. Furthermore, at least "half" of the instrument (PDA) can be used in other applications besides free flight. Examples are orientation after the flight, playing games or listen music just to name a few.

The use of PDA has also other implications since it opens new dimensions for the software features. Regarding the software, the FlyMaster is also very different from the competition. In fact it offers almost, not to say all, the features from other instruments but also provides some new features, particularly in terms of flight optimization. Examples are start optimization, final glide optimization, and thermal map by layer. Another great difference is some original forms of showing the information. Basically, the FlyMaster uses less numbers and more graphics in order to facilitate riding the information. Furthermore, the FlyMaster uses some context information to decide which information to show in order to avoid the pilot interference.

Flight tests made to date were very promising. The FlyMaster is already the choice of the 2006 Paragliding European Champion Luca Donini.

In spite of the FlyMaster features there is a lot of work to be done. Currently, we are developing testing wireless temperature sensors and wireless wind speed sensor. With the inclusion of these sensors a lot of new features can be added to the FlyMaster.

References

1. Harris, D.: Flight Instruments and Automatic Flight Control Systems. Blackwell Science
2. Chen, G., Kotz, D.: A Survey of Context-Aware Mobile Computing Research, Dartmouth College (2000)
3. Satyanarayanan, M.: Fundamental challenges in mobile computing. In: PODC 1996. Proceedings of the fifteenth annual ACM symposium on Principles of distributed computing (1996)
4. Weiser, M.: The Computer for the 21st Century. Scientific American 265, 94–104 (1991)
5. Ramos, C.: Ambient Intelligence → Artificial Intelligence. IADIS InternetWeb 2007 Plenary Session (2007)
6. Reinhart, R.O.: Basic Flight Physiology. McGrawHill, New York (1996)
7. Flytec GmbH, 4005 XL GPS - Manual (1999)
8. Brauniger GmbH, IQ Competition GPS - Manual (1999)
9. Aupetit, H.: Parapente - Tecnica Avanzada, Editorial Perfils (2001)
10. Brauniger GmbH, Compeo - Manual (2004)
11. National Marine Electronics Association. NMEA 0183
12. Fédération Aéronautique Internationale, Technical Specification for IGC-Approved GNSS Flight Recorders (2006)

Argumentation-Based Decision Making in Ambient Intelligence Environments

Goreti Marreiros[1,3], Ricardo Santos[1,2], Paulo Novais[4], José Machado[4],
Carlos Ramos[1,2], José Neves[4], and José Bula-Cruz[5]

[1] GECAD – Knowledge Engineering and Decision Support Group
Porto, Portugal
{goreti,csr}@dei.isep.ipp.pt
[2] College of Management and Technology– Polytechnic of Porto
Felgueiras, Portugal
rjs@estgf.ipp.pt
[3] Institute of Engineering – Polytechnic of Porto
Porto, Portugal
{goreti,csr}@dei.isep.ipp.pt
[4] University of Minho
Braga, Portugal
{pjon,jmac,jneves}@di.uminho.pt
[5] University of Trás-os-Montes e Alto Douro
Vila Real, Portugal
jcruz@utad.pt

Abstract. In Group Decision Making argumentation has a crucial role; we have
a set of participants, with different points of view that exchange ideas or engage
in a process of argumentation and counter-argumentation, negotiate, cooperate,
collaborate or even discuss techniques and/or methodologies for problem
solving. In this paper we propose an argumentation-based system, where
intelligent agents simulates the behaviour of individuals as members of a group
in a decision making process. Our agents operate in ambience intelligent
environments and behave depending on rational and emotional factors.

Keywords: Argumentation, Group Decision Making, Artificial Societies,
Multi-agent systems.

1 Introduction

Group decision making processes are omnipresent in several everyday activities. We
can have more formal decision groups like committees or management teams who
have to decide for instance if it is more advantageous to acquire a competitor
enterprise and their technology or invest in research to develop the technology
internally. Or we can have more informal groups like, for instance, a married couple
who have to decide what movie to watch.

Along the last 20 years, several Group Decision Support Systems (GDSS) were
developed, some dedicated to be used exclusively in decision rooms and other ones
with features to support ubiquitous group decision meetings. The main goal of GDSS
is to help a group that is responsible for decision making.

J. Neves, M. Santos, and J. Machado (Eds.): EPIA 2007, LNAI 4874, pp. 309–322, 2007.
© Springer-Verlag Berlin Heidelberg 2007

A more recent approach is to Group Decision support is the use of agent based systems to support or simulate group decision processes. The use of multi-agent systems is very suitable to simulate the behaviour of groups of people working together and, in particular, to group decision making modelling, once it caters for individual modelling, flexibility and data distribution. Multi-agent systems have been also used to support and promote the creation of Ambient Intelligence environments [1][2].

In this work we propose the simulation of group decision making through an agent based approach. The simulation of group decision making can be viewed as an automated negotiation process. In the multi-agent literature, various interaction and decision mechanisms for automated negotiation have been proposed and studied.

Approaches to automated negotiation can be classified in three categories [3], namely game theoretic, heuristic and argumentation based. We think that an argumentation-based approach is the most adequate for group decision making (simulation), since agents can justify possible choices and convince other elements of the group about the best or worst alternatives.

Argumentation is intuitively adversarial in nature, so it is natural to relate argumentation to a dialogue in which two parties attempt to persuade one another [4]. There is a great variety of dialogues ranging from exchanges of pre-formatted messages to argumentation-based dialogues [5]. In this latter category, in 1995, Douglas Walton and Erik Krabbe [6] presented a classification where they identify six major classes (although they recognized that others may exist, like for instance common dialogues):

- **Inquiry dialogues** – a lack of knowledge is in the origin of this type of dialogue, two or more participants will collaborate to establish the truth or falsity of some proposition. The dialogue ends when a proof is found, or when the participants agree that it is not possible to achieve the proof. An example can be the dialogues that are involved in a scientific research.
- **Persuasion** – this dialogue occurs when there is a conflict between two participants, and each one aims to persuade the other to endorse of some proposition or course of action.
- **Negotiation** – the aim of this dialogue is to reach a division of scare resources that can be acceptable to all. Each participant in the negotiation dialogue aims to maximize his part in the division. An example may be a typical seller/buyer dialogue
- **Deliberation** – a dialogue where two or more participants share the responsibility of forming a plan of action. The goal of this dialogue is to achieve an agreement on a plan of action, and each participant aims to agree on the plan that is more favourable for their intentions. Contrary to persuasion this dialogue does not start from a conflict but from the need to perform an action.
- **Information seeking dialogue** – one specific participant inquire other participant about the truthful of a proposition. This dialogue is similar to an inquiry dialogue, but differs in the initial knowledge. One agent thinks that the other has more knowledge in a specific topic. An example of this type of dialogue can be an expert consultation.

- **Eristic dialogue** – participants use this dialogue as an alternative for physical fighting, thus it is a very intense dialogue, with each one aiming to win.

It is certainly difficult to classify a real dialogue exactly as one, and just one, of the mentioned types, most of them are in fact a combination of different types. In this work we will focus on exploring the persuasion dialogues that seems to be the most adequate to phase of group decision making that we are working: the choice phase (at this phase there is no space for negotiation, because alternatives are already settled).

The work described in this paper is being implemented under the scope of ArgEmotionAgents (POSC/EIA/56259/2004 - Argumentative Agents with Emotional Behaviour Modelling for Participants' Support in Group Decision-Making Meetings) project.

2 Ambient Intelligence

Ambient Intelligence (AmI) deals with a new world where computing devices are spread everywhere (ubiquity), allowing the human being to interact in physical world environments in an intelligent and unobtrusive. AmI environments may be diverse (e.g. decision room, hospitals, stores), but the idea behind is to provide a better support to the human being and the access to the essential knowledge in order to make better decisions when interacting with these environments.

In this work it is proposed a multi-agent (simulator) argumentation based system whose aim is to simulate group decision making processes. The simulator considers and supports the emotional factors of participants (agents) and their associated processes of argumentation.

This system is integrated in an ubiquitous agent based group decision support system [7] and intended to be used for intelligent decision making, a part of an ambient intelligence environment where networks of computers, information and services are shared (Fig. 1) [7][8].

As an example of a potential scenario, it is considered a distributed meeting involving people in different locations (some in a meeting room, others in their offices, possibly in different countries) with access to different devices (e.g. computers, PDAs, mobile phones, or even embedded systems as part of the meeting room or of their clothes). Fig. 1 shows an Ambient Intelligent Decision Laboratory with several interactive Smartboards. The meeting is distributed but it is also asynchronous, so participants do not need to be involved at any time.

Fig. 1. Ambient Intelligent Decision Lab

3 Multi-agent Simulator

In our previous work we identified the main agents involved in a simulation of a group decision meeting [9]: Participant Agents; Facilitator Agent; Register Agent; Voting Agent and Information Agent. Fig. 2 illustrates the multi-agent model.

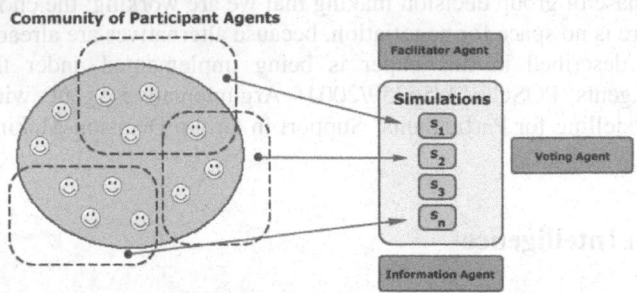

Fig. 2. Multi-agent model to simulate group decision

The Facilitator Agent caters for the simulation tasks in its organization (e.g. in decision problem and decision rules configuration). The Facilitator Agent also administrates the group formation process (i.e., the selection of the Participant Agents) and manages the inclusion of new Participant Agents into the agent community or agency. During the simulation, the facilitator agent coordinates and summarizes the simulation results.

Based on experience, it is also known that almost all the group decision making meetings have one or more voting rounds. The Voting Agent executes tasks related with the voting simulation process, according to the decision rules settled by the Facilitator Agent.

The Information Agent holds and retrieves information about the different proposals (or alternatives) that are evaluated by the agency during the group decision simulation process.

The Participant Agents also act according to the role usually fulfilled by the humans in the group decision making process. The set of Participant Agents set a community where the agents are created with social and emotional attributes that personalize their behaviour.

In the remain text of this section we will first present the architecture of participants agents [7], because they represent the main role in group decision making and then we will briefly detail the components of this architecture (Fig. 3).

In the knowledge layer the agent has information about the environment where he is situated, about the profile of the other participant's agents that compose the simulation group, and regarding its own preferences and goals (its own profile). The information in the knowledge layer is dotted of uncertainty and will be accurate along the time through interactions done by the agent.

The interaction layer is responsible for the communication with other agents and by the interface with the user of the group decision making simulator.

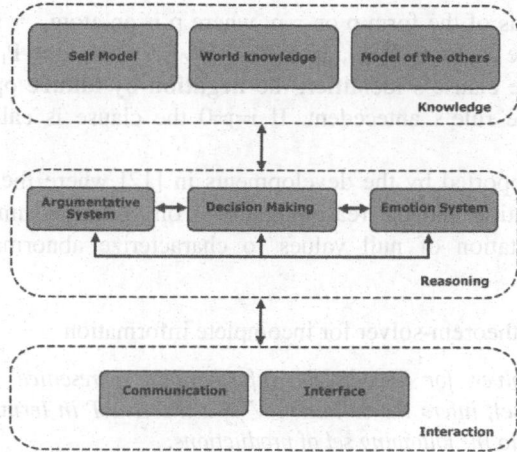

Fig. 3. Participant agent architecture

The reasoning layer contains three major modules:

- The argumentative system – that is responsible for the arguments generation. This component will generate explanatory arguments and persuasive arguments, which are more related with the internal agent emotional state and about what he thinks of the others agents (based on a profile, where is included the emotional state);

- The decision making module – will support agents in the choice of the preferred alternative and will classify all the set of alternatives in three classes: preferred, indifferent and inadmissible;

- The emotional system [10] – will generate emotions and moods, affecting the choice of the arguments to send to the others participants, the evaluation of the received arguments and the final decision.

4 Argumentation System

This component will generate persuasive arguments based on the information that exists in the Participant Agent knowledge base. We adopt the same ontology as in [11]. So, we have the following arguments: appeal to prevailing practice; a counter example; an appeal to past promise; an appeal to self-interest; a promise of future reward; and a threat.

In this section we start by presenting the participant agent knowledge base and its strategies, followed by the presentation of the argumentation protocol for two agents and the description of the persuasion process (i.e. selection of the opponent, arguments generation, selection and evaluation).

4.1 Participant Agent Knowledge Base

The Participant Agents Knowledge Bases (KB) are made of logic clauses of the form $r_k:P_{i+j+1} \leftarrow P_1 \wedge P_2 \wedge \dots \wedge P_{i-1} \wedge \text{not } P_i \wedge \dots \wedge P_{i+j}$, where $i, j, k \in N_0, P_1, \dots, P_{i+j}$ are literals; i.e., formulas of the form p or ¬p, where p is an atom, ¬ stands for explicit negation and where r_k, not, P_{i+j+1}, and $P_1 \wedge P_2 \wedge \ldots \wedge P_{i-1} \wedge$ not $P_i \wedge \ldots \wedge P_{i+j}$ stand, respectively, for the clause's identifier, the **negation-by-failure** operator, the rule's consequent, and the rule's antecedent. If i=j=0 the clause is called a **fact** and is represented as $r_k:P_1$.

This work is supported by the developments in [12] where the representation of incomplete information and the reasoning based on partial assumptions is studied, using the representation of null values to characterize abnormal or exceptional situations.

Definition 1. Meta theorem-solver for incomplete information

A meta theorem-solver for incomplete information, represented by the signature demo:T,V →{true,false}, infers the valuation V of a theorem T in terms of false, true and unknown according to the following set of productions:

demo(T,true) ← T.

demo(T,false) ← ¬T.

demo(T,unknown) ← not T,

  not ¬T.

The Knowledge Base of a Participant Agent KB_{AgP} is:

$KB_{AgP}= AgP_iO \ U \ AgP_iOO \ U \ AgP_iP \ U \ AgP_iPO \ U \ AgP_iW$, where AgP_iO are the goals of agent AgP_i, AgP_iOO are the set of goals that AgP_i believes the other agents hold, AgP_iP contains the model of its own profile, AgP_iPO contains what AgP_i believes about the other agents profile, and AgP_iW contains generic knowledge about the environment.

The AgP_iPO is defined according to a set of characteristics enumerated afterward:

- Emotional state – characterizes the mood of the agent, and can be positive, negative or neutral;
- Benevolence – an agent can be, or not, benevolent. If an agent is benevolent, and receive a request from another agent to change his preferred alternative to alternative X, he will accept the request without the need to support arguments, since the alternative X belong to his internal set of preferred alternatives. If a benevolent agent receives more than one request to change his preference, and all the requests are acceptable, he will accept the request of the most credible agent;
- More/less preferred argument – agents may have a specific preference about the arguments to send;
- Gratitude debts of the agent – result from previous interactions (simulations) in the community of participant agents;
- Enemies – agents that, for some particular reason, our participant agent does not like to interact;

- Credibility - each agent evaluates the credibility of other agents based on two key components: trustworthiness and expertise.

4.2 Participant Agent Strategies

During its lifetime and particularly in a specific simulation agents behave in order to accomplish their goals. The goals of each participant agent (AgP_iO) are defined according to the agent strategies. In the model that we are proposing participant agents have two distinct strategies: behaviour dependent and goal dependent.

In the goal related strategy two tactics are defined influencing the agent argumentation dialogues, namely:

- Defending the most preferred alternative;
- Avoiding that an unacceptable alternative becomes the group solution.

The goal related strategy is influenced by another strategy that is the time dependent behaviour. If an agent is time-dependent and supposing that he is arguing to defend his preferred, but he concludes that it is impossible to achieve an agreement on that alternative, he will change his tactic to avoid unacceptable alternatives.

The behaviour related strategy determines the method for the selection of the participant to persuade (opponent):

- Determined – the agent will consume all the available arguments with a specific agent beforehand to other agent;
- Anxious – an agent using this strategy will try to persuade a specific agent, but if that agent refuses the persuasion he will move forwards to try to persuade another agent.

Strategies will influence agent's behaviour, namely the argumentation flow and certainly the simulation results.

4.3 Argumentation Protocol

During a group decision making simulation, participants' agents may exchange the following locutions: request, refuse, accept and request with argument.

Request (AgP_i, AgP_j, α, arg) - in this case agent AgP_i is asking agent AgP_j to perform action α, the parameter *arg* may be void and in that case it is a request without argument or may have one of the arguments specified in the end of this section.

Accept (AgP_j, AgP_i, α) - in this case agent AgP_j is telling agent AgP_i that it accepts its request to perform α.

Refuse (AgP_j, AgP_i, α) - in this case agent AgP_j is telling agent AgP_i that it cannot accept its request to perform α.

In Fig. 4, it is possible to see the argumentation protocol for two agents. However, note that this is the simplest scenario, because in reality, group decision making involves more than two agents and, at the same time AgP_i is trying to persuade AgP_j that, this agent may be involved in other persuasion dialogues with other group members.

Begin
  request(AgP_i,AgP_j,Action)
  If refuse (Proponent, AgP_i, Action)
    do
      $Arg \leftarrow select\_arg(AgP_i,AgP_j,Action,Args\_List)$
      request(AgP_i,AgP_j,Action,Arg)
    While (refuse (Proponent, AgP_i, Action) and not timeout)
End

Fig. 4. Argumentation protocol for two agents

As we previously refer we adopt the same ontology as in [11]. So, we have the following arguments: appeal to prevailing practice; a counter example; an appeal to past promise; an appeal to self-interest; a promise of future reward; and a threat. These are the arguments that agents will use to persuade each other.

4.4 Opponent Selection

A first step in an argumentation dialogue is the opponent selection. In this process the participant agent considers three different aspects, namely its own strategies (behaviour and goal related), the other agents voting preferences and the quality of the information detained about the others group members.

For evaluating the quality of the information, the following quality operators are defined: $Q_{EmotionalState}$, $Q_{Gratitude}$, $Q_{Credibility}$, $Q_{Enemies}$, $Q_{Benevolent}$, $Q_{TimeDependent}$ and $Q_{PreferredArguments}$.

where the quality of the information about the property K is given by: $Q_K = 1/Card$, *Card* is the cardinality of the exception set for K, if the exception set is . If the exception set is disjoint the information quality is given by

$$Q_K = \frac{1}{C_1^{Card} + ... + C_{card}^{card}},$$ where C_{card}^{card} is a card-combination subset, with card elements.

The quality of the information that agent AgP_i detains about agent AgP_j is measured by the following formula:

$$Q^{AgP_i}(Profile_{AgP_j}) = \frac{\sum_{k=1}^{N} Q_k^{AgP_j} * W_k^{AgP_i}}{\sum_{k=1}^{N} W_k^{AgP_i}}$$

where N is the number of properties of the profile, $Q_k^{AgP_j}$ is the quality measure of K and $W_k^{AgP_i}$ (in interval [0,1]) represents the contribution of K (weight) in the agent profile construction. A property with weight 0 will not be considered in the profile construction process.

4.5 Arguments Generation

In this section we describe and specify the persuasive arguments that our agents may change. For each argument it is given a little example.

Threats

As previously referred, threats are very common in human negotiation, and they can assume two distinct forms: you should perform *action A* otherwise I will perform *action B*; and you should **not** perform *action A* otherwise I will perform *action B*.

In our model this type of arguments may be formalized as a triplet:

Threat(Justification, Conclusion, Threatened_goal)

Example 1: AgP_1 ask AgP_2 to vote on alternative A_i with the argument that if he refuse he will vote on alternative A_j that he believes is unacceptable for AgP_2.

Promises

Like threats we can also have two distinct forms of rewards: If you perform action *A* I can perform *B*; and If you do not perform action *A* I perform *B*.

In our model this type of arguments may be formalized as a triplet:

Promise(Justification, Conclusion, Promised_goal)

Example 2: AgP_1 ask AgP_2 to vote on alternative A_i with the argument that if he accept he will stay in debt of gratitude.

Appeal to counter-example

In this case, the participant agent that makes a request supported by this argument, expect to convey the opponent that there is a contradiction between what he says and his past actions.

The argument appeal to a counter-example is an explanatory argument and in our model is formalized as:

Appeal_counter_example (Justification, Conclusion)

Example 3: AgP1 asks AgP2 to vote on alternative Ai, if AgP2 refuses then AgP1 may counter argument with an appeal to a counter-example saying for instance that in the past he preferred alternative Ak and change to Al, so why not to do the same now?

Appeal to self-interest

In this case the participant agent that makes a request supported by this argument expects to convince his interlocutor that making action *A* is of his best interest.

The argument appeal to self-interest is an explanatory argument and in our model is formalized as:

Appeal_self_interest (Justification, Conclusion)

Example 4: Suppose that AgP1 ask AgP2 to avoid voting on alternative Ai, supported by the argument that AgP3 voted in the Ai alternative and AgP2 do not like AgP3.

Appeal to past reward
In this case the participant agent that sends such kind of argument expects that his interlocutor performs an action based on a past promise.

In our model the appeals to past rewards are formalized as:

Appeal_past_reward (Justification, Conclusion)

Example 5: If in some point in the past agent AgP1 sent a request to AgP2 to vote on a specific alternative with the promise that he will stay with a debt of gratitude, if AgP2 accept it, in a future decision AgP2 can send a request supported by this argument.

Appeal to prevailing practice
In this case, a participant agent believes that the opponent agent will refuse to perform a request action since it contradicts one of its own goals. For that reason the participant agent sends a request with a counter-example from third agent's actions or from past actions of the opponent.

In our model the appeal to prevailing practice is formalized as:

Appeal_prevailing_practice (Justification, Conclusion)

Example 6: Suppose that AgP_1 knows that AgP_2 had a strong preference for the alternative A_i and changed for alternative A_j. If he intends to ask AgP_3 to change his preference to alternative A_j he could support his request with the argument that another agent, who also prefers alternative A_i, to change his preference to alternative A_j.

The last four types of arguments are explanatory arguments and are formalized as tuple:

Argument_type(Justification, Conclusion).

4.6 Arguments Selection

In [11] it is used an existent pre-order for the selection of arguments to send, the strongest argument is a threat and the weakest argument is an appeal to prevailing practice. Another approach that can be finding in literature to govern the argument selection process is based on mixture of the alternatives utility and the trust in the interlocutor.

In our model it is proposed that the selection of arguments should be based on agent emotional state. We propose the following heuristic [9]:

- If the agent is in a good mood he will start with a weak argument;
- If the agent is in bad mood he will start with a strong argument.

We adopt the scale proposed in [11] for the definition of strong and weak arguments. We defined two distinct classes of arguments, namely a class for the weaker ones (i.e., appeals) and a class for the remainders (i.e., promises and threats). Inside each class the choice is conditionally defined by the existence in the opponent profile of a (un)preference by a specific argument. In case the agent does not detain information about that characteristic of the opponent, the selection inside each class follow the order defined in [11].

4.7 Arguments Evaluation

In each argumentation round the participant agents may receive requests from several partners, and probably the majority is incompatible (e.g., in the same round the agent AgP_i may receive a request to vote on alternative A_i and other to vote on alternative A_j). The agent should analyse all the requests based on several factors, namely the proposal utility, the credibility of proponent and the strength of the argument.

If the request does not contain an argument, the acceptance is conditioned by the utility of the request for the self, the credibility of the proponent and one of its profile characteristics, i.e., benevolence. Next is presented the algorithm for the evaluation of this type of requests (without arguments):

$$Req^t_{AgP_i} = \left\{ request^t_1(AgP, AgP_i, Action),....,request^t_n(AgP, AgP_i, Action) \right\}$$

where AgP represents the identity of the agent that perform the request, n the total number of request received in instant t and Action the request action (e.g., voting on alternative number 1).

> **Begin**
> **If** $\neg$ $profile_{AgPi}(benovolent)$ **then**
>> **Foreach** $request(Proponent, AgP_i, Action) \in Req^t_{AgP_i}$
>>> $refuse\ (Proponent,\ AgP_i,\ Action)$
>
> **Else**
>> **Foreach** $request(Proponent, AgP_i, Action) \in Req^t_{AgP_i}$
>>> **If** $AgPO_{AgP_i} \vdash Action$ **then**
>>>> $Requests \leftarrow Requests \cup request(Proponent, AgP_i, Action)$
>>>
>>> **Else**
>>>> $refuse\ (Proponent,\ AgP_i,\ Action)$
>
>> $(AgP, Requested\_Action) \leftarrow Select\_more\_credible(Requests)$
>> **Foreach** $request(Proponent, AgP_i, Action) \in Requests$
>>> **If** $(Proponent = AgP\ or\ Request\_Action = Action)$ **then**
>>>> $accept\ (Proponent,\ AgP_i,\ Action)$
>>>
>>> **Else**
>>>> $refuse\ (Proponent,\ AgP_i,\ Action)$
>
> **End**

5 Implementation

Some details of the implementation of the simulator are described here. The system was developed in Open Agent Architecture (OAA), Java and Prolog. OAA is structured in order to: minimize the effort involved in the creation of new agents, that can be written in different languages and operating on diverse platforms; encourage the reuse of existing agents; and allow for dynamism and flexibility in the formation of agent communities. More information about OAA can be found in www.ai.sri.com/~oaa/.

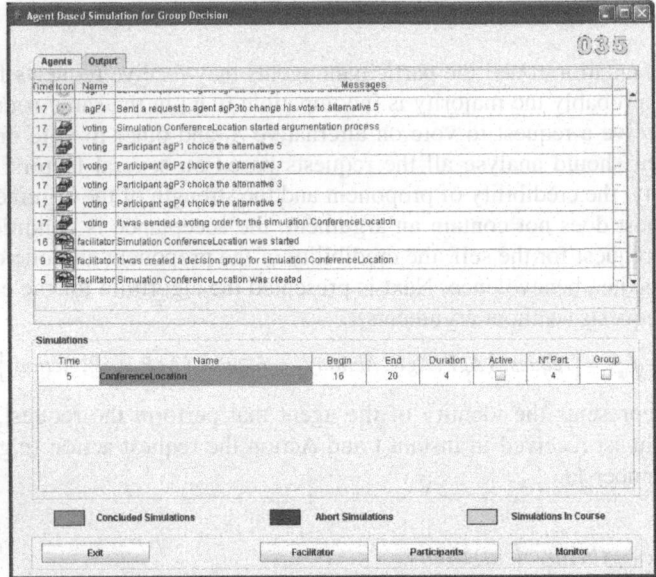

Fig. 5. Argumentation dialogues

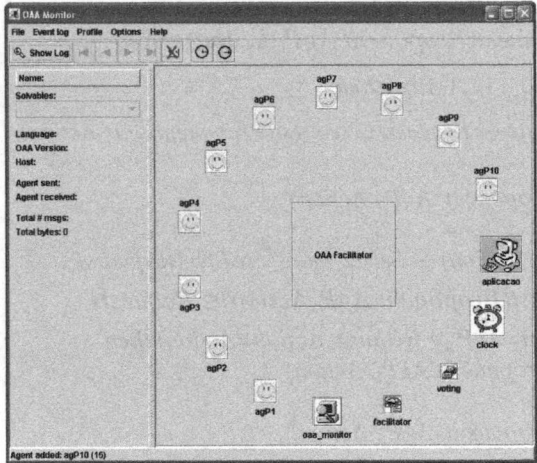

Fig. 6. Community of agents

Some screens of the prototype may be found in Figures 5 and 6.

Figure 5 shows an extract of the arguments exchanged by the participant agents. Once a simulation is accomplished, agents update the knowledge about the other agent's profile (e.g. agent credibility).

Fig. 6. shows the collection of agents that work at a particular moment in the simulator: ten participant agents, the facilitator agent (responsible for the follow-up of all simulations), the voting agent, the clock agent (OAA is not specially designed for simulation, for that reason it was necessary to introduce a clock agent to control the simulation), the *oaa_monitor* (i.e. an agent that belongs to the OAA platform, and is used to trace, debug and profile communication events for an OAA agent community) and the application agent (responsible for the communication between the community of agents and the simulator interface).

6 Conclusions

This work proposes a multi-agent simulator of group decision making processes, where the agents present themselves with different emotional states, being able to deal with incomplete information, either at the representation level, or at the reasoning one. Each agent represents a group decision member. This representation facilitates the simulation of persons with different behavioural characteristics.

The discussion process between group members (agents) is made through the exchange of persuasive arguments, built around the same premises stated to above.

The idea to create a multi-agent community to simulate argumentation process during group decision making envisages to create a tool to support one or more group members. With this kind of tool supported participants will be able to preview what may occur if a specific argumentation strategy is used and to test different argumentation strategies before deciding to adopt one. It can be seen as a kind of "what-if" performance dealing with the argumentation phase of group decision making. Future work will be concerned with the establishment of good participants' profiles and to the identification if the actuation of these participants during the meeting is in accordance with the profile (context-aware).

References

1. Hagras, H., Callaghan, V., Colley, M., Clarke, G., Pounds-Cornish, A., Duman, H.: Creating an Ambient-Intelligence Environment Using Embedded Agents. IEEE Intelligent Systems 19(6), 12–20 (2004)
2. Masthoff, J., Vasconcelos, W., Aitken, C., Correa da Silva, F.: Agent-Based Group Modelling for Ambient Intelligence. In: AISB Symposium on Affective Smart Environments, Newcastle, UK (2007)
3. Jennings, N., Faratin, P., Lomuscio, A., Parson, S., Sierra, C., Wooldridge, M.: Automated negotiation: Prospects, methods, and challenges. Journal of Group Decision and Negotiation 2(10), 199–215 (2001)
4. Bench-Capon, T., Dunne, D.: Argumentation and Dialogue in Artificial Intelligence. In: IJCAI 2005 tutorial notes, University of Liverpool, Liverpool, UK (2005)
5. Amgoud, L., Belabbès, S., Prade, H.: A Formal General Setting for Dialogue Protocols. In: Euzenat, J., Domingue, J. (eds.) AIMSA 2006. LNCS (LNAI), vol. 4183, pp. 13–23. Springer, Heidelberg (2006)
6. Walton, D., Krabbe, E.: Commitment in Dialogue. Basic Concepts of Interpersonal Reasoning. State University of New York Press, Albany, NY (1995)
7. Marreiros, G., Santos, R., Ramos, C., Neves, J., Novais, P., Machado, J., Bulas-Cruz, J.: Ambient Intelligence in Emotion Based Ubiquitous Decision Making. In: Proc. Artificial Intelligence Techniques for Ambient Intelligence, IJCAI 2007 – Twentieth International Joint Conference on Artificial Intelligence, Hyderabad, India (2007)

8. Marreiros, G., Santos, R., Freitas, C., Ramos, C., Neves, J., Bulas-Cruz, J.: Modeling Group Decision Making Processes with Artificial Societies considering Emotional Factors. In: Symposium on Artificial Societies for Ambient Intelligence, Newcastle, UK (2007)

9. Marreiros, G., Ramos, C., Neves, J.: Dealing with Emotional Factors in Agent Based Ubiquitous Group Decision. In: Enokido, T., Yan, L., Xiao, B., Kim, D., Dai, Y., Yang, L.T. (eds.) Embedded and Ubiquitous Computing – EUC 2005 Workshops. LNCS, vol. 3823, pp. 41–50. Springer, Heidelberg (2005)

10. Santos, R., Marreiros, G., Ramos, C., Neves, J., Bulas-Cruz, J.: Multi-agent Approach for Ubiquitous Group Decision Support Involving Emotions. In: Ma, J., Jin, H., Yang, L.T., Tsai, J.J.-P. (eds.) UIC 2006. LNCS, vol. 4159, pp. 1174–1185. Springer, Heidelberg (2006)

11. Kraus, S., Sycara, K., Evenchick, A.: Reaching agreements through argumentation: a logical model and implementation. Artificial Intelligence 104(1-2), 1–69 (1998)

12. Neves, J.: A Logic Interpreter to Handle Time and Negation in Logic Data Bases. In: Proceedings of ACM 1984, The Fifth Generation Challenge, pp. 50–54 (1984)

Intelligent Mixed Reality for the Creation of Ambient Assisted Living

Ricardo Costa[1], José Neves[2], Paulo Novais[2], José Machado[2], Luís Lima[1], and Carlos Alberto[3]

[1] College of Management and Technology - Polytechnic of Porto, Felgueiras, Portugal
{rfc,lcl}@estgf.ipp.pt
[2] DI-CCTC, Universidade do Minho, Braga, Portugal
{pjon,jmac,jneves}@di.uminho.pt
[3] Hospital Geral de Santo António, EPE, Porto, Portugal
calberto.admn@hgsa.min-saude.pt

Abstract. Demographical and social changes have an enormous effect on health care, emergency and welfare services. Indeed, as the average age continues to rise, it is set the mood to an exponential growth in assistance and care, resulting in higher service costs, a decrease in quality of service, or even both. On the other hand, as part of the evolution of traditional Virtual Reality Environments (or Intelligent Mixed Reality), a striving expression for Ambient Intelligence (AmI) it is possible to outline the role of AmI in healthcare, by focusing on its technological, logical (relational) and common sense nature. Our goal is to have in place an electronically-based monitoring system. This would reduce response time to adverse events, improve analytics and reporting, and will provide caregivers with the information they need to positively impact the care of individual patients.

Keywords: Ambient Intelligence, Collaborative Networks and Decision Trees.

1 Introduction

Nowadays, we are faced with two concurrent trends, namely the pervasive diffusion of intelligence in the space around us, through the development of wireless network technologies and intelligent sensors, and the increase of richness and completeness of communications, through the development of multimedia technologies, including an increased attention to the aspects of human perception and of machine-person-machine interaction. The possible result of these converging trends is *Ambient Intelligence (AmI)*. AmI is a new paradigm in information technology, in which people are empowered through a digital environment that is aware of their presence and context, and is sensitive, adaptive, and responsive to their needs, habits, gestures and emotions. *AmI* can be defined as the merger of *ubiquitous computing* and *social user interfaces*. It builds on advanced networking technologies, formed by a broad range of mobile devices and other objects. Adding adaptive user-system interaction methods, based on new insights in the way people interact with computing devices (social user interfaces), digital environments can be created which improve the quality

J. Neves, M. Santos, and J. Machado (Eds.): EPIA 2007, LNAI 4874, pp. 323–331, 2007.

of life of people by acting on their behalf [1]. These context aware systems combine ubiquitous information, communication, and entertainment, with enhanced personalization, natural interaction and intelligence. The path to pursue, in order to achieve this goal, relies on a mix of different receptiveness from Artificial Intelligence, Psychology or Mathematical Logic (just to name a few), coupled whit different computational paradigms and methodologies for problem solving, such as that of agent, and its social counterpart, in the form of Group Decision Support Systems [2].

1.1 Inter-organization Cooperation

In Inter-Organization Cooperation there are factors that tend to be surrounded in the local milieu, and may be seen as the social embedded processes that allow organizations to obtain outside complementary knowledge and be innovative in the course of interaction among different actors, i.e., the local or regional milieu needs to include not only the substances related to the service structure or economics terms, but also social, cultural and institutional ones [4,5]. Thus, in the interaction of the different actors, the cooperation elements can be found in a kind of common language, social relationship, norms, values and institutions, which in our work will be set in terms of an extension to the logic programming language, being their knowledge bases built as logical theories that found their foundations on this extension [6]. Conclusions are supported by deductive proofs, or by arguments that include conjectures and motivate new topics of inquiry, i.e., if deduction is fruitless the agent inference engine resorts to abduction, filling in missing pieces of logical arguments with plausible conjectures to obtain answers that are only partly supported by the facts available.

1.2 Ambient Intelligence in HealthCare

Ambient Intelligence is a relative new paradigm in the, so-called, information society. It envisages a digital environment that is aware of people presence and context, thus been able to adapt and response to their personal needs, habits, gestures and emotions [7,8]. These distinctive characteristics are a major advantage in HealthCare, where the goal is to make persons (patients) to feel as better as possible, despite their condition. Actually, *AmI* can be used to provide personalized, intelligent, assistive technology that can promote patient recovery, combined with a sustained independence, thus achieving a better quality of life.

2 Business Integration for Healthcare

Our objective is to present an intelligent multi-agent system that will be able to monitor, interact and serve its costumers, being those elderly people and/or their relatives. This system will be interconnected, not only to healthcare institutions, but also with leisure centers, training facilities, shops and relatives, just to name a few. The VirtualECare Architecture is a distributed one with different components interconnected through a network (e.g., LAN, MAN, WAN), each one with a different role (Figure 1). A top-level description of the roles of the architecture components is given below:

- **Supported User** – Elderly with special health care needs, being under unending supervision, thus allowing that collected data is sent on time to the *CallCareCenter* and forwarded to the *Group Decision Supported System* team;
- **Home** – *SupportedUser* which must be under constant supervision, being the collected data sent to the *Group Decision Supported System* team through the *CallCareCenter* or to *CallServiceCenter*, according to its source;
- **Group Decision** – It is in charge of all the decisions taken in the VirtualECare platform. Our work will be centered on this key component;
- **Call Service Center** – Entity with all the necessary computational and qualified personal resources, capable of receiving and analyze the diverse data, and to take the necessary actions according to it;
- **Call Care Center** – Entity with all the necessary computational and qualified personal resources (i.e., healthcare professionals and auxiliary), capable of receiving and analyze the diverse data and take the necessary actions according to it;
- **Relatives** – *SupportedUser* relatives which may and should have an active role in attending their love ones.

In order to take the right decisions, it is necessary to consider a digital profile of the *SupportedUser,* that allow for a better understand of his/her special needs, ranging from the patient Electronic Clinic Process to their own personal preferences and/or experiences. Such an approach will help healthcare providers to integrate,

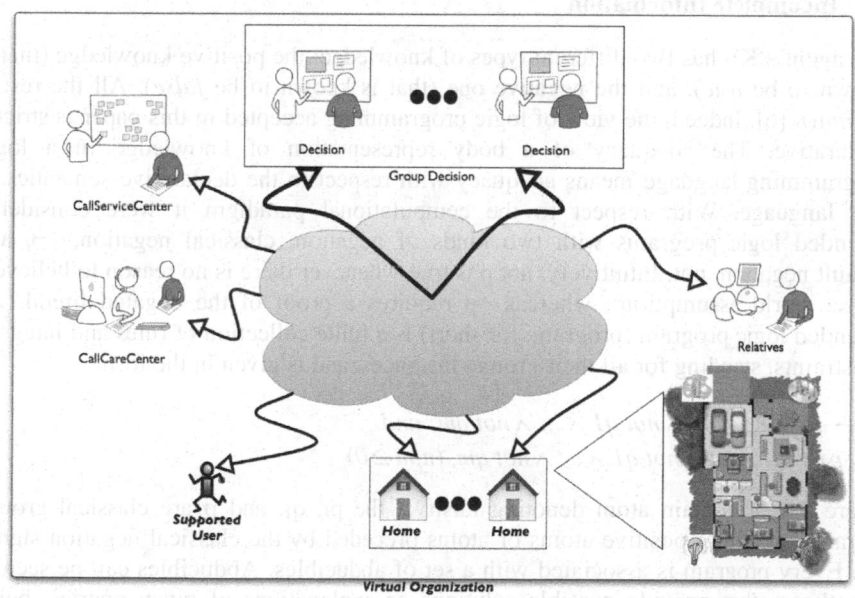

Fig. 1. The System Architecture for VirtualECare

analyze, and manage complex and incongruent clinical, research and administrative data. It will provide tools and methodologies for creating an information-on-demand environment that can improve quality-of-living, safety, and quality of patient care.

3 Group Decision Support Systems in HealthCare

In order to model the call centers, and according to the methodology for problem solving that is being subscribed (that is based on the concept of agent), and given in the form:

$$Agent(Id)::area\_of\_expertise \land organizational\_factors \land interest\_topics \land$$
$$disponibility \land credibility \land reputation \land availability.$$

the decision making process is based on a socialization process of such virtual entities, and here understtood as MultiAgent Systems, where *Id, area_of_expertise, organizational_factors, interest_topics, disponibility, credibility, reputation* and *availability* denote, respectively, the identification of the agent, the set of areas where the agent is an expert, the information about the institution where the agent is enrolled (e.g., employee numbers), the interests of the agent, its disponibility, credibility, reputation, and availability. The community of agents is given as a set of N agents, $\{AgP_1, AgP_2, \dots AgP_N\}$, denoted by AgP. The availability of each agent can be classified as **uncommitted**, **committed**, or **in action**. An **uncommitted** agent stands for someone that may or may not join the MAS. A **committed** agent has agreed to be part of the MAS, but the inclusion process does not started yet.

3.1 Incomplete Information

The agent`s KB has two different types of knowledge: the positive knowledge (that is known to be *true*), and the negative one (that is known to be *false*). All the rest is *unknown* [6]. Indeed, the view of logic programming accepted in this paper is strictly declarative. The adequacy of a body representation of knowledge in a logic programming language means adequacy with respect to the declarative semantics of that language. With respect to the computational paradigm it were considered extended logic programs with two kinds of negation, classical negation, ¬, and default negation, not. Intuitively, not p is true whenever there is no reason to believe p (close world assumption), whereas ¬p requires a proof of the negated literal. An extended logic program (program, for short) is a finite collection of rules and integrity constraints, standing for all their ground instances, and is given in the form:

$$p \leftarrow p1 \land \dots \land pn \land not\ q1 \land \dots \land not\ qm;\ and$$
$$?\ p1 \land \dots \land pn \land not\ q1 \land \dots \land not\ qm,\ (n,m \geq 0)$$

where ? is a domain atom denoting falsity, the pi, qj, and p are classical ground literals, i.e., either positive atoms or atoms preceded by the classical negation sign ¬ [6]. Every program is associated with a set of abducibles. Abducibles can be seen as hypotheses that provide possible solutions or explanations of given queries, being given here in the form of exceptions to the extensions of the predicates that make the program. The main idea here is to compute all consequences of the program, even those leading to contradiction, as well as those arising from contradiction. Suppose that in the KB of the AgR the information related to the areas of expertise of the AgP$_i$ identified as Peter, is represented in Program 1.

area_of_expertise('Peter', pediatrics).
¬area_of_expertise('Peter', oncologist).

Program 1 - It contains information related to the expertise areas of a specific agent.

If the KB is questioned if the area of expertise of Peter is Pharmacy the answer should be unknown, because there is no information related to that. On other hand, situations of incomplete information may involve different kinds of nulls. The ELP language will be used for the purpose of knowledge representation. One of the null types to be considered stands for an unknown value, a countable one (i.e., it is able to form a one-to-one correspondence with the positive integers). As an example, let us suppose that one of the agents that belong to the agent community AgP, at the registration phase, does not specify its interest topics; it just informs that it has interest topics. This means that the interest topics of the agent are unknown (Program 2).

¬skill(A,B) ← not skill(A,B) ∧ not exceptiont$_{skill}$(A,B).
exception$_{skill}$(A,B) ← skill(A, something).
skill('John', something).

Program 2 - Information related to the agent interest topics.

Another type of null value denotes information of an enumerated set. Following the previous example, suppose that an agent does not give information related to its availability, but its state of affairs is one of the three: **uncommitted**, **committed** or **in_action** (Program 3).

¬availability(A,B) ← not availability(A,B) ∧ not exception$_{availability}$(A,B).
exception$_{availability}$('John',committed).
exception$_{availability}$('John',uncommitted).
exception$_{availability}$('John',in_action).
¬((exception$_{availability}$('John´,A) ∨ exception$_{availability}$('John',B)) ∧ ¬
(exception$_{availability}$('John´,A) ∧ exception$_{availability}$('John',B)).

/ This invariant denotes that the agent states of **committed**, **uncommitted** and in_action are disjointed */*

Program 3 - Information related to the agent's availability.

where the set of duo (A,B) is given by ((**committed**, **uncommitted**), (**committed**, **in_action**), (**uncommitted**, **in_action**)).

3.2 The Decision Making Process

It is now possible, based on a set of (evolving) Decision Trees (DTs), to follow, day in, day out, the elderly. However, DTs are not simple representations of a decision making process, they may also apply to categorization problems, i.e., instead of saying that one wish to represent a DT to plan what to do, on a weekend, we could ask what kind of weekend is to be expected. This could easily be phrased as a question of learning a DT to decide in which category a weekend fits in (e.g., if it rains and it is windy then it is a weekend not to be remembered).

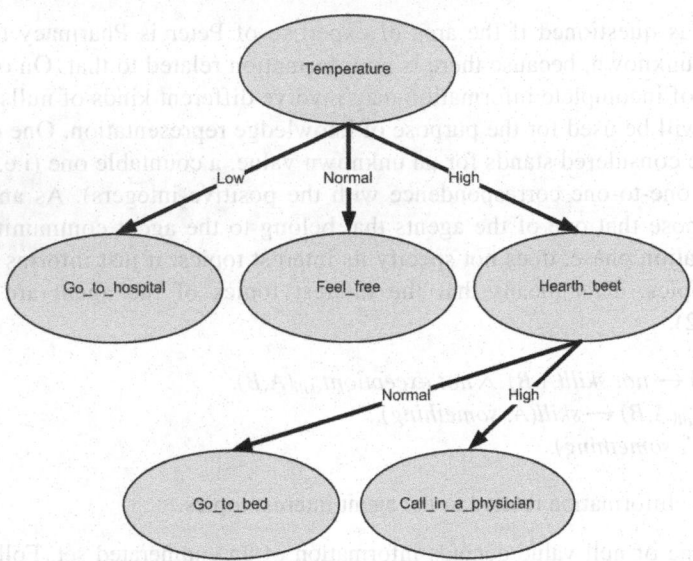

Fig. 2. Decision Tree to watch the Elderly State of Health

One may now look to the process of DTs construction (e.g., to decide what to do at the weekend). One may use some background information as axioms and deduce what to do (e.g. one can know that the family is in town and that they want to go to the cinema). Then, using, for example, Modus Ponens, we may decide to go to the cinema. Another way to stand around, it is by generalizing from previous experiences (e.g., let us consider all the times we had a really good weekend). If this is the case, one is using an inductive, rather than deductive method to construct the DTs (e.g., Modus Mistakens) (Figures 2 and 3).

On the other hand, there is a link between decision tree representations and logical representations, which can be exploited to make it easier to understand (to read) learned DTs. If we think better about it, every DT is actually a disjunction of implications (i.e., *if ... then* statements), and the implications are Horn clauses, i.e. a conjunction of terms implying a single term, which may be given in the form:

if family_is_visiting then go-to-the-cinema ∨
if ¬family_is_visiting ∧ weather_is_sunny then play_tenis ∧ visit_friends ∨
...

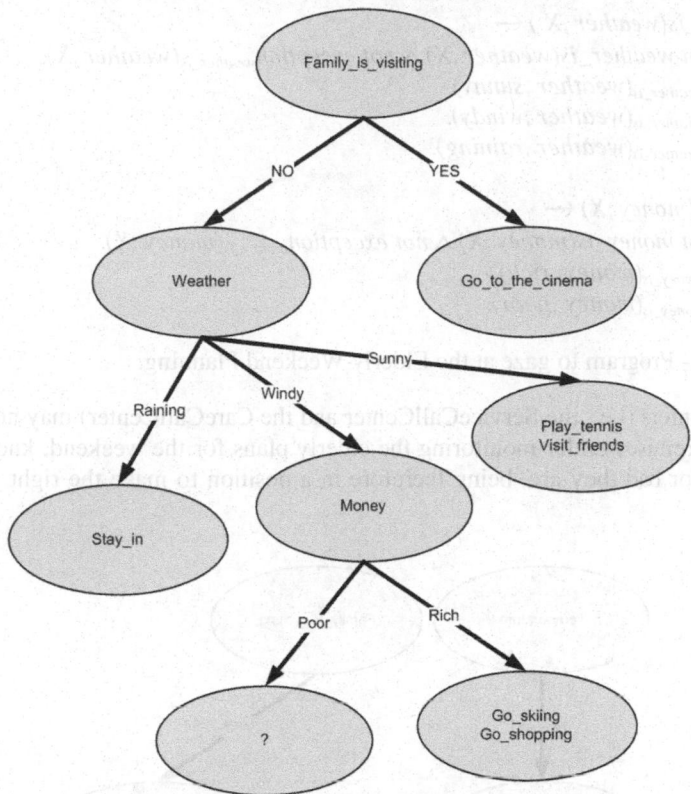

Fig. 3. Decision Tree to gaze at the Elderly Weekend Planning

The DTs depicted in Figures 2 and 3 may now be given in terms of logic programs or theories. For Figure 3 one may have (Program 4):

if family_is_visiting then go-to-the-cinema.
if ¬family_is_visiting ∧ weather_is(weather, sunny) then play_tenis.
if ¬family_is_visiting ∧ weather_is(weather, sunny) then visit_friends.
...
if ¬family_is_visiting ∧ weather_is(weather, windy) ∧ money_is(money, rich) then
go_shopping.
if ¬family_is_visiting ∧ weather_is(weather, windy) ∧ money_is(money, rich) then
go_skiing.
¬((go_shopping ∨ go_skiing) ∧ ¬(go_shopping ∧ go_skiing)). /* This invariant
denotes that the options of *go_shopping* and *go_skiing* are disjointed */
...
family_is_visiting.
¬ family_is_visiting.
¬((family_is_visiting ∨ (¬ family_is_visiting)) ∧ ¬(family_is_visiting ∧ (¬
family_is_visiting)). /* This invariant denotes that the occurrences *family_is_visiting*
and ¬ *family_is_visiting* are disjointed */

$\neg$ *weather_is(weather ,X)* ←
    not weather_is(weather ,X) ∧ *not exception$_{weather\_is}$(weather ,X).*
exception$_{weather\_is}$(weather ,sunny).
exception$_{weather\_is}$(weather ,windy).
exception$_{weather\_is}$(weather ,raining).

$\neg$ *mone_is(money ,X)* ←
    not money_is(money ,X) ∧ *not exception$_{money\_is}$(money ,X).*
exception$_{money\_is}$(money ,rich).
exception$_{money\_is}$(money ,poor).

Program 4 - Program to gaze at the Elderly Weekend Planning.

The call centers (i.e., the ServiceCallCenter and the CareCallCenter) may now receive from the premises under monitoring the elderly plans for the weekend, knowing also how good or bad they are, being therefore in a position to make the right judgments (Figure 4).

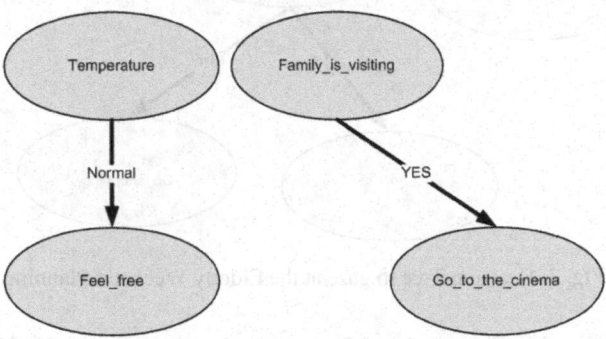

Fig. 4. Messages sent to the call centers

An advice is got in return (e.g., you may go with your family to the cinema). This advice is given not only in terms of the factual information gathered by the sensors, in terms of the DT or the corresponding logical formulae depicted by Program 4, but also attending to an evaluation of the quality of information of such formulae [9].

4   Conclusions

In this paper we describe a framework to handle context informaton in AmI environments. Based on a set of evolving DTs, and the figure of quality-of-information [9], it was also shown how user context information can be exploited to find and propose an advice. In future work, and attending to such environments, we may use collaborative networks as a support for different, but interconnected virtual organizations, that could provide to all the population in general, and the elderly, in particular, specialized services (e.g., healthcare, entertainment, learning), without delocalizing or messing up with their routine, in a more effective and intelligent way.

References

1. Giráldez, M., Casal, C.: The Role of Ambient Intelligent in the Social Integration of the Elderly. IOS Press, Amsterdam (2005)
2. Marreiros, G., Novais, P., Machado, J., Ramos, C., Neves, J.: An Agent-based Approach to Group Decision Simulation using Argumentation. In: ABC 2006. International MultiConference on Computer Science and Information Tecnology, Workshop Agent-Based Computing III, Wisla, Poland, pp. 225–232 (2006)
3. Augusto, J.C., McCullah, P., McClelland, V., Walden, J.-A.: Enhanced Healthcare Provision Through Assisted Decision-Making in a Smart Home Environment. In: 2nd Workshop on Artificial Inteligence Techniques for Ambient Inteligence (2007)
4. Dosi, G.: Sources, procedures and microeconomics effects of innovation. Economic Literature 26, 1120–1171 (1998)
5. Malmberg, A.: Industrial Geography: agglomeration and local milieu. Progress in Human Geography 20, 392–403 (1996)
6. Neves, J.: A Logic Interpreter to Handle Time and Negation in Logic Data Bases. In: The Fifth Generation Challenge, pp. 50–54. ACM Press, New York (1984)
7. Riva, G.: Ambient Intelligence in Health Care. CYBERPSYCHOLOGY & BEHAVIOR 6 (2003)
8. Marreiros, G., Santos, R., Ramos, C., Neves, J., Novais, P., Machado, J., Bulas-Cruz, J.: Ambient Intelligence in Emotion Based Ubiquitous Decision Making. In: Proceedings of the International Joint Conference on Artificial Intelligence (IJCAI 2007) - 2nd Workshop on Artificial Intelligence Techniques for Ambient Intelligence (AITAmI 2007) (2007)
9. Neves, J., Machado, J., Analide, C., Abelha, A., Brito, L.: The Halt Condition in Genetic Programming. In: Proceedings of EPIA 2007. LNCS (LNAI), Springer, Heidelberg (to appear, 2007)

Medical Imaging Environment – A Multi-Agent System for a Computer Clustering Based Multi-display

Victor Alves[1], Filipe Marreiros[1], Luís Nelas[2], Mourylise Heymer[1], and José Neves[1]

[1] Department of Informatics, University of Minho, Braga, Portugal
[2] Radiconsult, Braga, Portugal
{valves,jneves}@di.uminho.pt,
f_marreiros@yahoo.com.br,
m_heymer@yahoo.com,
luis.nelas@radiconsult.com

Abstract. This paper presents a solution to minimize a problem that normally arises from the huge amount of images that a radiologist usually has to interpret. A multi-agent system that implements a multi-display for medical imaging based on computer clustering of normal personal computers is therefore described, as well as the multi-agent architecture that caters for the system evolution. An evaluation study was performed and its results are presented.

Keywords: Medical Image Viewer; Multi-display Systems; Medical Imaging; DICOM; Ambient Intelligence, Computer Clustering, Observer Performance Evaluation, Radiologist Productivity, Screen Real Estate Problem, Ergonomics.

1 Introduction

The evolution of equipments used in the medical image practice, confronts the physicians with a new problem: the capacity to interpret a huge amount of image workload. The current workflow reading approaches are becoming inadequate for reviewing the 300 to 500 images of a routine Computer Tomography (CT) of the chest, abdomen, or pelvis, and even less for the 1500 to 2000 images of a CT angiography or functional Magnetic Resonance (MR) study. On the other hand, the image visualization computer programs continue to present the same procedures for image readings, i.e., imitating the manual process where the images films are viewed using a light screen (as it is found in most of the commercial medical imaging viewers). These insights have given us the motivation to overcome such shortcomings by increasing the amount of display area, in order to allow faster navigation and analysis of the medical images by the radiologist or the referring physician. To fulfill this objective a multi-display system with an intelligent viewing protocol feature, was created to allow for the visualization of selected medical images across the computer's displays in use. To support the overall multi-display system a multi-agent architecture was developed. It takes advantage of scalability and gives one control and interoperability over all the systems' components.

The research in this area has its main focus on the radiology field, mainly on image processing rather than image presentation, as it follows from most of the medical

J. Neves, M. Santos, and J. Machado (Eds.): EPIA 2007, LNAI 4874, pp. 332–343, 2007.

image viewers available. From the several studies on image presentation that were considered, it must be referred the work developed at the Simon Fraser University [1, 2]. Their focus is on the best way to present MR images in a single computer screen. In [2] a traditional light screen was emulated using several techniques to overcome the screen real estate problem. This can be described as the problem of presenting information within the space available on a computer screen. Van der Heyden et al. [3] explored several presentation issues in the development of medical imaging viewing systems to overcome computer screen size limitations. Mathie and Strickland, and Kim et al. [4,5] explored the stack mode solution where images are stacked all one in top of the other and are viewed in a user controlled cine mode. Reiner et al. [6] addressed several dynamic processes, using multi-planar reconstructions, volumetric navigation, and electronic decision support tools. They state that the result is optimization of the human-computer interface with improved productivity, diagnostic confidence, and interpretation accuracy. Our system presents an alternative approach to the real estate problem; a system with no limitations, for it is a multi-display system scalable, in terms of computer power, according to the needs and resources available. It also addresses the study presentation issue with the development of an intelligent hanging agent that implements an individual viewing protocol for each radiologist.

In the medical field, 3D reconstruction may be used to visualize anatomical volumes. Although these techniques have a great potential, given that we may directly view the entire data set, the radiologists are experts in performing 3D mental reconstructions using the 2D images. Not only due to this fact, but once it is much easier to parameterize, it results that in most of the cases, mainly 2D images, are preferred for medical diagnostic purposes.

1.1 Multi-Agent Systems

Multi-Agent Systems (MAS) set a new paradigm in problem-solving via theorem proving, i.e., agent-based computing has been hailed as a significant break-through in problem solving and/or a new revolution in software development and analysis[7]. Indeed, agents are the focus of intense interest on many sub-fields of Computer Science, being used in a wide variety of applications, ranging from small systems to large, open, complex and critical ones. Agents set not only a very promising technology, but are emerging as a new way of thinking, a conceptual paradigm for analysing problems and for designing systems, for dealing with complexity, distribution and interactivity. Indeed, it may be a new form of computing and "intelligence". To develop such systems, a standard specification method is required, and it is believed that one of the keywords for its wide acceptance is simplicity. Indeed, the use of intelligent agents to simulate human decision making in the medical arena offers the potential to set an appropriate software development and analysis practice and design methodology that do not distinguish among agents and humans, until implementation. Being pushed in this way, the design process, the construction of such systems, in which humans and agents can be interchanged, is simplified, i.e. the modification and development in a constructive way, of multi-agent systems with a human-in-the-loop potential aptitude is becoming central in the process of agent-oriented software development and analysis. These systems have provided a clear means of monitoring the agent's behavior with significant impact in their process of knowledge acquisition and validation. MAS are a natural connection to intelligent systems evolution, being elements for task substitution or delegation, usually performed by humans. However agent based systems have some restrictions, such as global system control and universal view absences, and some lack of confidence and fear of competence delegation by humans. To delegate tasks, bilateral confidence relations have to be established. Organizations may also mature their experience relatively to the use of autonomous software components.

Based on such a framework, a multi-agent system was developed that enables the multi-display of archived images. Indeed, multi-agent systems have proven to be extremely scalable and offer great flexibility to the overall system design (e.g. one may point out to a multi-agent system to aid the diagnostic process [8], or a web-based medical training system [9]).

1.2 Medical Digital Imaging Systems

The use of computers to processes radiology images began in the 70's. In 1983 the ACR (American College of Radiology) and NEMA (National Electrical Manufacturers Association) formed a committee to develop a standard that contemplates the transmission and storage of digital medical images. This standard was called DICOM (Digital Imaging and Communications in Medicine). Now, and based on the DICOM standard, it is possible to create medical images repositories designated by PACS (Picture Archiving and Communications Systems). PACS are used in conjunction with the DICOM standard, either to store studies sent by the modalities or to answer queries made by the viewers, returning the wanted stored studies.

There are still two other crucial information systems, the RIS (Radiology Information System) and the HIS (Hospital Information System). The RIS is the information system of the Imagiolagy/Radiology Service. It handles all the information regarding the radiological studies, ranging from the scheduling, and the execution to payment. The HIS is the hospital information system; it handles all the hospital information. The HIS and RIS are interconnected. These information systems have their own standard, known has HL7 (Health Level Seven). This standard caters for the identification of patients, orders of processes, stored reports, but it cannot handle DICOM data. The DICOM standard defines a hierarchical information model (Fig. 1).

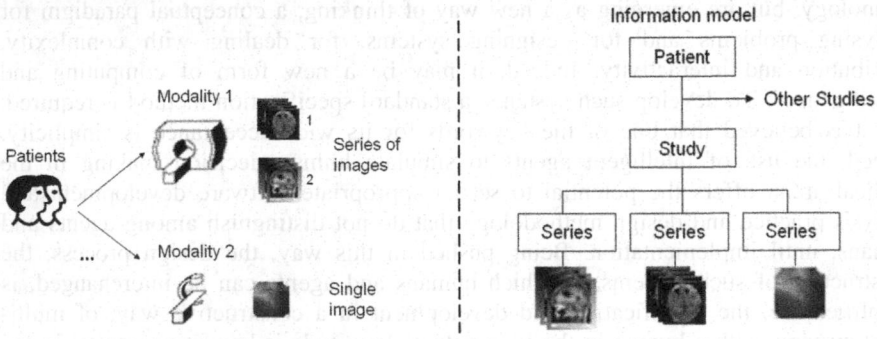

Fig. 1. DICOM Information Model

2 Developed Multi-display System

The developed system has two main functional components, namely the Control Station and the Visualization Terminals (Fig. 2). The Visualization Terminals are arranged in a grid that can grow in columns and rows. This grid may or may not be fully populated (e.g. one may only use two terminals (1, 1) and (2, 1)). This grid that is viewed in the control station simulates the real distribution of the displays. The person responsible for the configuration of the visualization terminals should have this into consideration, in order to fulfill the user expectations, since the user has to have the information in the correct position. The Control Station is responsible for user interface management, processing the user input and controlling the information displayed on the visualization terminals. It has several tools and functionalities for image analyses and processing.

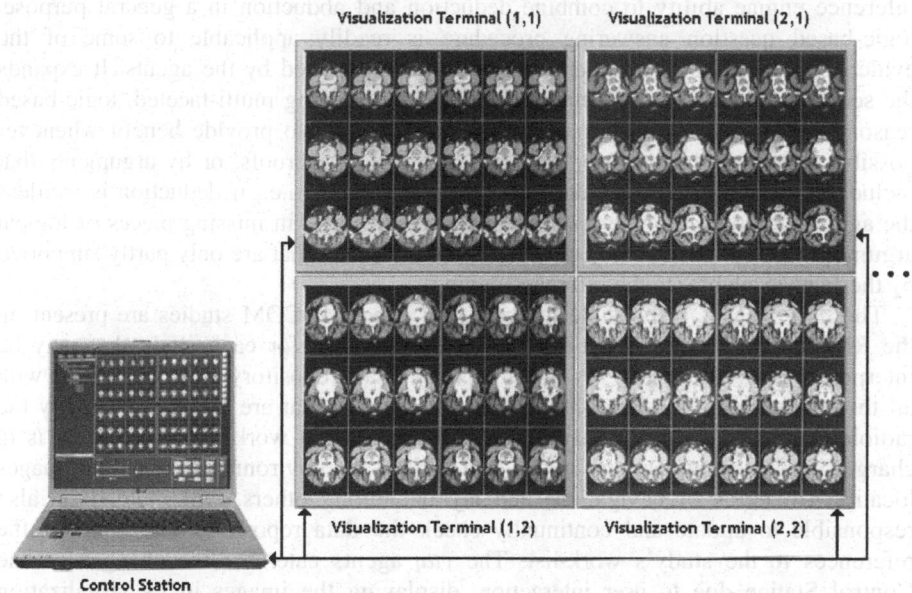

Visualization Terminal (1,1)   Visualization Terminal (2,1)

Visualization Terminal (1,2)   Visualization Terminal (2,2)

Control Station

Fig. 2. Multi-Display System with the control station and 2x2 visualization terminals

2.1 Multi-Agent Architecture

To support the overall multi-display system a multi-agent architecture was used, offering scalability in terms of the number of monitor displays and giving control and interoperability among the system's components (Fig. 3). The agents that were developed stand for the *dpa* (Data Prepare Agent), *csa* (Control Station Agent), *iha* (Intelligent Hanging Agent) and *vta$_i$* (Visualization Terminal Agents), being their knowledge bases built as logical theories that find their foundations on an extension to Horn clause logic, given in the form:

A rule: 1 positive literal, at least 1 negative literal. A rule has the form "~P1 ∨ ~P2 ∨ ... ∨ ~P$_k$ ∨ Q$_1$ ∨ ... ∨ Q$_m$". This is logically equivalent to "[P$_1$ ∧ P$_2$ ∧ ... ∧ P$_k$] => [Q$_1$

∧ ... ∧ Q$_m$]"; thus, an if-then implication. Examples: "~man(X) ∨ mortal(X)" (All men are mortal); "~parent(X,Y) ∨ ~ancestor(Y,Z) ∨ ancestor(X,Z)" (If X is a parent of Y and Y is an ancestor of Z then X is an ancestor of Z). A fact or unit: 1 positive literal, 0 negative literals, e.g. "man(socrates)", "parent(elizabeth, charles)", "ancestor(X,X)" (Everyone is an ancestor of themselves (in the trivial sense)). A negated goal: 0 positive literals, at least 1 negative literal.

The agents sense their environment (Fig. 3 - the medical imaging equipments, the messages on the blackboard and the radiologist) and act according to the changes that may occur on it. On the other hand, to generate opportune hypothetical statements, the agents' inference engine uses a form of reasoning called abduction. In simplest terms, abductive inference follows the pattern: if Q is known to be true and P => Q is also known to be true, then posit P is to be understood as a hypothetical support, or explanation, for Q. This mode of reasoning transforms every rule in an agents' knowledge base (at least potentially) into a template for hypothesis generation. The inference engine ability to combine deduction and abduction in a general purpose, logic-based question answering procedure is readily applicable to some of the evidence assembly and argument construction tasks faced by the agents. It expands the scope of automated question answering by combining multi-faceted, logic-based reasoning techniques with information retrieval search to provide benefit whenever possible [10]. Conclusions are supported by deductive proofs, or by arguments that include conjectures and motivate new topics of inquiry, i.e., if deduction is fruitless the agents' inference engine resorts to abduction, filling in missing pieces of logical arguments with plausible conjectures to obtain answers that are only partly supported by the facts available (to the inference engine).

The *dpa* agent is responsible for verifying if new DICOM studies are present on the RIS. If new studies are found, it creates a profile for each study that may be interpreted by the *csa* agent. It will also update a data repository in the hard drive with all the references to the work-list studies (i.e. studies that are to be analyzed by the radiologists). The *iha* agent learns how each radiologist works. The *csa* agent is in charge of all user interactions as this will change the environment, since the images location (in terms of navigation) and layout, among others, will mute. It is also responsible to update and continually check the data repository that contains the references to the study's work-list. The *vta$_i$* agents cater for the changes on the Control Station due to user interaction, displaying the images in its visualization terminals accordingly. Each of the *vta$_i$* agents controls its personal computer.

The blackboard implements a shared memory environment. It is a process that runs on the main memory and is responsible to manage the attributes of the active visualization terminals; in particular it maintains and updates the IP (Internet Protocol) address, the terminal position and the screen resolution. It is therefore possible for the control station agent to access this information, and to use it accordingly (e.g. to send messages to the visualization terminals agents).

The Knowledge Base (KB) is composed of several configuration data repositories present in the hard drives, which are used to maintain the root location, the work-list studies references, the locations of the visualization terminals, and IPs of the blackboard and visualization terminals. It also logs the radiologist behavior, which leads to an intelligent behavior of the viewing protocol (i.e. it takes into consideration and acts according to the physician practice) [11].

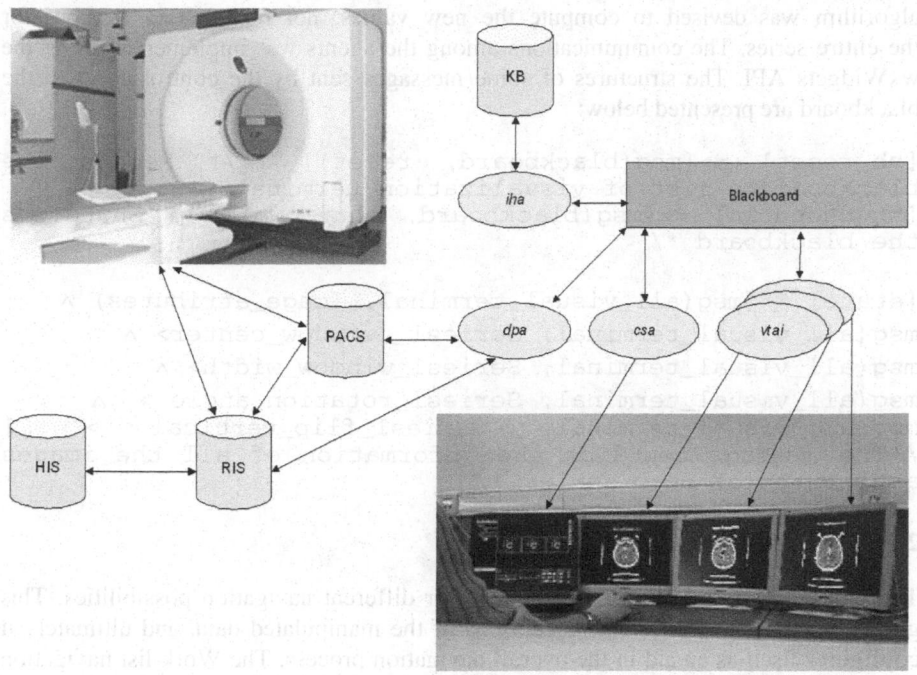

Fig. 3. Multi-agent architecture

2.2 System Implementation

The system was implemented using the programming languages Java and C/C++. The DICOM parser was implemented in Java and uses the PixelMed Java DICOM Toolkit. From the wxWidgets library we used the GUI toolkit and the wxWidgets sockets API for communication. The key aspect of the implementation relies on the introduction of a wxWidgets component that enables the usage of OpenGL. Indeed, OpenGL is used to render the medical images, and therefore based or supported on the use of texture mapping commands to generate and load textures to the graphics card. When a study series is loaded it is necessary to create a texture for each of its medical images, previously to be rendered. To create the textures we need the DICOM raw data, where the density values are found, for modalities such as MR and CT. These are the modalities that were considered in this work, as our system was designed around them. With the raw data it is possible to compute each texel color. In order to accomplish this goal we had to use the pseudo-code found in part 3, C.11.2.1.2, of the DICOM standard. When the user changes the image attributes like the window center/width, the texture (images) has to be updated and replaced. We replace the textures instead of generating new ones, given that in most cases this process is faster. When a series is unloaded, all correspondent textures have to be deleted. To fulfill this goal it is necessary to check if the texture is in the graphic card and, if that is the case, delete it. When the user sets the image parameters he/she is affecting the entire series, since the images are all co-related. Having this in mind, an algorithm was devised to compute the new values, not for each image, but for the entire series. The communications among the agents was implemented using the wxWidgets API. The structures of some messages sent by the control agent to the blackboard are presented below:

```
[bb_reset]  => [msg(blackboard, reset)]     /* Resets the
blackboard's list of visualization terminals */
[bb_shutdown] => [msg(blackboard, shutdown)] /* Shutdowns
the blackboard */
...
[state] => [msg(all_visual_terminal, Image_atributes) ∧
msg(all_visual_terminal, Series1_ window_center> ∧
msg(all_visual_terminal, Series1_window_width> ∧
msg(all_visual_terminal, Series1_rotation_angle > ∧
msg(all_visual_terminal,  Series1_flip_vertical  >  ...]
/*The message contains the information of all the images
attributes of both series */
```

2.3 The Radiologists' Interface

The Control Station User Interface caters for different navigation possibilities. This gives the radiologist different perceptions of the manipulated data, and ultimately it configures itself as an aid in the overall navigation process. The Work-list navigation allows the user to navigate throughout the studies present in the system. Figure 4 points out the system's work-list interface. The work-list has a hierarchical structure of five levels, namely L1-Patient, L2-Modality, L3-Study, L4-Series, and L5-Image. This structure is slightly different from the DICOM standard that was introduced earlier, since we added an additional level, the Modality one (it assumes itself as one of the most important search keys when radiologists search for studies/images). By selecting one series, the user can load it. The navigation throughout the images in a series is made using sliders. There are two sliders, one for each series and an additional slider to navigate in the series of the selected image. Figure 4 shows the navigation sliders, where there is an additional check box, the "Sync" one, to synchronize navigation in both series. In this way, when the user manipulates one slider, automatically the other slider will change accordingly.

The control station interface presents a virtual environment built around different navigation levels (Fig. 5). The lower level of this environment emulates the real visualization terminals, and gives an overview of the entire group of terminals used. The same content of the real visualization terminals can be seen in this environment, but scaled to the available screen area, as shown in figure 4. As it is seen on the figure referred to above, one of the visualization terminals is bigger than the others, which is due to its higher screen resolution. Proportionality and positioning is maintained, aiding the users in making a clearer identification of the real visualization terminals. Ideally, the real terminals should be placed according to the virtual ones, if not the user will be confused. The next level is the monitor selected level. To go to this level the user just has to double-click over the visualization terminal, and then he/she will have only that terminal in the available screen area. To go back to the lower level the

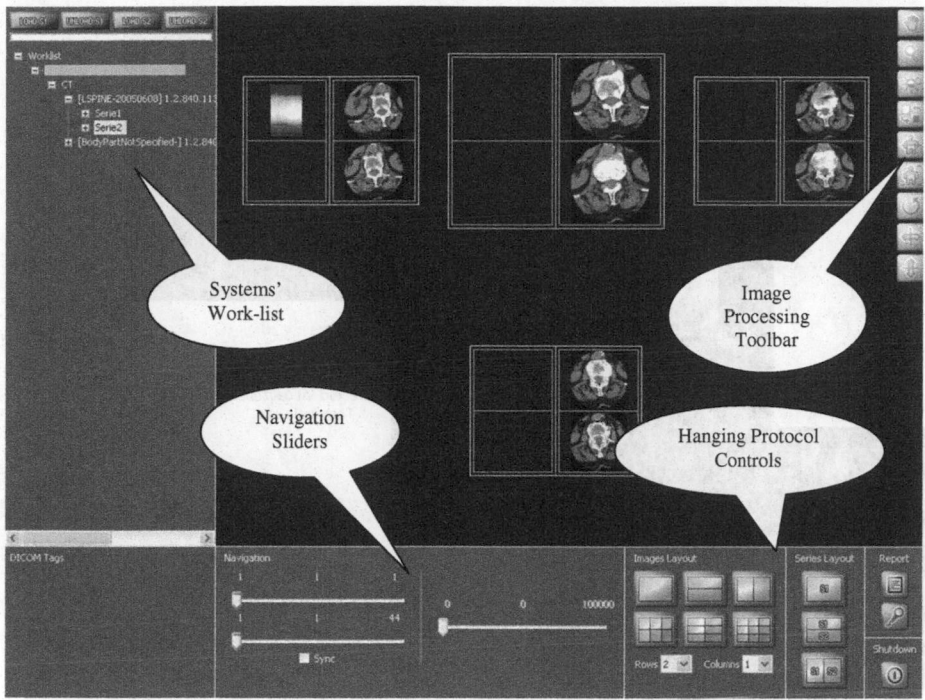

Fig. 4. Control Station Interface where the Systems' Work-list (presenting only one blanked patient with two CT studies), Navigation Sliders, Image Processing Toolbar and the Hanging Protocol Controls are pointed out

user just has to Ctrl+click. The last level is the selected image level. To go to this level there are two possibilities. The former, is used when the user is at the terminal overview level and he/she has a hanging protocol were only one series and one image per terminal is displayed. When the user double-clicks over one of the virtual terminal, control will go directly to this level. The other possibility is for the user (at the monitor selected level) to double-click over one image. When the user is at this level the image analysis tools and the slider to navigate the series of the selected image becomes available. The navigation using this slider is independent and will not affect the image positioning of the virtual and real visualization terminals. This mode of navigation is identical to the one found in traditional viewers (it seems important to support this function as the users are familiar with it). To go back to the lower level the user must use Ctrl+click. Figure 5 depicts the virtual environment levels.

In order for image viewing software to be valuable, it must display the images in a manner that is useful for the tasks that must be accomplished by a radiologist. The hanging protocol defines how the images and series are displayed (layout) and the predefined settings (WindowLevel/WindowWidth, zoom, rotation, arrangement of the images and series in the visualization terminal) in all the virtual and real visualization terminals. As pointed out by Moise and Atkins [12] the time spent preparing a study for review by radiologists can be considerable reduced. In figure 4 we

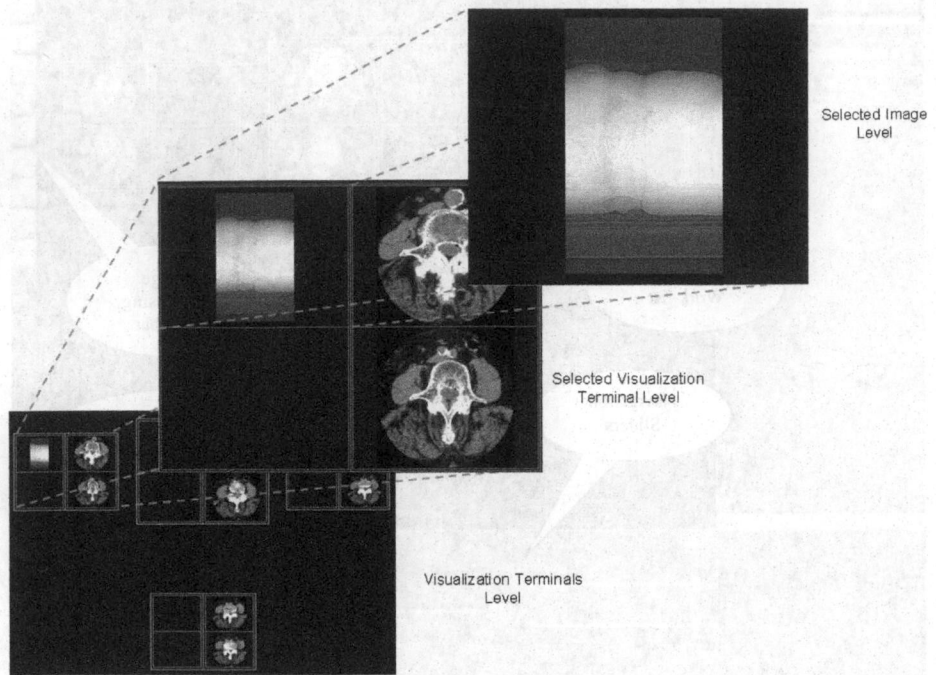

Selected Image
Level

Selected Visualization
Terminal Level

Visualization Terminals
Level

Fig. 5. Virtual Environment Levels

can see the hanging protocol controls. As it may be observed we have separate the buttons for the images from those of the series. The images lay within the area attributed to the series, and these series may only have three possible configurations (these configurations can be clearly identified by looking at the correspondent buttons). When the intelligent hanging agent (*iha*) is activated the initial viewing protocol for each study are by it initialized. In the same figure it is possible to view how the changes of the hanging protocol affect the virtual visualization terminals. In this particular case both series are separated horizontally. Inside the series we can view the images layout (in this case, 2 rows and 1 column), which gives two images per series. The first series contains only one image and this is why only that image is displayed. Using this tool, users can quickly and easily change the hanging of all the images and series across all the displays. Compared to hardcopy readings where the images are printed in a film and hanged on a light screen, (i.e. in a fixed grid format, containing each film a few images), our system offers the same features, plus the possibility to change the grid, changed image settings and, obviously, to discard the film.

3 System Evaluation

The evaluation of the system was done at CIT-Centro de Imagiologia da Trindade, a private medical imaging facility within Hospital da Trindade in OPorto, with the collaboration of a neuroradiologist and a radiologist. We used two four body light boxes for the film reading. For the multi-display system, we used a notebook as the control station, having sorted the viewing stations in pairs of two, four, six and eight. The viewing monitors were used in two rows (like the light boxes) for ergonomics sake. The physicians since they are seated, prefer to have their eyes travel sideways, keeping the images at the same distance, then to have to look up higher, straining their neck and their focus. We used mainly MR imaging of the brain and spine for neuroradiology, and MR imaging of the abdomen for radiology. We are conscious though that nowadays with the very fast multi-detector helical CT equipments; this modality produces an enormous amount of images, perhaps more difficult to follow than the MR itself and in this way justifying even more the use of multi-display systems to help the physician reporting his findings. The timing of the reporting on the multi-display system was done using the systems' logging feature. The radiologists stated that diagnostic confidence, and interpretation accuracy was maintaining. In this study we also did an evaluation of the intelligent hanging feature provided by the *iha* agent. After the first day we noticed that the time saved by this feature seemed to be constant along the day and will be around one minute per study. The physician after the first display of the images on the monitors tends to "play around" with the following boards when the necessity of a different approach, different display (zoom, measures, etc.), become necessary. Anyway, just from this feature alone we obtain a daily gain of nearly half hour of workload of physician work in this modality. We found out that the great gains in time were mainly at the advantages of using the multi-display digital system. The fact that the images are ordered, sorted by the programmed way the physician wants them (WL/WC, layouts of the monitors, etc.) saved a lot of time as comparing to film reading. As can be seen in the following tables the difference in time spent by the physician reading studies by the different methods and the multi-display system is substantial. It can be stated that the optimum number of monitors is six although in the brain studies little difference was noted between four and six monitors, mainly because the amount of images weren't so large and the image layout of the monitors (number of images per display) was different (table 1).

In the abdominal studies time was cut by almost in half from film reading as to using six monitors, and more than six minutes as to a two monitor configuration. In a twenty abdominal MR study physician period we can say that using six monitors instead of two can gain more than two hours workload at the end of the period. In comparison to film reading that time rises to three and a half hours. Comparing with the values obtained by Wideman and Gallent [13] where they evaluate a standard

Table 1. System evaluation by physician A (neuroradiologist) and physician B (radiologist). The values presented are average times expressed in minutes of one week workflow.

physician	Body part	Num. of Images (Average)	Hardcopy reading	Softcopy reading			
				2x1	2x2	2x3	2x4
A	Spine	50	15.6 m.	12.2 m.	10.1 m.	9.4 m.	10.4 m.
A	Brain	30	14.8 m.	11.9 m.	9.8 m.	10.2 m.	10.0 m.
B	Abdominal	100	21.1 m.	15.4 m.	9.7 m.	9.2 m.	9.8 m.

digital visualization system our system performs better. Their values are quite similar to our 2x1 configuration.

4 Conclusions

We have presented a multi-display visualization system used to support the medical image visualization process. Due to its scalability one can easily assemble a system that grows according to the user needs. The navigation facilities and the wider work area of this system support them in the image viewing process, thus improving the diagnostic efficiency in terms of average time spent with each study. On the other hand it can be built with conventional hardware (i.e., no special graphics cards or other specific hardware).

Results from the evaluation study support the feasibility of the proposed approaches and clearly indicate the positive impact of an augmented display area and an intelligent viewing protocol on the radiologist workflow. The result is optimization of the human-computer interface with improved productivity, maintaining diagnostic confidence, and interpretation accuracy.

Although our work is aimed for the medical field, it can be easily rewritten for other areas (e.g., in an advertising context, we can imagine a store with a network of computers where one wants to set dynamically the advertising images of those monitors using a control station).

Acknowledgments. We are indebted to *CIT- Centro de Imagiologia da Trindade*, for their help in terms of experts, technicians and machine time.

References

1. Van der Heyden, J.E., Atkins, M.S., Inkpen, K., Carpendale, M.S.T.: MR image viewing and the screen real estate problem. In: Proceedings of the SPIE-Medical Imaging 1999 (1999)
2. Van der Heyden, J.E., Atkins, M.S., Inkpen, K., Carpendale, M.S.T.: A User Centered Task Analysis of Interface Requirements for MRI Viewing. Graphics Interface (1999)
3. Van der Heyden, J., Inkpen, K., Atkins, M., Carpendale, M.: Exploring presentation methods for tomographic medical image viewing. Artificial Intelligence in Medicine 22 (2001)
4. Mathie, A.G., Strickland, N.H.: Interpretation of CT Scans with PACS Image Display in Stack Mode, Radiology 203 (1997)
5. Kim, Y.J., Han, J.K., Kim, S.H., Jeong, J.Y., An, S.K., Han, Son, Lee, Lee, Choi: Small Bowel Obstruction in a phantom Model of ex Vivo Porcine Intestine: Comparison of PACS Stack and Tile Modes for CT Interpretation. Radiology 236 (2005)
6. Reiner, B.I., Siegel, E.L., Siddiqui, K.: Evolution of the Digital Revolution: a Radiologist Perspective. Journal of Digital Imaging 16(4) (2003)
7. Wooldrige, M.: An Introduction to Multi-Agent Systems. John Wiley & Sons, Chichester (2002)
8. Alves, V.: Resolução de Problemas em Ambientes Distribuídos, Uma Contribuição nas Áreas da Inteligência Artificial e da Saúde. PhD Thesis, Braga, Portugal (2002)

9. Alves, V., Neves, J., Maia, M., Nelas, L., Marreiros, F.: Web-based Medical Teaching using a Multi-Agent System. In: Zhang, S., Jarvis, R. (eds.) AI 2005. LNCS (LNAI), vol. 3809, Springer, Heidelberg (2005)
10. Neves, J.: A Logic Interpreter to handle time and negation in logic data bases. In: Proceedings of the ACM 1984 Annual Conference, San Francisco, California, USA (1984)
11. Neves, J., Alves, V.: Intelligent Hanging of Medical Imaging Studies using Concept Learning. Technical Report, Universidade do Minho, Braga, Portugal (2007)
12. Moise, A., Atkins, M.S.: Design Requirements for Radiology Workstations. Journal of Digital Imaging 17(2) (2004)
13. Wideman, C., Gallet, J.: Analog to Digital Workflow Improvement: A Quantitative Study. Journal of Digital Imaging 19(suppl. 1) (2006)

Chapter 5 - Second Workshop on Building and Applying Ontologies for the Semantic Web (BAOSW 2007)

Partial and Dynamic Ontology Mapping Model in Dialogs of Agents

Ademir Roberto Freddo, Robison Cris Brito, Gustavo Gimenez-Lugo,
and Cesar Augusto Tacla

Universidade Tecnológica Federal do Paraná
Av. Sete de Setembro, 3165, Curitiba, Paraná, Brazil
{freddo,robison}@utfpr.edu.br, tacla@cpgei.cefetpr.br,
gustavo@dainf.cefetpr.br

Abstract. This paper describes a partial and dynamic ontology mapping model
for agents to achieve an agreement about meaning of concepts used in the
content part of messages during a dialog. These agents do not share an
ontology. The proposed model prescribes phases to cluster and to select the
clusters with background knowledge in an ontology, operations to interpret the
content of a message based on syntactic, semantic approaches, and the dialog
between agents according to the difficulty in finding similar concepts in both
ontologies. A case study is presented.

Keywords: Ontology, Ontology Mapping, Agent Communication.

1 Introduction

A multi-agent system is an open, dynamic and heterogeneous environment. It is open
because new agents can be added to or removed from the system without interrupting
its execution, and dynamic because of its openness and non-determinism. Agents are
not reactive meaning they not forced to respond to every request and they may
answer the same request in different ways given that they acquire new knowledge as
time goes by. It is heterogeneous because agents may have various knowledge
representations encoded in different ontologies.

The heterogeneity of representation is the focus of this work and affects system
openness and dynamics. In order for a system have such properties an incoming agent in
the system must be able to locate and communicate with other ones in a transparent way,
so it needs to address the semantic interoperability problem. When two or more agents
communicate, it is difficult to achieve an agreement about the meaning of concepts if
they do not have a common language and a common ontology. In this work we assume
that agents share a communication language (e.g. FIPA-ACL [1]), but the content of
messages is expressed by means of concepts coming from different ontologies.

The most straightforward way to solve the interoperability problem is developing a
common ontology to all agents, but this would be very unlikely in a multi-agent
system, because it would require all agents involved to reach a consensus on which
ontology to use. Besides, a common ontology forces an agent to abandon its own
world view and adopt one that is not specifically designed for its task [2].

J. Neves, M. Santos, and J. Machado (Eds.): EPIA 2007, LNAI 4874, pp. 347–356, 2007.

Ontology alignment is an alternative way to deal with the interoperability problem. This approach keeps agents with their particular ontologies, creating a set of correspondences between two or more ontologies. These correspondences are expressed as mappings that enable agents to interpret the content part of messages and to combine their knowledge. We defined mapping as [3]: "Given two ontologies A and B, mapping one ontology with another means that for each concept (node) in ontology A, we try to find a corresponding concept (node), which has the same or similar semantics, in ontology B and vice verse" The mapping between two ontologies results in a formal representation that contains expressions that link concepts from one ontology to the second [4]. In order to communicate efficiently using different ontologies, agents need to map their ontologies in runtime.

In literature there are several approaches to map ontologies: AnchorPrompt [4], QOM [5], NOM [6], COMA[7], OLA[8] and S-Match [9]. Such methods try to do a total mapping, i.e. identify all the correspondences among concepts. Nevertheless, in open and dynamic environments, the total mapping in runtime is expensive because of the computational complexity, so a partial mapping would be more adapted to this kind of environment. The partial mapping tries to map only concepts used in a dialog. In total mapping, each concept of an ontology is compared with all concepts of the other one. When two ontologies cover different domains and there are just some overlaps, the total mapping spends resources to find relationships between concepts that cannot be mapped. The partial mapping consists of mapping only those concepts present in the communication.

Besana and co-authors [10] proposed a framework that maps only the terms found in a dialog. However, the proposed solution is computational expensive because the terms in the message are compared with all the concepts of the target ontology to find correspondences. Besides, it is necessary to generate, filter and select the hypotheses creating an argument tree.

In this paper, we present a partial and dynamic model to map ontologies in a dialog between agents to address the semantic interoperability problem in open and dynamic environments. We believe that agents are unlikely to share same domain ontologies given that they are created by different teams with different purposes, and it is very likely that they have different knowledge representations. Nevertheless they share transport protocols. It is worthwhile to investigate how agents may learn to dialog with others. The model prescribes some activities: creating and finding clusters of concepts in the ontology, selecting clusters by means of syntactic and semantic approaches, and the dialog between agents to exchange information about source and target ontology.

In the following section we provide basic definitions and concepts. Section 3 presents the partial and dynamic model, and section 4 describes a case study. In section 5, the conclusion and presentation of future works are detailed.

2 Definitions

In this section, we provide some basic definitions about ontology mapping, background knowledge, and agent communication.

2.1 Approaches to Ontology Mapping

There are a lot of approaches to classify ontology matching techniques [11, 12]. Shvaiko and Euzenat [11] proposed the classification in accordance with the nature of its operations. The classification is based on general properties of matching techniques, interpretation of input information, and kind of input information. Thus, a method is classified as: string-based techniques (prefix, suffix, edit distance, n-gram), language-based techniques (tokenization, lemmatization, elimination), constraint based techniques (data type and multiplicity comparison), linguistic resources (thesauri, synonyms, hyponyms), alignment reuse (corpus), graph-based techniques (graph matching, children, leaves, relations), upper level formal ontologies, taxonomy-based techniques and model-based (propositional satisfiability, Description Logic - based techniques).

More generally, approaches can be subdivided into syntactic [4, 5, 6, 7, 8] and semantic methods [9, 10]. The syntactic approach fails when ontologies are syntactically different or when there is a little lexical overlay among the concept labels.

The automated semantic mapping is a difficult task because the intended meaning of a concept is given by the ontology designers who give one interpretation to it, amongst the possible multiple interpretations, fixing it to a context of use not explicitly represented in the system [13]. In semantic mapping, the similarity between concepts is normally given by their structural position in the ontology by computing semantic relation (equivalence, more general) and by analyzing the meaning of the concepts which is codified in the elements and the structures of ontologies [9].

Syntactic and semantic techniques only explore the linguistics evidences and rely on label and structure similarity presented in the ontologies. These evidences may not be enough to fill the semantic gap between ontologies, thus they can fail for not exploring all heterogeneity that exists between ontologies [14, 15, 16, 17]. In some cases, evidences of the local ontologies are not enough for the mapping and may be necessary to adopt another approach. The use of background knowledge can improve the matching results, because it brings new meanings to the concepts that are not represented in the local ontologies.

2.2 Approaches to Background Knowledge

A few approaches have considered the use of background knowledge coming from different sources: online textual resources [18], reference domain ontologies [19], online available semantic data [20], corpus of schemas and mappings [21], exploiting interaction contexts [10], and external resources [9].

Online textual resources can provide an important textual source but it is not structured. The drawback in runtime is that knowledge has to be extracted without human validation, and online texts cannot be considered as reliable semantic resources [18].

The reference domain ontologies are considered a semantic bridge between ontologies. A weakness is that the appropriate reference ontology needs to be manually selected prior to the mapping. In open and dynamic environments, it implies a dialog between agents to decide which reference ontology should be used. Another problem is that there is not reference ontology for all domains, and even when they exist, they are unlikely to cover all the intend mappings between the ontologies [19].

Online available semantic data in the semantic web is a source to provide background knowledge. The semantic web could be used to select reference ontologies or to search concepts [20].

A corpus offers large corpora of texts, alternative representations of concepts in the domain, and therefore can be leveraged for multiple purposes [21]. This paper focuses on using it to improve mapping for exploiting a corpus. Texts in a corpus need human validation to be given trustworthy semantics.

The use of external resources as WordNet [22] does not identify all the correspondences between two ontologies, because it is not a complete semantic resource. It is used by the S-Match as a lexical resource that has symbols, terms and its lexical relation (synonyms, hyponyms). The thesaurus is an external resource that could be used.

The use of background knowledge overrides some of the syntactic limitations and structural semantic approaches in getting a mapping between dissimilar ontologies, although separately it can not be enough for partial and dynamic mappings.

2.3 Agent Communication Language

In order to make possible the interaction between agents, it is necessary a common language and some basic understanding of ontology concepts. Agent Communication Language provides to the agents a way to exchange messages and knowledge.

Agent communication language is based on speech-acts theory [23]. In general speech-acts have two components: a performative and a propositional content. The nature of the required interaction determines the speech acts required. Request, assertions, answer the questions, inquire and inform are examples of performatives. The FIPA-ACL [1] performatives are used in this work.

The content of a message refers to whatever the communicative act applies to (if the communicative act is the sentence, the content is the grammatical object of it). FIPA-ACL imposes that the content language must be able to express propositions, objects and actions.

3 Proposed Model

The model prescribes functions that allow agents for communicating and exchanging knowledge with others who do not share the same ontologies. The messages between agents have two parts: performative and content. We assume that the agents are able to interpret the meaning of the performatives. The model is based on three hypotheses: i) it is unlikely to build a universal domain ontology that covers all applications' requirements; ii) the agents are able to exchange messages (transport); iii) the agents may not be able to interpret the concepts or symbols in the content part of a message.

The concepts used in the content part of a message are interpreted with partial and dynamic mapping (Figure 1). The agents have their tasks definitions and concepts to describe the domain in which they act.

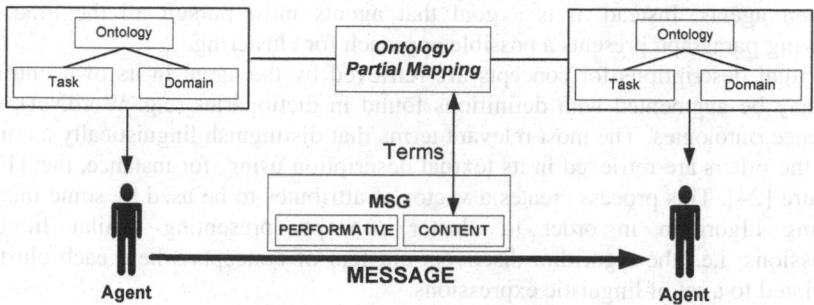

Fig. 1. Agent Communication

The partial mapping model has the following modules: clustering of ontology's concepts (partitioning the ontology) with background knowledge; selecting clusters; interpreting message contents using syntactic and semantic approaches, and the dialog between agents. The Figure 2 presents a schematic view of the model.

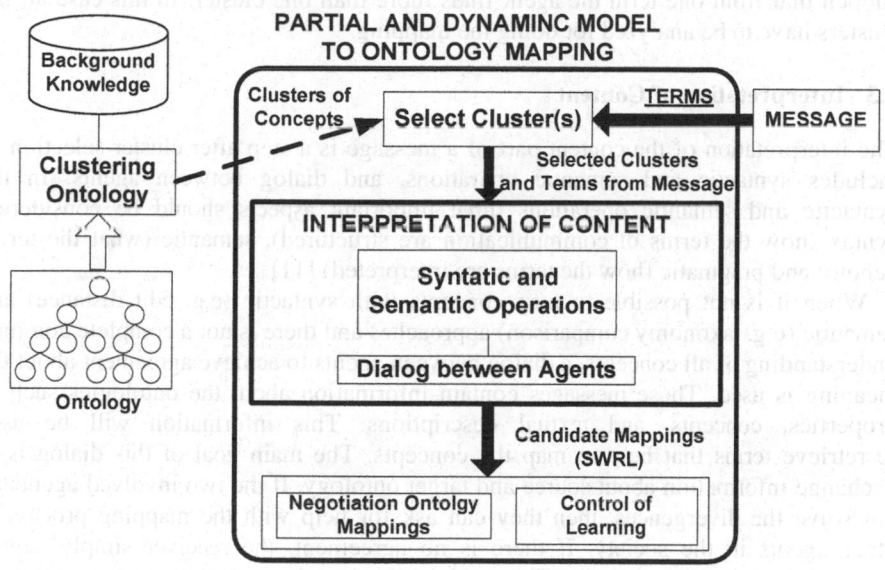

Fig. 2. Partial, Dynamic Model Mapping

3.1 Clustering Ontologies

When the ontologies to be mapped have thousands of concepts, the mapping process will take insufferably long time. Thus, the search space must be reduced to some areas, clusters or blocks of concepts. Clustering occurs when new concepts are included in the ontology or modified and it may be implemented by some machine learning algorithm using background knowledge. It is not a step in the communication between agents. Instead, it is a goal that agents must pursuit all the time. The following paragraph presents a possible approach for clustering.

Textual descriptions for concepts are retrieved by the agent in its own ontology, and may be augmented with definitions found in dictionaries (e.g. WordNet), or in reference ontologies. The most relevant terms that distinguish linguistically a concept from the others are retrieved in its textual description using, for instance, the TF/IDF measure [24]. This process creates a vector of attributes to be used by some machine learning algorithm in order to cluster concepts presenting similar linguistic expressions, i.e., the algorithm discovers clusters of concepts where each cluster is associated to a set of linguistic expressions.

3.2 Selecting Clusters

Cluster selection is done during the communication between agents. When an agent receives a message, it retrieves the terms in the content part of the message and tries to find some lexical and syntactic correspondences with the terms describing each cluster of concepts. Then, it selects the cluster of the ontology. Only the concepts in the selected cluster will be searched for similarity with the term in hands. It may happen that from one term the agent finds more than one cluster. In this case all the clusters have to be analyzed for doing the mapping.

3.3 Interpretation of Content

The interpretation of the content part of a message is a step after cluster selection. It includes syntactic and semantic operations, and dialog between agents. In the syntactic and semantic operations three important aspects should be considered: syntax (how the terms of communication are structured), semantic (what the terms denote) and pragmatic (how the terms are interpreted) [11].

When it is not possible to map concepts with syntactic (e.g. edit distance) and semantic (e.g. taxonomy comparison) approaches and there is not a complete common understanding of all concepts, a dialog between agents to achieve agreement about the meaning is used. These messages contain information about the ontologies such as properties, concepts, and textual descriptions. This information will be used to retrieve terms that help to map the concepts. The main goal of this dialog is to exchange information about source and target ontology. If the two involved agents do not solve the divergences, then they can ask for help with the mapping process to other agents in the society. If there is no agreement, the receiver simply cannot respond the incoming message. The result of this step is a set of mappings expressed in SWRL – Semantic Web Rule Language [25].

3.4 Negotiation Ontology Mapping

These mappings in SWRL may be not satisfactory to both agents, and thus additional negotiation to identify a mutually agreeable set of correspondences may be required. Laera [26] proposed an approach for exchanging different arguments that support or reject possible correspondences or mappings. Then each agent can decide according to its preferences and interests that support or reject possible candidate mappings.

3.5   Learning Control

After interpreting the content of a message, when the terms have been mapped, a learning mechanism can be used to store such mappings. This repository of mappings stores past experiences and can be useful to discover other mappings, to cluster and select clustres from ontologies. Mappings must be revised when new versions of ontologies are generated.

4   Case Study

Let us assume that two agents, named X and Y, need to interact using two independent but overlapping ontologies in the tourism domain. Figure 3 shows the ontology of the agent X used in case study.

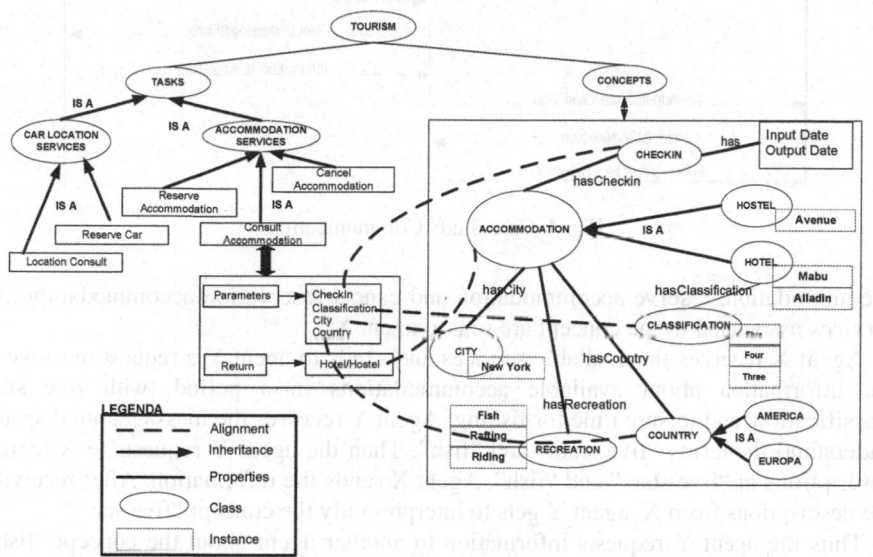

Fig. 3. Task/Domain Ontology of an Agent

As Figure 4 illustrates, agent X requests accommodation services to agent Y. Agent Y receives the message, checks that ontology used by X is different from yours, verify the performative, and retrieve the more relevant terms from the content of the message. These terms are used to select possible one cluster of the ontology in order to reduce the search space for the mapping task.

After cluster selection, the next step is interpretation of content. This interpretation is subdivided in syntactic and semantic operations; and a dialog between agents to exchange information about the concepts. This dialog can be made with other agents in the society.

In this example, the agent Y identifies the concept *accommodation* in the ontology and sends to agent X the follow services concerning accommodation: inform about

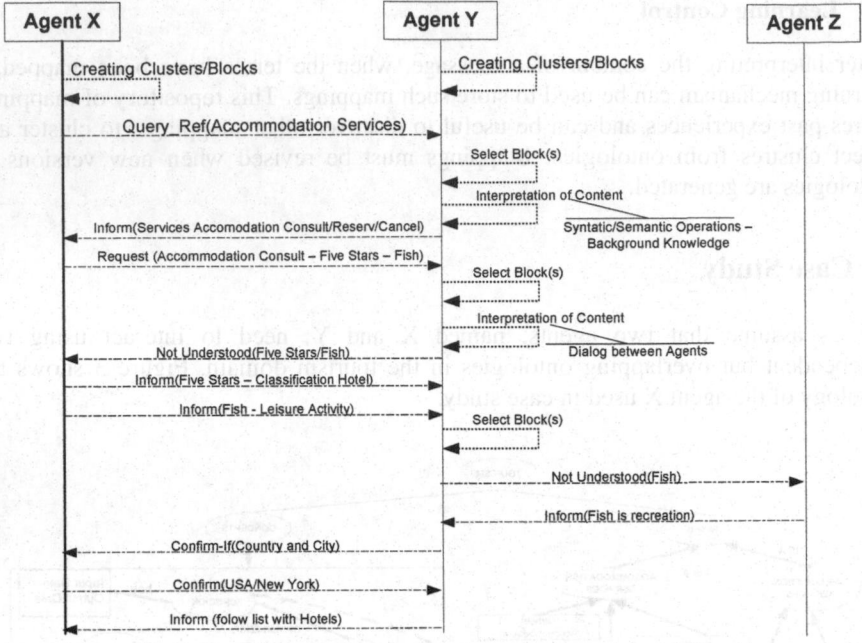

Fig. 4. Case Study Communication

accommodation, reserve accommodation and cancel reservation accommodation. All services associated to the concept are sent to agent X.

Agent X receives the available services and sends to agent Y a request message to get information about available accommodations in a period, with five stars classification, and leisure time for fishing. Agent Y receives the message, but does not understand the terms "five stars" and "fish". Then the agent Y requests to X textual descriptions of "five stars" and "fish". Agent X sends the information. After receiving the descriptions from X, agent Y gets to interpret only the concept "five star".

Thus the agent Y requests information to another agent about the concept "fish". The agent Z answers the request and sends its textual description about "fish" to the agent Y. Finally, it gets to interpret the content (mappings) and send the available hotels with the requirements. After interpreting content it is possible to negotiate or arguing the mappings. This negotiation identifies different preferences between agents and the mapping becomes more refined.

The learning control is used to store the mappings that can be used in other situations when it is necessary.

5   Conclusions

In this paper we have presented the problem of semantic interoperability or the interpretation of content in a dialog between agents in a dynamic and open environment. To lead with this problem, we propose a model to map the concepts involved in a communication between agents. The model prescribes functions for building partial and dynamic ontology mapping in order to agents achieve an agreement about the meaning of concepts. The proposed functions prescribe phases to cluster and to select clusters in an ontology, operations to interpret the content of a message based on syntactic, semantic approaches and the dialog between agents. The main contribution at the current state of the work is to give a general view of the mapping process trying to integrate several mapping approaches/techniques in order to build one adapted for runtime mapping.

First of all, we need to go deeper in the pragmatics of the mapping problem. So, the first action is to make a prototype to evaluate the usefulness and computational capacity of the model. We are going to assess our model participating in the benchmark ontology alignment evaluation initiative [27]. The domain of the ontology to be used is bibliographical references. Several hard problems remain to be tackled. For instance, how agents will dialog for explaining a concept to the other when syntactic and semantic approaches both fail. Besides exchanging concepts' properties and textual descriptions, agents could exchange some individuals. Currently, we do not envisage a way to check if the interpretation given by an agent to a concept is the one expected by the other agent (pragmatics). But, even with humans this is not easily perceived, and only after a lot of dialogs/exchanges one can be "sure" that the other has well interpreted his/her message. So, with agents we plan do to the same. When an agent receives expected answers, it will "know" that the other agent has understood its messages. Otherwise, it would not send more messages to the agent who does not understand it.

References

1. FIPA. Foundation for Intelligent Physical Agents (1996), available in http://www fipa org
2. van Diggelen, J., Beun, R., Dignum, F., van Eijk, R., Meyer, J.-J.: ANEMONE: An e_ective minimal ontology negotiation environment. In: Stone, P., Weiss, G. (eds.) AAMAS 2006. Proceedings of the Fifth International Conference on Autonomous Agents and Multi-agent Systems, pp. 899–906. ACM Press, New York (2006)
3. Xiaomeng, S.: A text categorization perspective for ontology mapping. Technical report, Department of Computer and Information Science, Norwegian University of Science and Technology, Norway (2002)
4. Noy, N., Musen, M.: The PROMPT Suite: Interactive Tools for Ontology Merging and Mapping. International Journal of Human-Computer Studies 56(6), 983–1024 (2003)
5. Ehrig, M., Staab, S.: QOM - Quick Ontology Mapping. In: Proceedings of the International Semantic Web Conference, pp. 683–697 (2004)
6. Ehrig, M., Sure, Y.: Ontology mapping - an integrated approach. In: Proceedings of the First European Semantic Web Symposium, ESWS (2004)
7. Do, H., Rahm, E.: Coma – a system for flexible combination of schema matching approaches. In: Bressan, S., Chaudhri, A.B., Lee, M.L., Yu, J.X., Lacroix, Z. (eds.) VLDB 2002. LNCS, vol. 2590, pp. 610–621. Springer, Heidelberg (2003)
8. Euzenat, J., Valtchev, P.: An integrative proximity measure for ontology alignment. In: Fensel, D., Sycara, K.P., Mylopoulos, J. (eds.) ISWC 2003. LNCS, vol. 2870, pp. 33–38. Springer, Heidelberg (2003)
9. Giunchiglia, F., Shvaiko, P., Yatskevich, M.: S-Match: An Algorithm and an Implementation of Semantic Matching. In: Proceeding of the 1st. European Semantic WebSymposium, pp. 61–75 (2004)

10. Besana, P., Robertson, D., Rovatsos, M.: Exploiting interaction contexts in P2Pontology mapping. In: P2PKM 2005 (2005)
11. Shvaiko, P., Euzenat, J.: A Survey of Schema-based Matching Approaches. J. Data Semantics IV, 146–171 (2005)
12. Rahm, E., Bernstein, P.A.: A survey of approaches to automatic schema matching. VLDB 10(4), 334–350 (2001)
13. Bachimont, B.: Engagement sémantique et engagement ontologique: conception et réalisation d'ontologies en Ingénierie des connaissances. In: Charlet, J., Zacklad, M., Kassel, G., Bourigault, D. (eds.) Ingénierie des connaissances, évolutions récentes et nouveaux défis, Eyrolles, Paris (2000)
14. Klein, M.: Combining and relating ontologies: an analysis of problems and solutions. In: IJCAI 2001. Workshop on Ontologies and Information Sharing, Seattle, USA (2001)
15. Euzenat, J.: Towards a principled approach to semantic interoperability. In: Pérez, A.G., Gruninger, M., Stuckenschmidt, H., Uschold, M. (eds.) Proc. IJCAI workshop on ontology and information sharing, Seattle (WA US), pp. 19–25 (2001)
16. Bouquet, P., Euzenat, J., Franconi, F., Serafini, L., Stamou, G., Tessaris, S.: D2.2.1 Specification of a common framework for characterizing alignment – Knowledge Web Project. Realizing the semantic web. IST-2004-507482. Program of the Comission of the European Communities (2004)
17. Visser, P.R.S., Jones, D.M, Bench-Capon, T.J.M., Shave, M.J.R.: Assessing Heterogeneity by Classifying Ontology Mismatches. In: F01S 1998. Proceedings International Conference on Formal Ontology in Information Systems, lOS Press, Amsterdam (1998)
18. Van Hage, W., Katrenko, S., Schreiber, G.: A Method to Combine Linguistic Ontology-Mapping Techniques. In: Gil, Y., Motta, E., Benjamins, V.R., Musen, M.A. (eds.) ISWC 2005. LNCS, vol. 3729, Springer, Heidelberg (2005)
19. Aleksovski, Z., Klein, M., Ten Katen, W., Van Harmelen, F.: Matching Unstructured Vocabularies using a Background Ontology. In: Staab, S., Svátek, V. (eds.) EKAW 2006. LNCS (LNAI), vol. 4248, Springer, Heidelberg (2006)
20. Sabou, M., D'Aquin, M., Motta, E.: Using the Semantic Web as Background Knowledge for Ontology Mapping. In: 2nd workshop on Ontology matching, at 5th Intl. Semantic Web Conf., November 6, 2006, Georgia, US (2006)
21. Madhavan, J., Bernstein, P.A., Doan, A., Halevy, A.: Corpus Based Schema Matching. ICDE, 57–68 (2005)
22. Miller, G.A.: WordNet: An on-line lexical database. International Journal of Lexicography 3(4), 235–312 (1990)
23. Searle, J.: Os Actos de Fala - Um Ensaio de Filosofia da Linguagem, Traduzido por: Carlos Vogt. Coimbra: Livraria Almedina. Tradução de: Speech Acts – An Essay in the Philosophy of Language (1981)
24. Salton, G.: Automatic Text Processing: The Transformations, Analysis, and Retrieval of Information by Computer. Addison-Wesley, Reading (1989)
25. Horrocks, I., Patel-Schneider, P., Boley, H., Tabet, S., Grosof, B., Dean, M.: SWRL: a semantic web rule language combining OWL and RuleML (2004), http://www.w3.org/Submission/SWRL/
26. Laera, L., Tamma, V., Euzenat, J., Bench-Capon, T., Payne, T.: Reaching agreements over ontology alignments. In: Cruz, I., Decker, S., Allemang, D., Preist, C., Schwabe, D., Mika, P., Uschold, M., Aroyo, L. (eds.) ISWC 2006. LNCS, vol. 4273, Springer, Heidelberg (2006)
27. Ontology Alignment Evaluation Initiative Test Library. Reference Ontology (2007), available in: http://oaei.ontologymatching.org/2007/benchmarks/

Using Ontologies for Software Development Knowledge Reuse

Bruno Antunes, Nuno Seco, and Paulo Gomes

Centro de Informatica e Sistemas da Universidade de Coimbra
Departamento de Engenharia Informatica, Universidade de Coimbra
bema@student.dei.uc.pt, {nseco,pgomes}@dei.uc.pt
http://ailab.dei.uc.pt

Abstract. As software systems become bigger and more complex, software developers need to cope with a growing amount of information and knowledge. The knowledge generated during the software development process can be a valuable asset for a software company. But in order to take advantage of this knowledge, the company must store and manage it for reuse. Ontologies are a powerful mechanism for representing knowledge and encoding its meaning. These structures can be used to model and represent the knowledge, stored in a knowledge management system, and classify it according to the knowledge domain that the system supports. This paper describes the Semantic Reuse System (SRS), which takes advantage of ontologies, represented using the knowledge representation languages of the Semantic Web, for software development knowledge reuse. We describe how this knowledge is stored and the reasoning mechanisms that support the reuse.

Keywords: Ontologies, Sematic Web, Software Reuse, Knowledge Management.

1 Introduction

The new vision of the web, the Semantic Web [1], aims to turn the web more suitable for machines, thus make it more useful for humans. It brings mechanisms that can be used to classify information and characterize it's context. This is mainly done using knowledge representation languages that create explicitly domain conceptualizations, such as ontologies [2]. These mechanisms enable the development of solutions that facilitate the access and exchange of relevant information.

The storage and access to relevant information is a central issue in knowledge management [3], which comprises a set of operations directed to the management of knowledge within an organization, helping the organization in the achievement of its objectives. The technologies that are behind the Semantic Web vision provide an opportunity to increase the efficiency of knowledge management systems, turning them more valuable. These technologies provide knowledge representation structures that can be used to improve the storage, search and retrieval functionalities of common knowledge management systems.

J. Neves, M. Santos, and J. Machado (Eds.): EPIA 2007, LNAI 4874, pp. 357–368, 2007.

One of the problems that software development companies face today is the increasing dimension of software systems. Software development projects have grown by complexity and size, as well as in the number of functionalities and technologies that are involved. The resources produced during the design and implementation phases of such projects, are an important source of knowledge inside software development companies. These resources contain the knowledge that has been used to solve various development problems of past projects. Some of these problems will certainly appear in the future, for this reason the software development resources produced in the past, represent an important source of knowledge that can be reused in the future. In order to make this knowledge useful and easily shareable, efficient knowledge management systems are needed. These knowledge management systems must be able to efficiently store, manage, search and retrieve all kinds of knowledge produced in the development of software systems. To accomplish this, the stored knowledge must be well described, classified and accessible from where it is needed.

As referred before, ontologies are one of the most important concepts present in the Semantic Web architecture. They are a powerful mechanism for representing knowledge and encoding its meaning, allowing the exchange of information that machines are able to process and concisely understand. These structures can be used to model and represent the knowledge stored in a knowledge managemente system and classify it according to the knowledge domain that the system supports. In this paper we describe the *Semantic Reuse System* (SRS), a system for the reuse of software development knowledge that we are developing at the Artificial Intelligence Laboratory (AILab) of the University of Coimbra.

The goal of SRS is the management and reuse of software development knowledge using the mechanisms provided by the Semantic Web, such as RDF, RDFS and OWL [4], to represent the ontologies that reflect the knowledge used by the system. We make use of a representation ontology, used to represent tne different types of artifacts that the system deals with, and a domain ontology, to classify this these artifacs. As software development knowledge, we consider the various elements that result from the software development process, such as specification documents, design diagrams, source code, etc. We will name each one of these elements as a *Software Development Knowledge Element* (SDKE). Our approach aims to provide efficient mechanisms for storing, searching, retrieving and managing the stored knowledge, using ontologies and the Semantic Web languages to represent them.

The next section describes the SRS architecture, as well as the main functionalities and requirements of the system. Section 3 describes the knowledge base used in the system, and its subcomponents. In section 4 the knowledge reuse mechanisms are described and illustrated. Section 5 presents similar works to SRS and section 6 concludes the paper with final remarks and some future work.

2   SRS Architecture

The SRS system works as a platform that provides a way to store the software development knowledge in the form of SDKE's. Besides being stored in the file

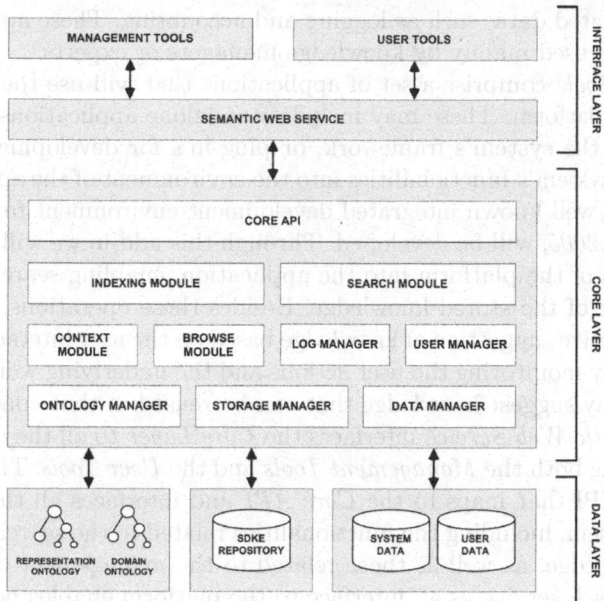

Fig. 1. The architecture of SRS

system and accessible through an interface framework, the different elements should be described using a *Representation Ontology* and classified through a *Domain Ontology*. These additional knowledge representation structures will be used to empower the search mechanisms and make easier the reuse of stored knowledge. To take advantage of the system functionalities, a set of client applications can be developed. These client applications can be directed to users and developers, implemented as standalone applications or as plug-in's for well known development environments, or to those responsible for the management of the stored knowledge and supporting structures of the platform. The system provides an API for these applications using a web service, for better portability and integration.

In figure 1, we present the system's architecture. The system can be structured in three different logical layers: data, core and interface. The data layer corresponds to the knowledge base and is described in section 3, it stores all the knowledge needed for the system reasoning, including the knowledge representation structures and the SDKE's. The core layer implements the reasoning mechanisms, including the search, suggestion and browsing facilities, which are described in section 4. The interface layer comprises the semantic web service and the applications that use the system's functionalities through this semantic web service.

The *Management Tools* comprise a set of applications that can be used to manage the system and its knowledge base. There is a *Knowledge Base Manager* that is used to manage the knowledge representation structures, both the *Domain Ontology* and the *Representation Ontology*, the *SDKE Repository* and the system related data, such as logging and accounting. These applications are intended to be used mainly by knowledge managers or experts.

The *User Tools* comprise a set of applications that will use the services provided by the platform. These may include standalone applications specially designed to use the system's framework, or plug-in's for development tools that integrate the system's functionalities into the environment of the application. An add-in for the well known integrated development environment from *Microsoft*, *Visual Studio 2005*, will be developed. Through this add-in we will integrate the functionalities of the platform into the application, enabling searching, submission and reuse of the stored knowledge. Besides these operations, we intend to provide pro-active suggestion of knowledge based on the user interaction with the application. By monitoring the user actions and the underlying working context, the system may suggest knowledge that can be reused in that context.

The *Semantic Web Service* interfaces the *Core Layer* to all the client applications, including both the *Management Tools* and the *User Tools*. The web service provides an API that maps to the *Core API* and interfaces all the functionalities of the system, including the functionalities related to the search, storage and reuse of knowledge, as well as those related to the management of the system. The use of a web service as an interface to the platform enables portability and integration. Specially, the Semantic Web Services [5] are an attempt to make web services prepared to be efficiently used by software agents. This is done by providing agent-independent mechanisms to attach metadata to a web service that describes its specification, capabilities, interface, execution, prerequisites and consequences. The OWL-S [6] language has been developed in order to support these mechanisms.

3   Knowledge Base

This section describes the knowledge base used in SRS. It comprises the *Representation Ontology*, the *Domain Ontology*, the *SDKE Repository*, and *System and User's Data*. The ontologies in the platform will be represented using languages from the Semantic Web, namely RDF, RDFS and OWL [4], and will be managed using the *Jena Semantic Framework* [7] (http://jena.sourceforge.net/), which is an open source Java framework for building Semantic Web applications, providing an API for manipulating RDF graphs, including support for RDFS and OWL. The *SDKE Repository* will be managed using *Apache Lucene* [8](http://lucene.apache.org/), which provides an open source high-performance engine for full-featured text search and document storage written in Java. The system and user data will be managed through a relational database. The next subsections describe each of these parts.

3.1   Representation Ontology

The *Representation Ontology* defines a set of classes, properties and relations between classes that are used to model and describe the different types of SDKE's.

This ontology is used to define every type of SDKE's that can be used in the system. The different properties that apply to each one of them and the specific relations that can exist between them.

This ontology is specified taking into account the different types of elements that we want to store and reuse. This is mainly done before the deployment of the system. But this ontology should be viewed as a dynamic structure, since it should be possible to do some adjustments to the model after it has been deployed, without affecting the system as a whole. The *Representation Ontology* will also include the metadata model developed by the *Dublin Core Metadata Initiative* (DCMI) [9], which is an organization engaged in the development and promotion of interoperable metadata standards for describing resources that enable more intelligent information systems. The DCMI is integrated in the representation ontology as annotation properties to SDKE objects. The adoption of these standards enables the system to easily exchange data with other platforms, or even with intelligent agents. As important as describing the components that are being stored for reusing, is the association, to these components, of what we call of reuse documentation. This reuse documentation will be defined in the *Representation Ontology*, using a metadata model based on the work of Sametinger [10], providing vital information for those who will reuse the stored knowledge in future development projects. In figure 2, we present an example of part of the *Representation Ontology*. Note that what appears in the figure is the ontology editor of the *Knowledge Base Manager*, which is one of the management tools of SRS. The figure represents part of the representation ontology taxonomy.

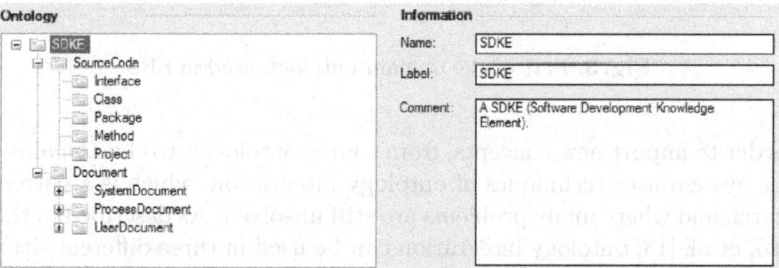

Fig. 2. Part of the representation ontology used in SRS

3.2   Domain Ontology

The *Domain Ontology* is used to represent the knowledge of the domain that the system supports. It is comprised of concepts and relations between concepts. The SDKE's stored in the platform are indexed to one or more concepts in this ontology, reflecting the knowledge stored in each one of them. Having relations that semantically relate concepts with other concepts, the system is able to retrieve not only knowledge that match a query, but also knowledge that is semantically related and eventually relevant to the user.

This ontology is dynamically built during the submission of new knowledge to the platform. To construct this ontology, the system uses other source ontologies, from which it retrieves the concepts and relations that are needed to index new knowledge. One of the source ontologies that is used is *WordNet* [11], which is a large lexical database of English. In *WordNet*, nouns, verbs, adjectives and adverbs are grouped into sets of synonyms, each expressing a distinct concept. The system retrieves the concepts from WordNet, that it needs too classify the knowledge supported, and import them to the domain ontology. But the system may also search, retrieve and import ontologies that contain the desired concepts, using tools such as the *Swoogle Semantic Web Search Engine* [12], which is a crawler-based indexing and retrieval system for documents containing knowledge representation structures of the Semantic Web. In figure 3, we present an example of part of the *Domain Ontology*. As in the previous figure, what appears in the figure is the ontology editor of the *Knowledge Base Manager*, which is one of the management tools of SRS. The figure represents part of the domain ontology taxonomy.

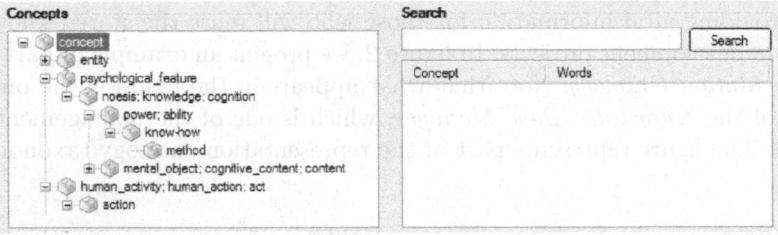

Fig. 3. Part of the domain ontology used in SRS

In order to import new concepts, from source ontologies to the *Domain Ontology*, the system uses techniques of ontology integration, which is a current field of research and where many problems are still unsolved. As described in the work of Pinto, et al, [13] ontology integration can be used in three different situations: integration of ontologies when building a new ontology reusing other available ontologies; integration of ontologies by merging different ontologies about the same subject into a single one that "unifies" all of them; and the integration of ontologies into applications. From these, the situation we face in our system is building a new ontology reusing other available ontologies. Specially, we want to build our *Domain Ontology* by importing the concepts and their relations, from source ontologies, as long as we need them to index new knowledge. By using *WordNet* as the main knowledge source, our *Domain Ontology* could be seen as a small part of *WordNet*, since we would only bring to our ontology the concepts needed to reflect the stored knowledge. But, having *WordNet* as the only source ontology creates some problems, such as domain vagueness, since it is general, and language specificness, since it is in English. Another alternative would be reusing ontologies available in large repositories or in the Internet.

This approach, although partially solving the problem of domain vagueness and language specific, brings even more complex challenges, such as ontology evaluation, ontology ranking, ontology segmentation and ontology merging, which have been described by Alani [14].

3.3 SDKE Repository

The *SDKE Repository* represents the knowledge repository in the platform. When submitted, the development knowledge is described and indexed, using the knowledge representation structures described before, and is stored in the repository in the same form it as been submitted. This way, the repository contains all the files that have been submitted to the platform, representing the SDKE's indexed by the system. These files are stored using *Apache Lucene*, as referred before, which also provides advanced text search capabilities over the stored documents.

3.4 System and User's Data

The *System and User's Data* is comprised of data related to system configuration, logging and accounting. This data comprises parameters and other configuration information used by the system; logging data generated by the user's interaction with the system, which can be used to analyze the user's behavior and eventually improve the system's efficiency and precision; and data for user accounting, which includes user accounts, user groups, access permissions, etc.

4 Knowledge Reuse

In the previous section we have described the knowledge that the system uses and stores, this section presents the reasoning mechanisms used for manipulating this knowledge. There are a set of modules that implement the basic operations on the platform: the *Core API* is an API to all the core functionalities, and acts as an interface between the core modules and the *Semantic Web Service*; the *Log Manager* implements an interface to all the operations related to system's logging; the *User Manager* implements an interface to all the operations related to users; the *Ontology Manager* implements an interface to all the operations over the *Representation Ontology* and the *Domain Ontology*; the *Storage Manager* implements an interface to all the operations over the *SDKE Repository*; and the *Data Manager* implements an interface to all the operations over the *System Data* and *User Data*.

In the next sub-sections, we describe in detail the most relevant modules present in the *Core Layer*: *Indexing Module*, *Context Module*, *Search Module* and *Browse Module*. These modules implement the operations related to the submission, search, retrieval and management of knowledge in the platform.

4.1   Indexing Module

The *Indexing Module* implements and provides an interface to submit and index new knowledge to the platform. The submitted knowledge, in the form of software development artifacts, must be described using the *Representation Ontology*, indexed to concepts in the *Domain Ontology* and stored in the *SDKE Repository*.

The description of the artifacts using the *Representation Ontology* is done taking in account their type and associated metadata. The *Representation Ontology* defines the different types of artifacts that can be stored in the platform, their relations and associated metadata. Given the artifacts and the corresponding metadata, the system creates instances of classes in the *Representation Ontology* that represents each one of the artifacts. The indexing of the artifacts in the *Domain Ontology* comprises two phases: first the system must extract from the artifacts the concepts used for indexing and then it must verify if those concepts exist in the domain ontology or if they must be imported from source ontologies. The extraction of concepts from the artifacts is done using linguistic tools from *Natural Language Processing* (NLP) [15]. To index the SDKE in the *Domain Ontology*, the system first searches the *WordNet* [11] for synsets containing the base forms of the relevant terms of the SDKE. The synsets that contain the relevant terms are then imported to the *Domain Ontology*. The synsets are imported in conjunction with the whole set of its hypernyms, which form one or more paths from the synset to the root of the synset's hierarchy. The hypernym of a concept is a concept that is more general that the first and that contains its definition. The terms not found in *WordNet* are added to the root of the concepts hierarchy in the *Domain Ontology*. These concepts are added with a special property indicating that they were not imported from the *WordNet* and are waiting the validation of the knowledge manager. With all the concepts that classify the SDKE added to the *Domain Ontology*, the system indexes the SDKE to these concepts. After being described and indexed, the artifacts must be stored in the *SDKE Repository*.

4.2   Context Module

The *Context Module* implements and provides an interface to retrieve knowledge that is relevant for a development context. As development context we consider a set of elements that can be found in a common IDE, such as packages, interfaces, classes, methods, etc, and an event, such as the creation of any of the referred elements. The process of interaction between the user and the *Context Module* is represented in figure 4.

The *Context Module* can be useful to suggest knowledge for reuse in a development environment, such as a programming IDE. By using this module, an IDE can pro-actively suggest knowledge for reuse upon certain actions of the developer. For instance, when the developer creates a new package, the IDE can use the *Context Module* to retrieve knowledge that is relevant to the package being created taking into account the context that surrounds that package. As relevant knowledge,

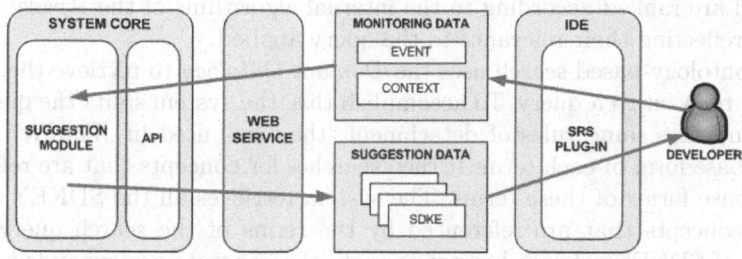

Fig. 4. The process of interaction between the user and the context module

we can consider two kinds of knowledge: suggested knowledge and related knowledge. The suggested knowledge represents knowledge suitable for replacing the element that fired up the event, so if the developer is creating a package the system will suggest packages that can be reused. The related knowledge represents knowledge that cannot replace the element that fired up the event, but is somehow related to that element or its context, and can also be useful to the developer. For instance, if a user is modifying a class, the system can suggest the piece of text from the specification documentat where the class is described. The retrieval of SDKE's, through keyword-based search or context-based search, is implemented by the *Search Module*, which we describe next.

4.3 Search Module

The *Search Module* implements and provides an interface to all the operations of searching and retrieving stored knowledge. Three different search methods were developed: keyword-based search, ontology-based search and full search. The process of interaction between the user and the *Search Module* is represented in figure 5.

The keyword-based search relies on the full-text search capabilities of the *Apache Lucene* [8] engine that supports the *SDKE Repository*. When SDKE's are stored in the repository, the *Apache Lucene* engine indexes their textual data so that keyword-based searches can be made in the repository. The results

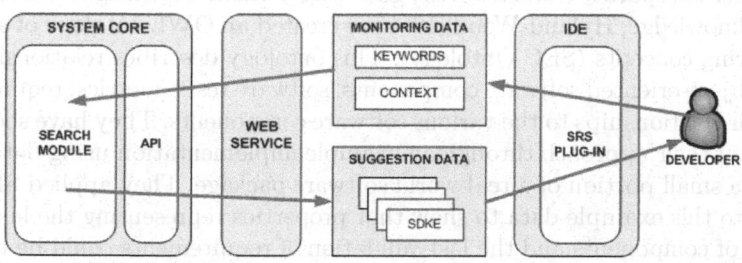

Fig. 5. The process of interaction between the user and the search module

returned are ranked according to the internal algorithms of the *Apache Lucene* engine, reflecting their relevance to the query applied.

The ontology-based search uses the *Domain Ontology* to retrieve the SDKEs that are relevant to a query. To accomplish this, the system splits the query into terms and uses some rules of detachment, the same used in *WordNet* [11], to get the base form of each term. It then searches for concepts that are referenced by the base forms of these terms. The system retrieves all the SDKE's indexed by the concepts that are referenced by the terms of the search query. If the number of SDKE's retrieved is not enough, the concepts are expanded to their hyponyms and hypernyms. Then the SDKE's indexed by the expanded concepts are retrieved and again, if they are not enough, the process is repeated until a defined maximum expansion length is reached. Finally, the retrieved SDKE's are ranked according to the number of concepts they are indexed to.

4.4   Browse Module

The *Browse Module* provides an interface to all the operations related to the browsing of the stored knowledge. This can be done using either the *Representation Ontology* or the *Domain Ontology*. This means that the knowledge stored in the platform can be organized by its type or by the concepts it is associated with.

5   Related Work

This section introduces some recent works in the field of knowledge management for software development using Semantic Web technologies.

In their work, Hyland-Wood, et al, [16] describe ongoing research to develop a methodology for software maintenance using Semantic Web techniques. They propose a collection of metadata for software systems that include functional and non-functional requirements documentation, metrics, success or failure of tests and the means by which various components interact. By recording and tracking the changes in the metadata it is possible to notify the developers of changes that may influence further development. The system metadata is encoded using Semantic Web techniques such as RDF, OWL and SPARQL Query Language [17].

In order to separate the software engineering domain knowledge from the operational knowledge, Hyland-Wood [18] have created an OWL ontology of software engineering concepts (SEC Ontology). This ontology describes relationships between object-oriented software components, software tests, metrics, requirements and their relationships to the various software components. They have shown the validity of their approach through an example implementation using data representing a small portion of a real-world software package. They applied SPARQL queries to this example data to show that properties representing the last modification of components and the last validation of requirements could be updated and that subsequent queries could be used to determine state changes. Based on this grounding work, they envision a software maintenance methodology where developers could manually input information about requirements through some kind of IDE which would store software system metadata in a RDF graph and accept SPARQL queries. Changes in metadata could be tracked and developers would be notified when these changes required actions.

Recent work by Thaddeus [19] has proposed a system similar to SRS, which uses formal logics for representing software engineering knowledge and reasoning. The proposed system, uses a multi-agent approach to implement several reasoning mechanisms, like: knowledge access, process workflow automation or automated code generation. They use three different ontologies: Project Ontology, Process Ontology and Product Ontology. This system has some common goals with SRS, but has a different approach to the overall architecture. While Thaddeus uses a multi-agent approach, SRS uses a Producer/Consumer approach based on several types of clients that can be connect to the main system through a Semantic Web Service.

6 Conclusions

In this paper we described the SRS system, which intends to explore new ways to store and reuse knowledge. We have used ontologies and the Semantic Web technologies to build a software development reuse platform, representing and storing knowledge in RDF, RDFS and OWL. The SRS platform provides several ways to reuse and share knowledge, including a pro-active way of suggesting relevant knowledge to the software developer using the user context.

Ontologies and the Semantic Web technologies enable the association of semantics to software artifacts, which can be used by inference engines to provide new functionalities such as semantic retrieval, or suggestion of relevant knowledge. The SRS provides easy integration with client applications through the semantic web service. Thus, tools for Software Engineering and MDA (Model Driven Architecture) that use metadata (for exemple, the EODM plug-in for Eclipse) can be integrated in SRS has knowledge providers. These tools can also be modified to act as knowledge retrievers.

References

1. Berners-Lee, T., Hendler, J., Lassila, O.: The semantic web. Scientific American 284(5), 34–43 (2001)
2. Zuniga, G.L.: Ontology: Its transformation from philosophy to information systems. In: Proceedings of the International Conference on Formal Ontology in Information Systems, pp. 187–197. ACM Press, New York (2001)
3. Liebowitz, J., Wilcox, L.C.: Knowledge Management and Its Integrative Elements. CRC Press, Boca Raton, USA (1997)
4. McGuinness, D.L., van Harmelen, F.: Owl web ontology language overview. Technical report, W3C (2004)
5. McIlraith, S.A., Son, T.C., Zeng, H.: Semantic web services. IEEE Intelligent Systems 16(2), 46–53 (2001)

6. Martin, D.L., Paolucci, M., McIlraith, S.A., Burstein, M.H., McDermott, D.V., McGuinness, D.L., Parsia, B., Payne, T.R., Sabou, M., Solanki, M., Srinivasan, N., Sycara, K.P.: Bringing semantics to web services: The owl-s approach. In: Cardoso, J., Sheth, A.P. (eds.) SWSWPC 2004. LNCS, vol. 3387, pp. 26–42. Springer, Heidelberg (2005)
7. Carroll, J.J., Dickinson, I., Dollin, C., Reynolds, D., Seaborne, A., Wilkinson, K.: Jena: Implementing the semantic web recommendations. Technical Report HPL-2003-146, HP Laboratories Bristol (2003)
8. Hatcher, E., Gospodnetic, O.: Lucene in Action. Manning Publications (2004)
9. Hillmann, D.: Using dublin core (2005)
10. Sametinger, J.: Software Engineering with Reusable Components. Springer, Heidelberg (1997)
11. Miller, G.A.: Wordnet: A lexical database for english. Communications Of The ACM 38(11), 39–41 (1995)
12. Ding, L., Finin, T.W., Joshi, A., Pan, R., Cost, R.S., Peng, Y., Reddivari, P., Doshi, V., Sachs, J.: Swoogle: A search and metadata engine for the semantic web. In: Grossman, D., Gravano, L., Zhai, C., Herzog, O., Evans, D.A. (eds.) Proceedings of the 2004 ACM CIKM International Conference on Information and Knowledge Management, November 8-13, 2004, pp. 652–659. ACM Press, Washington, DC, USA (2004)
13. Pinto, H.S., Gómez-Pérez, A., ao P. Martins, J.: Some issues on ontology integration. In: Proceedings of the Workshop on Ontologies and Problem Solving Methods (1999)
14. Alani, H.: Position paper: Ontology construction from online ontologies. In: Carr, L., Roure, D.D., Iyengar, A., Goble, C.A., Dahlin, M. (eds.) Proceedings of the 15th international conference on World Wide Web, WWW 2006, Edinburgh, Scotland, UK, May 23-26, 2006, pp. 491–495. ACM Press, New York (2006)
15. Jackson, P., Moulinier, I.: Natural Language Processing for Online Applications: Text Retrieval, Extraction and Categorization. John Benjamins, Amsterdam (2002)
16. Hyland-Wood, D., Carrington, D., Kaplan, S.: Toward a software maintenance methodology using semantic web techniques. In: Proceedings of Second International IEEE Workshop on Software Evolvability (2006)
17. Seaborne, A., Prud'hommeaux, E.: Sparql query language for rdf. Technical report, W3C (2006)
18. Hyland-Wood, D.: An owl-dl ontology of software engineering concepts (2006)
19. Thaddeus, S., S.V., K.R.: A semantic web tool for knowledge-based software engineering. In: SWESE 2006. 2nd International Workshop on Semantic Web Enabled Software Engineering, Springer, Athens, G.A., USA (2006)

Chapter 6 - First Workshop on Business Intelligence (BI 2007)

Analysis of the Day-of-the-Week Anomaly for the Case of Emerging Stock Market

Virgilijus Sakalauskas and Dalia Kriksciuniene

Department of Informatics, Vilnius University, Muitines 8, 44280 Kaunas, Lithuania
{virgilijus.sakalauskas,dalia.kriksciuniene}@vukhf.lt

Abstract. The aim of the article is to explore the day-of-the-week effect in emerging stock markets. This effect relates to the attempts to find statistically significant dependences of stock trading anomalies, which occur in particular days of the week (usually the first or the last trading day), and which could be important for creating profitable investment strategies. The main question of the research is to define, if this anomalies affects the entire market, or it is applicable only for the specific groups of stocks, which could be recognized by identifying particular features. The investigation of the day-of-the-week effect is performed by applying two methods: traditional statistical analysis and artificial neural networks. The experimental analysis is based on financial data of the Vilnius Stock Exchange, as of the case of emerging stock market with relatively low turnover and small number of players. Application of numerous tests and methods reveals better effectiveness of the artificial neural networks for indicating significance of day-of-the-week effect.

Keywords: day-of-the-week effect, neural network, statistical analysis, stock market.

1 Introduction

The day-of-the-week effect is one of the controversial questions for research. Yet it receives extensive interest in the scientific literature, because better understanding of this anomaly could lead to more profitable trading strategies by strengthening positive influence of the effect, identifying most affected portfolios of stocks, or avoiding losses by better knowledge of the negative influence of this effect.

The effect means that particular trading conditions occur quite regularly on specific days of the week. Published research for the United States and Canada finds that daily stock market returns tend to be lower on Mondays and higher on Fridays [4, 6]. In contrast, daily returns in Pacific Rim countries tend to be lowest on Tuesdays [3]. Using daily data of one hundred years of the Dow Jones and of about 25 years of the S&P500, confirmed Monday Effect as the best calendar rule [18].

On the other hand, there is quite big number of research works, which suggest the diminishing influence of the day-of-the-week effect. The results presented in Kohers et al [11] indicate that while the day-of-the-week effect was clearly prevalent in the vast majority of developed markets during the 1980s is now losing its importance.

J. Neves, M. Santos, and J. Machado (Eds.): EPIA 2007, LNAI 4874, pp. 371–382, 2007.
© Springer-Verlag Berlin Heidelberg 2007

In Korea and the United Kingdom the day-of-the-week effect has disappeared starting in the 1990s, [10, 17]. The attempts to analyze the effect by taking into account different markets, variables, and methods, showed it to be highly dependable on data and on the computational tools applied for its identification.

Most research works analyse mature and developed markets, which provide high volumes of historical data, covering long periods. The emerging markets differ by comparatively low turnover and small number of market players, which could require different approaches and methods for the analysis, and for making trading anomalies significantly evident.

Main variable, used for day-of-the-week effect investigation is mean return, and the most recognized finding about it states, that the average return for Monday is significantly negative for the countries like the United States, the United Kingdom, and Canada [1]. The other important indicators are the daily closing prices and indexes [2]. There are numerous statistical analyses of mean return and variance. Empirical evidence has shown departure from normality in stock returns [7, 19], which suggests that another aspect appropriate for study is testing higher moments of stock returns. This type of analysis is rare, but some already published experimental studies found out that both skewness and kurtosis are diversified away rapidly [19].

Basher in [18] has studied the day of week effect in 21 emerging stock market by using both unconditional and conditional risk analysis and applying different analysis models. He confirmed the evidence of day-of-the-week effect only for Philippines, Pakistan and Taiwan markets, even after adjusting research results for the market risk.

Application of the neural networks extended the traditional statistical analysis by including more variables for research, in addition to return and index data.

In the work of Kumar and Thenmozhi [12] the results of applying back propagation and recurrent neural network models outperformed the linear ARIMA models (in this research the NIFTY index futures were used as input data).

Gencay and Qi in [8, 14] have presented evidence in favor of nonlinear predictability of stock returns. Both authors used neural network models and positively evaluated the performance of their models in comparison to the conventional linear models. On the other hand, their research was limited because of type of variables used for analysis: Qi used not only financial but also economic variables, such as industrial output and inflation. The problem was that these data were not available at the time when the investors had to make their investment decisions. In contrast, Gencay used only past prices and technical indicators (differences between prices and moving average of the chosen number of the most recent prices).

The investigations of market inefficiency by Reschenhofer in [15] revealed, that there existed influence of days-of-the-week, and it made possible to earn by applying rules-related strategies. The analysis required all available information, including trading volumes, dividends, earning-price ratios, prices of the other assets, interest rates, exchange rates, and, subsequently, usage of powerful nonparametric regression techniques (e.g., artificial neural networks), and automated model selection criteria.

These findings made it evident, that the day-of-the-week effect and the other stock market anomalies depend on individual markets and economics, and their occurrence is more common in the emerging stock markets.

The goal of this research was to explore the possibility to recognize the day-of-the-week effect by analysing daily trade volume data of the emerging stock markets with low liquidity. The empirical research was performed by using data of the Vilnius Stock Exchange. We used the Vilnius OMX Index and 24 equities mean return (time interval 2003-01-01 to 2006-11-21). One of the goals was to explore if the day-of-the-week-effect affects the whole market, or it is valid only for some particular segments of stocks. For implementing the experiment, we composed three portfolios of stocks with the low, medium and high daily turnover, and explored the validity of hypothesis of presence of the day-of-the-week effect for each group of securities. The stock portfolios characterized by traditional variables of return and index were explored by analysing the first moments calculated for the days of the week, and by applying methods based on analysis of the higher moments. Application of the artificial neural networks for the expanded number of variables of the data set was used as well.

In the following chapter the research methodology and organization of research data set is defined. The research procedures and conclusions revealed better effectiveness of the neural networks for significant confirmation of the presence of day-of-the-week effect, than is was possible to achieve by traditional statistical analysis.

2 Data and Results

The data set was formed from the Vilnius Stock Exchange trading information [20]. The Vilnius Stock Exchange is the only regulated stock market in Lithuania providing trading, listing and information services. The Vilnius Stock Exchange is part of OMX, which offers access to approximately 80 percent of the Nordic and Baltic stock markets through its exchanges in Copenhagen, Stockholm, Helsinki, Riga, Tallinn and Vilnius. The main financial data of Vilnius stock exchange is characterized by its market value of 7 EUR billions, share trading value per business day of near 2 million EUR, approximately 600 trading transactions per business day, and 44 shares in the equity list [20].

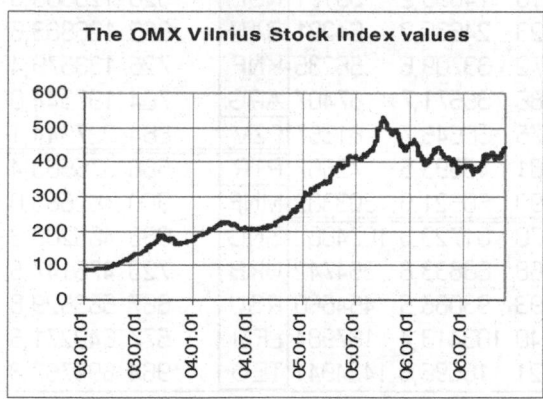

Fig. 1. The OMX Index values

According to these financial indicators, Vilnius Stock Exchange belongs to the category of small emerging securities markets. The OMX Vilnius Stock Index is a capitalization weighted chain linked total-return index, which is evaluated on a continuous basis using the most recent prices of all shares that are listed on the Vilnius Stock Exchange. For the calculations we use the OMX Vilnius Stock Index and 24 shares values of the time interval from 2003-01-01 to 2006-11-21 on daily basis (Fig. 1).

For ensuring validity of the research, mean return data of twenty four equities (out of total 44 listed in OMX Vilnius Stock Index) were selected for representing all variety of the equity list, according to the capitalization, number of shares, daily turnover, profitability and risk.

The traditional understanding of return is presented by expression (1), where return on the time moment t, R_t is evaluated by logarithmic difference of stock price over time interval (t-1,t],

$$R_t = \ln(\frac{P_t}{P_{t-1}}) = \ln(P_t) - \ln(P_{t-1}),$$ (1)

where P_t indicates the price of financial instrument at time moment t [16].

The return values of 24 equities were assigned to the variables, named correspondingly to their symbolic notation of Vilnius Stock Exchange. The initial analysis by summary statistics of the data set revealed quite big differences of average trading volume of stocks (Table 1). The following task of the research was to check, if differences of trade volume of stocks could influence occurrence of day-of-the-week effect. The shares were ranked by daily turnover and divided into three groups according to mean average of the daily trading volume as presented in the Figure 2:

Descriptive Statistics (VOLUME)						
Var	Valid N	Mean	Std.Dev.	Valid N	Mean	Std.Dev.
LEL	818	14800,2	35171	NDL	326	120808,3	1065115
KBL	723	24636,2	64291	SAN	650	126883,6	922287
LNS	872	33709,6	56235	KNF	725	138579,4	423522
LEN	886	35571,7	67407	APG	764	197344,0	546005
VBL	775	36645,7	61357	PZV	884	227944,1	2245328
LJL	881	45693,5	72567	PTR	566	375563,4	1263903
LLK	590	50521,9	303382	MNF	950	462588,0	868492
KJK	470	81723,3	1074062	SNG	883	463206,3	3138579
ZMP	768	88833,8	354742	UKB	723	476147,5	2048537
LDJ	893	93065,5	454560	RSU	867	563529,8	4198952
RST	940	103413,4	167984	LFO	577	649271,5	9034340
UTR	521	118596,5	1424941	TEO	964	699757,3	1104185

Fig. 2. Summary statistics of stocks daily turnover

In the first group, seven equities had the lowest daily trading volume, falling into the interval from 14 to 51 thousand LTL (1 EUR=3.45 LTL), with the average one day turnover equal to 34 511.3 LTL. The second group consisted of 10 equities with average trading volume from 81 to 228 thousand LTL per day, and mean value 129 719.2 LTL. The daily turnover of the third group varied from 375 to 700 thousand LTL, and had mean value within group of 527 152 LTL.

According to the average value of turnover, the differences between the groups were quite evident. Our goal was to check, if the equities, grouped according to different trading volumes, did not differ in their profitability, and if the day-of-the-week effect could be observed in any of groups.

Therefore, three stock portfolios were formed from the earlier described stock groups. For each portfolio, mean return was calculated and assigned correspondingly to the variables: I group, II group, III group.

The hypothesis for exploring the day-of-the week effect has been formulated as H_0: mean return distribution is identical for all days of the week.

The day-of the-week effect was analysed within the three portfolios by applying various statistical methods, starting from exploring differences of turnover and volatility, indicated by basic and higher moment's analysis.

The hypothesis was then checked by applying traditional t-test and one-way ANOVA statistical tests.

Var	Descriptive Statistics (Return)								
	N	Mean	Std. Dev.	Skew-ness	Kurto-sis	N	Mean	Std. Dev.	Skew-ness	Kurto-sis
	Mon day					Thurs day				
Return	194	0,055	0,905	-0,34	3,9	196	0,197	1,031	-0,52	3,2
I group	194	0,173	1,026	-0,04	1,1	196	0,054	1,113	-0,02	0,9
II group	194	0,001	0,684	0,41	1,7	196	0,142	0,659	0,34	1,0
III group	194	0,020	1,669	-8,71	103,5	196	0,010	2,655	-11,76	155,5
	Tues day					Fri day				
Return	199	0,151	0,943	0,15	2,6	194	0,205	0,891	0,88	4,4
I group	199	0,045	1,121	-0,33	1,2	194	0,084	0,980	-0,53	1,0
II group	199	0,160	0,742	0,09	1,2	194	0,177	0,660	-0,02	1,0
III group	199	0,093	0,885	0,01	1,2	194	-0,005	2,812	-12,54	168,7
	Wedn esday					All days				
Return	200	0,190	0,886	0,12	1,3	983	0,160	0,932	0,01	3,1
I group	200	0,137	1,036	0,11	0,4	983	0,098	1,056	-0,16	0,9
II group	200	0,144	1,353	-8,56	103,9	983	0,125	0,866	-6,47	124,0
III group	200	0,250	0,932	0,30	1,3	983	0,075	1,961	-14,09	252,4

Fig. 3. Day-of-the-week Summary Statistics by groups

In Fig. 3, the summary statistics for variable RETURN and group variables were evaluated for each day of the week. Here we can notice that only variables RETURN (market index) and II group show lowest trading average return for Monday and the highest trading average on Friday. These tendencies of the analysed emerging stock market are similar to the developed stock markets [5,9,13,18].

The portfolios with the smallest and highest daily turnover had hardly predictable mean return over the days of the week. According to the standard deviation analysis, the index variable RETURN, and mean return variables of I group and II group had no difference in volatility among the days of the week, except for the III group variable, which had the biggest volatility for almost all days of the week (Fig.3).

Similar tendencies could be observed by analyzing higher moments - skewness and kurtosis of mean return. Their values were most declined from zero (and therefore most different from the normal distribution) for the portfolios with high and medium daily turnover (II group, III group).

Firstly, we explored differences between mean returns of the days of the week with the help of traditional t-test and one-way ANOVA. The results of t-test revealed that statistically significant difference in profitability was indicated only for the stocks from II group, and only for Monday pair wise to other days of week. The difference between Monday and Friday mean returns shown in fig.4.

	T-tests; Grouping: Var2 (Return) Group 1: Monday Group 2: Friday				
Variable	Mean Monday	Mean Friday	t-value	df	p
RETURN	0,0550	0,2045	-1,6393	386	0,1020
I group	0,1727	0,0836	0,8745	386	0,3824
II group	0,0011	0,1768	-2,5767	386	0,0103
III group	0,0197	-0,0053	0,1066	386	0,9152

Fig. 4. T-test Results

The hypothesis of mean return identity was not rejected for any other pairs of days of the week of all portfolios. Application of t-test did not confirm day-of-the-week effect for OMX Vilnius Stock Index return (variable RETURN), for no pair of the days of the week. Similar results were obtained by applying analysis of variance (Fig. 5).

	Analysis of Variance (Return)					
Variable	df Effect	MS Effect	df Error	MS Error	F	p
RETURN	4	0,745	978	0,870	0,857	0,489
I group	4	0,591	978	1,117	0,529	0,715
II group	4	0,968	978	0,749	1,293	0,271
III group	4	2,222	978	3,851	0,577	0,679

Fig. 5. ANOVA Results

None of the portfolios indicated significant difference between the days of the week. Application of F criteria did not allow us reject the hypothesis for none of the four analyzed variables.

From the above presented results it was concluded that the daily trade volume had no significant influence for occurrence of day-of-the-week effect for the portfolios, formed of stocks with different levels of turnovers. The significant mean return difference was observed between Monday and other days of the week only for the case of variable II group. The turnover rate influenced the liquidity of the stock, and at the same time increased volatility and risk.

Further, Kolmogorov-Smirnov test was applied for testing hypothesis stating that two samples were drawn from the same population. This test is generally sensitive to the shapes of the distributions of the two samples, denoted by dispersion, skewness, kurtosis and other characteristics. It is generally applied for testing the influence of higher moments for the distribution [7,15]. Kolmogorov-Smirnov test was applied to investigate the day-of-the-week effect for the portfolios, composed of stocks with different daily turnovers. By applying test to all possible pairs of the days, it was defined, that only the variable II group had significant difference in mean return for Monday comparing to other days of the week.

Variable	Kolmogorov-Smirnov Test(Return in vol) By variable Var2 Marked tests are significant at p <,05000				
	Max Ne Differnc	Max Po Differnc	p-level	Mean Monday	Mean Wed-day
RETURN	-0,111	0,013	p > .10	0,0550	0,1897
I group	-0,026	0,074	p > .10	0,1727	0,1368
II group	-0,174	0,005	p < .01	0,0011	0,1437
III group	-0,118	0,019	p > .10	0,0197	0,2505

Fig. 6. Kolmogorov-Smirnov test results

In the Fig. 6, the Kolmogorov-Smirnov test values for illustrating difference between Monday and Wednesday mean return distributions are presented.

3 Application of Neural Networks Methods

In this part the results of application of artificial neural networks are presented. They were used to explore presence of the day-of-the-week effect for the same data set of the 24 equities from the list of Vilnius Stock Exchange, as described in Part 2. The computational tool STATISTICA (ST) Neural Networks standard module was used. The classification was performed by using nominal output variables-day of the week. In our research applied two standard types of neural networks, implemented in this software: MLP (Multilayer Perceptrons) and RBF (Radial Basis Function Networks), positively evaluated for their good performance of classification tasks [16].

These two types of networks have some different features. For the linear analysis the MLP units are defined by their weights and threshold values, which provide equation of the defining line, and the rate of falloff from that line [16]. A radial basis function network (RBF), has a hidden layer of radial units, each modelling a Gaussian response surface.

In this research both types of neural networks, MLP and RBF, were used to distinct Monday and Friday from the other trading days of the week. To make the results comparable to those, obtained by statistical analysis, mean return values were used as the main indicator. Moreover, additional input variables were used for classification, such as number of deals and shares, turnover, and H-L (high minus low price). For training of the network we used from 700 to 1000 items of data, accordingly to each investigated security. The non-trading days of the particular securities were eliminated from the data sets. Further, the detailed steps of the data processing procedure by applying neural networks for one of the securities (TEO, Fig.7) are presented. The research was aimed to distinct Monday from other days of the week. The same procedure was applied for each of the 24 securities of the data set.

	1 DEALS	2 NO_OF_SH	3 TURNOVER	4 DAY	5 RETURN	6 H_L
2003.01.06	1	3000	2790	Monday	-3,175	0,00
2003.01.07	11	46112	44268	Other	3,175	0,00
2003.01.08	7	452159	427048	Other	-1,047	0,02
2003.01.09	10	41673	39273	Other	-1,058	0,01
2003.01.13	17	131765	121224	Monday	-1,081	0,00
2003.01.14	4	123000	114390	Other	1,081	0,00
2003.01.15	4	1025	953	Other	0,000	0,00
2003.01.16	8	223455	210053	Other	2,128	0,02
2003.01.17	9	15580	14760	Other	-1,058	0,01
2003.01.21	23	421665	392208	Other	-2,174	0,02
2003.01.22	18	90652	81458	Other	-2,222	0,03
2003.01.23	21	366545	329560	Other	-1,130	0,00
2003.01.24	6	39969	34873	Other	0,000	0,01
2003.01.27	14	38537	33509	Monday	-2,299	0,01
2003.01.28	11	46119	39522	Other	-1,170	0,01
2003.01.29	16	314980	277513	Other	-2,381	0,03

Fig. 7. TEO securities data set segment used in calculations

The performance of the best-found network (according to the correct classification rate, error, area under ROC curve) is presented in the report, generated by the Intelligent Problem Solver (Fig. 8).

The report (Fig. 8) informed, that it took 3:43 min for finding the best classification algorithm, which resulted in the improved MLP (3 layer) network specified as

3:3-7-1:1 (3 input variables, 7 hidden units and 1 output variable), and the error level of 0.36817. The error was calculated as the root mean square (RMS) of the errors, measured for each individual case. While performing the ST Neural Networks training algorithms attempted to minimize the training error, the final performance measure indicated rate of correctly classified cases. This result depends on the Accept and Reject thresholds (confidence limits).

In the Fig. 8, the correct classification rate is 0.623377, therefore 62.33% cases were correctly classified. The Statistica (ST) Neural Networks software has a function of assigning the network performance to the categories 'poor' (rate is less than 0.6), 'O.K' and 'High' (over 0.8), according to the achieved correct classification rate.

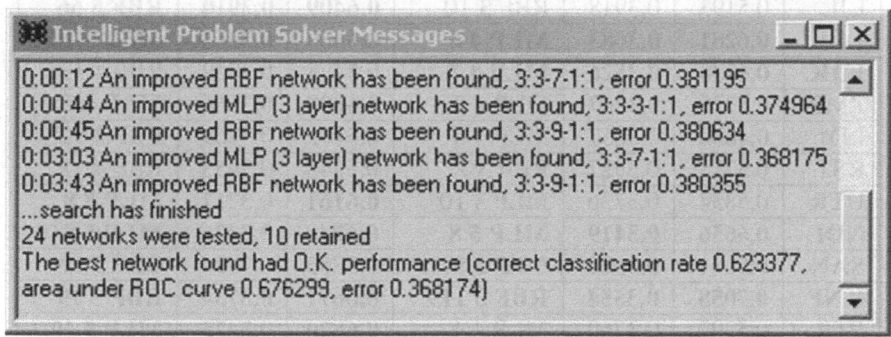

Fig. 8. Intelligent Problem Solver report of TEO equities Monday effect

The outcome of analysis for security (TEO) confirmed that it was influenced by the Monday effect. The constructed neural network distinct Mondays from the other trading days of the week with the sufficient reliability (we assumed classification performance as significant, if it was higher, than 0.6).

As the correct classification rate highly exceeds wrongly classified data, the Monday effect for TEO security is confirmed.

The above-described data processing and classification procedure for all the 24 securities is summarized in Table 1.

From the summary of analysis (Table 1) we concluded that Monday effect was present for only 11 equities, and Friday effect was present for 9 equities. Both effects were present for 4 equities. The analysis of the performance of neural networks (MLP versus RBF) showed, that the best results for indicating Monday effect were achieved by using MLP (18 times), and the RBF performed better for 6 times. The Friday effect was revealed with the same precision by both types of neural networks (12 times each - MLP and RBF). The neural networks MLP and RDF used similar number of input variables, but the number of elements in hidden layer for RBF was much bigger.

Comparing the performance of the methods of neural networks and general linear statistical analysis revealed the advantages of the neural computations, which confirmed the presence of the day-of-the-week effect in 20 cases, whereas application of Kolmogorov-Smirnov test could show significant indications only for 3 such cases.

Table 1. Results of application of neural network for stock trading data (the presence of Monday and Friday effect is marked by bold font)

	Monday			Friday		
	Perfor-mance	Error	Network Input Hidd.	Perfor-mance	Error	Network Input Hidd.
LEL	0,5623	0,3909	MLP 2 6	0,5770	0,3849	MLP 3 6
KBL	**0,6501**	**0,3590**	**MLP 5 10**	0,5214	0,3994	RBF 4 9
LNS	**0,6066**	**0,3656**	**MLP 5 7**	0,5436	0,3961	MLP 3 3
LEN	0,5576	0,3815	MLP 3 6	**0,7291**	**0,3590**	**RBF 5 180**
VBL	0,5943	0,3640	MLP 5 6	0,5814	0,3870	MLP 4 6
LJL	0,5193	0,3918	RBF 3 10	**0,6409**	**0,3910**	**RBF 5 66**
LLK	**0,6281**	**0,3683**	**MLP 4 6**	0,6638	0,3563	MLP 5 10
KJK	**0,6738**	**0,3526**	**MLP 4 7**	0,5778	0,3829	RBF 4 13
ZMP	0,5776	0,3807	MLP 4 7	**0,6007**	**0,3566**	**MLP 4 7**
LDJ	0,4850	0,3857	MLP 4 6	0,5795	0,3869	RBF 3 12
RST	0,5713	0,3927	RBF 3 7	0,5245	0,3953	RBF 4 12
UTR	0,5854	0,3736	MLP 3 10	**0,6161**	**0,3717**	**MLP 5 8**
NDL	**0,6656**	**0,3419**	**MLP 5 8**	**0,6748**	**0,3426**	**MLP 5 8**
SAN	0,5671	0,3682	MLP 5 8	0,5855	0,3924	RBF 3 19
KNF	**0,7058**	**0,3554**	**RBF 3 119**	**0,6671**	**0,3734**	**RBF 3 79**
PTR	0,5707	0,3760	MLP 4 4	**0,6820**	**0,3475**	**MLP 5 10**
PZV	**0,6282**	**0,3639**	**MLP 5 11**	0,5734	0,3953	RBF 3 18
APG	**0,6453**	**0,3633**	**MLP 5 11**	0,5668	0,3964	RBF 2 13
MNF	**0,6346**	**0,3622**	**MLP 4 8**	0,5569	0,3916	RBF3 13
SNG	0,5571	0,3936	RBF 2 7	0,5447	0,3951	RBF 2 9
UKB	0,5795	0,3818	RBF 3 17	0,5906	0,3767	MLP 4 11
RSU	0,5709	0,3828	MLP 3 5	0,5456	0,3918	MLP 4 5
LFO	**0,6583**	**0,3720**	**RBF 4 34**	**0,6761**	**0,3480**	**MLP 6 10**
TEO	**0,6234**	**0,3682**	**MLP 3 7**	0,5563	0,3912	MLP 3 4

4 Summary and Conclusion

In this research the day-of-the-week effect was explored for the case of emerging stock market. Although the analysis of the developed stock markets in the scientific literature substantiated the diminishing of this effect due to market development, still there is insufficient scientific research of the emerging markets from this point of view. The day-of-the-week effect was explored for the index variable RETURN, consisting from Vilnius Stock Exchange OMX Index daily returns values, as the index variable is the most common object for analysis in the scientific literature. Three data sets of mean return of 24 Vilnius Stock Exchange equities were formed by using daily turnover values. They were assigned to variables I group (low turnover), II group (medium turnover) and III group (high turnover).

By applying traditional statistical research methods and analysis of the higher moments, the statistically significant difference among Monday to the other days of the week was observed only for II group variable, which consisted of medium turnover stocks. Yet there was no significant influence of trading volume for occurrence of the day-of-the-week effect for the mixed stocks with different turnover and according to the index value. This can lead to the conclusion that the statistical research methods are less effective for exploring markets without grouping stocks according to their turnover. The day-of-the-week effect and possibly other anomalies, which could result in including them into profitable trading strategies are difficult to notice, unless we can identify groups of stocks, where the explored effect is stronger expressed.

The investigation of the day-of-the-week effect by applying the artificial neural network methods was based on analysis of numerous variables: return, number of deals and shares, turnover, H-L (high minus low price). By applying ST Neural Network software the best classifying neural networks were selected, yet no preference could be given for MLP and RBF neural networks due to their similar performance. Neural networks were more effective for revealing Monday and Friday effect, as they discovered more stocks, affected by day of the week influence, comparing to traditional statistical analysis.

The research results helped to conclude the effectiveness of application of neural networks, as compared to the traditional linear statistical methods for such type of classification problem, where the effect is vaguely expressed and its presence is difficult to confirm. The effectiveness of the method has been confirmed by exploring variables, influencing the day-of-the-week effect.

References

1. Balaban, E., Bayar, A., Kan, O.B.: Stock returns, seasonality and asymmetric conditional volatility in World Equity Markets. Applied Economics Letters 8, 263–268 (2001)
2. Basher, S.A., Sadorsky, P.: Day-of-the-week effects in emerging stock markets. Applied Economics Letters 13, 621–628 (2006)
3. Brooks, C., Persand, G.: Seasonality in Southeast Asian stock markets: some new evidence on day-of-the-week effects. Applied Economics Letters 8, 155–158 (2001)
4. Connolly, R.A.: An examination of the robustness of the weekend effect. Journal of Financial and Quantitative Analysis 24, 133–169 (1989)
5. Cross, F.: The behaviour of stock prices on Fridays and Mondays. Financ. Anal. J. 29, 67–69 (1973)
6. Flannery, M.J., Protopapadakis, A.A.: From T-bills to common stocks: investigating the generality of intra-week return seasonality. Journal of Finance 43, 431–450 (1988)
7. Galai, D., Kedar-Levy, H.: Day-of-the-week Effect in high Moments, Financial Markets. Institutions & Instruments 14(3), 169–186 (2005)
8. Gencay, R.: The predictability of security returns with simple technical trading. Journal of Empirical Finance 5, 347–359 (1998)
9. Jaffe, J.F., Westerfield, R., Ma, C.: A twist on the Monday effect in stock prices: evidence from the U.S. and foreign stock markets. Journal of Banking and Finance 13, 641–650 (1989)
10. Kamath, R., Chusanachoti, J.: An investigation of the day-of-the-week effect in Korea: has the anomalous effect vanished in the 1990s? International Journal of Business 7, 47–62 (2002)

11. Kohers, G., Kohers, N., Pandey, V., Kohers, T.: The disappearing day-of-the-week effect in the world's largest equity markets. Applied Economics Letters 11, 167–171 (2004)
12. Kumar, M., Thenmozhi, M.: Forecasting Nifty Index Futures Returns using Neural Network and ARIMA Models. Financial Engineering and Applications (2004)
13. Mills, T.C., Coutts, J.A.: Calendar effects in the London Stock Exchange FTSE indices. The European Journal of Finance 1, 79–93 (1995)
14. Qi, M.: Nonlinear predictability of stock returns using financial and economic variables. Journal of Business and Economic Statistics 17, 419–429 (1999)
15. Reschenhofer, E.: Unexpected Features of Financial Time Series: Higher-Order Anomalies and Predictability. Journal of Data Science 2, 1–15 (2004)
16. StatSoft Inc. Electronic Statistics Textbook. StatSoft, Tulsa, OK (2006), web: http://www.statsoft.com/textbook/stathome.html
17. Steeley, J.M.: A note on information seasonality and the disappearance of the weekend effect in the UK stock market. Journal of Banking and Finance 25, 1941–1956 (2001)
18. Syed, A.B., Sadorsky, P.: Day-of-the-week effects in emerging stock markets. Applied Economics Letters 13, 621–628 (2006)
19. Tang, G.Y.N.: Day-of-the-week effect on skewness and kurtosis: a direct test and portfolio effect. The European Journal of Finance 2, 333–351 (1998)
20. The Nordic Exchange (2006), http://www.baltic.omxgroup.com/

A Metamorphosis Algorithm for the Optimization of a Multi-node OLAP System

Jorge Loureiro[1] and Orlando Belo[2]

[1] Departamento de Informática, Escola Superior de Tecnologia, Instituto Politécnico de Viseu,
Campus Politécnico de Repeses, 3505-510 Viseu, Portugal
jloureiro@di.estv.ipv.pt
[2] Departamento de Informática, Escola de Engenharia, Universidade do Minho
Campus de Gualtar, 4710-057 Braga, Portugal
obelo@di.uminho.pt

Abstract. In OLAP, the materialization of multidimensional structures is a *sine qua non* condition of performance. Problems that come along with this need have triggered a huge variety of proposals: the picking of the optimal set of aggregation combinations, to materialize into centralized OLAP repositories, emerges among them. This selection is based on general purpose combinatorial optimization algorithms, such as greedy, evolutionary, swarm and randomizing approaches. Only recently, the distributed approach has come to stage, introducing another source of complexity: space. Now, it's not enough to select the appropriate data structures, but also to know where to locate them. To solve this extended problem, optimizing heuristics are faced with extra complexity, hardening its search for solutions. This paper presents a polymorphic algorithm, coined as metamorphosis algorithm that combines genetic, particle swarm and hill climbing metaheuristics. It is used to solve the extended cube selection and allocation problem generated in M-OLAP architectures.

1 Introduction

On-Line Analytical Processing (OLAP) addresses the special requirements of today's business user queries, allowing the multidimensional vision and aiming a fast query answering, independently of the aggregation level of the required information. In fact, speed is increasingly important in decision making: today it is not enough to deliver the required information and under adequate shape, it must be given fast. This latter characteristic implies the pre-computation and materialization of aggregated queries' answers, denoted as materialized views or subcubes. But the number of those structures is very huge (implying high storage space and maintenance strain), and the number of users also increases. All these factors impose a great stress over the OLAP server and some solutions must be intended. Firstly, the number of subcubes to materialize can be limited, as many of them are simply of no use. But, selecting the most beneficial is not easy, constituting a recognizably NP-hard [4] combinatorial problem, dealt by approximate solution, mainly using greedy, evolutionary, and recently, particle swarm approaches. Secondly, the old "divide ut imperes" maxim may also be

J. Neves, M. Santos, and J. Machado (Eds.): EPIA 2007, LNAI 4874, pp. 383–394, 2007.

used. We may distribute the materialized structures by several hardware platforms, trying to gain the known advantages of database distribution: a sustained growth of processing capacity (easy scalability) without an exponential increase of costs, and an increased availability of the system, as it eliminates the dependence from a single source and avoidance of bottlenecks. And this distribution may be achieved in different ways. Here, we focus in one of them: distributing the OLAP cube by several nodes, inhabiting in close or remote sites, interconnected by communication links, generating a multi-node OLAP approach (M-OLAP). The cube views selection problem is now extended, as we have a new dimension: space. It's not enough to select the subcubes; they also have to be conveniently located. In the distributed scenery we have a number n of storage and processing nodes, named OLAP server nodes (OSN), with a known processing power and storage space, interconnected by a network, being able to share data or redirecting queries to other nodes. The problem gets bigger if the distribution is disposed as an arbitrary network: not only a single level, but a topology that may be shaped as any graph, where nodes of great storage capacity, known as static (maintained incrementally), coexist with many others, eminently dynamic and smaller, whose maintenance may be done by complete recomputing of subcubes from scratch. All components are connected with a network and the services may be distributed, being configured by the system's administrator.

The distributed scenario increases the complexity of the cube selection problem: some solutions of the centralized approach might be intended, but now have to be extended, in order to be able to deal with the dimensional feature of the new architecture. Typical (greedy) approaches (e.g. [4], [3], [7]) were extended in [1] to the distributed data warehouse. The authors introduced the distributed aggregation lattice and proposed a distributed node set greedy algorithm that addressed the distributed view selection problem, being shown that this algorithms has a superior performance than the corresponding standard greedy algorithm, using a benefit per unit space metric. However they didn't include maintenance cost into the general optimization cost goal and also didn't include communication and node processing power parameters into the cost formulas. Moreover genetic approaches, common in centralized OLAP (e.g. in [14], [8]) were proposed. Also, a genetic co-evolutionary approach is applied to the selection and allocation of cubes in M-OLAP systems, where the genotype of each specie is mapped to the subcubes to materialize in each node. This work uses the distributed lattice framework, but introduces a cost model that includes real communication cost parameters, using time as cost unit and cost estimation algorithms which use the intrinsic parallel nature of the distributed OLAP architecture [10].

A recent optimization method, the discrete particle swarm (DiPSO) [5] was also used to centralized OLAP approaches [9]. This optimizing method was also extended to M-OLAP system, and the reported tests' results show that it has a good scalability, supporting easily an M-OLAP architecture with several nodes and also suited for OLAP medium or high dimensionalities.

The idea of hybrids is an old genetic approach: building a specie which combines the virtues of two others trying to limit their demerits is becoming successfully applied, in many different areas, e.g. production of new vegetables species (corn, wheat, soya). This same idea has been pursued in a new class of combinatorial optimization problems: hybrid genetic and memetic algorithms [11]. Genetic hybridization has been proposed also in cube selection domain, mainly using a genetic algorithm combined with a local search approach [14] But a similar yet different approach is also common in nature: many species assume different shapes in its phenotype in different life epochs or under different environmental conditions: between each phenotypic appearance many transformations occur, during the so called metamorphosis, which, many times, generates a totally different living being (recall the similarities between a caterpillar and a butterfly, passing through a puppet phase). Assuming totally different shapes seems to be a way for the living being to better globally adapt, changing its ways to a particular sub-purpose. E.g. for the butterflies, the larva state seems to be ideal for feeding and growing; the butterfly seems to be perfectly adapted to offspring generating, especially increasing the probability of diverse mating and colonization of completely new environments (recall the spatial range that a worm could run and the space that a butterfly could reach and search).

Having used genetic and swarm algorithms, we gained some insights about each one's virtues and limitations that were somewhat disjoint, and then, when combined, may generate a globally better algorithm. And this schema could be easily implemented as a unified algorithm because of the similarities between the solutions' evaluation scheme and the easy transposing of solutions' mapping. In [6], a life cycle algorithm model was proposed that is, in its core essence, the metamodel of the algorithm proposed here. The underlying idea is combining the features of several algorithms, selecting a mix which may counterbalance the restrictions of one with the virtues of some other into the mix. This way, it's expectable that the aggregate performance will surpass the intrinsic performance of any of the participant algorithms.

This work proposes an algorithm that we coined as MetaMorph M-OLAP, based on the combining of genetic, particle swarm and hill-climbing with simulated annealing, showing how this particular optimization approach was applied to the distributed cube selection problem. Now, the optimizing aim is minimizing the query and maintenance cost, observing a space constraint concerning to an M-OLAP architecture.

2 Hybrids, Memetic and Multi-shape Algorithms

All these algorithms, although different, have in common the fact that they are combinations or mixes of some others. The former two are somewhat similar, using the combination of two search strategies, being sometimes referred as genetic local search or even hybrid evolutionary algorithms, without explicitly distinguishing them from other methods of hybridization. They both use search strategies simultaneously (in the same main iteration): all search agents (SA) use one strategy and then the other, a sequence that was repeated over and over again.

The multi-shape algorithms try to mimic the lifecycle of individuals from birth to maturity and reproduction. And all theses morphologies and behaviors of the phenotype manifest within the same genome, trying to adapt to a particular environment or to a particular objective. The transposition of this strategy to optimizing algorithms creates a self-adaptive search heuristic in which each individual (containing the candidate solution) can decide whether it would prefer to assume different algorithms' shapes. In [6], the authors proposed a three algorithms life cycle (whose structure is shown in Algorithm 1) where each SA can be a genetic individual of a population of a genetic algorithm (GA), a swarm particle of a particle swarm optimization (PSO) algorithm, or a hill climber, as each of these search strategies has particular strengths but also individual weaknesses. But this is not a question of the number of algorithms that are used: the way how they operate is also different. The multi-shape algorithm has an operating granularity at a finer level: each component algorithm operates not for all population at one turn, but at SA level. Each participant algorithm (with its population) may be active in any iteration. Each SA may decide, by itself, when the metamorphose has to happen.

This is a model and focus shift: hybrids algorithms have a population of solutions that are the subject of action of the participant algorithms by turn; in multi-shape algorithms, the focus is the SA (not the whole population) and the participant algorithms don't act over the SA, but it assumes each new personality, and what is the most, each SA decides, by itself, when it has to suffer the metamorphosis.

Algorithm 1. Multi-Shape Algorithm (adapted from [6]).

```
Multi-Shape Algorithm
1. Initialization:
     Generate a swarm of particles //the SA begins as a particle
2. Main LifeCycling Iteration:
     While (not terminate-condition) Do:
       For Each search_agent In population, Do:
         Evaluate fitness(search_agent);
         Switch LifeCycle Stage if no recent improvement;
       For (PSO search _agents) Do:
         Move particles: apply part.'dynamics rules/eq.
       For (GA search_agents) Do:
         Select new population;
         Generate offspring;
         Mutate population;
       For (HillClimbers search_agents) Do:
         Find possible new neighboring solution;
         Evaluate fitness of new solution;
         Shift to new solution with probability p;
     End While
3. Return the result
```

3 M-OLAP Cube Selection and Metamorphosis Algorithm

The schema of an M-OLAP architecture is simple: we have several OLAP Server nodes (OSN) and a base node where the base relations inhabits, e.g. a data warehouse, connected through a communication network.

This way, beyond the traditional intra-node subcubes' dependencies (the arrows of the OLAP lattice [4]), we also have inter-node dependencies because a subcube, in a particular OSN, may be also computed using some other subcubes located into a different OSN, incurring into an additional cost, corresponding to the transport of the generated subcube between nodes. In this work we used a cost model and corresponding cost estimation formulas, corresponding serial and parallel query cost estimation algorithm (SQA and PQA) [10] and also a parallel maintenance cost estimation algorithm (2PHSMPP) [10].

3.1 Problem Definition

M-OLAP Cube selection problem is an extension of the traditional cube selection problem. As stated before, now we deal with a new factor: space. The communication facility allows to gain the advantages of database distribution but with additional costs (communications) and also an increase in the management of data complexity.

Definition 1. Selection and allocation of distributed M problem. Let $Q=\{q_1, ..., q_n\}$ be a set of queries with access frequencies $\{fq_1, ..., fq_n\}$ and query extension $\{qe_1,...,qe_n\}$; let update frequency and extension be $\{fu_1, ..., fu_n\}$ and $\{ue_1,...,ue_n\}$, respectively, and let S_{Ni} be the amount of materializing space by OLAP node i. A solution to the selection and allocation problem is a set of subcubes $M=\{s_1,...,s_n\}$ with a $\sum_j |s_{jN_i}| \leq S_{N_i}$, so that the total costs of answering all queries Q and maintaining M, $Cq(Q, M)+Cm(M)$ are minimal.

3.2 Metamorphosis Algorithm

The algorithm, whose acronym is MetaMorph M-OLAP, is derived directly from Algorithm 1. A pictorial scheme of metamorphosis algorithm is shown in Fig. 1. The circular arrows symbolize the metamorphoses that are triggered by an impasse condition: when the current form of the SA, for a specified number of iterations, does not achieve any further solution's improvement, it must suffer a metamorphosis, assuming another form. This way, the current population of SAs, in a moment, is a mix of any of the possible type: a swarm particle (SP-SA), a GA individual (GA-SA) or a hill climber (HC-SA).

Here, we used a normal version of Discrete Particle Swarm Algorithm (DiPSO) [5] and a standard and co-evolutionary version of GA (Coev-GA). Concerning to hill climbing (HC), we used a population of hill climbers with a simulated annealing metaheuristic, using a scheme for moving them, where, in each move, a random number of bits (of its position in the n-dimensional search space) is allowed to be reset (from 1 to 0) and a random number may be set (passing from 0 to 1).

The range of the number of bits to be reset or set is a parameter which is linearly decreased with the number of iterations. Also a simulated annealing move accepting is used, allowing the hill climber to escape from local optima, thus the final acronym HC-SA^2. If the current position of a HC-SA^2 was x_c and a new position x_n has been selected from its neighborhood, a move to a worst x_n will be accepted with a probability: $p = \exp(\frac{fitness(x_n) - fitness(x_c)}{T})$, not totally denying up-hill moves, allowing the escape from local minima.

Inevitably, the flight of the particles, the mating and mutation of individuals or the hill climber's rumble generate genomes or positions, whose M solution counterpart violates some imposed constraint. Some solutions may be intended: 1) simply rejecting the "bad" solution, reprocessing the task that generate the unviable state; 2) applying a penalty to the fitness of the invalid SA; 3) repair the SA, repositioning it in an allowed space region. Here we adopted a solution of type 3, simply picking randomly

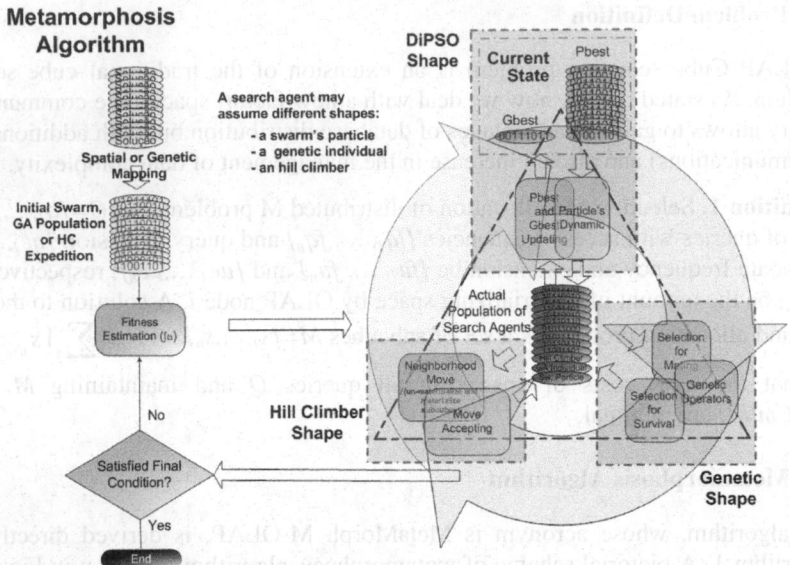

Fig. 1. Metamorphosis algorithm: the searching agent changes its appearance and behavior turning into a swarm's particle, a genetic individual or a hill climber

successive subcubes in M to un-materialize until the constraint has been met. Also a solution of type 1 was used, when the Hamming distance of the new solution was low (e.g. in mutation). Other solutions were tried (e.g. an inverse greedy heuristic to select the subcube to un-materialize) and also a type 2 solution, but their results weren't good enough.

3.3 Problem Mapping

In each OSN of M-OLAP architecture inhabits a cube lattice: the possible M (materialized set) has a number of subcubes $nS=n.Ls$, where n is the number of OSNs and Ls the number of subcubes into the lattice. The materialization of all these subcubes is, usually, neither possible nor convenient (recall the definition 1). Then, a given subcube may be materialized (1) or non-materialized (0).

The mapping to the standard GA is straightforward: the binary string that composes the genotype is mapped to the subcubes into all OSNs. E.g. for a 3 OSNs M-OLAP architecture ($n=3$), with a lattice with 8 subcubes ($Ls=8$), a genome (10011000 00011111 10010001) means that subcubes S0, S3 and S4 of OSN1, S3, S4, S5, S6 and S7 of OSN2 and S0, S3 and S7 of OSN3 are materialized. The mapping in the co-evolutionary version is similar: each specie maps to the materialized set of each OSN.

Concerning to DiPSO, we have a spatial mapping paradigm: each particle, in a given dimension, may be in a (1) or (0) position. Then, mapping each dimension to each possible subcube, we may consider that, when the particle is in a 0 position, the corresponding subcube in M is not materialized and reciprocally. We must have a DiPSO space of nS dimensions (where nS is computed as in GA mapping) and then, each particle, flying into this nS dimensional space, may propose any possible M materializing solution. E.g. for the same 3 OSNs M-OLAP architecture, if the particle was in a (0101101 10011000 01100010) position, this means that subcubes S1,S3,S4 and S7 of OSN1, S0,S3,S4 of OSN2, and S1,S2,S6 of OSN3 are materialized.

Finally, the same DiPSO space paradigm may be used in hill climbing: each hill climber may ramble into a nS dimensional space. But this ramble is neighborhood limited: as said in section 3.2, in each move, a usually low number of 1 positions may pass to 0, and reciprocally. Another vision, which resembles more the way how the algorithm runs, is, perhaps, more intuitive: each hill climber has a rucksack that contains the materialized subcubes M. Being M the materialized set of a given iteration (the ones contained into the rucksack), a number nOut=rnd(low_limit_D, up_limit_D) of subcubes is randomly selected to be taken off the rucksack, being substituted by a number nIn=rnd(low_limit_C, up_limit_C) of subcubes not yet into the rucksack. Low_limit and up_limit of charging and discharging are parameter settings. These values are decreased for a value dD and dC with frequency Ist (number of search steps). This way, the level of exploration may be higher in the beginning, being decreased progressively. Also T (the temperature control parameter), which controls the annealing process, is decreased of a ΔT amount after tfT steps, favoring exploration in initial search phases, and the exploitation in the final search steps.

Summarizing, in all three SAs states, a solution is coded as a binary string, which is interesting, as this facilitates the metamorphose process.

3.4 Metamorphose Description

When a SA decides to metamorphose, a paradigm shift happens: the support data and behavior change. All data that supports the state of a given trait must migrate to corresponding data structures of the next manifestation. The essence (its solution) must be retained, passing its legacy to the newer body. Although all three SAs states have a similar problem coding, there are some features that might be solved. Concerning to SP-SA $\rightarrow$ GA-SA metamorphosis, knowing the performance dependency of genetic algorithms with the size of the population, needing a higher number than PSO, we adopted a transition of one SP-SA into n GA-SAs, or naming that multiple relation as multiplicity, we adopted a multiplicity of n. In fact, in past experiments, we have found that values of n=5, 20 particles and 100 individuals, were good settings (this same 1-to-many solution was referred in [6], as future work proposal). The opposite problem happens when a GA-SA has to turn into a HC-SA^2 (the next cycle metamorphose). This time, n GA-SAs must origin a HC-SA^2. When each GA-SA decides to metamorphose we put it into a staging area of n places. When full, the conversion process is triggered that: 1) uses a selection scheme to elect the GA-SA to survive, the one to transform into a HC-SA^2; 2) adopts a majority rule: the HC-SA^2 string will be the result of a majority referring to the value of each bit, e.g. for n=5, if we have (1001) (1100) (1011) (0100) (0110) the result string will be (1100), object of a low probability possible repairing. Finally, concerning to the HC-SA^2 $\rightarrow$ SP-SA metamorphose, as they both use the same solution space paradigm, the initial position of the SP-SA is the last position of HC-SA^2. This way, we only have to generate the velocity vector (that is lost in SP-SA $\rightarrow$ GA-SA metamorphosis), generating one which is consistent with the particle's position.

4 Experimental Evaluation

This work has the main purpose of gaining some insights about the viability of applying the algorithm that is proposed here to the distributed OLAP cube selection problem, especially about its performance and scalability. To perform the experimental study of the algorithms we used the test set of Benchmark's TPC-R [12], selecting the smallest database (1 GB), from which we used 3 dimensions (customer, product and supplier). To broaden the variety of subcubes, we added some other attributes to each dimension, generating hierarchies, as follows: customer (c-n-r-all); product (p-t-all) and (p-s-all); supplier (s-n-r-all). With this, we generated a cube, whose subcubes and sizes may be found in [9]. Whenever the virtual subcube (base relations) is scanned, this has a cost three times the subcube of lower granularity. This is a simpler OLAP scheme, but a higher complexity case will be left to future work. We generated several query sets to simulate the query profile. Given the stochastic nature of the algorithm, each test was repeated 10 times, being taken its average.

The algorithm was developed in Java. A simplified class diagram is shown in Fig. 2. The Search_Agents class (a super-class) has four sub-types, corresponding to the three forms that the search agent may assume, plus the variation (in the GA). M_OLAP_State corresponds to the general distribution state and M allows to estimate the fitness of any M. Other classes were used to simulate the environment, parameter setting and input/output services, represented as subsystems in Fig. 2.

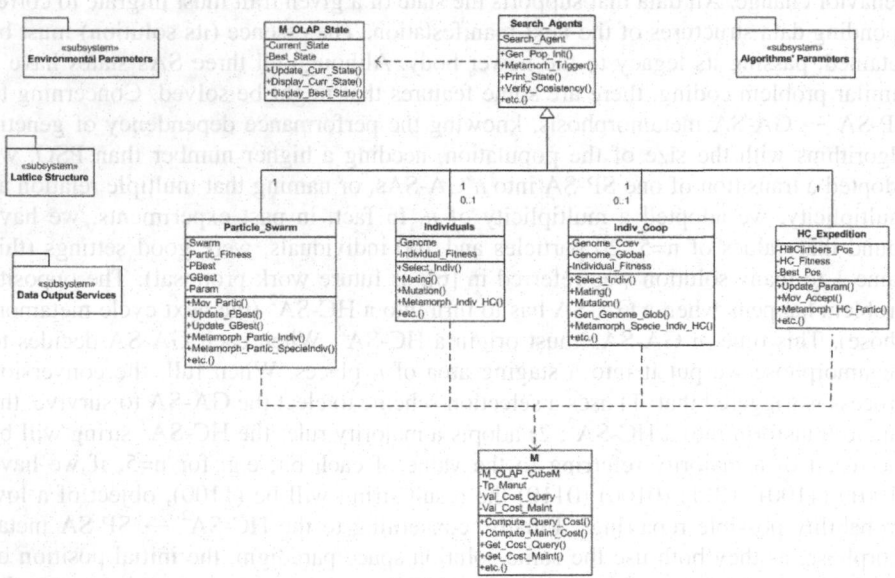

Fig. 2. MetaMorph M-OLAP algorithm's architecture

4.1 Algorithm's Parameters Settings

Table 1 shows the main settings of the parameters for the three component algorithms. Some of them were changed latter in particular tests; each situation and the new parameters setting is referred, when appropriate.

Table 1. Main parameter settings of metamorphosis compound algorithms

Alg.	Parameters	Settings
DiPSO	*Vmax, w, c1 and c2* (parameters in equations of particles' dynamic control [5])	Vmax=10, w: [0.99,1.00], varying linearly with the number of iterations; $c1=c2=1.74$.
GA	*Selection type (S), cross-over type(CoT) and cross-over probability (CoP), % mutation (%M) and mutation probability(MP)*	S: Competition, with 85% probability of selecting the fittest individual for crossing; CoT=one point, randomly generated by generation; CoP=90%; %M=5% and MP=1/8 bits.
HC-SA^2	*low_limit_D, up_limit_D, low_limit_C, up_limit_C, dD,dC,tpGerPInit* (parameters described in section 3.3)	4, 9, 3, 8, 1, 1, random.

4.2 Test Set and Results

The test set was designed intending to analyze the impact of some control parameters over the performance of MetaMorph M-OLAP algorithm:

- the initial shape of the SAs population (which of them is the most beneficial one);
- the number of SAs;
- the number of iterations without fitness improvement of the SA that triggers the metamorphosis process, (I);
- the scalability concerning to the queries number and M-OLAP OSNs number.

For the first test we used a query set with 90 queries, randomly generated, and the GA standard version component. We ran the algorithm with the three possible initial shapes of SA's population, using a multiplicity of 5 and allowing 500 iterations. The results are shown in Fig. 3, plot a). As we can see, the initial SA's shape has a reduced impact on the quality of the solutions. However, a HC-SA initial shape seems to be the most beneficial, once it achieves better solutions faster. It was also observed that the HC-SA initial shape has also a positive impact over the run-time of the algorithm. It is also evident that, for this particular problem and environmental conditions, 200 iterations will be enough for the tests: beyond that iteration number, the algorithm has not achieved any significant further benefits.

The second test tries to evaluate the impact of the number of SAs onto the performance of the algorithm. We used maximal populations of 50, 100 and 200 SAs, what means that, for a maximal population of 100 ABs, we will have an initial population of 20 HC-SA^{2}s, and we may have a maximal population of 20 SP-SAs, 100 GA-SAs or 20 HC-SA^{2}s (denoted as 20/100/20). Fig. 3, plot b) shows the obtained results.

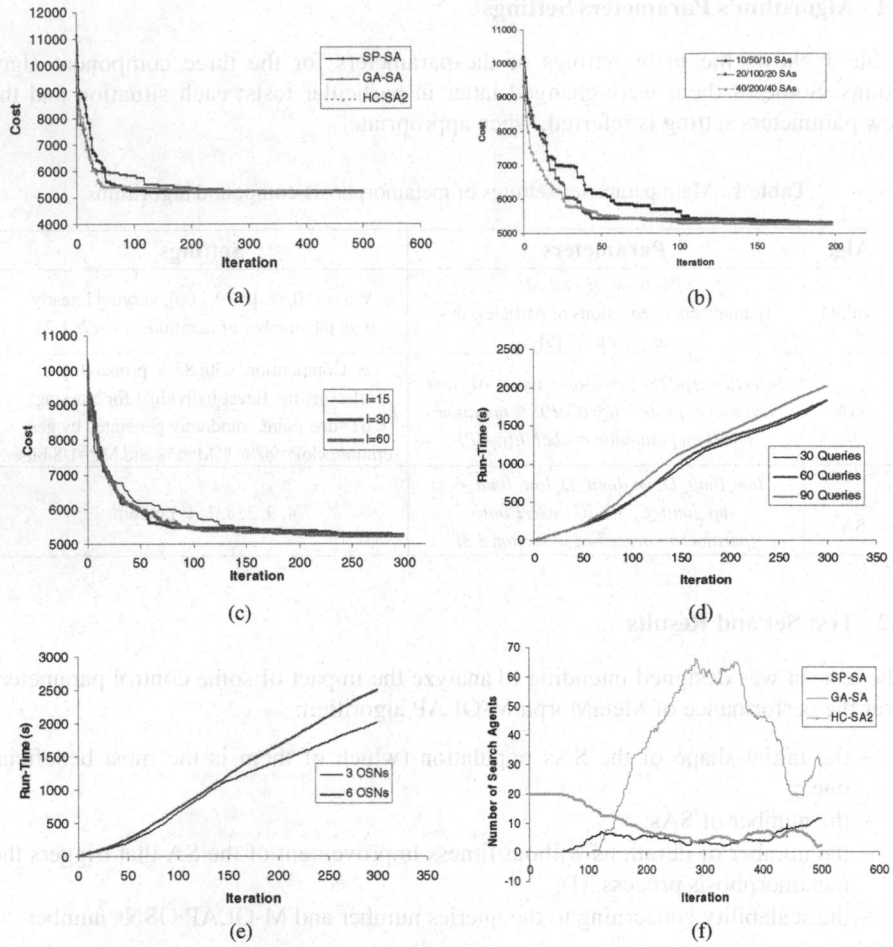

Fig. 3. Plots that show the results of the performed tests

As we can see, the number of the population of search agents has a limited impact on the quality of the solutions, although a larger population achieves good solutions faster. This behavior is possibly explained due to the positive impact that will have, in this particular issue, the forms SP-SA and HC-SA^2, whose quality of achieved solutions are relatively independent of the population size. It was observed also that the run-time of the algorithm grows at a lower rate than the ratio of the population's number for low populations (between 10/50/10 and 20/100/20); but the run-time has a high increase for the 40/200/40 population. This behavior may be explained possibly by the higher efficacy of GA-SA for higher individual populations, increasing the predominance of GA-SAs, and the genetic algorithm is also the most "expensive" of them all. Anyway, the quality *versus* population relation seems to bend to small populations, because the higher run-time (when using high populations) doesn't return an important improvement of the quality of the solutions. Populations of 20/100/20 seem to be a good trade-off to use for the rest of the tests.

The third test aims to evaluate the impact of I (the number of iterations without fitness improvement that triggers the metamorphosis of the SA). For I, we used the values 15, 30 and 60. The results of the performed tests are shown in the plot c) of Fig. 3. As we can see, this parameter seems to have a reduced impact on the algorithm. However, a detailed analysis showed that a value of I=30 or 60 seems to be a good choice for I.

The last test valuates the scalability of the MetaMorph M-OLAP concerning to the number of queries and the number of OSNs. We used 3 queries' set (with 30, 60 and 90 queries) and two OLAP architectures: the one with 3 OSNs and another with 6 OSNs. Plot d) and e) of Fig. 3 show the algorithm's run-time.

Analyzing plot d) we can see that the number of queries has a reduced impact (only a 16% increase when the queries number increases 3 times). Concerning to the number of OSNs, the plot e) shows that it also has a reduced impact on the run-time. A 2 times increase implied only a 20% increase of the run-time. Both evidences show that MetaMorph M-OLAP algorithm is scalable both for the queries' number and M-OLAP OSNs' number.

5 Conclusions and Future Work

This paper proposes an algorithm – the MetaMorph M-OLAP – for the selection and allocation of M-OLAP cubes. The algorithm not only provides a new method to solve this problem, but also constitutes a repository of metaheuristics, as each algorithm may be used by itself, being then the algorithm also a framework for M-OLAP cube selection metaheuristics. This is enforced by the design of the algorithm, which allows the easy inclusion of new heuristics. The preliminary tests' results seem to point out that MetaMorph M-OLAP seems to be a good choice for this problem. It has a great independence for the majority of the control parameters, showing a great adaptability. In fact, the population of each form of SAs seems to fluctuate according to the relative performance of each of the compound algorithms, as shown in plot f) of Fig. 3. This characteristic insures the adaptability of the algorithm in face of the diversity of the particular case that it has to solve.

For future work, we intend to pursue two research directions:

- We plan to perform a more exhaustive test of MetaMorph M-OLAP algorithm, particularly including a maintenance cost constraint (which is already supported by the component algorithms, as they only accept moves or offspring which do obey the constraint) comparing its performance with a greedy algorithm to be developed, possibly a version for M-OLAP architectures of two-phase greedy or integrated greedy [13], referred as having a good scalability.
- We also want to investigate more deeply the metamorphosis process, and to include in it: a) Diversification mechanisms, using a probabilistic way of transferring the solution from the SA present form to the next; changes to the metamorphosis cycle may also be attempted, substituting the inflexible cycle of Fig. 1 by a user defined or an adaptive mechanism. b) New heuristics into the Meta-Morph M-OLAP algorithm, as tabu search [2] (option already suggested in [6])

and ant colony optimization, another life inspired algorithm, which is turning
out to be a good optimization method.

Acknowledgements

The work of Jorge Loureiro was supported by a grant from PRODEP III, acção 5.3 –
Formação Avançada no Ensino Superior, Concurso N.° 02/PRODEP/2003.

References

1. Bauer, A., Lehner, W.: On Solving the View Selection Problem in Distributed Data Ware-house Architectures. In: SSDBM 2003. Proceedings of the 15th International Conference on Scientific and Statistical Database Management, pp. 43–51. IEEE, Los Alamitos (2003)
2. Glover, F., Laguna, M.: Tabu Search. Kluwer Academic Publishers, Dordrecht (1997)
3. Gupta, H., Mumick, I.S.: Selection of Views to Materialize under a Maintenance-Time Constraint. In: Proc. of the International Conference on Database Theory (1999)
4. Harinarayan, V., Rajaraman, A., Ullman, J.: Implementing Data Cubes Efficiently. In: Proceedings of ACM SIGMOD, Montreal, Canada, pp. 205–216 (1996)
5. Kennedy, J., Eberhart, R.C.: A Discrete Binary Version of the Particle Swarm Optimiza-tion Algorithm. In: SMC 1997. Proc. of the 1997 Conference on Systems, Man and Cyber-netics, pp. 4104–4109 (1997)
6. Krink, T., Løvbjer, M.: The LifeCycle Model: Combining Particle Swarm Optimization, Genetic Algorithms and HillClimbers. In: Guervós, J.J.M., Adamidis, P.A., Beyer, H.-G., Fernández-Villacañas, J.-L., Schwefel, H.-P. (eds.) PPSN VII. LNCS, vol. 2439, pp. 621–630. Springer, Heidelberg (2002)
7. Liang, W., Wang, H., Orlowska, M.E.: Materialized View Selection under the Mainte-nance Cost Constraint. Data and Knowledge Engineering 37(2), 203–216 (2001)
8. Lin, W.-Y., Kuo, I.-C.: A Genetic Selection Algorithm for OLAP Data Cubes. Knowledge and Information Systems 6(1), 83–102 (2004)
9. Loureiro, J., Belo, O.: A Discrete Particle Swarm Algorithm for OLAP Data Cube Selec-tion. In: ICEIS 2006. Proc. of the 8th International Conference on Enterprise Information Systems, Paphos – Cyprus, May 23-27, 2006, pp. 46–53 (2006)
10. Loureiro, J., Belo, O.: Establishing more Suitable Distributed Plans for MultiNode-OLAP Systems. In: Proceedings of the 2006 IEEE International Conference on Systems, Man, and Cybernetics, Taipei, Taiwan, October 8-11, 2006, pp. 3573–3578 (2006)
11. Moscato, P.: Memetic Algorithms: A Short Introduction. In: Corne, D., Dorigo, M., Glover, F. (eds.) New Ideas in Optimization, ch. 14, pp. 219–234. McGraw-Hill, London (1999)
12. Transaction Processing Performance Council (TPC): TPC Benchmark R (decision support) Standard Specification Revision 2.1.0. tpcr_2.1.0.pdf, available at http://www.tpc.org
13. Yu, J.X., Choi, C.-H., Gou, G., Lu, H.: Selecting Views with Maintenance Cost Con-straints: Issues, Heuristics and Performance. Journal of Research and Practice in Informa-tion Technology 36(2) (May 2004)
14. Zhang, C., Yao, X., Yang, J.: An Evolutionary Approach to Materialized Views Selection in a Data Warehouse Environment. IEEE Trans. on Systems, Man and Cybernetics, Part C 31(3) (2001)

Experiments for the Number of Clusters in K-Means

Mark Ming-Tso Chiang and Boris Mirkin

School of Computer Science & Information Systems, Birkbeck University of London,
London, UK
{mingtsoc,mirkin}@dcs.bbk.ac.uk

Abstract. K-means is one of the most popular data mining and unsupervised learning algorithms that solve the well known clustering problem. The procedure follows a simple and easy way to classify a given data set through a pre-specified number of clusters K, therefore the problem of determining "the right number of clusters" has attracted considerable interest. However, to the authors' knowledge, no experimental results of their comparison have been reported so far. This paper presents results of such a comparison involving eight selection options presenting four approaches. We generate data according to a Gaussian-mixture distribution with clusters' spread and spatial sizes variant. Most consistent results are shown by the least squares and least modules version of an intelligent version of the method, iK-Means by Mirkin [14]. However, the right K is reproduced best by the Hartigan's [5] method. This leads us to propose an adjusted iK-Means method, which performs well in the current experiment setting.

1 Introduction

K-Means, in its model-free version, arguably is the most popular clustering method now and in predictable future. This is why studying its properties is of interest not only to the classification, data mining and machine learning communities, but also to an increasing numbers of practitioners in business intelligence, bioinformatics, customer management, engineering and other application areas. The procedure follows a simple and easy way to classify a given data set through a certain number of clusters K, therefore the problem of determining "the right number of clusters" attracts considerable interest ([7], [14]). This paper focuses upon an experiment aiming at comparing of various options for selecting the number of clusters in K-Means (see [5], [7], [14]) and analysis of its results. The setting of our experiment is described in section 2. Section 3 presents all the clustering algorithms involved, including the Hartigan-adjusted iK-Means method. A Gaussian-mixture data generator is described in section 4. Our evaluation criteria are in section 5. The evaluation results are presented in section 6. The conclusion focuses on issues and future work.

2 Algorithm Descriptions

2.1 Generic K-Means

K-Means is an unsupervised clustering method that applies to a data set represented by the set of N entities, I, the set of M features, V, and the entity-to-feature matrix

J. Neves, M. Santos, and J. Machado (Eds.): EPIA 2007, LNAI 4874, pp. 395–405, 2007.

$Y=(y_{iv})$, where y_{iv} is the value of feature $v \in V$ for entity $i \in I$. The method produces a partition $S=\{S_1, S_2,..., S_K\}$ of I in K non-overlapping classes S_k, referred to as clusters, each with a specified centroid $c_k=(c_{kv})$, an M-dimensional vector in the feature space (k=1,2,...K). Centroids form set $C=\{c_1, c_2,..., c_K\}$. The criterion, minimised by the method, is the within-cluster summary distance to centroids:

$$W(S, C)= \sum_{k=1}^{K} \sum_{i \in S_k} d(i, c_k) \qquad (1)$$

where d is typically the Euclidean distance squared or the Manhattan distance. In the first case criterion (1) is referred to as the square error criterion and in the second, the absolute error criterion.

Given K M-dimensional vectors c_k as cluster centroids, the algorithm updates cluster lists S_k according to the Minimum distance rule. For each entity i in the data table, its distances to all centroids are calculated and the entity is assigned to the nearest centroid. Then centroids are updated according to the criterion used. This process is reiterated until clusters do not change. Before running the algorithm, the original data are pre-processed (standardised) by subtracting the grand mean from each feature wand dividing it by its range. The above algorithm is referred to as *Straight K-Means*.

We use either of two methods for calculating the centroids: one by averaging the entries within clusters and another by taking the within-cluster median. The first corresponds to the least-squares criterion and the second to the least-moduli criterion [14].

2.2   Selection of the Number of Clusters with the Straight K-Means

In K-Means, the number K of clusters is pre-specified (see [10], [6]). Currently, a most popular approach to selection of K involves multiple running K-Means at different K with the follow-up analysis of results according to a criterion of correspondence between a partition and a cluster structure. Such, "internal", criteria have been developed using various probabilistic hypotheses of the cluster structure by Hartigan [5], Calinski and Harabasz [2], Tibshirani, Walther and Hastie [18] (Gap criterion), Sugar and James [17] (Jump statistic), and Krzanowski and Lai [9]. We have selected three of the internal indexes as a representative sample.

There are some other approaches to choosing K, such as that based on the silhouette width index [8]. Another one can be referred to as the consensus approach [15]. Other methods utilise a data based preliminary search for the number of clusters. Such is the method iK-Means [14]. We consider two versions of this method – one utilising the least squares approach and the other the least moduli approach in fitting the corresponding data model.

We use six different internal indexes for scoring the numbers of clusters. These are: Hartigan's index [5], Calinski and Harabasz's index [2], Jump Statistic [17], Silhouette width [8], Consensus distribution's index [15] and the DD index [14], which involves the mean and variance of the consensus distribution.

Before applying these indexes, we run the straight K-Means algorithm for different values of K in a range from START value (typically 4, in our experiments) to END

K-Means Results Generation

For K = START: END

For diff_init=1: number of different K-Means initialisations
- randomly select K entities as initial centroids
- run Straight K-Means algorithm
- calculate W_K, the value of W(S, C) (1) at the found clustering
- for each K , take the clustering corresponding to the smallest W_K among different initialisations

end diff_init

end K

---

value (typically, 14). Given K, the smallest W(S, C) among those found at different K-Means initialisations, is denoted by W_K. The algorithm is in the box above.

In the following subsections, we describe the statistics used for selecting "the right" K at the clustering results.

2.2.1 Variance Based Approach

Of many indexes based on W_K to estimate the number of clusters, we choose the following three: Hartigan [5], Calinski & Harabasz [2] and Jump Statistic [17], as a representative set for our experiments. Jump Statistic is based on the extended W, according to the Gaussian mixture model. The threshold 10 in Hartigan's index of estimating the number of clusters is "a crude rule of thumb" suggested by Hartigan [5], who advised that the index is proper to use only when the K-cluster partition is obtained from a (K-1)-cluster partition by splitting one of the clusters. The three indexes are described in the box below.

Hartigan (HK):
- calculate $HK=(W_K/W_{K+1}-1)(N-K-1)$, where N is the number of entities
- find the very first K at which HK is less than 10

Calinski and Harabasz (CH):
- calculate $CH=((T-W_K)/(K-1))/(W_K/(N-K))$, where $T=\sum_{i\in I}\sum_{v\in V} y_{iv}^2$ is the data scatter
- find the K which maximises CH

Jump Statistic (JS):
- for each entity i, clustering $S=\{S_1,S_2,...,S_K\}$, and centroids $C=\{C_1,C_2,...,C_K\}$
- calculate $d(i, S_k)=(y_i-C_k)^T\Gamma^{-1}(y_i-C_k)$ and $d_k=(\sum_k\sum_{i\in S_k} d(i, S_k))/M*N$, where

M is the number of features, N is the number of rows and Γ is the covariance matrix of Y
- select a transformation power, typically M/2
- calculate the jumps $JS= d_K^{-M/2} - d_{K-1}^{-M/2}$ and $d_0^{-M/2}\equiv0$
- find K that maximises JS

2.2.2 Structural Approach

Instead of relying on the overall variance, the Silhouette Width index by Kaufman and Rousseeuw [8] evaluates the relative closeness of individual entities to their clusters. It calculates the silhouette width for each entity, the average silhouette width for each cluster and the overall average silhouette width for the total data set. Using this approach each cluster could be represented by the so-called silhouette, which is based on the comparison of its tightness and separation. The silhouette width s(i) for entity $i \in I$ is defined as

$$s(i) = \frac{b(i) - a(i)}{\max(a(i), b(i))} \tag{2}$$

where a(i) is the average dissimilarity between i and all other entities of the cluster to which i belongs and b(i) is the minimum of the average dissimilarity of i and all the entities in other clusters.

The silhouette width values lie in the range [−1, 1]. If the silhouette width value is close to 1, it means that sample is well clustered. If the silhouette width value for an entity is about zero, it means that that the entity could be assigned to another cluster as well. If the silhouette value is close to −1, it means that the entity is misclassified. The largest overall average silhouette width indicates the best number of clusters. Therefore, the number of clusters with the maximum overall average silhouette width is taken as the optimal number of the clusters.

2.2.3 Consensus Approach

We apply the following two consensus-based statistics for estimating the number of clusters: Consensus distribution area [15] and Average distance [14]. These two statistics represent the consensus over multiple runs of K-Means for different initialisations at a specified K. First of all, consensus matrix is calculated. The consensus matrix $C^{(K)}$ is an $N \times N$ matrix that stores, for each pair of entities, the proportion of clustering runs in which the two entities are clustered together.

An ideal situation is when the matrix contains 0's and 1's only: all runs lead to the same clustering. Consensus distribution is based on the assessment of how the entries in a consensus matrix are distributed within the 0-1 range. The cumulative distribution function (CDF) is defined over the range [0, 1] as follows:

$$CDF(x) = \frac{\sum_{i < j} 1\{C^{(K)}(i, j) \leq x\}}{N(N - 1)/2} \tag{3}$$

where 1{cond} denotes the indicator function that is equal to 1 when cond is true, and 0 otherwise. The difference between two cumulative distribution functions can be partially summarized by measuring the area under the two curves. The area under the CDF corresponding to $C^{(K)}$ is calculated using the following formula:

$$A(K)= \sum_{i=2}^{m} (x_i - x_{i-1})CDF(x_i) \qquad (4)$$

where set $\{x_1, x_2, ..., x_m\}$ is the sorted set of entries of $C^{(K)}$. We can calculate the proportion increase in the CDF area as K increases, computed as follows

$$\Delta(K+1)= \begin{cases} A(K), & K=1 \\ \dfrac{A(K+1) - A(K)}{A(K)}, & K \geq 2 \end{cases} \qquad (5)$$

The number of clusters is selected when a large enough increase in the area under the corresponding CDF, which is to find the K which maximises $\Delta(K)$. The index average distancing is based on the entries of the consensus matrix $C^{(K)}(i,j)$ obtained from the consensus distribution algorithm. The mean and the variance of these entries μ_K and σ_K^2 for each K can be calculated. We define $avdis(K)= \mu_K*(1- \mu_K)- \sigma_K^2$, which is proven to be equal to the average distance between the R clusterings

$$M(\{S^t\})= \frac{1}{m^2} \sum_{u,w=1}^{m} M(S^u, S^w) \text{ (Mirkin 2005, p.229), where } M=(|\Gamma_S|+|\Gamma_T|-2a)/\binom{N}{2},$$

$|\Gamma_S|= (\sum_{t=1}^{K} N_{t+}^2 - N)/2$, $|\Gamma_T|= (\sum_{u=1}^{L} N_{+u}^2 - N)/2$ in the contingency table of the two

partitions (see section 4.3). The index is $DD(K)=(avdis(K) - avdis(K+1))/avdis(K+1)$. The number of clusters is decided by the maximum value of $DD(K)$.

2.3 Choosing K with Intelligent K-Means

Another approach to selecting the number of clusters is proposed in Mirkin [14] as the so-called intelligent K-Means. It initialises K-Means with the so-called Anomalous pattern approach, which is described in the box below:

Anomalous Pattern (AP):
1. Find an entity in I, which is the farthest from the origin and put it as the AP centroid c.
2. Calculate distances $d(y_i,c)$ and $d(y_i,0)$ for each i in I, and if $d(y_i,c)<d(y_i,0)$, y_i is assigned to the AP cluster list S.
3. Calculate the centroid c' in the S. If c' differs from c, put c' as c, and go to step 2, otherwise go to step 4
4. Output S and its centroid as the Anomalous Pattern.

The distance and centroid in the AP with the Least Squares criterion are the Euclidean squared and the average of the within-cluster entries, respectively, whereas the iK-means with the Least Modules criterion are the Manhattan distance and the median of the within- cluster entries, respectively.

The intelligent K-Means algorithm iteratively applies the Anomalous Pattern procedure and after no unclustered entities remain, removes the singletons and takes the centroids of remaining clusters and their quantity to initialise K-Means. The algorithm is as follows:

Intelligent K-means:

0. Put t=1 and I_t the original entity set.
1. Apply AP to I_t to find S_t and C_t.
2. If there are unclustered entities left, put $I_t \leftarrow I_t - S_t$ and t=t+1 and go to step 1.
3. Remove all the found clusters with the cluster size 1. Denote the number of remaining clusters by K and their centroids by $c_1, c_2, ..., c_K$.
4. Do Straight K-means with $c_1, c_2, ..., c_K$ as initial centroids.

The iK-Means algorithm differs from those in the previous section by the following: (a) it uses just one run of the iterative AP algorithm, (b) it utilizes yet another parameter, the discarding threshold, which is taken to be DT=1 in the following up experiments.

To further enhance iK-Means methods LS and LM, one can make a notice that, according to Chiang and Mirkin [3], LS and, especially, LM may produce sometimes excessive numbers of clusters, which may contribute to their losing to other methods in these situations. We interpret this as an indication that the discarding threshold value DT, set to be always 1, can be overly restrictive. On the other hand, the tables show that HK results, on average, reasonably reproduce the number of clusters generated. This leads us to suggest that the HK number-of-cluster results should be taken as a reference to adjust the discarding threshold in iK-Means. An iterative procedure for such an adjustment is in the box below.

HK-adjusted iK-Means

0. HK-number: Find the number of clusters K_h by using R runs of Straight K-Means at each K with the Hartigan rule.
1. iK-Means number: Find the number of clusters by using iK-Means with the discarding threshold $DT=1$. Let it be K_ls for LS and K_lm for LM.
2. Adjust: If K_ls (or K_lm) is 1.15 times greater than K_h, increase the discarding threshold by 1 and go to step 1 with the updated DT. Otherwise, halt. (The adjustment factor value of 1.15 has been found experimentally.)

2.4 Selection

Here is the list of methods for finding the number of clusters in our experiment, with the acronyms assigned:

Table 1. Set of methods for selection of the number of clusters in K-Means under comparison

Method	Acronym
Hartigan	HK
Calinski & Harabasz	CH
Jump Statistic	JS
Silhouette Width	SW
Consensus Distribution area	CD
Davdis	DD
Least Squares	LS
Least Modules	LM
Adjusted Least Squares	ALS
Adjusted Least Modules	ALM

3 Data Generation for the Experiment

There is a popular distribution in the literature on computational intelligence, the mixture of Gaussian distributions, which can supply a great variability of cluster shapes, sizes and structures ([1] and [10]). Yet there is an intrinsic difficulty related to the huge number of parameters defining a Gaussian mixture distribution: (a) the cluster probabilities; (b) cluster centres; and (c) cluster covariance matrices, of which the latter involve $KM^2/2$ parameters, where M is the number of features, which is about a 1000 at K=10 and M=15 – by far too many for modelling in an experiment. However, there is a model involving the so-called Probabilistic Principal Components (PPCA) framework that uses an underlying simple structure covariance model ([16] and [19]).

The Gaussian mixture data are generated as implemented in a MATLAB Toolbox freely available on the web [4]. Our sampling functions are based on a modified version of that proposed in Wasito and Mirkin [20]. The mixture model type in the functions defines the covariance structure. We use either of two types: the spherical shape or the probabilistic principal component analysis (PPCA) shape [19]. The cluster spatial sizes are taken constant at the spherical shape, and variant at the PPCA shape. The cluster spatial size with the PPCA structure can be defined by multiplying its covariance matrix by a factor. We maintain two types of the cluster spatial size factors: the linear and quadratic distributions of the factors. To implement these, we take the factors to be proportional to the cluster's index k (the linear distribution being k-proportional) or k^2 (the quadratic distribution being k^2-proportional) (k=1,...,K).

Cluster centroids are generated randomly from a normal distribution with mean 0 and standard deviation 1 and then they are scaled by a factor expressing the spread of the centroids. Table 2 presents experimentally chosen spread values, which are used in the experiments. The PPCA model runs with the manifest number of features 15 and the dimension of the PPCA subspace equal to 6.

In the experiments, we generated Gaussian mixtures with 9 clusters. The cluster proportions (priors) we took were uniformly random.

Table 2. Cluster spread used in the experiments

Spread	Spherical	PPCA	
		k-proport.	k²-proport.
Large	2	10	10
Small	0.2	0.5	2

4 Evaluation Criteria

4.1 Number of Clusters

This criterion is based on the difference between the number of generated clusters (7 or 9) and that in the selected clustering.

The number of clusters measure is rather rough; it does not take into account the clusters' content, that is, similarity between generated clusters and those found with the algorithms.

4.2 Distance Between Centroids

This is not quite an obvious criterion when the number of clusters in a resulting partition is greater than the number of clusters generated. In our procedure, we use three steps to score the similarity between the real and obtained centroids: (a) assignment, (b) distancing and (c) averaging. These steps are described in the box below for both the weighted and unweighted distance cases in terms of centroids $e_1, e_2, ... e_L$ of found clusters $Q_1, Q_2, ... Q_L$, and generated centroids $g_1, g_2, ..., g_K$ of generated clusters $P_1, P_2, ..., P_K$.

Distance between two sets of centroids

1. *Assignment:* For each $k=1,....K$, assign g_k with that e_l which is the closest to it.

   If there remains any not assigned centroid e_i, find that g_k that is the nearest to it.

2. *Finding distances:* Denote by E_k the set of those e_l that have been assigned to g_K and take $\alpha_{lk} = q_l/|E_k|$ (weighted version) or $\alpha_{lk} = 1$ (unweighted version). Define, for each $k=1,...,K$

$$dis(k) = \sum_{e_l \in E_k} d(g_k, e_l) * \alpha_{lk} .$$ (The distance d here is Euclidean squared distance.)

3. *Averaging:*   Calculate $D = \sum_{k=1}^{K} p_k * dis(k)$ where $p_k = N_k = |N_k|$, in the weighted version, or $p_k = 1$, in the unweighted version.

4.3 Partition Confusion Measures

To measure the similarity between two partitions, their contingency (confusion) table is to be used. The entries in the contingency table are the co-occurrences of the generated partition clusters (row category) and the obtained clusters (column category), that is, counts of numbers of entities that fall simultaneously in both clusters. The generated cluster (row category) is denoted by $k \in T$, the obtained partition (column category) is denoted by $h \in U$ and the co-occurrences counts are denoted by N_{kh}. The frequencies of row and column categories usually are called marginals and denoted by N_{k+} and N_{+h}. The probabilities are defined accordingly: $p_{kh}=N_{kh}/N$, $p_{k+}=N_{k+}/N$, and $p_{+h}=N_{+h}/N$, where N is the total number of entities. Of the four used contingency-based measures (the relative distance, Tchouproff coefficient, the average overlap, and the adjusted Rand index), only the adjusted Rand index will be presented.

The adjusted Rand index ([6], [21]) is defined as follows:

$$Ari = \frac{\sum_{k \in T}\sum_{h \in U}\binom{N_{kh}}{2} - \left[\sum_{k \in T}\binom{N_{k+}}{2}\sum_{h \in U}\binom{N_{+h}}{2}\right]/\binom{N}{2}}{\frac{1}{2}\left[\sum_{k \in T}\binom{N_{k+}}{2}+\sum_{h \in U}\binom{N_{+h}}{2}\right]-\left[\sum_{k \in T}\binom{N_{k+}}{2}\sum_{h \in U}\binom{N_{+h}}{2}\right]/\binom{N}{2}} \tag{6}$$

where $\binom{N}{2} = \frac{N(N-1)}{2}$.

5   Evaluation Results

The evaluation results with the adjusted iK-means methods are shown in Table 3. The distance between centroids shown in Table 3 are rescaled according to Table 2 in such away that the distances at different spatial size distributions become comparable. Specifically, at the small spreads, the spread factor at k2-proportional distribution, 2, is four times greater than that at k-proportional, 0.5, and 10 times greater than that at the equal sizes, 0.2. By multiplying the distances between centroids at equal sizes by 100=102 and at k-proportional sizes by 16=42, we make them comparable with those at the k2-proportional distribution. (Note, the distance between centroids is Euclidean squared, which implies the need in the quadratic adjustment of the factors.) Similarly, at the large spreads, the spread factors at the variant size distributions are the same while that at the constant size is 5 times less. Multiplying the distances between centroids at equal sizes by 25, we make all the distances comparable. After having done this, we can see one more effect in the Table 3. All methods perform better when the cluster spatial sizes are less different: at the constant sizes the best and at the k2-proportional sizes the worst. This can be seen as conforming to the idea that K-Means best delivers at constant radius clusters indeed. The HK adjustment works too: ALS and ALM join HK to perform better on the number of clusters. On the other evaluation measures, the pattern supports the view that the iK-Means based methods perform well. LS, LM, ALS and ALM perform better over both the distance between centroids and cluster contents. Overall, the adjusted versions ALS and ALM win at most situations.

Table 3. The average values of evaluation criteria at 9-clusters data sets with NetLab Gaussian covariance matrix for the large and small spread values (LaS and SmS, respectively) in Table 2. The standard deviations are after slash, per cent. The three values in a cell refer to the three cluster structure models: the spherical on top, the PPCA with k-proportional cluster sizes in the middle, and the PPCA with k2-proportional cluster sizes in the bottom. Two winners of the ten methods (eight from Table 1 plus adjusted LS and LM which are denoted by ALS and ALM) and are highlighted using the bold font, for each of the options. Distances between centroids are rescaled according to Table 2, as explained in section 6.

	Estimated number of clusters		Adjusted distance between centroids		Adjusted Rand Index	
	LaS	SmS	LaS	SmS	LaS	SmS
HK	8.27/6	**7.6/10**	10310.00/13	38601.00/14	0.89 / 9	0.29/10
	8.55/7	**9.4 / 9**	11833.21/14*	**47448.96**/15	0.90 / 9	0.37/11
	9.35/7	9.12/10	12154.99/15	55286.55/14	0.84 / 9	0.28/12
CH	11.55/8	4.00 / 0	10096.25/12*	41927.00/12	0.82 / 9	0.25/12
	12.10/4	5.30 / 5	11788.38/14*	46924.64/19	0.81 / 8	0.21/12
	11.15/8	4.11 / 8	12146.83/13	53779.46/15	0.79 / 9	0.22/12
JS	12.12/8	4.50 / 0	10084.50/13*	41927.00/13	0.77/10	0.25/12
	12.75/9	6.15 / 8	11785.21/13*	46533.28/15	0.82 / 8	0.24/13
	12.10/8	4.45 / 5	12131.86/12	53699.24/14	0.80 / 8	0.22/11
SW	6.29/8	4.54/10	10456.50/12	41866.00/14	0.92/10	0.26/13
	6.95/7	4.95 / 4	11876.31/13*	45540.96/16	0.92 / 8	0.27/12
	7.15/8	4.28/11	12203.58/12	53583.12/16	0.85 / 6	0.22/13
CD	5.31/7	5.11 / 9	10749.00/12	37393.00/12	0.78/12	0.27/13
	5.30/6	5.10/10	11943.98/13	46361.76/18	0.78/12	0.28/14
	5.20/6	5.31 / 9	12265.98/12	55040.86/15	0.75/12	0.25/13
DD	5.67/3	6.42 / 8	10884.25/12	40997.00/13	0.75/12	0.27/12
	4.90/3	5.60 / 9	11979.30/13	47940.48/18	0.74/12	0.24/12
	5.3/3	5.83 / 8	12286.43/12	53912.13/13	0.71/12	0.27/10
LS	8.67/6	13.00/18	**10061.75/12**	**33591.00/23**	**0.99 / 9**	**0.48/12**
	8.80/6	10.80/16	**11771.70/12**	**42582.56/20**	**0.99/10**	**0.42/12**
	7.95/7	13.44/18	12031.13/11	54026.92/15	0.90 / 9	**0.45/12**
LM	**9.33/6**	25.00/18	**10004.50/12**	38112.00/25	0.92 / 9	0.38/12
	8.80/7	16.10/17	**11767.34/13**	**42377.60/20**	**0.99/10**	0.41/12
	10.00/6	23.11/18	12114.01/12	53507.21/16	0.84/10	**0.41/12**
ALS	8.50/5	**7.60 / 6**	10086.75/12*	33849.00/12	**0.99/11**	**0.50/11**
	8.70/7	**9.90 / 7**	11871.70/15*	43536.32/11	**0.99/11**	**0.42/12**
	8.70/9	**9.40 / 9**	**11031.13/12**	**52098.21/12**	**0.95/11**	0.38/12
ALM	**8.70/6**	7.50 / 6	10504.50/12	**30556.00/12**	**0.99/12**	0.44/10
	8.70/7	10.60 / 9	11867.34/15*	44298.88/11	**0.99/10**	0.38/11
	9.50/9	**9.60 / 9**	**10114.01/13**	53057.21/11	**0.92/13**	0.35 / 9

* within 1% of the best value.

6 Conclusions

Of the ten different procedures considered in these experiments, most consistent results are shown by the least-squares version of iK-Means LS [14] closely followed by its least-moduli counterpart LM. Rather unexpectedly, Hartigan's "rule of thumb" HK [5], appears to best reproduce the number of clusters, which can be used for the adjustment of the discarding coefficient in iK-Means methods as described above.

The future research should include, first of all, greater coverage of potential data distributions, in terms of greater freedom in covariance parameters of the Gaussians as well as involving other types of distributions.

References

1. Banfield, J.D., Raftery, A.E.: Model-based Gaussian and non-Gaussian clustering. Biometrics 49, 803–821 (1993)
2. Calinski, T., Harabasz, J.: A Dendrite method for cluster analysis. Communications in Statistics 3(1), 1–27 (1974)
3. Chiang Mark, M.T., Mirkin, B.: Determining the number of clusters in the Straight K-means: Experimental comparison of eight options. In: Proceeding of the 2006 UK workshop on Computational Intelligence, pp. 119–126 (2006)
4. Generation of Gaussian mixture distributed data, NETLAB neural network software (2006), http://www.ncrg.aston.ac.uk/netlab
5. Hartigan, J.A.: Clustering Algorithms. J. Wiley & Sons, New York (1975)
6. Hubert, L.J., Arabie, P.: Comparing partitions. Journal of Classification 2, 193–218 (1985)
7. Jain, A.K, Dubes, R.C.: Algorithms for Clustering Data. Prentice Hall, Englewood Cliffs (1988)
8. Kaufman, L., Rousseeuw, P.: Finding Groups in Data: An Introduction to Cluster Analysis. J. Wiley & Son, New York (1990)
9. Krzanowski, W., Lai, Y.: A criterion for determining the number of groups in a dataset using sum of squares clustering. Biometrics 44, 23–34 (1985)
10. McLachlan, G., Basford, K.: Mixture Models: Inference and Applications to Clustering. Marcel Dekker, New York (1988)
11. McQueen, J.: Some methods for classification and analysis of multivariate observations. In: Fifth Berkeley Symposium on Mathematical Statistics and Probability, vol. II, pp. 281–297 (1967)
12. Milligan, G.W., Cooper, M.C.: An examination of procedures for determining the number of clusters in a data set. Psychometrika 50, 159–179 (1985)
13. Mirkin, B.: Eleven ways to look at the Pearson chi squares coefficient at contingency tables. The American Statistician 55(2), 111–120 (2001)
14. Mirkin, B.: Clustering for Data Mining: A Data Recovery Approach. Chapman and Hall/CRC, Boca Raton Fl (2005)
15. Monti, S., Tamayo, P., Mesirov, J., Golub, T.: Consensus clustering: A resampling-based method for class discovery and visualization of gene expression microarray data. Machine Learning 52, 91–118 (2003)
16. Roweis, S.: EM algorithms for PCA and SPCA. In: Jordan, M., Kearns, M., Solla, S. (eds.) Advances in Neural Information Processing Systems, vol. 10, pp. 626–632. MIT Press, Cambridge (1998)
17. Sugar, C.A., James, G.M.: Finding the number of clusters in a data set: An information-theoretic approach. Journal of American Statistical Association 98(463), 750–778 (2003)
18. Tibshirani, R., Walther, G., Hastie, T.: Estimating the number of clusters in a dataset via the Gap statistics. Journal of the Royal Statistical Society B 63, 411–423 (2001)
19. Tipping, M.E., Bishop, C.M.: Probabilistic principal component analysis. J. Roy. Statist. Soc. Ser. B 61, 611–622 (1999)
20. Wasito, I., Mirkin, B.: Nearest neighbours in least-squares data imputation algorithms with different missing patterns. Computational Statistics & Data Analysis 50, 926–949 (2006)
21. Yeung, K.Y., Ruzzo, W.L.: Details of the Adjusted Rand index and clustering algorithms. Bioinformatics 17, 763–774 (2001)

A Network Algorithm to Discover Sequential Patterns

Luís Cavique

ESCS-IPL, Portugal
lcavique@escs.ipl.pt

Abstract. This paper addresses the discovery of sequential patterns in very large databases. Most of the existing algorithms use lattice structures in the space search that are very demanding computationally. The output of these algorithms generates a large number of rules. The aim of this work is to create a swift algorithm for the discovery of sequential patterns with a low time complexity. In this work, we also want to define tools that allow us to simplify the work of the final user, by offering a new visualization of the sequences, while bypassing the analysis of thousands of association rules.

Keywords: data mining, frequent sequence mining, graph theory.

1 Introduction

Sequential Pattern Discovery is a very important data mining subject with a wide range of applications. Sequence analysis is used to determine data patterns throughout a sequence of temporal states. Nowadays, sequence analysis is widely applied to find click stream data in the web sites. Several authors have been trying to find customers' purchase patterns of goods in the retail [3], people's careers [5] and in the financial sectors [11].

The Sequential Patterns Discovery, which extends itemset mining, is a complex task for the user. The user must specify a minimum support threshold to find the desired patterns. A useless output can be expected by pruning either too many or too few items. The process must be repeated interactively, which becomes very time consuming for large databases.

The association rules' systems that support the itemset and sequence mining, usually generate a huge number of rules, and therefore, it is difficult for the user to decide which rules to use.

Since most of the existing algorithms use a lattice structure in the search space and need to scan the database more than once, they are not compatible with very large databases.

These three handicaps are strong challenges to be overcome in this research area. In this work we present an algorithm that can discover sequence patterns with low time complexity and that can cut down on the work of the user.

In section 2 the related work with Frequent Pattern Mining is presented using a taxonomy of algorithms that distinguish between the Frequent Itemset Mining and the

J. Neves, M. Santos, and J. Machado (Eds.): EPIA 2007, LNAI 4874, pp. 406–414, 2007.

Frequent Sequence Mining. In section 3 the new network based algorithm, the Ramex algorithm, is presented. The sub-section 3.1 reports some network concepts in graph theory and the sub-section 3.2 describes the algorithm. In section 4, computational experience is reported using a web click stream dataset. Finally, in section 5, we draw some conclusions.

2 Related Work

Data Mining includes a wide range of procedures and algorithms, although, in the most recent literature three subjects belong to the core topics: the clustering (with strong origins in statistics), the classification and the frequent pattern mining. In frequent pattern mining we include the frequent itemset mining (or market basket analysis) and the frequent sequence mining.

In Frequent Pattern Mining two approaches can be used: The Frequent Itemset Mining (FIM) only needs two attributes as input: the transaction and the item. This subject is also known as the Market Basket Analysis. On the other hand the Frequent Sequential Mining (FSM) usually needs three attributes as input: the customer, the date and the item. The information retrieved by the FSM, that includes the variable time, offers a dynamic view of the purchasing process [12].

Most of the Frequent Pattern algorithms use lattice structures in the search space. In the search two methods can be used, the breadth-first search and the depth-first search. A third approach can be carried out by transforming the problem and condensing the data.

Table 1. Taxonomy of the Frequent Pattern Mining

	Frequent Itemset Mining (FIM)	Frequent Sequential Mining (FSM)
Lattice, breadth-first search	Apriori	AprioriAll GSP
Lattice, depth-first search	FP-growth	SPADE
Transformed Structures	Similis	Markov Chain Model Ramex

In table 1, a taxonomy of the Frequent Pattern Mining algorithms is presented. The Frequent Itemset Mining algorithms include for the breadth-first search, the Apriori algorithm [2] and for the depth-first, search the FP-growth [9]. The FIM deals with static itemsets, i.e., they don't change with time.

On the other hand, the Frequent Sequence Mining involves a sequence of steps in time. A typical purchasing sequence starts when a customer buys, for instance, a laptop. There are lots of other items that can be bought afterward, like a mouse, then a CD with software, a bag, a memory stick, a floppy disk drive, a printer/scanner, a modem/router and other multimedia items.

The Frequent Sequence Mining, FSM, algorithms include for the breadth-first search, the AprioriAll algorithm [1] and the GSP (Generalized Sequential Pattern)

[13] and for the depth-first search the SPADE (Sequential PAttern Discovery using Equivalence classes) [14]. They use a lattice structure in the search space resulting in high computation time when dealing with large databases.

The transformation method shrinks the data into new data structures, and afterward it uses known techniques to extract the patterns. The Similis algorithm [4] transforms the database into a weighted graph and heuristic search techniques discover complete sub-graphs that correspond to itemsets. For Sequential Pattern Mining the Markov Chain Models and the proposed algorithm Ramex use network structures to condense the data. With the Markov Chain Model a global view is possible since all items are taken into consideration. On the other hand, the Association Rules usually present a set of rules that only show the items with high support.

Most of the existing software for cross-selling, using association rules, generates thousands of rules, and therefore it is difficult for the marketeer (or user) to predict the next-item that each customer will buy. To implement cross-selling strategies we want an algorithm that can discover the next item, that reduces the work of the user and with a low time complexity algorithms.

3   Ramex Algorithm

The aim of this approach is to create a tree of sequences, with as many branches as needed to visit all the vertexes, starting by a vertex, called root. Ramex is the name of present algorithm that means branch in Latin. This term was chosen just like Apriori and Similis, since they are all Latin names.

3.1 Network Concepts

In this sub-section a short bibliographic overview in graph theory is given to be reused in the next sub-section.

Given a connected undirected graph, a spanning tree of the graph is a sub-graph that connects all the vertexes. A minimum weight spanning tree is the spanning tree with a weight that is lower than or equal to the weight of every other spanning tree. This problem is easily solved using a greedy algorithm. The algorithms proposed by Kruskal and Prim are well known examples. In Kruskal's algorithm, the edges are chosen without worrying about the connections to previous edges, but avoiding the cycles. In Prim's algorithm the tree grows from an arbitrary root.

In a connected directed graph G, an acyclic sub-graph B is a branch in G if the in-degree of each vertex is at the most 1. Let w(B) be the sum of the weighted arcs in branch B. The maximum weight branching problem is the optimization problem that finds the largest branch B possible. Edmonds' branching algorithm [6], proved by Karp [10] can be described in two steps, as follows:

i) The condensation process: the input is the weighted directed graph G, let G_1 be equal to G, repeat the process by condensing the cycles, G_k is the digraph which is acyclic.

ii) The unraveling process: Let $B_k=G_k$, then constructs B_{k-1} from B_k by expanding the condensed cycles, until a tree is obtained.

Fulkerson [7] presents the same problem with an additional constraint, the branch must start in a vertex called the root. His algorithm for the maximum packing rooted directed cuts in a graph is equal to the weight of the minimum spanning tree directed away from the root. The algorithm is also described in two steps: the condensation process, to remove the cycles, and the unraveling process where the branch is created.

3.2 Ramex Description

The Ramex approach has strong connections with the Markov Chain Models. In the Markov Chain Model each state is an item and in the transition matrix, each Pij is the probability of moving from state i to state j in one step. The Ramex, like the Markov Chain Models, also presents a global view, since all the items are taken into consideration. In the Ramex approach instead of the relative frequencies, the absolute frequencies are used. In each transition the number of times from an item to the next-item is reported.

The Ramex algorithm is presented in two main steps: the transformation of the problem and the search of the sequences.

Ramex Algorithm
Input: a database (customer, date, item);
Output: a sequence of items;
1) Network Transformation
1.1) Sort data by customer, date and item;
1.2) Create a new attribute next-item;
1.3) Build a state transition network G;
2) Find highly probable branch sequence B;
2.1) Condensation process;
2.2) Unraveling process.

In the transformation of a database into a network, the raw data (customer, date, item) must be sorted in such a way that each customer sequence can be identified. For each line in the table the new attribute next-item is created. Then a network, i.e. a graph G with a source and a sink is created, where each customer sequence has an initial node called source (or root) and a final node called sink.

The network, where cycles are allowed, condenses the information of the database by incorporating all the customer sequences. In the network G each state corresponds to an item and each transition represents the sequence from one item to the next-item. The weight of each arc corresponds to the number of times that one item precedes the next-item.

The second step is to find in the network G the highly probable branch sequence B, given a root vertex, where the Maximum Weight Rooted Branching Algorithm is applied, as defined in Fulkerson [7]. Like Edmonds' branching algorithm [6], Fulkerson's algorithm has $\Theta(N^2)$ in the worst case time complexity, where N is the number of vertexes.

In the original table a new attribute "next-item" must be included. Afterward, using the feature Cross Reference Query, that exists in most of the databases, the adjacency matrix can be obtained. The adjacency matrix of G indicates the number of times that a transition occurs from one item to the next-item.

Table 2. Adjacency matrix

	A	laptop	mouse	bag	CD	mem	Z
A	-	3					
laptop		-	2	1			
mouse			-	2	2		
bag				-		3	
CD					-		2
mem			2			-	1
Z							-

Numeric Example. Using the adjacency matrix presented in table 2, a network in figure 1-I can be built. In the network the source is represented by A and the sink is represented by Z. In this type of network, cycles may be present. A cycle can be identified: mouse, bag, memory and mouse again. Starting at the root A, all vertexes must be visited, except Z, avoiding cycles, where the sum of the weighted arcs is as heavy as possible.

In the condensation phase, figure 1-II, the existing cycle is shrunk, obtaining a directed acyclic graph. In the unraveling phase, figure 1-III, the nodes are expanded, obtaining the tree in figure 1, where the dotted line represents the forbidden arcs.

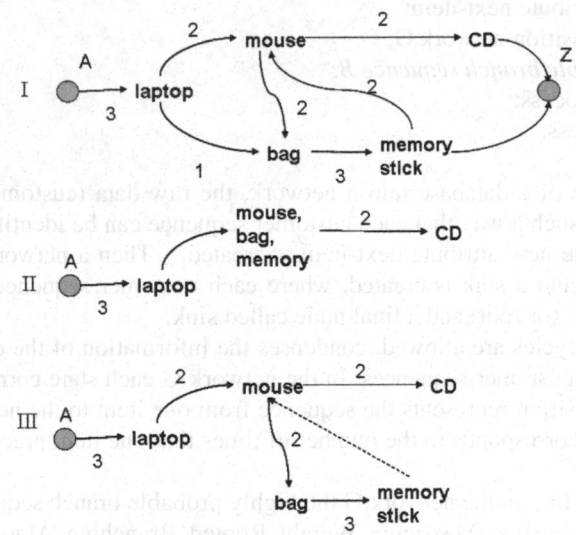

Fig. 1. Phases of the Maximum Weight Rooted Branching Algorithm

So, this optimization problem is solved by the Maximum Weight Rooted Branching Algorithm and the solution is a tree, as represented in fig.1-III. In this figure the sum of the weights is equal to 3+2+2+2+3=12. In comparison, the value of the Minimum Weight Rooted Branching is 3+1+3+2+2=11, which uses the arc (laptop, bag).

4 Computational Experience

To validate the Ramex algorithm, a dataset named Web Click-Stream [8] was used. This file has 250,711 observations, from 22,527 web visitors and 36 web pages. The dataset has three attributes: client identifier (cookie), date/ time and web page. In this case a sequence is given by the clicks a client produces on a particular date.

Table 3. Statistic Measure of the dataset

Statistic Measure	Number Clicks
Average	11.1
Mode	5
Minimum	1
First quartile	6
Median	8
Third quartile	13
Maximum	192
Skewness	4.7
Kurtosis	40.8

Before using the algorithm it is important to make a brief exploratory data analysis. Table 1 shows some statistic measures for the number of clicks. The average number of clicks per visit is 11, the mode is 5 and the median 6. The minimum number of clicks is 1 and the maximum is 192, which seems far-fetched. The distribution is heavy skewed to the right and presents a high kurtosis equal to 41.

The algorithm Ramex-phase-1 transforms the dataset and generates a network with 36 nodes, 539 arcs with 42% of density.

Afterward, in the Ramex-phase-2, the Maximum Weight Rooted Branching Algorithm is applied, as defined in Fulkerson [7]. The input of the algorithm needs an initial node (root), in this case it is number 1, and the Ramex output is a tree as shown in figure 2.

We think that a tree is the best way to represent the most frequent sequences. In this way, we can find many branches in the tree, that includes all the vertexes, with short branches and long branches. Each branch corresponds to a weighted sequence of clicks. By looking closely at the tree we can see different clusters, after node 12-home. The sequence 17, 8, 9 corresponds to news, feedback, feed-post that represents the users who are looking for news. The branch that starts in 13-login represents users who login, logout and register. The sequence 5 and 28 represents users of the catalog. The branch that starts with 22-program is divided into two sub-branches. Sub-branch 18, 23, 33 represents the users that want information and also want to test the

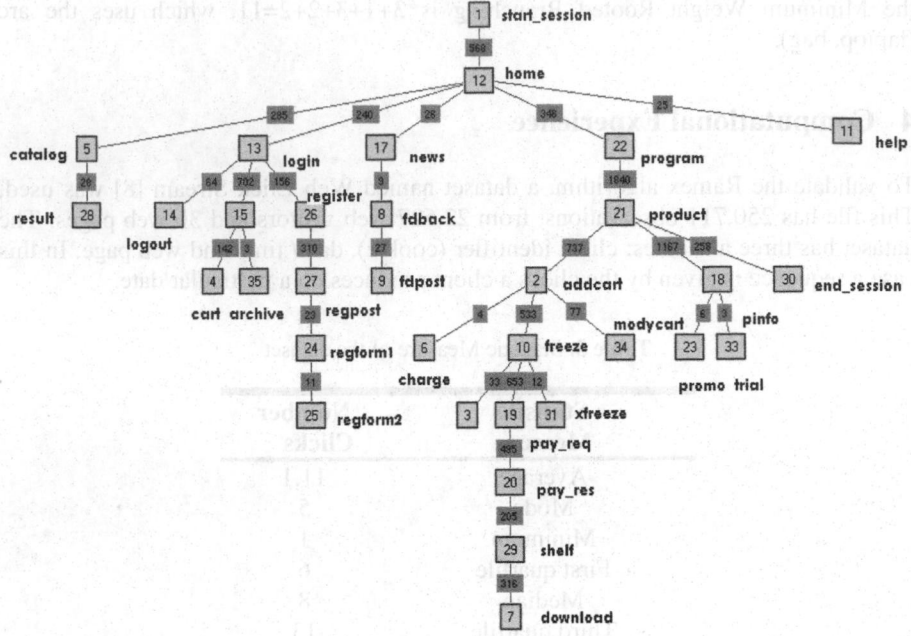

Fig. 2. Ramex Output

product. The second sub-branch, that starts with 2-add-cart, represents the users that add a product to the basket cart. Afterward, the users pay and finally download the product. The main advantage is that we can find long sequences of clicks and even clusters of clicks using a low complexity time algorithm.

It is hard to make comparisons among different methods that use completely different approaches and return different outputs. However, in this section we will report the advantages and the disadvantages of Ramex, compared to the Markov Chain Model, the Link Analysis and the Association Rules, keeping in mind that the sequence patterns must be found.

Markov Chain Models. Using a Markov Chain Model the path of the most likely transitions is given in table 4 [8]. These results are identical to the solution given by the Ramex algorithm. However, the Ramex algorithm returns a more complete solution, presenting the needed branches that cover all the vertexes.

Table 4. Markov Chain Model the forward path

Page i	Page j	Prob (i, j)
Start_session	Home	45.81%
Home	Program	17.80%
Program	Product	70.18%
Product	P_info	26.73%

Link Analysis. Link Analysis usually uses a circular graph, where the clicks between two pages can be shown. Strong links can be found between some pages, like 21-product, 18-pinfo and 21-product, 22-program. An extension of this approach is the search of complete sub-graphs, in order to find strong connected pages. The drawback is that the time sequence is lost. When studying the click-stream, the visualization is critical. The visualization with a circular graph, used in the link analysis, is recommended when the time is not relevant.

Association Rules.   In the market basket analysis, the Soul-Mate algorithm compares a k-itemset with an identical k-itemset of a single customer. On the other hand, Apriori-like algorithms find all the exact k-itemsets. In the sequence pattern discovery the AprioriAll counts the exact number of sequences. In a more general way, the Ramex algorithm, like the Markov Chain Models, mixes all the sequences in the digraph going further than a single comparison or an exact count, by creating a pattern. Emerging from the network pattern a weighted tree can be revealed, offering new directions for the decision maker.

5 Conclusions

Knowledge Discovery in Databases (KDD) and Data Mining appear in the literature so tied to each other that sometimes they seem to be the same. However, Data Mining is part of Knowledge Discovery. Knowledge Discovery in Databases can be divided in three sub-process: the data pre-processing, the data mining and the post-processing. The pre-processing includes the ETL, extraction, transformation and loading the data into the data warehouses, and involves subjects like data reduction, normalization and data quality. On the other hand, the data post-processing includes the visualization and patterns interpretation for decision making.

In this paper a new approach in Data Mining to discover Sequence Patterns with a low time complexity algorithm is presented. This work also includes the post-processing dimension of KDD, the visualization of the sequences using a tree seems easier than using association rules or using the link analysis. Once the branches of items are known the user can easily decide what is the next-item for each customer.

In Data Mining the time complexity of the algorithms is very important. To discover sequence patterns we propose the Ramex algorithm. To run the algorithm no parameter are needed. On the other hand, other algorithms use the min-support as a parameter to control the combinatorial expansion. This algorithm doesn't use the lattice structure in the search space, but it uses data condensation in a network. The procedure that returns the output is a low time complexity algorithm with $\Theta(N^2)$ in the worst case, where N is the number of items of the condensed network.

The Ramex algorithm finds large branch patterns that sometimes do not correspond to the exact sequence count, but express the combination of several sequences, as patterns tend to do.

References

1. Agrawal, R., Srikant, R.: Mining sequential patterns. In: Proceedings 11th International Conference Data Engineering, ICDE, pp. 3–14. IEEE Press, Los Alamitos (1995)
2. Agrawal, R., Srikan, R.: Fast algorithms for mining association rules. In: Proceedings of the 20th International Conference on Very Large Databases, pp. 478–499 (1994)
3. Berry, M., Linoff, G.: Data Mining Techniques for Marketing, Sales and Customer Support. John Wiley and Sons, Chichester (1997)
4. Cavique, L.: A Scalable Algorithm for the Market Basket Analysis. Journal of Retailing and Consumer Services, Special Issue on Data Mining in Retailing and Consumer Services (accepted paper, 2007)
5. Dunham, M.H.: Data Mining: Introductory and Advanced Topics. Prentice Hall, Pearson Education Inc. (2003)
6. Edmonds, J.: Optimum branchings. J. Research of the National Bureau of Standards 71B, 233–240 (1967)
7. Fulkerson, D.R.: Packing rooted directed cuts in a weighted directed graph. Mathematical Programming 6, 1–13 (1974)
8. Giudici, P.: Applied Data Mining: Statistical Methods for Business and Industry. John Wiley and Sons, Chichester (2003)
9. Han, J., Pei, J., Yin, Y.: Mining frequent patterns without candidate generation. In: Proceedings of the 2000 ACM SIGMOD International Conference on Management of Data, Dallas, Texas, United States, pp. 1–12 (2000)
10. Karp, R.M.: A simple derivation of Edmonds' algorithm for optimum branchings. Network 1, 265 (1971)
11. Prinzie, A., Van den Poel, D.: Investigating Purchasing Patterns for Financial Services using Markov, MTD and MTDg Models. In: Working Papers of Faculty of Economics and Business Administration, Ghent University, Belgium 03/213, Ghent University, Faculty of Economics and Business (2003)
12. Silvestri, C.: Distributed and Stream Data Mining Algorithms for Frequent Pattern Discovery, Universit'a Ca' Foscari di Venezia, Dipartimento di Informatica, Dottorato di Ricerca in Informatica, Ph.D. Thesis TD-2006-4 (2006)
13. Srikant, R., Agrawal, R.: Mining sequential patterns: Generalizations and performance improvements. In: Apers, P.M.G., Bouzeghoub, M., Gardarin, G. (eds.) EDBT 1996. LNCS, vol. 1057, pp. 3–17. Springer, Heidelberg (1996)
14. Zaki, M.J.: Spade: An efficient algorithm for mining frequent sequences. Machine Learning 42, 31–60 (2001)

Adaptive Decision Support for Intensive Care

Pedro Gago[1], Álvaro Silva[2], and Manuel Filipe Santos[3]

[1] Escola Superior de Tencologia e Gestão do Instituto Politécnico de Leiria, Portugal
[2] Instituto de Ciências Biomédicas Abel Salazar, Porto, Portugal
[3] Universidade of Minho, Portugal

Abstract. The condition of patients admitted to an Intensive Care Unit is complex to the point that it is often very difficult for physicians to accurately determine the most adequate course of action. However, an ICU is a data rich environment where patients are continuously connected to sensors that allow data collection. Datasets containing such data may hide invaluable information regarding the patients' prognosis. Previous work on intensive care data, produced prediction models that were integrated into a decision support system called INTCare. Although presenting interesting results, INTCare uses static models that are expected to become less accurate over time. As an alternative, this paper presents the results of a set of experiments using an ensemble approach to the prediction of the final outcome of ICU patients, given the data collected during the first 24 hours after ICU admission. Results for both the static and dynamic ensembles (where model weights are updated after each prediction) are presented.

1 Introduction

Since the 1960's computer applications whose purpose was that of supporting the decision making process have been designed [22]. Even though the first computer applications in business environments were intended to ease the operational activities like order processing, billing or inventory control, the need arose for tools that could ease the tasks related to decision support [1].

In the medical area several expert systems were built and deployed [6,14,21]. Nevertheless, the failure rate was high as the effort required to update the knowledge base was excessive and the scope of the expert systems was very limited. Researchers started shifting their attention to the automation of the knowledge acquisition process by using methods from several areas of expertise (e.g. machine learning, statistics). Knowledge Discovery from Databases (KDD) [9] is well suited for this task. In fact, given that there is enough data KDD techniques make knowledge acquisition easier, simplifying the task of building decision support tools. Despite KDD being a semi-automatic process, the predictive models still need to be re-evaluated on a regular basis to detect any loss of predictive accuracy. In fact, model performance is known to degrade over time as the world does not remain in a stationary state [13] (e.g. in the medical area new drugs and therapeutic procedures are constantly being developed). Whenever performance drops bellow acceptable values it is necessary to repeat the KDD process

J. Neves, M. Santos, and J. Machado (Eds.): EPIA 2007, LNAI 4874, pp. 415–425, 2007.
© Springer-Verlag Berlin Heidelberg 2007

or, at the very least, retrain the models using the latest data. Thus, an adaptive DSS must include mechanisms to detect this drop in performance and to act accordingly in order to maintain the needed performance levels [19]. Moreover, it is now well established that prediction accuracy can usually be improved by using ensembles of prediction models instead of a single model [7,15]. Despite ensemble performance is being usually better than that of a single model, the quality of ensemble predictions also degrades with the passing of time [13].

In this paper we present a decision support system aiming at predicting the final outcome for patients staying in an Intensive Care Unit (ICU). The system is connected to the hospital's computer network allowing for continuously assessment of its performance. The prediction is the result of an ensemble of classifiers, composed of both neural networks and decision trees as it has been shown that the different model types contribute to a lower number of coincident failures, thus increasing the ensemble performance [26]. Whenever necessary the system automatically alters the ensemble, either by changing the models weights or by deleting poor performing models.

In section 2 we present an overview of ensemble learning and some techniques used to promote base model diversity. Next, the problem we are addressing is presented. In section four we describe data used in this work and in section five we present the experimental results. Finally section six includes the discussion of the results and pointers to future work.

2   Ensemble Learning

By combining classifiers we accept an increase in complexity hoping that it leads to better results [15]. Instead of a single classifier we train a group of somewhat diverse classifiers (as a group of similar classifiers does not offer an increase in performance). It has been shown that multiple classifier systems are able to improve prediction results if each of the base models has an accuracy above 50% with somewhat uncorrelated errors. Ensembles are capable of producing better results due to their ability to overcome (at least partially) some problems faced by single model predictors. Dieterich [7] calls them computational, statistical and representational problems. Using one dataset one can induce several different classifiers having the same accuracy on the training data but with different generalisation capabilities. The statistical reason for ensembles to perform better than single classifiers is due to the fact that by averaging the results of several base models we are reducing the risk of choosing a bad one. Computational problems refers to the fact that single classifier may converge to a sub-optimal solution. Again, combining several classifiers may provide a better answer. Finally, the representational problem deals with the fact that, given limited time and data, whatever the training algorithm used it will be unable to completely explore the space of possible classifiers. Several methods for ensemble creation try to force base model diversity using different techniques [8]:

- Different initializations - e.g. varying the initial weights in neural networks;
- Different parameter choices - e.g. amount of pruning in a decision tree;

- Different architectures - e.g. number of internal nodes of a neural network;
- Different classifiers - e.g. using decision trees and neural networks;
- Different training sets - Bagging[4], boosting and variants [10] are probably the most widespread approaches. Both attempt to promote model diversity by selecting different subsets of records for the training of each base model;
- Different feature sets - using different subsets of attributes [12].

Extensive surveys on ensemble building methods can be found in [5] and in [20].

Although there are several diversity measures used for ensemble construction, the correct type of diversity measure to be used in each situation is not known [15]. Also, in terms of ensemble architecture we are still far from a consensus. Generally speaking, ensemble architecture may be static, with fixed structure and combination function or it can allow for dynamic ensembles in that the ensemble base models and combination function can be altered whenever necessary. This approach, even if more complex, is potentially of easier generalization as the dynamic nature of these ensembles optimizes their functioning for the current data. Moreover, in a dynamic environment performance starts to lower as new data arrives. In effect, it can be expected that models (or ensembles) perform better when using the latest data [13].

The problem of detecting when and how to retrain the models in dynamic environments in order to achieve good results has been extensively studied. This is related with the detection of *Concept Drift* [23] (detecting when there are changes in the hidden context that induce changes in the target concept). Several strategies have been proposed to deal with the concept drift issue with most work being focused on the detection of sudden concept drift. However, concept drift in many real world problems is much slower and difficult to detect [2]. Medical knowledge is relatively stable and thus we expect the concept drift present in our data to be very slow.

3 Problem Description

We are currently developing an intelligent Decision Support System called INTCare [11]. INTCare predicts the patient's outcome (the patient status at the time of hospital discharge: dead or alive) and also predicts organ failure for six organ or systems (cardiovascular, respiratory, hepatic, renal, central nervous system and hematologic). At the present time, models included in INTCare were obtained via batch off-line training even though the INTCare's architecture allows for integration with the Hospital's electronic records. This makes it possible for INTCare to collect information both for making predictions and to the appraisal of its predictive performance. In this paper we present the results of several experiments on the use of this information to build a predictive model that maintains an interesting predictive performance in the ICU. In particular we are going to address the problem of creating a system capable of predicting hospital mortality for ICU patients using data collected during the first 24 hours after ICU admission. Moreover, such system must be able to function without human intervention, i.e. it must automatically adjust to new data. Other constraints relate to the relative lack of data as typical 10-bed ICU has less than 500 patients per year. For this study we have only four months of ICU data (collected in three different ICUs).

3.1 Data Description

Data to be used was collected in three ICUs in the late 1990's. The patient's data was manually collected and registered by the nursing staff. Data available includes 216 records for the first ICU (ICU 1), 312 records for ICU 2 and 352 records for ICU 3. Data is not balanced, with the number of survivors being larger than the number of deaths (163 survivors and 53 deaths for ICU 1, 261 survivors and 51 deaths for ICU 2 and 287 survivors and 65 deaths for ICU 3).

Four variables contain the information that remains unchanged during the patient's stay, including where the patient came from, the type of admission, the patient's age and the Simplified Acute Physiology Score (SAPS II) score [17] (SAPS II is a severity of disease classification system). The remaining variables (except for the outcome) contain values collected during the patient's first day in the ICU. There are six variables containing the Sequential Organ Failure Assessment (SOFA) score values [25]. The SOFA score is a scoring system to determine the extent of a person's organ function or rate of failure (a SOFA score of 3 or 4 indicates organ failure). After consultation with the ICU doctors three new variables were created: the sum of the SOFA values (total SOFA), the sum of the renal and cardiovascular SOFA values and the sum of the renal, cardiovascular and respiratory SOFA values, as doctors believe these new variables may be

Table 1. The attributes of the intensive care data

Attribute	Description	Domain Values
SAPS II	SAPS II score	$\{0, 1, \ldots, 163\}$
age	Patients' age	$\{18, \ldots, 100\}$
admtype	Admission type	$\{1, 2, 3\}^a$
admfrom	Admission origin	$\{1, 2, \ldots, 7\}^b$
sResp	Respiratory SOFA score	$[0, \ldots, 4]$
sCard	Cardiovascular SOFA score	$[0, \ldots, 4]$
sCoag	Coagulation SOFA score	$[0, \ldots, 4]$
sHepa	Hepatic SOFA score	$[0, \ldots, 4]$
sCNS	Central Nervous System SOFA score	$[0, \ldots, 4]$
sRenal	Renal SOFA score	$[0, \ldots, 4]$
sTotal	Total SOFA score	$[0, \ldots, 24]$
sReCa	Renal + Cardio SOFA score	$[0, \ldots, 8]$
sRCR	Renal + Cardio + Resp. SOFA score	$[0, \ldots, 12]$
NFail	Number of organ Failures	$[0, \ldots, 6]$
NEV	Total number of events	$[0.0, \ldots, 19.0]$
NCR	Total number of critical events	$[0.0, \ldots, 8.0]$
PCTCR	Percentage of critical events	$[0.0, \ldots, 1.0]^c$
death	The occurrence of death	$\{0, 1\}^d$

^{a} 1 - Non scheduled surgery, 2 - Scheduled surgery, 3 - Medical.
^{b} 1 - Operating Theater, 2 - Recovery room, 3 - Emergency room, 4 - Ward,
5 - Other ICU, 6 - Other hospital, 7 - Other sources.
^{c} Or '-1' when the total number of events is zero.
^{d} 0 - No death, 1 - Death.

relevant to predict hospital outcome. Also, another variable contains the number of organ failures. Another group of variables was derived from the information available on the intermediate outcomes, which are defined from four monitored biometric measurements: the systolic blood pressure, the heart rate, the oxygen saturation and the urine output (UR). Finally, the last attribute denotes the patients' final outcome (status at the time of hospital discharge). For each of the ICUs considered the data is ordered by date, allowing us to simulate the temporal evolution of the system. Information regarding the definition of events and critical events may be found in [18]. The main features of the clinical data are described in Table 1.

4   INTCare

In previous work [11] an agent-based intelligent decision support system (INT-Care) was designed to predict hospital outcome and organ failure for patients admitted to an ICU. INTCare automatically collects data from the bed side monitors and from the medical team. As it may have access to the patients electronic records INTCare is able to check out if the predictions made were correct, enabling a supervised learning approach. The architecture of the system is presented in Figure 1. It shows the different components, from data acquisition to model induction to the delivery of results to physicians. The experiments presented in this paper are related to the design of the ensemble agent that has the responsibility of updating the prediction models in INTCare in order to improve its performance.

4.1   Ensemble Creation

Ensembles have been shown outperform single classifier systems in many different problems [15,7]. Moreover, hybrid ensembles seem lead to an even greater improvement [26]. Nevertheless, such promising results are usually obtained in off-line learning experiments. In order to try and replicate as much as possible the functioning of the INTCare system in a real setting we did not modify the order of the records in the available data. In doing so we opted not to merge data from different ICUs into a big dataset as we would possibly loose potentially valuable information. This limited us to relatively small datasets and so we decided not to use techniques that rely on record sampling, like boosting or bagging. In fact, boosting has been shown to perform poorly in the presence of noise and outliers, especially when used on small datasets [3,10]. We chose an approach similar to the Random Subspace Method [12] as it can be used on small datasets as it introduces diversity by selecting attributes, not by sampling of the available records. Alternative methods for inducing base model diversity may be explored in the future. Each model in the ensemble is trained on a randomly selected subset of the training dataset. This dataset contains all the records available for training but only a subset of the attributes. In effect, each attribute as a 50% chance of being selected for inclusion. This enables an

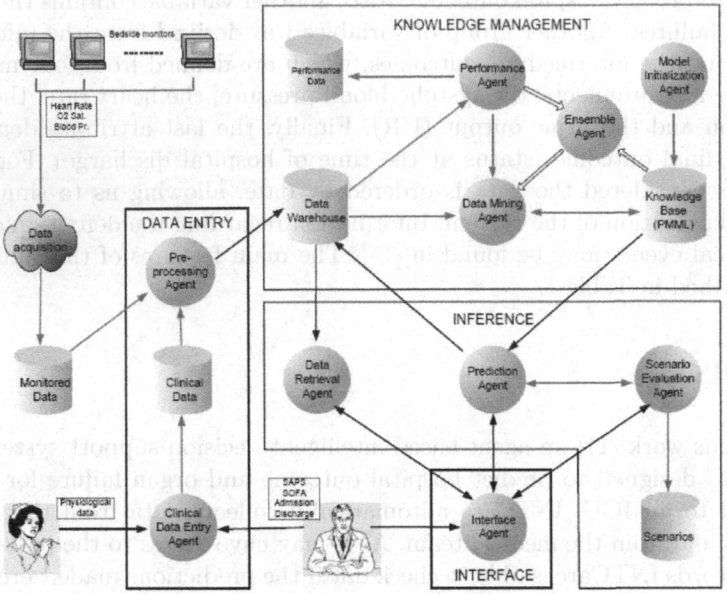

Fig. 1. INTCare's architecture

increase in base model diversity even if we are using small datasets as it ensures that base models are trained on different data. Moreover, the ensemble is composed of two types of base models (neural networks and decision trees) thus increasing its representational power [7]. Base models were trained to produce a binary output and the final prediction is a numeric value corresponding to the weighted average of the predictions made by each individual model:

$$outcome = \sum_{i=1}^{n} w_i o_i$$

where n represents the number of models in the ensemble, w_i represents the weight of the model i and o_i stands for the outcome of that model i.

In this first step, the ensemble outcome is the average of the individual predictions as all the models weights are set to $\frac{1}{n}$.

4.2 Ensemble Evolution

Decision Support Systems must be able to adapt to changes in their environment. Ideally, such systems should be autonomous in the sense that they are able to keep producing good results even when faced with changes in their environment [24]. When using multi-classifier systems, the adaptation may be achieved by modifying the combination function so that the best performing models contribute more to the solution. Other possible actions in this step include deleting poor performing models and creating new ones or modifying the combination function. Initially all models have the same weights. The second step in the training procedure involves dynamic adjustments of the weights in accordance to the models performance. The evolution step develops as follows: after each prediction the outcome predicted by each model is compared with the correct one. Whenever a model makes a correct prediction its weight is increased. Similarly, the weight is decreased (by the same value) if the prediction is wrong. When a model's weight reaches a negative value that model is deleted from the ensemble as it is performing poorly.

5   Experimental Results

With the available data from the chosen ICUs we created three subsets as illustrated in Figure 2. Dataset A was used for initial training, dataset B was used for ensemble evolution (weight updates and eventual elimination of models) and dataset C (containing 30% of the available data) was used for testing. Temporally, data in dataset A is the oldest with dataset C being composed of the latest data. In the present work we intend to evaluate the usefulness of a dynamic hybrid ensemble in maintaining or improving predictive performance levels in such context. To evaluate the results we used the average of the values of the area under the Receiver Operating Characteristic curve (AUC ROC) obtained after 10 runs of each experiment. The ROC curves are often used in the medical area to evaluate computational models for decision support, diagnosis and prognosis [16,27]. A model presenting an AUC of 1 has perfect discriminative power (perfect predictive ability) while a value of 0.5 corresponds to random guessing.

Our first experiment served to determine the number of models to be included in the ensemble. We evaluated ensembles both with 25 and with 100 base models. At the same time we tested four different degrees of evolution by reserving different percentages of the total data for training (dataset A) and ensemble evolution (dataset B):

- Configuration 1 – 20% for training and 50% for evolution
- Configuration 2 – 30% for training and 40% for evolution
- Configuration 3 – 40% for training and 30% for evolution
- Configuration 4 – 50% for training and 20% for evolution

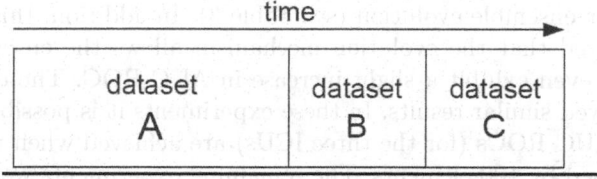

Fig. 2. Dataset creation

Table 2. Results for the three configurations after the initial training (% of AUC ROC)

	ICU 1		ICU 2		ICU 3	
Models	25	100	25	100	25	100
Config.						
1	58.6±3.4	58.3±2.1	75.5±2.0	79.0±0.8	60.7±3.5	59.9±3.5
2	58.6±2.3	55.8±5.6	69.1±2.1	74.6±0.8	58.7±5.4	64.8±1.2
3	59.0±2.9	59.2±1.1	68.4±1.7	72.4±1.6	57.4±3.9	63.4±3.8
4	58.3±2.5	58.8±1.5	70.4±2.0	76.8±1.1	58.0±4.3	65.4±2.9
Average	58.6	58.0	70.8	75.7	58.7	63.4

Table 3. Results after ensemble evolution (% of AUC ROC)

	ICU 1		ICU 2		ICU 3	
Models	25	100	25	100	25	100
Config.						
1	56.1±4.2	55.5±1.7	75.7±2.1	78.8±0.8	62.2±2.4	62.1±2.1
2	57.6±2.3	55.6±1.4	69.4±2.0	74.7±0.8	59.1±6.3	64.9±1.4
3	59.4±2.9	59.9±5.5	68.4±1.6	72.5±1.4	57.6±4.0	63.5±3.5
4	58.2±2.4	58.2±1.6	70.7±1.8	76.8±1.0	58.8±3.8	64.4±2.7
Average	57.8	56.3	71.0	75.7	59.4	63.7

Table 4. Results for the balanced training datasets (% of AUC ROC).

	ICU 1		ICU 2		ICU 3	
Config.	In.	Ev.	In.	Ev.	In.	Ev.
1	56.9±1.7	42.4±4.4	78.9±1.1	80.6±0.6	48.3±4.0	63.7±3.3
2	59.9±1.0	45.7±1.6	75.7±2.2	79.5±0.5	65.8±2.6	65.9±2.6
3	56.3±1.2	56.8±1.1	70.8±3.8	76.2±1.0	64.1±3.5	68.9±1.8
4	59.2±1.3	60.0±1.3	70.9±2.2	75.7±1.0	65.1±2.4	67.3±1.8
Average	58.1	51.2	74.1	78.0	60.8	66.4

In. - After initial training; Ev. - After ensemble evolution

In Table 2 we present the results obtained in each configuration and the global averaged results. As can be seen, better results are achieved when using ensembles of 100 base models in opposition to ensembles composed of 25 base models. In fact, after the initial training we observed an increase close to 2% in the value of the area under the ROC. These models also have an higher average AUC ROC after ensemble evolution (see Table 3). In addition, this first experiment also showed that the evolution mechanism allows the ensemble to stay up-to-date and even exhibit a slight increase in AUC ROC. The different configurations showed similar results. In these experiments it is possible to observe that the best AUC ROCs (for the three ICUs) are achieved when using the ensembles composed by 100 elements. The remaining experiments were conducted using ensembles of 100 base models.

As the data is unbalanced (most records are for outcome 0) we investigated the effect of data balancing on the ensemble performance. New ensembles were built using balanced training datasets. No changes were made to the evolution and evaluation datasets. In Table 4 we can observe that when a balanced training dataset is used the discriminative power of the ensembles as measured by the AUC ROC is greater than it was when using unbalanced data.

Except for ICU 1 (the one with only 216 records), results from Table 4 compare favorably to those in Table 2.

6 Conclusions and Future Work

Future decision supports systems must be capable of adapting to changes in their environment [19,24]. In the medical area this will allow for an easier integration of these tools in everyday use as its reliability tends to increase. In this paper we presented the results of a set of experiments in building the adaptive module of a decision support system (INTCare). Considering the available data, both in the present time and in the foreseeable future, we tested dynamic hybrid ensemble architectures allowing for unassisted operation while maintaining acceptable performance. We concluded that, for our problem, one should use ensembles of 100 base elements with weights updates independently of the data distribution. Moreover, data balancing for training of the base models was found to lead to better results and must be included in the architecture demands. The results seem to be promising and we expect to be able to confirm them when more data is available. For ICUs 2 and 3 one has observed increased AUC ROC values (more than 2%), despite the evolution of the ensembles being done over small datasets. Less encouraging results regarding ICU 1 are likely to be related to the fact that the dataset used was too small. Interesting research issues regard the automatic inclusion of new models, giving rise to such question as to when to train and how to include the new models in the ensemble (the value of initial weight for the new model). Future work includes extending the architecture to include organ failure prediction. Other necessary developments will consider the need to incorporate all the available data as it is being registered. For patients staying several days in the ICU, the prediction models must take into account not only data from the initial 24 hours but also all data stored during the stay. Moreover, relevant clinical information is possibly hidden in the sequence in witch clinical adverse events occur. As the INTCare system has access to data collected via bed side monitoring this seems to be an interesting path to be explored.

Acknowledgments

We thank FRICE and the BIOMED project BMH4-CT96-0817 for the provision of part of the EURICUS II data and support for this study, which is integrated in a PhD program, developed at Instituto de Ciências Biomédicas Abel-Salazar from University of Porto and the Departments of Computer Science/Information Systems from the University of Minho. Part of this work was supported by the Fundação para a Ciência e Tecnologia (grant SFRH BD 28840 2006).

References

1. Arnott, D., Pervan, G.: A critical analysis of decision support systems research. Journal of Information Technology 20(2), 67–87 (2005)
2. Baena-García, M., del Campo-Ávila, J., Fidalgo, R., Bifet, A., Gavaldá, R., Morales-Bueno, R.: Early drift detection method. In: ECML PKDD Workshop on Knowledge Discovery from Data Streams (2006)
3. Breiman, L.: Arcing classifiers. The Annals of Statistics 26(3), 801–849 (1998)
4. Breiman, L.: Bagging predictors. Machine Learning 24(2), 123–140 (1996)
5. Brown, G., Wyatt, J.L., Harris, R., Yao, X.: Diversity creation methods: A survey and categorisation. Journal of Information Fusion 6(1), 5–20 (2005)
6. Buchanan, B., Shortliffe, E.: Rule-Based Expert Systems: The MYCIN experiments of the Stanford Heuristic Programming Project. Addison-Wesley, Reading, MA (1984)
7. Dietterich, T.G.: Ensemble methods in machine learning. In: Kittler, J., Roli, F. (eds.) MCS 2000. LNCS, vol. 1857, pp. 1–15. Springer, Heidelberg (2000)
8. Duin, R.P.W.: The combining classifier: To train or not to train? In: Kasturi, R., Laurendeau, D., Suen, C. (eds.) Proceedings 16th International Conference on Pattern Recognition, vol. II, pp. 765–770. IEEE Computer Society Press, Los Alamitos (2002)
9. Fayyad, U.M., Piatetsky-Shapiro, G., Smyth, P.: From data mining to knowledge discovery: an overview. In: Advances in knowledge discovery and data mining, pp. 1–34. American Association for Artificial Intelligence, Menlo Park, CA, USA (1996)
10. Freund, Y., Schapire, R.E.: A decision-theoretic generalization of on-line learning and an application to boosting. In: European Conference on Computational Learning Theory, pp. 23–37 (1995)
11. Gago, P., Santos, M.F., Silva, Á., Cortez, P., Neves, J., Gomes, L.: Intcare: a knowledge discovery based intelligent decision support system for intensive care medicine. Journal of Decision Systems 14(3), 241–259 (2005)
12. Ho, T.K.: The random subspace method for constructing decision forests. IEEE Transactions on Pattern Analysis and Machine Intelligence 20(8), 832–844 (1998)
13. Klinkenberg, R., Ruping, S.: Concept drift and the importance of examples. In: Text Mining – Theoretical Aspects and Applications. Physica-Verlag (2003)
14. Kulikowski, C.A., Weis, S.M.: Artificial Intelligence in Medicine. In: Representation of expert knowledge for consultation: the CASNET and EXPERT projects, pp. 21–56. Westview Press, Boulder (1982)
15. Kuncheva, L.I.: Combining Pattern Classifiers: Methods and Algorithms. Wiley-Interscience, Chichester (2004)
16. Lasko, T.A., Bhagwat, J.G., Zou, K.H., Ohno-Machado, L.: The use of receiver operating characteristic curves in biomedical informatics. J. of Biomedical Informatics 38(5), 404–415 (2005)
17. Le Gall, J.R., Lemeshow, S., Saulnier, F.: A new simplified acute physiology score (saps ii) based on a european/north american multicenter study. JAMA 270(24), 2957–2963 (1993)
18. Silva, Á., Cortez, P., Santos, M.F., Gomes, L., Neves, J.: Mortality assessment in intensive care units via adverse events using artificial neural networks. Artificial Intelligence in Medicine 36(3), 223–234 (2006)
19. Michalewicz, Z., Schmidt, M., Michalewicz, M., Chiriac, C.: Adaptive Business Intelligence. Springer, Heidelberg (2006)

20. Polikar, R.: Ensemble based systems in decision making. IEEE Circuits and Systems Magazine 6(3), 21–45 (2006)
21. Pople, H.E.: Evolution of an Expert System: from INTERNIST to CADUCEUS. In: Artificial Intelligence in Medicine, pp. 179–208. Elsevier Science, Amsterdam (1985)
22. Shim, J.P., Warkentin, M., Courtney, J.F., Power, D.J., Sharda, R., Carlsson, C.: Past, present, and future of decision support technology. Decis. Support Syst. 33(2), 111–126 (2002)
23. Tsymbal, A.: The problem of concept drift: definitions and related work. Technical Report TCD-CS-2004-15, Department of Computer Science, Trinity College Dublin, Ireland (2004)
24. Vahidov, R., Kersten, G.E.: Decision station: situating decision support systems. Decision Support Systems 38, 283–303 (2004)
25. Vincent, J., Moreno, R., Takala, J., et al.: The sofa (sepsis-related organ failure assessment) score to describe organ dysfunction/failure. Intensive Care Med. 1996 22, 707–710 (1996)
26. Wang, W., Partridge, D., Etherington, J.: Hybrid ensembles and coincident-failure diversity. In: Proceedings of the International Joint Conference on Neural Networks, vol. 4, pp. 2376–2381. IEEE Press, Washington, USA (2001)
27. Zweig, M.H., Campbell, G.: Receiver-operating characteristic (roc) plots: a fundamental evaluation tool in clinical medicine. Clin. Chem. 39(4), 561–577 (1993)

A Tool for Interactive Subgroup Discovery
Using Distribution Rules*

Joel P. Lucas[1], Alípio M. Jorge[2,3], Fernando Pereira[2,3],
Ana M. Pernas[4], and Amauri A. Machado[4]

[1] Departamento de Informática y Automática – Universidad de Salamanca – Salamanca, Spain
[2] Faculdade de Economia – Universidade do Porto, Portugal
[3] LIAAD, INESC Porto L.A., Portugal
[4] Instituto de Física e Matemática – Universidade Federal do Pelotas (UFPel), Pelotas – RS –
Brazil

Abstract. We describe an approach and a tool for the discovery of subgroups
within the framework of distribution rule mining. Distribution rules are a kind of
association rules particularly suited for the exploratory study of numerical
variables of interest. Being an exploratory technique, the result of a distribution
mining process is typically a very large number of patterns. Exploring such results
is thus a complex task and limits the use of the technique. To overcome this
shortcoming we developed a tool, written in Java, which supports subgroup
discovery in a post-processing step. The tool engages the analyst in an interactive
process of subgroup discovery by means of a graphical interface with well defined
statistical grounds, where domain knowledge can be used during the identification
of such subgroups amid the population. We show a case study to analyze the
results of students in a large scale university admission examination.

Keywords: Data Mining, Subgroup Discovery, Post-processing, Visualization,
Association Rules, Distributions.

1 Introduction

Data mining is the discovery of non-trivial, implicit, previously unknown, and
potentially useful information from data [5]. Typically, data mining tasks are divided
in two broad categories: directed data mining, where the data analyst has a very well
defined and measurable objective, and undirected data mining, where the objective is
very broadly defined and the success of the task relies on aspects which are subjective
and hard (or impossible) to formalize. For this last category of tasks, it is very
important to have tools that provide a general view of the obtained knowledge –
especially if the number of patterns resulting from the data mining step is very large –
and at the same time allow the data analyst to spot particularly interesting patterns and
to examine them more carefully. Visualization plays naturally a strong role here.

---

* Supported by POCI/TRA/61001/2004 Triana Project (Fundação Ciência e Tecnologia),
FEDER e Programa de Financiamento Plurianual de Unidades de I & D.

J. Neves, M. Santos, and J. Machado (Eds.): EPIA 2007, LNAI 4874, pp. 426–436, 2007.

However, it is also important that the concepts behind the visualization techniques are well founded and well defined.

In this paper we describe an approach to subgroup discovery, which is mostly a non-directed task [14]. The aim is to find well characterized and expressive groups of individuals that have an interesting distribution for a particular variable or property of interest. Although in general this property of interest can have any type, we developed a method suited for numerical properties. Our approach is based on distribution rule mining [10], which is related to association rule mining [1]. We propose a tool that visualizes the mining results and enables the interactive exploration of interesting subgroups. The exploration space is based on solid statistical grounds.

2 Subgroup Discovery

Subgroup Discovery is typically an undirected task, which was first suggested by Klösgen [13]. Its general aim is to find frequent patterns corresponding to subsets of individuals from a given population which have some unusual, unexpected or deviating feature. Such patterns do not state a general behavior of the whole dataset, but in specific subsets. Despite their locality, such patterns are often useful, either because the interest of the analyst is very focused (as it is the case when a marketer is looking for market niches), or the available data does not allow to derive a complete description of the behavior for the whole population [14].

The unusual, unexpected or deviating feature mentioned above may correspond to a specific dataset variable, which we refer to as the "Property of Interest". Such property can drive the subgroup discovery process, where relations between the property of interest (regarded as a dependent variable) with the rest of the dataset variables (regarded as independent variables) are sought. Independent variables are used to produce the description of subgroups, so that their meaning can be understood by the data analyst [14].

Subgroup discovery has some similarities with clustering. However, these are distinct tasks. In clustering, we want to place each individual in its own group, according to criteria which are global to the dataset. A single cluster cannot be evaluated without considering all the other clusters, whereas in subgroup discovery we can identify groups that are interesting regardless of other potential groups around. Association rule mining can also be often regarded as subgroup discovery, since each rule/itemset represents a subgroup, and the interest of the subgroup can be measured using objective association rule interest measures such as lift or conviction. However, AR mining can be used for other purposes, and subgroup discovery can potentially be done using other techniques.

3 The Proposed Approach

The approach we employed in this work was proposed by Jorge et. al. [11] and Pereira [17]. The approach suggests the use of visualization techniques for interactive subgroup discovery. The data analyst is not directly confronted with the, possibly very large, output of the subgroup discovery algorithm. The visual exploration is driven, not only by objective measures of the discovered patterns, but also by the analyst's domain knowledge which can used in the selection of each exploration step.

The approach finds subgroups expressed as Distribution Rules (DR) [10], a kind of association rules with a statistical distribution on the consequent [11]. Each rule corresponds to a subgroup of potential interest for the population under study. A DR may be formally defined as follows: $A \rightarrow y=Dy|A$, where A is a set of conditions corresponding to the antecedent part of a DR, y is a property of interest and $Dy|A$ is an empirical distribution of y when A is observed. $Dy|A$ is composed by a set of pairs $<y_i/\text{freq}(y_i)>$, where y_i is one particular value of y found when A is observed and $\text{freq}(y_i)$ is the frequency of y_i when the items from A are observed.

We now show a simple example. Table 1 shows a hypothetical dataset with data from patients suffering from cold, flu or pneumonia.

Table 1. Hypothetical dataset

Patient	City	Diagnosis	Age
1	Pelotas	Cold	35
2	Bagé	Cold	18
3	Bagé	Pneumonia	45
4	Canguçu	Flu	58
5	Pelotas	Pneumonia	60
6	Bagé	Flu	14
7	Pelotas	Cold	32
8	Canguçu	Pneumonia	38
9	Pelotas	Pneumonia	74
10	Pelotas	Pneumonia	60

One observable pattern is the following DR:

$City=Pelotas$ & $Diagnosis=pneumonia \rightarrow age = \{60/2, 74/1\}$.

This corresponds to the subgroup of people coming from the city of Pelotas and having a diagnosis of pneumonia. Patterns like this can be automatically found using the data mining method described in [10]. One important advantage of this method is that the numerical property of interest does not have to be previously discretized, because instead of dealing with crisp values or intervals, the method works with raw distributions. Another advantage is that the objective interest of each subgroup is related to the Kolmogorov-Smirnov statistic for the comparison of two distributions. For each subgroup, we perform a KS test comparing the distribution of y within the subgroup with the distribution of y for the whole population. Interesting groups (from an objective point of view) are the ones which have distributions that significantly differ from the expected.

The number of subgroups discovered can be very large. To support the data analyst we represent graphically the set of subgroups in 2 dimensional plots. The x and y axes of these plots correspond to statistical measures (such as mean, median, mode, standard deviation, kurtosis and skewness) that characterize the distribution of each subgroup [11].

These 2D plots provide an overview of the results. After that, the data analyst can pinpoint (by clicking on the 2D plot), and view, the distribution of the property of interest of a particular subgroup, visually compare its distribution with the whole population, observe in detail other statistical features of the subgroup (such as its support), and also its definition in terms of variables in the dataset.

4   The Developed Tool

The main aim of the tool developed in this work is to provide support for the analyst in a decision making processes. The tool may be seen as an interactive visualization environment, basically composed of a graphical description of subgroups by means of statistical measures that characterize the distributions of such subgroups.

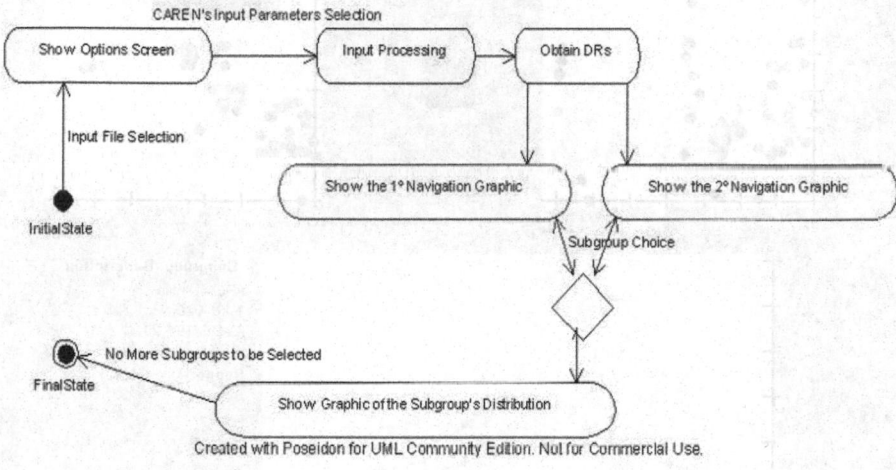

Fig. 1. Subgroup discovery process using the tool

The subgroups shown by the tool's graphical user interface correspond to distribution rules. These are obtained with CAREN-DR [10], which is based on CAREN [3] association rule engine. CAREN-DR is transparently executed for the user by the subgroup exploration tool.

Before running CAREN-DR, the analyst sets the value of parameters needed for obtaining the DRs. This is done through the graphical interface. One of these parameters sets the attribute to be used as the property of interest.

The DRs are obtained from a dataset stored in CSV (*Comma Separated Value*) files. Afterwards, post-processing and analysis take place. A general overview of the obtained subgroups is described by two 2D alternate navigation graphics (figure 2). Then, the data analyst may choose a particular subgroup to visualize (in a third graphic) its distribution and other features. Figure 1 shows a diagram representing the discovery process.

For the x and y axis of the navigation charts 1 and 2, the data analyst can choose two statistical measures among six available: median, mean, mode, kurtosis, skewness and standard deviation. The choice of the measures can be done according to each population, to the analysis purpose and to the analyst preferences. Figure 2 shows the tool's main screen, where the two navigation graphics can be seen at the top.

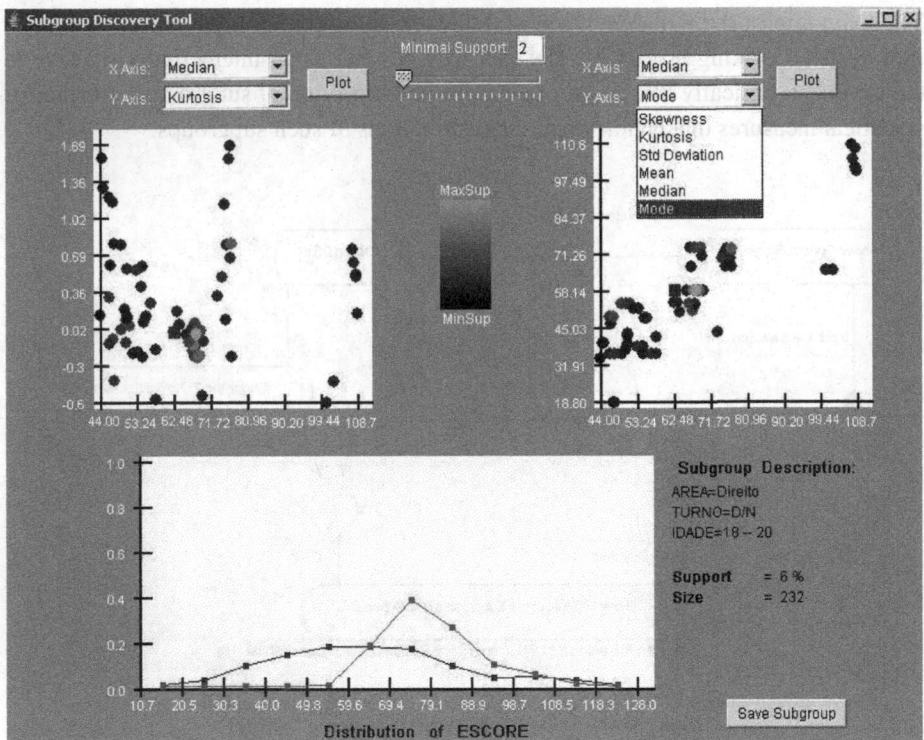

Fig. 2. The Tool's Main Screen

The dots on the navigational plots correspond to the subgroups obtained. The blue square corresponds to the distribution of the whole population. The support of each subgroup is expressed by different color grades: the greener a dot is, the higher is its support. The minimum support threshold value, which was previously defined by the data analyst for distribution rule discovery, is presented on a selection field between the two navigation graphics. This threshold can always be changed at run time.

At the lower part of figure 2 we can see the graphical display of the distribution of the property of interest for a selected subgroup (green line). On the right, we can see the subgroup description, which is very important for the data analyst, and the support and size (the absolute support) of the subgroup. For guidance, the distribution of the property of interest for the whole population is also shown (blue line). If the subgroup is considered interesting, it can be saved for later presentation.

At any time the analyst can go back to the navigation charts, change the statistical measures that serve as coordinates, change minimal support and focus on other subgroups.

5 Implementation Details

The Java platform was chosen to implement the tool. Therefore, it is platform-independent and may be easily placed in Web environments as a *Java Applet*. Furthermore, the Java Platform provides a vast collection of classes and methods, supporting the development of interactive and user-friendly interfaces.

In addition to standard libraries provided by the Java Platform, we employed some extra libraries on the implementation: the *Jakarta Mathematics* and the *GenJava-CSV*. The *Jakarta Mathematics* is a library developed within the *Jakarta Commons* project [8] and it was released under the Apache License, version 2.0. *Jakarta Mathematics* provides a vast library of classes that may be used to solve computing problems from diverse domains, including the mathematical domain too. However, in this work we used only the package that provides algorithms for descriptive statistics.

The other library we employed, the *GenJava-CSV*, supports read and write operations of CSV files. *GenJava-CSV* is available under the BSD license and is part of a set of open source libraries implemented by a group of developers named OSJava (*Open Sourced Java*) [15].

Additionaly, in the tool we can select one of two languages: English and Portuguese. Other languages can be easily included.

6 Case Study

As a final step of this work, we show a case study that illustrates some features of the developed tool. For that, we have used a dataset from the 2005 *Vestibular* (competitive examination for students admission) of the *Universidade Federal de Pelotas* (UFPel), located in Brazil. The aim of the data analysis is to understand the characteristics of the students with respect to the outcome in the admission's exam. Therefore, our study focused on identifying associations between students' performance in the exams and their profile as provided by the vestibular enrolment form. Among the information provided by enrolment forms, we used the following attributes in the analysis: AGE, the DOMAIN area of the course chosen, CITY of origin and the preferred SHIFT (morning or night shift). Subsequently, we added an attribute corresponding to the final SCORE yielded by the student at the admission examination. This attribute is the property of interest for subgroup discovery.

In the pre-processing phase, we considered only the students who completed the examination process. Moreover, we discretized the attribute AGE in the following intervals: equal or lower than 17, between 18 and 20, between 21 and 23, between 24 and 27, between 28 and 30, between 31 and 35, between 36 and 42, between 43 and 50, greater than 50. The values of the CITY attribute were reduced to: Pelotas; Porto Alegre; Rio Grande; Cities out of the Rio Grande do Sul state; Small cities that belong to the region of Pelotas; Rest of non-capital cities in Rio Grande do Sul state. The values of the DOMAIN area attribute were adjusted in order to be restricted to the following ones: literature and arts, human science, exact and technological science, agrarian science, life science, law and medicine. The courses of law and medicine were separated from their domain area because they are the courses that have the highest number of candidate students.

6.1 Obtained Results

After building the subgroups and analyzing the navigation charts (2 dimensional plots in figure 2) we have observed that subgroups related to candidate students for the course of medicine stand out. Such subgroups present the lowest value of the standard deviation measure. Therefore we could presume that, in general, candidate students for the course of medicine reached similar scores. This also indicates the likelihood of score ties and a source of difficulties in student selection.

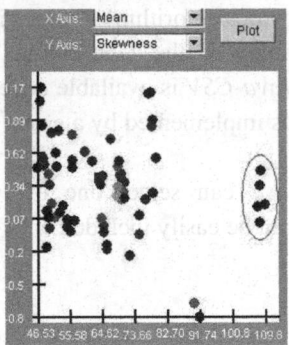

Fig. 3. Subgroups of students from the medicine course (within ellipse)

Among the previous mentioned subgroups, one had candidate students for medicine living outside Rio Grande do Sul. This subgroup has the highest mean among discovered subgroups. This sort of insight provided by subgroup discovery with the help of a graphical tool provides profiling information to the admissions office and other university services. This information can be used in course marketing (by targeting to areas where good students can be found) and also in course design.

Figure 3 represents the navigational chart for the skewness (Y axis) and mean (X axis) measures. There, we immediately identify five subgroups having relatively high score values. By clicking on each of the five subgroups on the navigation graphic, we can analyze, individually, the description of each one through the descriptive graphic (as the one showed on the bottom part of figure 2). These are the subgroups with candidate students for the course of medicine.

On the other hand, there are subgroups with low mean values. What are the features of these students? The S2 subgroup, for example, has a mean value of 46.6 (lower than the average). S2 is composed of candidate students for night courses of literature and arts domain area. Taking into account that S2 has a low standard deviation value (11.1), we may presume that the low score is a persistent feature of this subgroup. The data analyst can also take a closer look at the distribution of the values for the property of interest within this subgroup and compare this with the distribution of the whole population. That can be obtained by clicking on S2' dot on any of the navigation charts. The distribution of the subgroup (more to the left) clearly shows the concentration of lower scores, although showing some students with relative success.

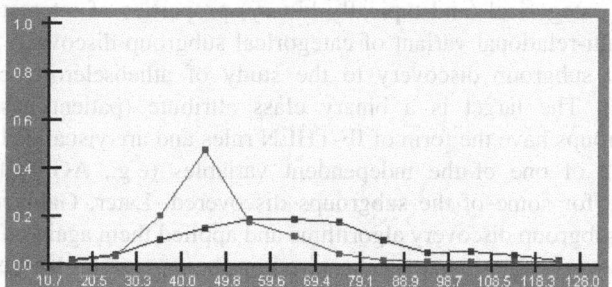

Fig. 4. Subgroup S2 (distribution to the left) when compared to the whole population (lower distribution)

Taking into account age, we observe that the subgroup composed of students who are between 28 and 30 years old (S5) has a mean value lower than the one composed of students who are between 24 and 27 years old (S6), which, subsequently, has a mean value lower than the one composed of students who are between 21 and 23 years old (S7). Not surprisingly, the subgroup composed of students who are between 18 and 20 years old (S8) has the highest mean value. We can observe this by clicking first on the subgroup represented by the uppermost red dot placed on the navigation chart presented on figure 5 and then analyzing the distribution of the property of interest. Subsequently, we repeated the same process for the subgroups represented by the red dots placed under the uppermost one.

Taking into account the city attribute, we observe that students who live in small cities in Pelotas region, have the lowest mean when compared to other subgroups from other cities. S9 has a mean value of 56, while the whole population has a mean value of 64. Additionally, the standard deviation value of S9 (15.0) is lower than the majority of the subgroups. Again, this indicates a prevalent feature.

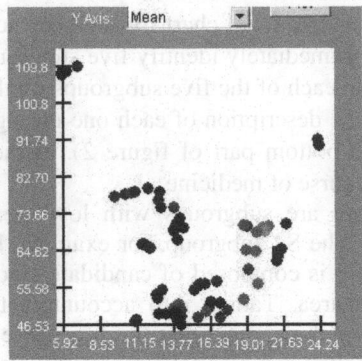

Fig. 5. Subgroups S5, S6, S7 and S8 highlighted in a navigation graphic

7 Related Work

Klösgen [13] identified the subgroup discovery data mining problem. The examples shown are for categorical (and typically binary) properties of interest. Wrobel [18] proposed a multi-relational variant of categorical subgroup discovery. Gamberger et al. [6] applied subgroup discovery to the study of atheosclerotic coronary heart disease (CHD). The target is a binary class attribute (patient has/doesn't have disease). Subgroups have the form of IF-THEN rules and are visualized by displaying the distribution of one of the independent variables (e.g., AGE), for the whole population and for some of the subgroups discovered. Later, Gamberger et al. [7] proposed two subgroup discovery algorithms and applied them again on a CHD study. The rule discovery algorithms are inspired in beam-search. Kavsek et al. [12] adapted the APRIORI association rule discovery algorithm for subgroup discovery with categorical properties of interest. Browsing and post-processing environments for association rules discovery include PEAR [9] and Ma et al.'s work [16]. These works propose browsing large sets of association rules structured as a generality lattice of itemsets. Distribution rules are related to Quantitative Association Rules (QAR) [1], but take advantage of the whole distribution instead of specific distribution measures. However, the two dimensional visual browsing space approach we present, can also be used with QAR. In this case, the representation of the subgroup would be merely textual since we would not have a subgroup distribution to display.

8 Conclusions

Subgroup discovery is an undirected data mining task that can be accomplished by applying a distribution rule mining algorithm to a dataset. Each distribution rule corresponds to a subgroup, with a logical definition and a characteristic distribution for one numerical property of interest. An interesting subgroup is one with a distribution that is significantly different from the distribution of the whole population. The difference between the distributions is established using the Kolmogorov-Smirnov statistic.

The large number of subgroups produced motivated the development of a Java based tool that provides a visual interactive environment for exploratory data analysis. A broad view of the subgroup space is given by 2D plots defined by statistical measures of the distributions. These plots serve as navigational charts that enable the analyst to identify interesting subgroups. Each group can be further explored by viewing its definition and distribution.

A case study on a population of university admission examinations illustrated the main features of this visual data mining environment. We saw how outstanding groups are identified and how the inspection of these groups enables the discovery of characteristics of the different groups of candidate students. The extracted knowledge can potentially be used for marketing purposes (student targeting), course design or even social studies on the underlying population.

As limitations of this work we should point out the difficulty in objectively validating the discovery process. This would always be difficult in an interactive process. However, controlled experiments with artificial tasks could be designed in order to measure the usability of the tool.

References

1. Aumann, Y., Lindell, Y.: A statistical theory for quantitative association rules. Journal of Intelligent Information Systems (2003)
2. Agrawal, R., Srikant, R.: Fast algorithms for mining association rules in large databases. In: Proceedings of the 20th International Conference on Very Large Databases, pp. 487–499 (1994)
3. Azevedo, P.J.: Caren - A Java Based Apriori Implementation for Classification Purposes, Technical Report, Universidade do Minho, Portugal (2003)
4. Fayyad, U.M., Piatetsky-Shapiro, G., Smyth, P.: From Data Mining to Knowledge Discovery: An Overview. In: Advances in Knowledge Discovery and Data Mining, pp. 11–34 (1996)
5. Frawley, W.J., Piatetsky-Shapiro, G., Matheus, C.J.: Knowledge discovery in databases: An overview. In: Advances in Knowledge Discovery and Data Mining, pp. 57–70 (1992)
6. Gamberger, D., Lavrac, N.: Active subgroup mining: a case study in coronary heart disease risk group detection. Artificial Intelligence in Medicine 28(1), 27–57 (2003)
7. Gamberger, D., Lavrac, N., Wettschereck, D.: Subgroup visualization: A method and application in population screening. In: Proceedings of the International Workshop on intelligent Data Analysis in Medicine and Pharmacology, IDAMAP (2002)
8. JAKARTA-Commons (Webpage accessed in January 2007), http://jakarta.apache.org/commons/
9. Jorge, A., Poças, J., Azevedo, P.J.: Post-processing operators for browsing large sets of association rules. In: Lange, S., Satoh, K., Smith, C.H. (eds.) DS 2002. LNCS, vol. 2534, pp. 414–421. Springer, Heidelberg (2002)
10. Jorge, A.M., Azevedo, P.J., Pereira, F.: Distribution rules with numerical properties of interest. In: 10th European Conference on Principles and Practice of Knowledge Discovery in Databases. LNCS (LNAI), Springer, Berlin (2006)

11. Jorge, A.M., Pereira, F., Azevedo, P.J.: Visual interactive subgroup discovery with numerical properties of interest. In: Discovery Science 2006. LNCS (LNAI), Springer, Barcelona (2006)
12. Kavsek, B., Lavrac, N., Jovanoski, V.: Apriori-sd: Adapting association rule learning to subgroup discovery. In: Proceedings of the fifth International Symposium on Inteligent Data Analysis, pp. 230–241. Springer, Heidelberg (2003)
13. Klösgen, W.: Exploration of simulation experiments by discovery. In: AAAI 1994 Workshop on Knowledge Discovery in Databases. LNCS (LNAI), pp. 251–262. Springer, Barcelona (1994)
14. Klösgen, W.: Applications and Research Problems of Subgroup Mining. In: 11th International Symposium on Foundations of Intelligent Systems, pp. 1–15 (1999)
15. OSJava. Open Sourced Java (Webpage accessed in November 2006), http://www.osjava.org/
16. Ma, Y., Liu, B., Wong, C.K.: Web for data mining: organizing and interpreting the discovered rules using the web. SIGKDD Explor. Newsl. 2(1), 16–23 (2000)
17. Pereira, F.: Descoberta de subgrupos com regras de associação. MSc dissertation on Data Analysis and Decision Support Systems, Faculdade de Economia do Porto, Universidade do Porto (2006)
18. Wrobel, S.: An algorithm for multi-relational discovery of subgroups. In: Komorowski, J., Żytkow, J.M. (eds.) PKDD 1997. LNCS, vol. 1263, pp. 78–87. Springer, Heidelberg (1997)

Quantitative Evaluation of Clusterings for Marketing Applications: A Web Portal Case Study

Carmen Rebelo[1], Pedro Quelhas Brito[2,3], Carlos Soares[2,3], Alípio Jorge[2,3], and Rui Brandão[4]

[1] FCUP - Faculty of Sciences, University of Porto, Portugal
[2] LIAAD/INESC Porto LA, Rua de Ceuta 118 6o andar, 4050-180 Porto
[3] Faculty of Economics, University of Porto, Portugalo
[4] Infinivista
mrebelo@fc.up.pt, {pbrito,csoares,amjorge}@fep.up.pt

Abstract. The potential value of a market segmentation for a company is usually assessed in terms of six criteria: identifiability, substantiality, accessibility, responsiveness, stability and actionability. These are widely accepted as essential criteria, but they are difficult to quantify. Quantification is particularly important in early stages of the segmentation process, especially when automatic clustering methods are employed. With such methods it is easy to produce a large number of segmentations but only the most interesting ones should be selected for further analysis. In this paper, we address the problem of how to quantify the value of a segmentation according to the criteria above. We propose several measures and test them on a case study, consisting of a segmentation of portal users.

1 Introduction

The ultimate goal of marketing is customer satisfaction and is therefore based on knowing and respecting his/her needs [10]. On the other hand, it should enable actions that have a positive impact on business results. Marketing represents an important dimension in the management of commercial web portals as in any other business. In this context, the main concerns are to improve current user experience and act in a timely fashion to diminish churn.

The heterogeneity of customer needs in markets makes it impossible for companies to meet all of them. Attempting to do so usually leads to diminished profitability. It is necessary to divide customers into sub-groups, each one containing customers with similar needs. This approach has first been called *market segmentation* by Smith [14]. Segmentation should be the first step in the definition of any marketing strategy [7].

A common approach to segmentation uses data analysis methods on customer profile data. In e-business, data generated by the activity of users can be easily obtained. This data can be used to observe user behaviour, build profiles of users, which can then be used as data for segmentation [9].

J. Neves, M. Santos, and J. Machado (Eds.): EPIA 2007, LNAI 4874, pp. 437–448, 2007.

To be successful, the results of a market segmentation must be carefully evaluated from a business point of view. A largely accepted existing proposal include the following set of criteria [16]: identifiability, substantiality, accessibility, responsiveness, stability and actionability. These criteria, however, have been defined in very abstract terms. Therefore, the evaluation of a market segmentation is essentially a subjective process which must be done by a human expert. Given that market segmentations are typically obtained by a trial-and-error process, in which different methods and variables are considered, the costs required to evaluate them may be significant in terms of human resources.

Many measures have been proposed for the objective assessment of clusterings (e.g., [4] and references therein). Although such measures can be useful to evaluate market segmentations, they do not necessarily represent criteria that are relevant for this particular application area. In this paper we propose a set of measures taht can be used to quantify the marketing-specific criteria given above. The goal of these measures is to provide support in the development process, by guiding the selection of segmentations. However, they do not replace human expert evaluation, which is necessary for final validation and selection of the segmentation(s) to be used for further marketing operations.

The paper is organized as follows. In Section 2 we describe the case study which was addressed in this work. In Section 3 we give an example of a manual evaluation of a market segmentation. In Section 4 we investigate how to quantify the six dimensions of market segmentation quality mentioned earlier [16]. We propose a set of measures for that purpose and apply them on some of the segmentations obtained in our case study. In Section 5 we illustrate how they can be used to reduce the number of segmentations to be manually analysed. The business perspective of how the results of this work can be useful is given in Section 6. Finally we present some final remarks, including some ideas for future work (Section 7).

2   Description of the Case Study

Our case study is concerned with PortalExecutivo.com (PE), a web portal for the analysis, management and sharing of knowledge in Management and Economics [3].[1] It is a subscription-based service which is targeted to business executives. The goal is to provide its customers a portal that fulfills most of their needs in terms of information (e.g., reports on the Portuguese and international economy) and electronic services (e.g., trading in the stock exchange), both in terms of their professional as well as individual interests.

The access logs used are generated by the Content Management System of the portal. Although the underlying information is similar to the logs generated by the web server, the former have some advantages: they are cleaner (e.g., requests for figures are not logged) and contain more detailed information, including user and

---

[1] As part of the restructuring of the company owning PE, the site has been offline as of April 2006.

Table 1. Basic statistics describing the data before and after the pre-processing steps

	Before	After
Users	3,236	1,381
Accesses	458,394	267,749
Sessions	71,546	46,794
Content items	21,185	14,701

Table 2. Sample of measures containing information about users

Variable	Description
cliques	number of accesses
sessoes	number of sessions
temputil	number of days between first and last access
ultaces	number of days since last access
nclv	number of different classes of content accessed
ndsv	number of dossiers accessed
nartv	number of types of articles visited

session ID, the ID and title of the content accessed, and the template, which indicates the type of action (e.g., accessing content, navigation and printing). The data analyzed is from 29 May 2002 to 11 January 2005. This represents a total of 959 days, which is long enough to enable the detection of patterns in the behavior of users.

Although cleaner than the logs generated by the web server, several pre-processing operations were required in order to maximize the quality of the data: elimination of special users who are not customers; identification and correction of different aliases for a single user; and the elimination of short-lived users. Table 1 describes the effect of these operations on some basic descriptive statistics of the data.

An additional pre-processing operation consisted of reducing the number of different values in the template field (type of action). Originally, 612 different values existed, which we grouped into 16 general categories, such as article, report, stock exchange, navigation and registration.

We have computed a set of variables from the access logs to characterize the behavior of the users of PE. Some of the measures are based on the survey of related work which was carried out earlier, while others are specific to the case study at hand and were based on an exploratory analysis of the data (more details can be found in Rebelo [13]). The measures can be divided in two groups, corresponding to the object which they provide information about: users and sessions. For illustration purposes, we present some of the measures describing users in Table 2. More information about data preparation can be found in [13].

3   Analysis of a Market Segmentation

One of the methods used to automatically generate user segments from data was the Ward hierarchical clustering method [16]. For illustration purposes, we analyse a partition of the users into five clusters. The interpretation of the segments obtained, which is essential in marketing, is quite difficult when it is based on more than very few variables, as is the case. To reduce the number of variables to be analyzed, we have used a Factor Analysis (FA) approach [12,6]. It is a statistical data reduction technique which enables the replacemente of the set of original variables by a smaller set of linear combinations of those variables, called factors, with minimal loss in the information contained in the data. The smaller the set of variables obtained, the easier will be to interpret the data. We will illustrate the application of the method on the case study addressed in this work.

3.1   Generation of Principal Components

FA is only advantageous if there are significant correlations between the original variables. The Kaiser-Meyer-Olkin (KMO) test [12] may be used to test the significance of correlations between a set of variables. The values obtained with this test are in the range of 0 and 1 and it only makes sense to perform a FA for values of this test larger than 0.5. In our case, the result of the KMO test was 0.701, which means that FA is applicable.

Next, we compute the Principal Components (PC). These are a set of ordered linear combinations of the original variables, which have mean 0 and are not correlated with each other [6]. This means that they are orthogonal to each other. The first PC (PC1) is the one representing more variance in the data, the second PC (PC2) represents most of the remaining variance and so on.

The analysis of the eigen values associated with each PC is essential, not only to evaluate the descriminative power of each of them but also to decide how many components should be analysed such that the loss in information is minimized. Two criteria are used commonly for this purpose, Pearson's and Kaiser's. According to the former, the PCs analyzed should account for 80% of the total variance. According to the latter, which is suitable only for normalized data, all PCs with eigen value larger than 1 should be considered. In our case, the first five PCs have eigen values larger than 1 and explain 73.42% of the data. Given that they nearly satisfy both criteria, we could select them for further analysis. Given that our purpose is merely illustrative, we will consider only the first two PCs.

The interpretation of the PCs is based on the analysis of correlation circles (Figure 1). Each point represents the coordinates of the corresponding variable in the first (x-axis) and second (y-axis) PCs. The larger the coordinate value, the larger the weight of the variable. In other words, the more important the variable is in the value of the corresponding PC. Additionally, the variable may have a positive or negative influence on the value of the PC, depending on its sign. We observe that PC1 (horizontal axis) is mostly defined by number of accesses and sessions (`cliques` and `sessoes`), maximum duration of the session (`maxdur`), lifetime (`temputil`), most recent access (`ultaces`) and diversity of type of content (`nclv`, `ndsv` and `nartv`). Therefore, we can say that PC1 represents usage level.

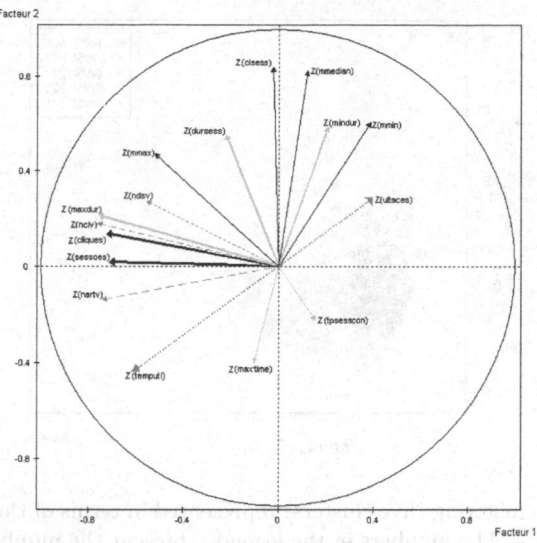

Fig. 1. Coordinates of the variables on the first and second PCs

As for PC2 (vertical axis), it is mostly defined by accesses per session (`clsess`, `mmedian`, `mmin` and `mmax`) and average and medium duration of sessions (`dursess` and `mindur`). It can be interpreted as a representation of the level of activity of the sessions of the corresponding user.

3.2   Visualization of Clusters

We now analyze the clusters by plotting them in a 2-dimensional space defined by the PCs obtained earlier. If we visualize the five clusters using PC1 and PC2 (Figure 2), we observe that three of them are clearly defined, occupying the center of the plot, and the other two are satellite groups. If we only consider four or three clusters, the latter groups are integrated in the others.

We observe that cluster 4W5f represents the users who profit most from the portal, because they have relatively high usage level and activity level within sessions. Cluster 2W5f includes individuals with a reasonable usage level, i.e., close to the average level, but with sessions with the lowest activity of all users. Finally, cluster 1W5f contains individuals with a usage level below average but with very active sessions. We also observe that the cluster with the most active users is the smallest of the three.

Our results are consistent with earlier ones, which indicate that the behavior of users evolves with time [11]. We observe that users in cluster 1W5f have sessions with a high activity level but with a long time since their latest access, which probably means that they have abandoned the portal. The users in cluster 4W5f have longer liftime and shorter interval between sessions but their sessions are not particularly active. This is consistent with the work of Bucklin and Sismeiro [2] who state that efficiency increases with experience.

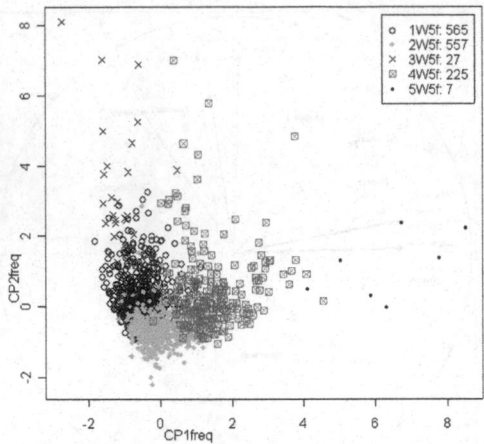

Fig. 2. Results of clustering (five clusters) represented in terms of the first (x-axis) and second (y-axis) PCs. The numbers in the legend represent the number of users in each cluster.

This example illustrates the effort required to analyse a segmentation from a marketing point of view. Therefore, it becomes clear that, although it is easy to obtain many segmentations with different automatic clustering methods, the cost of evaluating them may be quite high. Alternatively, it is possible to try only a few different segmentations. However, this reduces the probability of finding one which is satisfactory from the business point of view. In the following section, we will describe measures to quantify the quality of a segmentation that may be useful in the process, especially by enabling the elimination of unsuitable ones and, thus, reduce the cost of finding a good one.

4   Quantitative Evaluation of a Market Segmentation

A set of six criteria commonly used to assess the quality of market segmentations are [16]: identifiability, substantiality, accessibility, responsiveness, stability and actionability. However, these are defined in an informal way. In this section we propose some measures that can be used to assess most of them in a quantitative way. We test them on six different segmentations, briefly described in Table 3. Besides two well-known methods, hierarchical clustering method (using Ward distance) [16] and k-means clustering, we have used a fuzzy k-means clustering method [1].

The *identifiability* of a segmentation represents how well it makes it possible to identify distinct groups of customers. In other words, this criterion depends on how clearly defined the borders between the segments are. Therefore, we may quantify identifiability using the accuracy of a classification algorithm trained on the problem of classifying different customers in the different segments.[2] For

---

[2] Given that the model is not used for predictive purposes, we use training error.

Table 3. Summary of the segmentations analysed. Three types of customer characterization were used: *Frequency* includes measures of the type listed in Table 2; the variables in *Action* describe the distribution of accesses in terms of type of action, as defined earlier; in *Session* we have variables that describe the most common sessions of the user. More details can be found in [13].

Id	Segments	Method	User profile
W3f	3	Hierarchical clustering	Frequency
K3f	3	K-means clustering	Frequency
W3p	3	Hierarchical clustering	Action
K3p	3	K-means clustering	Action
fz4	4	fuzzy k-means clustering	Action
manual	3	Manual	Session

marketing purposes, it makes sense to choose methods that are commonly used and/or generate models that are easy to interpret. In this work, we have used a linear discriminant analysis and a decision tree learner [8].

The results in Table 4 show that in all cases, the classification algorithms obtain higher accuracies than a naive classifier that always predicts the majority class. This indicates that all segmentation achieve minimum levels of identifiability. Additionally, we can say that the most identifiable segmentations are the ones obtained with the absolute values of session activity (W3f and K3f), which achieve high accuracy with both methods. However, we can say that the remaining segmentations also achieve a high level of identifiability because a high accuracy is obtained with one of the algorithms, linear discriminant (W3p, K3p and fz4) and decision tree (manual). We note that, although a high accuracy means that the segmentation has high identifiability, the opposite is not necessarily true. A low accuracy of a classification method means that that particular method is not able to discriminate the segments, but there may be another method that can.

The second criterion, *substantiality*, refers to the size of the segments, which must be sufficiently large to enable a cost-effective targetting. Sufficiency naturally depends on the application. In extreme cases (e.g., e-commerce) it is possible to target individual customers. However, we may say that, in general, a balanced distribution of customers across segments is desirable. Therefore, we quantify substantiality with the two measures. The first is given by $S_1 = (max(n_i) - min(n_i))/\bar{n}$, where n_i is the size in segment i (i.e., number of customers in the segment) and $\bar{n}$ is the mean segment size. The lower the value of S_1, the more balanced the distribution is and, thus, the higher the substantiality of the segmentation. The second measure is given by $S_2 = min(n_i)/\bar{n}$, which provides a perspective of the minimum segment size relative to the mean cluster size. The higher the value of S_2, the higher the substantiality.

Analysing the values of S_1 and S_2 in Table 4 we observe that segmentations K3p and fz4 have the highest substantiality. Concerning manual, although it has a high value of S_2, it's value of S_1 is also high, indicating that it is not very

Table 4. Quantitative assessment of four market segmentation evaluation criteria on six examples

Measure	W3f	K3f	W3p	K3p	fz4	manual
Identifiability						
Linear discriminant (% accuracy)	88	95	89	93	90	70
Decision tree (% accuracy)	90	94	54	49	42	93
Majority class (% accuracy)	43	70	45	39	29	47
Substantiality						
S_1	0.8	1.8	0.7	0.3	0.4	0.7
S_2	0.5	0.3	0.6	0.9	0.7	0.7
n_1	592	965	622	545	390	647
n_2	557	275	481	435	382	395
n_3	232	141	278	401	330	337
n_4	-	-	-	-	255	-
Accessibility						
λ_1	2.2	2.1	2.9	2.5	5.3	0.8
λ_2	1.1	1.5	0.4	1.2	1.1	0.2
variables	5	5	7	4	4	4
leaves	8	11	11	7	6	8
Responsiveness						
$\langle d \rangle$	4.0	3.5	3.5	3.6	3.5	3.9
weighted covariance	11.6	13.0	1.1	0.1	1.0	0.13
d_1	2.3	2.1	2.6	3.9	2.9	3.8
d_2	3.3	5.0	4.5	3.7	2.7	4.4
d_3	9.9	9.9	3.8	3.2	4.2	3.7
d_4	-	-	-	-	4.9	-

balanced. These observations are confirmed by the individual n_i values. Finally, K3f is clearly the segmentation with the lowest substantiality.

The *accessibility* of a segmentation indicates how easily reachable each of the segments is expected to be. In other words, it assesses the possibility to develop clearly diverse marketing strategies for the different segments. To measure identifiability, we were concerned with the borders separating the segments. Here we try to assess how clearly different the centers of the segments are. We quantify this criterion using two measures. First we analyse the eigen values, λ_i, of the discriminating function obtained by the linear discriminant: the higher the value, the larger the amount of variation between groups is described by the discriminant [12]. We also analyse the compactness of the decision tree, in terms of the number of variables used and leaves: the smaller the number of variables and leaves, the simpler the description of the segments.

We observe that the λ_1 for fz4 has the highest value, indicating that this segmentation has higher accessibility than the others (Table 4). On the other hand, the λ values for the manual segmentation are quite low, which indicate a low accessibility. The analysis based on the compactness of the decision tree also confirms the observations regarding fz4. However, the number of variables used and leaves for manual is also quite low, which is somewhat contradictory with the previous observation that this segmentation has low accessibility. This can be explained because these measures are based on different models and depend on how well they fit the data. This means that they should be used in combination.

The fourth criterion in *responsiveness*, which refers to how well the segments are expected to react to the marketing strategies targeted to them. This is determined by the homogeneity of the profiles of the customers in each of the segments. To quantify this property we use the weighted mean dispersion, $\langle d \rangle$, where the weights are given by the cluster sizes. This measure provides information about the intra-cluster homogeneity and inter-cluster heterogeneity: the smaller the dispersion, the more compact the segments, and, thus, the higher the potential responsiveness.

The values of indicate that all segmentations have similar values of $\langle d \rangle$, with slightly higher valuer for W3f and K3f. However, on closer analysis of the segmentations based on frequency variables (Table 3), we we observe that two of the segments actually have very low dispersion (d_i). However, the remaining segment has very high dispersion. Therefore, these segmentations generate two clusters with high potential responsiveness and one with very low responsiveness. This example illustrates the need to complement the information provided by the mean some information about the variance, such as, the (weighted) covariance. The values of weighted covariance (Table 4) are, in fact much higher for those two segmentations in comparison to the others. This indicates that there is a large variation in the degree of responsiveness of the different segments in those segmentations. On the other hand, the covariance of K3p is quite low, which, combined with a reasonable value of the weighted mean, indicates that this segmentation has the highest responsiveness of the six.

The next criterion is concerned with the *stability* of the segmentation with time. A segmentation is only useful if its segments remain as unchanged as possible for a period which is long enough to identify them and to develop, implement and deploy a suitable marketing strategy. Recently, there has been some work on the analysis of the stability of clusterings [15,5], so we will not address the issue here. In those approaches, a few measures are proposed that could be used in this work.

The final criterion is *actionability*, which represents the alignment of the segmentation with the goals and the competences of the company. In other words, it assesses the extent to which it enables effective marketing strategies to be developed. It differs from accessibility because it depends on the specific goals of the company, rather than simply on the variables which are used to build the segmentation. The dependence on the application-specific properties makes it very difficult to quantify actionability in an abstract way. Therefore, we do not address it in this work.

5 Selection of Segmentations for Manual Analysis

In Table 5, we present a summary of the results, based on the analysis of the values in Table 4. This table indicates that fz4 is the best segmentation of the

Table 5. Summary of the results obtained on six example market segmentations. Segmentations are classified as good ('+'), average (blank) or bad ('-'), based on a comparative analysis of the values in Table 4.

Criterion	W3f	K3f	W3p	K3p	fz4	manual
Identifiability	+	+				
Substantiality	-			+	+	
Accessibility				+		
Responsiveness	-	-		+		-

set. However, these results cannot be used blindly. For instance, it may also be worthwhile to consider for further manual assessment K3p, although it obtains mostly average results, because it was generated using a different method (K-means clustering). For similar reasons, it may be worthwhile to analyse W3f, although it has one of the worst responsiveness values in the group. It is not only generated with a different method (hierarchical clustering) but it is also based on a different set of variables.

6   The Business Perspective

From the point of view of the owners of the portal, our results provide interesting information. The segmentations provided can be used in, at least, three different ways. First, they can be used for monitoring purposes, in particular concerning usability issues. They provide the ability to quickly identify "problem areas" or whether there was a change in usage patterns. This is a key business concern in web portals. While the often used general numbers, such as page views and sessions, are important, they sometimes hide emerging trends. These can be perceived through the correct use of a segmentation, which allows for faster and more effective problem identification and solving. Of particular importance are changes in the relative weight of the different clusters, although this information should be used in combination with other types of business information. As an example, an increase in the click pattern of new users may indicate a need to improve usability, or to provide better guidance.

Secondly, segmentations can also improve the perception of different "life-cycles" of users. The main point is to understand where the transition between different clusters are occurring and how they can be improved upon in order to prevent churn and increase usage. As an example, further investigation on what distinguishes between users who move from the high usage level clusters (1W5F and 3W5F) to the efficient usage cluster (4W5f and 5W5f) and those who abandon the portal (Section 3). This information could be used to help the former to move faster and to design strategies to avoid users from abandoning.

Thirdly, suitable segmentations can significantly improve the planning of marketing activities (development of actions targeted to specific segments) as well as their assessment (measuring their impact). For instance, we have observed in Section 3 that users who abandon have sessions with many clicks, possibly indicating that they have difficulty in finding the information they are looking for. To address this issue, new improved FAQ session with clear communication to new users could be developed. The effect of this measure should then be assessed only on the corresponding clusters. Additionally, we observed that users who access a more diverse set of features from the portal have longer lifetime. This indicates that users should be exposed to as many of those features as soon as possible.

7   Conclusions

The use of automatic clustering methods is becoming increasingly popular in marketing. It is now possible to generate many different customer segmentations. The downside of this trend is that the mount of effort required to analyse a segmentation from a marketing perspective is significant. Therefore, when many segmentations are available, the costs of analysing them may become too high. Therefore, it is necessary to provide the user with tools to select only a few segmentations for manual evaluation.

In this paper we address the problem of quantifying the quality of market segmentations, which can be used for that purpose. We propose measures that represent six criteria which are commonly used for market segmentation evaluation [16]. We show how these measures can be used to support the segmentation process, by enabling an easier selection of the most promising segmentations. Therefore, the amount of manual effort required to obtain a suitable market segmentation may be significantly reduced, thus reducing costs and enabling a more efficient use of human resources.

Although the measures proposed enabled us to satisfactorily achieve our goals on data from a real problem, a few issues have been identified. Further validation using more data and, preferably, on real marketing scenarios, will enable us to improve the set of measures proposed. In practise, it is essential to keep in mind what the differences between the methods used to generate the clusterings under comparison and carefully assess and interpret the information derived from those clusterings.

From the business point of view, there are several tools that can be used to achieve the goals of marketing, in e-business as in other areas. Segmentation and factor analysis are of particular importance but the best results will be obtained by the wise combination of several tools, including other statistical and BI methods, satisfaction and customer surveys, etc.

Acknowledgements

The authors are grateful to PortalExecutivo.com for their support, and, in particular, to Carlos Sampaio for his collaboration. The financial support of the POSC/EIA/58367/2004/Site-o-Matic Project (Fundação Ciência e Tecnologia) co-financed by FEDER is gratefully acknowledged.

References

1. Brochado, A.: A segmentação de mercado: bases de segmentação e métodos de classificação. aplicação ao mercado de vinho verde. Master's thesis, Escola de Gestão do Porto (2002) (in Portuguese)
2. Bucklin, R.E., Sismeiro, C.: A model of web site browsing behavior estimated on clickstream data. Journal of Marketing Research XL, 249–267 (2003)
3. Cabrita, S., Valentim, S.: Portal executivo - construção de um mercado. In: Celeste, P. (ed.) Estratégias de Marketing - colectânea de casos portugueses, pp. 127–143, Escolar Editora (2005) (in Portuguese)
4. Campello, R.J.G.B.: A fuzzy extension of the rand index and other related indexes for clustering and classification assessment. Pattern Recognition Letters 28(7), 833–841 (2007)
5. Carvalho, C., Jorge, A.M., Soares, C.: Personalization of e-newsletters based on web log analysis and clustering. In: WI 2006. Proceedings of the 2006 IEEE/WIC/ACM International Conference on Web Intelligence, pp. 724–727. IEEE Computer Society, Washington, DC, USA (2006)
6. Gorsuch, R.L.: Factor Analysis. Lawrence Erlbaum, Mahwah (1983)
7. Hooley, G., Saunders, J., Piercy, N.F.: Marketing Strategy and Competitive Positioning, 2nd edn. Prentice-Hall, Englewood Cliffs (1998)
8. Ihaka, R., Gentleman, R.: R: A language for data analysis and graphics. Journal of Computational and Graphical Statistics 5(3), 299–314 (1996)
9. Kosala, R., Blockeel, H.: Web mining research: A survey. SIGKDD Explorations 2(1), 1–15 (2000)
10. Kotler, P.: Marketing Management, 11th edn. Prentice-Hall, Englewood Cliffs (2003)
11. Moe, W.W., Fader, P.S.: Capturing evolving visit behavior in clickstream data. Technical report, Wharton School (2002)
12. Pestana, M.H., Gageiro, J.N.: Análise de dados para as ciências sociais. A. complementaridade do SPSS. Sílabo, 3rd edn. (2003)
13. Rebelo, C.: Segmentação do comportamento online utilizando clickstream data. Master's thesis, Faculty of Economics of Porto (2006) (in Portuguese)
14. Smith, W.: Product differentiation and market segmentation as alternative marketing strategies. Journal of Marketing (48), 32–45 (1956)
15. Spiliopoulou, M., Ntoutsi, I., Theodoridis, Y., Schult, R.: Monic: modeling and monitoring cluster transitions. In: KDD 2006. Proceedings of the 12th ACM SIGKDD international conference on Knowledge discovery and data mining, pp. 706–711. ACM Press, New York (2006)
16. Wedel, M., Kamakura, W.A.: Market Segmentation - Conceptual and Methodological Foundations. Kluwer, Dordrecht (1998)

Resource-Bounded Fraud Detection

Luis Torgo

LIAAD-INESC Porto LA / FEP, University of Porto
R. de Ceuta, 118, 6., 4050-190 Porto, Portugal
ltorgo@liaad.up.pt
http://www.liaad.up.pt/~ltorgo

Abstract. This paper describes an approach to fraud detection targeted at applications where this task is followed by a posterior human analysis of the signaled frauds. This is a frequent setup on fraud detection applications (e.g. credit card misuse, telecom fraud, etc.). In real world applications this human inspection is usually constrained by limited resources. In this context, standard fraud detection methods that simply tag each case as being (or not) a possible fraud are not very useful if the number of tagged cases surpasses the available resources. A much more useful approach is to produce a ranking of fraud that can be used to optimize the available inspection resources by first addressing the cases with higher rank. In this paper we propose a method that produces such ranking. The method is based on the output of standard agglomerative hierarchical clustering algorithms, resulting in no significant additional computational costs. Our comparisons with a state of the art method provide convincing evidence of the competitiveness of our proposal.

1 Introduction

Fraud detection is a hot topic in several research areas (e.g. [5,13,15]). Due to the intrinsic characteristics of fraudulent events it is often associated with other research topics like outlier detection, anomaly detection or change detection. The connecting feature among these topics is the interest on deviations from "normal" behavior. Depending on the characteristics of the fraud data available different data mining methodologies can be applied. Namely, when the available data includes information regards each observation being (or not) fraudulent, supervised classification techniques are typically used (e.g. [4]). On the contrary, in several domains such classifications do not exist and thus unsupervised techniques are required (e.g. [3]). Finally, we may have a mix of both types of data with a few labelled observations (e.g. resulting from past inspection activities) and a large set of unlabeled cases. These situations are often handled with semi-supervised approaches (e.g. [12]).

Several authors (e.g. [13]) have criticized the use of labelled data for fraud detection. These authors have noted that in most real world applications it is difficult to have reliable labels due to several factors like the cost of obtaining them, among others. In this paper, we assume the available data is not labelled. The method we propose produces a ranking of fraud probability for a set of

J. Neves, M. Santos, and J. Machado (Eds.): EPIA 2007, LNAI 4874, pp. 449–460, 2007.
© Springer-Verlag Berlin Heidelberg 2007

unlabeled observations. Many detection methods provide yes/no answers to this task. We claim that for real world applications of fraud detection systems this type of answer may lead to sub-optimal decisions. In effect, in most applications the detection systems are used to help in planning posterior inspection activities. These activities are typically constrained by a limited amount of resources (human or other). In this context, it is preferable to have a ranking of fraud instead of a set of cases predicted as fraudulent. Such ranking is much more flexible for the correct use of the available resources and will most probably lead to better results. Without these ranks and if the cases labeled as fraudulent are more than what the available resources allow to inspect, the user is left with the unguided task of deciding which ones to inspect. By providing a rank of fraud, the resources can be used on the cases that have a higher probability of fraud.

2   Outlier Ranking

The approach we propose for obtaining a ranking of fraud probability assumes that frauds are rare and that can be regarded as outliers from the bulk of "normal" data. Outlier detection is a well studied topic (e.g. [2]). Different approaches have been taken to this task. Distribution-based approaches (e.g. [6]) assume a certain parametric distribution of the data and signal outliers as observations that deviate from this distribution. The main drawbacks of these approaches lie on the constraints of the assumed distributions. Depth-based methods (e.g. [14]) are based on computational geometry and compute different layers of k-d convex hulls and then represent each data point in this space together with an assigned depth. In practice these methods are too inefficient for dealing with large data sets. Knorr and Ng [10] introduced distance-based outlier detection methods. These approaches generalize several notions of distribution-based methods but still suffer from several problems, namely when the density of the data points varies (e.g. [17]). Density-based local outliers [17] are able to find this type of outliers and are the appropriate setup whenever we have a data set with a complex distribution structure.

Clustering algorithms can also be used to identify outliers as a side effect of the clustering process (e.g. [11]). Most clustering methods rely on a distance metric and thus can be seen as distance-based approaches. However, iterative methods like hierarchical clustering algorithms (e.g. [7]) can also handle different density regions. In effect, if we take for instance agglomerative hierarchical clustering methods, they proceed in an iterative fashion by merging two of the current groups (which initially are formed by single observations) based on some criterion that is related to their proximity. This decision is taken locally, i.e. for each pair of groups, and takes into account the density of these two groups only. It is based on this observation that we plan to explore hierarchical clustering methods as a form of producing outlier rankings that are able to handle applications with both global and local outlier types.

2.1   Height-Based Outlier Factors

Our proposal is based on an agglomerative hierarchical clustering method (e.g. [7]). These methods start with as many clusters as there are training observations and then go through an iterative process where at each stage two of the current clusters are merged to form a new grouping of the data. This merging process results in a tree-based structure usually known as a dendogram. The merging step is guided by the information contained on the distance matrix of all available data. Several methods can be used to select the two groups to be merged at each stage. For instance, the single linkage method selects the pair of groups which has the smallest distance between any of their members. One of our goals is to have an outlier detection method that can handle local outliers, i.e. being able to capture areas of high local density and spot nearby points that somehow break this density, though from a global perspective they could look near to these areas and thus would not be regarded as outliers. In this context, we have selected to work with the Ward's [19] agglomeration method. Merging according to this method is carried out by selecting the pair of groups which would result in a new group with minimal variance, i.e. maximally compact. This means that outliers tend to be selected later on this iterative process because, by definition, they are clearly separated from their neighborhood and thus will increase the variance of a group when joining it. Informally, the idea behind our proposal is to use the height (in the dendogram) at which any observation is merged into a group of observations as an indicator of its outlyingness. If an observation is really an outlier this should only occur at later stages of the merging process, i.e. the observation should be merged at a higher level than "normal" observations. More formally, we set the outlyingness factor of any observation as,

$$OF_H(x) = \frac{h}{N} \tag{1}$$

where h is the level of the hierarchy H at which the case is merged[1], and N is the number of training cases (which is also the maximum level of the hierarchy by definition of the hierarchical clustering process).

One of the main advantages of our proposal is that we can use a standard hierarchical clustering algorithm to obtain the OF_H values without any additional computational cost. This means our proposal as a time complexity of $O(N^2)$ and a space complexity of $O(N)$ [8]. We use the *hclust()* function of the statistical software environment R [16], which is based on Fortran code by F. Murtagh [9]. This function includes in its output a matrix (**merge**) that can be used to easily obtain the necessary values for calculating directly the value of OF_H according to Equation 1.

Figure 1.(a) shows an artificial data set with two marked clusters of observations with very different density. As it can be observed there are two clear outliers: observations 1 and 12. While the former can be seen as a global outlier, the latter is clearly a local outlier. In effect, it is only regarded as an outlier because of the high

---

[1] Counting from bottom up.

density of its neighbors, as it is in effect nearer observation 2 than, say the 14th
from the 15th. However, as these two latter are in a less compact region their distance is not regarded as a signal of outlyingness. This is a clear example of a data
set with both global and local outliers and we would like our method to clearly
signal both 1 and 12 as observations with a high probability of being outliers.

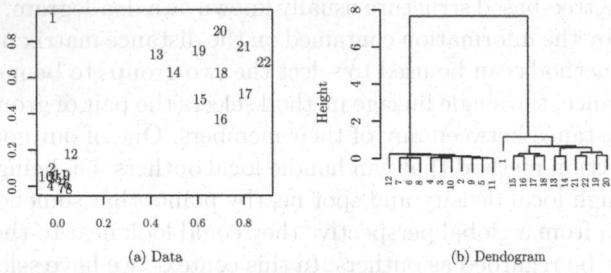

| (a) Data | (b) Dendogram |

Fig. 1. An artificial example

Figure 1.(b) shows the dendogram obtained by using an agglomerative (Ward)
hierarchical clustering algorithm. As it can be seen, both 1 and 12 are the last
observations to be individually merged into some cluster. As such, it does not
come as a surprise that when running our method on this data we get the top 5
outliers shown on Table 1.

Table 1. Outlier ranking for the example of Figure 1.

Rank	CaseID	OF_H
1	1	0.9091
2	12	0.6818
3	17	0.5909
4	18	0.5909
5	19	0.5455

In spite of the success, this method has serious problems when facing compact
groups of outliers. In effect, if we have a data set where there are a few outliers
that are very similar to each other, they will be merged with each other very
quickly (i.e. at a low level of the hierarchy) and thus will have a very low OF_H
value in spite of being outliers. Figure 2 illustrates this problem. For this data
set, the method ranks observations 9 and 10, which are clear outliers, as the
least probable outliers (they are in effect the first to be merged by the Ward
method).

The example of Figure 2 shows a clear failure of our initial proposal. The
failure results from considering only the height at which individual observations
are merged and not groups of observations. However, if there is a small group[2]

---

[2] Such that it could make sense to talk about a set of outliers.

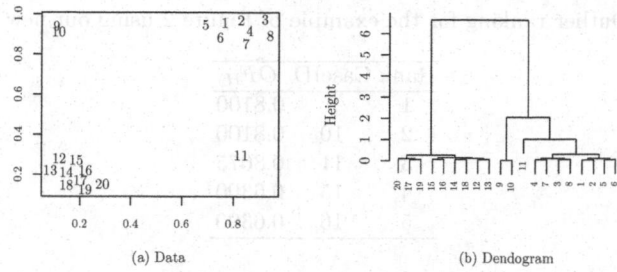

(a) Data   (b) Dendogram

Fig. 2. A problematic artificial example for our initial proposal

of similar observations that is quite different from others, and thus will only be merged with other groups at later stages, our proposal will not consider this as a signal of outlyingness of the members of that group. Still, we should remark that the general idea of our proposal remains valid as long as we generalize it for these situations. We can do this by assigning a value similar to that of Equation 1 to all members of the smallest group of any merge that occurs along the hierarchical clustering process. However, we should reinforce this value with some size-dependent factor (i.e. the smallest the most probable that we are facing outliers). Formally, for each merge of a group g_s with a group g_l, where $|g_s| < |g_l|$, we set the outlier factor of the members of g_s as,

$$OF(g_s) = \left(1 - \frac{|g_s|}{N}\right) \times \frac{h}{N} \qquad (2)$$

where $|g_s|$ is the cardinality of the smallest group, g_s, and h is the level of the hierarchy where the merge occurs. The OF value of the larger group g_l is set to zero. The value of OF ranges from zero to one, and it is maximum when a single observation is merged at the last level of the hierarchy.

Any observation can belong to several groups along its upwards path through the dendogram. As such, it will probably get several of these scores at different levels. We set the outlyingness factor of any observation as the maximum OF score it got along its path through the dendogram. By proceeding this way we are in effect taking care of local outliers, which at some merging stage might have got a very high score of OF because they are clear outliers with respect to some group that they have merged with, even though at higher levels of the hierarchy (i.e. seen more globally), they might not get such high OF values. This means that the outlyingness factor of an observation is given by,

$$OF_H(x) = \max_{g \in G_x} OF(g) \qquad (3)$$

where G_x is the set of groups in the dendogram to which x belongs.

Applying this method to the problematic example of Figure 2, we get the outlier ranking shown in Table 2. This is the expected result for this problem,

Table 2. Outlier ranking for the example of Figure 2 using our new proposal

Rank	CaseID	OF_H
1	9	0.8100
2	10	0.8100
3	11	0.8075
4	15	0.6300
5	16	0.6300

which means that this new formulation is able to handle compact and small groups of outlier observations, like for instance observations 9 and 10 of this problem.

We should remark that although we have always used our proposal in the context of a hierarchical clustering process using the Ward's agglomerative criterion, our proposal is not dependent on this criterion. In effect, the method could be applicable to any agglomerative criterion and/or distance metric. Still, we think the Ward's method is more adequate for finding local outliers.

3   Experimental Evaluation

In this section we present several experiments providing some insights on the effective behavior of our proposal. We compare our method with the state of the art in terms of obtaining degrees of outlyingness: the LOF method [17].

As we did not have access to real world data sets of fraud detection we decided to use some supervised regression data sets obtained from Torgo's repository [18]. These data sets can be used to provide an idea of how well does our method captures the most deviating observations and at the same time, as these are supervised regression problems with a continuous target variable, we can check whether the top ranked outliers correspond to unusual values of the target variable. This is somehow similar to the process it would be followed if the method were to be used in real world applications of fraud detection: first the method would provide a ranking and then, after human inspection, we would confirm the degree of fraud of the observations. The assumption we are making here is that observations that have unusual values on the input variables (so as to make them stand as outliers), will also have unusual values on the target variable. This assumption is reasonable provided there is some relationship between the target and input variables and the unknown regression surface is reasonably smooth. The data sets we have used are the following:

- *Boston Housing* (BH) - 506 cases described by 14 variables. The data concerns the task of predicting the median price of houses in different areas of Boston. The input variables describe some socio-economical features of the areas. The distribution of the prices has a normal-like shape around the value 22, with a few extremely high or low prices for some areas.

- *Abalone* (AB) - 4177 cases described by 9 variables. The data concerns the prediction of the number of rings in an abalone, which is supposed to be directly related to their age (the goal of the original application). The distribution of the number of rings also follows a normal-like distribution around the value 9, with a few small values and also some unusually high values.
- *Alga 1* (A1) - 200 cases described by 12 variables. This data concerns the prediction of the concentration values of a rare harmful alga species in 200 water samples drawn from different European rivers. The distribution is concentrated around values very near zero, with a few unusually high concentration values (known as algae blooms).
- *Machine-CPU* (MC) - 209 observations described by 7 variables. This data set concerns the task of predicting the relative CPU performance based on some hardware features. The target variable has a distribution centered on small values (around 50) with a few extreme values of performance.

Regards the compared methods, we have based our implementation of OF_H on the *hclust()* function of the statistical software environment R [16], as mentioned before. With respect to LOF we have used the implementation of this method available on the package *dprep* [1] of the same software environment. For the outlier ranking tasks we have eliminated the target variable information. We have used the two principal components to obtain a 2-D plot of all data points of each domain. In all graphs we report the proportion of variance of the original data that is explained by these 2-D summarization. For all data sets it is above 95% which means that the plots are a good spatial representation of the original data. For the larger data sets of have signalled the top 20 outliers according to each method, while for smaller we have only used the top 10.

Figure 3 shows the results for the Boston Housing domain. With a few exceptions in the case of LOF, most of the signalled points can easily be accepted as outliers. The solutions obtained by both methods are quite similar, with LOF being able to identify a few outliers at the bottom right corner that OF_H

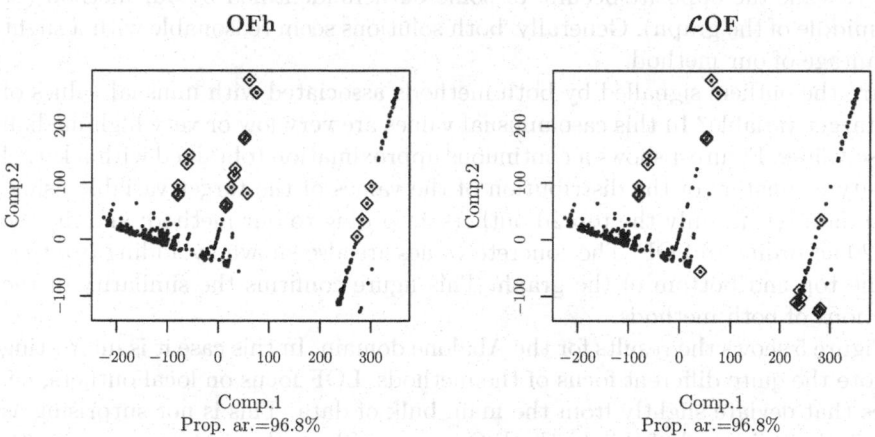

Fig. 3. The top 20 outliers for the Boston Housing domain

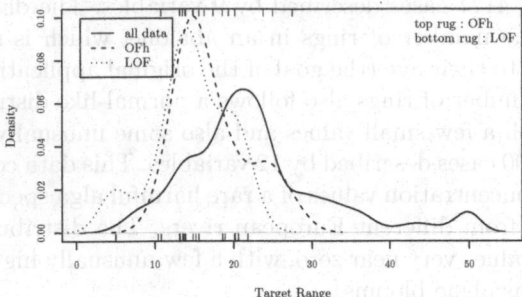

Fig. 4. Boston target variable distribution for all data set and for the outliers

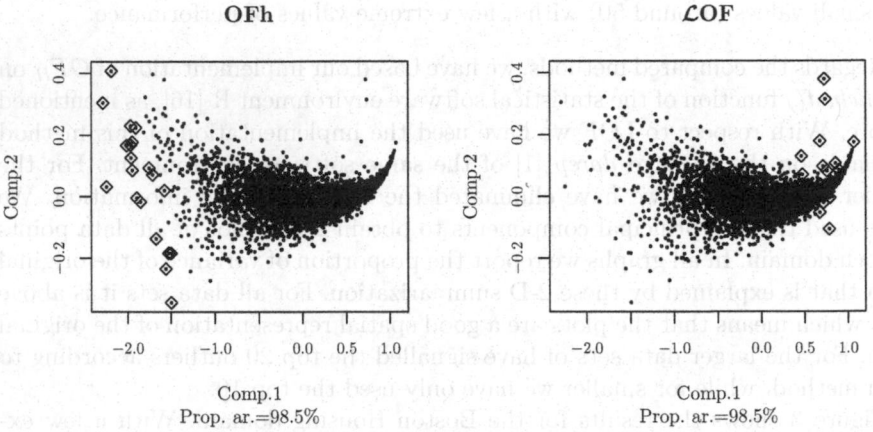

Fig. 5. The top 20 outliers for the Abalone domain

is not, while the opposite occurs for some outliers identified by our method (at the middle of the graph). Generally, both solutions seem reasonable with a slight advantage of our method.

Are the outliers signalled by both methods associated with unusual values of the target variable? In this case unusual values are very low or very high median house values. Figure 4 shows a continuous approximation (obtained with a kernel density estimator) of the distribution of the values of the target variable using: i) all data set; ii) only the top 20 outliers according to our method; and iii) the top 20 according to LOF. The concrete values are also shown by adding two rugs at the top and bottom of the graph. This figure confirms the similarity of the solutions of both methods.

Figure 5 shows the results for the Abalone domain. In this case it is interesting to note the quite different focus of the methods. LOF focus on local outliers, i.e. cases that deviate slightly from the main bulk of data. This is not surprising as that is the main goal of this method. Our proposal, on the contrary seems to be focused on other type of outliers that deviate largely from the main bulk of data

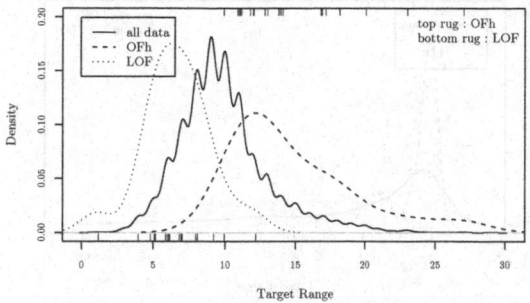

Fig. 6. Abalone target variable distribution for all data set and for the outliers

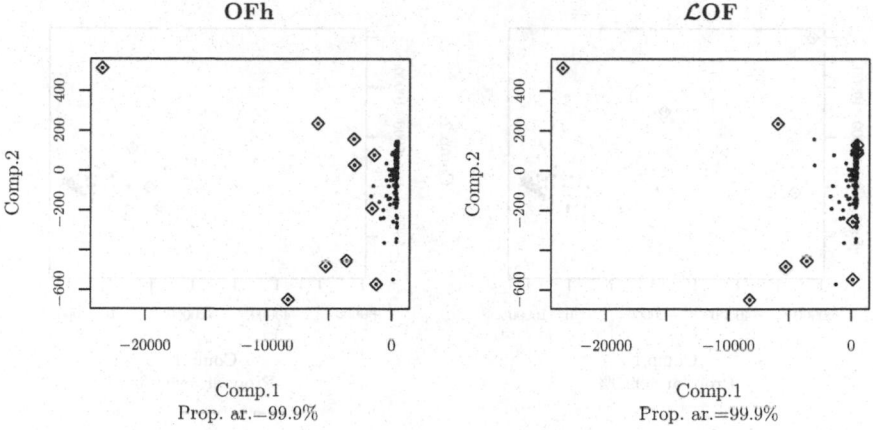

Fig. 7. The top 10 outliers for the Alga 1 domain

and are located in less dense regions. Both methods provide correct indications with respect to finding outliers (though LOF seems to be making a few mistakes). We have checked that our method also ranks high the outliers signalled by LOF, though not including them on the top 20. It is interesting to check whether these two different groups of outliers also correspond to different target variable values. Figure 6 confirms this. The outliers signalled by our method have unusually high number of rings, whilst LOF seems to be more focused on extreme low values.

Figure 7 shows the solutions for the Alga 1 domain. Once again both methods do a good job at spotting the most deviating observations. Still, we can claim a slight advantage of our proposal as a few outliers signalled by LOF can hardly be regard as such. With respect to the corresponding target variable distribution (c.f. Figure 8) the advantage of our method is not so clear, as LOF includes two larger algae blooms on its top 10 outliers.

Finally, Figure 9 shows the rankings for the Machine CPU domain. We should start by referring that due to the fact that the two principal components only

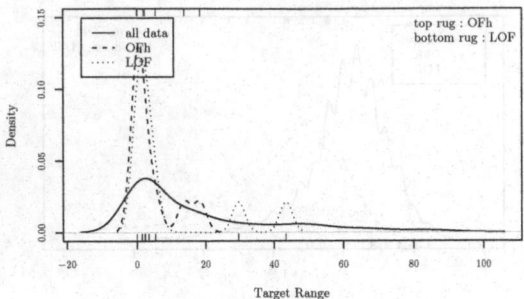

Fig. 8. Alga 1 target variable distribution for all data set and for the outliers

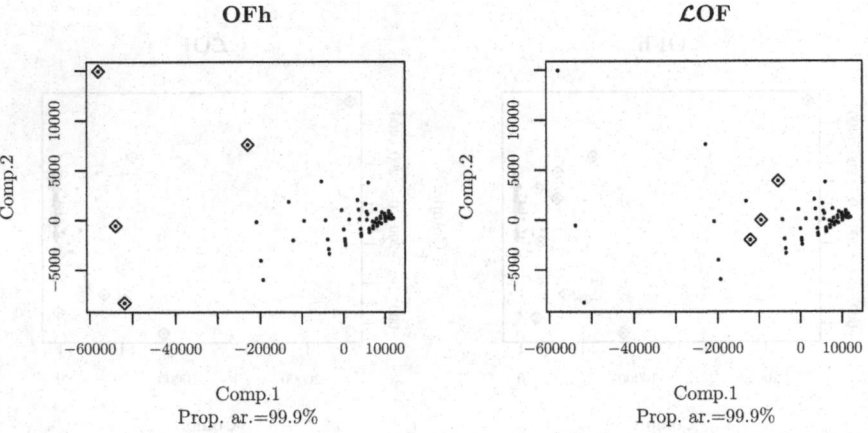

Fig. 9. The top 10 outliers for the Machine CPU domain

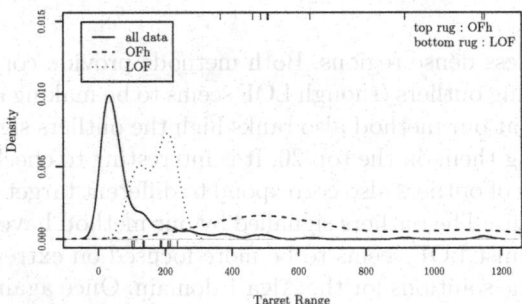

Fig. 10. Machine CPU target variable distribution for all data set and for the outliers

use two of the original variables of the domain, several data points get plotted on the same place as they have equal values on these variables. That is the explanation for apparently less than 10 outliers being plotted in the graphs. In this domain we observe a clear advantage of our proposal as LOF misses the most obvious outliers. This is confirmed when looking at the distribution data (Figure 10), where we see that the outliers ranked higher by our method effectively correspond to the most extreme values of CPU performance.

4 Conclusions

In this paper we have presented an outlier ranking method that can be applied to fraud detection problems allowing a resource-aware planning of any posterior inspection activities, which is a key requirement of several business activities.

Compared to other existing approaches the most distinguishing feature of our work is the fact that it relies on the output of common hierarchical clustering algorithms, thus not requiring additional computational efforts to obtain a ranking of outliers.

The initial set of experiments that we have presented, comparing our method to a state of the art alternative (LOF), confirm the validity of our proposal. In effect, we have always found the results of our method to be equivalent and in some cases even slightly superior to the results of LOF.

Further work should extend the analysis of the method both experimentally as well as theoretically. We also plan to make a more deep analysis of the computational requirements of our proposal.

References

1. Acuna, E., Rodriguez, C.: dprep: Data preprocessing and visualization functions for classification. R package version 1.0
2. Hodge, V., Austin, J.: A survey of outlier detection methodologies. Artificial Intelligence Review 22, 85–126 (2004)
3. Bolton, R.J., Hand, D.J.: Unsupervised profiling methods for fraud detection. In: Credit Scoring and Credit Control VII (2001)
4. Ghosh, S., Reilly, D.: Credit card fraud detection with a neural network. In: Proc. of the 27th Annual Hawaii Intern. Conf. on System Science. DSS/Knowledg-Based Systems, vol. 3, pp. 621–630 (1994)
5. Bolton, R.J., Hand, D.J.: Statistical fraud detection: A review. Statistical Science 17(3), 235–255 (2002)
6. Hawkins, D.: Identification of Outliers. Chapman & Hall, Sydney, Australia (1980)
7. Kaufman, L., Rousseeuw, P.: Finding Groups in Data: an introduction to cluster analysis. Wiley Series in Probability and Mathematical Statistics. Wiley, Chichester (1990)
8. Murtagh, F.: Complexities of hierarchic clustering algorithms: state of the art. Computational Statistics Quarterly 1, 101–113 (1984)
9. Murtagh, F.: Multidimensional clustering algorithms. In: COMPSTAT Lectures 4, Physica-Verlag, Wuerzburg (1985)
10. Knorr, E., Ng, R.: Algorithms for mining distance-based outliers in large datasets. In: Proc. 24th Int. Conf. Very Large Data Bases VLDB, pp. 392–403 (1998)
11. Ng, R., Han, J.: Efficient and efective clustering method for spatial data mining. In: Proc. of VLDB 1994 (1994)

12. Nigam, K., McCallum, A., Thrun, S., Mitchell, T.: Text classication from labeled and unlabeled documents using em. Machine Learning 39, 103–134 (2000)
13. Phua, C., Lee, V., Smith, K., Gayler, R.: A comprehendive survey of data mining-based fraud detection research. Artificial Intelligence Review (submitted)
14. Preparata, F., Shamos, M.: Computational Geometry: an introduction. Springer, Heidelberg (1988)
15. Fawcett, T., Provost, F.: Adaptive fraud detection. Data Mining and Knowledge Discovery 1(3), 291–316 (1997)
16. R Development Core Team R: A Language and Environment for Statistical Computing. R Foundation for Statistical Computing (2007) ISBN 3-900051-07-0
17. Breunig, M., Kriegel, H., Ng, R., Sander, J.: Lof: identifying density-based local outliers. In: ACM Int. Conf. on Management of Data, pp. 93–104 (2000)
18. Torgo, L.: Repository of regression data sets, http://www.liaad.up.pt/~ltorgo/Regression/DataSets.html
19. Ward, J.: Hierarchical grouping to optimize an objective function. Journal of the American Statistical Association 58(236) (1963)

Chapter 7 - First Workshop on Computational Methods in Bioinformatics and Systems Biology (CMBSB 2007)

System Stability Via Stepping Optimal Control: Theory and Applications

Binhua Tang[1], Li He[2], Sushing Chen[3], and Bairong Shen[1,4]

[1] Dept. of Biomedical Engineering, Tongji University, Shanghai, China
[2] Dept. of Electronics & Communication, Sun Yat-sen University, Guangzhou, China
[3] CAS-MPG Partner Institute of Computational Biology, Shanghai 200031 China
[4] Institute of Medical Technology, University of Tampere, FIN-33014, Finland
bairong.shen@uta.fi

Abstract. A novel stepping optimal control algorithm is proposed for system stability with concrete applications to physical and biological systems. According to the Vanecek-Celikovsky classification criteria and pertinent control objectives, the anterior parameter-set control law is first introduced to stabilize the parameter-corrupted system, and then the posterior target-set control law pilots the stabilized system to the preconcerted target region. By these means, the concerned problem evolves into a twin subordinate constructive subproblem. Relevant applications to system stability, within numerical simulations, have proved the efficiency and robustness of the proposed algorithm as a potentially significant tool for systems biology and particularly pathological analysis.

Keywords: Stepping optimal control algorithm, Vanecek-Celikovsky classification criteria, System stability, Systems biology.

1 Introduction

Decades of development in system science has facilitated the expanding of control techniques with applications to many complex dynamical systems (*e.g.*, electronics engineering, biochemical systems, economics, *etc.*). This has actuated the efficient and robust controlling tactics to be the focus of such scientific fields. At present, many algorithms have been proposed, such as self-controlling feedback [1], continuous control [2], and the variable structure [3] as well as the robust and optimal control [4], [7], [8], therefore, accelerated the concrete applications in those fields.

Provided with the generalization, this paper will deal with cases with the systematic parameters under interior and/or exterior corruptions. For instances, the relevant parameters could be influenced by the systematic or external noises. In biological systems, kinetic parameters are often changed by internal mutations or environmental perturbations and these kinds of changes may make the systems behavior abnormally and therefore cause diseases. It becomes very important to drive such generalized parameter-corrupted systems to any preconcerted target set. In the present paper, we developed an algorithm to decompose those problems into twin

J. Neves, M. Santos, and J. Machado (Eds.): EPIA 2007, LNAI 4874, pp. 463–472, 2007.
© Springer-Verlag Berlin Heidelberg 2007

subordinate constructive problems according to the Vanecek-Celikovsky classification criteria and factual controlling objectives, *i.e.*, one concerning the stabilizing of the relevant parameters and the other for tracking the terminal target set.

Pertinent deductions and simulations on a revised Rucklidge-like system and a protein-protein reaction network have proved the efficiency and reliability of the recommended algorithm for such stability analysis purposes under the condition of the intrinsic and/or exterior interferences which impact correlative system parameters.

2 The Proposed Parameter-Target Stepping Optimal Control Algorithm

Consider such a kind of systems as follows:

$$\dot{x} = Ax + Bg(x,t), \quad x(0) \in R^n \tag{1}$$

$$\dot{\overline{x}} = A\overline{x} + B(g(\overline{x},t) + u(t)), \quad \overline{x}(0) \in R^n \tag{2}$$

where $g: R^n \times R \to R^m$, x, $\overline{x} \in R^n$, $u(t) \in R^m$, $B \in R^{n \times m}$ and controllable (A, B). Denote $e(t) = \overline{x}(t) - x(t)$, the synchronization between the driving system (1) and its response system (2), can be converted into addressing the control law, $u(t) = f(x, \overline{x}, t)$, s.t. $\lim\limits_{t \to \infty} \|\overline{x}(t) - x(t)\| = \lim\limits_{t \to \infty} \|e(t)\| = 0$.

Denote the error system as:

$$\dot{e} = Ae(t) + B(g(\overline{x},t) - g(x,t) + u(t)) \tag{3}$$

Then the synchronization issue could be substituted by investigating the above equation.

As described in Section 3, one kind of optimal control law is introduced to stabilize the parameter-corrupted system, (13), also namely the target system, and then its corresponding response system is listed below:

$$\begin{cases} \dot{x}_1 = -a_1 x_1 + b_1 y_1 - y_1 z_1 + \mu_1 \\ \dot{y}_1 = x_1 - c_1 y_1 + \mu_2 \\ \dot{z}_1 = -d_1 z_1 + y_1^2 + \mu_3 \end{cases} \tag{4}$$

where a_1, b_1, c_1, d_1 are the parameters with interfering items; μ_1, μ_2, μ_3 for the control laws. According to the Vanecek-Celikovsky classification criteria [6] and the control objectives [7], the adaptive feedback law should be decomposed into twin subordinate constructive subjects, the anterior parameter-set control law μ_i^*, and the posterior target-set control law μ_i', for $i = 1, 2, 3$.

$$\begin{cases} \mu_1 = \mu_1^* + \mu_1' \\ \mu_2 = \mu_2^* + \mu_2' \\ \mu_3 = \mu_3^* + \mu_3' \end{cases} \quad (5)$$

2.1 Anterior Parameter-Set Control Law: μ_i^*

Such kind of control law is for homing utilities to the parameter-corrupted system, normally leading those ill-posed statuses to controllable regions via the anterior parameter-set control law, to some extent, as dependable tactics improving the systematic performance and robustness under such conditions of undesirable interferences.

A nonlinear function describing the diversification of parameter sets is illustrated as:

$$\dot{x} = [f(x) + \Delta f(x)] + [g(x) + \Delta g(x)]u \quad (6)$$

where $\Delta f(x)$ and $\Delta g(x)$ for the interference factors in the parameter-set system. One kind of parameter-set control law can be devised as:

$$u^* = \frac{1}{b(x)}[-R^{-1}B^T Pe - k_0 \, \text{sgn}(e^T PB)]\pi \quad (7)$$

where $k_0 \in R^+$, P, R for the positive symmetric matrixes satisfying the Riccati equation:

$$A^T P + PA - PBR^{-1}B^T P = -Q \quad (8)$$

For any initial condition, $x_0 \in R^n$ and $e(0) \in R^r$, there exists:

$$\lim_{t \to \infty} e(t) = \lim_{t \to \infty} \Delta f(x(t)) = 0 \quad (9)$$

i.e., via the law, the parameter-perturbed system can be redirected to the initially-anticipated set.

2.2 Posterior Target-Set Control Law: μ_i'

Denote the previously-stabilized system, after exerting the anterior parameter-set control law, as:

$$\dot{x} = f(x) + g(x)u, \quad f(x^*) = 0 \quad (10)$$

where $x \in R^n$ for the state variable, $u \in R^m$ the control law to be devised, $f(x)$: $R^n \to R^n$ and $g(x)$: $R^n \to R^{n \times m}$ are the continuous functions. The optimal control tactics for impelling (10) from any initial conditions to the anticipated target-set should satisfy the objective function below:

$$O(u) = \int_0^\infty [q(x) + u^T Ru]dt \tag{11}$$

which will attain the minimum, where $q(x)$ one positive, continuous and differentiable function. With the dynamic programming principles, such controlling issues are ascribed to minimum-solving of one class of partial differential equations below:

$$\min_{u \in U} | \frac{ds}{dt} + w | = | \frac{ds}{dt} + w |_{u=u^o} = 0 \tag{12}$$

where $s(x(t)) = \min_{u \in U} O(u) = \min_{u \in U} \int_0^\infty [q(x) + u^T Ru]dt$, $w = q(x) + u^T Ru$, and U is for the control law set. In the undermentioned application section, such characteristic laws are derived on the base of specific issues under discussion.

3   Applications to Physical and Biological Systems

3.1   The Rucklidge System

Firstly, we consider such an ill-posed Rucklidge system below [5], with aberrancy of parameters, ax, by, cy and dz, where $a = a_1 + a_2$, $d = d_1 + d_2$:

$$\begin{cases} \dot{x} = -ax + by - yz \\ \dot{y} = x - cy \\ \dot{z} = -dz + y^2 \end{cases} \tag{13}$$

The noise-corrupted one, or the Rucklidge-like system with the parameter set, *e.g.*, $a=2.2$, $b=6.7$, $c=0.02$, $d=1.03$, is illustrated within Fig.1. Such case of simulation is derived under the initial condition of independent variables, $[x_0, y_0, z_0] = [1, 0, 4.5]$.

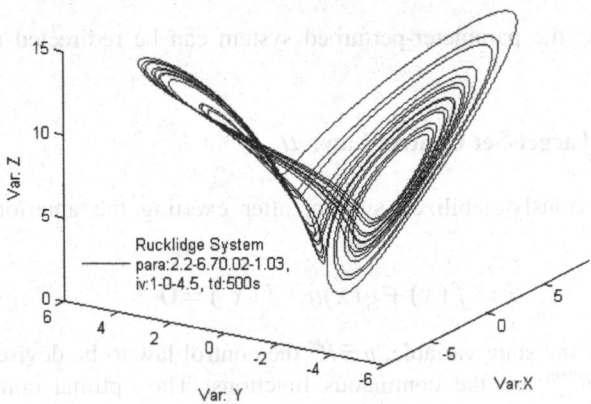

Fig. 1. The phase plane portrait of the parameter-corrupted Rucklidge-like system

The system (13) also has the properties listed below:

A. Symmetry of an equivalence relation (invariability): the system conformably satisfies the transform: $(-x, -y, z) \rightarrow (x, y, z)$;

B. Dissipation: For (13), there exists: $\nabla V = \dfrac{\partial \dot{x}}{\partial x} + \dfrac{\partial \dot{y}}{\partial y} + \dfrac{\partial \dot{z}}{\partial z} = -a - c - d$, i.e., the system has the dissipation for $a+c+d > 0$; it shall converge to the initial V_0 with the exponential rate of $\dfrac{dV}{dt} = -(a+c+d)V$, i.e., for $t \rightarrow \infty$, the system trajectory shall converge into the spatial set of zero cubage with the exponential rate of $-(a+c+d)$, the system dynamics will then be fixed in one single attractor;

C. Equilibrium points and its stabilities: (13) has one equilibrium point $S_1(0, 0, 0)$; for $d(b-ac) > 0$, there exists $S_{2,3}(\pm c\sqrt{d(b-ac)}, \pm\sqrt{d(b-ac)}, b-ac)$, the systematic Jacobian matrix is depicted as:

$$\begin{bmatrix} -a & b-z & -y \\ 1 & -c & 0 \\ 0 & -2y & -d \end{bmatrix}$$

(1) For $S_1(0, 0, 0)$, following the eigenfunction: $f(\lambda)=(\lambda+d)\,[\lambda^2+(a+c)\lambda+ac-d]$ of the above Jacobian matrix, one can get its eigenvalues as below:

$$\lambda_1 = d, \quad \lambda_{2,3} = [-(a+c) \pm \sqrt{(a+c)^2 - 4(ac-b)}]/2.$$

1) For $ac-b < 0$, λ_2, λ_3 are two eigenvalues of contrary signs, $S_1(0, 0, 0)$ is an unstable saddle point;

2) For $ac-b > 0$, $(a+c)^2-4(ac-b) \geq 0$ and $a+c > 0$, (3) only has negative eigenvalues, $S_1(0, 0, 0)$ is a stable saddle point; while for $a+c < 0$, (3) has the positive eigenvalues, and $S_1(0, 0, 0)$ is an unstable one;

3) For $(a+c)^2-4(ac-b) < 0$, (3) only has imaginary eigenvalues, $S_1(0, 0, 0)$ is also an unstable one.

(2) As to $S_{2,3}(\pm c\sqrt{d(b-ac)}, \pm\sqrt{d(b-ac)}, b-ac)$, its corresponding Jacobian matrix is depicted as:

$$\begin{bmatrix} -a & ac & \mp\sqrt{d(b-ac)} \\ 1 & -c & 0 \\ 0 & \mp 2\sqrt{d(b-ac)} & -d \end{bmatrix}$$

the eigenfunction of which is namely, $f(\lambda)=\lambda^3+(a+c+d)\lambda^2+(a+c)d\lambda+2(ac-b)d$, and according to Routh-Hurwits stability criteria, for $a+c+d>0$, $(a+c)d>0$, $(ac-b)d>0$ and $(a+c+d)(a+c)<2(ac-b)$, it only has negative eigenvalues.

The anterior parameter-set control law is adopted as:

$$U_f = \begin{bmatrix} ax_1^* - bx_2^* + x_2x_3 \\ -x_1^* \\ x_3^* - x_2^2 \end{bmatrix} \tag{14}$$

Then the relevant Lyapunov function $L(x)=\gamma^T S\gamma + c(z-z^*)^2$, and one positive symmetric matrix $S^{2\times2}$ are adopted, s.t. $\left.\dfrac{\partial}{\partial u_i}\left|\dfrac{ds}{dt}\right.+w\right|=0$ $(i=1,2,3)$, the posterior target set control law can be deduced as:

$$\begin{cases} u_1^o = -\frac{s_{11}}{r_1}(x-x^*)-\frac{s_{12}}{r_1}(y-y^*) \\ u_2^o = -\frac{s_{12}}{r_2}(x-x^*)-\frac{s_{22}}{r_2}(y-y^*) \\ u_1^o = -\frac{c}{r_3}(z-z^*) \end{cases} \tag{15}$$

Consequently, the proposed algorithms have been verified on its reliability and efficiency with the demonstration in Fig.2.

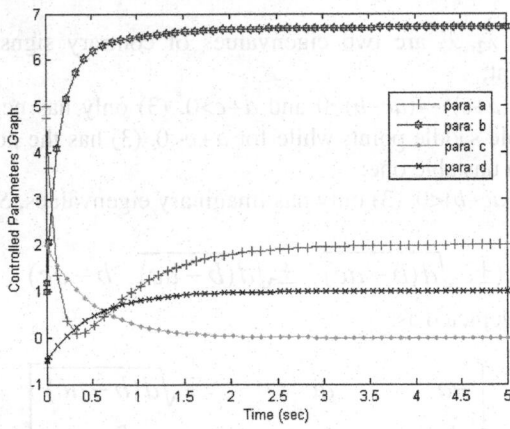

(a) Anterior parameter-set control law: After 3 seconds, the system reaches the original parameter-set with the deviation being rectified to $a = 2$, $b = 6.7$, $c = 0$, $d = 1$;

Fig. 2. Numerical simulation of the stepping optimal control law on a Rucklidge-like system

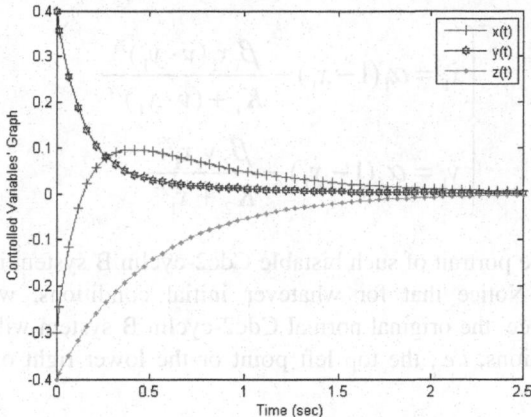

(b) Posterior target-set control law: The initial condition of the previously-stabilized system was set at (-0.3, 0.4, -0.4), the law aims to lead the system to the terminal target-set, (0, 0, 0).

Fig. 2. (*continued*)

3.2 Protein-Protein Reaction Networks

In this section, this paper considers further applications of the proposed algorithms in biological systems, in particular, the protein-protein reaction networks, one typical instance by D. Angeli, *et al.*[9], listed as below:

$$\begin{cases} \dot{x}_1 = \alpha_1 x_2 - \dfrac{\beta_1 x_1 (v \cdot y_1)^{\gamma_1}}{K_1 + (v \cdot y_1)^{\gamma_1}} \\[3mm] \dot{x}_2 = -\alpha_1 x_2 + \dfrac{\beta_1 x_1 (v \cdot y_1)^{\gamma_1}}{K_1 + (v \cdot y_1)^{\gamma_1}} \\[3mm] \dot{y}_1 = \alpha_2 y_2 - \dfrac{\beta_2 y_1 x_1^{\gamma_2}}{K_2 + x_1^{\gamma_2}} \\[3mm] \dot{y}_2 = -\alpha_2 y_2 + \dfrac{\beta_2 y_1 x_1^{\gamma_2}}{K_2 + x_1^{\gamma_2}} \end{cases} \qquad (16)$$

where an active form (with x_1 denoting active Cdc2 and y_1 denoting active Wee1) and an inactive form (x_2 and y_2 denoting inactive Cdc2 and Wee1, respectively). Take $x_2 = 1 - x_1$, $y_2 = 1 - y_1$, *i.e.*, the total concentrations of Wee1 and Cdc2-cyclin B are assumed constants and the concentrations of Cdc2 and Wee1 (active or inactive) are measured in fractional terms, we can then eliminate two variables from these equations and obtain system (17):

$$\begin{cases} \dot{x}_1 = \alpha_1(1-x_1) - \dfrac{\beta_1 x_1 (v \cdot y_1)^{\gamma_1}}{K_1 + (v \cdot y_1)^{\gamma_1}} \\[4mm] \dot{y}_1 = \alpha_2(1-y_1) - \dfrac{\beta_2 y_1 x_1^{\gamma_2}}{K_2 + x_1^{\gamma_2}} \end{cases} \tag{17}$$

The phase plane portrait of such bistable Cdc2-cyclin B system is illustrated as the following Fig.3. Notice that for whatever initial conditions, without interior or exterior interference, the original normal Cdc2-cyclin B system will evolve to one of the bistable situations, *i.e.*, the top left point or the lower right one, as depicted in Fig.3.

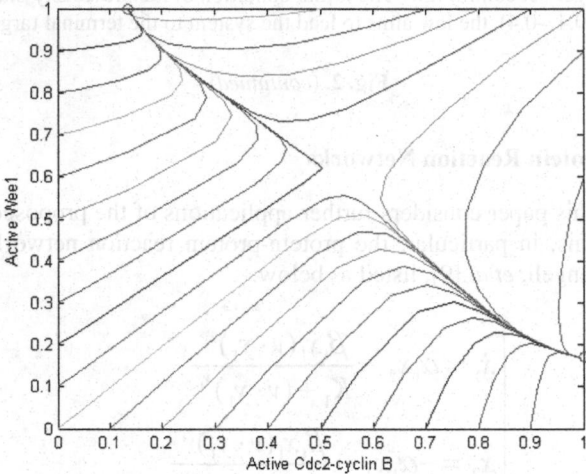

Fig. 3. The phase plane portrait of the bistable Cdc2-cyclin B system, where $\alpha_1=\alpha_2=1$, $\beta_1=200$, $\beta_2=10$, $\gamma_1=\gamma_2=4$, $K_1=30$, $K_2=1$, $v=1$

On the term of generality, this paper considers such conditions when systematic noise and/or exterior interruptions cause the reaction process to become unstable, *i.e.*, it can not be redirected to one of the two points, which has concrete relations appertaining to pathological changes in biological systems.

By comparing coefficients of the proposed posterior target-set control law, the requisite positive symmetric matrices can be obtained, listed as below:

$$Q = \begin{bmatrix} 4 & -9.6 \\ -9.6 & 25 \end{bmatrix}, \quad S = \begin{bmatrix} 1 & 0 \\ 0 & 3 \end{bmatrix} \tag{18}$$

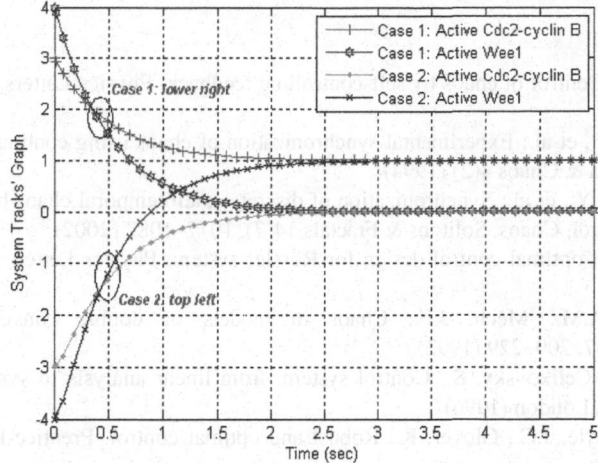

Fig. 4. The illustrative simulation on the Cdc2-cyclin B system. After 2.5 seconds or so, the ill-posed system reaches the original target-set, with the final states of active Cdc2-cyclin B and active Wee1 being rectified to 1 and 0.13 for *Case 1* (corresponding to the lower right corner in Fig.3), for *Case 2* (corresponding to the top left corner in Fig.3), those terminal values are 0.13 and 1 respectively.

With such control law, the previously ill-posed system can be readjusted to the normal conditions. Within Fig.4, two cases of bistable occurrences are illustrated to verify the initially-proposed algorithms.

4 Discussions and Conclusions

A novel stepping optimal control algorithm has been proposed, composed of the anterior parameter-set and the posterior target-set control laws. This paper verifies such control laws in one Rucklidge-like system with the generalized form of parameter-corrupted conditions, and the other application is for analyzing the stability of a bistable protein-protein reaction network, engendering interrelated factors to pathological changes in the concrete biological systems. Analysis and simulations has proved the efficiency and robustness of the proposed parameter-target stepping control algorithms.

Our future work will be emphasized particularly on stability analysis and its constructive objectives for those large-scale biological reaction networks, including such monostable and bistable ones. As the case stands, associative stability analysis and its reconstruction conjecture on large-scale biological reaction networks may act as an operable and efficacious pathway for the pathological analysis-related applications.

References

1. Pyragas, K.: Control of chaos by self-controlling feedback. Physics Letters A 170, 421–428 (1992)
2. Kapitaniak, T., et al.: Experimental synchronization of chaos using continuous control. Int. J. Bifurcations & Chaos 4(2) (1994)
3. Yin, X., Ren, Y., et al.: Synchronization of discrete spatiotemporal chaos by using variable structure control. Chaos, Solitons & Fractals 14(7), 1077–1082 (2002)
4. Marat, R.: On optimal control design for Rössler system. Physics Letters A 333, 241–245 (2004)
5. Rucklidge, A.M., Mech, J.F.: Chaos in models of double convection. J. Fluid Mechanics 237, 209–229 (1992)
6. Vanecek, A., Celikovsky, S.: Control system: from linear analysis to synthesis of chaos. Prentice-Hall, London (1996)
7. Zhou, K., Doyle, J.C., Glover, K.: Robust and optimal control. Prentice-Hall, Englewood Cliffs (1996)
8. Zhou, K., Doyle, J.C.: Essentials of robust control. Prentice-Hall, Upper Saddle River, New Jersey (1998)
9. Angeli, D., et al.: Detection of multistability, bifurcations, and hysteresis in a large class of biological positive-feedback systems. Proc. Natl. Acad. Sci. USA 101(7), 1822–1827 (2004)

Evaluating Simulated Annealing Algorithms in the Optimization of Bacterial Strains

Miguel Rocha[1], Rui Mendes[1], Paulo Maia[1], José P. Pinto[1], Isabel Rocha[2], and Eugénio C. Ferreira[2]

[1] Departament of Informatics / CCTC - University of Minho
Campus de Gualtar, 4710-057 Braga - Portugal
mrocha@di.uminho.pt, rcm@di.uminho.pt, paulo.maia@di.uminho.pt
[2] IBB - Institute for Biotechnology and Bioengineering
Center of Biological Engineering - University of Minho
Campus de Gualtar, 4710-057 Braga - Portugal
irocha@deb.uminho.pt, ecferreira@deb.uminho.pt

Abstract. In this work, a Simulated Annealing (SA) algorithm is proposed for a Metabolic Engineering task: the optimization of the set of gene deletions to apply to a microbial strain to achieve a desired production goal. Each mutant strain is evaluated by simulating its phenotype using the Flux-Balance Analysis approach, under the premise that microorganisms have maximized their growth along natural evolution. A set based representation is used in the SA to encode variable sized solutions, enabling the automatic discovery of the ideal number of gene deletions. The approach was compared to the use of Evolutionary Algorithms (EAs) to solve the same task. Two case studies are presented considering the production of succinic and lactic acid as the target, with the bacterium *E. coli*. The variable sized SA seems to be the best alternative, outperforming the EAs, showing a fast convergence and low variability among the several runs and also enabing the automatic discovery of the ideal number of knockouts.

Keywords: Simulated Annealing, Set based representations, Variable size chromosomes, Metabolic Engineering, Flux-Balance Analysis.

1 Introduction

Metabolic Engineering has been generating tools appropriate to introduce directed genetic modifications in microorganisms, to make them fit to comply with industrial purposes, i.e. to be able to synthesize some desired compounds, rather than to follow their natural aims (e.g. the maximization of growth) [15][10]. The importance of those approaches has been increasing as many traditional chemical processes are being replaced by biotechnology for the production of valuable products, such as pharmaceuticals, fuels or food ingredients.

Most often, these processes imply that the microorganism's metabolism needs to be modified, a task that can be complex. Current methods are still based mostly on intuitive design principles and scarcely on effective mathematical

J. Neves, M. Santos, and J. Machado (Eds.): EPIA 2007, LNAI 4874, pp. 473–484, 2007.

models that can predict cellular behaviour. Nevertheless, and although whole cell models are still far away, it is possible to predict cellular metabolism under some simplifying assumptions, using existing mathematical models of metabolism.

One of the most important approaches in this direction considers the cell to be in a steady-state, i.e., the concentrations of all the metabolites is considered constant throughout time. This imposes a number of constraints over the fluxes of all reactions, that can be used to predict cellular behavior. This is the basis of the Flux Balance Analysis approach [7], where a particular flux is typically optimized using linear programming. In the most usual case, a flux for biomass production is defined, whose maximization is taken as the objective function, thus assuming that the microbes have evolved towards optimal growth [6]. Solving this optimization problem for genome-scale models results in getting the values for all the fluxes of the reactions occurring in the cell.

In this way, it is possible to predict the behavior of a microorganism, both in its wild type and also in mutant forms. This allows the definition of a bi-level optimization problem, adding a layer that searches for the best mutant that can be obtained from the wild type by applying a selected set of genetic modifications. In this work, this set will be restricted to the possibility of deleting genes from the wild type. The idea is to force the microorganisms to synthesize a desired product by selected gene deletions. Therefore, the underlying optimization problem consists in reaching an optimal subset of gene deletions to optimize an objective function related with the production of a given compound.

A first approach to this problem was proposed by the *OptKnock* algorithm [2], where mixed integer linear programming methods are used to reach a guaranteed optimum solution. This algorithm suffered from two important drawbacks: the impossibility of considering nonlinear objective functions and the considerable computation time required that only allowed the problem to be solved for a relatively small number of reactions.

An alternative approach was proposed by the *OptGene* algorithm [11] that considers the application of Evolutionary Algorithms (EAs). EAs are capable of providing near optimal solutions in a reasonable amount of time and also allow the optimization of nonlinear objective functions. *OptGene* proposes EAs with two alternative representation schemes: binary or integer. The first is closer to the natural evolution, but is more complex and leads to solutions with a large number of knockouts. The latter allowed for a more compact encoding scheme, representing only the gene deletions. One of its major limitations is the need to define *a priori* the number of gene knockouts, that remains fixed. In [13] an EA with a set-based representation is proposed to extend *OptGene*. The use of variable-sized chromosomes to encode the sets was a major improvement, allowing the automatic definition of the ideal number of gene deletions.

In this work, an alternative optimization strategy is proposed based on the use of a Simulated Annealing (SA) algorithm. The proposed algorithm encodes solutions using a variable size set-based representation, making use of mutation operators similar to the ones used by the EAs. The proposed algorithm will be

tested, using as a benchmark the two case studies proposed in [13] and comparing
the results obtained with the ones achieved by the EA's.

2 Simulation Algorithms for the Prediction of Metabolic Behavior

One of the many potential applications of the recently sequenced and annotated
genomes of microorganisms is the reconstruction of genome-scale metabolic networks. The set of metabolic reactions obtained can therefore be used to simulate
the phenotypic behaviour of microorganisms. One approach is to write dynamic
mass balances for each metabolite in the network, generating a set of ordinary
differential equations that may be used to simulate the dynamic behavior of
metabolite concentrations. However, there is still insufficient data on kinetic expressions and parameters, and it is only possible to simulate dynamic conditions
for a few pathways[3].

Therefore, a steady state approximation is generally applied, where for each
metabolite in the network, the sum of all productions and consumptions will be
zero, weighted by the stoichiometric coefficients. Thus, for metabolite i, where
$i = 1, \ldots, M$ (M is the number of metabolites) the following constraint is defined:

$$\sum_{j=1}^{N} S_{ij} v_j = 0 \qquad (1)$$

where S_{ij} is the stoichiometric coefficient for metabolite i in reaction j and v_j
is the reaction rate or flux over the reaction j. It is possible to define a matrix
S, composed of the S_{ij} values, $j = 1, \ldots, N$ (N is the number of reactions); v is
the N-dimensional vector of the fluxes of the reactions.

The mass balances are therefore reduced to a set of linear homogeneous equations. The maximum/minimum values of the fluxes can be set by additional
constraints in the form $\alpha_j \leq v_j \leq \beta_j$, that are also used to specify both thermodynamic and environmental conditions (e.g. availability of nutrients).

For most of the metabolic networks, and because the number of fluxes is
greater than the number of metabolites, the set of linear equations obtained
from the application of Equation 1 to the M metabolites usually leads to an
under-determined system, for which there exists an infinite number of feasible
flux distributions that satisfy the constraints. However, if a given linear function
over the fluxes is chosen to be maximized, it is possible to obtain a single solution
by applying standard algorithms (e.g. *simplex*) for linear programming problems.
This methodology is known as Flux Balance Analysis (FBA) [7].

The combination of this technique with the existence of validated genome-
scale stoichiometric models [4][1] allows to simulate the phenotypic behaviour
of a microorganism under defined environmental conditions without performing
any experiments. The most common flux chosen for maximization is the biomass,
based on the premise that microorganisms have maximized their growth along
natural evolution, a premise that has been confirmed experimentally in some
cases [6].

3   Simulated Annealing

Simulated Annealing (SA) is an optimization algorithm inspired in the annealing process used in metallurgy, where a melt, initially at high temperature, is slowly cooled so that the system at any time is approximately in thermodynamic equilibrium. As the cooling proceeds, the system becomes more ordered, approaching a minimal energy state when the temperature reaches zero. If the initial temperature of the system is too low or the cooling process is not sufficiently slow the system may become trapped in a local minimum energy state.

In the original Metropolis scheme, an initial state of a thermodynamic system is chosen at energy E and holding temperature T constant. The initial configuration is perturbed and the change in energy ΔE is computed. A better configuration is always accepted, while a worse configuration is only accepted with a probability given by the Boltzmann factor

$$p[accept] = e^{-\frac{\Delta E}{T}} \tag{2}$$

This process is then repeated a number of trials, sufficient to give good sampling statistics for the current temperature, and then the temperature is decreased according to a given cooling schedule. The entire process is repeated until the temperature is sufficiently low.

This Monte Carlo approach may be used to optimize real or combinatorial problems [8]. The current state is a solution to the optimization problem and the energy represents its objective function value. The solution is perturbed by any process that will generate a new solution from the current one, also denoted as a *mutation* operator. This perturbation may depend on the temperature, thus allowing larger steps to be taken when the temperature is high and fine-tuning when the temperature is low. The implementation used in this work allows the use of any kind of mutation or combination thereof to perturb the current solution and generate a new one. Any number of mutations may be applied, each with a given probability (that must sum 1). The solution encoding is also very flexible and may use for instance a binary representation, real values, sets, permutations, trees, etc.

The configuration parameters for the algorithm are the initial and final temperatures, the number of iterations performed at each temperature and the cooling schedule used. The choice of these parameters is of paramount importance to the performance of the algorithm. If the initial temperature is too low or the cooling schedule is not slow enough, the optimization process may become stuck in a local optimum. On the other hand, if the initial temperature is too high, the cooling is too slow or the number of iterations per temperature is too high, the algorithm wastes a potentially large amount of computational time while searching for solutions. The cooling schedule used in this paper is among the most popular ones, where the temperature decrease is exponential, defined according to the following equation:

$$T_{n+1} = \alpha T_n \tag{3}$$

where $0 < \alpha \leq 1$. To ensure that the cooling schedule is sufficiently slow, the parameter α should be given values close to the unity.

As the choice of initial (T_0) and final temperatures (T_f) is problem dependent, it was decided to use the following configuration parameters:

ΔE_0 – The difference in energy that corresponds to an acceptance probability of 50% of worse solutions at the beginning of the run;

ΔE_f – The difference in energy that corresponds to an acceptance probability of 50% of worse solutions at the end of the run;

trials – The number of iterations per temperature;

NFEs – The number of function evaluations.

Using these parameters, the initial temperature, the final temperature and the scale parameter were computed using the following equations

$$T_0 = -\frac{\Delta E_0}{\log 0.5} \tag{4}$$

$$T_f = -\frac{\Delta E_f}{\log 0.5} \tag{5}$$

$$\alpha = \exp\left(\frac{\log T_f - \log T_0}{\left[\frac{\text{NFEs}}{\text{trials}}\right]}\right) \tag{6}$$

The advantage of using ΔE_0 and ΔE_f is that it allows the user who knows the fitness landscape of the optimization problem to automatically define the temperatures by reasoning over the values of the objective function. Supplying the number of function evaluations instead of the scale parameter α allows the user to accurately define the number of function evaluations the optimization algorithm will use, enabling a simpler comparison with other approaches.

4   The Proposed Algorithm

4.1   Representation Scheme and Mutation Operator

The problem addressed in this work consists in selecting, from a set of genes in a microbe's genome, a subset to be deleted in order to maximize a given objective function related to the microorganism's metabolism. The encoding of a solution is achieved by a set-based representation, where only gene deletions are represented. Each solution consists of a set of integer values representing the genes that will be deleted. Therefore, if the value i is in the set, this means the i-th gene in the microbe's genome is knocked out. Each value in the set is an integer with a value between 1 and N.

Two variants of this representation can be defined, considering fixed or variable sized sets. In the fixed-size alternative, the mutation operator creates solutions always of the same size. A random mutation operator is used that replaces a gene by a random value in the allowed range, avoiding duplicates in the set. In variable-sized representations, sets with distinct cardinalities can be encoded and compete in the search process. In this case, two additional mutation operators are defined to be able to create solutions with a distinct size:

– *Grow*: consists in the introduction of a new gene into the chromosome, whose value is randomly generated in the available range (avoiding duplicates in the set).
– *Shrink*: a randomly selected gene is removed from the genome.

In the SA, the *Grow* and *Shrink* mutations are each used with a probability of 25% each, meaning that half of the new individuals are created in this way. The remaining are created by the aforementioned random mutation operator. In the experiments reported in this work, when a variable size is used, the minimum size is set to 1 and the maximum size is set to the number of genes (N), thus not restricting the possible range of solutions.

4.2   Decoding and Evaluating

The principle considered is a correspondence between the values in the set and metabolic reactions, i.e., each value represented in the set represents a particular enzyme that catalyzes a metabolic reaction. That enzyme is associated with a particular gene (or genes) that should be deleted for that reaction to be eliminated. The decoding process works by taking each value in the set and forcing the flux it indexes to the value 0, therefore disabling that reaction from the metabolic model. The process proceeds with the simulation of the mutant using FBA. The output is the set of values for the fluxes of all reactions, that are then used to compute the fitness value, given by an appropriate objective function.

One possible objective function is the Biomass-Product Coupled Yield (BPCY) [11], given by:

$$BPCY = \frac{PG}{S} \tag{7}$$

where P stands for the flux representing the excreted product; G for the organism's growth rate (biomass flux) and S for the substrate intake flux. Besides optimizing for the production of the desired product, this function also allows to select for mutants that exhibit high growth rates, i.e., that are likely to exhibit a higher productivity, an important industrial aim.

An alternative is to maximize only the value of the product's flux (P), but imposing a minimum threshold to the value of the biomass (G_{min}). Therefore, the objective function (denoted as Product Flux with Minimum Biomass (PFMB)) will be defined as: $PFMB = P$, if $G > G_{min}$; otherwise $PFMB = 0$.

4.3   Initialization

The initial solution is a set with randomly generated elements. In the variable size variant, the size of the individual is randomly created in the range [1,12]. The same process is used in the EAs to initilize each individual in the population.

4.4 Pre-processing and Post-processing

In genome-scale models the number of variables (fluxes over metabolic reactions) is in the order of hundreds or a few thousands and therefore the search space is very hard to address. Thus, every operation that gives a contribution to reduce this number, greatly improves the convergence of the algorithms. In this work, a number of operations is implemented to reduce the search space:

- Removal of fluxes that, given the constraints of the linear programming problem, cannot exhibit values different from 0.
- Equivalent variables, i.e. pairs of variables that are constrained to have the same value by the model. Each group of equivalent variables is replaced by a single variable.
- Discovery of essential genes that can not be deleted from the microorganism genome. As these genes should not be considered as targets for deletion, the search space for optimization is reduced. This list can be manually edited to include genes that are known to be essential, although that information can not be reached from the mathematical model.
- Identification of artificial fluxes that are associated with external metabolites and exchange fluxes that represent transport reactions. These are not allowed to be knocked out, since generally this would not have a biological meaning.

The best solution in each run goes through a simplification process, by identifying all gene deletions that contribute to the fitness of the solution, removing all deletions that keep the objective function unaltered. The aim is to keep only the necessary knockouts, given that the practical implementation of a gene deletion is both time consuming and costly.

4.5 Implementation Issues

The implementation of the proposed algorithms was performed by the authors in the *Java* programming language. In the implementation of FBA, the *GNU linear programming package (GLPK)*[1] was used to run the *simplex* algorithm.

5 Experiments

5.1 Experimental Setup

Two case studies were used to test the aforementioned algorithms. Both consider the microorganism *Escherichia coli* and the aim is to produce succinic and lactic acid (case studies I and II, respectively), with glucose as the limiting substrate. The genome-scale model for this microorganism used in the simulations was developed by Reed et al [12]. This model considers the *E. coli* metabolic network, including a total of $N = 1075$ fluxes and $M = 761$ metabolites. After the pre-processing stages, the simplified model remains with $N = 550$ and $M = 332$

---

[1] http://www.gnu.org/software/glpk/

metabolites. Furthermore, 227 essential genes are identified, which leaves 323 variables to be considered by the optimization algorithms.

The proposed SA is compared to the EAs proposed in [13]. Both algorithms were implemented in its fixed and variable size versions. In the first case, the cardinality of the set (k) took a number of distinct values. In the EA the population size was set to 100. The SA used $\Delta E_0 = 0.005$, $\Delta E_f = 5E - 5$ and $trials = 50$. In both cases, the termination criteria was defined based on a maximum of 50000 fitness evaluations. For each experimental setup, the process was repeated for 30 runs and the mean and 95% confidence intervals were calculated.

5.2   Case Study I: Succinic Acid

Succinic acid is one of the key intermediates in cellular metabolism and therefore an important case study for metabolic engineering[9]. The knockout solutions that lead to an improved phenotype regarding its production are not straightforward to identify since they involve a large number of interacting reactions. Succinic acid and its derivatives have been used as common chemicals to synthesize polymers, as additives and flavoring agents in foods, supplements for pharmaceuticals, or surfactants. Currently, it is produced through petrochemical processes that can be expensive and have significant environmental impacts.

In Table 1, the results for the EAs and SA, both fixed (the number of knockouts k is given) and variable sized (last row) are given, taking the BPCY as the objective function. The results show the mean, the 95% confidence interval and the maximum value of the BPCY for each configuration. In the last column, the mean of the number of gene deletions (after the simplification process) is shown.

In Table 1 it is possible to observe that, when using set-based representations with fixed size chromosomes, the results improve with the increase on the number of gene deletions. The improvement obtained when increasing from 6 to 20 gene deletions is essentially visible in the increase of the mean, since the best solution suffers minor improvements. Furthermore, there is less variability in the results, given by the smaller confidence intervals. The variable size alternatives seem to be able to automatically find the appropriate number of gene deletions. They also present a very low variability, given the small confidence interval.

Table 1. Results obtained for the case study I - production of succinic acid

k	EA Mean	Conf. int.	Best	Knockouts	SA Mean	Conf. int.	Best	Knockouts
2	0.0458	±0.0288	0.0752	2.0	0.0475	±0.0290	0.0752	2.0
4	0.1172	±0.0769	0.3366	3.6	0.1096	±0.0674	0.3440	3.7
6	0.2458	±0.1108	0.3573	5.8	0.3184	±0.1121	0.3573	5.9
8	0.2963	±0.0969	0.3577	7.1	0.3401	±0.0781	0.3576	7.2
10	0.3218	±0.0739	0.3578	8.1	0.3566	±0.0554	0.3578	8.1
12	0.3496	±0.0370	0.3578	8.8	0.3573	±0.0015	0.3578	8.7
20	0.3575	±0.0012	0.3578	10.8	0.3577	±0.0001	0.3578	10.2
VS	0.3507	±0.0372	0.3579	11.8	0.3577	±0.0001	0.3579	11.4

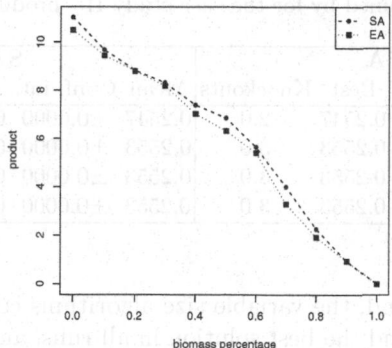

Fig. 1. Convergence plots for the variable-sized EA and SA in case study I

A comparison between EAs and SA show that the latter seems to obtain better results. When k is small (2 or 4) the results are comparable; when k increases the mean of the SA improves faster. Although some of the confidence intervals are overlapping, the standard deviations are smaller in the SA showing less variability in the results. Regarding the variable size versions, the SA has a better mean and a very reduced confidence interval, denoting the capacity to consistently find high quality results.

A distinct set of results was obtained by running both algorithms (variable sized variants) with the PYMB as the objective function. The minimum biomass was varied as a percentage of the wild type's value, in 10% intervals. The mean of the results for each algorithm is plotted in Figure 1. It is noticeable that the SA has slightly higher mean values for almost every point in the graph although the confidence intervals are overlapping in most cases (these are not shown for improved visualization). Nevertheless, the fact that SA achieves higher means and lower standard deviation values is an aditional indicator of its performance.

5.3   Case Study II - Lactic Acid

Lactic acid and its derivatives have been used in a wide range of food-processing and industrial applications like meat preservation, cosmetics, oral and health care products and baked goods. Additionally, and because lactate can be easily converted to readily biodegradable polyesters, it is emerging as a potential material for producing environmentally friendly plastics from sugars [5]. Several microorganisms have been used to produce lactic acid, such as *Lactobacillus* strains. However, those bacteria have undesirable traits, such as a requirement for complex nutrients which complicates acid recovery. *E. coli* has many advantageous characteristics, such as rapid growth and simple nutritional requirements.

In Table 2, the results for the case study II are given. The first conclusion that can be drawn is that this case study seems to be less challenging than the one presented on the previous section. In fact, 3 gene deletions seem to be enough to obtain the best solution, and therefore the results with k larger than 6 are not

Table 2. Results obtained by for the case study II - production of lactic acid

k	EA				SA			
	Mean	Conf. int.	Best	Knockouts	Mean	Conf. int.	Best	Knockouts
2	0.2547	±0.0000	0.2547	2.0	0.2547	±0.0000	0.2547	2.0
4	0.2553	±0.0000	0.2553	3.0	0.2553	±0.0000	0.2553	3.0
6	0.2553	±0.0000	0.2553	3.0	0.2553	±0.0000	0.2553	3.0
VS	0.2553	±0.0000	0.2553	3.0	0.2553	±0.0000	0.2553	3.0

shown. On the other hand, the variable size algorithms confirm its merits once they are again able to find the best solution in all runs, and automatically finds the adequate number of knockouts.

5.4   Discussion

Two features that are important when comparing meta-heuristic optimization algorithms are the computational effort required and the convergence of the algorithm to a good solution. The computational burden of the alternatives compared is approximately the same, since the major computational effort is devoted to fitness evaluation and the same number of solutions is evaluated in every case. A typical run of each algorithm for the case studies presented will take approximately two or three hours in a regular PC.

Regarding the convergence of the algorithms, a plot of the evolution of the objective function along the generations of the SA and EA is given in Figure 2 (the mean of the 30 runs is plotted). Only the variable sized versions were selected to allow a better visualization. It is clear from this plot that the SA converges faster than the EA, obtaining high quality results early in the runs. Both

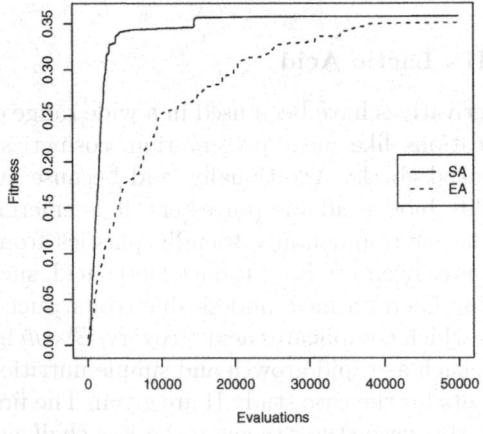

Fig. 2. Convergence plots for the variable-sized EA and SA in case study I

algorithms are similar in terms of computational effort, since most of the time is spent evaluating the solutions.

It is also important to refer that the approach followed is to solve the problem *in silico*, using computer models. Therefore, the results may or may not be biologically feasible. Nevertheless, a validation of the results was conducted since the best solutions obtained were analyzed by experts in Biotechnology using Bioinformatics databases (e.g. *EcoCyc*[2]).

6   Conclusions and Further Work

In this work, a contribution to Metabolic Engineering was provided by the development of a variant of Simulated Annealing that is able of reaching a near optimal set of gene deletions in a microbial strain, to maximize the production of a given product. This algorithm was able to improve previous results from the use of Evolutionary Algorithms. These were tested in a case study that dealt with the production of succinic acid by the *E. coli* bacterium. Important contributions of this work were the introduction of a set-based representation, that made use of variable size solutions, an uncommon feature in SA algorithms.

There are still a number of features that need to be introduced. These include other algorithms for simulation and distinct objective functions. Regarding the former, an alternative algorithm for simulating mutants' phenotype is the MOMA algorithm, that was proposed by Segre et al [14], where it is assumed that knockout metabolic fluxes undergo a minimal redistribution with respect to the flux configuration of the wild type. This implies solving a quadratic programming problem, whose aim is to minimize the differences between the fluxes in the mutant and the ones in the wild type. It would also be interesting to consider an objective function capable of taking into account the number of knockouts of a given solution and the cost of its experimental implementation.

Acknowledgments

The authors thank the Portuguese Foundation for Science and Technology (FCT) for their support through project ref. POSC/EIA/59899/2004. partially funded by FEDER.

References

1. Borodina, I., Nielsen, J.: From genomes to in silico cells via metabolic networks. Current Opinion in Biotechnology 16(3), 350–355 (2005)
2. Burgard, A.P., Pharya, P., Maranas, C.D.: Optknock: A bilevel programming framework for identifying gene knockout strategies for microbial strain optimization. Biotechnol. Bioeng. 84, 647–657 (2003)

---

[2] http://www.ecocyc.org

3. Chassagnole, C., Noisommit-Rizzi, N., Schmid, J.W., Mauch, K., Reuss, M.: Dynamic modeling of the central carbon metabolism of escherichia coli. Biotechnology and Bioengineering 79(1), 53–73 (2002)
4. Covert, M.W., Schilling, C.H., Famili, I., Edwards, J.S., Goryanin, I.I., Selkov, E., Palsson, B.O.: Metabolic modeling of microbial strains in silico. Trends in Biochemical Sciences 26(3), 179–186 (2001)
5. Hofvendahl, K., Hahn-Hagerdal, B.: Factors affecting the fermentative lactic acid production from renewable resources. Enzyme Microbial Technology 26, 87–107 (2000)
6. Ibarra, R.U., Edwards, J.S., Palsson, B.G.: Escherichia coli k-12 undergoes adaptive evolution to achieve in silico predicted optimal growth. Nature 420, 186–189 (2002)
7. Kauffman, K.J., Prakash, P., Edwards, J.S.: Advances in flux balance analysis. Curr. Opin. Biotechnol. 14, 491–496 (2003)
8. Kirkpatrick, S., Gelatt Jr., C.D., Vecchi, M.P.: Optimization by simulated annealing. Science 220(4598), 671–680 (1983)
9. Lee, S.Y., Hong, S.H., Moon, S.Y.: In silico metabolic pathway analysis and design: succinic acid production by metabolically engineered escherichia coli as an example. Genome Informatics 13, 214–223 (2002)
10. Nielsen, J.: Metabolic engineering. Appl. Microbiol. Biotechnol. 55, 263–283 (2001)
11. Patil, K., Rocha, I., Forster, J., Nielsen, J.: Evolutionary programming as a platform for in silico metabolic engineering. BMC Bioinformatics 6(308) (2005)
12. Reed, J.L., Vo, T.D., Schilling, C.H., Palsson, B.O.: An expanded genome-scale model of escherichia coli k-12 (ijr904 gsm/gpr). Genome Biology 4(9) R54.1–R54.12 (2003)
13. Rocha, M., Pinto, J.P., Rocha, I., Ferreira, E.C.: Optimization of Bacterial Strains with Variable-Sized Evolutionary Algorithms. In: Proceedings of the IEEE Symposium on Computational Intelligence in Bioinformatics and Computational Biology, pp. 331–337. IEEE Press, Honolulu, USA (2007)
14. Segre, D., Vitkup, D., Church, G.M.: Analysis of optimality in natural and perturbed metabolic networks. Proc. National Acad. Sciences USA 99, 15112–15117 (2002)
15. Stephanopoulos, G., Aristidou, A.A., Nielsen, J.: Metabolic engineering principles and methodologies. Academic Press, San Diego (1998)

Feature Extraction from Tumor Gene Expression Profiles Using DCT and DFT

Shulin Wang[1,2], Huowang Chen[1], Shutao Li[3], and Dingxing Zhang[1]

[1] School of Computer Science, National University of Defense Technology,
Changsha, Hunan 410073, China
jt_slwang@hnu.cn
[2] School of Computer and Communication, Hunan University,
Changsha, Hunan 410082, China
[3] College of Electrical and Information Engineering, Hunan University,
Changsha, Hunan 410082, China

Abstract. Feature extraction plays a key role in tumor classification based on gene expression profiles, which can improve the performance of classifier. We design two novel feature extraction methods to extract tumor-related features. One is combining gene ranking and discrete cosine transform (DCT) with principal component analysis (PCA), and another is combining gene ranking and discrete Fourier transform (DFT) with PCA. The proposed feature extraction methods are proved successfully and effectively to classify tumor dataset. Experiments show that the obtained classification performance are very steady, which are evaluated by support vector machines (SVM) and K-nearest neighbor (K-NN) classifier on two well-known tumor datasets. Experiment results also show that the 4-fold cross-validated accuracy rate of 100% is obtained for the leukemia dataset and 96.77% for the colon tumor dataset. Compared with other related works, the proposed method not only has higher classification accuracy rate but also is steadier in classification performance.

Keywords: gene expression profiles, tumor classification, feature extraction, support vector machines, discrete cosine transform.

1 Introduction

Machine learning have been broadly used in many fields of bioinformatics, especially in tumor classification based on gene expression profiles, which is a novel molecular diagnostic method compared with traditional tumor diagnostic methods based on the morphology appearance of tumor. However, feature extraction which is a dimensionality reduction technique plays a key role in tumor classification when using machine learning based classification methods, because there are too many redundant genes and too much noise in gene expression profiles. In tumor classification, feature extraction can be viewed as a searching process among all possible transformation of the gene expression profiles for the best one which retains

J. Neves, M. Santos, and J. Machado (Eds.): EPIA 2007, LNAI 4874, pp. 485–496, 2007.
© Springer-Verlag Berlin Heidelberg 2007

as much classification information as possible in the novel feature space with the lowest possible dimensionality. In pattern recognition, feature extraction yields three advantages: (1) improving classification performance; (2) reducing the time of classification; (3) visualizing classification results when mapping gene expression profiles into 2 or 3-dimensional space. Therefore, in this paper we focus on the feature extraction problem and our aim is to explore a novel dimensionality reduction method which can improve the performance of tumor classification.

2   Related Works

Due to the characteristics of gene expression profiles such as high dimensionality and small sample size, how to select tumor-related genes and to extract integrated features to drastically reduce the dimensionality of tumor samples constitutes a challenging problem. Feature extraction is an important aspect of pattern classification. The goal of feature extraction is to eliminate redundancies and noisiness in gene expression profiles to obtain many integrated attributes which can correctly classify the tumor dataset. For example, Zhang [1] adopted independent component analysis (ICA) to extract independent components (ICs) from gene expression profiles to be used as the classification information. We also adopted factor analysis to extract latent factors to be used as the input of support vector machines (SVM) [2], which has the meaning of biology. Li [3] proposed a feature extraction method which combines gene ranking based on T-test method with kernel partial least squares to extract features with the classification ability. Nguyen [4] proposed an analysis procedure which involves dimension reduction using partial least squares and classification using logistic discrimination and quadratic discriminant analysis. Although the enumerative methods are effective for tumor classification, the obtained accuracy rate based on feature extraction is not higher than the classification based on gene selection.

In this paper we propose two similar and novel feature extraction methods to extract tumor-related features. One is combining gene ranking and discrete cosine transform (DCT) with principal component analysis (PCA), and another is combining gene ranking and discrete Fourier transform (DFT) with PCA. The proposed feature extraction methods are proved successfully in our experiments which are conducted using SVM classifier and K-nearest neighbor (K-NN) classifier on two well-known tumor datasets.

3   Tumor Classification Model

3.1   Gene Expression Profiles and Problem Description

Let $G = \{g_1, \cdots g_n\}$ be a set of genes and $S = \{s_1, \cdots s_m\}$ be a set of samples. Here, m is the number of samples and n is the number of genes measured. The corresponding gene expression matrix can be represented as $X = (x_{i,j})$, $1 \leq i \leq m, 1 \leq j \leq n$. The matrix X is composed of m row vectors $s_i \in R^n$, $i = 1, 2, \cdots, m$.

$$X = \begin{bmatrix} \overbrace{\begin{matrix} & n & genes & \end{matrix}} & & \overbrace{Class} \\ x_{1,1} & x_{1,2} & \cdots & x_{1,n} & l_1 \\ x_{2,1} & x_{2,2} & \cdots & x_{2,n} & l_2 \\ \vdots & \vdots & \cdots & \vdots & \vdots \\ x_{m,1} & x_{m,2} & \cdots & x_{m,n} & l_m \end{bmatrix}$$

where $x_{i,j}$ is the expression level value of sample s_i on gene g_j, and l_i denotes the class label that the sample s_i belongs to. Each vector $s_i \in S$ may be thought of as a point in n-dimensional space. Each of the n columns consists of an m-element expression vector for a single gene.

Suppose $Acc(T)$ denotes the classification ability of feature subset T, which is usually measured by accuracy rate. Let W denotes a transform to gene expression profiles X. Feature transform can be represented as $Y = W(X)$, our task is to find the optimal transform W^* among many transforms, which can be denoted as

$$Acc(W^*(X)) = \max_W (Acc(W(X))) \tag{1}$$

However, finding the optimal transform is not easy task, because we have less prior knowledge about gene expression profiles.

3.2 The Algorithm Model

Many tumor classification methods based on feature extraction have been proposed, but their performance of classification need to be improved further, which can be seen in Table 4. Therefore, to further improve the classification accuracy rate, we design a novel feature extraction model which mainly consists of five steps and which is shown in Fig. 1.

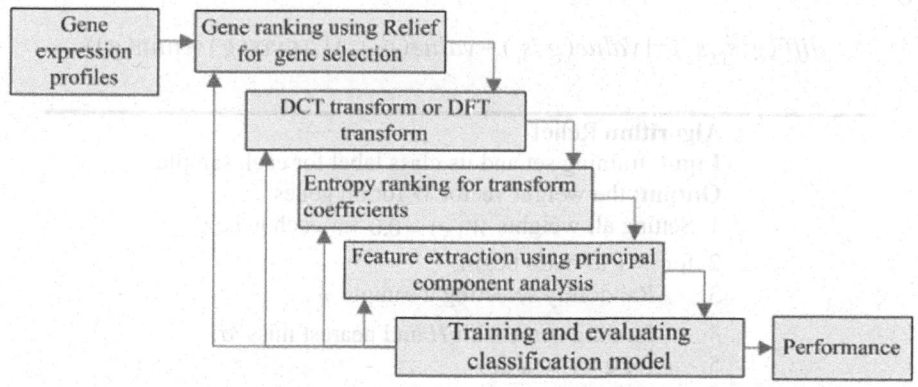

Fig. 1. The basis procedure of our classification algorithm model

The function of every step in algorithm model is described as follows.

Step 1: Gene ranking and selection: for each gene $g_i \in G$, we firstly calculate its weight value using Relief algorithm [5] and then rank all genes by their weights. After gene ranking, we simply take the top-ranked genes with the highest weight value as the informative gene subset G_{top}.

Step 2: DCT or DFT transform: applying DCT or DFT transform to the top-ranked gene subset G_{top} to remove the redundancies between genes and obtaining transform coefficients.

Step 3: Entropy ranking: computing relative entropy for each transform coefficient and ranking all coefficients according to their relative entropy.

Step 4: Feature extraction: extracting principal components (PCs) using PCA from the top-ranked coefficients.

Step 5: Training classifier using training set based on the extracted PCs and evaluating the classifier using testing set to obtain predictive accuracy rate.

When implementing this algorithm model, there are many feedbacks from step 5 to step 1, 2, 3, 4, respectively, the objective of which is to find the best combination of parameters in every step. However, the procedure of finding best combination of parameters is very time-consuming, and no theory guides us selecting those parameters.

3.3 Gene Selection

Gene selection is necessary for performing tumor classification, for there are a large number of genes not related to tumor in gene expression profiles, which can decrease the classification accuracy rate. We adopt Relief algorithm [5] described in Fig. 2 to select the tumor-related genes.

Function $diff(g, s_1, s_2)$ calculates the difference between the values of the gene g for the two samples s_1 and s_2. It was defined as:

$$diff(g, s_1, s_2) = | value(g, s_1) - value(g, s_2) | /(\max(g) - \min(g))$$

Algorithm Relief
Input: training set and its class label for each sample
Output: the weight vector W for all genes
1. Setting all weights $W(g) := 0.0$ for each gene g;
2. for $i := 1$ to m do begin
3.   Randomly selecting a sample R_i;
4.   Finding nearest hit H and nearest miss M;
5.   for $g := 1$ to n do
6. $W[g] := W[g] - diff(g, R_i, H)/m + diff(g, R_i, M)/m$;
7. end;

Fig. 2. The basic Relief algorithm

3.4 Discrete Cosine Transform and Discrete Fourier Transform

The DCT transform [6] is a Fourier-related transform and similar to the discrete Fourier transform, and has a strong energy compaction property and the next best performance in compaction efficiency, so DCT is often used in signal and image processing. The mathematics for the 1-D DCT algorithm is defined by equation (2).

$$X(k) = \alpha(k) \sum_{n=0}^{N-1} x(n) \cos[\frac{\pi(2n+1)k}{2N}] \quad 0 \le k \le N-1 \tag{2}$$

The defined equation for the inverse DCT is

$$x(n) = \sum_{k=0}^{N-1} \alpha(k) X(k) \cos[\frac{\pi(2n+1)k}{2N}] \quad 0 \le n \le N-1 \tag{3}$$

In both equation (1) and (2), $\alpha(0) = \sqrt{1/N}$, $\alpha(k) = \sqrt{2/N}$, $1 \le k \le N-1$. Obviously, for $k = 0$, $X(0) = \sqrt{1/N} \sum_{k=0}^{N-1} x(k)$. Hence, if the transformed sequence is gene expression profiles, the first coefficient, which is called as DC coefficient, is the average value of gene expression in a sample, and the other coefficients are referred to as AC coefficients.

We also apply DFT transform to gene expression profiles in contrast with DCT. The formula of DFT is defined as:

$$X(k) = \sum_{n=0}^{N-1} x(n) e^{-\frac{2\pi i}{N}kn} \quad 0 \le k \le N-1 \tag{4}$$

The inverse discrete Fourier transform (IDFT) is given by:

$$x(n) = \frac{1}{N} \sum_{k=0}^{N-1} X(k) e^{\frac{2\pi i}{N}kn} \quad 0 \le n \le N-1 \tag{5}$$

In our algorithm we only apply DCT and DFT transform to the selected gene set, and their inverse transform is not necessary in our algorithm.

3.5 Entropy Ranking for DCT Coefficients

After DCT transform, we apply relative entropy [7], also known as Kullback-Lieber distance, to measuring the separability of coefficients between the two classes ω_1 and ω_2. Relative entropy for each coefficient is defined as:

$$d_{12} = \frac{1}{2}\left(\frac{\sigma_1^2}{\sigma_2^2} + \frac{\sigma_2^2}{\sigma_1^2} - 2\right) + \frac{1}{2}(\mu_1 - \mu_2)^2\left(\frac{1}{\sigma_1^2} + \frac{1}{\sigma_2^2}\right)$$

where μ_1 (resp. μ_2) denotes the mean value of the coefficient in class ω_1 (resp. ω_2) and σ_1 (resp. σ_2) denotes the standard deviation of the coefficient in class ω_1 (resp. ω_2). Then all coefficients are ranked according to their relative entropy which reflects the ability of classification.

3.6 Support Vector Machines and K-Nearest Neighbor

SVM is a relatively new type of statistic learning theory, originally introduced by Vapnik [8]. SVM builds up a hyper-plane as the decision surface to maximize the margin of separation between positive and negative samples. Given a labeled set of m training samples $S = \{(F_i, y_i) \mid (F_i, y_i) \in R^n \times \{\pm 1\}, i = 1, 2, \cdots m\}$, where $F_i \in R^n, y_i \in \{\pm 1\}$ is a label of sample eigenvector F_i, and the discriminant hyper-plane is defined by formula

$$f(x) = \sum_{i=1}^{m} \alpha_i y_i K(F_i, x) + b \tag{6}$$

where $K(F_i, x)$ is a kernel function and the sign of $f(x)$ determines which class the unknown sample eigenvector x belongs to. Constructing an optimal hyper-plane is equivalent to finding all the support vectors α_i and a bias b.

In contrast with SVM classifier we also adopt K-nearest neighbor [9] to be used as our classifier. The K-NN classifier is the most common and non-parametric method. To classify an unknown sample x, the K-NN extracts the k closest vectors from the training set using similarity measures, and makes decision for the label of the unknown sample x using the majority class label of the k nearest neighbors. We adopt Euclidean distance to measure the similarity of samples.

4 Experiments

4.1 The Descriptions of Two Well-Known Tumor Datasets

We have experimented with two well-known datasets: the leukemia dataset [10] and the colon tumor dataset [11]. The descriptions of the two datasets are shown in Table 1. The leukemia dataset consists of 72 bone marrow or peripheral blood samples including 47 acute lymphoblastic leukemia (ALL) samples and 25 acute myeloid leukemia (AML) samples. The colon tumor dataset consists of 62 samples of colon epithelial cells including 40 colon cancer samples and 22 normal samples.

Table 1. Descriptions of two tumor datasets in our experiments

Tumor Dataset	#Gene	#Sample	Subtype 1	Subtype 2
Leukemia dataset	7,129	72	47(ALL)	25(AML)
Colon tumor dataset	2,000	62	40(Tumor)	22(Normal)

4.2 Experiment Setup

Due to many parameters to be set in our algorithm, making a good experiment scheme becomes difficulty. Firstly, after using Relief algorithm to rank all genes, how to determine the number of the selected informative genes is not ease task, owing to lack of prior knowledge about gene expression profiles. Fig. 3 shows the distribution of gene weight which is computed using Relief algorithm on the colon tumor dataset and

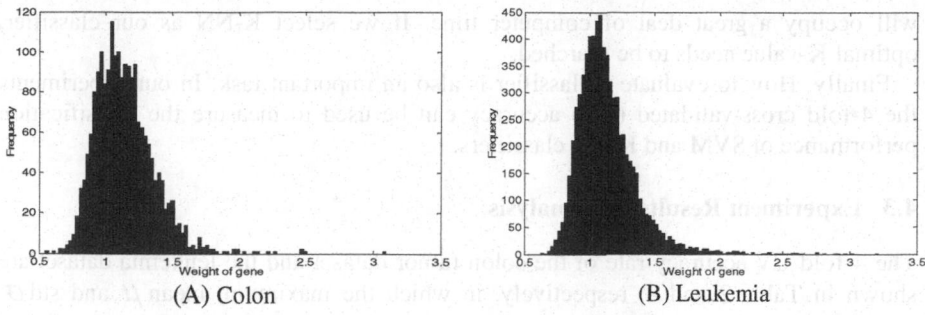

(A) Colon   (B) Leukemia

Fig. 3. The distribution of gene weight

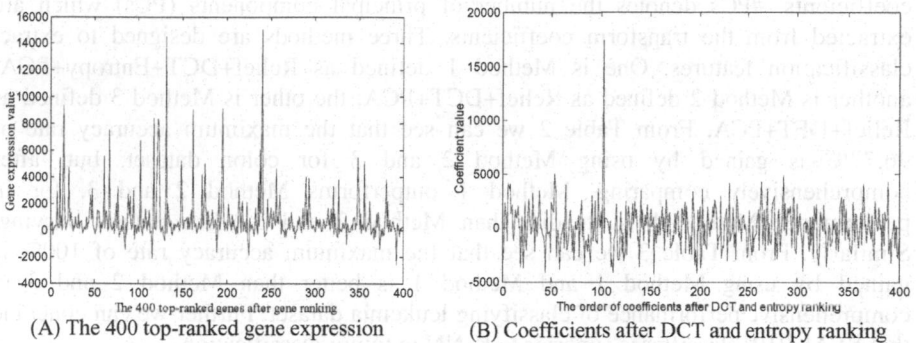

(A) The 400 top-ranked gene expression   (B) Coefficients after DCT and entropy ranking

Fig. 4. DCT transform of a leukemia sample

the leukemia dataset, respectively, from which we can see that the number of gene with higher weight is less, so we will try to select 400 top-ranked genes as initial informative genes to be used as tumor classification.

Secondly, after DCT transform and the entropy ranking of coefficients, there exists difficulty in how to determine the number of coefficients to be used as the input of PCA. Fig. 4 shows the DCT transform of a leukemia sample, in which (A) shows the original expression value of the 400 top-ranked genes after gene ranking with Relief algorithm and (B) shows the coefficients after DCT transform and entropy ranking. We will try to extract principal components from 8 to 100 top-ranked coefficients.

Thirdly, how many principal components should be extracted from those coefficients also need to be seriously considered. It is inevitable for too many principal components to decrease the accuracy rate of classifier. We will try to extract 3 to 5 principal components to be used as the input of SVM or K-NN classifier.

Fourthly, the parameter selection of classifier also greatly affects the classification performance. Usually, different classifier parameters are sensitive to different datasets. In experiment, we select LIBSVM software [12] to be used as our classifier, which requires specifying the type of kernel and the regularization parameter C. Generally, the recommended kernel for nonlinear problems is the Gaussian radial basis function (RBF) kernel $K(x, y) = \exp(-\gamma \|x - y\|^2)$ that is also used in our experiments. However, finding the best combination for the parameter pairs (C, γ)

will occupy a great deal of computer time. If we select K-NN as our classifier, optimal K-value needs to be searched.

Finally, How to evaluate a classifier is also an important task. In our experiments the 4-fold cross-validated (CV) accuracy can be used to measure the classification performance of SVM and K-NN classifiers.

4.3   Experiment Results and Analysis

The 4-fold CV accuracy rate of the colon tumor dataset and the leukemia dataset are shown in Table 2 and 3, respectively, in which the maximum, mean μ and std σ denote the maximum, mean value μ and standard deviation σ of the accuracy rate which are computed from 92 accuracy rates obtained from 8 to 100 top-ranked coefficients. #PCs denotes the number of principal components (PCs) which are extracted from the transform coefficients. Three methods are designed to extract classification features. One is Method 1 defined as Relief+DCT+Entropy+PCA; another is Method 2 defined as Relief+DCT+PCA; the other is Method 3 defined as Relief+DFT+PCA. From Table 2 we can see that the maximum accuracy rate of 96.77% is gained by using Method 2 and 3 for colon dataset, but, after comprehensively comparing, Method 1 outperforms Method 2 and 3, for in performance Method 1 is steadier than Method 2 and 3 when dataset varying. Similarly, From Table 3 we can see that the maximum accuracy rate of 100% is gained by using Method 1 and Method 1 is better than Method 2 and 3 in comprehensive performance of classifying leukemia dataset. Further we can conclude that SVM-RBF classifier is superior to K-NN in tumor classification.

Fig. 5(A) and (B) respectively show the comparison of the classification accuracy rate using Method 1 and Method 2 which are evaluated by SVM classifier on the two datasets, from which we can obviously see that Method 1 is superior to Method 2 when extracting 5 PCs from 8 to 100 top-ranked coefficients. When extracting PCs

Table 2. The accuracy rate of the colon tumor dataset

Feature extraction	Classifier	#PC	Accuracy rate (%)		
			maximum	mean μ	std σ
Method 1 (Relief+DCT+ Entropy+PCA)	SVM-RBF	3	95.16	91.90	1.41
		4	93.55	90.96	0.96
		5	93.55	90.79	0.97
	K-NN	3	93.55	90.05	1.61
		4	90.32	87.67	1.20
		5	91.94	87.63	1.35
Method 2 (Relief+DCT+PCA)	SVM-RBF	3	96.77	90.01	4.03
		4	90.32	88.17	1.95
		5	91.94	88.81	1.42
	K-NN	3	91.94	87.04	2.53
		4	91.94	88.33	1.60
		5	93.55	88.03	1.97
Method 3 (Relief+DFT+PCA)	SVM-RBF	3	96.77	89.58	4.07
		4	93.55	89.13	1.58
		5	96.77	89.72	1.84
	K-NN	3	93.55	85.22	2.95
		4	91.94	87.93	1.66
		5	91.94	84.60	3.88

Table 3. The accuracy rate of the leukemia dataset

Feature extraction	Classifier	#PC	Accuracy rate (%)		
			maximum	mean μ	std σ
Method 1 (Relief+DCT+ Entropy+PCA)	SVM-RBF	3	100	98.40	0.74
		4	100	98.69	0.69
		5	100	98.75	0.68
	K-NN	3	100	97.64	1.15
		4	100	97.36	1.45
		5	100	96.82	1.50
Method 2 (Relief+DCT+PCA)	SVM-RBF	3	98.61	96.31	0.78
		4	98.61	96.92	1.24
		5	98.61	97.76	1.16
	K-NN	3	97.22	95.53	0.91
		4	95.83	94.79	1.22
		5	97.22	94.52	1.45
Method 3 (Relief+DFT+PCA)	SVM-RBF	3	98.61	97.19	1.08
		4	98.61	98.01	0.99
		5	98.61	98.15	0.90
	K-NN	3	97.22	94.46	2.47
		4	98.61	95.06	3.45
		5	97.22	94.06	3.57

from 40 to 100 top-ranked coefficients, the obtained classifiers are steadier in classification performance.

Fig. 6(A) and (B) respectively show the comparison of the classification accuracy rate using Method 1 and Method 2 which are evaluated by K-NN classifier on the two datasets, from which we can obviously see that for colon dataset Method 1 is not always superior to Method 2 in accuracy rate when extracting 5 PCs from 8 to 100 top-ranked coefficients, but for leukemia dataset Method 1 is always superior to Method 2.

The comparison of the classification accuracy rate based on feature extraction with DFT transform on the two datasets using SVM-RBF and K-NN classifiers are shown in Fig. 7(A) and (B), respectively. Note that the number of extracted PCs is 5. After comparing tumor classification performance we find that SVM-RBF is superior to K-NN classifier based on feature extraction of Method 3 in classification accuracy rate and stability for the two datasets.

From these experiments we can conclude that every step in our tumor classification model contributes to the classification performance, so every step is necessary in improving classification performance. The reason that adopting DCT or DFT can improve the classification performance is that DCT and DFT can remove the noise in gene expression profiles and especially DCT has the strong energy compaction property.

Moreover, the advantage of our method is that our method can attain very high accuracy rate when extracting only 3 PCs from transform coefficients, so the classification results can be visualized in the form of 3D scatter plot, which is convenient for doctor to diagnose tumor from vision in clinical application.

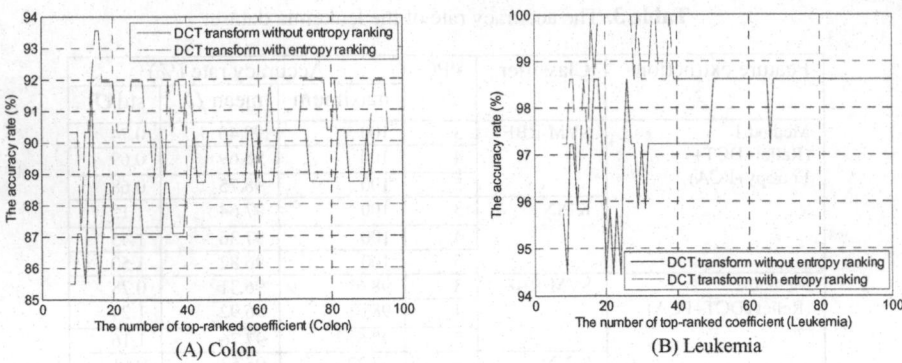

Fig. 5. Classification comparison using SVM classifier for the two datasets

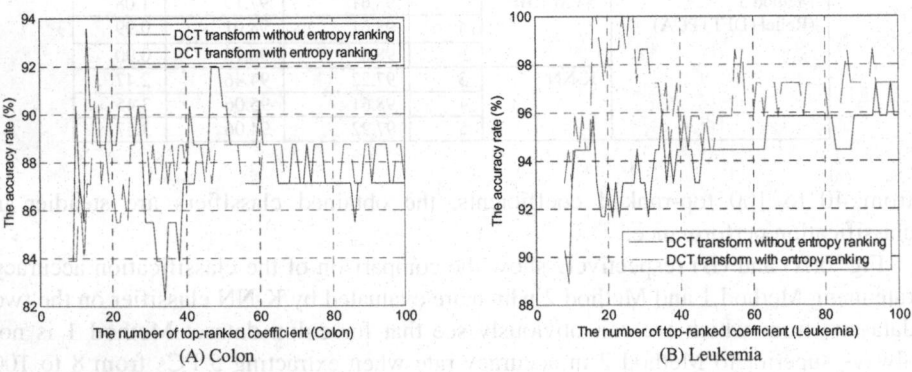

Fig. 6. Classification comparison using K-NN classifier for the two datasets

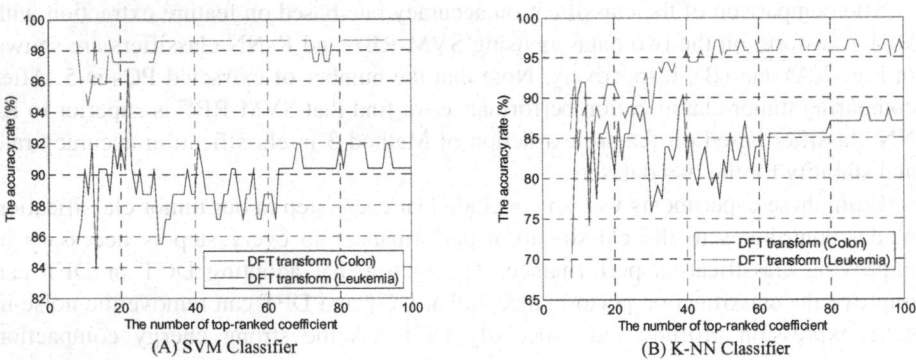

Fig. 7. The comparison of classification based on feature extraction with DFT transform

5 Comparison with Other Related Works

Tumor classification has a promising future in clinical application, so a great deal of research has been done. Many pattern recognition approaches have been successfully

Table 4. The classification accuracy comparison among different classification approaches

Feature Extraction	Classifier	Dataset	CV Acc.%	Ref.
T-test + kernel partial least square	SVM	Colon	91.90	[3]
		Leukemia	100	
Feature score criterion + factor analysis	SVM with RBF	Colon	91.94	[2]
		Leukemia	100	
Joint classifier and feature optimization (JCFO) Linear kernel		Colon	**96.8**	[13]
		Leukemia	**100**	
Principal component analysis (PCA)	Logistic discriminent	Colon	87.10	[4]
		Leukemia	94.20	
Partial least square	Logistic discriminant	Colon	93.50	[4]
		Leukemia	95.90	
	Quadratic discriminant analysis	Colon	91.90	[4]
		Leukemia	96.40	
Independent component analysis (ICA)	Calculating the ratio of tumor and normal ICs	Colon	91.90	[1]
TNoM Scores	SVM with linear kernel	Colon	77.4	[14]
		Leukemia	94.4	
Relief+DCT+Entropy+PCA	**SVM with RBF kernel**	**Leukemia**	**100**	**This**
Relief+DCT+PCA	**SVM with RBF kernel**	**Colon**	**96.77**	**paper**

applied to the tumor classification based on gene expression data. Table 4 shows the comparison of the classification accuracy rate among different tumor classification approaches, from which we can conclude that our approach outperforms others in accuracy rate except for literature [13] and obtains almost the same accuracy rate as literature [13].

6   Conclusions and Future Work

A five-step feature extraction model from high dimensional gene expression profiles is proposed, which are examined on two well-known tumor datasets to assess the classification performance using SVM and K-NN classifiers. Our experiment results show that the 4-fold CV accuracy rate of 100% is obtained for the leukemia dataset and 96.77% for the colon tumor dataset. Compared with other related works, the designed classifiers not only have a higher accuracy rate but also are steadier relatively, which are two important factors to evaluate a classifier. Another advantage is that the classification results can be visualized when extracted only 3 PCs with high accuracy rate. However, there are too many parameters to be set in our feature extraction method, which directly leads to time-consuming run to find the best combination of those parameters, which is our future work to be solved. Moreover, we will integrate more different classifiers into our methods to improve tumor classification performance to found the basis of the clinical application for tumor diagnosing.

Acknowledgement

This research is supported by the Program for New Century Excellent Talents in University and the Excellent Youth Foundation of Hunan Province (NO. 06JJ1010).

References

1. Zhang, X., Yap, Y.L., Wei, D., Chen, F., Danchin, A.: Molecular diagnosis of human cancer type by gene expression profiles and independent component analysis. European Journal of Human Genetics 05(9), 1018–4813 (2005)
2. Wang, S., Wang, J., Chen, H., Tang, W.: The classification of tumor using gene expression profile based on support vector machines and factor analysis. In: Intelligent Systems Design and Applications, Jinan, pp. 471–476. IEEE Computer Society Press, Los Alamitos (2006)
3. Li, S., Liao, C., Kwok, J.T.: Gene feature extraction using T-test statistics and kernel partial least squares. In: King, I., Wang, J., Chan, L., Wang, D. (eds.) ICONIP 2006. LNCS, vol. 4234, pp. 11–20. Springer, Heidelberg (2006)
4. Nguyen, D.V., Rocke, D.M.: Tumor classification by partial least squares using microarray gene expression data. Bioinformatics 18(1), 39–45 (2002)
5. Kira, K., Rendell, L.A.: The feature selection problem: traditional methods and a new algorithm. In: Swartout, W. (ed.) Proceedings of the 10th National Conference on Aritficial Inteligence, pp. 129–134. AAAI Press/The MIT Press, Cambridge, MA (1992)
6. Ahmed, N., Natarajan, T., Rao, K.R.: Discrete Cosine Transform. IEEE Trans. Computers C-23, 90–94 (1974)
7. Theodoridis, S., Koutroumbas, K.: Pattern recognition, pp. 341–342. Academic Press, London (1999)
8. Vapnik, V.N.: Statistical learning theory. Springer, New York (1998)
9. Dasarathy, B.: Nearest Neighbor Norms: NN Pattern Classification Techniques. IEEE Computer Society Press, Los Alamitos (1991)
10. Golub, T.R., Slonim, D.K., Tamayo, P., Huard, C., Gaasenbeek, M., Mesirov, J.P., Coller, H., Loh, M.L., Downing, J.R., Caligiuri, M.A., Bloomfield, C.D., Lander, E.S.: Molecular classification of cancer: class discovery and class prediction by gene expression monitoring. Science 286, 531–537 (1999)
11. Alon, U., Barkai, N., Notterman, D.A., Gish, K., Ybarra, S., Mack, D., Levine, A.: Broad patterns of gene expression revealed by clustering analysis of tumor and normal colon tissues by oligonucleotide arrays. Proc. Nat. Acad. Sci. USA 96, 6745–6750 (1999)
12. Chang, C.-C., Lin, C.-J.: LIBSVM: A library for support vector machines (2001), Software available at http://www.csie.ntu.edu.tw/ cjlin/libsvm
13. Krishnapuram, B., Carin, L., Hartemink, A.: Gene expression analysis: Joint feature selection and classifier design. In: Scholkopf, B., Tsuda, K., Vert, J.-P. (eds.) Kernel Methods in Computational Biology, pp. 299–318. MIT, Cambridge (2004)
14. Ben-Dor, A., Bruhn, L., Friedman, N., Nachman, I., Schummer, M., Yakhini, Z.: Tissue classification with gene expression profiles. Journal of Computational Biology 7, 559–584 (2000)

Chapter 8 - Second Workshop on Intelligent Robotics (IROBOT 2007)

An Omnidirectional Vision System for Soccer Robots

António J.R. Neves*, Gustavo A. Corrente, and Armando J. Pinho

Dept. de Electrónica e Telecomunicações / IEETA
Universidade de Aveiro, 3810-193 Aveiro, Portugal
{an,gustavo,ap}@ua.pt

Abstract. This paper describes a complete and efficient vision system developed for the robotic soccer team of the University of Aveiro, CAMBADA (Cooperative Autonomous Mobile roBots with Advanced Distributed Architecture). The system consists on a firewire camera mounted vertically on the top of the robots. A hyperbolic mirror placed above the camera reflects the 360 degrees of the field around the robot. The omnidirectional system is used to find the ball, the goals, detect the presence of obstacles and the white lines, used by our localization algorithm. In this paper we present a set of algorithms to extract efficiently the color information of the acquired images and, in a second phase, extract the information of all objects of interest. Our vision system architecture uses a distributed paradigm where the main tasks, namely image acquisition, color extraction, object detection and image visualization, are separated in several processes that can run at the same time. We developed an efficient color extraction algorithm based on lookup tables and a radial model for object detection. Our participation in the last national robotic contest, ROBOTICA 2007, where we have obtained the first place in the Medium Size League of robotic soccer, shows the effectiveness of our algorithms. Moreover, our experiments show that the system is fast and accurate having a maximum processing time independently of the robot position and the number of objects found in the field.

Keywords: Robotics, robotic soccer, computer vision, object recognition, omnidirectional vision, color classification.

1 Introduction

After Garry Kasparov was defeated by IBM's Deep Blue Supercomputer in May 1997, forty years of challenge in the artificial intelligence (AI) community came to a successful conclusion. But it also was clear that a new challenge had to be found.

"By mid-21st century, a team of fully autonomous humanoid robot soccer players shall win the soccer game, complying with the official rules of the FIFA, against the winner of the most recent World Cup." This is how the ultimate goal was stated by the RoboCup Initiative, founded in 1997, with the aim to foster the development of artificial intelligence and related field by providing a standard problem: robots that play soccer.

It will take decades of efforts, if not centuries, to accomplish this goal. It is not feasible, with the current technologies, to reach this goal in any near term. However,

---

* This work was supported in part by the Fundação para a Ciência e a Tecnologia (FCT).

J. Neves, M. Santos, and J. Machado (Eds.): EPIA 2007, LNAI 4874, pp. 499–507, 2007.
© Springer-Verlag Berlin Heidelberg 2007

this goal can easily create a series of well directed subgoals. The first subgoal to be accomplished in RoboCup is to build real and simulated robot soccer teams which play reasonably well with modified rules. Even to accomplish this goal will undoubtedly generate technologies with impact on broad range of industries.

One problem domain in RoboCup is the field of Computer Vision. Its task is to provide basic information that is needed to calculate the behavior of the robots. Especially omnidirectional vision systems have become interesting in the last years, allowing a robot to see in all directions at the same time without moving itself or its camera [13,12,10,11]. Omnidirectional vision is the method used by most teams in the Middle Size League.

The main goal of this paper is to present an efficient vision system for processing the video acquired by an omnidirectional camera. The system finds the white lines of the playing field, the ball, goals and obstacles. The lines of the playing field are needed because knowing the placement of the playing field from the robot's point of view is equal to know the position and orientation of the robot.

For finding the goals, the ball and the obstacles, lines are stretched out radially from the center of the image and, if some defined number of pixels of the respective colors are found, the system saves that position associated to the respective color. For finding the white lines, color transitions from green to white are searched for.

For color classification, the first step of our system, a lookup table (LUT) is used. Our system is prepared to acquire images in RGB 24-bit, YUV 4:2:2 or YUV 4:1:1 format, being necessary only to choose the appropriated LUT. We use the HSV color space for color calibration and classification due to its special characteristics [1].

This paper is organized as follows. In Section 2 we describe the design of our robots. Section 3 presents our vision system architecture, explaining the several modules developed and how they are connected. In Section 4 we present the algorithms used to collect the color information of the image using radial search lines. In Section 5 we describe how the object features are extracted. Finally, in Section 6, we draw some conclusions and propose future work.

2   Robot Overview

CAMBADA players were designed and completely built in-house. The baseline for robot construction is a cylindrical envelope, with 485 mm in diameter. The mechanical structure of the players is layered and modular. The components in the lower layer are the motors, wheels, batteries and an electromechanical kicker. The second layer contains the control electronics. The third layer contains a computer. The players are capable of holonomic motion, based on three omni-directional roller wheels [2].

The vision system consists on a firewire camera mounted vertically on the top of the robots. A hyperbolic mirror placed above the camera reflects the 360 degrees of the field around the robot. This is the main sensor of the robot and it is used to find the ball, the goals, detect the presence of obstacles and the white lines.

The robots computing system architecture follows the fine-grain distributed model [7] where most of the elementary functions are encapsulated in small microcontroller-based nodes, connected through a network. A PC-based node is used to execute higher-level control functions and to facilitate the interconnection of off-the-shelf devices, e.g.

cameras, through standard interfaces, e.g. Firewire. For this purpose, Controller Area Network (CAN) has been chosen [3]. The communication among robots uses the standard wireless LAN protocol IEEE 802.11x profiting from large availability of complying equipment.

The software system in each player is distributed among the various computational units. High level functions run on the computer, in Linux operating system with RTAI (Real-Time Application Interface). Low level functions run partly on the microcontrollers. A cooperative sensing approach based on a Real-Time Database (RTDB) [4,5,6] has been adopted. The RTDB is a data structure where players share their world models. It is updated and replicated in all players in real-time.

3   Vision System Architecture

A modular multi-process architecture was adopted for our vision system (see Fig. 1).

When a new frame is ready to download, one process is automatically triggered and the frame is placed in a shared memory buffer. After that, another process analyzes the acquired image for color classification, creating a new image with "color labels" (an 8 bpp image). This image is also placed in the shared image buffer, which is afterward analyzed by the object detection processes, generically designated by $Proc[x]$, $x = 1, 2, \ldots N$. These applications are encapsulated in separate Linux processes. Once started, each process gets a pointer to the most recent image frame available and starts tracking the respective object. Once finished, the resulting information (e.g. object detected or not and position) is placed in the real-time database. This database may be accessed by any other processes in the system, particularly for world state update.

The activation of the distinct image-processing activities is carried out by a process manager. Scheduling of vision related processes relies on the real-time features of the Linux kernel, namely the FIFO scheduler and priorities. At this level, Linux executes each process to completion, unless the process blocks or is preempted by other process with higher real-time priority. This ensures that the processes are executed strictly according to their priority with full preemption.

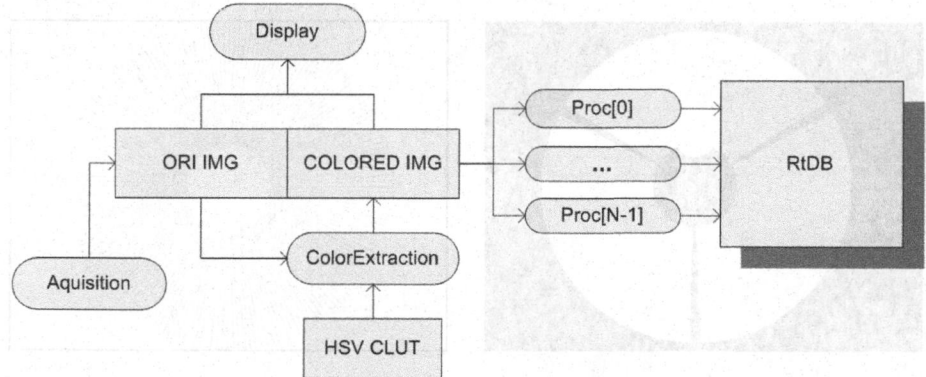

Fig. 1. The architecture of our vision system. It is based on a multi-process system being each process responsible for a specific task.

4 Color Extraction

Image analysis in the RoboCup domain is simplified, since objects are color coded. Black robots play with an orange ball on a green field that has yellow and blue goals and white lines. Thus, a pixel's color is a strong hint for object segmentation. We exploit this fact by defining color classes, using a look-up table (LUT) for fast color classification. The table consists of 16777216 entries (2^{24}, 8 bits for red, 8 bits for green and 8 bits for blue), each 8 bits wide, occupying 16 MB in total. If another color space is used, the table size is the same, changing only the "meaning" of each component. Each bit expresses whether the color is within the corresponding class or not. This means that a certain color can be assigned to several classes at the same time. To classify a pixel, we first read the pixel's color and then use the color as an index into the table. The value (8 bits) read from the table will be called "color mask" of the pixel.

The color calibration is done in HSV (Hue, Saturation and Value) color space due to its special characteristics. In our system, the image is acquired in RGB or YUV format and then is converted to HSV using an appropriate conversion routine.

There are certain regions in the received image that have to be excluded from analysis. One of them is the part in the image that reflects the robot itself. Other regions are the sticks that hold the mirror and the areas outside the mirror. For that, we have an image with this configuration that is used by our software. An example is presented in Fig. 2. The white pixels are the area that will be processed, black pixels will not. With this approach we can reduce the time spent in the conversion and searching phases and we eliminate the problem of finding erroneous objects in that areas.

To extract the color information of the image we use radial search lines to analyze the color information instead of processing all the image. This approach has two main advantages. First, that of accelerating the process due to the fact that we only process about 30% of the valid pixels. Second, the use of omnidirectional vision difficults the detection of the objects using, for example, their bounding box. In this case, it is more desirable to use the distance and angle. The proposed approach has a processing time almost constant, independently of the information around the robot, being a desirable

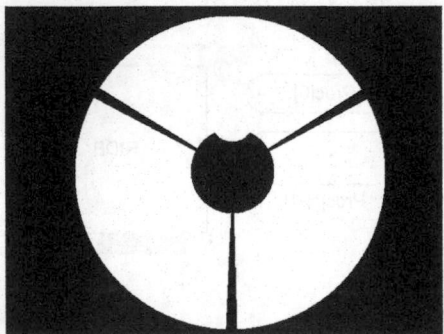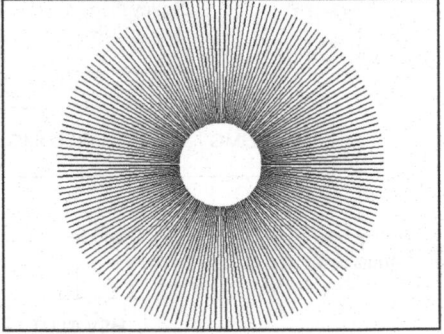

Fig. 2. On the left, an example of a robot mask. White points represent the area that will be processed. On the right, the position of the radial search lines.

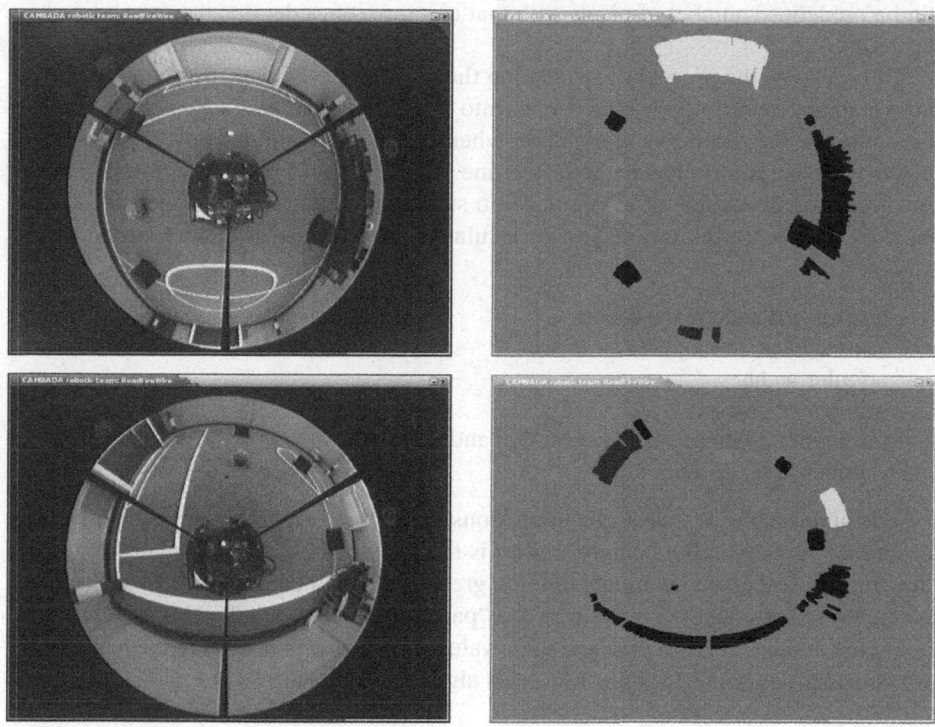

Fig. 3. An example of the blobs found in two images. On the left, the original images. On the right, the blobs found. For each blob, we calculate useful information that is used later to calculate the position of each object.

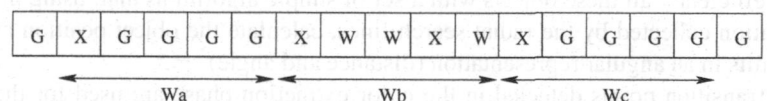

Fig. 4. An example of a transition. "G" means green pixels, "W" means white pixels and "X" means pixels with a color different from green or white.

property in Real-Time Systems. This is due to the fact that the system processes almost the same number of pixels in each frame.

A radial search line is a line that starts in the center of the robot with some angle and ends in the limit of the image (see the image on the right of Fig. 2). They are constructed based on the Bresenham line algorithm [8,9]. For each search line, we iterate through its pixels to search for two things: transitions between two colors and areas with specific colors.

We developed an algorithm to detect areas of a specific color which eliminates the possible noise that could appear in the image. Each time that a pixel is found with a color of interest, we analyze the pixels that follows (a predefined number) and if we don't find more pixels of that color we "forget" the pixel found and continue. When we find a predefined number of pixels with that color, we consider that the search line has this color.

To accelerate the process of calculating the position of the objects, we put the color information found, in each search line, into a list of colors. We are interested in the first pixel (in the respective search line) where the color was found and the number of pixels with that color found in the search line. Then, using the previous information, we separate the information of each color into sets that we named blobs (see Fig. 3). For each blob some useful information is calculated that will help in the detection of each object:

- average distance to the robot;
- mass center;
- angular width;
- number of pixels;
- number of green pixels between blob and the robot;
- number of pixels after blob.

The algorithm to search for the transitions between green pixels and white pixels is described as follows. If a non green pixel is found, we will look for a small window in the "future" and count the number of non green pixels and the number of white pixels. Next, we look for a small window in the "past" and a small window in the future and count the number of green pixels. If these values are greater than a predefined threshold, we consider this point as a transition. This algorithm is illustrated in Fig. 4.

5   Object Detection

The objects of interest that are present in the RoboCup environment are: a ball, two goals, obstacles (other robots) and the green field with white lines. Currently, our system detects efficiently all these objects with a set of simple algorithms that, using the color information collected by the radial search lines, calculate the object position and / or their limits in an angular representation (distance and angle).

The transition points detected in the color extraction phase are used for the robot localization. All the points detected are sent to the Real-time Database, afterward used by the localization process.

To detect the ball, we use the following algorithm:

1. Separate the orange information into blobs.
2. For each blob, calculate the information described in the previous section.
3. Sort the orange blobs that have some green pixels before or after the blob by descending order, using their number of orange pixels as measure.
4. Choose the first blob as candidate. The position of the ball is the mass center of the blob.

Regarding the goals, we are interested in three points: the center, the right post and the left post. To do that, we use the following algorithm:

1. Ignore the information related to radial search lines which have both blue and yellow information (they correspond to the land marks).

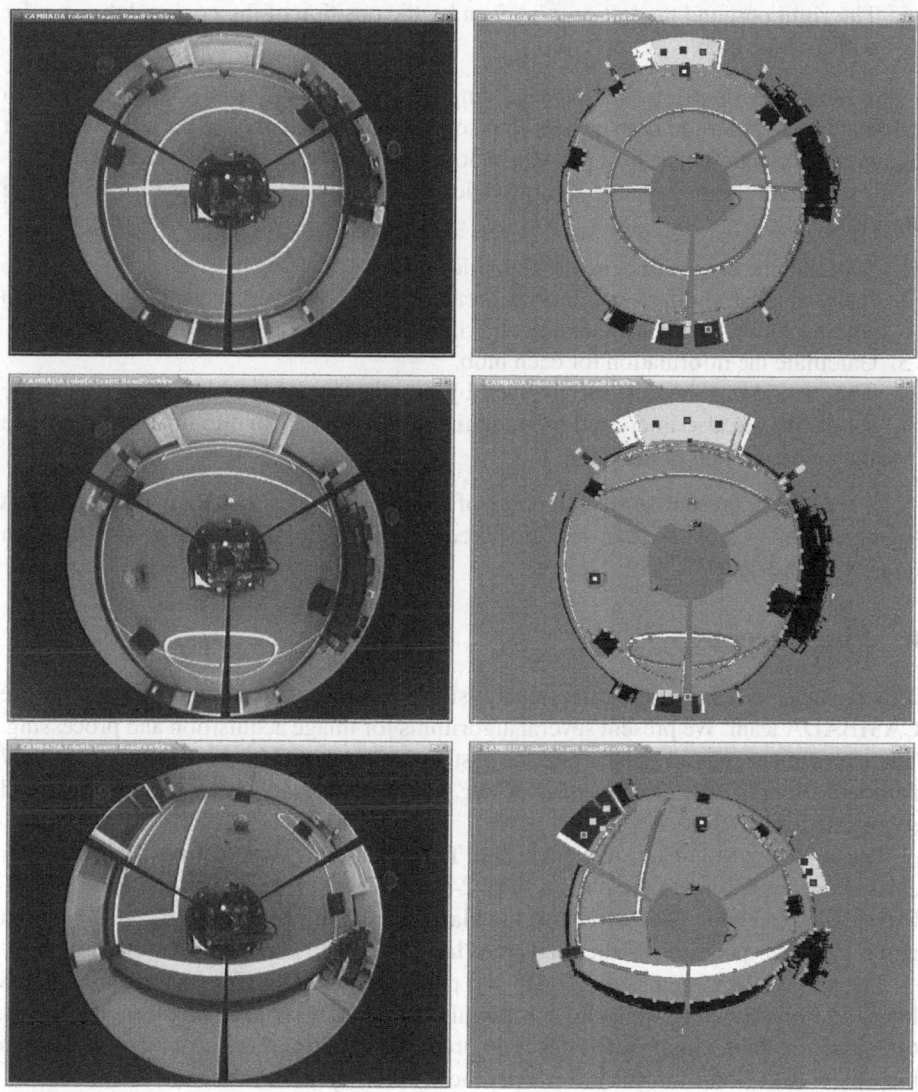

Fig. 5. On the left, examples of original images. On the right, the corresponding processed images. Marks in the blue and yellow goals mean the position of the goal (center) and the possible points to shoot. The mark over the ball points to the mass center. The several marks near the white lines (magenta) are the position of the white lines. The cyan marks are the position of the obstacles.

2. Separate the valid blue and yellow information into blobs.
3. Calculate the information for each blob.
4. Sort the yellow and the blue blobs that have some green pixels before the blob by descending order, using their angular width as measure.
5. Choose the first blob as candidate for each goal. Their position is given by the distance of the blob relatively to the robot.

6. The right post and the left post is given by the position of the goal and the angular width of the blob.

Another important information regarding the goals, is the best point to shoot. To calculate it, we split the blob chosen into several slices, and choose the one with most pixels blue or yellow. The best point to shoot is the mass center of the slice chosen.

To calculate the position of the obstacles around the robot, we use the following algorithm:

1. Separate the orange information into blobs.
2. If the angular width of one blob is greater than 10 degrees, we split the blob into smaller blobs, in order to obtain better information about obstacles.
3. Calculate the information for each blob.
4. The position of the obstacle is given by the distance of the blob relatively to the robot. We are also interested in the limits of the obstacle and to obtain that we use the angular width of the blob.

In Fig. 5 we present some examples of acquired images and their correspondent segmented images. As we can see, the objects are correctly detected (see the marks in images on the right).

6 Final Remarks

This paper presents the omnidirectional vision system that has been developed for the CAMBADA team. We present several algorithms for image acquisition and processing. The experiments already made and the last results obtained in the ROBOTICA 2007 competition prove the effectiveness of our system regarding the object detection and robot self-localization.

The objects in RoboCup are color coded. Therefore, our system defines different color classes corresponding to the objects. The 24 bit pixel color is used as an index to a 16 MBytes lookup table which contains the classification of each possible color in a 8 bit entry. Each bit specifies whether that color lays within the corresponding color class.

The processing system is divided in two phases: color extraction, using radial search lines, and object detection, using specific algorithms. The objects involved are: a ball, two goals, obstacles and white lines. The processing time and the accuracy obtained in the object detection confirms the effectiveness of our system.

As future work, we are developing new algorithms for camera and color calibration, in particular autonomous algorithms. Moreover, we are improving the presented algorithms in order to use the shape of the objects instead of using only the color information to improve the object recognition.

References

1. Caleiro, P.M.R., Neves, A.J.R., Pinho, A.J.: Color-spaces and color segmentation for real-time object recognition in robotic applications. Revista do DETUA 4(8), 940–945 (2007)
2. Carter, B.: *EST LA*.: Mechanical Design and Modeling of an Omni-directional RoboCup Player. In: Birk, A., Coradeschi, S., Tadokoro, S. (eds.) RoboCup 2001. LNCS (LNAI), vol. 2377, Springer, Heidelberg (2002)

3. Almeida, L., Pedreiras, P., Fonseca, J.A.: FTT-CAN: Why and How. IEEE Trans. Industrial Electronics (2002)
4. Almeida, L., Santos, F., Facchinetti, T., Pedreira, P., Silva, V., Lopes, L.S.: Coordinating Distributed Autonomous Agents with a Real-Time Database: The CAMBADA Project. In: Aykanat, C., Dayar, T., Körpeoğlu, İ. (eds.) ISCIS 2004. LNCS, vol. 3280, pp. 876–886. Springer, Heidelberg (2004)
5. Pedreiras, P., Teixeira, F., Ferreira, N., Almeida, L., Pinho, A., Santos, F.: Enhancing the reactivity of the vision subsystem in autonomous mobile robots using real-time techniques. In: Bredenfeld, A., Jacoff, A., Noda, I., Takahashi, Y. (eds.) RoboCup 2005. LNCS (LNAI), vol. 4020, Springer, Heidelberg (2006)
6. Santos, F., Almeida, L., Pedreiras, P., Lopes, L.S., Facchinetti, T.: An Adaptive TDMA Protocol for Soft Real-Time Wireless Communication among, Mobile Autonomous Agents. In: Santos, F. (ed.) Proc. WACERTS 2004, Int. Workshop on Architecture for Cooperative Embedded Real-Time Systems (in conjunction with RTSS 2004), Lisboa, Portugal (2004)
7. Kopets, H.: Real-Time Systems Design Principles for Distributed Embedded Applications. Kluwer, Dordrecht
8. Bresenham, J.E.: A linear algorithm for incremental digital display of circular arcs. CA CM 20(2), 100–106 (1977)
9. Bresenham, J.E.: Algorithm for computer control of a digital plotter. IBM Systems J. 4(1), 25–30 (1965)
10. Heinemann, P., et al.: Fast and Accurate Environment Modelling using Omnidirectional Vision. Dynamic Perception, 9–14, Infix (2004)
11. Heinemann, P., et al.: Tracking Dynamic Objects in a RoboCup Environment - The Attempto Tübingen Robot Soccer Team. In: Polani, D., Browning, B., Bonarini, A., Yoshida, K. (eds.) RoboCup 2003. LNCS (LNAI), vol. 3020, Springer, Heidelberg (2004)
12. Gaspar, J., Winters, N., Grossmann, E., Santos-Victor, J.: Toward Robot Perception using Omnidirectional Vision. In: Innovations in Machine Intelligence and Robot Perception, Springer, Heidelberg (2004)
13. Hoffmann, J., et al.: A vision based system for goal-directed obstacle avoidance. In: Nardi, D., Riedmiller, M., Sammut, C., Santos-Victor, J. (eds.) RoboCup 2004. LNCS (LNAI), vol. 3276, Springer, Heidelberg (2005)

Generalization and Transfer Learning in Noise-Affected Robot Navigation Tasks

Lutz Frommberger

SFB/TR 8 Spatial Cognition
Project R3-[Q-Shape]
Universität Bremen
Enrique-Schmidt-Str. 5, 28359 Bremen, Germany
lutz@sfbtr8.uni-bremen.de

Abstract. When a robot learns to solve a goal-directed navigation task with reinforcement learning, the acquired strategy can usually exclusively be applied to the
task that has been learned. Knowledge transfer to other tasks and environments is
a great challenge, and the transfer learning ability crucially depends on the chosen state space representation. This work shows how an agent-centered qualitative
spatial representation can be used for generalization and knowledge transfer in a
simulated robot navigation scenario. Learned strategies using this representation
are very robust to environmental noise and imprecise world knowledge and can
easily be applied to new scenarios, offering a good foundation for further learning
tasks and application of the learned policy in different contexts.

1 Introduction

In goal-directed navigation tasks, an autonomous moving agent has found a solution
when having reached a certain location in space. Reinforcement Learning (RL) [1]
is a frequently applied method to solve such tasks, because it allows an agent to autonomously adapt its behavior to a given environment. In general, however, this solution does only apply to the problem that the system was trained on and does not work
in other environments, even if they offer a similar structure or are partly identical. The
agent lacks an intelligent *understanding* of the general structure of geometrical spaces.

The ability to transfer knowledge gained in previous learning tasks into different
contexts is one of the most important mechanisms of human learning. Despite this,
the question whether and how a learned solution can be reused in partially similar settings is still an important issue in current machine learning research. Recently the term
transfer learning was used for this research field. In contrast to *generalization*, which
describes the ability to apply a learned strategy to unknown instances within the same
task, transfer learning tackles generalization ability across different tasks.

In real world applications we frequently experience changes in the environment that
modify the characteristics of a task. If the system input is provided by sensors, it is
subject to environmental noise, and the amount and type of noise is subject to external
conditions like, e.g., the weather. Consequently, applying a learned strategy to the same
task under different conditions can also be seen as a process of transfer learning, because
the task is not anymore the same as before. The complexity of the transfer process is

J. Neves, M. Santos, and J. Machado (Eds.): EPIA 2007, LNAI 4874, pp. 508–519, 2007.

heavily influenced by the agent's perception of the world. *Abstraction* of the state space is a key issue within this context.

In this work we investigate a simulated indoor robotics scenario where an autonomous robot has to learn a goal-directed wayfinding strategy in an unknown office environment. We will show that the choice of an appropriate spatial abstraction for the state space can minimize transfer efforts and enables the agent to learn a strategy that reaches the goal in the same environment under different conditions as well as showing a sensible navigation behavior in totally unknown environments. In particular, we will investigate if the chosen abstraction mechanism can be a means to enable a strategy transfer from an abstract simulation to a real robotics system in the future.

After an overview of related work, Sect. 3 introduces the domain we are investigating and shortly summarizes the RLPR representation used in this work. A study on the influence of environmental noise in the perception of the agent follows in Sect. 4. Section 5 investigates the properties of knowledge transfer under noisy conditions, presents a new algorithm to create a new policy for a target environment, and discusses the remaining challenges. The paper closes with a summary and an outlook.

2   Related Work

While most approaches on robot navigation with RL concentrate on the discrete state space of grid-worlds, there is also work being done in continuous domains. Lane and Wilson describe relational policies for spatial environments and demonstrate significant learning improvements [2]. However, their approach runs into problems when non-homogeneities such as walls and obstacles appear. A landmark based qualitative representation was also used in [3]. For navigation the authors rely on rather complex actions, and obstacle avoidance is handled by a separate component. Transfer to other environments has not been performed within these approaches.

Various work has been done in the domain of transfer learning: Thrun and Schwartz tried to to find reusable structural information that is valid in multiple tasks [4]. They introduced so-called skills, which collapse a sequence of actions into one single operation. Konidaris and Barto use the distinction between agent-space and problem-space to build options that can be transferred across tasks [5]. Taylor and Stone present an algorithm that summarizes a learned policy into rules that are leveraged to achieve a faster learning in a target task in another domain [6]. Torrey et al. extract transfer rules out of a source task based on a mapping specifying problem similarities [7]. The latter two approaches have in common that they both rely on external knowledge about the similarities between source and target task. The work we present here focuses on the abstraction method of RLPR [8] that uses this external knowledge to create identical parts of the state space representation for problems sharing the same structure.

3   Learning with Qualitative State Space Abstraction

3.1   A Goal Directed Robot Navigation Task

The scenario considered in this work is an indoor navigation task where an autonomous robot learns to find a specified location in an unknown environment, the goal state. This

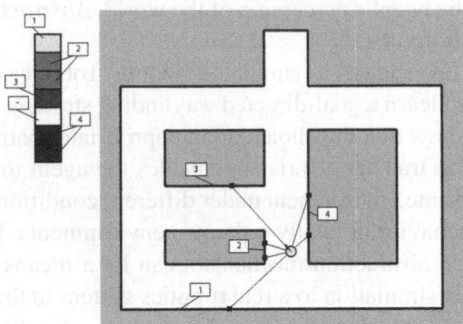

Fig. 1. The navigation task: a robot in a simplified simulated office environment with uniquely distinguishable walls. The lines departing from the robot visualize landmark scans. Detected landmarks are depicted in the upper left. The label "–" means that nothing was perceived within the agent's scanning range. The target location is the dead end.

can be formalized as a Markov Decision Process (MDP) $\langle \mathcal{S}, \mathcal{A}, T, R \rangle$ with a continuous state space $\mathcal{S} = \{(x, y, \theta), x, y \in \mathbb{R}, \ \theta \in [0, 2\pi)\}$ where each system state is given by the robot's position (x, y) and an orientation θ, an action space $\mathcal{A}$ of navigational actions the agent can perform, a transition function $T : \mathcal{S} \times \mathcal{A} \times \mathcal{S} \to [0, 1]$ denoting a probability distribution that the invocation of an action a at a state s will result in a state s', and a reward function $R : \mathcal{S} \to \mathbb{R}$, where a positive reward will be given when a goal state $s^* \in \mathcal{S}$ is reached and a negative one if the agent collides with a wall. The goal of the learning process within this MDP is to find an optimal policy $\pi : \mathcal{S} \to \mathcal{A}$ that maximizes the reward the agent receives over time.

To avoid the problems of a continuous state space, we consider the agent's *observation* instead of $\mathcal{S}$, using a function $\psi : \mathcal{S} \to \mathcal{O}$ that assigns an observation o to every state s. This results in a Partially Observable Markov Decision Process (POMDP) $\langle \mathcal{S}, \mathcal{A}, \mathcal{O}, T, R \rangle$ with $\mathcal{O} = \{\psi(s), s \in \mathcal{S}\}$ being the set of all possible observations in $\mathcal{S}$. We now use this POMDP to approximate the underlying MDP, i.e., we solve the POMDP as if it was an MDP. The used function ψ is introduced in Sect. 3.2.

The robot is assumed to be able to perceive walls around it within a certain maximum range. It is capable of performing three different actions: moving forward and turning a few degrees both to the left and to the right. Both turns include a small forward movement; and some noise is added to all actions. There is no built-in collision avoidance or any other navigational intelligence provided. See Fig. 1 for a look on the simulation testbed.

3.2 State Space Abstraction with RLPR

This section shortly introduces the *Relative Line Position Representation* (RLPR) as presented in [8]. RLPR is specifically designed for robot navigation tasks in indoor environments.

The idea behind RLPR is to divide each system state into two separate parts which represent two different aspects of goal-directed agent navigation. *Goal-directed behavior* towards a certain target location depends on the position of the agent within the

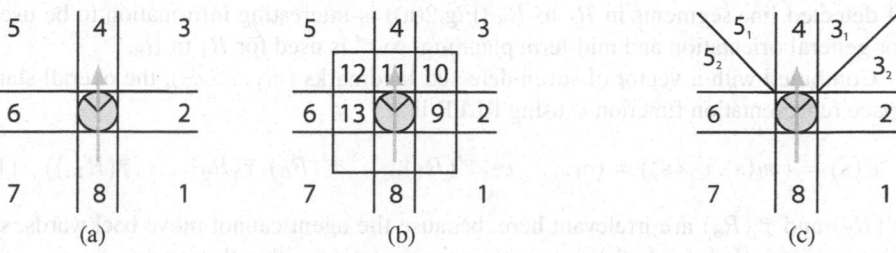

Fig. 2. RLPR: Neighboring regions around the robot in relation to its moving direction. Note that the regions in the immediate surroundings (b) are proper subsets of $R_1, \ldots, R_8$ (a). We present a variation of the representation that divides regions R_3 and R_5 into two parts to assure a higher resolution in this area (c).

world. It can, for example, be encoded by a circular ordering of detected landmarks. (e.g. cf. [9]). In this work, we assume that every wall in the environment can be identified, and recognized walls at n discrete angles around the agent $\psi_l(s) = (c_1, \ldots, c_n)$ serve as a representation for goal directed behavior (see Fig. 1). *Generally sensible behavior* regarding the structure of the world is the same at structurally similar parts within and across environments. In contrast to goal-directed behavior it is independent of the goal-finding task to solve. This distinction closely relates to the concepts of *problem-space* and *agent-space* [5]. RLPR is designed to encode the agent-space for generally sensible behavior. It benefits from the background knowledge that the structuring elements for navigation inside of buildings are the walls, which induce sensible paths inside the world which the agent is supposed to follow.

RLPR is a qualitative spatial abstraction of real world sensory data. It encodes the position of line segments perceived by the the agent's sensory system relative to its moving direction. The representation scheme of RLPR is based on the Direction Relation Matrix [10]. The space around the agent is partitioned into bounded and unbounded regions R_i (see Fig. 2). Two functions $\bar{\tau} : \mathbb{N} \to \{0, 1\}$ and $\bar{\tau}' : \mathbb{N} \to \{0, 1\}$ are defined: $\bar{\tau}(i)$ denotes whether there is a line segment detected within a sector R_i and $\bar{\tau}'(i)$ denotes whether a line spans from a neighboring sector R_{i+1} to R_i. See Fig. 3 for an example.

$\bar{\tau}_i$ is used for bounded sectors in the immediate vicinity of the agent (R_9 to R_{13} in Fig. 2(b)). Objects that appear there have to be avoided in the first place. The position

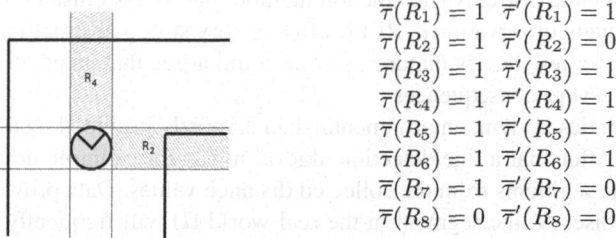

$$\bar{\tau}(R_1) = 1 \quad \bar{\tau}'(R_1) = 1$$
$$\bar{\tau}(R_2) = 1 \quad \bar{\tau}'(R_2) = 0$$
$$\bar{\tau}(R_3) = 1 \quad \bar{\tau}'(R_3) = 1$$
$$\bar{\tau}(R_4) = 1 \quad \bar{\tau}'(R_4) = 1$$
$$\bar{\tau}(R_5) = 1 \quad \bar{\tau}'(R_5) = 1$$
$$\bar{\tau}(R_6) = 1 \quad \bar{\tau}'(R_6) = 1$$
$$\bar{\tau}(R_7) = 1 \quad \bar{\tau}'(R_7) = 0$$
$$\bar{\tau}(R_8) = 0 \quad \bar{\tau}'(R_8) = 0$$

Fig. 3. RLPR values in an example situation. Region R_2 (right) and R_4 (front) are marked.

of detected line segments in R_1 to R_8 (Fig.2(a)) is interesting information to be used for general orientation and mid-term planning, so $\bar{\tau}'$ is used for R_1 to R_8.

Combined with a vector of seven detected landmarks $(c_1, \ldots, c_7)$, the overall state space representation function ψ using RLPR is

$$\psi(s) = (\psi_l(s), \psi_r(s)) = (c_1, \ldots, c_7, \bar{\tau}'(R_1), \ldots, \bar{\tau}'(R_6), \bar{\tau}(R_9), \ldots, \bar{\tau}(R_{13})) \quad (1)$$

$\bar{\tau}'(R_7)$ and $\bar{\tau}'(R_8)$ are irrelevant here, because the agent cannot move backwards, so they are not included into the state space representation within this work.

Figure 2(c) shows a new variant of the RLPR partition that divides the regions in the front left and right of the agent into two parts and increases the resolution in this area. This variant proved to be beneficial especially for transfer tasks into environments lacking landmark information. All the experiments in Sect. 5 have been conducted using it, the others use the representation given in (1).

Because RLPR performs abstraction directly and only on the state space representation, it can be combined with every underlying learning approach. RLPR creates a very small and discrete state space and is therefore suitable for easy to handle table based value function representations. It has been shown that this representation outperforms coordinate- and distance based representations by robustness and learning speed [8].

GRLPR (Generalizing RLPR) is a technique to enable applicability of RLPR-based representations to structurally similar regions within the same or across different learning tasks. The function approximation method of tile coding [11] (which is also known as CMACs) is applied to the landmark part $\psi_l(s)$. A tile size big enough that the whole landmark space of N landmarks can fit within one tile ensures that each update of the policy affects all system states with the same RLPR representation. As an effect, the agent can reuse structural knowledge acquired within the same learning task and the learned policy is immediately applicable to different learning tasks in the same environment and even to new, unknown worlds (ad-hoc strategy transfer). A discussion of the transfer properties of GRLPR based strategies follows in Sect. 5.2.

4   RLPR and Environmental Noise

4.1   Unreliable Line Detection

The question how RLPR-based strategies behave under different conditions and if they can also be transferred from simulation to a real system puts a focus on the behavior under environmental noise. An abstraction method like RLPR clusters several different percepts to a single observation o. RLPR offers a very strong abstraction from detected line segments to a few binary features, so one could argue that an erroneous classification may have severe consequences.

In a real robotics system, environmental data is mostly provided by a 2D laser range finder (LRF). After that a line detection algorithm (as for example described in [12]) will extract line segments from the collected distance values. Data provided by an LRF is subject to noise: A line segment in the real world (l) will frequently be detected as two or more separate, shorter line segments $(l_1, \ldots, l_n)$. However, RLPR is very robust towards such errors: $\bar{\tau}$ (used for $R_9, \ldots$) regards the existence of features in an area, so

Fig. 4. The noise model: The real walls (thin lines) are detected as several small chunks (thick lines). Depicted is a noise level of $\rho = 20$.

misclassifications are only to be expected in borderline cases. $\bar{\tau}'$ delivers a wrong result only if the "hole" between l_i and l_{i+1} is exactly at one of the region boundaries.

To evaluate this, we applied a noise model to the line detection in the simulator. Depending on a parameter ρ, $k \in \{1, \ldots, \rho + 1\}$ holes of 20 cm each are inserted into each line segment l of a length of 5 meters, i.e., a wall with a length of 5 meters will in the average be detected as $\frac{1}{2}(\rho + 1) + 1$ line segments (for comparison, the robot's diameter is 0.5 meters). Additionally, the start and end points of the resulting line segments $l_1, \ldots, l_k$ are then relocated into a random direction. The result is a rather realistic approximation of line detection in noisy LRF data (see Fig. 4).

For this experiment and all subsequent ones described in this work we used Watkins' $Q(\lambda)$ algorithm [13]. During training, the agent uses an ϵ-greedy policy with $\epsilon = 0.15$ for exploration. A positive reward is given when the agent reaches the target location, a negative reward is given when the agent collides with a wall. Test runs without random actions (and without learning) have been performed after every 100 training episodes. A step size of $\alpha = 0.05$, a discount factor of $\gamma = 0.98$, and $\lambda = 0.9$ was used in all the trials. A learning episode ends when the agent reaches the goal state, collides with a wall, or after a certain number of actions. Within this section, the state space representation from (1) was used with GRLPR generalization. In all the experiments in this work, the world in Fig. 1 was used for training.

Figure 5 (upper graphs) shows how the learning success is affected by different noise values ρ. Of course, as the noisy perception influences the transition probabilities T, learning time increases with the value of ρ and learning success decreases somewhat. But up to $\rho = 20$ a success rate of 95% in tests can be reached after 40,000 learning runs. Regarding collisions, the differences between the noise levels are even smaller compared to the differences with regard to learning speed. Even with $\rho = 20$ (which means more than 11 detected line segments per 5 meter in the average), the agent performs comparably well.

The sensibility of the line detection to group points together is a parameter to the algorithm and can be appropriately adjusted. For RLPR, it is better to aberrantly detect two adjacent line segments as a single one, because it does not matter to the navigation behavior as long as the gap is not that big that the robot could pass through it. Thus, a simple trick can reduce the impact of noise: Each line can be prolonged by a few centimeters to reduce the probability of holes. Figure 5 (lower graphs) shows that for $\rho = 20$ such prolongation leads to a significant improvement and a result comparable to very low noise levels.

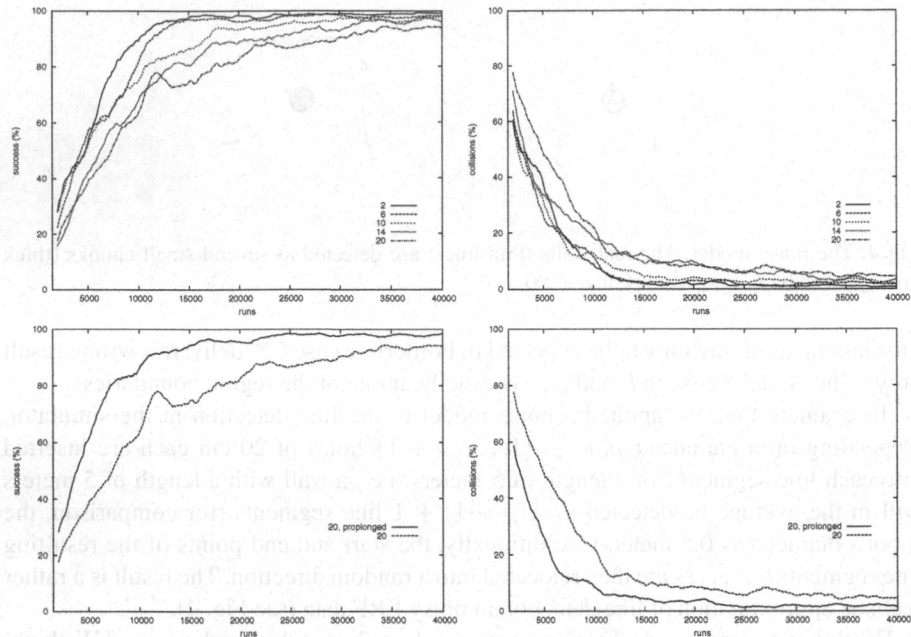

Fig. 5. Learning with unreliable line detection. Upper row: Learning success (left) and collisions (right) with different noise levels ρ. Even under heavy noise, the system learns successfully (but somewhat slower) with a low number of collisions. Lower row: Line segment prolongation significantly improves the learning behavior at a noise level of $\rho = 20$.

4.2   Unreliable Landmark Detection

Also landmark detection can be (and usually is) influenced by distortion. In another experiment we tested how the structural information of RLPR helps the system to cope with very unreliable landmark detection. The noise model implied here is the following: Each single landmark scan c_i returns the correct landmark detection with a probability of σ, so the probability that $(c_1, \ldots, c_7)$ is detected correctly is σ^7. For $\sigma = 95\%$ the overall probability to detect the whole vector correctly is around 70%, 3% for $\sigma = 60\%$, and only 0.2% for $\sigma = 40\%$. In case of a failure, no landmark is detected.

Figure 6 shows the learning success for values of $\sigma = 95, 85, 60, 40$ and 20%. The higher the failure in landmark detection, the lower is the rate at which the agent reaches the goal. However, even with $\sigma = 60\%$, the goal is reached in about two third of the test runs after 40,000 episodes of training. This is still a very good value, because the goal area is located behind a crossing, and reaching it is subject to a decision, which is hard to draw when the agent is that uncertain of where it is. Furthermore, the number of collisions is more or less negligible after very few runs, almost independent of σ. This shows that GRLPR does its job in enabling the agent to learn a good general navigation behavior even in the absence of utilizable landmark information.

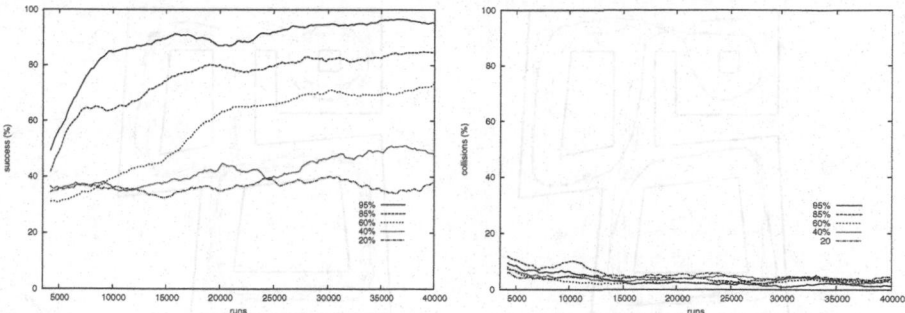

Fig. 6. Learning with unreliable landmark information for different levels of input distortion σ: The learning success of reaching the goal (left) decreases with higher values of σ, because the agent is uncertain of where it is; but is still comparably high even for very high distortion levels. The number of collisions (right) is hardly affected by landmark distortion.

Overall, the results in noisy environmental condition show a very robust behavior of the learned strategies and are a strong hint that strategies learned in simulation may also be successfully applied to a real robot system.

5   Knowledge Transfer to Unknown Environments

5.1   Ad-Hoc Strategy Transfer with Noisy Training Data

It has been shown in [8] that strategies learned with GRLPR can be transferred to different scenarios than the one the system was trained in without any further effort (ad-hoc strategy transfer), even if the new environments are differently scaled, contain different corridor angles or even lack any landmark information. This is because the RLPR representation of the new world is identical and the tile coding approximation of the value function returns a reasonable Q-value also for landmarks that have never been perceived before.

As the results so far provided transferred strategies that were trained under optimal environmental conditions we tested the transfer of a policy that was trained with noisy line detection. We trained the system in the world depicted in Fig. 1 with GRLPR both with a perfect line detection and with a noise parameter $\rho = 10$. The learned strategies have then been applied to another world (Fig. 7). Without environmental noise in training, the robot moves much closer to the walls compared to the noise-affected strategy which keeps the agent predominantly in the middle of the corridors and creates movement patterns that look well from a human perspective. The strategy of the noise-free learning task uses features in the bounded RLPR regions ($R_9...$) for navigation, while the noise-affected agent learned to avoid them: the agent plays for safety. Thus, noise in training does not only not disturb the agent in learning a transferable strategy, it proves being beneficial in keeping it away from possible danger and provides sensible trajectories within unknown worlds.

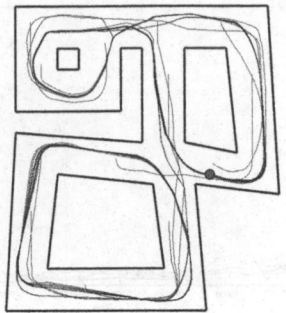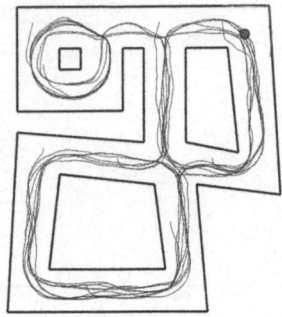

Fig. 7. Ad-hoc strategy transfer under noisy line detection: Trajectories in an unknown environment, learned with GRLPR with perfect line detection (left) and with distorted line detection with a noise parameter $\rho = 10$ (right). The noise-affected strategy shows a reasonable navigation behavior and keeps the robot safely in the middle of the corridors.

5.2   A Strategy Transfer Algorithm

GRLPR based strategies, however, need to be carefully checked when being transferred to a target scenario. Not all policies perform well in new worlds, even if they showed a perfect behavior in the source problem. We identified two drawbacks in the ad-hoc concept of GRLPR knowledge transfer, which are caused by the way generalization is achieved: First, in the target environment, the system is confronted with completely unknown landmark vectors and usually returns a reasonable reward. However, the reward is not independent of the new landmark vector, and different landmark inputs may result in different actions, even if the RLPR representation is the same. Second, because the system is trained with discounted rewards, Q-values are high for states near the goal state and decrease with distance from it. As tile coding sums up values in different tiles, the states in the vicinity of the goal have the biggest impact on the generalized strategy.

To get rid of these problems, we present a new method to generate a strategy for generally sensible navigation behavior in an observation space $\mathcal{O}'$ without landmark information with $\mathcal{O}' = \{\psi_r(s)\}$, $s \in \mathcal{S}$. It works both on policies learned with and without GRLPR generalization; but in all the experiments conducted we used non-GRLPR based policies in the source task. Given a learned policy π with a value function $Q_\pi(o, a)$ ($o \in \mathcal{O}$, $a \in \mathcal{A}$), we achieve a new policy π' with $Q_{\pi'}(o', a)$ ($o' \in \mathcal{O}'$, $a \in \mathcal{A}$) with the following function:

$$Q_{\pi'}(o', a) = \frac{\sum_{c \in \{\psi_l(s)\}} (\max_{b \in \mathcal{A}}(|Q_\pi((c, o'), b)|))^{-1} \, Q((c, o'), a)}{\#\{((c, o'), a) | Q_\pi((c, o'), a) \neq 0\}} \tag{2}$$

This is just a weighted sum over all possible landmark observations (in reality, of course, only the visited states have to be considered, because $Q(o, a) = 0$ for the others, so the computational effort is very low). Is is averaged over all cases where the Q-value is not zero (# describes a cardinality operator over a set here). The weighting factor scales all values according to the maximum reward over all actions to solve the mentioned problem of states far from the goal having lower Q-values.

A nice property of strategies created like this is that, in contrast to the GRLPR based strategies, they can be analyzed before applying them. It is easy to count the observation states in $\mathcal{O}$ that contributed to $Q_{\pi'}(o', a)$ to get an idea how big the data basis for an observation is. It is also possible to define a confidence measure on the new strategy π'.

$$\text{conf}(o') = \begin{cases} \frac{1}{\#\mathcal{A}} \sum_{a \in \mathcal{A}} (\max_{b \in \mathcal{A}} (Q_{\pi'}(o', b)) - Q_{\pi'}(o', a)) & Q_{\pi'}(o', a) \neq 0 \; \forall a \in \mathcal{A} \\ 0 & \text{else} \end{cases}$$

(3)

gives a measure for an observation o' of how certain the strategy is in choosing the best action according to the Q-value at this state over another one. If there is no information collected for an action a yet, the result is defined as 0. Of course there are situations where two or more actions are equally appropriate, so this measure only makes sense when summed up over the whole observation space for the whole strategy:

$$\text{Conf}(\pi') = \frac{\sum_{o' \in \mathcal{O}'} \text{conf}(o')}{\#\{o' | \text{conf}(o') \neq 0\}}$$

(4)

$\text{Conf}(\pi)$ can be used to compare strategies before applying them to another world. When comparing the confidence values of a task learned with and without line detection distortion, for example, $\text{Conf}(\pi')$ converged to 0.434 for the undistorted case and to 0.232 in the same world with a distortion of $\rho = 10$.

However, strategies that are created with (2) don't prove to be completely successful and even show similar results as GRLPR based strategies: Some of them work perfectly well in the target environment, and some don't and lead to collisions frequently. Even after introducing the weight factor, the differences of the output of the value function Q still seem to be too high.

A solution to this problem is to totally abstract from the rewards and only regard which action the agent will greedily be chosen in a particular state and which one has the least reward expectation. We then sum up over the decisions, not over the expected reward. We modify (2) accordingly:

$$Q_{\pi'}(o', a) = \frac{\sum_{c \in \{\psi_l(s)\}} \phi_b((c, o'), a)}{\#\{Q((c, o'), a) \neq 0\}}$$

(5)

$$\phi_b(s, a) = \begin{cases} 1 & \text{if } Q(o, a) = \max_{b \in \mathcal{A}} Q(o, b) \\ -1 & \text{if } Q(o, a) = \min_{b \in \mathcal{A}} Q(o, b) \\ 0 & \text{else} \end{cases}$$

This mechanism proves to create strategies that are robustly applicable to new environments. Figure 8 shows that the strategies perform very well even under line detection distortion in the target world. In contrast to GRLPR based strategies or those created with (2), experiments show that you can pick any strategy after an arbitrary learning time and will not experience outliers that don't succeed at all. Strategies with higher confidence values perform better, and $\text{Conf}(\pi')$ asymptotically increases over learning time. In contrast to that, GRLPR based strategies show a tendency to perform worse after longer training. Furthermore, the transferred strategies prove to be robust against

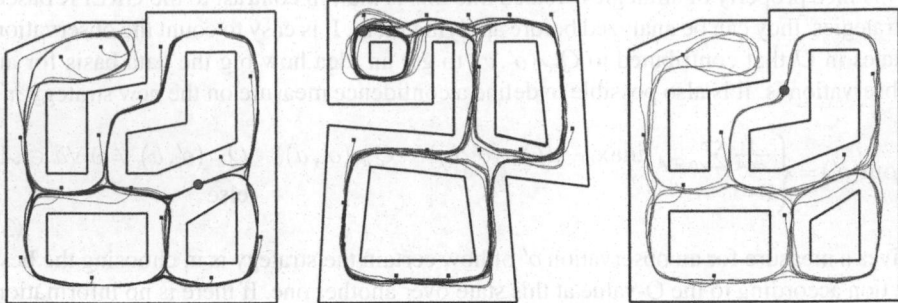

Fig. 8. Trajectories of policies transferred with (5) to unknown environments: The agents keeps in the middle of corridors and shows a sensible navigation behavior even at line detection noise levels in the target environment of $\rho = 1$ (left), $\rho = 10$ (middle) or $\rho = 20$ (right).

noise and do not, like frequently observable in GRLPR based ones, show a behavior that navigates closely to walls. Without noise, the agent navigates collision-free, and also for high values of ρ, collisions rarely occur.

Strategies $Q_{\pi'}$ that were transferred to $\mathcal{O}'$ and only rely on the RLPR part of the representation can then be used as a basis for new goal-directed learning tasks in unknown worlds: If the value $Q_{\pi''}((c, o'), a)$ of the policy to learn is undefined at an observation $(c, o') \in \mathcal{O}$ (because this observation has not been visited yet), it can be initialized with of $Q_{\pi'}(o', a)$, which provides the learning agent with a good general navigation behavior from scratch.

5.3 Remaining Challenges and Limitations

The algorithm presented in (5) performs very well to create generally sensible navigation strategies for different indoor environments. However, the learned policies do not completely prevent the agent from collisions when environmental noise is applied. Collisions may occur in two situations: First, when corridors are getting narrower and the agent tries to perform an U-turn or the agents is driving around a very steep curve, collisions may occur. The training has been performed in a very regular environment, so the learned turning strategy may bring the robot closer to a wall than it was used to in training. Second, and more frequently, the agent crashes when it is directly heading into a corner. In this case, the RLPR representation signals obstacles everywhere in front of the robot. Due to the agent's motion dynamics that don't include turning in place or even driving backwards, the robot is doomed in a dead-end it cannot escape anymore.

Usually, the transferred policy hinders the agent to reach such situations, but under conditions of noise this cannot be prevented completely. To avoid such trouble, it is suggested to provide (possibly negatively rewarded) actions that allow for escaping from such situations, especially when planning to apply learned strategies from simulation to a real robotics platform.

6   Conclusions and Future Work

In this work we investigated into generalization and transfer properties of robot navigation strategies learned in simulation with RL while using the abstraction method of RLPR. We showed that RLPR is a sensible and robust abstraction and enables learning both with distorted line segment and landmark detection. A new algorithm was presented for transferring goal-directed navigation strategies to generally sensible navigation strategies that can be transferred to arbitrary environments. The results prove to be robust and consistent, and the navigation policies enable successful navigation in unknown worlds even under heavy environmental noise.

Future work will have to provide a thorough analysis of the impact of the confidence measure for transferred strategies. Regarding the promising results, a next step will be to show that strategies learned in simulation can be transferred to a real robot.

Acknowledgments

This work was supported by the DFG Transregional Collaborative Research Center SFB/TR 8 "Spatial Cognition". Funding by the German Research Foundation (DFG) is gratefully acknowledged.

References

1. Sutton, R.S., Barto, A.G.: Reinforcement learning: an introduction. In: Adaptive Computation and Machine Learning, MIT Press, Cambridge, MA (1998)
2. Lane, T., Wilson, A.: Toward a topological theory of relational reinforcement learning for navigation tasks. In: Proceedings of FLAIRS 2005 (2005)
3. Busquets, D., de Mántaras, R.L., Sierra, C., Dietterich, T.G.: Reinforcement learning for landmark-based robot navigation. In: Alonso, E., Kudenko, D., Kazakov, D. (eds.) AAAMAS 2002. LNCS (LNAI), vol. 2636, Springer, Heidelberg (2003)
4. Thrun, S., Schwartz, A.: Finding structure in reinforcement learning. In: Advances in Neural Information Processing Systems, vol. 7 (1995)
5. Konidaris, G.D., Barto, A.G.: Building portable options: Skill transfer in reinforcement learning. In: Proceedings of IJCAI 2007 (January 2007)
6. Taylor, M.E., Stone, P.: Cross-domain transfer for reinforcement learning. In: Proceedings of ICML 2007, Corvallis, Oregon (June 2007)
7. Torrey, L., Shavlik, J., Walker, T., Maclin, R.: Skill acquisition via transfer learning and advice taking. In: Fürnkranz, J., Scheffer, T., Spiliopoulou, M. (eds.) ECML 2006. LNCS (LNAI), vol. 4212, pp. 425–436. Springer, Heidelberg (2006)
8. Frommberger, L.: A generalizing spatial representation for robot navigation with reinforcement learning. In: Proceedings of FLAIRS 2007, Key West, FL (May 2007)
9. Schlieder, C.: Reasoning about ordering. In: Kuhn, W., Frank, A.U. (eds.) COSIT 1995. LNCS, vol. 988, pp. 341–349. Springer, Heidelberg (1995)
10. Goyal, R.K., Egenhofer, M.J.: Consistent queries over cardinal directions across different levels of detail. In: Tjoa, A.M., Wagner, R., Al-Zobaidie, A. (eds.) Proceedings of the 11th International Workshop on Database and Expert System Applications, pp. 867–880 (2000)
11. Sutton, R.S.: Generalization in reinforcement learning: Successful examples using sparse tile coding. In: Advances in Neural Information Processing Systems, vol. 8 (1996)
12. Lu, F., Milios, E.: Robot pose estimation in unknown environments by matching 2D range scans. Journal of Intelligent and Robotic Systems (1997)
13. Watkins, C.: Learning from Delayed Rewards. PhD thesis, Cambridge University (1989)

Heuristic Q-Learning Soccer Players: A New Reinforcement Learning Approach to RoboCup Simulation

Luiz A. Celiberto Jr.[1,2], Jackson Matsuura[2], and Reinaldo A.C. Bianchi[1]

[1] Centro Universitário da FEI
Av. Humberto de Alencar Castelo Branco, 3972.
09850-901 – São Bernardo do Campo – SP, Brazil
[2] Instituto Tecnológico de Aeronáutica
Praça Mal. Eduardo Gomes, 50.
12228-900 – São José dos Campos – SP, Brazil
celibertojr@uol.com.br, jackson@ita.br, rbianchi@fei.edu.br

Abstract. This paper describes the design and implementation of a 4 player RoboCup Simulation 2D team, which was build by adding Heuristic Accelerated Reinforcement Learning capabilities to basic players of the well-known UvA Trilearn team. The implemented agents learn by using a recently proposed Heuristic Reinforcement Learning algorithm, the Heuristically Accelerated Q–Learning (HAQL), which allows the use of heuristics to speed up the well-known Reinforcement Learning algorithm Q–Learning. A set of empirical evaluations was conducted in the RoboCup 2D Simulator, and experimental results obtained while playing with other teams shows that the approach adopted here is very promising.

Keywords: Reinforcement Learning, Cognitive Robotics, RoboCup Simulation 2D.

1 Introduction

Modern Artificial Intelligence textbooks such as AIMA [11] introduced a unified picture of the field, proposing that the typical problems of the AI should be approached by multiple techniques, and where different methods are appropriate, depending on the nature of the task. This is the result of the belief that AI must not be seen as a segmented domain. According to this tendency, the applications domains probed by this field also changed: chess player programs that are better than a human champion, a traditional AI domain is a reality; new domains became a necessity.

The RoboCup Robotic Soccer Cup domain was proposed by several researchers [6] in order to provide a new long-term challenge for Artificial Intelligence research. The development of soccer teams involves a wide range of technologies, including: design of autonomous agents, multiagent collaboration, strategy definition and acquisition, real-time reasoning, robotics, and sensor-fusion.

J. Neves, M. Santos, and J. Machado (Eds.): EPIA 2007, LNAI 4874, pp. 520–529, 2007.

Soccer games between robots constitute real experimentation and testing activities for the development of intelligent, autonomous robots, which cooperate among each one to achieve a goal. This domain has become of great relevance in Artificial Intelligence since it possesses several characteristics found in other complex real problems; examples of such problems are: robotic automation systems, that can be seen as a group of robots in an assembly task, and space missions with multiple robots, to mention but a few.

The RoboCup Simulation League has the goal to provide an environment where teams can be created in order to compete against each other in a simulated soccer championship. Since the simulator provides the entire environment of the players, teams have only one task: to develop the strategies of their players.

However, the task is not trivial. To solve this problem, several researchers have been applying Reinforcement Learning (RL) techniques that been attracting a great deal of attention in the context of multiagent robotic systems. The reasons frequently cited for such attractiveness are: the existence of strong theoretical guarantees on convergence [14], they are easy to use, and they provide model-free learning of adequate control strategies. Besides that, they also have been successfully applied to solve a wide variety of control and planning problems.

One of the main problems with RL algorithms is that they typically suffer from very slow learning rates, requiring a huge number of iterations to converge to a good solution. This problem becomes worse in tasks with high dimensional or continuous state spaces and when the system is given sparse rewards. One of the reasons for the slow learning rates is that most RL algorithms assumes that neither an analytical model nor a sampling model of the problem is available *a priori*. However, in some cases, there is domain knowledge that could be used to speed up the learning process: "Without an environment model or additional guidance from the programmer, the agent may literally have to keep falling off the edge of a cliff in order to learn that this is bad behavior" [4].

As a way to add domain knowledge to help in the solution of the RL problem, a recently proposed Heuristic Reinforcement Learning algorithm – the Heuristically Accelerated Q–Learning (HAQL) [1] – uses a heuristic function that influences the choice of the action to speed up the well-known RL algorithm Q–Learning. This paper investigates the use of HAQL to speed up the learning process of teams of mobile autonomous robotic agents acting in a concurrent multiagent environment, the RoboCup 2D Simulator. It is organized as follows: section 2 describes the Q–learning algorithm. Section 3 describes the HAQL and its formalization using a heuristic function. Section 4 describes the robotic soccer domain used in the experiments, presents the experiments performed, and shows the results obtained. Finally, Section 5 summarizes some important points learned from this research and outlines future work.

2   Reinforcement Learning and the Q–Learning Algorithm

Reinforcement Learning is the area of Machine Learning [9] that is concerned with an autonomous agent interacting with its environment via perception and action. On each interaction step the agent senses the current state s of the environment, and chooses an action a to perform. The action a alters the state s of the environment, and a scalar reinforcement signal r (a reward or penalty) is provided to the agent to indicate the desirability of the resulting state. In this way, "The RL problem is meant to be a straightforward framing of the problem of learning from interaction to achieve a goal" [13].

Formally, the RL problem can be formulated as a discrete time, finite state, finite action Markov Decision Process (MDP) [9]. Given:

- A finite set of possible actions $a \in \mathcal{A}$ that the agent can perform;
- finite set of states $s \in \mathcal{S}$ that the agent can achieve;
- A state transition function $\mathcal{T} : \mathcal{S} \times \mathcal{A} \to \Pi(\mathcal{S})$, where $\Pi(\mathcal{S})$ is a probability distribution over $\mathcal{S}$;
- A finite set of bounded reinforcements (payoffs) $\mathcal{R} : \mathcal{S} \times \mathcal{A} \to \Re$,

the task of a RL agent is to find out a stationary policy of actions $\pi^* : \mathcal{S} \to \mathcal{A}$ that maps the current state s into an optimal action(s) a to be performed in s, maximizing the expected long term sum of values of the reinforcement signal, from any starting state.

In this way, the policy π is some function that tells the agent which actions should be chosen, under which circumstances [8]. In RL, the policy π should be learned through trial-and-error interactions of the agent with its environment, that is, the RL learner must explicitly explore its environment.

The Q–learning algorithm was proposed by Watkins [15] as a strategy to learn an optimal policy π^* when the model ($\mathcal{T}$ and $\mathcal{R}$) is not known in advance. Let $Q^*(s,a)$ be the reward received upon performing action a in state s, plus the discounted value of following the optimal policy thereafter:

$$Q^*(s,a) \equiv R(s,a) + \gamma \sum_{s' \in S} T(s,a,s')V^*(s'). \tag{1}$$

The optimal policy π^* is $\pi^* \equiv \arg\max_a Q^*(s,a)$. Rewriting $Q^*(s,a)$ in a recursive form:

$$Q^*(s,a) \equiv R(s,a) + \gamma \sum_{s' \in S} T(s,a,s') \max_{a'} Q^*(s',a'). \tag{2}$$

Let $\hat{Q}$ be the learner's estimate of $Q^*(s,a)$. The Q–learning algorithm iteratively approximates $\hat{Q}$, i.e., the $\hat{Q}$ values will converge with probability 1 to Q^*, provided the system can be modeled as a MDP, the reward function is bounded ($\exists c \in \mathcal{R}; (\forall s,a), |R(s,a)| < c$), and actions are chosen so that every state-action pair is visited an infinite number of times. The Q learning update rule is:

$$\hat{Q}(s,a) \leftarrow \hat{Q}(s,a) + \alpha \left[r + \gamma \max_{a'} \hat{Q}(s',a') - \hat{Q}(s,a) \right], \tag{3}$$

where s is the current state; a is the action performed in s; r is the reward received; s' is the new state; γ is the discount factor ($0 \leq \gamma < 1$); $\alpha = 1/(1 +$

$visits(s, a)$), where $visits(s, a)$ is the total number of times this state-action pair has been visited up to and including the current iteration.

An interesting property of Q–learning is that, although the exploration-exploitation tradeoff must be addressed, the $\hat{Q}$ values will converge to Q^*, independently of the exploration strategy employed (provided all state-action pairs are visited often enough) [9].

3   The Heuristically Accelerated Q–Learning Algorithm

The Heuristically Accelerated Q–Learning algorithm [1] was proposed as a way of solving the RL problem which makes explicit use of a heuristic function $\mathcal{H} : \mathcal{S} \times \mathcal{A} \to \mathfrak{R}$ to influence the choice of actions during the learning process. $H_t(s_t, a_t)$ defines the heuristic, which indicates the importance of performing the action a_t when in state s_t.

The heuristic function is strongly associated with the policy: every heuristic indicates that an action must be taken regardless of others. This way, it can be said that the heuristic function defines a "Heuristic Policy", that is, a tentative policy used to accelerate the learning process. It appears in the context of this paper as a way to use the knowledge about the policy of an agent to accelerate the learning process. This knowledge can be derived directly from the domain (prior knowledge) or from existing clues in the learning process itself.

The heuristic function is used only in the action choice rule, which defines which action a_t must be executed when the agent is in state s_t. The action choice rule used in the HAQL is a modification of the standard $\epsilon - Greedy$ rule used in Q–learning, but with the heuristic function included:

$$\pi(s_t) = \begin{cases} \arg\max_{a_t} \left[\hat{Q}(s_t, a_t) + \xi H_t(s_t, a_t) \right] & \text{if } q \leq p, \\ a_{random} & \text{otherwise,} \end{cases} \quad (4)$$

where:

- $\mathcal{H} : \mathcal{S} \times \mathcal{A} \to \mathfrak{R}$: is the heuristic function, which influences the action choice. The subscript t indicates that it can be non-stationary.
- ξ: is a real variable used to weight the influence of the heuristic function.
- q is a random value with uniform probability in [0,1] and p $(0 \leq p \leq 1)$ is the parameter which defines the exploration/exploitation trade-off: the greater the value of p, the smaller is the probability of a random choice.
- a_{random} is a random action selected among the possible actions in state s_t.

As a general rule, the value of the heuristic $H_t(s_t, a_t)$ used in the HAQL must be higher than the variation among the $\hat{Q}(s_t, a_t)$ for a similar $s_t \in S$, so it can influence the choice of actions, and it must be as low as possible in order to minimize the error. It can be defined as:

$$H(s_t, a_t) = \begin{cases} \max_a \hat{Q}(s_t, a) - \hat{Q}(s_t, a_t) + \eta & \text{if } a_t = \pi^H(s_t), \\ 0 & \text{otherwise.} \end{cases} \quad (5)$$

where η is a small real value and $\pi^H(s_t)$ is the action suggested by the heuristic.

Table 1. The HAQL algorithm

Initialize $Q(s, a)$.

Repeat:

Visit the s state.

Select an action a using the action choice rule (equation 4).

Receive the reinforcement $r(s, a)$ and observe the next state s'.

Update the values of $H_t(s, a)$.

Update the values of $Q(s, a)$ according to:

$\quad Q(s, a) \leftarrow Q(s, a) + \alpha[r(s, a) + \gamma \max_{a'} Q(s', a') - Q(s, a)]$.

Update the $s \leftarrow s'$ state.

Until some stop criteria is reached.

where: $s = s_t$, $s' = s_{t+1}$, $a = a_t$ e $a' = a_{t+1}$.

For instance, if the agent can execute 4 different actions when in state s_t, the values of $\hat{Q}(s_t, a)$ for the actions are $[1.0 \quad 1.1 \quad 1.2 \quad 0.9]$, the action that the heuristic suggests is the first one. If $\eta = 0.01$, the values to be used are $H(s_t, 1) = 0.21$, and zero for the other actions.

As the heuristic is used only in the choice of the action to be taken, the proposed algorithm is different from the original Q–learning only in the way exploration is carried out. The RL algorithm operation is not modified (i.e., updates of the function Q are as in Q–learning), this proposal allows that many of the conclusions obtained for Q–learning to remain valid for HAQL.

The use of a heuristic function made by HAQL explores an important characteristic of some RL algorithms: the free choice of training actions. The consequence of this is that a suitable heuristic speeds up the learning process, and if the heuristic is not suitable, the result is a delay which does not stop the system from converging to a optimal value.

The idea of using heuristics with a learning algorithm has already been considered by other authors, as in the Ant Colony Optimization presented in [2] or the use of initial Q-Values [7]. However, the possibilities of this use were not properly explored yet. The complete HAQL algorithm is presented on table 1. It can be noticed that the only difference to the Q–learning algorithm is the action choice rule and the existence of a step for updating the function $H_t(s_t, a_t)$.

4   Experiment in the RoboCup 2D Simulation Domain

One experiment was carried out using the RoboCup 2D Soccer Server [10]: the implementation of a four player team, with a goalkeeper, a first defender (fullback) and two forward players (strikers) that have to learn how to maximize the number of goals they score, minimizing the number of goals scored by the opponent. In this experiment, the implemented team have to learn while playing against a team composed of one goalkeeper, one defender and two striker agents from the UvA Trilearn 2001 Team [3].

The Soccer Server is a system that enables autonomous agents programs to play a match of soccer against each other. A match is carried out in a client/server style: a server provides a virtual field and calculates movements of players and a ball, according to several rules. Each client is an autonomous agent that controls movements of one player. Communication between the server and each client is done via TCP/IP sockets. The Soccer Server system works in real-time with discrete time intervals, called cycles. Usually, each cycle takes 100 milliseconds and in the end of cycle, the server executes the actions requested by the clients and update the state of world.

The space state of the defending agents is composed by its position in a discrete grid with N x M positions the agent can occupy, the position of the ball in the same grid and the direction the agent is facing. This grid is different for the goalkeeper and the defender: each agent has a different area where it can move, which they cannot leave. These grids, shown in figure 1, are partially overlapping, allowing both agents to work together in some situations. The space state of the attacking agents is also composed by its position in a discrete grid (shown in figure 2) the direction the agent is facing, the distance between the agents and the distance to the ball. The direction that the agent can be facing is also discrete, and reduced to four: north, south, east or west.

The defender can execute six actions: turnBodyToObject, that keeps the agent at the same position, but always facing the ball; interceptBall, that moves the agent in the direction of the ball; driveBallFoward, that allows the agent to move with the ball; directPass, that execute a pass to the goalkeeper; kickBall, that kick the ball away from the goal and; markOpponent, that moves the defender close to one of the opponents.

The goalkeeper can also perform six actions: turnBodyToObject, intercept-Ball, driveBallForward, kickBall, which are the same actions that the defender can execute, and two specific actions: holdBall, that holds the ball and move-ToDefensePosition, that moves the agent to a position between the ball and the goal. Finally, the strikers can execute six actions: turnBodyToObject, intercept-Ball and markOpponent, which are the same as described for the defender, and kickBall, that kick the ball in the direction goal; dribble, that allows the agent to dribble with the ball, and; holdBall, that holds the ball.

All these actions are implemented using pre-defined C++ methods defined in the BasicPlayer class of the UvA Trilearn 2001 Team. "The BasicPlayer class contains all the necessary information for performing the agents individual skills such as intercepting the ball or kicking the ball to a desired position on the field" [3, p. 50].

The reinforcement given to the agents were inspired on the definitions of rewards presented in [5], and are different for the agents. For the goalkeeper, the rewards consists of: ball caught, kicked or driven by goalie $= 100$; ball with any opponent player $= -50$; goal scored by the opponent $= -100$. For the defender, the rewards are: ball with any opponent player $= -100$; goal scored by the opponent $= -100$. The rewards of the strikers consists of: ball with any opponent player $= -50$; goal scored by the agent $= 100$; goal scored by the agent's teammate $= 50$.

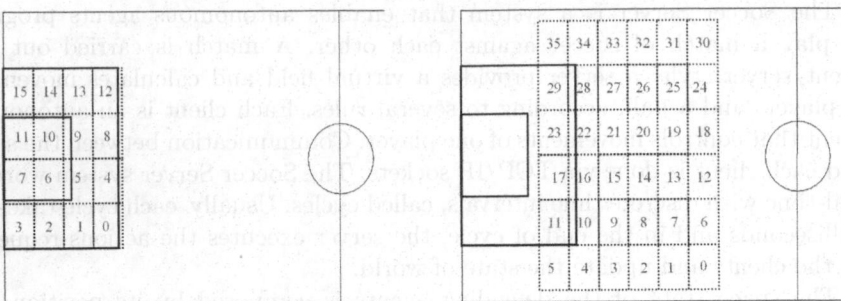

Fig. 1. Discrete grids that compose the space state of the goalkeeper (left) and the defender (right)

62	61	60	59	58	57	56	55	54
53	52	51	50	49	48	47	46	45
44	43	42	41	40	39	38	37	36
35	34	33	32	31	30	29	28	27
26	25	24	23	22	21	20	19	18
17	16	15	14	13	12	11	10	9
8	7	6	5	4	3	2	1	0

Fig. 2. Discrete grid that compose the space state of the strikers

The heuristic policy used for the all the agents is described by two rules: if the agent is not near the ball, run in the direction of the ball, and; if the agent is close to the ball, do something with it. Note that the heuristic policy is very simple, leaving the task of learning what to do with the ball and how to deviate from the opponent to the learning process. The values associated with the heuristic function are defined using equation 5, with the value of η set to 500. This value is computed only once, at the beginning of the game. In all the following episodes, the value of the heuristic is maintained fixed, allowing the learning process to overcome bad indications.

In order to evaluate the performance of the HAQL algorithm, this experiment was performed with teams of agents that learns using the Q–learning algorithm, the HAQL algorithm. The results presented are based on the average of 10 training sessions for each algorithm. Each session is composed of 100 episodes consisting of matches taking 3000 cycles each. During the simulation, when a teams scores a goal all agents are transferred back to a pre-defined start position, presented in figure 3.

Fig. 3. Position of all the agents at the beginning of an episode

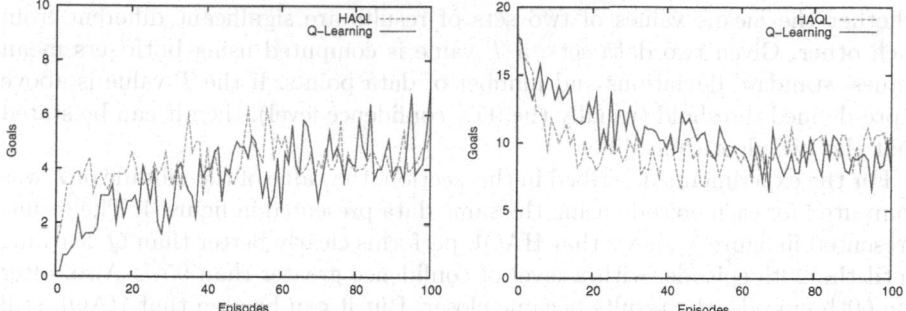

Fig. 4. Average goals scored by the learning team (left) and scored against it (right), using the Q Learning and the HAQL algorithms, for training sessions against UvA Trilearn agents

The parameters used in the experiments were the same for the two algorithms, Q–learning and HAQL: the learning rate is $\alpha = 1.25$, the exploration/ exploitation rate $p = 0.05$ and the discount factor $\gamma = 0.9$. Values in the Q table were randomly initiated, with $0 \leq Q(s, a, o) \leq 1$. The experiments were programmed in C++ and executed in a Pentium IV 2.8GHz, with 1GB of RAM on a Linux platform.

Figure 4 shows the learning curves for both algorithms when the agents learn how to play against a team composed of one goalkeeper, one defender and two strikers from the UvA Trilearn Team 2001 [3]. It presents the average goals scored by the learning team in each episode (left) and the average goals scored by the opponent team (left), using the Q–Learning and the HAQL algorithm. It is possible to verify that Q–learning has worse performance than HAQL at the initial learning phase, and that as the matches proceed, the performance of both algorithms become more similar, as expected.

Student's t–test [12] was used to verify the hypothesis that the use of heuristics speeds up the learning process. The t-test is a statistical test used to compute

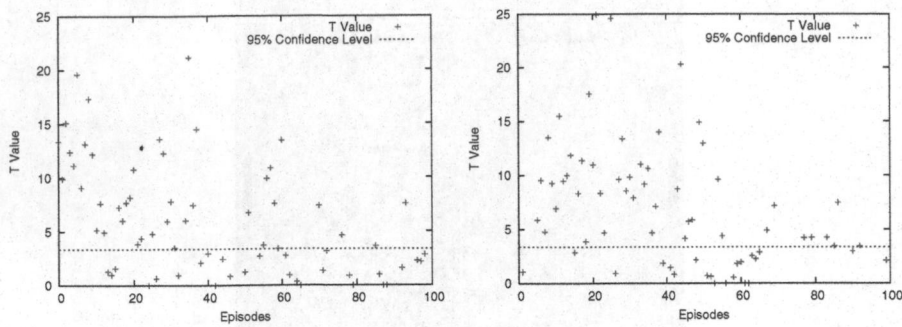

Fig. 5. Results from Student's t test between Q–learning and HAQL algorithms, for the number of goals scored (left) and conceded (right)

whether the means values of two sets of results are significant different from each other. Given two data sets, a T value is computed using both sets mean values, standard deviations and number of data points. If the T value is above a pre-defined threshold (usually the 95% confidence level), then it can be stated that the two algorithms differ.

For the experiments described in this section, the value of the module of T was computed for each episode using the same data presented in figure 4. The result, presented in figure 5, shows that HAQL performs clearly better than Q–learning until the 20th episode, with a level of confidence greater than 95%. Also, after the 60th episode, the results became closer. But it can be seen that HAQL still performs better than Q–learning in some cases.

5   Conclusion and Future Works

This paper presented the use of the Heuristically Accelerated Q–Learning (HAQL) algorithm to speed up the learning process of teams of mobile autonomous robotic agents acting in the RoboCup 2D Simulator.

The experimental results obtained in this domain showed that agents using the HAQL algorithm learned faster than ones using the Q–learning, when they were trained against the same opponent. These results are strong indications that the performance of the learning algorithm can be improved using very simple heuristic functions.

Due to the fact that the reinforcement learning requires a large amount of training episodes, the HAQL algorithm has been evaluated, so far, only in simulated domains. Among the actions that need to be taken for a better evaluation of this algorithm, the more important ones include:

- The development of teams composed of agents with more complex space state representation and with a larger number of players.
- Working on obtaining results in more complex domains, such as RoboCup 3D Simulation and Small Size League robots [6].

– Comparing the use of more convenient heuristics in these domains.
– Validate the HAQL by applying it to other the domains, such as the "car on the hill" and the "cart-pole".

Future works also include the incorporation of heuristics into other well known RL algorithms, such as SARSA, $Q(\lambda)$, Minimax-Q and Minimax-QS, and conceiving ways of obtaining the heuristic function automatically.

References

[1] Bianchi, R.A.C., Ribeiro, C.H.C., Costa, A.H.R.: Heuristically Accelerated Q-Learning: a new approach to speed up reinforcement learning. In: Bazzan, A.L.C., Labidi, S. (eds.) SBIA 2004. LNCS (LNAI), vol. 3171, pp. 245–254. Springer, Heidelberg (2004)

[2] Bonabeau, E., Dorigo, M., Theraulaz, G.: Inspiration for optimization from social insect behaviour. Nature 406(6791) (2000)

[3] de Boer, R., Kok, J.: The Incremental Development of a Synthetic Multi-Agent System: The UvA Trilearn 2001 Robotic Soccer Simulation Team. Master's Thesis, University of Amsterdam (2002)

[4] Hasinoff, S.W.: Reinforcement learning for problems with hidden state. Technical report, University of Toronto (2003)

[5] Kalyanakrishnan, S., Liu, Y., Stone, P.: Half field offense in RoboCup soccer: A multiagent reinforcement learning case study. In: Lakemeyer, G., Sklar, E., Sorenti, D., Takahashi, T. (eds.) RoboCup-2006: Robot Soccer World Cup X, Springer, Berlin (2007)

[6] Kitano, H., Minoro, A., Kuniyoshi, Y., Noda, I., Osawa, E.: Robocup: A challenge problem for ai. AI Magazine 18(1), 73–85 (1997)

[7] Koenig, S., Simmons, R.G.: The effect of representation and knowledge on goal–directed exploration with reinforcement–learning algorithms. Machine Learning 22, 227–250 (1996)

[8] Littman, M.L., Szepesvári, C.: A generalized reinforcement learning model: Convergence and applications. In: ICML 1996. Procs. of the Thirteenth International Conf. on Machine Learning, pp. 310–318 (1996)

[9] Mitchell, T.: Machine Learning. McGraw Hill, New York (1997)

[10] Noda, I.: Soccer server: a simulator of robocup. In: Proceedings of AI symposium of the Japanese Society for Artificial Intelligence, pp. 29–34 (1995)

[11] Russell, S., Norvig, P.: Artificial Intelligence: A Modern Approach. Prentice-Hall, Upper Saddle River, NJ (1995)

[12] Spiegel, M.R.: Statistics. McGraw-Hill, New York (1998)

[13] Sutton, R.S., Barto, A.G.: Reinforcement Learning: An Introduction. MIT Press, Cambridge (1998)

[14] Szepesvári, C., Littman, M.L.: Generalized markov decision processes: Dynamic-programming and reinforcement-learning algorithms. Technical report, Brown University, Department of Computer Science, Brown University, Providence, Rhode Island 02912 (1996) CS-96-11

[15] Watkins, C.J.C.H.: Learning from Delayed Rewards. PhD thesis, University of Cambridge (1989)

Human Robot Interaction Based on Bayesian Analysis of Human Movements

Jörg Rett and Jorge Dias

Institute of Systems and Robotics
University of Coimbra
Polo II, 3030-290 Coimbra, Portugal
{jrett,jorge}@isr.uc.pt

Abstract. We present as a contribution to the field of human-machine interaction a system that analyzes human movements online, based on the concept of Laban Movement Analysis (LMA). The implementation uses a Bayesian model for learning and classification, while the results are presented for the application to gesture recognition. Nowadays technology offers an incredible number of applications to be used in human-machine interaction. Still, it is difficult to find implemented cognitive processes that benefit from those possibilities. Future approaches must offer to the user an effortless and intuitive way of interaction. We present the Laban Movement Analysis as a concept to identify useful features of human movements to classify human actions. The movements are extracted using both, vision and magnetic tracker. The descriptor opens possibilities towards expressiveness and emotional content. To solve the problem of classification we use the Bayesian framework as it offers an intuitive approach to learning and classification. It also provides the possibility to anticipate the performed action given the observed features. We present results of our system through its embodiment in the social robot 'Nicole' in the context of a person performing gestures and 'Nicole' reacting by means of audio output and robot movement.

1 Introduction

Nowadays, robotics has reached a technological level that provides a huge number of input and output modalities. Apart from industrial robots, also social robots have emerged from the universities to companies as products to be sold. The commercial success of social robots implies that the available technology can be both, reliable and cost efficient. Surprisingly, higher level cognitive systems that could benefit from the technological advances in the context of human-robot interaction are still rare. Future approaches must offer an effortless and intuitive way of interacting with a robot to its human counterpart. One can think of the problem as a scenario where a robot is observing the movement of a human and is acting according to the extracted information (see Fig. 1). To achieve this interaction we need to extract the information contained in the observed movement and relate a appropriate robot action to it.

J. Neves, M. Santos, and J. Machado (Eds.): EPIA 2007, LNAI 4874, pp. 530–541, 2007.

Fig. 1. Nicole in position to interact

Our ultimate goal is to provide the robot with a cognitive system that mimics human perception in terms of anticipation and empathy. Towards the latter requirement this article will present the concept of Laban Movement Analysis (LMA) [1] as a way to describe intentional content and expressiveness of a human body movement. Two major components of LMA (i.e. *Space* and *Effort*) are described in detail. We show the technical realization of LMA for the cognitive system of the embodied agent which is based on a probabilistic (Bayesian) framework and a system for tracking of human movements. The system uses both a magnetic tracker and a visual tracker. The visual tracker extracts the movement-features of a human actor from a series of images taken by a single camera. The hands and the face of the actor are detected and tracked automatically without using a special device (markers) [2]. This work presents the Bayesian approach to LMA through the problem of learning and classification, also treating the system's online characteristic of anticipation. The probabilistic model anticipates the gesture given the observed features using the Bayesian framework. The system has been implemented in our social robot, 'Nicole' to test several human-robot interaction scenarios (e.g. playing).

If the perceptual system of a robot is based on vision, interaction will involve *visual human motion analysis*. The ability to recognize humans and their activities by vision is key for a machine to interact intelligently and effortlessly with a human-inhabited environment [3]. Several surveys on visual analysis of human movement have already presented a general framework to tackle this problem [4], [3], [5] and [6] usually emphasizing the three main problems: 1. Feature Extraction, 2. Feature Correspondence and 3. High Level Processing. One area of high level analysis is that of gesture recognition applied to control some sort of devices. In [7] DBNs were used to recognize a set of eleven hand gestures to manipulate a virtual display shown on a projection screen . Surveys specialized on gesture interfaces along the last ten years reflect the development and achievements [8], [9]. The most recent survey [10] is once more included in the broader context of human motion analysis emphasizing, once more the dependencies between low level features and high level analysis.

Section 2 presents the concept of LMA and its two major components (i.e. *Space*and *Effort*). Section 3 presents the system for tracking of human movements. Section 4 describes the Bayesian framework that is used to learn and classify human movements and presents the. Section 5 presents the results. Section 6 closes with a discussion and an outlook for future works.

2   Laban Movement Analysis (LMA)

Laban Movement Analysis (LMA) is a method for observing, describing, notating, and interpreting human movement. It was developed by a German named Rudolf Laban (1879 to 1958), who is widely regarded as a pioneer of European modern dance and theorist of movement education [11]. While being widely applied to studies of dance and application to physical and mental therapy [1], it has found little application in the engineering domain. Most notably the group of Norman Badler, who recently proposed a computational model of gesture acquisition and synthesis to learn motion qualities from live performance [12]. Also recently, researchers from neuroscience stated that LMA is quite useful to describe certain effects on the movements of animals and humans. In [13] LMA was adapted to capture the kinematic and non-kinematic aspects of movement in a reach-for-food task by human patients whose movements had been affected by stroke.

The theory of LMA treats five major components shown in Fig. 2 of which we adopted three. *Space* treats the spatial extent of the mover's *Kinesphere* (often interpreted as reach-space) and what form is being revealed by the spatial pathways of the movement. *Effort* deals with the dynamic qualities of the movement and the inner attitude towards using energy. Like suggested in [13] we have grouped *Body* and *Space* as kinematic features describing changes in the spatial-temporal body relations, while *Shape* and *Effort* are part of the non-kinematic features contributing to the qualitative aspects of the movement.

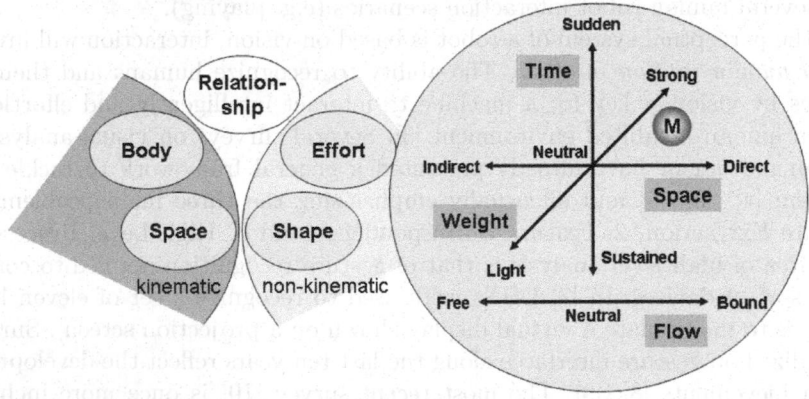

Fig. 2. Major components of LMA with the bipolar Effort factors as a 4-D space

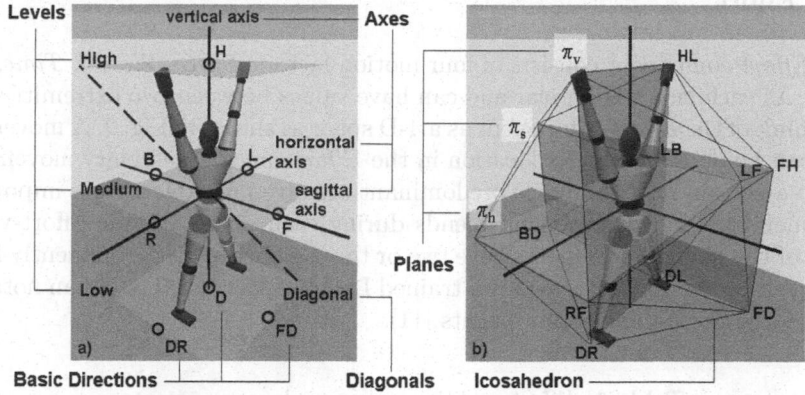

Fig. 3. The concepts of a) Levels of Space, Basic Directions, Three Axes, and b) Three Planes and Icosahedron

2.1 Space

The *Space* component addresses what form is being revealed by the spatial pathways of the movement. The actor is actually "carving shapes in space" [1]. *Space* specifies different entities to express movements in a frame of reference determined by the body of the actor. Thus, all of the presented measures are relative to the anthropometry of the actor. The concepts differ in the complexity of expressiveness and dimensionality but are all of them reproducible in the 3-D Cartesian system. The most important ones shown in Fig. 3 are: 1) The *Levels of Space* - referring to the height of a position, 2) The *Basic Directions* - 26 target points where the movement is aiming at, 3) The *Three Axes* - Vertical, horizontal and sagittal axis, 4) The *Three Planes* - Door Plane π_v, Table plane π_h, and the *Wheel Plane* π_s each one lying in two of the axes, and 5) The *Icosahedron* - used as *Kinespheric Scaffolding*. The *Kinesphere* describes the space of farthest reaches in which the movements take place. Levels and Directions can also be found as symbols in modern-day Labanotation [1]. Labanotation direction symbols encode a position-based concept of space. Recently, Longstaff [14] has translated an earlier concept of Laban which is based on lines of motion rather than points in space into modern-day Labanotation. Longstaff coined the expression *Vector Symbols* to emphasize that they are not attached to a certain point in space. The 38 *Vector Symbols* are organized according to *Prototypes* and *Deflections*. The 14 *Prototypes* divide the Cartesian coordinate system into movements along only one dimension (*Pure Dimensional Movements*) and movements along lines that are equally stressed in all three dimensions (*Pure Diagonal Movements*) as shown in Fig. 3 a). Longstaff suggests that the *Prototypes* give idealized concepts for labeling and remembering spatial orientations. The *Vector Symbols* are reminiscent of a popular concept from neuroscience, named *preferred directions*, which are the directions that trigger the strongest response from motion encoding cells in visual area MT of a monkey [15].

2.2   Effort

The *Effort* component consists of four motion factors: *Space*, *Weight*, *Time*, and *Flow*. As each factor is bipolar and can have values between two extremities one can think of the *Effort* component as a 4-D space as shown in Fig. 2. A movement (M) can be described by its location in the *Effort*-space. Exemplary movements where a certain *Effort*-value is predominant are given in table 1. It is important to remember, that a movement blends during each phase all four Effort-value. Most of the human movements have two or three *Effort*-values prominently high. In fact, it seems difficult even for a trained Laban performer (i.e. Laban notator) to perform single-quality movements [11].

Table 1. *Effort* qualities and exemplary movements

Effort	Movement
Space Direct	Pointing gesture
- Indirect	Waving away bugs
Weight Strong	Punching,
- Light	Dabbing paint on a canvas
Time Sudden	Swatting a fly
- Sustained	Stretching to yawn
Flow Bound	Moving in slow motion
- Free	Waving wildly

3   Tracking of Human Movements

For the tracking of human movements we use sensory data from a camera, which is mounted on our social robot, Nicole and a magnetic tracker as shown in Fig. 4. From the camera we collect 2-D position data of the hands and head with 15Hz. The magnetic tracker produces 3-D position and orientation data with 50Hz for each sensor. The number of sensors and their location depends on the performed action (e.g three sensors on hands and head for gestures). We have created a database of human movements, called HID-Human Interaction Da ase which is publicly accessible through the internet [16]. HID is organized in three categories of movements: 1. Gestures (e.g waving bye-bye), 2. Expressive movements in terms of LMA as presented in Table 1 (e.g. performing a punch) and 3. Manipulatory movements performing reaching, grasping and placing of objects (e.g. drinking from a cup). Figure 4 indicates some of the frames of references involved: The camera referential C in which the image is defined, the inertial referential I allowing us to register the image data in the vertical and the robot referential R which defines the position and orientation of the visual system relative to some world coordinate system W. In the current situation the frame of reference of the world W coincides with the on of the magnetic tracker M and the one we contribute to the human H.

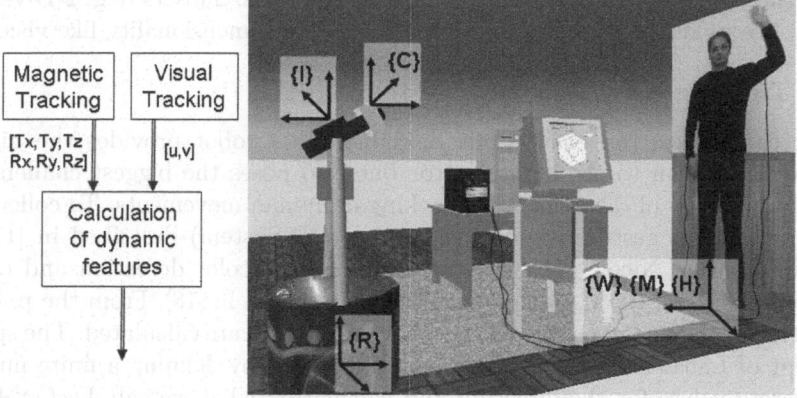

Fig. 4. The components and the frames of reference for tracking human movements

3.1 Tracking Using 6-DoF Magnetic Tracker

Using a 6-DoF magnetic tracker provides 3-D position data with a sufficiently high accuracy and speed. We use a Polhemus Liberty system with sensors attached to several body parts and objects. From the tracker data set of features is calculated and related to the Laban Movement Parameters (LMP). Figure 5 a) shows some sample images from the expressive movement "Stretching to yawn" and in Fig. 5 b) the trajectories for both hands. The tracker data is used to

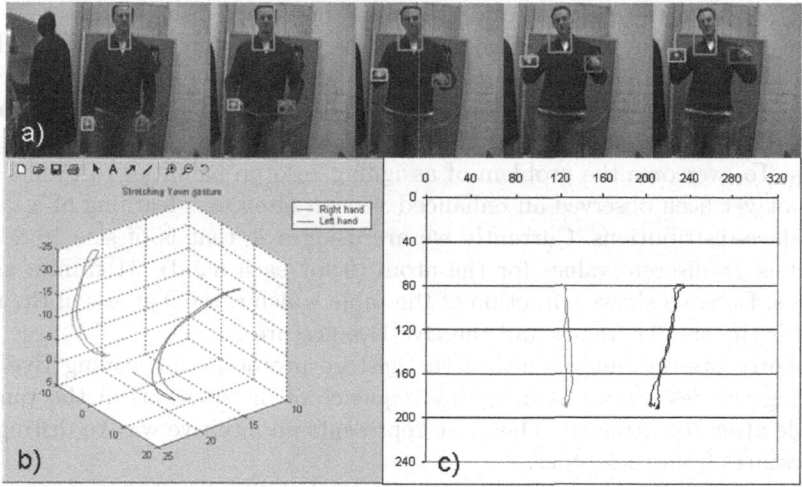

Fig. 5. Tracking of hands movement. a) Sample images b) Data from the magnetic tracker c) ... and the vision tracker.

learn the dependencies of the features from the LMPs. Subsets (e.g. 2-D vertical plane) are used to test the expressiveness in lower dimensionality like vision.

3.2 Tracking Using Vision

Using cameras as the basic input modality for a robot provides the highest degree of freedom to the human actor but also poses the biggest challenge to the functionality of detecting and tracking of human movements. To collect the data we use the gesture perception system (GP-System) described in [17] for our social robot Nicole. The system performs skin-color detection and object tracking based on the CAMshift algorithm presented in [18]. From the position data the displacement vectors dP between each frame are calculated. The spatial concept of Laban's *Vector Symbols* is implemented by defining a finite number of discrete values for the direction and calculating what we call *Vector Atoms* or simply *Atoms A*.

4 Bayesian Framework for Movement Classification

The classification of human movements is done with a probabilistic model using a Bayesian framework. The Bayesian framework offers a broad range of possible models (HMMs etc.) and has proven successful in building computational theories for perception and sensorimotor control of the human brain [19]. These models have already shown their usability in gesture recognition [20, 7].

The model for Laban *Space* uses as input (evidences) the *Atoms A*. Our solution assumes that the probability distribution for all possible values of atom A given all possible gestures G and frames I ,which is $\mathbf{P}(A|G, I)$ can be determined. As both, the gestures and the frame index are discrete values we can express $P(A|G, I)$ in form of a conditional probability table. The probabilities can be learned from training data using a certain number of atom-sequences for each gesture. A simple approach is the one known as Histogram-learning. It counts the number of different atom-values that appear for a gestures along the frames. To overcome the problem of assigning zero probabilities to events that have not yet been observed an enhanced version often uses learning of a family of Laplace-distributions. Currently we are using a le that is of size 18 x 31 x 6, that is 18 discrete values for the atom (9 for each hand), 31 frames and 6 gestures. Figure 6 shows a fraction of the table which is the 9 atoms of the right hand for the first 11 frames and the Bye-Bye gesture.

It represents the 'fingerprint' of the gesture prototype for waving Bye-Bye. Knowing the gesture we assume this sequence of distributions of the random variable atom to extracted. The table represents an intuitive way to distinguish two gestures from each other.

Applying Bayes rule we can compute the probability distribution for the gestures G given the frame I and the atom A expressed as $\mathbf{P}(G|I, A)$, which is the question the classification as based upon. $\mathbf{P}(G)$ represents the prior probabilities for the gestures. Assuming the the observed atoms are independently and identically distributed (i.i.d.) we can compute the probability that a certain gesture

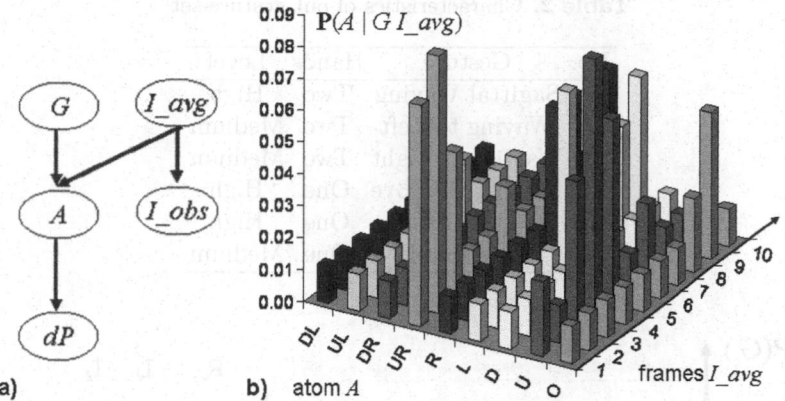

Fig. 6. a) Bayesian Net for the gesture model b) Learned Table $P(A|GI\_avg)$ for gesture 'Bye-Bye'

has caused the whole sequence of atoms $P(a_{1:n}|g, i_{1:n})$ by the product of the probabilities for each frame. Where $a_{1:n}$ represents the sequence of n observed values for atom and g a certain gesture from all gestures G. The jth frame of a sequence of n frames is represented by i_j. We are able to express the probability of a gesture g that might have caused the observed sequence of atoms $a_{1:n}$ in a recursive way. Assuming that each frame a new observed atom arrives we can state and expressing the real-time behavior by using the index t. We model the variance and mean speed of a performance by a Gaussian distribution $N(i\_obs, \sigma)$ expressed the probability that an observed frame i_obs maps to an average frame $iavg P(i_obs|i_avg)$.

Our Bayesian model is shown in (1). We see that the probability distribution of the gestures G at time $t + 1$ knowing the observed atoms a until $t + 1$ is equal to the probability distribution of G at time t times the probabilities of the current observed atom given the gestures G and frame i at $t+1$. The probability distribution of G for $t = 0$ is the prior.

$$\mathbf{P}(G_{t+1}|i_{1:t+1}, a_{1:t+1})$$
$$= \mathbf{P}(G_t)P(i\_obs_{t+1}|i\_avg_{t+1})\mathbf{P}(a_{t+1}|G, i_{t+1}) \tag{1}$$

We can likewise express our model in a *Bayesian Net* shown in Fig. 6. It shows the dependencies of the above mentioned variables including the displacement dP from the previous section. The rule for classification is based on the highest probability value above a minimum threshold, also known as maximum a posteriori estimation (MAP).

5   Results and Discussion

For this experiment we have used 15 video sequences from each human actor for each of 6 distinct gestures as shown in table 2. Figure 7 illustrates how

Table 2. Characteristics of out gesture-set

No.	Gesture	Hands	Level
1	Sagittal Waving	Two	High
2	Waving to Left	Two	Medium
3	Waving to Right	Two	Medium
4	Waving Bye-Bye	One	High
5	Pointing	One	High
6	Draw Circle	One	Medium

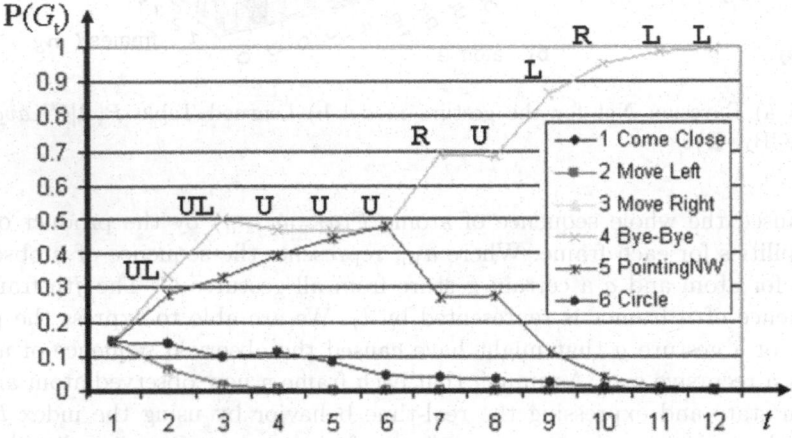

Fig. 7. Probability evolution for a Bye-Bye gesture input

the gesture-hypothesizes, evolve as new evidences (atoms) arrive taken from the performance of a Bye-Bye gesture. After twelve frames the probabilities have converged to the correct gesture-hypothesis (No. 4). After four frames the probabilities of the two-hand gesture-hypothesis have reached nearly zero. (No. 1, 2, and 3). Until the sixth frame the probabilities of both *High-Level* gestures grow (No. 4 and 5) indicating what is called pre-stroke phase in gesture analysis [21]. Conversely the probability of the *Medium-Level* gesture (No. 6) drops slowly towards zero. After the sixth frame the oscillating left-right movement (and its associated atoms) makes the probability of the Bye-Bye-gesture hypothesis rise and the Pointing-NW-gesture hypothesis drop. A similar behavior was revealed when the remaining five gestures were performed. An unknown gesture, i.e. an unknown sequence of atoms produced more than one gesture-hypothesizes with a significant probability.

For the Bye-Bye gesture (see Fig. 6) we can see, that during the first frames the most likely atom to be expected is the one that goes Up-Right (UR). This is similar for the Pointing gesture (see fig. 8) reflecting the already mentioned *Pre-Stroke* phase. The number of atoms during *Pre-Stroke* also reflect the *Levels of Space* in which the following *Stroke* [21] will take place. In our example we can

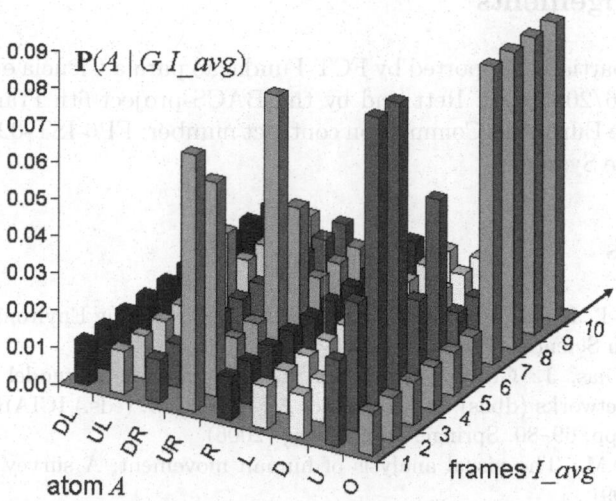

Fig. 8. Learned Table $P(A|GI\_avg)$ for gesture 'Pointing NW'

distinguish the two gestures during *Stroke* as the Bye-Bye gesture has a roughly equal distribution along the line of oscillation (e.g. left-right), while the Pointing gesture produces mainly zero-motion atoms (O).

6 Conclusions and Future Works

This work presented the application of the *Space* component of Laban Movement Analysis (LMA) to the Human-Robot Interface of the social robot, Nicole. It showed that trajectories of human movements can be learned and recognized using the concept of *Vector Symbols*. This work demonstrates that the *Bayesian approach for movement classification* provides a robust and reliable way to classify gestures in real-time. Using naive Bayesian classification we are able to anticipate a gesture from its beginning and can take decisions long before the performance has ended. We have shown that through *Bayesian Learning* the system memorizes learned data in an intuitive way which gives the possibility to draw conclusions directly from the look-up tables. In several trials the system was successfully performing Human-Robot Interaction with guests and visitors.

We are currently implementing the Bayesian models for the *Effort* and *Shape* component of the LMA. With a growing database (HID) we can evaluated classification and anticipation of expressive movements. Once evaluated, we want to put our attention to manipulatory movements and the use of LMA as a cue to describe objects properties. A parallel path follows the goal to improve visual tracking by high level knowledge derived from the LMA *Space* component. For the application we aim at shifting the scope of Nicole towards socially assistive robots that can be used in rehabilitation.

Acknowledgements

This work is partially supported by FCT-Fundação para a Ciência e a Tecnologia Grant #12956/2003 to J. Rett and by the BACS-project-6th Framework Programme of the European Commission contract number: FP6-IST-027140, Action line: Cognitive Systems.

References

1. Bartenieff, I., Lewis, D.: Body Movement: Coping with the Environment. Gordon and Breach Science, New York (1980)
2. Rett, J., Dias, J.: Gesture recognition using a marionette model and dynamic bayesian networks (dbns). In: Campilho, A., Kamel, M. (eds.) ICIAR 2006. LNCS, vol. 4141, pp. 69–80. Springer, Heidelberg (2006)
3. Gavrila, D.M.: The visual analysis of human movement: A survey. CVIU 73(1), 82–98 (1999)
4. Aggarwal, J.K., Cai, Q.: Human motion analysis: A review. CVIU 73(3), 428–440 (1999)
5. Pentland, A.: Looking at people: Sensing for ubiquitous and wearable computing. IEEE Transactions on PAMI 22(1), 107–119 (2000)
6. Moeslund, T.B., Granum, E.: A survey of computer vision-based human motion capture. CVIU 81(3), 231–268 (2001)
7. Pavlovic, V.I.: Dynamic Bayesian Networks for Information Fusion with Applications to Human-Computer Interfaces. PhD thesis, Graduate College of the University of Illinois (1999)
8. Pavlovic, V., Sharma, R., Huang, T.S.: Visual interpretation of hand gestures for human-computer interaction: A review. IEEE Transactions on Pattern Analysis and Machine Intelligence 19(7), 677–695 (1997)
9. Moeslund, T.B., Norgard, L.: A brief overview of hand gestures used in wearable human computer interfaces. Technical report, Computer Vision and Media Technology Lab, Aalborg University, DK (2003)
10. Moeslund, T., Hilton, A., Kruger, V.: A survey of advances in vision-based human motion capture and analysis. CVIU 103(2-3), 90–126 (2006)
11. Zhao, L.: Synthesis and Acquisition of Laban Movement Analysis Qualitative Parameters for Communicative Gestures. PhD thesis, University of Pennsylvania (2002)
12. Zhao, L., Badler, N.I.: Acquiring and validating motion qualities from live limb gestures. Graphical Models 67(1), 1–16 (2005)
13. Foroud, A., Whishaw, I.Q.: Changes in the kinematic structure and non-kinematic features of movements during skilled reaching after stroke: A laban movement analysis in two case studies. Journal of Neuroscience Methods 158, 137–149 (2006)
14. Longstaff, J.S.: Translating vector symbols from laban's (1926) choreographie. In: 26. Biennial Conference of the International Council of Kinetography Laban, ICKL, Ohio, USA, pp. 70–86 (2001)
15. Pouget, A., Dayan, R., Zemel, R.: Informa- information processing with population codes. Nature Reviews Neuroscience 1, 125–132 (2000)
16. Rett, J., Neves, A., Dias, J.: Hid-human interaction database (2007), http://paloma.isr.uc.pt/hid/

17. Rett, J., Dias, J.: Visual based human motion analysis: Mapping gestures using a puppet model. In: Bento, C., Cardoso, A., Dias, G. (eds.) EPIA 2005. LNCS (LNAI), vol. 3808, Springer, Heidelberg (2005)
18. Bradski, G.R.: Computer vision face tracking for use in a perceptual user interface. Intel Technology Journal (Q2), 15 (1998)
19. Knill, D.C., Pouget, A.: The bayesian brain: the role of uncertainty in neural coding and computation. TRENDS in Neurosciences 27, 712–719 (2004)
20. Starner, T.: Visual recognition of american sign language using hidden markov models. Master's thesis, MIT (February 1995)
21. Rossini, N.: The analysis of gesture: Establishing a set of parameters. In: Camurri, A., Volpe, G. (eds.) GW 2003. LNCS (LNAI), vol. 2915, pp. 124–131. Springer, Heidelberg (2004)

Understanding Dynamic Agent's Reasoning

Nuno Lau[1], Luís Paulo Reis[2], and João Certo[1,2]

[1] IEETA – Informatics Electronics and Telecommunications Department,
University of Aveiro, Portugal
[2] LIACC – Artificial Intelligence and Computer Science Lab., University of Porto, Portugal
`nunolau@ua.pt, lpreis@fe.up.pt, joao.certo@fe.up.pt`
`http://www.ieeta.pt/robocup`

Abstract. Heterogeneous agents that execute in dynamic, uncertain, partially cooperative, partially adversely environments have to take their decisions rapidly with an incomplete knowledge of actual environment conditions. This paper discusses different level of abstractions in agent's development for this type of domain, explains the principles of offline debugging and employs these principles in robotic agent's teams through the use of a new visual debugging tool. We argue that during the development of such complex agents, understanding agent reasoning is of crucial importance to the developer and that such understanding can only be done, in most of the cases, using an offline analysis of the agent's decisions. In order for the developer to rapidly perceive agent's reasoning we advocate visual debugging of the agent knowledge and reasoning at several levels of abstraction and at several different functional views. These principles have been applied with success in the context of a robotic soccer 2D simulation league team through the development of tools and extensive use of these analysis principles.

1 Introduction

Agents are software computational systems that have the following characteristics: autonomy, reactivity, pro-activity and social ability [1]. Although agents take their decisions autonomously, the developer should have some control over which type of actions are taken by the agent, motivating the use of "good" actions and deprecating the use of "bad" actions. Agents can execute in several different kinds of environments. In this paper we are interested in environments that are uncertain, dynamic, real-time, heterogeneous, partially cooperative, partially adversely like the soccerserver simulator of the RoboCup Simulation League [2]. Uncertainty is considered because the agent perceptions cannot give him a perfect and complete view of the actual environment. The environment is dynamic because the agent cannot assume that environment characteristics don't change while he is reasoning [3]. The real-time aspect comes from the obligation of sending an action with a predefined frequency external to the agent's control. Soccerserver environment is obviously partial cooperative and partially adversely for the agent is part of a multi agent system, together with 10 teammates, that constitutes his soccer team, all of which share the same goal, and he has to play against other muti-agent system with opposite goals. Heterogeneity is present in the agent's different physical capabilities

J. Neves, M. Santos, and J. Machado (Eds.): EPIA 2007, LNAI 4874, pp. 542–551, 2007.

so, each agent must take advantage of its strong points considering every opponent's physical attributes.

This type of agents take a large amount of decisions in very short time and understanding their decisions at the same time they are executing is a very complex task, because the developer doesn't have the necessary time to follow the agent reasoning. Some attempts to do agent's online debugging have been reported [7] but they imply changing the simulator and cannot guarantee that the behavior of the opponent team is not affected. Uncertain environments introduce a new level of complexity in the agent understanding process for the visualization normally provided to the developer may be substantially different than the knowledge in which the agent based its decisions. The difference between agents' knowledge and real environment should be minimized for it reinforces agent cooperation with its teammates. However, the minimization of the error can only be done by intelligent use of agent sensors [4] and that means taking "good" decisions that should be subject to developer debugging and understanding, reinforcing the need for good debugging practices and tools. Having these considerations in mind we believe that offline analysis of agent's decisions, meaning examining agent reasoning after execution and without real-time pressure, is the better way of making good and in-depth analysis of agent's reasoning.

Our second point related to debugging of agent's decisions is that every feature that may be expressed in a graphical way should be presented to the developer graphically. The time taken to perceive some item of information graphically is much less than the time to perceive exactly the same information reading a text based report. Graphical debugging of agents should show the superimposed information of real conditions, agent knowledge and agent reasoning. Text based reports should not be excluded but their analysis should be complementary and not the main developer's focus of attention.

Finally we will show that it is possible to define independent views of the agent's knowledge that enable the developer to focus on the relevant information at each debugging cycle. For example, it is not so important to see communication related information if the developer is tuning the ball possession behaviour of the team. The information shown to the developer at all debugging stages should be relevant and informative for too much information makes the analysis much harder.

In order to validate this approach we have developed a complete offline debugging tool that enables multi-level functional debugging of agents' teams performing in complex environments. Our debug tools allows visual, feature-based, offline debugging of agent's decisions using several layers of abstraction. The debugger was successfully applied in our FC Portugal RoboCup Simulation 2D and Simulation 3D teams that won more than twenty awards in RoboCup International Competitions.

This paper is organized as follows. After this introduction, the next section explains the multi-level world-state representation of the FC Portugal agent physical model. Section 3 discusses the principles of offline debugging. Section 4 presents the implementation of debugging tools that follow the principles explained before. Section 5 gives the development flow of the FC Portugal agent, focusing on the utilization of the debugging development tools. Finally section 6 concludes this paper.

2 Multi-lever World State

As stated in the introduction several layers of abstraction should be presented to the developer for debugging. Low-level layers give the developer detailed information on current situation and reasoning process, while high-level layers focus on high-level decisions of agents [11]. Maintaining internally a multi-level information structure is an important asset for the structuring of the debugging information in different abstraction levels. In FC Portugal agents we use a multi-level structure with data at four levels of abstraction:

• **Global Situation Information** – High-level information like result, time, game statistics (shoots, successful passes, etc.) and opponent behavior, used to decide the team tactic at a given moment [6];
• **Situation Information** – Information relevant to the selection of the appropriate formation and for the Strategic Positioning, Dynamic Role Exchange, Intelligent Vision and Advanced Communication mechanisms [5];
• **Action Selection Information** – Set of high-level parameters used to identify active situations and to select appropriate ball possession or ball recovery behaviors;
• **World State** – Low-level information, including positions and velocities of players and ball, player's orientations, looking directions and stamina.

3 Following Agent's Reasoning

As stated in the introduction our debugging methodology is based on the following principles:

• Offline debugging;
• Visual debugging;
• Superimposed real environment and agent physical knowledge;
• Feature-focused debugging;
• Information structured in layers of abstraction with different detail levels.

Offline debugging is needed not has an essential feature of our debugging principles but because the time necessary for in-depth analysis of agents reasoning is not available while the agent is executing, in the environments we are interested in, at least not without affecting the result of execution.

Visual debugging allows a much faster interpretation of agent's knowledge and reasoning and should be used whenever possible. The superimposition of real environment and agent beliefs about environment is a very important informational tool to understand agents that operate in uncertain environments. Not all features of agents are object of improvement at the same time, the selection of relevant information to provide to the developer is very important for too much (not so relevant) information is harder and more tiring to interpret that concise and relevant information. The level of detail should be selectable, just like in layered disclosure [8, 10], because an in depth analysis is not always necessary or advised.

In the context of the development of FC Portugal agents we have identified the following features that should be object of independent selection by the developer during a debugging session:

- Communication;
- Ball Possession;
- Ball Recovery;
- Synchronization;
- Low-level skills;
- Opponent Modeling.

The selection of each of these features, although independent, is not exclusive. The developer can activate one or more of the above listed features at each moment providing him with maximum flexibility in the selection of the information he finds relevant.

4 Implementation of Debugging Tools

4.1 Visual Debugger

The main debugging tool of FC Portugal is called Visual Debugger (Figure 1). Its implementation is based on CMUnited99 layered disclosure tool [8, 10] and the soccerserver logplayer application. CMUnited layered disclosure tool included the possibility of synchronous visualization of the game (using soccermonitor) and of one of the players reasoning (at several levels of abstraction) saved in action logfiles. We have integrated the two applications (logplayer with layered disclosure and soccermonitor) in a powerful team debugging tool and added the visual debugging capabilities and real and believed world-states superposition.

Visual information is much more easily handled than text information. Soccermonitor [2] includes the possibility of drawing points, lines and circles over the field, but this functionality is not reported as being used by other teams. The possibility of drawing over the field provided by soccermonitor was exploited, modified and extended (possibility of drawing text over the field).

Soccermonitor capabilities were limited relating the analysis of a particular situation so an amplification mode was developed (Fig. 2). In this mode, the area around the ball is magnified with a desired zoom level and offset to the ball. When in amplification mode the ball is drawn with a vector indicating its direction and velocity. Additionally all players are drawn with orientation lines (showing were the player is facing), a moving vector (showing the direction and velocity of the player), a view zone (that indicates where the player is looking and what he can see) and with a kickable area circle adapted to each player's physical characteristics. Whenever a player is able to kick the ball he is automatically highlighted.

On the other hand, on the team side, we extended the log action files syntax so that they could include draw commands [9]. Draw commands can be inserted at different reasoning levels of abstraction. A programming interface allows the player to save visual logs by combining simple primitives as points, lines, circles, crosses, squares and text over the field.

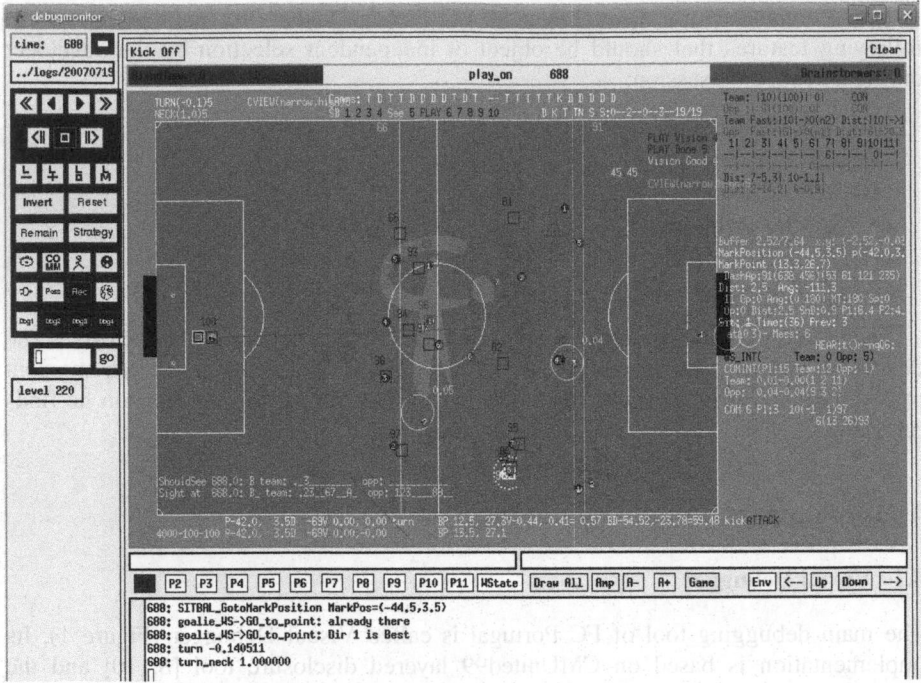

Fig. 1. The Visual Debugger

Pure textual action debug messages are shown in scrollable debug text windows. There is one of these windows for each of the players and one window with the real situation without errors derived from the simulator logfile. Finally two extra textual windows with only logfile information were created: an event board and a world state board. The event board displays the initial team selection of heterogeneous agents, and time stamped major environment events like goals, free kicks, offsides and substitutions. The world state board keeps track of the ball's and all players' information (stamina, recovery rate, neck angle…).

The debugger determines the read world state based on the visualization information saved in server record log files. The real positions and velocities of all objects are calculated and last player commands determined either by direct inspection of log (version 3) or are deduced from a set of simple rules that reason on world state changes (version 2 logs). This internal view of the real world state is shown in a text window. Some features (ball position/velocity, selected player position/velocity and ball distance) are shown directly over the field.

We used the possibility of drawing text over the field quite extensively to show player action mode, synchronization, basic action, lost cycles, evaluation metrics, position confidences and to compare players beliefs on its (and ball) position/velocities with real values. Drawing circles and lines over the field was used to show player beliefs on object positions, player view area, best pass, shoot or forward, communication events, evaluation parameters of each type of action, etc (Figure 3).

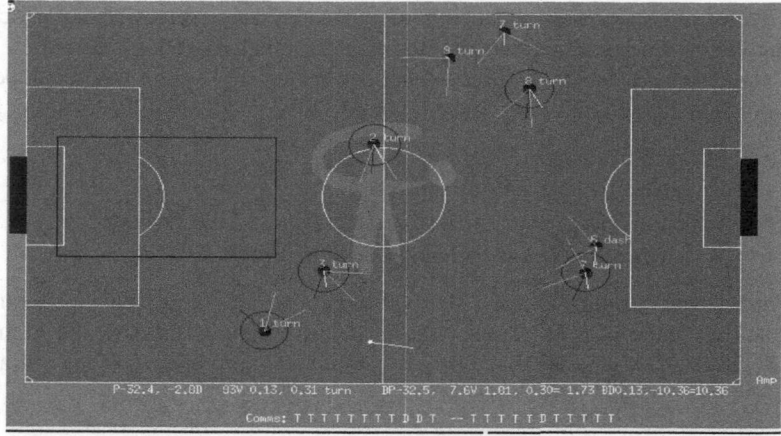

Fig. 2. Amplification mode of the Visual Debugger

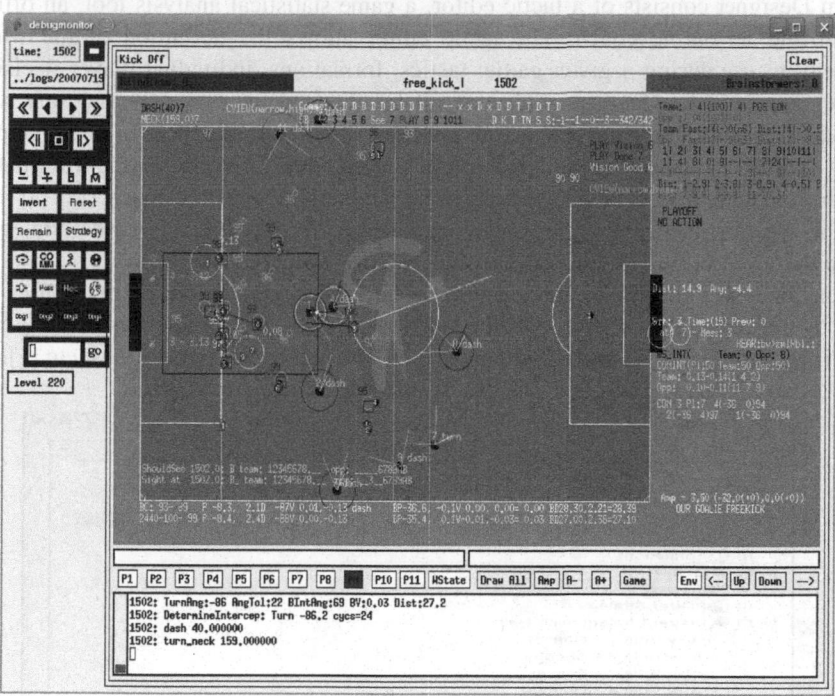

Fig. 3. Debugging for Player 9

Using Figure 3 as an example where player 9's point of view is shown, one can see both normal views and amplified views overlapped. This allows the user to directly follow the immediate action (amplified on the ball) and at the same time accompany the global player's view. For instance small colored balls are other players' positions and the near squares represent the position where player 9 thinks they are with a certain degree of confidence (attached number). At the same time with amplification, two large yellow circles are shown, highlighting players that can kick the ball.

This tool can be seen as a 4 dimensional debugging tool where the developer can rapidly position himself and receive the required visual and textual information:

- For each cycle time - dimension 1 -;
- In a given field space - dimension 2 -;
- With a given area focus (eg. Communication) - dimension 3 -;
- With a specified information depth – dimension 4 -.

We have used this tool to tune strategic positions, test new behaviors, tune the importance and precision of each of the evaluation metrics used in the high-level decision module, test world update, communication and looking strategies empirically, analyze previous games of other teams, etc.

4.2 Team Designer

Team Designer consists of a tactic editor, a game statistical analysis tool, an offline coach and an online coach. The tactic editor allows the definition of the whole strategy to use during a given game: tactics, formations, individual player decision, strategic positioning features, etc. may all be changed in a friendly and safe way. Some of the tactic parameters are defined by direct manipulation of their graphic view. This is the case for the definition of player home positions inside a formation and for the definition of new situations using the integrated offline coach.

The creation of a new strategy may use features from previously saved strategies, through the selection of which items are interesting to merge (Fig. 4).

The analysis tools gather game statistical information that is shown to the user and sent to the online coach (Fig. 5). The statistics include ball position in several field matrixes, shoots and passes by field areas, ball loss and recovery positions, etc [1].

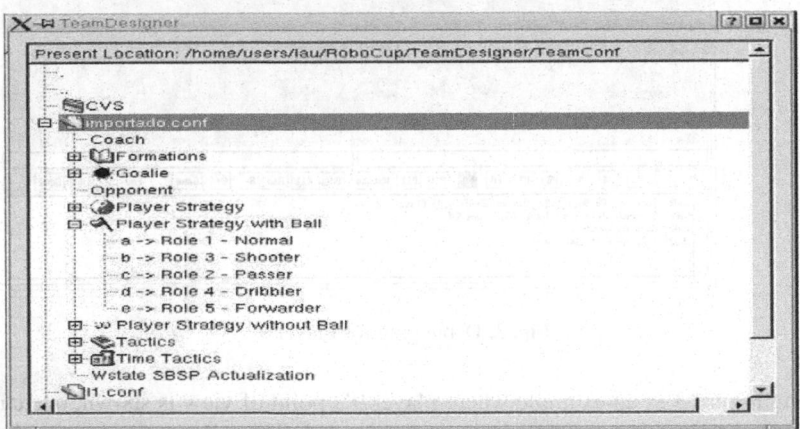

Fig. 4. Importing tactic features

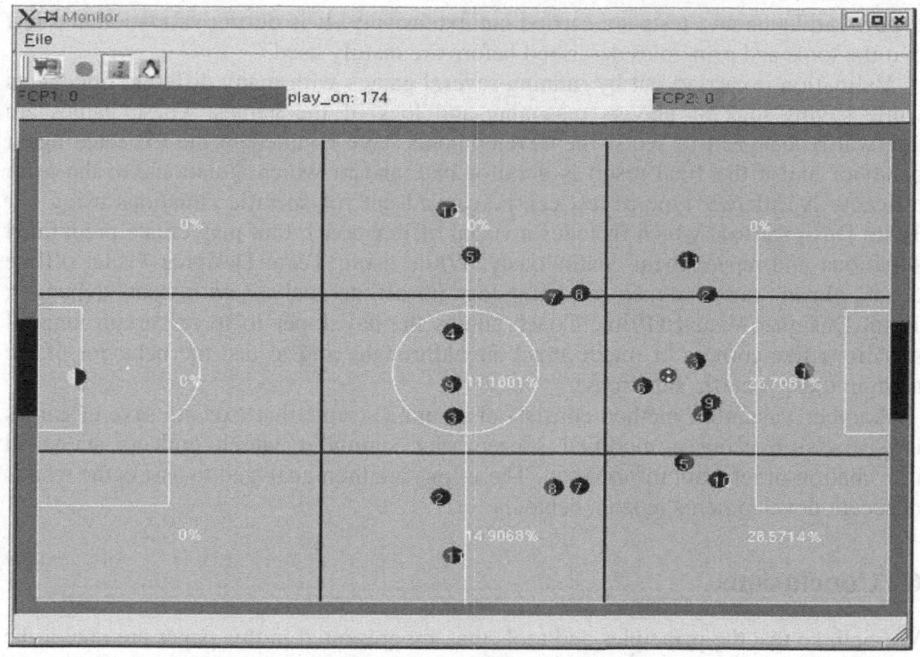

Fig. 5. Visualizing ball position in several areas

4.3 Offline Client

In some situations the action log files created by our players were not enough to understand what was really happening in a certain situation. A much finer debugging degree can be achieved by employing our offline client tool.

The principle of the offline client is that we can repeat the execution of a player over exactly the same setting of a previous game without the intervention of the server and without real-time constraints. Then we can use a normal debugger (like gdb) to examine the contents of all variables, set breakpoints, etc at the execution situations we want to analyze. A special log file, that records every interaction with the server and the occurrence of timer signals, must be generated to use the offline client. If an agent has probabilistic behavior some more information might be needed (the only probabilistic behavior of our agents is the optimum kick).

The offline execution of an agent is achieved through a stub routine that reads previously saved server messages from the log file instead of reading network messages. Player's behavior is maintained, as it is not affected by the substitution of this stub routine. The execution of a normal debugger over this offline client allows a very fine tracking of what happened.

5 Development Flow of FC Portugal Agent

The development of FC Portugal agent is a recursive process in which after the development of improvements or new features, issues that are not the topic of this paper, validation and tests are carried out extensively. It is during validation and test that the tools and principles described before are mainly used.

Validation is carried out by running several games with many different opponents while saving logs of players reasoning and logs of the games. These games are afterwards analyzed to see if the developments have resulted in the expected agent behavior and if the final result is good or bad, and in which situations, to the team efficacy. A different type of test can be carried out for specific situations using our Team Designer tool which includes a visual offline coach that may create predefined situations and repeat them continuously. While using Team Designer visual offline coach, players can save their reasoning logs for offline analysis with visual debugger. Team Designer Visual Offline Coach allows the developer to force certain unusual situations like corners or major attack breakthroughs and to test the behavior of the team in these specific situations.

Another validation method consists of running scripts that execute several games in succession using a modified soccerserver simulator which collects statistical information of relevant information. These logs are then analyzed to assess the results of recent developments in team behavior.

6 Conclusions

We believe that the principles and tools that are presented in this paper are one of the main assets of the FC Portugal RoboCup Simulation League team. This team has been one of the top teams over the last seven years wining five European and three World Championships in several soccer simulation leagues.

As the visual debugger is based on the official and supported logplayer and monitor, its continuous update in face of new simulator versions is relatively simple. The implemented visual debugger can be very useful for any team as some developed features only use the simulation logfiles. Even without a correspondent debugging on agent side these features could be sufficient to identify flaws. The four dimensional visual debugger enables a quick access with a selected content to a specific part of the game resulting in faster bug finding.

This paper presented the guideline principles of FC Portugal debug methodology. These principles have been integrated in several debugging tools and through it some of our team's flaws were detected. Detection of these flaws in positioning, passing, dribbling and shooting algorithms enabled a fine tuning of FC Portugal agents' behavior (feed backed into the visual debugger) resulting in an increased efficiency.

The continuous introduction of novel features in the team and the improvement of older features have always fitted nicely with our debugging principles and tools. The use of debugging tools has been essential in team development as it has enabled the fast identification and correction of some unwanted emergent behaviors. The utilization of these tools in the development flow of our agents is one of the aspects that have kept our team as one of the bests throughout the years in RoboCup.

Acknowledgements

This work was partially supported by project FC Portugal Rescue: Coordination of Heterogeneous Teams in Search and Rescue Scenarios - FCT/POSI/EIA/63240/2004.

Our thanks goes also to Peter Stone, Patrick Riley and Manuela Veloso for making available the CMUnited99 low-level source code that saved us a huge amount of time at the beginning of the project and to Johnny Santos, Claudio Teixeira and Rui Ferreira that contributed in the development of our debugging tools.

References

1. Badjonski, M., Schroeter, K., Wendler, J., Burkhard, H.-D.: Learning of Kick in Artificial Soccer. In: Proceedings of Fourth Int. Workshop on RoboCup, Melbourne (August 2000)
2. Chen, M., Foroughi, E., Heintz, F., Huang, Z., Kapetanakis, S., Kostiadis, K., Kummeneje, J., Noda, I., Obst, O., Riley, P., Steffens, T., Wang, Y., Yin, X.: RoboCup Soccer Server Manual (2002), http://downloads.sourceforge.net/sserver/manual.pdf
3. Kitano, H.: Robocup: The Robot World Cup Initiative. In: Agents 1997. Proceedings of 1st International Conference on Autonomous Agent, Marina del Ray, ACM Press, New York (1997)
4. Kostiadis, K., Hu, H.: A Multi-threaded Approach to Simulated Soccer Agents for the RoboCup Competition. In: Veloso, M.M., Pagello, E., Kitano, H. (eds.) RoboCup 1999. LNCS (LNAI), vol. 1856, pp. 366–377. Springer, Heidelberg (2000)
5. Reis, L.P., Lau, N., Oliveira, E.C.: Situation Based Strategic Positioning for Coordinating a Team of Homogeneous Agents. In: Hannebauer, M., Wendler, J., Pagello, E. (eds.) Balancing Reactivity and Social Deliberation in Multi-Agent Systems. LNCS (LNAI), vol. 2103, Springer, Heidelberg (2001)
6. Reis, L.P., Lau, N.: FC Portugal Team Description: RoboCup 2000 Simulation League Champion. In: Stone, P., Balch, T., Kraetzschmar, G.K. (eds.) RoboCup 2000. LNCS (LNAI), vol. 2019, pp. 29–40. Springer, Heidelberg (2001)
7. Riedmiller, M., Merke, A., Meier, D., Hoffman, A., Sinner, A., Thate, O., Ehrmann, R.: Karlsruhe Brainstormers - A Reinforcement Learning Approach to Robotic Soccer. Lecture Notes In Computer Science, pp. 367–372. Springer, Heidelberg (2001)
8. Stone, P.: Layered Learning in Multiagent Systems: A Winning Approach to Robotic Soccer. MIT Press, Cambridge (2000)
9. Stone, P., Riley, P., Veloso, M.: CMUnited-99 source code, Date Accessed: July 2007 (1999), accessible from http://www.cs.cmu.edu/ pstone/RoboCup/CMUnited99-sim.html
10. Stone, P., Riley, P., Veloso, M.: Layered Disclosure: Why is the agent doing what it's doing? In: Agents 2000, Fourth Int. Conf. on Autonomous Agents. Barcelona (June 2000)
11. Stone, P., Veloso, M.: Task Decomposition, Dynamic Role Assignment, and Low-Bandwidth Communication for Real-Time Strategic Teamwork. Artificial Intelligence 110(2), 241–273 (1999)

Chapter 9 - Fourth Workshop on Multi-agent Systems: Theory and Applications (MASTA 2007)

Convergence of Independent Adaptive Learners*

Francisco S. Melo and Manuel C. Lopes

Institute for Systems and Robotics,
Instituto Superior Técnico,
Lisboa, Portugal
{fmelo,macl}@isr.ist.utl.pt

Abstract. In this paper we analyze the convergence of independent adaptive learners in repeated games. We show that, in this class of games, independent adaptive learners converge to pure Nash equilibria in self play, if they exist, and to a best response strategy against stationary opponents. We discuss the relation between our result and convergence results of adaptive play [1]. The importance of our result stems from the fact that, unlike adaptive play, no communication/action observability is assumed. We also relate this result to recent results on the convergence of weakened ficticious play processes for independent learners [2,3]. Finally we present experimental results illustrating the main ideas of the paper.

1 Introduction

Game theory is traditionally used in economics, where it provides powerful models to describe interactions of economical agents. However, recent years have witnessed an increasing interest from the computer science and robotic communities in applying game theoretic models to multi-agent systems. For example, the interaction of a group of robots moving in a common environment can be naturally captured using a game theoretic model and their observed behavior suitably interpreted using game theoretic concepts.

When addressing game theory from a learning perspective, Boutilier [4] distinguishes two fundamental classes of learning agents: *independent learners* (IL) and *joint-action learners* (JAL). The former have no knowledge on the other agents, interacting with the environment as if no other decision-makers existed. In particular, they are unable to observe the rewards and actions of the other agents. Joint action leaners, on the contrary, are aware of the existence of other agents and are capable of perceiving (*a posteriori*) their actions and rewards.

Learning algorithms considering JALs are easily implementable from standard single-agent reinforcement learning algorithms [5]. Action observability allows a learning agent to build statistics on the other agents' behavior-rules and act in a best-response sense. This is the underlying principle of standard methods such as fictitious play [6] or adaptive play [1]. Joint action observability is also commonly assumed in several domains studied in the economic literature (*e.g.*, auctions or

---

* Work partially supported by POS_C that includes FEDER funds. The first author acknowledges the PhD grant SFRH/BD/3074/2000.

J. Neves, M. Santos, and J. Machado (Eds.): EPIA 2007, LNAI 4874, pp. 555–567, 2007.

exchanges[1]) and several learning algorithms are available that make use of such assumption [7, 8].

However, in many practical applications it is not reasonable to assume the observability of other agents' actions. Most agents interact with their surroundings by relying on sensory information and action recognition is often far from trivial. With no knowledge on the other agents' actions and payoffs, the problem becomes more difficult. In [9, 4] some empirical evidence is gathered that describes the convergence properties of reinforcement learning methods in multi-agent settings. In [10], the authors study independent learners in deterministic settings. Posterior works [11, 12] address non-deterministic settings. Recent results have established the convergence of a variation of fictitious play for independent learners [3]. In a different approach, Verbeeck et al. [13] propose an independent learning algorithm for repeated games that converges to a *fair periodical policy* that periodically alternates between several Nash equilibria.

In this paper, we propose and analyze the performance of *independent adaptive learning*, a variation of adaptive play for independent learners. This algorithm has an obvious advantage over the original adaptive learning algorithm [1], since it does not require each player to be able to observe the plays by the other agents. Furthermore, no *a priori* knowledge of the payoff function is required. Our results show that a very simple learning approach, requiring no communication or knowledge on the other agents, is still able to exhibit a *convergent* and *rational* behavior, in the sense of [14]. This means that independent adaptive learning is able to attain a Nash equilibrium in self-play and converge to a best-response strategy against stationary opponents. We show that, in weakly acyclic repeated games, independent adaptive learners converge to pure Nash equilibria, if they exist. This convergence is attained in both *beliefs* and *behavior*. We experimentally validate our results in several simple games.

2   Background

In this section we introduce some background material that will be used throughout the paper.

2.1   Strategic and Repeated Games

A strategic game is a tuple $(N, (\mathcal{A}_k), (r_k))$, where N is the number of players, $\mathcal{A}_k$ is the set of *individual actions* of player k, $k = 1, \ldots, N$ and $\mathcal{A} = \times_{k=1}^{N} \mathcal{A}_k$ is the set of *joint actions* for the group. Each function $r^k : \mathcal{A} \to \mathbb{R}$ is a *reward function* or *payoff function*, defining a preference relation on the set $\mathcal{A}$.

We represent an element $a \in \mathcal{A}$ as a N-tuple $a = (a_1, \ldots, a_N)$ and refer it as a *joint action* or *action profile*. The tuple $a_{-k} = (a_1, \ldots, a_{k-1}, a_{k+1}, \ldots, a_N)$ is a *reduced joint action*, and we write $a = (a_{-k}, a_k)$ to denote that the individual action of player k in the joint action a is a_k.

---

[1] Exchanges are also known as double actions.

In strategic games it is not possible to have memory effects in the players. If memory of past plays is possible, we refer to such a game as a *repeated game*. In a repeated game, N players repeatedly engage in a strategic game defined as usual as a tuple $(N, (\mathcal{A}_k), (r_k))$. The repeated interaction allows the players to maintain, for example, statistics describing the strategies of the other players and use these statistics to play accordingly.

A strategic game is *zero-sum* or *strictly competitive* if it has 2 players and $r_1 = -r_2$, and *general-sum* otherwise. A general sum game is *fully cooperative* if $r_1 = \ldots = r_N$.

2.2 Nash Equilibria

A *Nash equilibrium* of a strategic game $(N, (\mathcal{A}_k), (r_k))$ is an action profile $a^* \in \mathcal{A}$ such that, for every player $k = 1, \ldots, N$, $r_k(a^*) \geq r_k(a^*_{-k}, a_k)$, for all $a_k \in \mathcal{A}_k$. In a Nash equilibrium no player benefits from individually deviating its play from a^*. We emphasize that not every strategic game has a Nash equilibrium.

A *strategy* for player k is a probability distribution over the set $\mathcal{A}_k$. A strategy σ_k assigns a probability $\sigma_k(a_k)$ to each action $a_k \in \mathcal{A}_k$. We say that player k follows strategy σ_k when playing the game $(N, (\mathcal{A}_k), (r_k))$ if it chooses each action $a_k \in \mathcal{A}_k$ with probability $\sigma_k(a_k)$. If a strategy σ_k assigns probability 1 to some action $a_k \in \mathcal{A}_k$, then σ_k is a *pure strategy*. Otherwise, it is called a *mixed strategy*. We define the concepts of *joint strategy* or *strategy profile* and *reduced joint strategy* in a similar manner as defined for actions. The *support* of a strategy σ_k is the set of all actions $a_k \in \mathcal{A}_k$ such that $\sigma_k(a_k) > 0$.

A *mixed strategy Nash equilibrium* of a strategic game $(N, (\mathcal{A}_k), (r_k))$ is a strategy profile σ^* such that, for any strategy σ and for every player $k = 1, \ldots, N$,

$$\mathbb{E}_{\sigma^*}[R_k] \geq \mathbb{E}_{(\sigma^*_{-k}, \sigma_k)}[R_k] \tag{1}$$

where $\mathbb{E}_{\sigma^*}[\cdot]$ is the expectation conditioned on the strategy σ^* and R_k is the random variable denoting the outcome of the game for player k. The Nash equilibrium is *strict* if (1) holds with a strict inequality. Every strategic game $(N, (\mathcal{A}_k), (r_k))$ with finite $\mathcal{A}$ has a mixed strategy Nash equilibrium.

2.3 Fictitious Play

Fictitious play is an iterative procedure originally proposed by Brown [6] to determine the solution for a strictly competitive game. This procedure was shown to converge in this class of games in [15] and later extended to other classes of games by several authors (see, for example, [3]).

In its original formulation, two players repeatedly engage in a strictly competitive game. Each player maintains an estimate of the other player's strategy as follows: let $N_t(a)$ denote the number of times that the individual action a was played up to (and including) the tth play. At play t, player k estimates the other player's strategy to be

$$\hat{\sigma}_{-k}(a_{-k}) = \frac{N_t(a_{-k})}{t},$$

for each $a_{-k} \in \mathcal{A}_{-k}$. The expected payoff associated with each individual action of player k is then

$$EP(a_k) = \sum_{a_{-k} \in \mathcal{A}_{-k}} r_k(a_{-k}, a_k) \hat{\sigma}_{-k}(a_{-k}).$$

Player k can now choose its action from the set of best responses,

$$BR = \left\{ a_k \in \mathcal{A}_k \mid a_k = \arg\max_{u_k \in \mathcal{A}_k} EP(u_k) \right\}.$$

Robinson [15] showed that this methodology yields two sequences $\{\hat{\sigma}_1\}_t$ and $\{\hat{\sigma}_2\}_t$ converging respectively to σ_1^* and σ_2^* such that (σ_1^*, σ_2^*) is a Nash equilibrium for the game $(\{1,2\}, (\mathcal{A}_k), (r_k))$.

2.4 Adaptive Play

Adaptive play was first proposed by Young [1] as an alternative method to fictitious play. The basic underlying idea is similar to fictitious play, but the actual method works differently from fictitious play. For games which are *weakly acyclic*, adaptive play converges with probability 1 (w.p.1) to a pure strategy Nash equilibrium, both in *beliefs* and in *behavior*.

Let h be a vector of length m. We refer to any set of K samples randomly drawn from h without replacement as a K-*sample* and denote it generically by $K(h)$, where K and m are any two integers such that $1 \leq K \leq m$

Let $\Gamma = (N, (\mathcal{A}_k), (r_k))$ be a repeated game played at discrete instants of time $t = 1, 2, \ldots$. At each play, each player $k = 1, \ldots, N$ chooses an action $a_k(t) \in \mathcal{A}_k$ as described below, and the action profile $a(t) = (a_1(t), \ldots, a_N(t))$ is referred to as the *play at time* t. The history of plays up to time t is a vector $(a(1), \ldots, a(t))$.

Let K and m be two given integers as described above. At each time instant $t = 1, 2, \ldots$, each player $k = 1, \ldots, N$ chooses its action $a_k(t)$ as follows. For $t \leq m$, $a_k(t)$ is chosen randomly from $\mathcal{A}_k$; for $t \geq m + 1$, player k inspects K plays drawn without replacement from the most recent m plays. We denote by H_t the m most recent plays at time t. Let $N_K(a_{-k})$ be the number of times that the reduced action a_{-k} appears in the K-sample $K(H_t)$. Player k then uses $K(H_t)$ and determines the expected payoff $EP(a_k)$ for each $a_k \in \mathcal{A}_k$ as

$$EP(a_k) = \sum_{a_{-k} \in \mathcal{A}_{-k}} r_k(a^{-k}, a_k) \frac{N_K(a_{-k})}{K}$$

It then randomly chooses its action from the set of best responses,

$$BR = \left\{ a_k \in \mathcal{A}_k \mid a_k = \arg\max_{u_k \in \mathcal{A}_k} EP(u_k) \right\}.$$

Notice that this procedure is similar to fictitious play in that it chooses the best response action to the estimated reduced strategy $\hat{\sigma}_{-k}$. The only difference lies in the fact that adaptive play uses *incomplete history sampling*, while fictitious play uses the complete history.

Young [1] established the convergence of adaptive play for repeated games that are *weakly acyclic*. To properly introduce such result, let $\Gamma = \left(N, (\mathcal{A}_k), (r_k)\right)$ be a strategic game with finite action-space $\mathcal{A} = \times_{k=1}^{N} \mathcal{A}_k$. The *best response graph* for Γ is a directed graph $\mathcal{G} = (V, E)$, where each vertex corresponds to a joint action (*i.e.*, $V = \mathcal{A}$) and any two actions $a, b \in \mathcal{A}$, are connected by a directed edge $(a, b) \in E$ if and only if $a \neq b$ and there is exactly one player k for which b_k is a best-response to the pure strategy a_{-k} and $a_{-k} = b_{-k}$. A strategic game $\Gamma = \left(N, (\mathcal{A}_k), (r_k)\right)$ is *weakly acyclic* if, given any vertex in its best response graph there is a directed path to a vertex from which there is no exiting edge (a sink).

A sink as described in the previous definition corresponds necessarily to a strict Nash equilibrium. Given a weakly acyclic strategic game Γ, we denote by $L(a)$ the shortest path from the vertex a to a strict Nash equilibrium in the best response graph of Γ and by $L(\Gamma) = \max_{a \in \mathcal{A}} L(a)$. Young [1] showed that for any weakly acyclic strategic game, adaptive play converges w.p.1 to a strict Nash equilibrium as long as $K \leq \frac{m}{L(\Gamma)+2}$.

3   Independent Adaptive Leaning

In this section we describe *independent adaptive learning*, a variation of adaptive learning relying on independent learners. This algorithm has an obvious advantage over the original adaptive learning algorithm [1], since it does not require each player to be able to observe the plays by the other agents. Furthermore, no *a priori* knowledge of the payoff function is required.

3.1   Independent Adaptive Learning Process

Let $\Gamma = \left(N, (\mathcal{A}_k), (r_k)\right)$ be a repeated game played at discrete instants of time $t = 1, 2, \ldots$. At each play, each player $k = 1, \ldots, N$ chooses an action $a_k(t) \in \mathcal{A}_k$ and receives a reward $r_k(t)$. We are interested in developing a learning algorithm for independent players, *i.e.*, players that are not able to observe the plays of the others. Therefore, we consider that all plays and rewards referred henceforth concern a particular player k in Γ, except if explicitly stated otherwise. We refer to the pair $(a(t), r(t))$ as the play (of player k) at time t. The history of plays up to time t is a set $\mathcal{H}_t = \{(a(1), r(1)), (a(2), r(2)), \ldots, (a(t), r(t))\}$.

Let K and m be two integers $1 \leq K \leq m$. At each time instant $t = 1, 2, \ldots$, the player chooses its action $a(t)$ as follows. For $t \leq m$, $a(t)$ is chosen randomly from the corresponding action set $\mathcal{A}_k$; for $t \geq m+1$, the player inspects K plays drawn without replacement from its most recent m plays. Suppose, for definiteness, that the selected plays corresponded to times $t_1, \ldots, t_k$. The expected payoff associated with each action $u \in \mathcal{A}_k$ is

$$EP(u) = \frac{\sum_{i=1}^{K} r(t_i) \mathbb{I}_u(a(t_i))}{\sum_{i=1}^{K} \mathbb{I}_u(a(t_i))},$$

where $\mathbb{I}_u(\cdot)$ is the indicator function for the set $\{u\}$ with $u \in \mathcal{A}_k$.[2] Given $EP(u)$ for all $u \in \mathcal{A}_k$, the player now randomly chooses its action from the set

$$BR = \left\{ a \in \mathcal{A}_k \mid a = \arg\max_{u \in \mathcal{A}_k} EP(u) \right\}.$$

If one particular action $u \in \mathcal{A}_k$ is never played in the selected K plays, then the expected payoff should be taken as any sufficiently large *negative number* (we henceforth take it to be $-\infty$).

3.2   Convergence of the Independent Adaptive Learning Process

In this section we establish the convergence of our method by casting it as a variation of adaptive play as described in [1].

The main differences between our algorithm and the standard adaptive play lie on the fact that we do not assume any *knowledge of the payoff function* or any *observability of the actions of the other players*. Instead, we rely on the sampling process to implicitly provide this information.

Before introducing our main result, we need the following definitions.

Definition 1 (Greedy strategy). *An individual strategy σ_k is greedy with respect to (w.r.t.) a payoff function r if it assigns probability 1 to the action $a^* = \arg\max_{a \in \mathcal{A}^k} r(a)$.*

Definition 2 (GLIE strategy [16]). *An individual strategy σ_k is greedy in the limit with infinite exploration (GLIE) if (i) each action is visited infinitely often and (ii) in the limit, the policy is greedy with respect to some payoff function r w.p.1.*

A well-known example of GLIE policy is Boltzmann exploration:

$$\mathbb{P}\left[A_t = a \mid r\right] = \frac{e^{r(a)/T_t}}{\sum_{u \in \mathcal{A}} e^{r(u)/T_t}},$$

where T_t is a temperature parameter that decays at an adequate rate (see [16] for further details).

Theorem 1. *Let $\Gamma = \big(N, (\mathcal{A}_k), (r_k)\big)$ be a weakly acyclic N-player game. If*

$$K \leq \frac{m}{L(\Gamma) + 2},$$

then every independent adaptive learner following a GLIE policy will converge to a best response strategy to the other players' strategies with probability 1.

---

[2] The indicator function for a set A, $\mathbb{I}_A$, takes the value 1 when the argument is in A and 0 otherwise.

Proof. To prove our result we make use of two results from [1]. In this paper, Young showed that in weakly acyclic games, if $K \leq \frac{m}{L(\Gamma)+2}$, then as the *experimentation probability* approaches to zero, the limiting distribution "narrows" around the Nash equilibria in the game. This implies the convergence of the joint strategy to one such equilibrium w.p.1. The experimentation probability in [1] defines the probability of a player choosing non-greedy actions (*i.e.*, making a "mistake").

To prove our result, we make use of the results from [1] by first considering a fixed, positive exploration rate. The exploration rate in our algorithm plays the role of the "experimentation probability" in [1]. The independent adaptive learning process described in Subsection 3.1 yields an irreducible and aperiodic finite-state Markov chain whose state-space consists on the set of all m-long sequences of joint actions. This means that the sequence of histories provided by independent adaptive learning converges at an exponential rate to a stationary distribution. The conclusions of our theorem now follow from the results in [1] as long as we show that the probability of making "mistakes" in our algorithm goes to zero at a suitable rate, *i.e.*, slower than the aforementioned Markov chain converges to stationarity.

In our algorithm, if a particular action $u \in \mathcal{A}_k$ is never played in the selected K plays, then the associated expected payoff is $-\infty$. This means that, in our algorithm, "mistakes" can arise either due to the exploration or to the subestimation of action-values.

Two important observations are now in order. First of all, infinite exploration ensures that the probability of all players converging to a strategy other than a Nash equilibrium is 0. On the other hand, our assumption of a GLIE policy guarantees that the probability of exploration goes to zero as $t \to \infty$, while always ensuring sufficient exploration. This naturally implies that the probability of making exploration "mistakes" decreases to zero. Furthermore, it also implies that Nash equilibria will be sampled with increasing probability—as the exploration decreases, Nash equilibria will be played more frequently and consequently more frequently sampled, and consequently more frequently played, and so on. But this finally implies that, as $t \to \infty$, the probability of making "mistakes" due to sub-evaluation also decreases to zero.

These two remarks lead to the conclusion that the probability of making "mistakes" goes to zero at a slower rate than the GLIE policy becomes greedy which, by construction, is slower than the rate of convergence of the above Markov chain to stationarity. This allows us to apply the desired result from [1] and the proof is complete. ☐

4 Experimental Results

In this section we present the results of our method for several simple games. In each game, we applied our algorithm by running 1000 independent Monte-Carlo trials, each trial consisting of 900 plays of the same game. We used Boltzmann exploration with decaying temperature factor to ensure sufficient exploration of all actions. We present in Figures 3.a), 6.a), 9.a) and 12.a) the average evolution of the received payoff for each game (solid line) and the corresponding standard deviation (in dashed line) for each game. We also present in Figures 3.b), 6.b), 9.b) and 12.b) the percentage of trials that the algorithm converged to each joint strategy in each game.

Prisoner's dilemma. The prisoner's dilemma is a well-known game from game theory whose payoff function is represented in Fig. 1. In this game, two criminal prisoners are persuaded to confess/rat on the other by being offered immunity. If none of the prisoners confess, they will be sentenced for a minor felony. If one of the prisoners confesses and the other remains silent, the one confessesing is released while the other serves the full sentence. If both prisoners confess, they do not serve the full sentence, but still remain in jail for a long time.

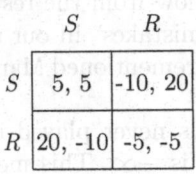

	S	R
S	5, 5	-10, 20
R	20, -10	-5, -5

Fig. 1. Payoff for the prisoner's dilemma. Each prisoner may opt by remaining silent (S) or by ratting on the other prisoner (R)

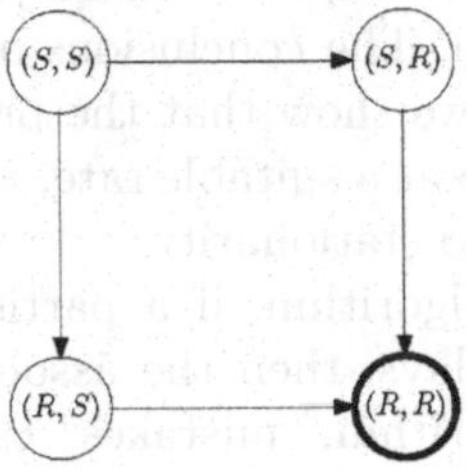

Fig. 2. Best-response graph for the prisoner's dilemma

This game is very interesting from a game theoretic point-of-view. In fact, both players would be better off by remaining silent, since they would both serve a short sentence. However, each player profits by confessing, no matter what the other player does. Therefore, both players will confess and therefore serve a long sentence. The joint action (R, R) is, therefore, a Nash equilibrium. This is clear from the best-response graph, depicted in Fig. 2, where it is also clear that the game is weakly acyclic.

As mentioned, this game has a single Nash equilibrium, consisting of the pure strategy (R, R). To this joint strategy corresponds a payoff of $(-5, -5)$. By observing Fig. 3.a) we can see that the average payoff received by each player converged to -5, indicating that the algorithm converged to the Nash equilibrium as expected. This is also clearly observable in Fig. 3.b): the algorithm converged to the joint strategy (R, R) 100% of the 1000 runs.

Diagonal game. We next considered a 2-player, fully cooperative game described by the payoff function in Fig. 4. Notice that the diagonal elements corresponding to the joint actions $(1, 1)$, $(2, 2)$, $(3, 3)$ and $(4, 4)$ yield higher payoff than the remaining joint actions, as if rewarding the two players for "agreeing" upon their individual actions.

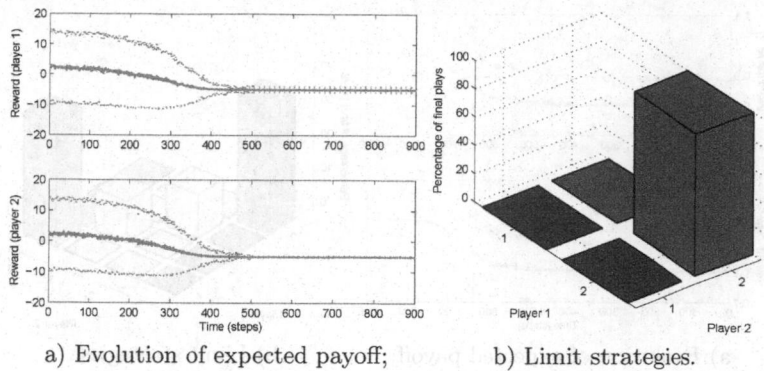

a) Evolution of expected payoff;   b) Limit strategies.

Fig. 3. Learning performance in the prisoner's dilemma

	1	2	3	4
1	1	0.75	0.75	0.75
2	0.75	0.9	0.75	0.75
3	0.75	0.75	0.9	0.75
4	0.75	0.75	0.75	1

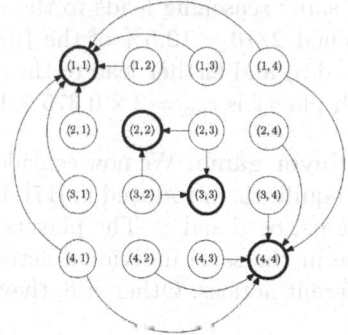

Fig. 4. Payoff for the fully cooperative, diagonal game

Fig. 5. Best-response graphs for the diagonal game

This game presents *four* pure Nash equilibria, corresponding to the diagonal
elements in the payoff matrix (Fig. 4). This motivates the naming of the game as
the "diagonal game". The four Nash equilibria are evident from the best-response
graph in Fig. 5. Notice, furthermore, that the game is weakly acyclic.

We applied our algorithm to both stances of the game and depicted the results
in Fig. 6.

Notice that the four equilibria do not yield similar payoffs and this will affect
the convergence pattern of the algorithm. We start by noticing in Fig. 6.a) that
the expected payoff for both players converges to 0.975. This value has a precise
interpretation that we provide next.

By close observation of the best-response graph in Fig. 5.b) we notice, for
example, that the equilibrium $(1, 1)$ can be reached from 7 different joint actions.
Out of the 16 possible joint actions, 5 lead to $(1, 1)$ and 2 other lead to $(1, 1)$
half of the times. This reasoning allows to conclude that we expect $(1, 1)$ to be
the limit point of our algorithm $6/16 = 37.5\%$ of the times. The same reasoning
can be applied to the equilibrium $(4, 4)$. As for the equilibria $(2, 2)$ and $(3, 3)$,

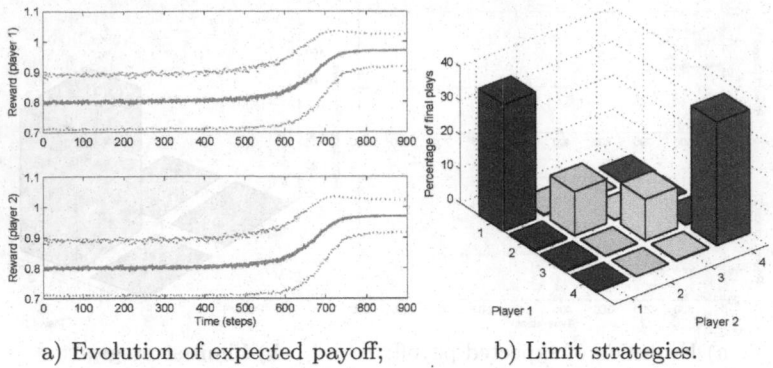

a) Evolution of expected payoff;   b) Limit strategies.

Fig. 6. Learning performance in the diagonal game when $\psi = 0.1$

the same reasoning leads to the conclusion that each of these equilibria will be reached $2/16 = 12.5\%$ of the times. These are, indeed, the results depicted in Fig. 6.b) and further lead to the conclusion that the average expected payoff for each player is $r_{av} = 2 \times 0.375 \times 1 + 2 \times 0.125 \times 0.9 = 0.975$.

3-Player game. We now consider a fully cooperative 3-player game with multiple equilibria introduced in [17]. In this game, 3 players have available 3 possible actions, α, β and γ. The players are rewarded maximum payoff if all 3 coordinate in the same individual action; they are rewarded a small payoff if all play different actions. Otherwise, they are penalized with a negative payoff.

	$\alpha\alpha$	$\alpha\beta$	$\alpha\gamma$	$\beta\alpha$	$\beta\beta$	$\beta\gamma$	$\gamma\alpha$	$\gamma\beta$	$\gamma\gamma$
α	10	-20	-20	-20	-20	5	-20	5	-20
β	-20	-20	5	-20	10	-20	5	-20	-20
γ	-20	5	-20	5	-20	-20	-20	-20	10

Fig. 7. Payoff for the 3-player game from [17]

The game has several Nash equilibria, marked in bold in the best-response graph in Fig. 5. Clearly, the game is weakly acyclic.

We applied our algorithm to the game. The results are depicted in Fig. 9.

Once again conducting an analysis similar to the one in the previous games, we expect the algorithm to converge to the optimal equilibria about 25.9% of the times and to the suboptimal equilibria about 3.7% of the times. The use of Boltzmann exploration leads to a slight increase in the number of runs converging to the optimal equilibria and consequent decrease in the number of runs converging to the suboptimal equilibria (Fig. 9.b)). This is also noticeable since the average payoff per player actually converges to 20 (Fig. 9.a)), which indicates that each optimal equilibrium is actually reached about 1/3 of the times.

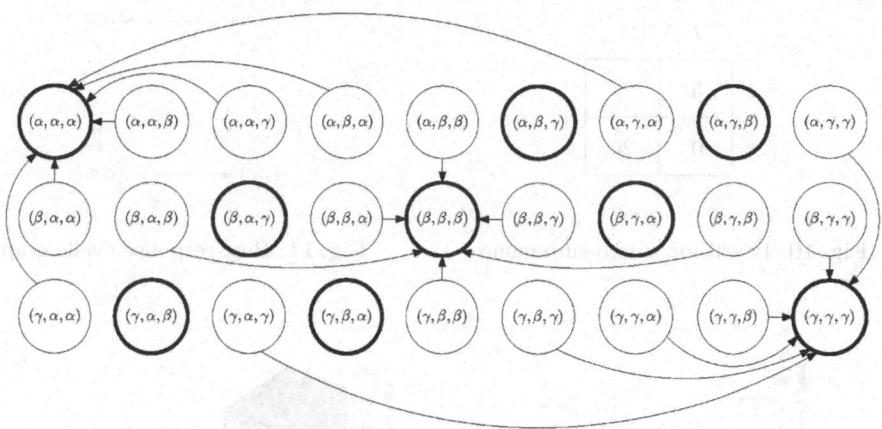

Fig. 8. Best-response graph for the 3-player game from [17]

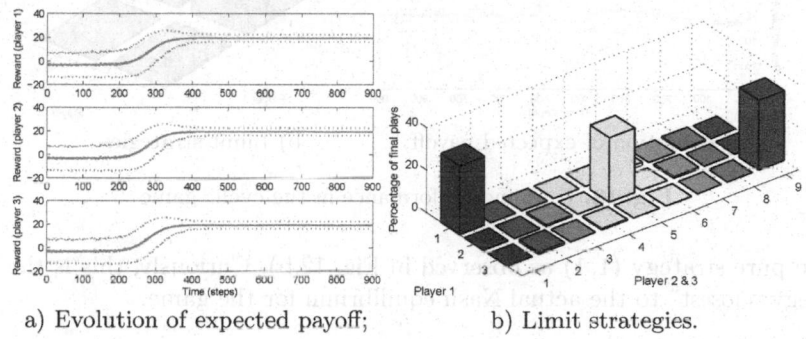

a) Evolution of expected payoff;   b) Limit strategies.

Fig. 9. Learning performance in the 3-player game from [17]

Cyclic game. Finally, we present a two-player, zero-sum game with no pure Nash equilibrium. The payoff function for the game is presented in Fig. 10. Since this game has no pure Nash equilibrium, it cannot be weakly acyclic, as verified from the best-response graph in Fig. 11. Therefore, it is not expected that our algorithm converges to an equilibrium, since the algorithm can only converge to pure strategies (and the equilibrium for this game is a mixed one).[3] We remark, however, that the Nash equilibrium for this game corresponds to an expected reward of 8 for player 1 and of −8 for player 2.

We applied our algorithm to the game, running 1000 independent Monte-Carlo runs, each consisting of 900 plays of the game. The results are depicted in Fig. 9.

Notice in Fig. 12.a) that the average payoff received by player 1 converged to about 5 (and to −5 for player 2). This means that the algorithm converged

---

[3] The Nash equilibrium for this game consists on the mixed strategy that plays action 1 with a probability 0.8 and action 2 with probability 0.2.

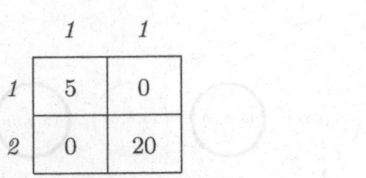

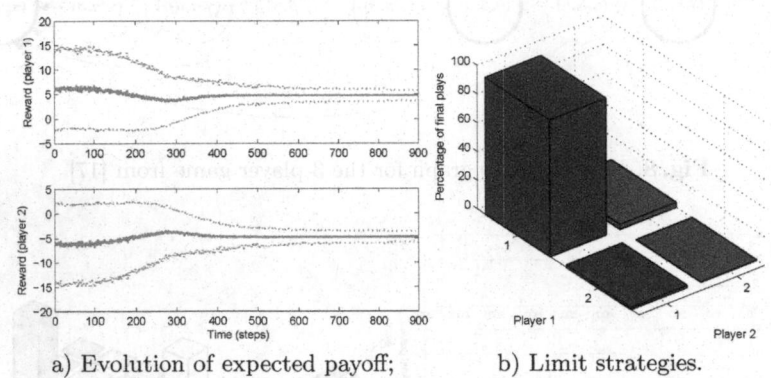

Fig. 10. Payoff for a zero-sum game

Fig. 11. Best-response cyclic graph

a) Evolution of expected payoff;   b) Limit strategies.

Fig. 12. Learning performance in the cyclic game

to the pure strategy $(1,1)$ as observed in Fig. 12.b). Curiously, this is the pure strategy "closest" to the actual Nash equilibrium for the game.

5   Conclusions

In this work we generalized adaptive play [1] to situations where actions and payoffs are not observable. We showed that our algorithm converges with probability 1 to a (pure) Nash equilibrium if it exists. However, if no (pure) Nash equilibrium exists, and as seen in the example of the cyclic game, the algorithm may eventually converge to the pure strategy which is "closest" to a mixed strategy Nash equilibrium for the game. Our algorithm, independent adaptive learning, proceeds as in standard adaptive play by using incomplete sampling of finite length history of past actions/payoffs. To handle the lack of action observability, the algorithm requires infinite exploration to avoid getting "stuck" in non-equilibrium strategies. We provided a formal proof of convergence and some experimental results obtained with our algorithm in several games with different properties. Further experimental results can be found in [18].

We are interested in extending the independent adaptive learning algorithm (or a variation thereof) to multi-state problems, such as Markov games. We are also interested in applying the algorithm to real world situations with a large number of agents with large action repertoires.

References

1. Young, H.P.: The evolution of conventions. Econometrica 61(1), 57–84 (1993)
2. Van der Genugten, B.: A weakened form of fictitious play in two-person zero-sum games. International Game Theory Review 2(4), 307–328 (2000)
3. Leslie, D.S., Collins, E.J.: Generalised weakened fictitious play. Games and Economic Behavior 56(2), 285–298 (2006)
4. Claus, C., Boutilier, C.: The dynamics of reinforcement learning in cooperative multiagent systems. In: AAAI, pp. 746–752 (1998)
5. Littman, M.L.: Value-function reinforcement learning in Markov games. Journal of Cognitive Systems Research 2(1), 55–66 (2001)
6. Brown, G.W.: Some notes on computation of games solutions. Research Memoranda RM-125-PR, RAND Corporation, Santa Monica, California (1949)
7. He, M., Leung, H.F., Jennings, N.R.: A fuzzy logic based bidding strategy for autonomous agents in continuous double auctions. IEEE Trans. Knowledge and Data Engineering 15(6), 1345–1363 (2002)
8. Bagnall, A.J., Toft, I.E.: Zero intelligence plus and Gjerstad-Dickhaut agents for sealed bid auctions. In: Kudenko, D., Kazakov, D., Alonso, E. (eds.) Adaptive Agents and Multi-Agent Systems II. LNCS (LNAI), vol. 3394, Springer, Heidelberg (2005)
9. Tan, M.: Multi-agent reinforcement learning: Independent vs. cooperative agents. In: Readings in Agents, pp. 487–494 (1997)
10. Lauer, M., Riedmiller, M.: An algorithm for distributed reinforcement learning in cooperative multi-agent systems. In: ICML, pp. 535–542 (2000)
11. Kapetanakis, S., Kudenko, D.: Improving on the reinforcement learning of coordination in cooperative multi-agent systems. In: Symp. AAMAS, pp. 89–94 (2002)
12. Lauer, M., Riedmiller, M.: Reinforcement learning for stochastic cooperative multi-agent-systems. In: AAMAS, pp. 1516–1517 (2004)
13. Verbeeck, K., Nowé, A., Parent, J., Tuyls, K.: Exploring selfish reinforcement learning in repeated games with stochastic rewards. JAAMAS 14, 239–269 (2006)
14. Bowling, M., Veloso, M.: Rational and convergent learning in stochastic games. In: IJCAI 2001. Proceedings of the 17th International Joint Conference on Artificial Intelligence, pp. 1021–1026 (2001)
15. Robinson, J.: An iterative method of solving a game. Annals of Mathematics 54, 296–301 (1951)
16. Singh, S., Jaakkola, T., Littman, M., Szepesvari, C.: Convergence results for single-step on-policy reinforcement-learning algorithms. Machine Learning 38(3) (2000)
17. Wang, X., Sandholm, T.: Reinforcement learning to play an optimal Nash equilibrium in team Markov games. In: NIPS, vol. 15, pp. 1571–1578 (2003)
18. Melo, F., Lopes, M.: Convergence of independent adaptive learners. Technical Report RT-603-07, Institute for Systems and Robotics (2007)

Multi-agent Learning: How to Interact to Improve Collective Results

Pedro Rafael and João Pedro Neto

Computer Science Department, University of Lisbon
Edifício C6, Campo Grande, 1749-016 Lisboa, Portugal
prafael.phd@netcabo.pt, jpn@di.fc.ul.pt

Abstract. The evolution from individual to collective learning opens a new dimension of solutions to address problems that appeal for gradual adaptation in dynamic and unpredictable environments. A team of individuals has the potential to outperform any sum of isolated efforts, and that potential is materialized when a good system of interaction is considered. In this paper, we describe two forms of cooperation that allow multi-agent learning: the sharing of partial results obtained during the learning activity, and the social adaptation to the stages of collective learning. We consider different ways of sharing information and different options for social reconfiguration, and apply them to the same learning problem. The results show the effects of cooperation and help to put in perspective important properties of the collective learning activity.

1 Introduction

When agents are involved in a common learning task and have similar goals, mutual interaction creates a vast field of possibilities to improve global performance. Sharing partial results and dynamically organizing the group are two forms of cooperation that can lead to such improvement. By sharing information that concerns intermediate steps of the learning task, not only can agents benefit directly from each other's experiences, but also combine different opinions and options, understand the validity of their individual learning paths, and be aware of the collective progress. By addressing the learning task as a flexible team, the group can adjust to specific stages and optimize their results: for example, when global experience in a certain task reaches a certain threshold or becomes abundant, the nomination of experts or the diminution of the number of agents assigned to that task are social reconfigurations capable of freeing useful resources.

To study the effects of these forms of cooperation on group learning, we consider a classical machine learning case-based reasoning (CBR) mechanism, and integrate different processes of interaction that allow collective learning, materializing a diversity of options for information sharing and social adaptation. We then apply the different systems of interaction to a collective learning task that involves a foraging situation, and evaluate their influence on global performance. Furthermore, we test the tolerance of these systems to communication problems and to situations in which some of the agents autonomously decide not to learn.

J. Neves, M. Santos, and J. Machado (Eds.): EPIA 2007, LNAI 4874, pp. 568–579, 2007.
© Springer-Verlag Berlin Heidelberg 2007

2 Related Work

The first important studies on multi-agent learning were centered on collective versions of reinforcement learning. According to Tan [10], collective learning can be improved at the cost of information exchange and social adaptation. Whitehead [13] proposes an architecture for learning from mutual observation that is equivalent to the circulation of <state, action, quality> triples.

Numerous other works have since then proposed collective versions of Q learning, some of which introducing ideas directly related to the characteristics and potential of group learning: Clouse [3] suggests that agents can ask for help as a strategy for collective improvement; Chalkiadakis and Boutilier [2] propose a cooperative model to explore the space of solutions; Szer and Charpillet [9] define an algorithm to broadcast results that concern intermediate learning stages, and study the effects of the circulation of different quantities of information; Vu, Powers and Shoham [11] investigate agent coordination, stating that having minimum levels of individual performance is good for the group.

As for investigations that concern other machine learning techniques, Modi and Shen [6] propose two distributed classification algorithms that allow collective learning in situations where some of the information is to be kept private, showing that some learning results are only possible when isolation gives place to cooperation. Ontañon and Plaza [8] investigate cooperation techniques for case-based reasoning, adopting a bartering system inspired on market transactions. Searching answers for fundamental questions of multi-agent learning, Nunes and Oliveira [7] propose an advice exchange system that allows the circulation of information between agents that have different learning mechanisms. Graça and Gaspar [4] investigate the performance of opportunistic non-learning agents that receive information from learning agents, concluding that the existence of agents with different tasks and roles can improve global performance.

Weiß and Dillenbourg have an important contribution [12] that offers a general perspective over the multi-agent learning area. According to them, the true potential for multi-agent learning resides in dynamic forms of interactivity.

3 Multi-agent Learning

Different environments and learning situations appeal for different kinds of multi-agent interaction. Considering the degree of agent interactivity involved, we distinguish three important categories for multi-agent learning: with no interactivity, with static interactivity, and with dynamic interactivity. We focus our investigation on a context of full cooperation: all agents share true data and perform their roles within the group at the best of their capacity.

3.1 Three Categories for Multi-agent Learning

In a group of agents that do not interact, learning remains an individual activity. This is a solution for situations in which communication is too difficult (for example, too expensive or too unreliable), or unsuitable (for example, when there is a need for privacy policies or in presence of hostility). In this individual multi-agent learning, the group evolves as a result of the isolated evolution of its individuals.

When communication is available and the learning problem is such that it can be divided into sub-problems, the global task can be addressed in a collective way using a system of static interactivity. In such a system, learning sub-tasks are assigned to the agents in an organized way (for example, an equal partition of the search space to each agent), and individual results are integrated by the collective through well-defined patterns of relationship and communication. In this divided multi-agent learning, the evolution of the group derives from the propagation of the results produced by organized individual evolution.

Instead of simply following the preset details of static interaction, the collective work can be executed in a more dynamic way. The assignment of sub-tasks and the definition of collective patterns of relationship and communication can be set and modified during the learning process, and follow the natural flow of circumstances. When such flexibility is feasible, the details of interaction can be tuned to both match the evolution of the team throughout the different stages of collective learning and also the succession of environmental states, even when they change in unpredictable ways. In this interactive multi-agent learning, the team evolves as a result of the dynamic organization and circulation of information.

3.2 Joint Perspective of the Three Categories

In a group of isolated agents that address the same learning problem, the work is multiplied by all the individuals involved. From the collective perspective, this leads to a high degree of redundancy. In situations where divided multi-agent learning is possible, this redundancy can be greatly reduced. For that to happen, collective learning has to be planned: learning sub-tasks have to be defined and distributed in ways that facilitate teamwork; ways of sharing and integrating information have to be created and organized; a communication system has to be available so that partial results become available to the group. When these conditions are met, and especially when the problem addressed has a degree of predictability that allows a well-defined sequence of collective learning steps, divided multi-agent learning has the potential to improve global learning performance at the cost of information exchange.

This potential can be extended to more unpredictable problems with an interactive multi-agent learning system. However, learning in a more dynamic way also extends the costs involved: planning a flexible and adaptive group learning activity is generally more complex than simply dividing the learning task; and the communication system has to respond to a wider scope of situations and fluctuations that result from the higher degree of freedom associated to agent interaction. Considering this, what are the advantages in increasing group interactivity? To find answers to this question, consider the following properties of interactive multi-agent learning:

- Information availability – Instead of sharing partial results at pre-defined moments, the circulation of information can happen at any stage of collective learning. Because of this, partial results are more readily available. This is especially relevant whenever important learning results are achieved.
- Diversity of learning paths – The existence of free patterns of agent interaction leads to a diversity of perspectives that can benefit the discovery of good

solutions. If a static division of the learning task may lead to over-specialization, a dynamic evolution leads to the healthy proliferation of different approaches and opinions on the same subject. The existence of different learning paths is especially important when local minimums are an issue.

- Adaptability – The possibility of reassigning sub-tasks during the learning process allows beneficial migrations of agents to match each situation. For example, if the completion of a sub-task becomes critical to the group, more resources can be dynamically allocated to it.
- Independence – When the learning task is divided, agents become mutually dependent. If a sub-task is postponed or not executed, the collective work can be delayed or even compromised. This limits the application of static interactivity to environments in which, for example: agents have autonomy on when to execute their tasks; agents can choose to stop sharing data; the communication system has periods of low performance. With dynamic interactivity, cooperation becomes more an option than a necessity. The group can readjust to better tolerate faults, or other suboptimal situations.

4 Cooperative Case-Based Reasoning

On investigating the effects of different levels and ways of interactivity on multi-agent learning, we consider the traditional CBR learning mechanism and extend it to allow agent interaction. Four generic group learning systems are described: one for individual multi-agent learning (no interactivity in the group), another for divided learning (static interactivity), and two for interactive learning (dynamic interactivity).

4.1 Cases and Storage

A case describes a learning experience and has the following generic format: < *problem, solution, result* >. The *result* is an expression of the quality of the *solution* used to solve the *problem*. In traditional CBR, cases are stored in a single case base, but in multi-agent CBR other options are possible.

We address collective CBR with the use of separate case bases, an option that makes information exchange crucial for the group learning process, leading to a deeper and more complete discussion concerning the introduction of interactivity. The use of separate case bases is a solution for situations that involve individual privacy (for example, when part of the stored data is to be kept in secrecy), or the need for efficiency on data access or storage (for example, to avoid traffic bottleneck situations that can result from having a centralized case base).

4.2 Traditional CBR Mechanism

The traditional individual CBR generic learning process is based on the following four basic steps [1]:

1. Information retrieval – Retrieval of previously gathered cases that concern problems somehow similar to the current;
2. Selection of a solution – Analysis of the available information and selection of a solution for the current problem;

3. Resolution of the problem – Resolution of the problem and evaluation of the solution selected;
4. Storage of a new case – Creation and storage of a new case.

When a group of agents practices individual learning this traditional mechanism needs no adaptation: each agent executes the learning algorithm and the collective learning task ends when all agents complete their isolated tasks. To allow other forms of multi-agent learning some of the steps have to be modified. Moreover, the execution of the algorithm may vary in consideration to agents that perform distinct social roles.

4.3 Divided Multi-agent CBR Learning

To illustrate a divided CBR learning system we consider the most straightforward solution: the learning task is partitioned and each sub-task is assigned to an agent. Each agent gathers information that concerns his area of specialization, and that information is shared with the others when pre-defined levels of experience are reached. When solving problems that do not belong to their partitions, agents do not gather information: they use the solutions provided by the specialist on that partition, or, when those solutions are not yet available, a random solution. To respond to this, steps 3 and 4 of the CBR generic learning process have to be modified, becoming:

3. Resolution of the problem – Resolution of the problem and evaluation of the solution selected. If the problem solved belongs to the assigned partition then execute step 4, otherwise solve next problem;
4. Storage of a new case – Creation and storage of a new case. If a pre-defined level of experience on the current problem is reached, broadcast best known solution.

4.4 Interactive Multi-agent CBR Learning

We exemplify interactive CBR learning with two different systems. The first system is based on free communication: agents ask each other for help and exchange advices whenever that is considered benefic. To contemplate this form of interaction, the first step of the CBR generic learning process needs to be extended: instead of considering his individual case base as the only source of information, each agent can emit help requests to his peers and acquire additional external information. This form of interaction leads to the formation of temporary groups of discussion whose purpose is the utilization of collective experience to improve individual decisions.

The second system is based on social adaptation: after a period of individual learning, and upon reaching a certain threshold of experience on a problem, the group elects a team of experts to address that problem. When elected, the experts receive all the information available about the problem in question, and, from that moment on, assume the responsibility for that learning sub-task. When they find a solution that is better than those previously known, they share it with the group. This idea has two essential purposes: to combine the results of different learning paths, and to free learning resources when experience is abundant. To materialize this form of interactivity, the forth step of the CBR generic learning process has to be extended:

4. Storage of a new case – Creation and storage of a new case. If a pre-defined level of experience on the current problem is reached, proceed to the election of experts.

If the new case belongs to the area of expertise and describes a new and better solution, broadcast the new solution.

5 Example

To evaluate the effects of applying the four different systems, we adopted a learning problem that involves a foraging situation. After presenting the foraging environment and the generic idea of the learning process, we specify the details for each of the four multi-agent learning systems used.

5.1 Foraging Environment

The foraging activity takes place inside foraging areas, and is performed by a set of foraging machines. During each foraging task, the machines carry to a container all the objects located in the area. The various components of the environment have the following characteristics: foraging areas have two dimensions and can be of three different sizes – small, medium and large; the container is always placed at the same relative position – on the left of the map, at half height; there are two types of objects – small and big; there are three types of foraging machines – light machines are fast and only carry one small object; heavy machines are slow and carry one object of any type; hyper-machines are very slow but have a limited ability of instantaneous hyperspace jump that leaves them at a fixed proximity of their target, and they can carry one object of any type.

All foraging tasks are composed of a foraging area of specific size and 25 objects (for example, 20 small plus 5 large); the tasks are performed by 7 machines (for example, 4 light, 1 heavy and 2 hyper). The location of the objects is known to the machines, and their transportation to the container is a purely mechanic activity. The configuration of the set of machines influences the duration of the tasks. For example, if the goal is to minimize the duration of tasks: it is a good idea to have a predominance of hyper-machines when the area size is large[1]; it is a bad idea to have a high number of light machines when the number of small objects is low.

5.2 Generic Learning Process

The purpose of the learning activity is the improvement of the foraging process by finding answers to the following question: for each task, which configuration of seven machines minimizes the foraging duration? The learning group chooses configurations of machines to successively solve a list of foraging tasks, gathering experiences that progressively reveal which are the best configurations for each task.

The CBR mechanism is based on a generic case $C_i = <T_i, M_i, D_i>$, where the *problem* is the foraging task T_i, the *solution* is the set of machines M_i, and the *result* is the foraging duration D_i. To clarify this definition, consider the following example:

---

[1] Their hyperspace jump ability makes them the best option when objects are distant from the container; their very low speed makes them the worst when objects are near.

- C_1 = <(large, 10, 15), (2, 1, 4), 455> – Case C_1 describes the foraging of 10 small objects and 15 big objects in an area of large size, performed by a set of 2 light machines, 1 heavy machine and 4 hyper-machines, with the duration of 455 turns.

The experience level of an agent on a foraging task T_i is measured according to the number of stored cases that describe different solutions for T_i, or for tasks sufficiently similar[2] to T_i. The measurement is such that cases that include T_i are considered twice more relevant than those including tasks similar to T_i.

The generic individual exploration process is based on random experimentation and evolves according to the experience level. When experience is low, exploration is very frequent; with the rising of experience, exploration is progressively replaced by the reutilization of the best solutions known; when experience is abundant, random experimentation ceases.

The individual retrieval of stored cases to find a good solution for a task T_i considers cases that include T_i and similar tasks. Whenever a set of machines M_j used in T_j similar to T_i has a better result than the solutions tried for T_i, then M_j is considered a promising solution and is selected.

5.3 Multi-agent Learning Process

Multi-agent learning is performed by a group of six agents. At each learning cycle the group receives a list of foraging tasks that are equally divided among the agents. Each agent provides a solution for each problem received and the cycle ends when all foraging tasks are completed. The collective performance is measured considering the sum of the foraging duration of all tasks: the lower the sum, the better the performance. As a result of the learning activity, this sum tends to decrease cycle after a cycle, expressing the improvement of collective performance.

If LT is the list of m tasks and $D(LT) = \{d_1, d_2, ..., d_m\}$ is the list of results (duration of tasks) achieved during a learning cycle n, the collective performance on that cycle is given by expression (1).

$$Performance(LT, n) = \sum_{x=1}^{m} d_x \in D(LT). \tag{1}$$

5.4 Four Systems for Multi-agent Learning

Individual. In the first system, the individual multi-agent learning system, each agent simply applies the generic learning process. The learning results achieved without interactivity serve as a control reference for the other three systems.

Divided. In the divided multi-agent learning system we consider six partitions of the space of possible tasks, and assign one for each agent. There are two partitions for each of the three different area sizes, and, of those two, one concerns tasks where there are more small objects than big, and the other one concerns the opposite (for example, (large, 10, 15), (large, 14, 11) and (medium, 14, 11) belong to different

---

[2] The criterion of similarity between two tasks is based on the quantification of the differences between area sizes, and between the proportions of small and big objects.

partitions). When solving cases that belong to his partition, each specialist follows the generic learning process. For each task T_i he is responsible for, and at four predefined moments, he shares with all other agents the best solution known for T_i. These moments are defined according to the experience level achieved in T_i: the first broadcast takes place at very early stages of learning, providing a quick answer that allows non-specialists to abandon random selection of solutions; the second broadcast takes place at a medium stage of learning, when there is a good probability that a solution of at least average quality is known; the third broadcast takes place when experience is abundant (it coincides with the moment when the specialist ceases the random experimentation of solutions); the forth broadcast takes place when a better solution is unlikely to be found.

Free Communication. In the interactive multi-agent learning system based on free communication, whenever an agent decides to explore new solutions for the current task T_i, two options are available (with equal selection probability): random experimentation or emission of a help request. When an agent chooses to emit a help request, he randomly chooses another agent as the receiver. The help request includes the case that describes his best known solution for T_i (when it exists). When the receiver knows a better solution, he responds with an advice that includes the case that describes that solution. When the receiver does not know a better solution, he stores the case described in the help request and informs the sender that he cannot help him. This simple interaction stimulates the free circulation of cases in an unpredictable and mutually benefic way: when an agent requests help, he sometimes provides it.

Election of Experts. In the system based on the election of experts, all agents learn in an individual way about each task T_i until one of them reaches a predefined level of experience (the same level considered for the second broadcast described in the divided learning system). At that instant, that agent is elected has the expert in T_i, and the other agents send all cases that describe solutions for that task to him. From that moment on, the expert becomes the only one responsible for learning about T_i, and whenever he finds a better solution he shares it with the others. This system combines characteristics of individual learning, divided learning (static definition of the moment of interaction) and interactive learning (dynamic social adaptation based on the assignation of roles and sub-tasks, free circulation of information with unpredictable characteristics both in quantity and quality).

6   Tests and Results

With the help of a simulation workbench specifically developed for this effect, we applied the four multi-agent learning systems to the foraging environment described. We made a significant number of tests, using different sets of tasks under various circumstances. In this section we present some of the most relevant tests, and discuss their results.

The tests describe the evolution of the collective performance (the sum of the duration of tasks as shown before in expression (1), from hereon simply referred as *duration* or performance) during successive learning cycles. Each numeric series presented expresses the average values of 20 simulations. All agents start with no experience.

Test 1 – A situation very suitable for divided learning. In the first test we consider a list of six tasks that have a high degree of similarity, being each one from a different partition: for example, the tasks (small, 12, 13), (small, 13, 12) and (medium, 12, 13) have high similarity and belong to three different partitions. This list of tasks simulates a predictable situation (known before the learning activity begins), and is especially suitable for divided learning because: the distribution of sub-tasks perfectly matches the six specialists; the alikeness between tasks makes the broadcasted experiences a useful source of promising solutions.

The results (Fig. 1) show that all systems with interactivity have a similar performance that is better than the individual system, and that the divided system is the fastest to get the best *duration* levels.

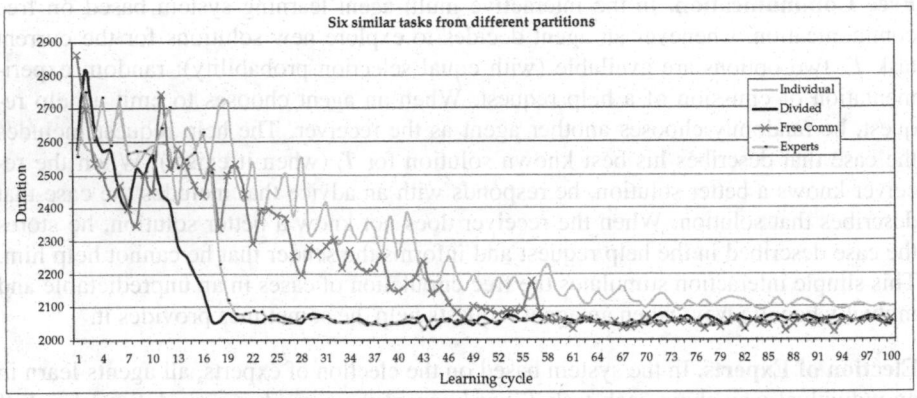

Fig. 1. Multi-agent learning in a situation very suitable for divided learning

Test 2 – A heterogeneous situation. In the second test we consider a list of twelve tasks, asymmetrically distributed by the six different partitions. Some of the tasks are

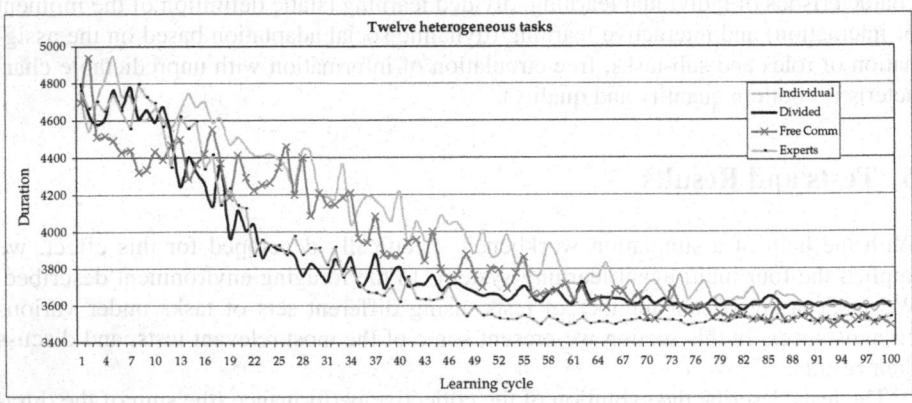

Fig. 2. Multi-agent learning in an unpredictable situation

equal, some similar, and others different. This list of tasks simulates an unpredictable situation, where the specific conditions of the learning activity are unknown before the beginning.

The results (Fig. 2) show that the two systems with dynamic interactivity have the best performance, with the experts system being the fastest. The divided system obtains a performance that is similar to the individual system.

Test 3 – Tolerance to communication problems. In the third test we consider the same list of twelve tasks and a communication system where 50% of the messages are permanently lost.

The results (Fig. 3) show that the two systems with dynamic interactivity have the best tolerance to this situation: compared to the previous test, their performance decreases slightly. The divided system shows less tolerance to difficult communication conditions. In another test concerning communication problems, where three of the agents were unable to communicate (simulating their isolation from the communication system), the divided system showed a clearly low tolerance level.

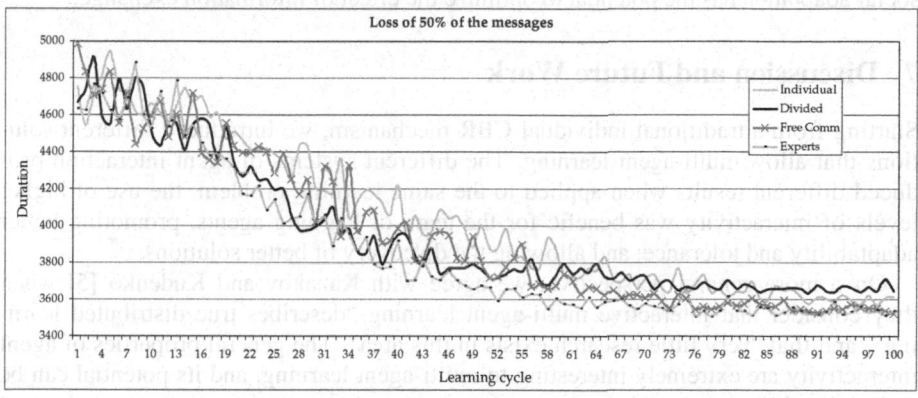

Fig. 3. Multi-agent learning when 50% of the messages are lost

Test 4 – Tolerance to autonomous decisions. In the fourth test we consider the same list of twelve tasks, and a situation where three of the agents do not to learn (simulating the appearance of other individual priorities). They still communicate, but they do not create learning cases.

The results (Fig. 4) again show that the two systems with dynamic interactivity have a good tolerance level. The divided and individual systems have little tolerance to this situation.

Commentary on the results. The system of static interaction motivated good results when circumstances were suitable. On less favorable grounds, and especially when tolerance was required, the divided system showed little adequacy.

The use of dynamic forms of interaction during multi-agent learning motivated a better collective performance, especially when circumstances appealed for adaptability and tolerance to suboptimal situations. The experts system was consistently able to achieve

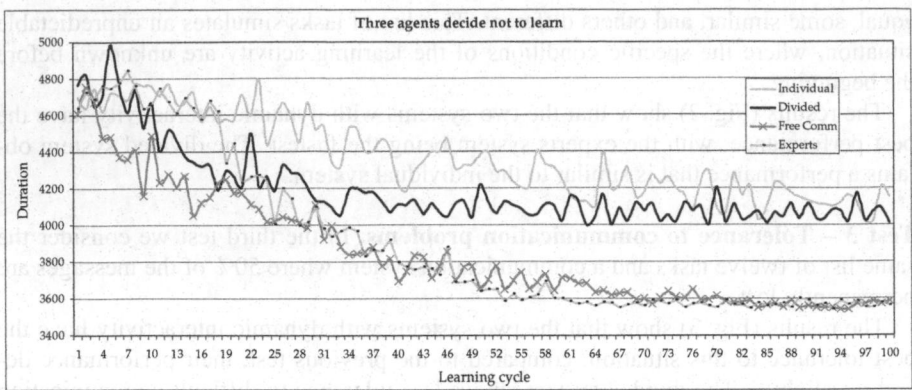

Fig. 4. Multi-agent learning when half of the agents decide not to learn

best performance levels earlier than the free communication system, confirming that social adaptation has the potential to optimize the effect of information exchange.

7 Discussion and Future Work

Starting from a traditional individual CBR mechanism, we introduced different solutions that allow multi-agent learning. The different systems of agent interaction produced different results when applied to the same learning problem: the use of higher levels of interactivity was benefic for the team of learning agents, promoting better adaptability and tolerance, and allowing the discovery of better solutions.

On a more general perspective, we agree with Kazakov and Kudenko [5] when they consider that interactive multi-agent learning "describes true distributed learning", and that "very little research exists in this area". The general properties of agent interactivity are extremely interesting to multi-agent learning, and its potential can be materialized in a vast number of ways.

According to the same authors, "it is an interesting open research problem to gain a clearer and more detailed understanding of the relationship between agent diversity and performance and learning in general". The results of our research show that the diversity of learning paths and social roles has the potential to improve collective learning performance. We want to go deeper on this subject, and study the impact of having structural diversity on agent attributes (for example, different innate levels of intelligence) and behavior (for example, different personalities).

References

1. Aamodt, A., Plaza, E.: Case based reasoning: foundational issues, methodological variations, and system approaches. In: Artificial Intelligence Communications (AICom), vol. 7, pp. 39–59. IOS Press, Amsterdam (1994)
2. Chalkiadakis, G., Boutilier, C.: Coordination in multiagent reinforcement learning: a bayesian approach. In: AAMAS 2003. Proceedings of the 2nd Conference on Autonomous Agents and Multiagent Systems (2003)

3. Clouse, J.A.: Learning from an automated training agent. In: Proceedings of the International Machine Learning Conference (IMLC) (1995)
4. Graça, P.R., Gaspar, G.: Using cognition and learning to improve agents' reactions. In: Alonso, E., Kudenko, D., Kazakov, D. (eds.) Adaptive Agents and Multi-Agent Systems. LNCS (LNAI), vol. 2636, pp. 239–259. Springer, Heidelberg (2003)
5. Kazakov, D., Kudenko, D.: Machine learning and inductive logic programming for multi-agent systems. In: Luck, M., Mařík, V., Štěpánková, O., Trappl, R. (eds.) ACAI 2001 and EASSS 2001. LNCS (LNAI), vol. 2086, pp. 246–270. Springer, Heidelberg (2001)
6. Modi, P.J., Shen, W.: Collaborative multiagent learning for classification tasks. In: Agents 2001. Proceedings of the 5th Conference on Autonomous Agents (2001)
7. Nunes, L., Oliveira, E.: Cooperative learning using advice exchange. In: Alonso, E., Kudenko, D., Kazakov, D. (eds.) Adaptive Agents and Multi-Agent Systems. LNCS (LNAI), vol. 2636, pp. 33–48. Springer, Heidelberg (2003)
8. Ontañon, S., Plaza, E.: A bartering approach to improve multiagent learning. In: Alonso, E., Kudenko, D., Kazakov, D. (eds.) Adaptive Agents and Multi-Agent Systems. LNCS (LNAI), vol. 2636, Springer, Heidelberg (2003)
9. Szer, D., Charpillet, F.: Coordination through mutual notification in cooperative multi-agent reinforcement learning. In: Kudenko, D., Kazakov, D., Alonso, E. (eds.) Adaptive Agents and Multi-Agent Systems II. LNCS (LNAI), vol. 3394, Springer, Heidelberg (2005)
10. Tan, M.: Multi agent reinforcement learning: independent vs cooperative agents. In: Proceedings of the 10th International Conference on Machine Learning, Amherst, MA, pp. 330–337 (1993)
11. Vu, T., Powers, R., Shoham, Y.: Learning against multiple opponents. In: AAMAS 2006. Proceedings of the 5th Conference on Autonomous Agents and Multiagent Systems (2006)
12. Weiß, G., Dillenbourg, P.: What is 'multi' in multi-agent learning. In: Dillenbourg, P. (ed.) Collaborative Learning, pp. 64–80. Pergamon Press, Oxford (1999)
13. Whitehead, S.D.: A complexity analysis of cooperative mechanisms in reinforcement learning. In: AAAI 1991. Proceedings of the 9th National Conference on Artificial Intelligence, pp. 607–613 (1991)

A Basis for an Exchange Value-Based Operational Notion of Morality for Multiagent Systems*

Antônio Carlos da Rocha Costa and Graçaliz Pereira Dimuro

Escola de Informática–PPGINF, Universidade Católica de Pelotas
96.010-000 Pelotas, RS, Brazil
{rocha,liz}@atlas.ucpel.tche.br

Abstract. In this paper, we sketch the basic elements of an operational notion of morality for multiagent systems. We build such notion on the qualitative economy of exchange values that arises in an artificial social system when its agents are able to assess the quality of services that they exchange between each other during their interactions. We present tentative formal definitions for some foundational concepts, and we suggest how the operational notion of morality introduced here can be useful in enhancing the stability and other features of multiagent systems.

1 Introduction

Ethics and Morality are tightly coupled subjects. Often, they are considered to be different: Ethics is concerned with universal values and the rules of behavior that derive from them, while Morality is concerned with the relationships that people establish between them, the value that people assign to them, and the rules of behavior for the conservation and improvement of the valued relationships.

In this paper, we attempt a transposition to the world of the artificial agents and multiagent systems of the main concepts involved in Morality. The Morality that we get as the result of this first attempt is *operational*, in the sense that its working is conceived to be embedded in the social organization of multiagent systems.[1]

The paper is structured as follows. We summarize in Sect. 2, in a way that surely does not makes justice to its original presentation, the analysis of Morality and Ethics in human beings put forward by Dwight Furrow in [5], which adopts a relational approach to morality. In Sec. 3 we show how the system of exchange values that arises from qualitatively assessed interactions [7] is able to support an operational notion of morality. In Sec. 4 we summarize the population-organization model of multiagent systems [2] on which we strongly build. In Sect. 5 we introduce in a tentative way our operational notion of morality, to be adopted by artificial agents and multiagent systems. Section 6 concludes the paper and presents the main issues that are not solved yet.

---

* Work partially supported by FAPERGS and CNPq.

[1] The authors are deeply aware that the discussion of Morality and Ethics presented in the paper makes no justice to the long tradition of Moral Studies in Philosophy and Human and Social Sciences, many crucial issues having not even been touched.

J. Neves, M. Santos, and J. Machado (Eds.): EPIA 2007, LNAI 4874, pp. 580–592, 2007.
© Springer-Verlag Berlin Heidelberg 2007

2 Morality and Ethics in Human Society

In [5], Dwight Furrow gives a textbook presentation of the philosophical areas of Ethics and Morality, but at the same time makes many points of his own, in particular, on the relational nature of morality.

The main ideas presented in the book, and the ones which seem to be the most important ones for the attempt of transposing Morality to the domain of artificial agents, are the following (summarized in a very sketchy way, possibly with some very strong personal biases):

Morality: Morality is a system of rules and values that help people keep sound the social relationships they establish between them. Moral rules determine criteria for judging if behaviors are acceptable in social relationships. Moral values operate as symbols of how social relationships should be maintained. Morality also determines what are the morally acceptable reactions that other agents may have concerning certain behaviors of a given agent. Such moral reactions are expressed in terms of praise or blame, and are assumed to operate respectively as prize or punishment for the agent that performed such behaviors.

Moral agency: Moral agency is the capacity that individuals have both to behave according to the moral rules of the society or social group in which they establish social relationships, and to justify their actions on the basis of the moral rules and of the concrete context in which they acted. The central element of moral agency is the capacity for moral deliberations, that is, the capacity of evaluating social actions according to the established morality and of deciding to perform them or not taking that evaluation into account.

Moral reasons: Moral reasons are the moral beliefs that moral agents have about what is morally correct or wrong, together with the moral rules to which they may appeal when required to justify some behavior that has been morally questioned by somebody else.

Moral obligations: Moral obligations are general rules that determine compulsory ways of behaving in social relationships, so that the general system of social relationships is kept stable and operational at an acceptable level of performance and efficacy, in accordance with the overall cultural setting of the society or social group to which they refer.

Moral character: Moral character is the set of personality traits that individuals have that are connected to the way they take moral decisions, positively valuating social relationships being, in general, the main one.

Moral responsibility: As a consequence of the capacity of moral deliberation, moral agents become morally responsible for their actions, that is, they become committed to the consequences of their actions for the social relationships in which they are involved, and thus become the target of the moral reactions of the other agents, expressed in terms of praise or blame.

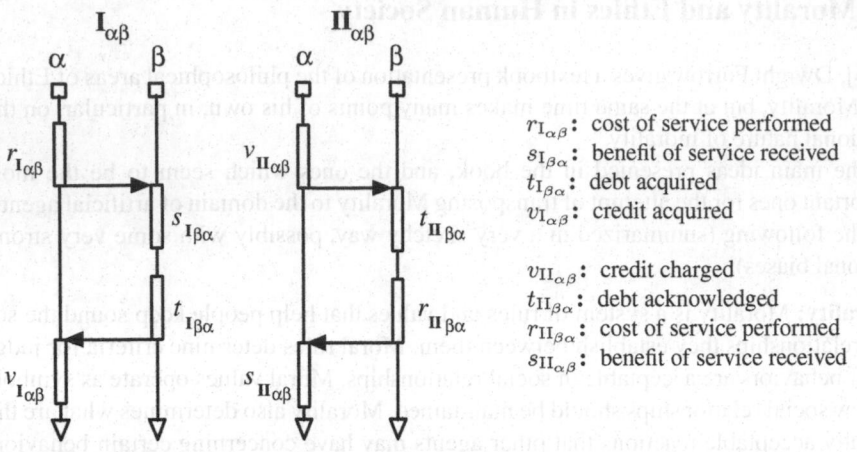

$r_{I_{\alpha\beta}}$: cost of service performed
$s_{I_{\beta\alpha}}$: benefit of service received
$t_{I_{\beta\alpha}}$: debt acquired
$v_{I_{\alpha\beta}}$: credit acquired

$v_{II_{\alpha\beta}}$: credit charged
$t_{II_{\beta\alpha}}$: debt acknowledged
$r_{II_{\beta\alpha}}$: cost of service performed
$s_{II_{\alpha\beta}}$: benefit of service received

Fig. 1. The two kinds of stages of social exchanges

3 Exchange Values and the Assessment of Social Relationships

Let's now assume that the agents of the system are able to qualitatively, and subjectively, assess the costs and benefits of the services that they exchange during their social interactions, and let's classify the social interactions according to the different ways the exchanges of services may happen within them.

In [7], the two basic kinds of stages of social exchanges are modeled as shown in Fig. 1. Agents α and β interact by exchanging services: in exchange stages of kind $I_{\alpha\beta}$, agent α performs a service for agent β. After assessing the service performed, the agents define the *exchange values* associated with that exchange stage: the cost ($r_{I_{\alpha\beta}}$) of performing the service, the benefit ($s_{I_{\beta\alpha}}$) acquired from receiving it, the debt ($t_{I_{\beta\alpha}}$) that β acquired with respect to α from getting the service performed, the credit ($v_{I_{\alpha\beta}}$) over β that α acquired from having performed the service. In exchange stages of kind $II_{\alpha\beta}$, agent α charges β some credit ($v_{II_{\alpha\beta}}$) that it has accumulated from services previously performed. Agent β performs a service to α in return for those services, with certain cost ($r_{II_{\beta\alpha}}$), but only after acknowledging that debt ($t_{II_{\beta\alpha}}$), and possibly just in the measure that the debt is acknowledged. Agent α gets, then, a final benefit ($s_{II_{\alpha\beta}}$).

4 A Relational Model of Artificial Social Systems

In this section, we sketch a relational model of artificial social systems. Artificial social systems are viewed as as relational systems of interacting agents, where agents play social roles and are connected to each other according to the different kinds of social relationships that exist between the roles that they play.

In [2], such a model was called the *Population-Organization* model. It emphasizes the modeling of social exchanges in small groups of individuals, adopting the point of view summarized above, according to which social interactions are exchanges of services between agents, the so called, *social exchanges* approach [7,6].

From such point of view, organizational structures are implemented by populations of individuals through two main mechanisms: first, the assignment of social roles to individuals (respectively, the individuals' adoptions of social roles); and second, the realization of social links through exchanges of services between agents.

In this section, we build heavily on the *time-invariant* version of the model introduced in [2] (leaving for future work the consideration of the time-varying version), but we depart somewhat from that presentation in order to allow for the introduction of the notion of *exchange value-based dynamics*, that was not introduced there.

The time-invariant Population-Organization model, $PopOrg = (Pop, Org; imp)$, is construed as a pair of structures, the population structure and the organization structure, together with an implementation relation. In the following subsections, we present each structure in turn. We end by presenting the implementation relation.

4.1 The Time-Invariant Population Structure

The population of a multiagent system consists of the set of agents that inhabit it. The population structure of a multiagent system is its population set together with the set of all behaviors that the agents are able to perform, and the set of all exchange processes that they can establish between them (for simplicity, we consider here only pairwise exchanges).

Let T be the time structure where a multiagent system is being analyzed (e.g., T is a discrete sequence of time instants).

The *population structure* of a time-invariant multiagent system is a tuple $Pop = (Ag, Act, Bh, Ep; bc, ec)$ where:

- Ag is the set of agents, called the *population set*;
- Act is the set of all *actions* (communication actions and actions on concrete objects) that may be performed by the agents of the system;
- $Bh \subseteq [T \to \wp(Act)]$ is the set containing any behavior (sequence of sets of actions) that any agent is able to perform;
- $Ep \subseteq [T \to \wp(Act) \times \wp(Act)]$ is the set containing any *exchange process* that any two agents may perform by executing together and/or interleaving appropriately their behaviors;
- $bc : Ag \to \wp(Bh)$ is the *behavioral capability* function, such that for all $a \in Ag$, $bc(a)$ is the set of behaviors that agent a is able to perform;
- $ec : Ag \times Ag \to \wp(Ep)$ is the *exchange capability* function, such that for all $a_1, a_2 \in Ag$, $ec(a_1, a_2)$ is the set of exchange processes that agents a_1 and a_2 may perform between them;
- $\forall a_1, a_2 \in Ag \; \forall e \in ec(a_1, a_2) \; \forall t \in T$:

$$Prj_1(e(t)) \subseteq \bigcup \{b(t) \mid b \in bc(a_1)\} \; \wedge \; Prj_2(e(t)) \subseteq \bigcup \{b(t) \mid b \in bc(a_2)\},$$

that is, the exchange capability is constrained by the behavioral capability function.

Given $t \in T$, we note that $bc(a)(t) = \{b(t) \mid b \in bc(a)\}$ is the family of sets of actions that agent a may perform at time t, given its behavioral capability $bc(a)$.

We also note that, in general, $ec(a_1, a_2)$ should be deducible from $bc(a_1)$ and $bc(a_2)$ and from any kind of restriction that limit the set of possible exchanges (e.g., limitations in the interaction medium, social norms, etc.), but since we are presenting a bare model where such restrictions are not present, it is sensible to include ec explicitly in the description of the population structure.

By the same token, $bc(a)$ should be deducible from any *internal description* of agent a, where its behavioral capability is constructively expressed, but since we are taking an external (observational) point of view, we include bc explicitly in the model.

Finally, we note that the definition is given in time-invariant terms. However, any of the sets Ag, Act, Bh, Ep included in the population structure, and both the behavioral capability bc and exchange capability ec, are time-variant, in general. This is formalized in [2], but not taken into account here.

The Time-Invariant Organization Structure. The *time-invariant organization structure* of a time-invariant population structure $Pop = (Ag, Act, Bh, Ep, bc, ec)$ is a structure $Org = (Ro, Li, lc)$, where

- $Ro \subseteq \wp(Bh)$ is the set of *social roles* existing in the organization, a role given by a set of behaviors that an agent playing the role may have to perform;
- $Li \subseteq Ro \times Ep \times Ro$ is the set of *social links* that exist in the organization between pairs of social roles, each social link specifying an exchange process that the agents performing the linked roles may have to perform;
- $lc : Ro \times Ro \rightarrow \wp(Li)$ is the link capability of the pairs of social roles, that is, the set of social links that the pairs of roles may establish between them;
- $\forall l \in Li \, \exists r_1, r_2 \in Ro : l \in lc(r_1, r_2)$, that is, every link has to be in the link capability of the two roles that it links.

The Time-Invariant Implementation Relation. The fact that a given organization structure is operating over a population structure, influencing the set of possible exchanges that the agents may have between them, is represented by an *implementation relation* $imp \subseteq (Ro \times Ag) \cup (Li \times Ep)$, where

- $Ro \times Ag$ is the set of all possible *role assignments*, i.e., the set of all possible ways of assigning roles to agents, so that if $(r, a) \in imp$, then the social role r is assigned to agent a, so that a is said to play role r (possibly in a shared, non-exclusive way) in the given organization;
- $Li \times Ep$ is the set of all possible *link supports*, i.e., the set of all possible ways of supporting social links, so that if $(l, e) \in imp$, then the social link l is said to be supported (in a possibly shared, non-exclusive way) by the exchange process e, and so indirectly supported by the agents that participate in e and that implement the roles linked by l.

We note that an organization implementation relation imp does not need to be a bijection: many roles may be assigned to the same agent, many agents may support a given role, many links may be supported by a given exchange process, many exchange processes may support a given link. Moreover, this relation may be partial: some roles may be assigned to no agent, some agents may be have no roles assigned to them, some links may be unsupported, some exchange processes may be supporting no link at all.

A *proper implementation relation* is an implementation relation that respects organizational roles and organizational links by correctly translating them in terms of agents, behaviors and exchange processes. Given an implementation relation $imp \subseteq (Ro \times Ag) \cup (Li \times Ep)$, a social role $r \in Ro$ is said to be *properly implemented* by a subset $A \subseteq Ag$ of agents whenever the following conditions hold:

(i) $\forall a \in A : (r, a) \in imp$, i.e., all agents in A participate in the implementation of r;
(ii) $\forall t \in T : \bigcup\{b(t) \mid b \in r\} \subseteq \bigcup\{b'(t) \mid b' \in bc(a), a \in A\}$, i.e., the set of behaviors required by r may be performed by the agents of A (in a possibly shared, non-exclusive way).

Also, a social link $l = (r_1, e, r_2) \in Li$ is said to be *properly implemented* by a subset $E \subseteq ec(a_1, a_2)$ of the exchange processes determined by the exchange capabilities of two agents a_1, a_2, whenever the following conditions hold:

(i) $\forall e \in E : (l, e) \in imp$, so that every exchange process helps to support the link;
(ii) $r_1 \, e \, r_2$ are properly implemented by the agents a_1 and a_2, respectively; and
(iii) $\forall t \in T : e(t) \subseteq \bigcup\{e'(t) \mid e' \in E\}$, i.e., the exchange process required by l may be performed by the ones in E (in a possibly shared, non-exclusive way).

That is, imp is a *proper implementation relation* if and only there is a subset $A \subseteq Ag$ of agents and a subset $E \subseteq ec(a_1, a_2)$ of the exchange processes determined by the exchange capabilities of any two agents $a_1, a_2 \in A$ such that each social role $r \in Ro$ is properly implemented by A and each social link $l \in Li$ is properly implemented by E.

A time-invariant population-organization structure $PopOrg = (Pop, Org, imp)$ is *properly implemented* if and only imp is a proper implementation relation.

5 Towards an Exchange Value-Based Notion of Morality for Artificial Social Systems

5.1 An Informal Account

First, we remark that the exchange values used to assess the exchanged services are of a subjective nature (e.g.: good, bad, regular, insufficient, excellent, etc.) and should thus be expressed in a qualitative way [4,3].

Values of the kind *cost* (r) and *benefit* (s) are called *material values* in [7], because they refer to the assessment of an actual service. Values of the kind *debt* (t) and *credit* (v) are called *virtual values* (or, *potential values*) because they refer to services that have yet to be performed.

Notwithstanding this qualitative nature of the exchange values, the fact is that by the agents continuously assigning such values to the services exchanged, and by their use of those values in their decisions about the relationships supported by such exchanges (e.g.: to discontinue a relationship that is giving no benefit to the agent) a truly *economy of exchange values* emerges in the social system, tightly connected to the *operational dynamics* of the system (e.g., to the variations in the efficiency, readiness or quality with which the agents perform their services) and to the *dynamics of the organization* of the system itself (when the organization is allowed to be time-variant).

Next, we notice that in [7] such economy of exchange values is proposed as the basis for an operatory model of moral behaviors, whose aims are the conservation of the extant social relationships that are positively valued by the agents.

Also, we notice that such view is completely coincident with that expressed in [5], where the caring about the continuity of social relationships is taken as the foundation of morality.

Finally, we notice that that view is also completely coincident with the approach proposed by the theory of social dependence [1], where agents are lead to care about others because of the objective dependence that arises between them, due to each agent's usual lack of capability to fully achieve its goals.

Thus, we are lead to conclude that a circle of ideas exists, linking the various approaches sketched above, so that an attempt can be made to establish an operational notion of morality for multiagent systems. Namely:

- the basic fact of the objective social dependences [1] leads agents to look for other agents, in order to exchange services that will help each other to achieve the goals required by the roles that they perform in the system;
- such exchange of services constitutes the operational basis for the social relationships that the agents establish with one another [7], supporting the organizational links between their respective roles [2];
- morality concerns essentially the caring about social relationships and the aim of conserving them when they are beneficial to the agents involved in them [5];
- the assessment and valuation of a social relationship existing between two agents can be neatly performed on the basis of the exchange values that emerge from the exchange of services promoted by such relationship (see below);
- thus, moral rules and moral values may possibly be expressed in terms of rules that help agents to decide about the correctness or wrongness of each other actions, basing their decisions on the achievement or not of equilibrated balances of exchange values, and taking as one of the basic moral values the conservation of social relationships.

Such notion of morality, that is, morality based on the assessment of balances of exchange values, requires three kinds of commitments from the part of the agents:

1. the commitment to assign material values in a realistic way, thus producing values of cost and benefit that do not overestimate neither underestimate the services performed;
2. the commitment to assign virtual values also in a realistic way, so that services expected to be performed in return in the future, to settle debts and credits acquired previously, are neither over- nor underestimated concerning the expectation of their results;
3. and finally, the commitment to respect in the future the virtual values of credit and debit that arose from past exchanges, so that services performed early in time are properly payed back later.

An *exchange value-based moral system* can thus be defined for a multiagent system, as a set of values and rules that enable agents to:

1. qualitatively evaluate the services they exchange between each other;
2. commit to debts and credits acquired from services performed earlier;
3. reason about the moral rules that aim the conservation of social relationships that are beneficial to the agents involved in them;
4. express praise (and blame) for actions that fit (do not fit) the moral rules adopted in the social system;
5. take moral rules, praises and blames into consideration in their social reasoning, that is, when they are deciding how to behave with respect to other agents in the social system.

5.2   Formal Conditions for Equilibrated Social Exchanges

Let $PopOrg = (Pop, Org; imp)$ be the (time-invariant) population-organization of a social system, where $Pop = (Ag, Act, Bh, Ep; bc, ec)$ is the population structure, $Org = (Ro, Li, lc)$ is the organization structure, and $imp \subseteq (Ro \times Ag) \cup (Li \times Ep)$ is the organization implementation relation. Let T be the time structure.

We assume that the way agents valuate their interactions (services exchanged) is modeled by a function $val : T \times Ag \times Act \rightarrow V$, so that $val(t, a, \sigma) = v$ denotes that, at time t, agent a assigned a value v to the service σ performed at that time by some agent in the system[2].

We then use such function to define the exchange values assigned to production and consumption of services in the different steps of the exchange stages. Let $I^t_{\alpha\beta}$ denote that an exchange stage of the kind $I_{\alpha\beta}$ occurred at some time t, with agent α performing some service to agent β. Let $\sigma(I^t_{\alpha\beta})$ be such service. Then we may define that $r_{I^t_{\alpha\beta}} = val(t, \alpha, \sigma(I^t_{\alpha\beta}))$ and $s_{I^t_{\alpha\beta}} = val(t, \beta, \sigma(I^t_{\alpha\beta}))$.

Analogously, let $II^t_{\alpha\beta}$ be an exchange stage of the kind $II_{\alpha\beta}$ that involved a service $\sigma(II^t_{\alpha\beta})$ and that occurred at some time t. Then, we also may state that $r_{II^t_{\alpha\beta}} = val(t, \beta, \sigma(II^t_{\alpha\beta}))$ and that $s_{II^t_{\alpha\beta}} = val(t, \alpha, \sigma(II^t_{\alpha\beta}))$.

The structure of the stages of service exchanges (Fig. 1) may then be used to support a simple definition of the condition of *equilibrated exchange*.

Let $V = (V, \leq)$ be a partially ordered *scale of exchange values*, assumed for simplicity to be common to all agents in the social system[3]. V is the structure from where the eight exchange values are taken.

Let $B_{\alpha\beta} : T \rightarrow V^8$ be the *balance* function of the interaction between agents α and β (see [4]), such that

$$B_{\alpha\beta}(t) = (R^t_{I_{\alpha\beta}}, S^t_{I_{\beta\alpha}}, T^t_{I_{\beta\alpha}}, V^t_{I_{\alpha\beta}}, V^t_{II_{\alpha\beta}}, T^t_{II_{\beta\alpha}}, R^t_{II_{\beta\alpha}}, S^t_{II_{\alpha\beta}})$$

aggregates the sum total of each of the eight exchange values:

---

[2] For simplicity, we consider here that services are instantaneous.

[3] Note that with several scales of exchange values being used in the society, a correspondence relation between such scales would be required. Otherwise, the agents would have no means to reach consensus about the correctness of their respective moral reasonings.

$$R^t_{\text{I}_{\alpha\beta}} = \sum_{0 \le \tau \le t} r_{\text{I}^\tau_{\alpha\beta}}$$

and similarly for the other exchange values.

A series of exchanges stages of the kinds $\text{I}_{\alpha\beta}$ and $\text{II}_{\alpha\beta}$ is equilibrated at time t if and only if the balance of the exchanges, given by the sum total of each kind of exchange value $B_{\alpha\beta}(t)$, satisfies the three conditions:

$$\mathcal{C}^t_{\text{I}_{\alpha\beta}} : (R^t_{\text{I}_{\alpha\beta}} = S^t_{\text{I}_{\beta\alpha}}) \wedge (S^t_{\text{I}_{\beta\alpha}} = T^t_{\text{I}_{\beta\alpha}}) \wedge (T^t_{\text{I}_{\beta\alpha}} = V^t_{\text{I}_{\alpha\beta}})$$
$$\mathcal{C}^t_{\text{II}_{\alpha\beta}} : (V^t_{\text{II}_{\alpha\beta}} = T^t_{\text{II}_{\beta\alpha}}) \wedge (T^t_{\text{II}_{\beta\alpha}} = R^t_{\text{II}_{\beta\alpha}}) \wedge (R^t_{\text{II}_{\beta\alpha}} = S^t_{\text{II}_{\alpha\beta}})$$
$$\mathcal{C}^t_{\text{I}_{\alpha\beta}\text{II}_{\beta\alpha}} : (V^t_{\text{I}_{\alpha\beta}} = V^t_{\text{II}_{\alpha\beta}})$$

The conditions guarantee that, if the relationship is equilibrated at time t, all investment made in services were fully credited to the agent that performed them (condition $\mathcal{C}^t_{\text{I}_{\alpha\beta}}$), every credit charged was fully paid (condition $\mathcal{C}^t_{\text{II}_{\alpha\beta}}$), and agents that charged other agents charged just the correct values that were owed (condition $\mathcal{C}^t_{\text{I}_{\alpha\beta}\text{II}_{\beta\alpha}}$).

5.3   A Tentative Formalization of Some Simple Moral Rules

Let's turn now to the operational morality that builds on the above conditions for equilibrated social exchanges. The core of morality being the set of its moral rules, we must give then the general structure of the moral rules that can make operational the proposed exchange value-based notion of morality.

The most basic exchange value-based moral rule concerns the *equilibrium of the material exchanges*. It is given here in an informal way, assuming that a notion of material profit has been defined (see below):

$\text{ME}_{\beta\alpha}$:

> If agents α and β have been interacting according to a certain social relationship, and it happens that agent β got greater material profit from the interaction than agent α got from it, then it is expected that at some time in the future agent β performs a service for agent α, in return for the greater profit received, and such that the final profit received by α from the return service equals the final profit received by β.

An action of performing a return service is thus a *moral action* if it respects such rule: keeps the profits of each agent equal to each other.

Such rule is of a moral nature, not of a juridical one, because it is not expected that an institutional authority monitors the agents in order to check that the rule is respected. On the contrary, it is expected that the agents help each other in checking the application of such rule, and that they manifest to each other their assessments (praise or blame) of the agents that do follow or do not follow it.

Also, it is not expected that agents that do not follow the rule will be officially sanctioned by an institutionalized authority. What is expected is that the agents will take into account in their social reasonings the manifestations that other agents expressed about the morality of the actions of each other, and deliberate according to such manifestations.

We note that if rule $ME_{\beta\alpha}$ is not followed by agent β when performing the return service for α, such pair of agents will not achieve an equilibrated state: either one of the agents will result with a balance more positive then it needs, or will result with a balance more negative then it deserves.

In any case, if the agents have assumed the exchange value-based morality we are talking about, and thus committed to an equilibrated relationship between them (which is the case we are assuming here), a *justification* for the failure in achieving the equilibrium is required from agent β when equilibrium is not achieved. Also, agent β should be prepared to present the reasons why the rule was not followed when it acted, and why, eventhough a moral rule was broken, β should not be blamed for it[4].

Formally, the moral rule underlying such situation may be expressed as follows. Let $B_{ab}(t) = (R^t_{I_{ab}}, S^t_{I_{ba}}, T^t_{I_{ba}}, V^t_{I_{ab}}, V^t_{II_{ab}}, T^t_{II_{ba}}, R^t_{II_{ba}}, S^t_{II_{ab}})$ be the current balance of exchange values in the relationship between agents a and b when a and b respectively take the places of α and β in the exchange schema. Correspondingly, let $B_{ba}(t) = (R^t_{I_{ba}}, S^t_{I_{ab}}, T^t_{I_{ab}}, V^t_{I_{ba}}, V^t_{II_{ba}}, T^t_{II_{ab}}, R^t_{II_{ab}}, S^t_{II_{ba}})$ be the current balance of exchange values in the relationship between a and b when a and b respectively take the places of β and α in the exchange schema. The *full current balance* of the relationship can then be expressed as follows, by adding exchange values that concern the same agent:

$$B(t) = (R^t_{I_{ab}} + R^t_{II_{ab}}, S^t_{II_{ab}} + S^t_{I_{ab}}, T^t_{I_{ab}} + T^t_{II_{ab}}, V^t_{I_{ab}} + V^t_{II_{ab}},$$
$$R^t_{I_{ba}} + R^t_{II_{ba}}, S^t_{I_{ba}} + S^t_{II_{ba}}, T^t_{I_{ba}} + T^t_{II_{ba}}, V^t_{I_{ba}} + V^t_{II_{ba}})$$

Applying to $B(t)$ conditions of equilibrium analogous to those in $C^t_{I_{\alpha\beta}}$, $C^t_{II_{\alpha\beta}}$ and $C^t_{I_{\alpha\beta}II_{\beta\alpha}}$, we get the following set of overall conditions of equilibrium for the social relationship between a and b at time t:

$$OC^t_{ab} : (R^t_{I_{ab}} + R^t_{II_{ab}} = S^t_{II_{ab}} + S^t_{I_{ab}}) \wedge (S^t_{II_{ab}} + S^t_{I_{ab}} = T^t_{I_{ab}} + T^t_{II_{ab}})$$
$$\wedge (T^t_{I_{ab}} + T^t_{II_{ab}} = V^t_{I_{ab}} + V^t_{II_{ab}})$$
$$OC^t_{ba} : (R^t_{I_{ba}} + R^t_{II_{ba}} = S^t_{I_{ba}} + S^t_{II_{ba}}) \wedge (S^t_{I_{ba}} + S^t_{II_{ba}} = T^t_{I_{ba}} + T^t_{II_{ba}})$$
$$\wedge (T^t_{I_{ba}} + T^t_{II_{ba}} = V^t_{I_{ba}} + V^t_{II_{ba}})$$
$$OC^t_{ab}OC^t_{ba} : (V^t_{I_{ab}} + V^t_{II_{ab}} = V^t_{I_{ba}} + V^t_{II_{ba}})$$

Building on those conditions of equilibrium, we can formally state the basic moral rule concerning the equilibrium of material profits.

First, we define the notion of *qualitative material profit* of agent α with respect to agent β at time t. Let $P = (\{pos, null, neg\}, <)$ be the strictly ordered set of possible values of qualitative material profits, ordered according to $neg < null < pos$.

---

[4] Moral rules are not supposed to be unbreakable. Actions, however, are asked to be justified on the basis of higher-level moral rules, when they do not follow moral rules of a certain level of generality [5].

Then, the material profit function $P_{\alpha\beta} : T \rightarrow \{pos, null, neg\}$ can be defined as:

$$P_{\alpha\beta}(t) = \begin{cases} pos & \text{if } (S^t_{\text{II}_{\alpha\beta}} + S^t_{\text{I}_{\alpha\beta}}) > (R^t_{\text{I}_{\alpha\beta}} + R^t_{\text{II}_{\alpha\beta}}), \\ null & \text{if } (S^t_{\text{II}_{\alpha\beta}} + S^t_{\text{I}_{\alpha\beta}}) = (R^t_{\text{I}_{\alpha\beta}} + R^t_{\text{II}_{\alpha\beta}}), \\ neg & \text{if } (S^t_{\text{II}_{\alpha\beta}} + S^t_{\text{I}_{\alpha\beta}}) < (R^t_{\text{I}_{\alpha\beta}} + R^t_{\text{II}_{\alpha\beta}}), \end{cases}$$

Thus, the moral rule $\text{ME}_{\beta\alpha}$ can be crudely expressed, in a tentative way, using first order predicate calculus, as follows:

$\text{ME}_{\beta\alpha}$:

$\forall t \in T : (P_{\beta\alpha}(t) > P_{\alpha\beta}(t)) \Rightarrow (prfm(\beta, \sigma, \alpha, t+1) \wedge (P_{\beta\alpha}(t+1) = P_{\alpha\beta}(t+1)))$

where $prfm(\beta, \alpha, \sigma, t+1)$ indicates that at time $t+1$ agent β performs a service σ for agent α, with the intended meaning that if at any time $t \in T$ it happens that agent β has a profited more than agent α from the interaction, then in the next time $t+1$ it should perform a service for α in order to equilibrate their respective profits[5].

5.4   Tentative Formulation of Procedures for Moral Judgements and Sanctions

Moral rules, supposed to be enforced by the population of agents on themselves through direct expressions of moral judgements on each other actions, not by institutionalized authorities, necessarily need to be supported by generally accepted procedures of moral judgement and sanction.

With respect to the basic exchange value-based moral rule $\text{ME}_{\beta\alpha}$ defined above, two simple such procedures may be specified.

First, for the moral judgement of the behavior of agent β in situations where β is supposed to re-install the equilibrium in its relationship with agent α, a formal definition of such judgement can be given thus.

Let $MJ = \{blame, praise\}$ be the set of possible results of moral judgements. Then, the moral judgement, at time t, of the behavior of agent β toward agent α, concerning their exchanges, is given by the function $MJ_{\beta\alpha} : T \rightarrow MJ$, defined as:

$$MJ^t_{\beta\alpha} = \begin{cases} praise & \text{if moral rule } \text{ME}_{\beta\alpha} \text{ was respected at time } t, \\ blame & \text{otherwise.} \end{cases}$$

Notice that a moral judgement like $MJ^t_{\beta\alpha}$ can be made by any agent in the system which is able of moral reasonings, that is, capable to reason with moral rules.

Second, with respect to sanctions, a very simple procedure can specified thus: agent α continues to interact with agent β as long as agent β respects the moral rule, and sanctions β with the ceasing of the interaction if the moral rule is disrespected.

This can be formally expressed in a preliminary way with the help of the imp relation between the organization and population. Let $e \in Ep$ be the exchange process that is

---

[5] We notice, immediately, the unnecessarily strong requirement imposed by such tentative formulation, that the return service be realized immediately. A formulation like the Role-Norms-Sanctions schema [9] for the rule should solve such problem.

occurring at time t between agents α and β. Let $i \in Li$ be the organizational link between the roles played by α and β in the organizational structure of the system. Let $(i, e) \in imp$ at time t, denoted by $imp^t(i, e) = true$. Then

$$imp^{t+1}(i, e) = \begin{cases} true & \text{if } MJ^t_{\beta\alpha} = praise, \\ false & \text{if } MJ^t_{\beta\alpha} = blame. \end{cases}$$

Note that the this preliminary formal specification of the sanction procedure does not tell much: it only tells that the interaction will not happen anymore in the system and that, thus, the organizational link will no more be implemented by imp. It does not even require that the moral judgement be accepted by agent α, which is clearly a must. On the other hand, sanctions like this clearly imply that the system has a time-varying organization structure, going beyond the formal organization structure considered in this paper.

6   Conclusion and Open Problems

We have shown that, given a multiagent system where agents are able to assess the quality of the services that they exchange with each other during their interactions, a natural notion of morality arises, based on the qualitative economy of exchange values that emerges from such interactions.

The operational nature of that morality may be directed to help the agents conserve the social relationships that exist in the system, and to improve the quality of the exchanges that implement them, thus contributing to the stability of the system.

We have, however, just scratched the basic ideas of such moral system. Much more details have to be worked out: the variety of moral rules and moral beliefs that can capture the dynamics of the economy of exchange values, the social procedures that allow moral judgements and moral sanctions to become effective, the details of the moral reasoning that the agents are required to do while taking into account the various aspects involved in the decision of continuing or discontinuing an interactions on the basis of moral judgements, etc. In particular, attention has to be given to the real impact of the availability of a moral system on the different properties (stability, effectiveness, adaptability, etc.) of the multiagent system where such moral system is installed.

In summary, in this paper we aimed only to briefly sketch the basic ideas concerning the possibilities open by the availability of an operational notion of morality in multi-agent system. But we have to emphasize, before we finish, the essential need of procedures that allow agents to assess the quality of the services exchanged in the system. An attempt to reach such result is, for instance, [8].

References

1. Castelfranchi, M.M.C., Cesta, A.: Dependence relations among autonomous agents. In: Decentralized A.I.-3, pp. 215–227. Elsevier, Amsterdam (1992)
2. Demazeau, Y., Costa, A.C.R.: Populations and organizations in open multi-agent systems. In: PDAI 1996. 1st National Symposium on Parallel and Distributed AI, Hyderabad, India (1996)

3. Dimuro, G.P., Costa, A.C.R.: Exchange values and self-regulation of exchanges in multi-agent systems: the provisory, centralized model. In: Brueckner, S.A., Serugendo, G.D.M., Karageorgos, A., Nagpal, R. (eds.) ESOA 2005. LNCS (LNAI), vol. 3464, pp. 92–106. Springer, Heidelberg (2005)
4. Dimuro, G.P., Costa, A.C.R., Palazzo, L.A.M.: Systems of exchange values as tools for multi-agent organizations. Journal of the Brazilian Computer Society 11(1), 31–50 (2005) (Special Issue on Agents' Organizations)
5. Furrow, D.: Ethics. Continuum, London (2005)
6. Homans, G.: Social Behavior – Its Elementary Forms. Addison-Wesley, New York (1961)
7. Piaget, J.: Sociological Studies. Routlege, London
8. Rodrigues, M.R., Luck, M.: Cooperative interactions: An exchange values model. In: COIN@ECAI 2006, Riva del Garda (August 28, 2006)
9. Nickles, M., Rovatsos, M., Weiss, G.: A Schema for Specifying Computational Autonomy. In: Petta, P., Tolksdorf, R., Zambonelli, F. (eds.) ESAW 2002. LNCS (LNAI), vol. 2577, pp. 92–103. Springer, Heidelberg (2003)

Intelligent Farmer Agent for Multi-agent Ecological Simulations Optimization

Filipe Cruz[1], António Pereira[1], Pedro Valente[1], Pedro Duarte[2], and Luís Paulo Reis[1]

[1] LIACC – Faculdade de Engenharia da Universidade do Porto
Rua Dr. Roberto Frias s/n, 4200-465 Porto, Portugal
Tel.: +351 22 508 14 00
[2] CEMAS – Universidade Fernando Pessoa
Praça 9 de Abril, 349, 4249-004 Porto, Portugal
Tel.: +351 22 507 13 00
{filipe.cruz,amcp,pedro.valente,lpreis}@fe.up.pt, pduarte@ufp.pt

Abstract. This paper presents the development of a bivalve farmer agent interacting with a realistic ecological simulation system. The purpose of the farmer agent is to determine the best combinations of bivalve seeding areas in a large region, maximizing the production without exceeding the total allowed seeding area. A system based on simulated annealing, tabu search, genetic algorithms and reinforcement learning, was developed to minimize the number of iterations required to unravel a semi-optimum solution by using customizable tactics. The farmer agent is part of a multi-agent system where several agents, representing human interaction with the coastal ecosystems, communicate with a realistic simulator developed especially for aquatic ecological simulations. The experiments performed revealed promising results in the field of optimization techniques and multi-agent systems applied to ecological simulations. The results obtained open many other possible uses of the simulation architecture with applications in industrial and ecological management problems, towards sustainable development.

Keywords: Ecological simulations, agents, optimization, simulated annealing, tabu search, genetic algorithms, reinforcement learning.

1 Introduction

Coastal ecosystems are used for multiple purposes (fishing, tourism, aquaculture, harbor activities, sports, etc.) and are the final destination of many pollutants generated by agriculture and other human activities. Considering that over 60% of the world's population lives within 60 km from the sea, the correct management of these ecosystems is very important towards world sustainable development [1] [2].

The diversity of usages and the opposite interests of stakeholders and some institutional authorities, coupled with the slowness of the decision process due to the huge number of possible decisions generated by the different management policies, make the implementation of efficient automatic management algorithms very difficult to achieve [3].

In this context, the use of intelligent agents [4] [5] [6] seems to be very promising. Each institutional authority and stakeholder may be modeled as an agent, interacting

J. Neves, M. Santos, and J. Machado (Eds.): EPIA 2007, LNAI 4874, pp. 593–604, 2007.

with simulation tools - able to predict the outcome of different decisions - getting results and configuring new conditions for further simulation experiments.

The human factors can be represented by agents with specific goals and visions for the coastal area that can sometimes contradictory or of unsure effect upon the ecosystem. Examples of such agents are tourism managers, bivalve farmers, and representatives of civil authorities. Each has different goals that they want to see maximized or minimized. Including them in a multi-agent system surrounding the ecological simulator allows them to be concurrently represented and enabling the prediction, to some extent, of the outcome through time of their interactions with the ecosystem [7].

Realistic simulations of ecosystems require describing several physical, chemical and biological processes in mathematical terms. Physical processes include flow and circulation patterns, mixing and dispersion of mass and heat, settling and resuspension of planktonic organisms and suspended matter, insulation and light penetration. Chemical processes include biogeochemical cycles of important elements, such as nitrogen and phosphorus. Biological processes include growth and death rate of any organisms that may alter the concentration of different elements. The accurate simulation of these processes is very important for setting up a reliable and realistic model of the whole ecosystem. Detailed representations of these processes can be found in the literature since the 1980's [8].

This paper refers, in particular, to the development of a bivalve farmer agent which interacts with a realistic ecological simulator (EcoDynamo [2]) to find out the best combination of bivalve seeding areas within an allowed area of exploitation and a maximum seeding occupied area.

The developed bivalve farmer agent has implementations and adaptations of known multi-objective optimization algorithms, well documented in the literature such as Simulated Annealing [9], Tabu Search [10], Genetic Algorithms [11] and Reinforcement Learning [12]. It attempts to take advantage of each algorithm positive aspects, minimizing their limitations. The bivalve farmer agent has the objective of finding an optimum combination of locations providing better shellfish growth and production within a bay.

The paper is organized as follows: section 2 describes the problem in analysis; section 3 presents the architecture and implementation; section 4 contains a description and analysis of obtained results; section 5 mentions the conclusions and indicates future work.

2 Problem Statement

As a proof of concept, a bivalve farmer agent was developed to discover by itself the best combinations of locations to seed and harvest bivalve species. The bivalve farmer agent interacts with the simulator application in order to run series of test simulations seeking to find the optimum, or very near optimum, combination of lagoon bivalve farming regions (cells) where bivalve production would be maximized.

The tests were carried out using one validated model for Sungo Bay, People's Republic of China [1]. Sungo Bay is modeled as a 2D vertically integrated, coupled hydrodynamic-biogeochemical model, based on a finite difference bathymetric staggered grid with 1120 cells (32 columns x 35 lines) and a spatial resolution of 500m

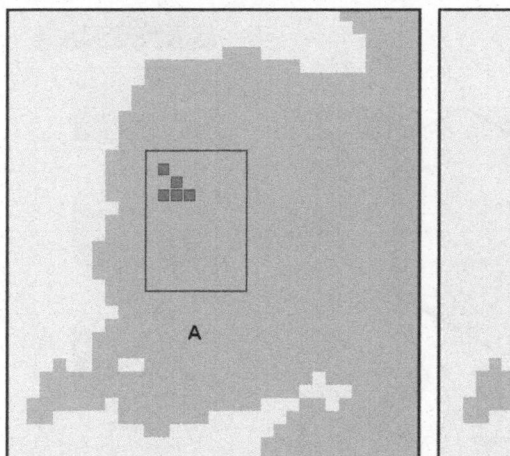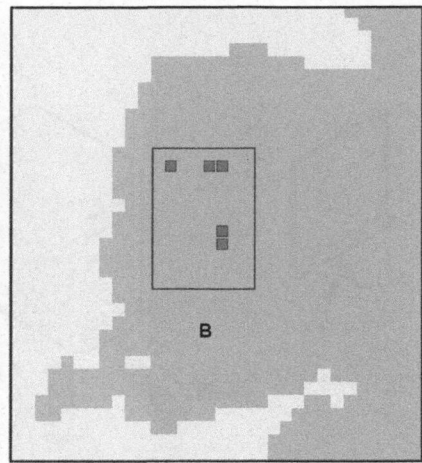

Fig. 1. Two different configurations (A and B) of farming regions: five farming cells are selected from the wide available area in each simulation

[1]. The idea is to find the best solution - five regions within a rectangular area of 88 cells (square regions) that maximize bivalve production. Hereafter, "solution" will be used to refer to any combination of five regions within the mentioned area. It is important to refer that due to the realistic characteristics of the ecological simulation, the existence of bivalves in one location will affect the growth of bivalves in the neighborhood. Simulated bivalves feed on suspended particles such as detritus and phytoplankton cells, with the potential to cause significant local depletion of these food items. Therefore, placing many regions together could negatively affect the potential yield of those regions. The extent of the influence depends on many factors such as tidal flux, quantity and quality of suspended food particles, and water quality, substantially increasing the complexity of the problem. Figure 1 shows two different possible combinations of regions (cells).

Taking into account the heavy time and processor power required to perform full length realistic simulations (complete bivalve culture cycle is approximately 1.5 years – about 1 576 800 simulation time steps of 30 seconds) it was a requirement of the bivalve farmer agent to intelligently choose its test combinations of cells with simulations of only 1000 time steps.

3 Implementation

3.1 Architecture

The implementation is based on a multi-agent architecture following the principles of Weiss [5] and Wooldridge [6], where agents communicate with the simulator application via TCP/IP packets. The simulation tool (EcoDynamo [2]) was developed for the simulation of aquatic ecosystems, and is able to communicate with several agents representing the human interests over the simulated ecosystem, namely the stakeholders, the national authorities and the municipality. This communication is

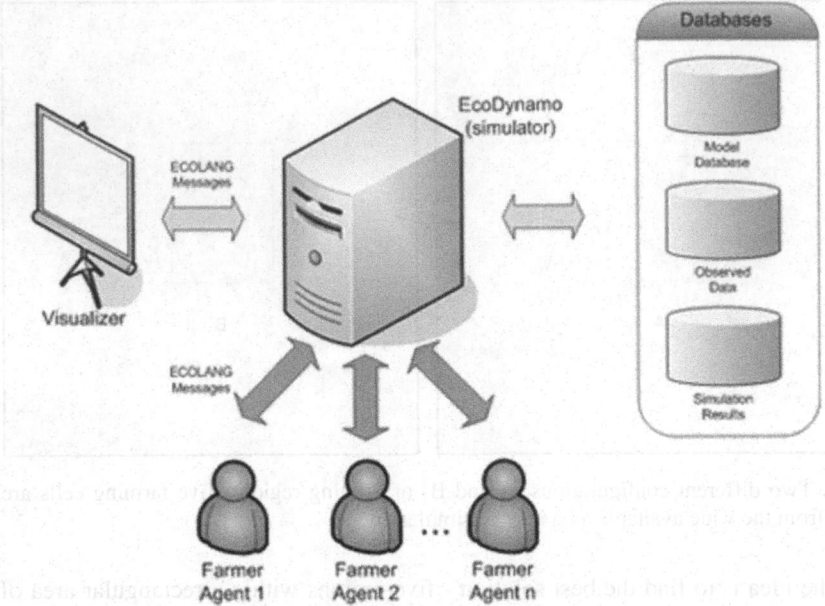

Fig. 2. Experimental system architecture

supported by ECOLANG messages [2] - ECOLANG is a high-level language that describes ecological systems in terms of regional characteristics and translates living agent's actions and perceptions. The format of the messages enables easy readability, simplicity and expandability, and is independent from any computational platform or operating system. Figure 2 shows the architecture used in the experiment with the simulator, the farmer agent and one visualizer application.

The architecture of the multi-agent system was structured to allow several agents to interact with the simulator at the same time with different purposes in mind. There is much space for diverse applications of machine learning in ecological modeling which can be better exploited with the proper architecture [7] [8] [13].

3.2 Implemented Algorithms

To implement the sense of intelligence in the choice of region combinations of the bivalve farmer agent, a system of customizable tactics was developed. The system allows different multi-objective optimum solution finder techniques to be applied at the same time. The base of the program is a simple hill-climbing optimization algorithm based on Simulated Annealing with Monte Carlo probability [9] [14], iteratively seeking a new possible solution and accepting it as the best found so far if its quality is considered higher than the previous best. The program, however, allows for several other configurable optimizations to be activated influencing the selection logic. These optimizations are based and adapted from documented implementations of Tabu Search [10] [14] [15], Genetic Algorithms [16] [17] and Reinforcement Learning [12]. A simple example: one configuration may allow the iterations to start functioning with a random search algorithm, trigger a genetic algorithm based crossbreeding at 45% of the search process, and switch to a low dispersion local neighbor solution selection near the end.

It is important to also notice that depending on the algorithms that are used, the selection of the initial combination to test can either greatly assist or greatly hinder the overall performance of the algorithm. Thus, a decision was made to create an architecture that would always start from a random seed but could then easily be configurable to alternate, in real time, between the different implemented algorithms. The selection of what and when to change is described through the so called tactics files, which are nothing more than simple text files with listed algorithms and their correspondent parameters. The fine tune of the parameters in these files to achieve better results can be compared to a meta-heuristic problem.

The algorithms implemented in the farmer agent were called FarmerSA, FarmerTabu, FarmerGA and FarmerRL. A small description of each one follows.

FarmerSA. The implemented Simulated Annealing (SA) algorithm follows the usual guidelines of Kirkpatrick [9], allowing the user to define the typical parameters of initial temperature, final temperature and descent rate.

As documented widely in literature, SA is based in the natural process of slower cooling giving higher resistance to certain materials. A threshold formula slowly increases the probability of accepting a given test solution of superior quality as the temperature lowers by each passing iteration. The overall concept behind the algorithm is allowing the system to have enough time to explore most part of its universe of solutions whilst the entropy is high enough to prevent the algorithm to follow only one path of the best solutions. By slowly restricting its entropy, the algorithm will eventually constrain itself to the best local solutions and hopefully this slow process will prevent it from getting stuck in local maximums instead of finding the global maximum. In critic terms, for some cases it can work very well, but for others it requires a great number of iterations to assure a high probability of finding a good solution and without certainty of being the optimum. It depends on the complexity/linearity of the solution search area itself, on how the neighbor solutions are defined, on the speed of the temperature drop and on a hefty amount of luck with the Monte Carlo probabilities. In synthesis - too many factors that make the exclusive use of this algorithm not recommendable for complex problems.

A special parameter exists in this implementation that allows the algorithm to ignore the Monte Carlo probability. This parameter is used to guarantee using a new solution when its quality is higher than the one considered as the best.

FarmerTabu. The adaptation of Tabu Search [10] was called FarmerTabu and its implementation is based on maintaining a hash list of all the previously tested solutions, so that when it is toggled on, it simply prevents the simulation from choosing a previously tested solution as the next solution to test, also keeping a counter of how many times the algorithm has revoked the choice of a new and previously untested option from its choice of possible next (referred to as neighbor) solution [14]. The user can define a threshold value for the counter of refused solutions to activate a special mutation factor that serves to stir the process into other solution search paths once the current path has apparently already been overly explored without much success.

FarmerGA. The FarmerGA implementation maintains a list of best solutions found so far and crossbreeds them to form new combinations of regions. The user can define the number of solutions to take into account as parents and the number of new breeds. Unlike common Genetic Algorithms [16], this FarmerGA doesn't account for any mutation factor, making imperative that the list of solutions used as parents contains combinations of regions as spread out through the solution universe as possible, in order to achieve results that guarantee a solution not restricted to a local maximum.

The way the FarmerGA crossbreeds the parents list is based on the quality hierarchy and in one parameter configurable by the user. In simple terms, it attempts to breed the best solution with all the others in the list of top candidates, the second best with half of all the other members of the list, the third best with a third of all the others and so on. The child genes are selected as the best individual locations amongst both parents. For each pair of two good solutions (A and B) as seen on figure 3, the best locations are genes of each parent chosen according to their individual performance (measured in tons of bivalve harvested in previous simulation). The child derived (C) is either registered in a list of children to be tested or disregarded if it is present on the FarmerTabu list.

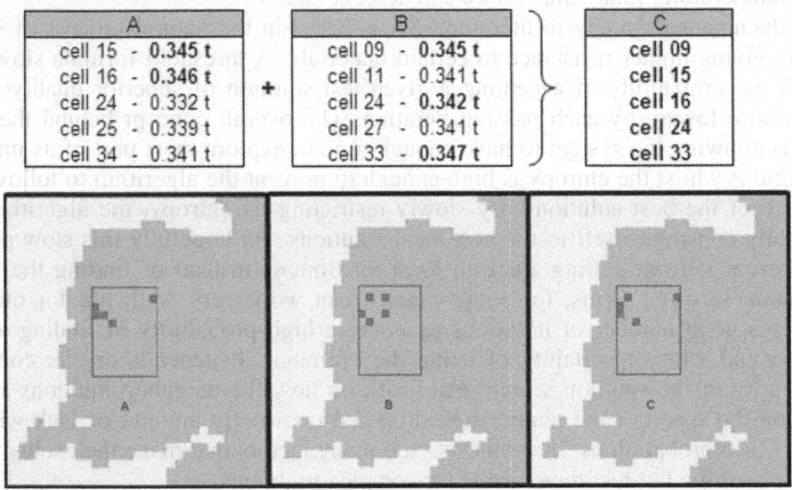

Fig. 3. FarmerGA breeding result: A and B - parent configurations; C - child configuration

FarmerRL. The FarmerRL optimization is somewhat far from what is usually referred to as Reinforcement Learning [12]. This implementation maintains a neighbors list for each possible region, containing information on its neighbor locations quality for farming. As the test iterations occur, calculated weights are summed to the quality of the areas selected in the tested solution, increasing or decreasing its value. The value of the weight depends on the geometric distance between the farming result of an individual zone and the average amongst all the zones of the tested solution.

The farming quality value of each region is taken into account when the next selection of other neighboring solutions for testing is performed. As more iterations occur, the quality values of good bivalve farming regions gets higher and the quality values of bad bivalve farming regions gets lower, thus restraining the scope of search to the better regions. The strength of the weight is configurable by the user and can vary in real-time. The definition of neighbor solutions also influences greatly on how this optimization will perform. There are 8 implemented types of algorithms for choosing the next neighbor solution of FarmerRL:

1. Choosing the best neighbors of each position, using Monte Carlo probability factor.
2. Choosing the best neighbors of each position, only using Monte Carlo probability factor for the positions below average quality.
3. Choosing the best neighbors of each position, without using Monte Carlo probability factor.
4. Choosing the best neighbors of each position, without using Monte Carlo probability factor, but changing only one of the positions.
5. Randomly shift all positions to their neighbors.
6. Randomly shift positions to their neighbors of only one of the positions.
7. Random search new position in positions below average quality.
8. All new positions completely randomly selected.

3.3 Tactics System

A tactics system was developed to allow the users to configure the real-time parameter reconfiguration of the implemented algorithms.

The base algorithm FarmerSA is always being applied and can be configured/parameterized through a text file (*config.cfg*). The parameters on this file will define the number of iterations that the process will take to finish (extracted from initial value, final values and descent rate of the temperature), the number of simulation steps and the tactics file to use during the process.

The remaining algorithms can be configured through the specific tactics file used. This tactics file contains text lines which are referred to as toggles lines. These lines contain the parameterization of the algorithms described previously. If these algorithms are not toggled in this file to start at a given time frame of the process, they will not be used during the process.

FarmerTabu and FarmerRL can be toggled on, re-parameterized or toggled off at any point in time of the process. FarmerGA functions in a different manner: as each toggle line for FarmerGA will trigger the process of running the genetic crossing itself, it is expected to only be toggled a few steps into the process so that it has time to build a large and wide spread enough list of good results for its cross breeding.

Toggles for the same algorithm can be defined more then once per tactics file, typically representing reconfigurations of the parameters at different time frames of the process. As seen in table 1 each toggle line (*tog*) contains several parameters, the first one refers to the algorithm it is configuring (*0 for FarmerTabu, 1 for FarmerGA, 2 for FarmerRL*), the second one to the time frame in percentage during which it shall be activated (e.g: *0.0 will be triggered right from the first iteration, 0.5 will be triggered after half the iterations have occurred*). The other parameters depend on the specific algorithm.

Table 1. Example of a tactics file: 'toggles_rand.tgl'

Keyword	Toggle Number	Param1	Param2	Param3	Param4
init					
tog	0 (TA)	0.0	1	0.8	2
tog	1 (GA)	0.5	1	15	8
tog	1 (GA)	0.8	1	15	8
tog	2 (RL)	0.0	1	0.05	
tog	2 (RL)	0.5	8	0.05	
tog	2 (RL)	0.7	7	0.05	
eof					

Table 1 exemplifies a tactics file with a total of 6 toggles defined:

- The first refers to FarmerTabu (*tog 0*), defining it to actuate from the beginning of the process (*0.0*), turned on (*1*) using a *0.8* threshold for repeated selections in tabu list before applying mutation of type *2* (distribute one random number of solution members for neighboring regions with good qualitative indicators).
- The second type of toggles refers to the FarmerGA (*tog 1*), it is used twice during this process, one at half of the iterations and again after 80% of the iterations have been processed. It is turned on (*1*) and will create a list of maximum 15 children members out of the top 8 members of the best found results so far.
- The third type of toggles refers to the FarmerRL (*tog 2*), which starts with neighbor selection algorithm type *1* (best neighbor with Monte Carlo probability factor), using quality weight update of *0.05*, then changes to neighbor selection type *8* (completely random) at half of the process with weight update of *0.05*, and finally changes to neighbor selection type *7* (half random) at 70% of the process and using weight update of *0.05*.

Different tactics may easily be defined and different syntaxes for the tactics configuration file are also very easy to define, including new parameters for the available algorithms.

4   Results and Discussion

Several different tactics were tested on the Sungo Bay model to assist on determining the optimum combination of 5 bivalve farming regions out of the allowed bivalve farming area. Table 2 shows the results of the best solution encountered by each tactic.

The different tactics were intuitively determined whilst undergoing the development of the application. Some were uniquely written to show the performance following traditional simplistic methods for terms of comparison - *hillclimber* performs a simple choice of the best close neighbor (only one position from the possible five is changed from each solution at a time) and its refined version named *neigh* which allowed for up to 3 positions to shift from each solution at a time.

Table 2. Results of tested tactics

Tactic	Result (Ton)
hillclimber 01	8019.142
hillclimber 02	8027.930
neigh 01	8051.824
neigh 02	8053.852
3GA 01	8049.961
3GA 02	8050.868
neighrand 01	8056.120
neighrand 02	8055.063
neighrand 03	8053.148
totrand 01	8051.824
totrand 02	8051.824

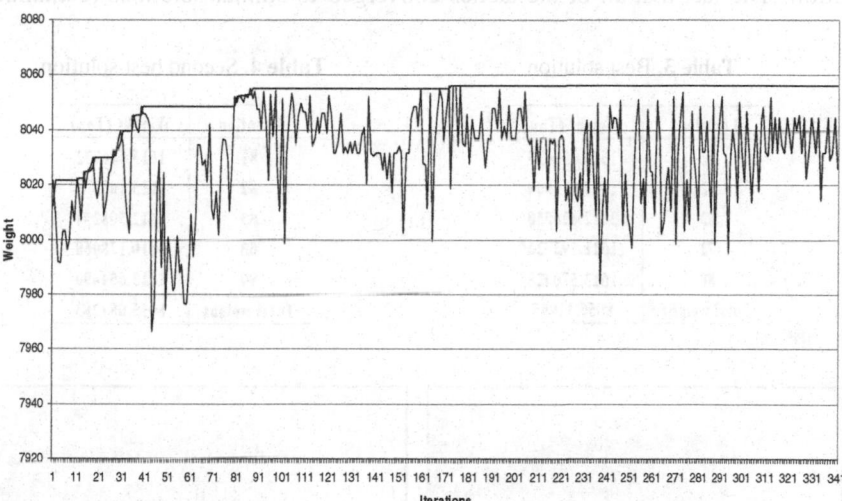

Fig. 4. Results of the *neighrand 01* development

Other tactics were aimed at abusing a certain capability of the application beyond reason to test its impact on the outcome result. One such is the *3GA* tactic which repeats the genetic algorithm variance three times without *a priori* having a proper scope of the whole possible universe. In similar fashion, *totrand* tactic had a more random search orientation (admits bursts of best neighbor searches) in an attempt to measure the real value of scoping the entire universe prior to applying the remaining available methods. At last, *neighrand* is the more balanced of all the tactics devised, despite not being fully fine tuned to always unravel the optimum or very near optimum solution.

Figure 4 shows the development throughout the iterations of tactic *neighrand 01*. This tactic was scripted by the file showed in table 1 (*toggles_rand.tgl*) - had FarmerTabu on during all the process, had 2 instances of FarmerGA occurring once at half way through the simulation and again at 80%, and its FarmerRL changed three times through the process to alter its neighbor selection parameters. The weight refers the amount of bivalves harvested in each iteration of the simulation. Each iteration represents the same simulation period of a new combination/solution of bivalve zones. So the first line on the graph shows the weight result for the current iteration, whilst the second line shows the weight result for the best solution found so far. Testing the different tactics under the same conditions in short time period simulation provided several different combinations of locations which could be tested later in longer time period simulations for the validation of results.

The best results from the tests carried out are presented in tables 3 and 4 and can be seen in figure 5. Analyzing the test results leads to conclude that the best farming regions of the area are located on the lower bottom of the global search area of 88 zones. There is a strong possibility that other, more optimum combinations could be found by re-analyzing the problem concentrating on that area alone.

A reminder must be added that the program is based on a random initial search solution. The fact that all of the tactics converged to similar solutions (containing a

Table 3. Best solution

Location	Weight (Ton)
81	1613.490035
82	1612.912944
83	1612.423538
72	1611.792726
80	1613.570428
Total weight	8056.11995

Table 4. Second best solution

Location	Weight (Ton)
81	1613.646772
82	1613.165278
83	1612.508254
63	1610.178968
80	1613.654490
Total weight	8055.084283

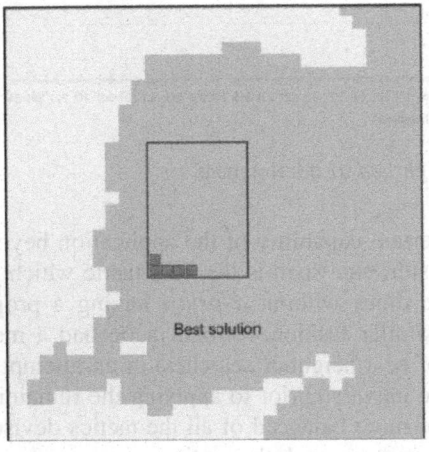 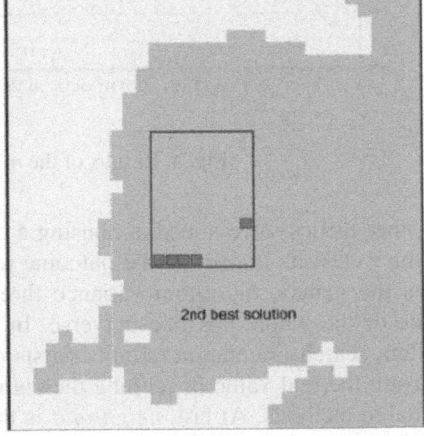

Fig. 5. Visualization of the two best solutions obtained by *neighrand 01* development

majority of elements in early 80s locations) reassures the quality of the location as a most profitable zone to explore bivalve farming.

5 Conclusions and Future Work

Generally, the management of coastal ecosystems may be done in many different ways and there is hardly one optimal solution, but most likely a "family" of "good" management options. Giving the large complexity of these systems and the numerous synergies between environmental conditions and management options, finding "good" choices cannot be reduced to simple optimization algorithms, assuming linear or some defined form of non-linear relationship between a set of parameters, variables and goal seeking functions. Mathematical models may be very useful in finding "good" management solutions. However, finding these may require many trial and error simulations and this is why using agents that may look automatically for the mentioned solutions may be advantageous. This requires the *a priori* definition of "good" solutions and constraints. For example, one may wish to increase aquaculture production but keeping water quality within certain limits for other uses.

This paper presented a different approach for the problem of bivalve aquaculture optimization in coastal ecosystems. The approach is based on the development of a bivalve farmer agent interacting with a realistic ecological simulation system. The farmer agent optimizes bivalve production by using known optimization techniques for solving the problem of selecting the best farming regions combinations, maximizing the production of the zone. The approach enabled also to achieve better results than using hand-tuned techniques.

From the comparison between distinct optimization methodologies, it was also concluded that better results are achieved by using a combination of different optimization methodologies by the use of our tactics configuration file. However more experiences with different scenarios and longer growing times must be carried out in order to fully validate this conclusion.

There is still plenty of research to be accomplished within the combined fields of ecological simulation and multi-agent systems. There are many scenarios where intelligent programs acting as agents could emulate and enhance the realistic behavior of many different factors within simulation systems. Decision support systems based on realistic ecological simulator have much to gain with the inclusion of multi-agent systems interaction in their architecture.

The methodology experienced in this work will be extended to test more combinations with benthic species and regions: one region/several benthic species, several regions/one benthic species, several regions/several benthic species, restricted farming areas/unrestricted farming areas, etc.

Also the optimization methodologies must be improved in order to allow the simulation period to grow in order to verify the behavior of the best tactics in one complete farming cycle (about 1.5 years).

Acknowledgements. This work is supported by the ABSES project – "Agent-Based Simulation of ecological Systems", (FCT/POSC/EIA/57671/2004). António Pereira is supported by the FCT research scholarship SFRH/BD/16337/2004. Filipe Cruz was supported by a POCI2010 program grant.

References

1. Duarte, P., Meneses, R., Hawkins, A.J.S., Zhu, M., Fang, J., Grant, J.: Mathematical modelling to assess the carrying capacity for multi-species culture within coastal waters. Ecological Modelling 168, 109–143 (2003)
2. Pereira, A., Duarte, P., Reis, L.P.: ECOLANG – A Communication Language for Simulations of Complex Ecological Systems. In: Merkuryev, Y., Zobel, R., Kerckhoffs, E. (eds.) Proceedings of the 19th European Conference on Modelling and Simulation, Riga, pp. 493–500 (2005)
3. Pereira, A., Duarte, P., Reis, L.P.: An Integrated Ecological Modelling and Decision Support Methodology. In: Zelinka, I., Oplatková, Z., Orsoni, A. (eds.) 21st European Conference on Modelling and Simulation, ECMS, Prague, pp. 497–502 (2007)
4. Russel, S., Norvig, P.: Artificial Intelligence: A modern approach, 2nd edn. Prentice-Hall, Englewood Cliffs (2003)
5. Weiss, G.: Multiagent Systems. MIT Press, Cambridge (2000)
6. Wooldridge, M.: An Introduction to Multi-Agent Systems. John Wiley & Sons, Ltd., Chichester (2002)
7. Pereira, A., Duarte, P., Reis, L.P.: Agent-Based Simulation of Ecological Models. In: Coelho, H., Espinasse, B. (eds.) Proceedings of the 5th Workshop on Agent-Based Simulation, Lisbon, pp. 135–140 (2004)
8. Jørgensen, S.E., Bendoricchio, G.: Fundamentals of Ecological Modelling. Elsevier, Amsterdam (2001)
9. Kirkpatrick, S., Gelatt, C.D., Vecchi, M.P.: Optimizing by Simulated Annealing. Science 220(4598) (1983)
10. Glover, F., Laguna, M.: Tabu Search. Kluwer Academic Publishers, Dordrecht (1997)
11. Michalewicz, Z.: Genetic Algorithms + Data Structures = Evolution Programs. Springer, Heidelberg (1999)
12. Sutton, R.S., Barto, A.G.: Reinforcement Learning: An Introduction. MIT Press, Cambridge (1998)
13. Dzeroski, S.: Applications of symbolic machine learning to ecological modelling. Ecological Modelling 146 (2001)
14. Mishra, N., Prakash, M.K., Tiwari, R., Shankar, F., Chan, T.S.: Hybrid tabu-simulated annealing based approach to solve multi-constraint product mix decision problem. Expert Systems with Applications 29 (2005)
15. Youssef, H., Sait, S.M., Adiche, H.: Evolutionary algorithms, simulated annealing and tabu search: a comparative study. Engineering Applications of Artificial Intelligence 14 (2001)
16. Amirjanov, A.: The development of a changing range genetic algorithm. Computer Methods in Applied Mechanics and Engineering 195 (2006)
17. Sait, S.M., El-Maleh, A.H., Al-Abaji, R.H.: Evolutionary algorithms for VLSI multiobjective netlist partitioning. Engineering Applications of Artificial Intelligence 19 (2006)

Tax Compliance Through MABS: The Case of Indirect Taxes

Luis Antunes, João Balsa, and Helder Coelho

GUESS/Universidade de Lisboa, Portugal
{xarax,jbalsa,hcoelho}@di.fc.ul.pt

Abstract. Multi-agent systems can be used in social simulation to get a deeper understanding of complex phenomena, in such a way that predictions can be provided and tested, and policies can be designed in a solid individually grounded manner. Our aim is explore this link between micro-level motivations leading to and being influenced by macro-level outcomes in an economic setting where to study the complex issue of tax evasion. While it is obvious why there is a benefit for people who evade taxes, it is less obvious why people would pay any taxes at all, given the small probability of being caught, and the small penalties involved. Our research program uses exploratory simulation and progressively deepening models of agents and of simulations to study the reasons behind tax evasion. We have unveiled some relatively simple social mechanisms that can explain the compliance numbers observed in real economies. We claim that simulation with multiple agents provides a strong methodological tool with which to support the design of public policies.

1 Introduction

In most societies, tax evasion is a serious economical problem that only recently has started to receive scientific attention [18,1]. Tax evasion creates in the individual person an idea of social unfairness that can only invite more evasion. Especially in developing countries, the perception of tax evasion, together with the notion of widespread corruption leads to social disturbance, and undermines both individual and group motivation towards better social mechanisms and fairer societies. People have the generalised idea that it is individually compensating to evade taxes, as penalties are low, and the probability of being caught rather small. And in fact this idea is supported by the models that most theories of tax evasion endorse. In such models, tax payers are represented by agents whose utilitarian rationality would inevitably lead to free-riding and evasion. However, the real picture is quite different. Real people pay far more than utilitarian theories predict. For instance, in the USA, although the fine value (or rate) can be neglected, and even though less than 2% of households were audited, the Internal Revenue Service (IRS) estimates that 91.7% of all income that should have been reported was in fact reported (numbers from 1988-1992-1995, cited from [2]). In fact, the whole scientific endeavour dealing with tax evasion is known as "tax compliance."

J. Neves, M. Santos, and J. Machado (Eds.): EPIA 2007, LNAI 4874, pp. 605–617, 2007.

Our claim is that multi-agent systems (MAS) provide a more effective approach to the study of tax compliance. The main reason behind this belief is that we can consider *individual* agents endowed with appropriate motivational models to provide an empirically rooted account of their decisions, whereas in traditional economical models the focus is on general laws that can be used to describe the behaviour of any agent. People are different from one another, and from their own motivations and decisions, together with the interactions they engage in, a rich and complex macro behaviour emerges. Such macro-level phenomena then influence individual decisions, as much as individual decisions build up the overall behaviour.

Multi-agent-based simulations (MABS) can explore the space of individual mental models, agent interactions and societal mechanisms and provide a deep understanding of the problem that can be used to predict future behaviours of the social system. With such strong explanatory power, policies to reduce evasion can be designed and rehearsed before deploying them in the real world.

In this paper, we present the application of the e*plore methodology to conduct social simulations to study the tax compliance problem. We introduce *indirect* taxes, which call for even more complex socialisation abilities, since in a transaction all agents must agree about whether to evade. Both our conclusions and the series of models they are based on will be used together with the National Tax Authority (DGCI) in a project to design tax policies through social simulation (DEEPO).

2 Tax Compliance Studies

Tax compliance is an interesting problem from a series of standpoints [2]. As a theoretical problem, it poses a challenge to individual rationality. As a social science issue, it can be studied as a complex instance of a micro-macro problem, in which perceptions, decisions and behaviours in both levels of consideration are mutually influenced in an intricate fashion. As a political science problem, there is a need to consider policy decisions and their consequences, as well as the practical enforcement mechanisms to be used. Finally, as an ethics problem, there are issues about the meaning of communities and the role of their individuals, the contrast between self-interest and moral obligations.

In most economical accounts, agents are supposed to take actions to pursue their own interest. In tax compliance the issue at stake is money and its possession. It would be natural to assume that most agents want to have as much money as possible, and so would have a natural inclination to evade taxes as much as possible. However, as we mentioned before, reality shows that this is not the case, and most taxes end up getting paid. One important goal of this scientific project is to uncover the kind of motivations that can support this kind of "awkward" behaviour. Could it be that people are afraid of getting caught and be exposed to public shame? Or do they fear fines and the heavy hand of justice? Or do they expect some added value from the tax-financed state-provided services? What kind of combinations of these motivations with social interaction mechanisms exist that can explain the numbers coming from available statistics?

Some recent theories go beyond the standard economic account and consider the individual motivations of individuals when deliberating on the compliance decision. One of these theories is Kahneman and Tversky's prospect theory [16], in which the individuals do not use expected utility (the so-called von Neumann-Morgenstern utility [22]) but a rather different choice function. In particular, this function is asymmetric and with different steepness for positive and negative values, meaning that people are risk averse for gains, risk seeking for losses and generally loss averse. Moreover, decision weight substitute probabilities and are usually lower in value. Prospect theory concerns itself with how decisions are *actually* made, whereas expected utility theory concerns itself with how decisions under uncertainty *should* be made [19].

Another recent account is strong expectations theory [15]. According to this approach, in a participative situation agents react strongly and immediately to the perceptions of the global (not average) behaviour of the other participants, in a sort of tit-for-tat attitude [7]. This new account can explain some compliance figures that cannot be supported by the traditional theories. However, this new approach indicates paths, but does not yet walk them. In our view, the most proper way to use these ideas is to deploy multi-agent-based simulation systems, and explore alternative models through social simulation.

3   Principled Exploration of MABS

The main problem with simulations are complexity and validity of results. Simulations are used to tackle complex problems, and models should always be simpler than the original phenomena to be studied. So there is the risk of designing models that are so simple that become useless, or so complex that are as hard to study as the target phenomenon. In this strive for the right standpoint, we find the risk of overpressuring for results, designing the model and tuning the simulation in order to get results that, while interesting, show little resemblance to reality. To avoid an *ad hoc* approach to simulation, and strengthen the confidence in the results and conclusions, it is important to follow a principled methodology.

Purposes of MABS. Social simulation serves several purposes: (i) By building computational models, scientists are forced to *operationalise* the concepts and mechanisms they use for their formulations. This point is very important as we are in cross-cultural field, and terminology and approaches can differ a lot from one area to another; (ii) to better *understand* some complex phenomenon. In MABS, 'understand' means to describe, to model, to program, to manipulate, to explore, to have a hands-on approach to the definition of a phenomenon or process; (iii) to *experiment* with the models, formulate conjectures, test theories, explore alternatives of design but also of definitions, rehearse different approaches to design, development, carry out explorations of different relevance of perceived features, compare consequences of possible designs, test different initial conditions and simulation parameters, explore 'what-if' alternatives; (iv) to *explain* a given phenomenon, usually from the real social world. The sense of explaining is linked to causality more than to correlation. As Gilbert [13] says, we need explanation not only at the macro level, but also at the individual level. Our explanation of the phenomena we observe in simulation is solid because we must make the effort of creating and validating the mechanisms at the micro level, by providing solid and valid reasons for individual behaviours; (v) to *predict* how our models react to change, and this prediction is verifiable in the real phenomenon, through empirical observations. It is important to stress that even empirical observations presuppose a model (which data were collected, which questionnaires were used, etc.); and (vi) when we have enough confidence in the validity and prediction capability of our simulation system, we are ready to help rehearse new policies and *prescribe* measures to be applied to the real phenomenon with real actors.

The e*plore Methodology. To accomplish these increasingly hard purposes, some methodologies have been proposed. Our recent revision of Gilbert's methodology [12] goes somewhat deeper in taking on the MABS purposes, and considers and explores not only the design of intervening agents, but also of societies and their mechanisms and even of the simulation experiments themselves. The inspiration for this approach comes from Sloman's explorations of the design space [20,21], and the idea of a 'broad but shallow' agent design extrapolated to the design of societies and experiments. We have introduced the idea of 'deepening' a design feature in a recent paper [3]. Here is a summary of the steps of e*plore methodology:

i. *identify the subject* to be investigated, by stating specific items, features or marks;
ii. *unveil state-of-the-art* across the several scientific areas involved to provide context. The idea is to enlarge coverage before narrowing the focus, to focus prematurely on solutions may prevent the in-depth understanding of problems;
iii. *propose definition* of the target phenomenon. Pay attention to its operationality;
iv. *identify relevant aspects* in the target phenomenon, in particular, *list individual and collective measures* with which to characterise it;
v. if available, *collect observations* of the relevant features and measures;
vi. *develop the appropriate models* to simulate the phenomenon. Use the features you uncovered and program adequate mechanisms for individual agents, for interactions among agents, for probing and observing the simulation. Be careful to base behaviours in reasons that can be supported on appropriate individual motivations. Develop visualisation and data recording tools. Document every design option thoroughly. *Run the simulations*, collect results, compute selected measures;
vii. return to step iii, and *calibrate everything*: your definition of the target, of adequate measures, of all the models, verify your designs, validate your mod-

els by using the selected measures. Watch individual trajectories of selected
agents, as well as collective behaviours;

viii. *introduce variation* in your models: in initial conditions and parameters, in
individual and collective mechanisms, in measures. Return to step v;

ix. After enough exploration of design space is performed, use your best models
to *propose predictions*. Confirm it with past data, or collect data and validate
predictions. Go back to the appropriate step to ensure rigour;

x. Make a generalisation effort and *propose theories and/or policies*. Apply to
the target phenomenon. Watch global and individual behaviours. Recali-
brate.

Deepening the Design. When building up experimental designs, it is usual to
defend and adopt the so-called KISS ("keep it simple, stupid!") principle [9]. In
some sense, Sloman's "broad but shallow" design principle starts off from this
principle. Still, models must never be simpler than they should. The solution
for this tension is to take the shallow design and increasingly deepen it while
gaining insight and understanding about the problem at hand. The idea is to
explore the design of agents, (interactions), (institutions), societies and finally
experiments (including simulations and analysis of their outcomes) by making
the initially simple (and simplistic) particular notion used increasingly more
complex, dynamic, and rooted in consubstantiated facts. As Moss argued in his
WCSS'06 plenary presentation, "Arbitrary assumptions must be relaxed in a way
that reflects some evidence." This complex movement involves the experimenter
him/herself, and, according to Moss, includes "qualitative micro validation and
verification (V&V), numerical macro V&V, top-down verification, bottom-up
validation," all of this whereas facing that "equation models are not possible,
due to finite precision of computers."

A possible sequence of deepening a concept representing some agent feature
(say parameter c, standing for honesty, income, or whatever) could be to consider
it initially a constant, then a variable, then assign it some random distribution,
then some empirically validated random distribution, then include a dedicated
mechanism for calculating c, then an adaptive mechanism for calculating c, then
to substitute c altogether for a mechanism, and so on and so forth.

4   The Ec^* Model Series

In a recent research project, we have used the e*plore methodology to propose
the exploration of the tax compliance problem through a series of models [4,5,10].
This hierarchical structure of models aims at transversing the space of possible
designs by progressively removing simplistic assumptions and criticisms posed
to the original model Ec_0, which represents the standard economics account of
the problem [1,23].

This strategic exploration of the experiment design space uses a set of techniques and their combinations to evolve models from simpler models, in such
as way as to progressively deepen the designs (of agents, of societies, even of

experiments) starting from broad but shallow, to finally build up theory from this exploration of models. These techniques include: refining, tiling, adding up, choosing, enlarging, etc.

So, we successively introduced new models with specific characteristics, either at the micro (individual) or at the macro (societal) levels, with some reasons, conjectures or intuitions. Ec_0^τ introduced expanded history in the individual decision; Ec_1 proposed agent *individuality*, whereas Ec_2 postulated individual *adaptivity*; Ec_3^* introduced sociality, it is the first model where the individual decision depends on social perceptions; Ec_3^{*i} explored one particular type of interaction, imitation; and finally Ec_4^* postulated social heterogeneity, different agent breeds in a conflictual relation. Other models are still being shaped, such as $Ec_?^{*k}$ a model where perception is limited to a k-sized neighborhood. This coverage of our problem and model space uses several combined techniques [6].

5   Results of Simulations

Along the series of simulations we have run [4,5,10], we could unveil some features and mechanisms that can contribute to explain the numbers we find in empirical studies. Remember that the standard theoretical model predicts that 0% of agents would pay, which is the result our model Ec_0 also yields. In this section we present the most important results and conclusions from the studies we conducted. To usefully tackle such a complex problem we have to include several of these mechanisms, and appropriately tune them to the different national and contextual realities we face.

Memory. In model Ec_2, we introduced a memory effect: agents caught evading would no longer pay only the fine and the due tax, they would also *remember* for some time that they had been caught. In this way, when facing a new decision to pay or evade taxes, the agent would take into account the recent past. We experienced with memories ranging from short to long term and found a decisive influence of this feature in the overall societal behaviour: with enough tax return declarations being audited (10%) and a substantial (but not unrealistic) fine (50%) and a very moderate memory degradation rate (0.01) we could reach compliance levels of 77%, which are very much in tune with real numbers.

History. In Ec_0^τ, we pointed out that when the central tax authority would find an agent evading there would be no sense in acting only on the present year. It would surely investigate the recent past, in order to uncover other possible evasions. The probabilistic calculations the agents in Ec_0 use to decide change substantially, but the points in which individual decision changes are not dramatically different. For instance, for an agent who evades taxes, if the fine is higher than 570%, it is not very important to change this to 540%. However, by systematically exploring the space of parameters, we could find that we could bring the evasion numbers down to 5% or 6%. To accomplish this we would have to act simultaneously on the fine, the probability of auditing and the number of past years to be investigated. For instance, when investigating 10 years, we can lower evasion to only 5% with a fine of 400% and a probability of auditing of 2%, whereas with a fine of 50% and a probability of auditing of 1% we would have 85% of evasion.

Imitation. In Ec_3^{*i}, we followed Frédéric Kaplan's ideas about imitation and stubbornness [17] as a social mechanism to enhance compliance figures. Some resilient agents would follow their tendency to pay their due taxes indenpendently of the financial inconvenience or utility calculations. And other agents would imitate the behaviours of their neighbours or other agents they would meet. With such a simple social mechanism (which has deeply rooted psychological support) we could dramatically increase the number of complying agents. In special circumstances (crowded environments, in which imitation opportunities are more frequent) a small percentage of stubborn compliers will yield a very high percentage of overall compliance. For instance, from 20% of stubborn compliers we could reach more than 95% of overall compliers, given adequate parameters. With this series of simulations, we concluded that when designing policies for fighting tax evasion, it could be more rewarding to disseminate information to the proper individuals than to advertise big measures of fine increasing or enhanced auditing methods. This is in line with Kahan's findings about strong reciprocity: publicising crime fighting measures will only make people imitate the bad behaviours, not the good [15].

Autonomous Inspectors. We further removed some of the classic criticisms to the standard approach in Ec_4^*. Here we introduced different breeds of agents, one for tax payers, and other for tax inspectors. Each inspector has autonomy to decide whether to perform an audit on a given agent or not. The inspector makes this decision by taking into account his/her personal constraints and expectations about the tax to be collected from the given individual. So, audits are no longer determined by a probability, the probability of being audited is no longer independent of the past or of the other agents being audited, the cost of an audit is not irrelevant anymore, and there is now a limit (a budget for the tax authority and for each inspector) for the number of audits that can be carried out. Since some of the parameters that inform the inspector decision can only be known after the audit is performed, we had to use estimates to fill in the values. For instance, the expected value of tax to be collected from an agent is estimated by his/her wealth times the probability that he/she is an evader, given by the Laplace rule. This assumes a lot of publicly known information, but we will experiment later with personal information and communication networks between inspectors.

The results of simulations emphasise the question of inspector mobility. When inspectors move step by step, we get the overall behaviour of a central authority. By allowing inspectors to take wider steps, we could reduce evasion by *circa* a quarter. Density of population remains an important feature that severely changes the evasion results. It is possible to obtain evasion values as little as 5%, provided that population density was very high. However, the inspectors had to perform an unrealistically high relative number os audits. This model still has to be tuned up in terms of the decision model the inspectors use to audit the tax payers.

6   Introducing Indirect Taxes

There have been some attempts at simulation with indirect taxes for policy recommendations purposes. For instance, Jenkins and Kuo [14] propose to study the introduction of a Value-Added Tax (VAT) in a country (Nepal) where there was a highly distorted indirect tax system, consisting mostly of excise taxes. The authors sustain that an European-style VAT is equivalent to a retail sales tax on final consumption, and conclude that to introduce a VAT and enhance government reliability on the VAT over time it is necessary to "move aggressively to broaden the base and enhance compliance." [14] They also conclude that "compliance will come about only if the tax laws are designed to reduce the incentive for evasion and the tax administration is strengthened to provide both service and enforcement" [14]. It remains to be determined *how* these laws (and corresponding mechanisms) should be designed and implemented.

On another case, Decoster [11] proposes a microsimulation model for policy rehearsal on a particular subset of indirect taxes (excise taxes over energy). The models used are static, and the calculation of the consequences of policy measures at the level of the individual actor is made by using a representative sample of the population. The conclusions indicate that the individual behavioural reactions are decisively important to calculate the average losses determined by the introduction of the new tax. Environmental taxes are regressive in terms of net expenditures and more heavily felt by small consumers.

Both these examples indicate the need and the benefits of conducting simulations to observe "in silica" the consequences of the deployment of new tax policies. However, the MABS approach can bring about a new vision. In MABS we are not worried about *general equilibrium*, rather we prefer to observe individual trajectories, and have some reasons behind the individual decisions to be made. The modelling effort is bigger and more prone to pitfalls, but the confidence on the simulation results is enhanced.

Direct vs. Indirect Taxes. In previous models we have only considered direct taxes: taxes that incide over the personal income of the agent. In our strategy of increasingly considering more complex scenarios, these taxes were the easiest to approach. The decision to comply or evade is entirely up to the individual agent being taxed. But with indirect taxes inciding over transactions, the panorama changes completely.

Now, to perform a tax evasion episode, the agent must find an adequate partner who explicitly agrees to this decision. As a consequence, the whole society gets much more complex. Starting from the individual, we have to consider which individual motivation leads to evasion. If this is obvious for the buyer, since he/she will pay less by evading taxes, it is less obvious why the seller would agree. This leads to the need of considering a complex tax structure involving all the participants in the society: the seller also gains from evasion since the evasion episode will necessarily be made outside his/her accounting records, so resulting in a lower revenue, which in turn would subject to direct taxes. Every participant gains, except of course the tax authority. So, in our models, we need not only direct taxes over the individual agents (income taxes) but also over the agents that perform sales (revenue taxes). In simple terms, we need direct taxes over revenues from work as well as direct taxes over profits from sales.

This more complex picture of the tax evasion episodes implies drastic modifications to our models. Agents have to decide whether to *try* to evade or not, and after this they have to decide whether to *propose* evasion to the other party or not. It could be the case that the other agent is a stubborn complier, or even if he is an evader, he would not be available to evade in this particular episode. Another possibility is that the agent is available to take the chance of evading given that enough money is at stake. But this would imply that our agents should have some sense (or reference) about absolute money quantities, which recognisably is a hard endeavour to take on[1]. Or, it could be even the case that he gets offended with the proposal and reports it to the tax authorities. We are in a complex world of possibilities and risk.

A Canonical Transaction Model. As before in our research, we opted for taking on the simplest possible model and building up from it. We consider only indirect taxes. Agents live and move in a simulated grid world. Each time they meet they have a chance for a transaction. They decide (for now by rolling a dice) whether or not to engage into this transaction. For the moment, it consists only in a transfer of money from one agent to another. If both parties agree to the transaction, they have to decide whether to propose to evade the due taxes. An agent can opt for doing this or not, and the other agent can freely decide to accept or decline the offer. According to their decisions they proceed with the transactions, pay or not their taxes, money changes hands and they get on with their lives. According with the "broad but shallow principle" most of the quantities involved are constant. The ones that are not are listed below:

- $\rho \in [0, 1]$ represents the propension to evade tax. An agent rolls a dice and opts for evasion if the number is less than ρ;
- $\sigma \in [0, 1]$ represents the individual probability that a transaction occurs.

We should note that these very simple mechanisms are a poor substitute for the real motivations involved in such complex decisions. However, at this stage we are interested in getting to know the mechanisms and dynamics of the society that results from such simple-minded agents. Later we can deepen these mechanisms while maintaining control over the simulation.

In a given encounter between two agents, we can easily calculate the probabilities with which transactions or collusions to evade taxes take place. However, we are more interested in observing the overall measures that different transaction designs lead to.

---

[1] We should thank Tarek el Sehity for pointing out this possibility to us.

There are no inputs of money except for the initial amount available in the population. The simulation runs until there are no longer agents capable of performing a transaction, or a certain number of cycles is reached. By that time, most of the money has been collected to the central authority as tax revenues. Buying and selling agents have no diversity of products to choose from, no different prices, no added value in their activities. This is the simplest transaction model we could conceive.

Now we study different strategies for the individual decisions involved with the transaction:

a. Buyer agent use its probability (σ) to decide to perform a transaction. The target agent has to comply, and taxes are duly paid;
b. Both agents use σ to decide to perform a transaction. They always pay the due taxes;
c. Both agents use σ to decide to perform a transaction, and their propension to evade (ρ), but are mutually frank about their intentions;
d. Agents decide independently (using σ, ρ) on whether to perform the transaction and evade or not, by using public information (tags) about the *a priori* intentions of the other agent;
e. Agents do not know anything about the intentions of others; they can propose an evasion episode and be reported to the central tax authority. We try out a scope of financial penalties.

7   Analysis of Results

We have programmed these models in NetLogo 3.0. The first experiments were used to debug, tune up and refine the several components of the system, as well as guide the construction of the models.

The use of e*plore methodology involves a series of techniques that introduce variations in the models of agents, of societies and of experiments, in order to completely travel through the space of possible designs, while gaining insights into the complex problems at hand. These variations start off from simple mechanisms in broad but shallow designs, and proceed by progressive deepening of those mechanisms.

The measures we took from these first experiments deal with the number of the active agents in the society and the number of collusions to evade taxes. Since there are no injections of capital, the society has no long term sustainability. We could see this happening in all the experiments we run, with strategies a. to e. mentioned above. The differences in duration of the society in terms of active agents have only to do with the pace with which the encounters promote more or less transactions.

One important note to make is that to take averages over series of experiments yields some sort of equalisation phenomenon between experiments with different strategies. Each simulation shows its own path and behaviour, according to the series of initial parameters and random events that determine its course. However, when we aggregate these individual measures over a series of runs, these differences get blurred, and the different strategies display seemingly similar figures.

The difference between the measures taken over runs of several strategies does not correspond to the difference between typical runs of the simulations. This is a strong argument to not consider aggregate measures but instead to have field specialists looking at the simulation runs in detail, so that the specific behaviour that prevails does not get mingled into opaque statistics. A similar argument was used by Axelrod [8], when he defended the analysis of typical and even of atypical runs of his series of experiments.

Even though there are not structural differences in the measures for agent survival, there are some quantitative differences. Societies where agents use strategies c. and d. behave as if every bandit is a shameless bandit, without any compunction at all, nor any dismay preventing its abusive deed. So, only the natural tendencies agents have towards compliance and evasion lead to smaller rate of collusions. However, in societies where agents follow strategy e., there is an additional filter (in fact one for each agent) opposing the idea of collusion to evade, which explains why the number of collusions is far lower, and the central authority ends up gathering a significantly larger amount of collected taxes. In a seemingly surprising fashion, the set of active agents (capable of engaging in transactions) does not survive longer. This is surely due to a decrease in the amount of money available for transactions, as funds are collected earlier by the central authority.

Evidently, this experimental and simulation setting is only ready to commence exploration, and most questions still remained unanswered. In the Future Work section we indicate where this exploration of the indirect taxation scenario will likely take us.

8 Conclusions

In this paper we argue that MAS and MABS are two new scientific approaches to the old problem of tax compliance. Most standard approaches to this complex and prominent problem either deal with the discovery of individual evaders through statistical or classification methods, or focus on the discovery of general laws that can be applied to a general (average) individual. Obviously, no individual is average, and this explains why such an old problem (reporting back to the ancient Egypt [18]) has had scientific attention only for the last 25 years. Within economics, the most promising approach is economic psychology, an interdisciplinary field of research concerned with how people behave when facing decision problems, and not really how some ideal person should behave.

We claim and have shown through extensive experimentation that social simulation with rationality-heterogeneous individualised multiple agents can provide the necessary concepts, mechanisms, and even tools and methodological principles with which to complement this new visions about the deciding person, and explore the space of possible designs to gain decisive insight into this complex problem. It is our goal to propose policy rehearsal by public entities as a new way of attacking political problems such as tax evasion, so that to build trust and confidence in the efficiency of such policies. We have unveiled some societal and individual mechanisms that display a far more realistic performance than the one of classical models based on probabilities and utilities. After dealing with the tax compliance problem, we have laid the basis of a canonical model of transactions that sets the ground for exploration of this far more complex taxation scenario, in which agents have to collude to evade, with all implied social interactions and their possible consequences.

8.1   Future Work

Future work will incide mainly in deepening the agent mental models in what concerns motivations for decisions. Our agents are intentionally simplistic, and we need deeper reasons for their behaviours. These could be found especially in a special attitude tax payers bear toward taxes and their central authority. As argued by Kirchler and Hoelzl [18], a tax climate of trust will be determinant in achieving voluntary compliance. People will have to find value in the taxes they pay, but this whole vision has only been taken into account until now in the so-called "public good theory," which is severely overly utilitarian. Redistribution of the tax revenues through public services (and the corresponding central authority view of treating tax payers not as robbers but as clients [18]), and each agent individual perception on the value (not necessarily monetary) of these services will be central in future approaches to the tax compliance simulations.

Another area in which we will also invest is the acquisition of a grounding for an absolute notion of money. From our extensive experimentation, this seems very important to explain the behaviour of people in real decision settings. One might be ready to steal an apple but reluctant to steal a car. A project recently proposed by el Sehity is taking a simple bartering economy and trying to isolate which of the exchanged commodities are good canditates for monetary units. This grounding of money in real commodities can help absolutise its value, for instance in terms of what is needed for the survival of individuals.

Finally, we will use all these results and explorations to run experiments in which we have direct and indirect taxes. A complete picture of the agent's attitude towards taxes can only be considered with both types of taxes and agents built and tuned up from premises coming from sets of real data. This strive towards real data is already undergoing.

References

1. Allingham, M.G., Sandmo, A.: Income tax evasion: A theoretical analysis. Journal of Public Economics 1(3/4), 323–338 (1972)
2. Andreoni, J., Erard, B., Feinstein, J.: Tax compliance. Journal of Economic Literature 36(2) (June 1998)
3. Antunes, L., Balsa, J., Coelho, H.: e*plore-ing the simulation design space. In: Proc. IV European Workshop on Multi-Agent Systems (2006)
4. Antunes, L., Balsa, J., Moniz, L., Urbano, P., Palma, C.R.: Tax compliance in a simulated heterogeneous multi-agent society. In: Sichman, J.S., Antunes, L. (eds.) MABS 2005. LNCS (LNAI), vol. 3891, Springer, Heidelberg (2006)

5. Antunes, L., Balsa, J., Respício, A., Coelho, H.: Tactical exploration of tax compliance decisions in multi-agent based simulation. In: Antunes, L., Takadama, K. (eds.) Proc. MABS 2006 (2006)
6. Antunes, L., Coelho, H., Balsa, J., Respício, A.: e*plore v.0: Principia for strategic exploration of social simulation experiments design space. In: Takahashi, S., Sallach, D., Rouchier, J. (eds.) Advancing Social Simulation: The First World Congress, Springer, Japan (2007)
7. Axelrod, R.: The Evolution of Cooperation. Basic Books, New York (1984)
8. Axelrod, R.: A model of the emergence of new political actors. In: Artificial societies: the computer simulation of social life, UCL Press, London (1995)
9. Axelrod, R.: Advancing the art of simulation in the social sciences. In: Conte, R., Hegselmann, R., Terna, P. (eds.) Simulating Social Phenomena. LNEMS, vol. 456, Springer, Heidelberg (1997)
10. Balsa, J., Antunes, L., Respício, A., Coelho, H.: Autonomous inspectors in tax compliance simulation. In: Proc. 18th European Meeting on Cybernetics and Systems Research (2006)
11. Decoster, A.: A microsimulation model for Belgian indirect taxes with a carbon/energy tax illustration for Belgium. Tijdschrift voor Economie en Management 40(2) (1995)
12. Gilbert, N.: Models, processes and algorithms: Towards a simulation toolkit. In: Suleiman, R., Troitzsch, K.G., Gilbert, N. (eds.) Tools and Techniques for Social Science Simulation, Physica-Verlag, Heidelberg (2000)
13. Gilbert, N.: Varieties of emergence. In: Proc. Agent 2002: Social agents: ecology, exchange, and evolution, Chicago (2002)
14. Jenkins, G.P., Kuo, C.-Y.: A VAT revenue simulation model for tax reform in developing countries. World Development 28(2), 763–774 (2000)
15. Kahan, D.M.: The logic of reciprocity: Trust, collective action and law. In: Gintis, H., Bowles, S., Boyd, R., Fehr, E. (eds.) Moral Sentiments and Material Interests – The Foundations of Cooperation in Economic Life, MIT Press, Cambridge (2005)
16. Kahneman, D., Tversky, A.: Prospect theory: An analysis of decision under risk. Econometrica 47(2), 263–291 (1979)
17. Kaplan, F.: The emergence of a lexicon in a population of autonomous agents (in French). PhD thesis, Université de Paris 6 (2000)
18. Kirchler, E., Hoelzl, E.: Modelling taxpayers behaviour as a function of interaction between tax authorities and taxpayers. In: Managing and Maitaining Compliance, Boom Legal Publishers (in press, 2006)
19. Montier, J.: Behavioural Finance: Insights into Irrational Minds and Markets. John Wiley and Sons, Chichester (2002)
20. Sloman, A.: Prospects for AI as the general science of intelligence. In: Prospects for Artificial Intelligence - Proceedings of AISB 1993, IOS Press, Amsterdam (1993)
21. Sloman, A.: Explorations in design space. In: Proc. of the 11th European Conference on Artificial Intelligence (1994)
22. von Neumann, J., Morgenstern, O.: Theory of Games and Economic Behavior. Princeton University Press, Princeton (1944)
23. Yizthaki, S.: A note on income tax evasion: A theoretical analysis. Journal of Public Economics 3(2), 201–202 (1974)

Chapter 10 - First Workshop on Search Techniques for Constraint Satisfaction (STCS 2007)

Efficient and Tight Upper Bounds
for Haplotype Inference by Pure Parsimony
Using Delayed Haplotype Selection

João Marques-Silva[1], Inês Lynce[2], Ana Graça[2], and Arlindo L. Oliveira[2]

[1] School of Electronics and Computer Science, University of Southampton, UK
jpms@ecs.soton.ac.uk
[2] IST/INESC-ID, Technical University of Lisbon, Portugal
{ines,assg}@sat.inesc-id.pt, aml@inesc-id.pt

Abstract. Haplotype inference from genotype data is a key step towards a better understanding of the role played by genetic variations on inherited diseases. One of the most promising approaches uses the pure parsimony criterion. This approach is called Haplotype Inference by Pure Parsimony (HIPP) and is NP-hard as it aims at minimising the number of haplotypes required to explain a given set of genotypes. The HIPP problem is often solved using constraint satisfaction techniques, for which the upper bound on the number of required haplotypes is a key issue. Another very well-known approach is Clark's method, which resolves genotypes by greedily selecting an explaining pair of haplotypes. In this work, we combine the basic idea of Clark's method with a more sophisticated method for the selection of explaining haplotypes, in order to explicitly introduce a bias towards parsimonious explanations. This new algorithm can be used either to obtain an approximated solution to the HIPP problem or to obtain an upper bound on the size of the pure parsimony solution. This upper bound can then used to efficiently encode the problem as a constraint satisfaction problem. The experimental evaluation, conducted using a large set of real and artificially generated examples, shows that the new method is much more effective than Clark's method at obtaining parsimonious solutions, while keeping the advantages of simplicity and speed of Clark's method.

1 Introduction

Over the last few years, an emphasis in human genomics has been on identifying genetic variations among different people. A comprehensive search for genetic influences on disease involves examining all genetic differences in a large number of affected individuals. This allows the systematic test of common genetic variants for their role in disease. These variants explain much of the genetic diversity in our species, a consequence of the historically small size and shared ancestry of the human population. One significant effort in this direction is represented by the HapMap Project[23], a project that aims at developing a haplotype map of the human genome and represents the best known effort to develop a public resource that will help finding genetic variants associated with specific human diseases.

For a number of reasons, these studies have focused on the tracking of the inheritance of Single Nucleotide Polymorphisms (SNPs), point mutations found with only

J. Neves, M. Santos, and J. Machado (Eds.): EPIA 2007, LNAI 4874, pp. 621–632, 2007.

two common values in the population. This process is made more difficult because of technological limitations. Current methods can directly sequence only short lengths of DNA at a time. Since the sequences of the chromosomes inherited from the parents are very similar over long stretches of DNA, it is not possible to reconstruct accurately the sequence of each chromosome. Therefore, at a genomic site for which an individual inherited two different values, it is currently difficult to identify from which parent each value was inherited. Instead, currently available sequencing methods can only determine that the individual is ambiguous at that site.

Most diseases are due to very complex processes, where the values of many SNPs affect, directly and indirectly, the risk. Due to a phenomenon known as linkage disequilibrium, the values of SNPs in the same chromosome are correlated with each other. This leads to the conservation, through generations, of large haplotype blocks. These blocks have a fundamental role in the risk of any particular individual for a given disease. If we could identify maternal and paternal inheritance precisely, it would be possible to trace the structure of the human population more accurately and improve our ability to map disease genes. This process of going from genotypes (which may be ambiguous at specific sites) to haplotypes (where we know from which parent each SNP is inherited) is called haplotype inference.

This paper introduces a greedy algorithm for the haplotype inference problem called Delayed Haplotype Selection (DS) that extends and improves the well-known Clark's method[5]. We should note that recent work on Clark's method studied a number of variations and improvements, none similar to DS, and all performing similarly to Clark's method. This new algorithm takes advantage of new ideas that have appeared recently, such as those of pure parsimony[10]. A solution to the haplotype inference by pure parsimony (HIPP) problem provides the smallest number of haplotypes required to explain a set of genotypes. This algorithm can then be used in two different ways: (1) as a standalone procedure for giving an approximate solution to the HIPP problem or (2) as an upper bound to the HIPP solution to be subsequently used by pure parsimony algorithms which use upper bounds on their formulation. Experimental results, obtained on a comprehensive set of examples, show that, for the vast majority of the examples, the new approach provides a very accurate approximation to the pure parsimony solution.

This paper is organised as follows. The next section introduces key concepts, describes the problem from a computational point of view, and points to related work, including Clark's method and pure parsimony approaches. Based on Clark's method, section 3 describes a new algorithm called *Delayed Haplotype Selection*. Afterwards section 4 gives the experimental results obtained with the new algorithm, which are compared with other methods and evaluated from the point of view of a parsimonious solution. Finally, section 5 presents the conclusions and points directions for future research work, including the integration of the greedy algorithm in pure parsimonious algorithms.

2   Problem Formulation and Related Work

2.1   Haplotype Inference

A *haplotype* represents the genetic constitution of an individual chromosome. The underlying data that forms a haplotype is generally viewed as the set of SNPs in a given region of a chromosome. Normal cells of diploid organisms contain two haplotypes, one inherited from each parent. The *genotype* represents the conflated data of the two haplotypes. The value of a particular SNP is usually represented by X, Y or X/Y, depending on whether the organism is homozygous with value X, homozygous with value Y or heterozygous. The particular base that the symbols X and Y represent varies with the SNP in question. For instance, the most common value in a particular location may be the guanine (G) and the less common variation cytosine (C). In this case, X will mean that both parents have guanine in this particular site, Y that both parents have cytosine at this particular site, and X/Y that the parents have different bases at this particular site. Since mutations are relatively rare, the assumption that at a particular site only two bases are possible does not represent a strong restriction. This assumption is supported by the so called *infinite sites model*[14], that states that only one mutation has occurred in each site, for the population of a given species.

Starting from a set of genotypes, the haplotype inference task consists in finding the set of haplotypes that gave rise to that set of genotypes. The variable n denotes the number of individuals in the sample, and m denotes the number of SNP sites. Without loss of generality, we may assume that the two values of each SNP are either 0 or 1. Value 0 represents the wild type and value 1 represents the mutant. A haplotype is then a string over the alphabet $\{0,1\}$. Genotypes may be represented by extending the alphabet used for representing haplotypes to $\{0,1,2\}$. A specific genotype is denoted by g_i, with $1 \leq i \leq n$. Furthermore, g_{ij} denotes a specific site j in genotype g_i, with $1 \leq j \leq m$. We say that a genotype g_i can be explained by haplotypes h_k and h_l iff for each site g_{ij}:

$$g_{ij} = \begin{cases} h_{kj} & \text{if } h_{kj} = h_{lj} \\ 2 & \text{if } h_{kj} \neq h_{lj} \end{cases}$$

In general, if a genotype g_i has $r \geq 1$ heterozygous sites, then there are 2^{r-1} pairs that can explain g_i. The objective is to find the set $\mathcal{H}$ of haplotypes that is most likely to have originated the set of genotypes in $\mathcal{G}$.

Definition 1. *(Haplotype Inference) Given a set $\mathcal{G}$ of n genotypes, each of length m, the haplotype inference problem consists in finding a set $\mathcal{H}$ of $2 \cdot n$ haplotypes, not necessarily different, such that for each genotype $g_i \in \mathcal{G}$ there is at least one pair of haplotypes (h_k, h_l), with h_k and $h_l \in \mathcal{H}$ such that the pair (h_k, h_l) explains g_i.*

Example 1. (Haplotype Inference) Consider genotype 02122 having 5 SNPs, of which 1 SNP is homozygous with value 0, 1 SNP is homozygous with value 1, and the 3 remaining SNPs are heterozygous (thus having value 2). Genotype 02122 may then be explained by four different pairs of haplotypes: (00100, 01111), (01100, 00111), (00110, 01101) and (01110, 00101).

We may distinguish between a number of approaches that are usually used for solving the haplotype inference problem: the statistical, the heuristic and the combinatorial approaches. The statistical approaches[19,22] use specific assumptions about the underlying haplotype distribution to approximate different genetic models, and may obtain highly accurate results. The heuristic approaches include, among others, Clark's method[5]. Finally, most combinatorial approaches are based on the pure parsimony criterion[10]. The later has shown to be one of the most promising alternative approaches to statistical models[3,17].

2.2 Clark's Method

Clark's method is a well-known algorithm that has been proposed to solve the haplotype inference problem[5]. Clark's algorithm has been widely used and is still useful today. This method considers both haplotypes and genotypes as vectors. The method starts by identifying genotype vectors with zero or one ambiguous sites. These vectors can be resolved in only one way, and they define the initially resolved haplotypes. Then, the method attempts to resolve the remaining genotypes by starting with the resolved haplotypes. The following step infers a new resolved vector NR from an ambiguous vector A and an already resolved genotype vector R.

Suppose A is an ambiguous genotype vector with r ambiguous sites and R is a resolved vector that is a haplotype in one of the 2^{r-1} potential resolutions of vector A. Then the method infers that A is the conflation of the resolved vector R and another unique vector NR. All of the ambiguous positions in A are set in NR to the opposite value of the position in R. Once inferred, this vector is added to the set of known resolved vectors, and vector A is removed from the set of unresolved vectors.

The key point to note is that there are many ways to apply the resolution rule, since for an ambiguous vector A there may be many choices for vector R. A wrong choice may lead to different solutions, or even leave orphan vectors, in the future, i.e., vectors that cannot be resolved with any already resolved vector R.

The Maximum Resolution (MR) problem[9] aims at finding the solution of the Clark's algorithm with the fewest orphans, i.e. with the maximum number of genotypes resolved. This problem is NP-hard as shown by Gusfield[9], who also proposed an integer linear programming approach to the MR problem.

2.3 Pure Parsimony

Chromosomes in the child genome are formed by combination of the corresponding chromosomes from the parents. Long stretches of DNA are copied from each parent, spliced together at recombination points. Since recombination is relatively infrequent, large segments of DNA are passed intact from parent to child. This leads to the well known fact that the actual number of haplotypes in a given population is much smaller than the number of observed different genotypes. The haplotype inference by pure parsimony approach was proposed by Hubbel but only described by Gusfield[9].

Definition 2. *(Haplotype Inference by Pure Parsimony) The haplotype inference by pure parsimony (HIPP) approach aims at finding a solution for the haplotype inference problem that minimises the total number of distinct haplotypes used. The problem of finding such a parsimonious solution is APX-hard (and, therefore, NP-hard)[16].*

Example 2. (Haplotype Inference by Pure Parsimony) Consider the following example, taken from a recent survey on the topic[11], where the set of genotypes is: 02120, 22110, and 20120. There are solutions that use five different haplotypes[1], but the solu-

---

[1] In general, up to $2 \cdot n$ distinct haplotypes may be required to explain n genotypes. However, in this particular case, there is no solution with six distinct haplotypes.

tion (00100, 01110), (01110, 10110), (00100, 10110) uses only three different haplotypes.

It is known that the most accurate solutions based on Clark's method are those that infer a small number of distinct haplotypes[10,20]. Although Clark's method has sometimes been described as using the pure parsimony criterion[19,1,22], this criterion is not explicitly used and an arbitrary choice of the resolving haplotype does not lead to a pure parsimony solution. The present paper proposes a method that, while still based on Clark's method, explicitly uses the pure parsimony criterion, leading to more precise results.

Several approaches, have been proposed to solve the HIPP problem. The first algorithms are based on integer linear programming[10,2,24], whereas the most recent and competitive encode the HIPP problem as a constraint satisfaction problem (either using propositional satisfiability[17,18] or pseudo-Boolean optimization[4]).

One should note that the implementation of exact algorithms for the HIPP problem often requires computing either lower or upper bounds on the value of the HIPP solution[24,18]. Clearly, Clark's method can be used for providing upper bounds on the solution of the HIPP problem. Besides Clark's method, which is efficient but in general not accurate, existing approaches for computing upper bounds to the HIPP problem require worst-case exponential space, due to the enumeration of candidate pairs of haplotypes[12,24]. Albeit impractical for large examples, one of these approaches is used in Hapar[24], a fairly competitive HIPP solver when the number of possible haplotype pairs is manageable.

The lack of approaches both accurate and efficient for computing upper bounds, prevented their utilization in recent HIPP solvers, for instance, in SHIPs[18]. Algorithm 1 summarizes the top-level operation of SHIPs. This SAT-based algorithm iteratively determines whether there exists a set $\mathcal{H}$ of distinct haplotypes, with $r = |\mathcal{H}|$ such that each genotype $g \in \mathcal{G}$ is explained by a pair of haplotypes in $\mathcal{H}$. The algorithm considers increasing sizes for $\mathcal{H}$, from a lower bound lb to an upper bound ub. Trivial lower and upper bounds are, respectively, 1 and $2 \cdot n$. For each value of r considered, a CNF formula φ^r is created, and a SAT solver is invoked (identified by the function call SAT(φ^r)). The algorithm terminates for a size of $\mathcal{H}$ for which there exist $r = |\mathcal{H}|$ haplotypes such that every genotype in $\mathcal{G}$ is explained by a pair of haplotypes in $\mathcal{H}$, i.e. when the constraint problem is satisfiable. (Observe that an alternative would be to use binary search.)

Algorithm 1. Top-level SHIPs algorithm

SHIPs($\mathcal{G}, lb$)

1   $r \leftarrow lb$
2   **while (true)**
3   **do** Generate φ^r given $\mathcal{G}$ and r
4   **if** SAT(φ^r) $= true$
5   **then return** r
6   **else** $r \leftarrow r + 1$

This paper develops an efficient and accurate approach for haplotype inference, inspired by pure parsimony, and which can be used to compute tight upper bounds to the HIPP problem. Hence, the proposed approach can be integrated in any HIPP approach, including Hapar[24] and SHIPs[18].

3   Delayed Haplotype Selection

A key drawback of haplotype inference algorithms based on Clark's method is that these algorithms are often too greedy, at each step seeking to explain *each* non-explained genotype with the most recently chosen haplotype. As a result, given a newly selected haplotype h_a, which can explain a genotype g_t, a new haplotype h_b is generated that only serves to explain g_t. If the objective is to minimize the number of haplotypes, then the selection of h_b may often be inadequate.

This section develops an alternative algorithm which addresses the main drawback of Clark's method. The main motivation is to avoid the excessive greediness of Clark's method in selecting new haplotypes. Therefore a *delayed* greedy algorithm for haplotype *selection* (DS) is used instead.

In contrast to Clark's method, where identified haplotypes are included in the set of chosen haplotypes, the DS algorithm maintains two sets of haplotypes. The first set, the *selected* haplotypes, represents haplotypes which have been chosen to be included in the target solution. A second set, the *candidate* haplotypes, represents haplotypes which can explain one or more genotypes not yet explained by a pair of selected haplotypes.

The initial set of selected haplotypes corresponds to all haplotypes which are required to explain the genotypes with no more than one heterozygous sites, i.e. genotypes which are explained with either one or exactly two haplotypes. Clearly, all these haplotypes must be included in the final solution.

At each step, the DS algorithm chooses the candidate haplotype h_c which can explain the largest number of genotypes. The chosen haplotype h_c is then used to identify additional candidate haplotypes. Moreover, h_c is added to the set of selected haplotypes, and all genotypes which can be explained by a pair of selected haplotypes are removed from the set of unexplained genotypes. The algorithm terminates when all genotypes have been explained.

Each time the set of candidate haplotypes becomes empty, and there are still more genotypes to explain, a new candidate haplotype is generated. The new haplotype is selected greedily as the haplotype which can explain the largest number of genotypes not yet explained. Clearly, other alternatives could be considered, but the experimental differences, obtained on a large set of examples, were not significant.

Observe that the proposed organization allows selecting haplotypes which will not be used in the final solution. As a result, the last step of the algorithm is to remove from the set of selected haplotypes all haplotypes which are not used for explaining any genotypes.

The overall delayed haplotype selection algorithm is shown in Algorithm 2 and summarizes the ideas outlined above. Line 2 computes the set of haplotypes $\mathcal{H}_S$ associated with genotypes $\mathcal{G}$ with one or zero heterozygous sites, since these haplotypes must be included in the final solution. Line 3 removes from $\mathcal{G}$ all genotypes that can be explained

---

Algorithm 2. Delayed Haplotype Selection

DELAYEDHAPLOTYPESELECTION($\mathcal{G}$)

```
1   ▷ Hₛ is the set of selected haplotypes; H_C is the set of candidate haplotypes
2   Hₛ ← CALCINITIALHAPLOTYPES(G)
3   G ← REMOVEEXPLAINEDGENOTYPES(G, Hₛ)
4   for each h ∈ Hₛ
5      do
6         for each g ∈ G
7            do if CANEXPLAIN(h, g)
8               then h_c ← CALCEXPLAINPAIR(h, g)
9                     H_C ← H_C ∪ {h_c}
10                    Associate h_c with g
11  while (G ≠ ∅)
12     do if (H_C = ∅)
13        then
14              h_c ← PICKCANDHAPLOTYPE(G)
15              H_C ← {h_c}
16        h ← h_c ∈ H_C associated with largest number of genotypes
17        H_C ← H_C − {h}
18        Hₛ ← Hₛ ∪ {h}
19        G ← REMOVEEXPLAINEDGENOTYPES(G, Hₛ)
20        for each g ∈ G
21           do if CANEXPLAIN(h, g)
22              then h_c ← CALCEXPLAINPAIR(h, g)
23                    H_C ← H_C ∪ {h_c}
24                    Associate h_c with g
25     Hₛ ← REMOVENONUSEDHAPLOTYPES(Hₛ)
26  return Hₛ
```

---

by a pair of haplotypes in $\mathcal{H}_S$. The same holds true for line 19. Lines 6 to 10 and 20 to 24 correspond to the candidate haplotype generation phase, given newly selected haplotypes. The DS algorithm runs in polynomial time in the number of genotypes and sites, a straightforward analysis yielding a run time complexity in $\mathcal{O}(n^2 m)$.

In practice, the delayed haplotype selection algorithm is executed multiple times, as in other recent implementations of Clark's method[20]. At each step, ties in picking the next candidate haplotype (see line 16) are randomly broken. The run producing the smallest number of haplotypes is selected.

Results in the next section suggest that delayed haplotype selection is a very effective approach. Nonetheless, it is straightforward to conclude that there are instances for which delayed haplotype selection will yield the same solution as Clark's method. In fact, it is possible for DS to yield solutions with more haplotypes than Clark's method. The results in the next section show that this happens very rarely. Indeed, for most examples considered (out of a comprehensive set of examples) DS is extremely unlikely to compute a larger number of haplotypes than Clark's method, and most often computes solutions with a significantly smaller number of haplotypes.

Table 1. Classes of problem instances evaluated

Class	#Instances	minSNPs	maxSNPs	minGENs	maxGENs
uniform	245	10	100	30	100
nonuniform	135	10	100	30	100
hapmap	24	30	75	7	68
biological	450	13	103	5	50
Total	854	10	103	5	100

4 Experimental Results

This section compares the delayed haplotype selection (DS) algorithm described in the previous section with a recent implementation of Clark's method (CM)[8]. In addition, the section also compares the HIPP solutions, computed with a recent tool[18], with the results of DS and CM. As motivated earlier, the objectives of the DS algorithm are twofold: first to replace Clark's method as an approximation of the HIPP solution, and second to provide tight upper bounds to HIPP algorithms.

Recent HIPP algorithms are iterative[18], at each step solving a Boolean Satisfiability problem instance. The objective of using tight upper bounds is to reduce the number of iterations of these algorithms. As a result, the main focus of this section is to analyze the absolute difference, in the number of haplotypes, between the computed upper bound and the HIPP solution.

4.1 Experimental Setup

The instances used for evaluating the two algorithms have been obtained from a number of sources[18], and can be organized into four classes shown in Table 1. For each class, Table 1 gives the number of instances, and the minimum and maximum number of SNPs and genotypes, respectively[2]. The uniform and nonuniform classes of instances are the ones used by other authors[3], but extended with additional, more complex, problem instances. The hapmap class of instances is also used by the same other authors[3]. Finally, the instances for the biological class are generated from data publicly available[13,21,7,6,15]. To the best of our knowledge, this is the most comprehensive set of examples used for evaluating haplotype inference solutions.

All results shown were obtained on a 1.9 GHz AMD Athlon XP with 1GB of RAM running RedHat Linux. The run times of both algorithms (CM and DS) were always a few seconds at most, and no significant differences in run times were observed between CM and DS. As a result, no run time information is included below.

4.2 Experimental Evaluation

The experimental evaluation of the delayed haplotype selection (DS) algorithm is organized in two parts. The first part compares DS with a publicly available recent

---

[2] Table 1 shows data for the original non-simplified instances. However, all instances were simplified using well-known techniques[3] before running any of the haplotype inference algorithms.

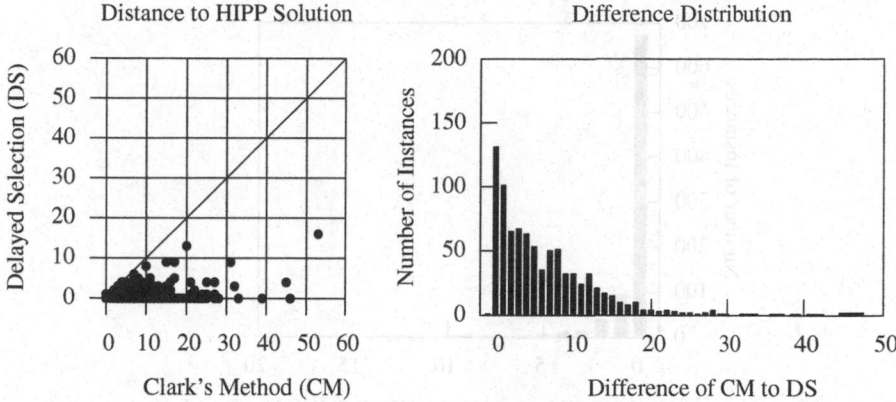

Fig. 1. Comparison of Clark's Method (CM) with Delayed Haplotype Selection (DS)

implementation of Clark's method (CM)[8], whereas the second part compares DS with an exact solution to the Haplotype Inference by Pure Parsimony (HIPP) problem[18]. In all cases, for both CM and DS, we select the best solution out of 10 runs. Other implementations of Clark's method could have been considered[20]. However, no significant differences were observed among these implementations when the objective is to minimize the number of computed haplotypes.

The results for the first part are shown in Figure 1. The scatter plot shows the difference of CM and of DS with respect to the exact HIPP solution for the examples considered. The results are conclusive. DS is often quite close to the HIPP solution, whereas the difference of CM with respect to the HIPP solution can be significant. While the distance of DS to the HIPP solution never exceeds 16 haplotypes, the distance of CM can exceed 50 haplotypes. Moreover, for a large number of examples, the distance of DS to the HIPP solution is 0, and for the vast majority of the examples the distance does not exceed 5 haplotypes. In contrast, the distance of CM to the HIPP solution often exceeds 10 haplotypes.

The second plot in Figure 1 provides the distribution of the difference between the number of haplotypes computed with DS and with CM. A bar associated with a value k represents the number of examples for which CM exceeds DS by k haplotypes. With one exception, DS always computes a number of haplotypes no larger than the number of haplotypes computed with CM. For the single exception, DS exceeds CM in 1 haplotype (hence -1 is shown in the plot). Observe that for 85% of the examples, DS outperforms CM. Moreover, observe that for a reasonable number of examples (40.1%, or 347 out of 854) the number of haplotypes computed with CM exceeds DS in more than 5 haplotypes. Finally, for a few examples (3 out of 854), CM can exceed DS by more than 40 haplotypes, the largest value being 46 haplotypes.

It should also be noted that, if the objective is to use either DS or CM as an upper bound for an exact HIPP algorithm, then a larger number of computed haplotypes represents a less tight, and therefore less effective, upper bound. Hence, DS is clearly preferable as an upper bound solution.

The results for the second part, comparing DS to the HIPP solution, are shown in Figure 2. As can be observed for the majority of examples (78.7%, or 672 out of 854),

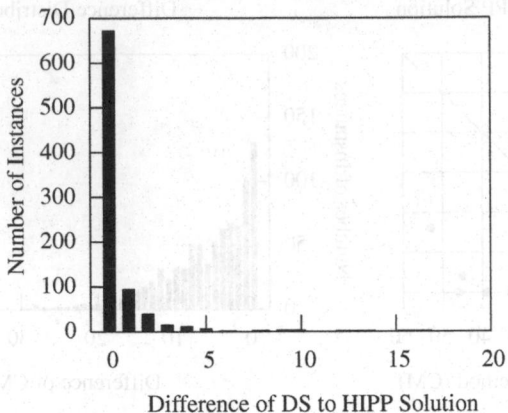

Difference of DS to HIPP Solution

Fig. 2. Comparison of Delayed Haplotype Selection (DS) with HIPP solution

DS computes the HIPP solution. This is particularly significant when DS is used as an upper bound for recent HIPP algorithms[18]. For examples where DS computes the HIPP solution, exact HIPP algorithms are only required to prove the solution to be optimum. For a negligible number of examples (0.9%, or 8 out of 854) the difference of DS to the HIPP solution exceeds 5 haplotypes. Hence, for the vast majority of examples considered, DS provides a tight upper bound to the HIPP solution.

The results allow drawing the following conclusions. First, DS is a very effective alternative to CM when the objective is to minimize the total number of computed haplotypes. Second, DS is extremely effective as an upper bound for exact HIPP algorithms. For most examples (99.1%, or 846 out of 854) the number of haplotypes identified by DS is within 5 haplotypes of the target HIPP solution.

5   Conclusions and Future Work

This paper proposes a novel approach for haplotype selection, which addresses one of the main drawbacks of Clark's method[5]: its excessive greediness. This is achieved by delaying haplotype selection, one of the greedy steps of Clark's method. This approach leads to a tight upper bound that can be used when modelling this problem as a constraint satisfaction problem. The main context for the work is the development of efficient and accurate upper bounding procedures for exact algorithms for the Haplotype Inference by Pure Parsimony (HIPP) problem. Nevertheless, the proposed approach can also serve as a standalone haplotype inference algorithm. Experimental results, obtained on a comprehensive set of examples, are clear and conclusive. In practice, the new *delayed haplotype selection* (DS) algorithm provides quite tight upper bounds, of far superior quality than a recent implementation of Clark's method. For the vast majority of the examples considered, the results for DS are comparable to those for HIPP, and for a large percentage of the examples, DS computes the actual HIPP solution.

As mentioned earlier, recent approaches for the HIPP problem iterate through increasingly higher lower bounds[18]. This implies that solutions to the haplotype inference problem are only obtained when the actual solution to the HIPP problem is identified. Thus, these recent approaches to the HIPP problem[18] *cannot* be used for computing approximate HIPP solutions. The work described in this paper provides an efficient and remarkably tight approach for computing upper bounds. This allows recent HIPP based algorithms[18] to compute the exact solution by iterating through decreasing upper bounds. Hence, at each step a solution to the haplotype inference problem is identified, and, therefore, these methods can be used for approximating the exact HIPP solution. The integration of the DS algorithm in recent solutions to the HIPP problem is the next natural step of this work.

Acknowledgments. This work is partially supported by Fundação para a Ciência e Tecnologia under research project POSC/EIA/61852/2004 and PhD grant SFRH/BD/28599/2006, by INESC-ID under research project SHIPs, and by Microsoft under contract 2007-017 of the Microsoft Research PhD Scholarship Programme.

References

1. Adkins, R.M.: Comparison of the accuracy of methods of computational haplotype inference using a large empirical dataset. BMC Genet. 5(1), 22 (2004)
2. Brown, D., Harrower, I.: A new integer programming formulation for the pure parsimony problem in haplotype analysis. In: Workshop on Algorithms in Bioinformatics (2004)
3. Brown, D., Harrower, I.: Integer programming approaches to haplotype inference by pure parsimony. IEEE/ACM Transactions on Computational Biology and Bioinformatics 3(2), 141–154 (2006)
4. Graça, A., Marques-Silva, J., Lynce, I., Oliveira, A.: Efficient haplotype inference with pseudo-Boolean optimization. Algebraic Biology 2007, 125–139 (July 2007)
5. Clark, A.G.: Inference of haplotypes from pcr-amplified samples of diploid populations. Molecular Biology and Evolution 7(2), 111–122 (1990)
6. Daly, M.J., Rioux, J.D., Schaffner, S.F., Hudson, T.J., Lander, E.S.: High-resolution haplotype structure in the human genome. Nature Genetics 29, 229–232 (2001)
7. Drysdale, C.M., McGraw, D.W., Stack, C.B., Stephens, J.C., Judson, R.S., Nandabalan, K., Arnold, K., Ruano, G., Liggett, S.B.: Complex promoter and coding region β_2-adrenergic receptor haplotypes alter receptor expression and predict in vivo responsiveness. National Academy of Sciences 97, 10483–10488 (2000)
8. Greenspan, G., Geiger, D.: High density linkage disequilibrium mapping using models of haplotype block variation. Bioinformatics 20(supp. 1) (2004)
9. Gusfield, D.: Inference of haplotypes from samples of diploid populations: complexity and algorithms. Journal of Computational Biology 8(3), 305–324 (2001)
10. Gusfield, D.: Haplotype inference by pure parsimony. In: Baeza-Yates, R.A., Chávez, E., Crochemore, M. (eds.) CPM 2003. LNCS, vol. 2676, pp. 144–155. Springer, Heidelberg (2003)
11. Gusfield, D., Orzach, S.H.: Haplotype Inference. In: Handbook on Computational Molecular Biology. Chapman and Hall/CRC Computer and Information Science Series, vol. 9, CRC Press, Boca Raton, USA (2005)
12. Huang, Y.-T., Chao, K.-M., Chen, T.: An approximation algorithm for haplotype inference by maximum parsimony. Journal of Computational Biology 12(10), 1261–1274 (2005)

13. Kerem, B., Rommens, J., Buchanan, J., Markiewicz, D., Cox, T., Chakravarti, A., Buchwald, M., Tsui, L.C.: Identification of the cystic fibrosis gene: Genetic analysis. Science 245, 1073–1080 (1989)
14. Kimura, M., Crow, J.F.: The number of alleles that can be maintained in a finite population. Genetics 49(4), 725–738 (1964)
15. Kroetz, D.L., Pauli-Magnus, C., Hodges, L.M., Huang, C.C., Kawamoto, M., Johns, S.J., Stryke, D., Ferrin, T.E., DeYoung, J., Taylor, T., Carlson, E.J., Herskowitz, I., Giacomini, K.M., Clark, A.G.: Sequence diversity and haplotype structure in the human abcd1 (mdr1, multidrug resistance transporter). Pharmacogenetics 13, 481–494 (2003)
16. Lancia, G., Pinotti, C.M., Rizzi, R.: Haplotyping populations by pure parsimony: complexity of exact and approximation algorithms. INFORMS Journal on Computing 16(4), 348–359 (2004)
17. Lynce, I., Marques-Silva, J.: Efficient haplotype inference with Boolean satisfiability. In: National Conference on Artificial Intelligence (AAAI) (July 2006)
18. Lynce, I., Marques-Silva, J.: SAT in bioinformatics: Making the case with haplotype inference. In: Biere, A., Gomes, C.P. (eds.) SAT 2006. LNCS, vol. 4121, Springer, Heidelberg (2006)
19. Niu, T., Qin, Z., Xu, X., Liu, J.: Bayesian haplotype inference for multiple linked single-nucleotide polymorphisms. American Journal of Human Genetics 70, 157–169 (2002)
20. Orzack, S.H., Gusfield, D., Olson, J., Nesbitt, S., Subrahmanyan, L., Stanton Jr., V.P.: Analysis and exploration of the use of rule-based algorithms and consensus methods for the inferral of haplotypes. Genetics 165, 915–928 (2003)
21. Rieder, M.J., Taylor, S.T., Clark, A.G., Nickerson, D.A.: Sequence variation in the human angiotensin converting enzyme. Nature Genetics 22, 481–494 (2001)
22. Stephens, M., Smith, N., Donelly, P.: A new statistical method for haplotype reconstruction. American Journal of Human Genetics 68, 978–989 (2001)
23. The International HapMap Consortium: A haplotype map of the human genome. Nature 437, 1299–1320 (October 27, 2005)
24. Wang, L., Xu, Y.: Haplotype inference by maximum parsimony. Bioinformatics 19(14), 1773–1780 (2003)

GRASPER
(A Framework for Graph Constraint Satisfaction Problems)

Ruben Viegas* and Francisco Azevedo

CENTRIA, Departamento de Informática
Universidade Nova de Lisboa
{rviegas,fa}@di.fct.unl.pt

Abstract. In this paper we present GRASPER, a graph constraint solver, based on set constraints, that shows promising results when compared to an existing similar solver at this early stage of development.

Keywords: Constraint Programming, Graphs, Sets.

1 Introduction

Constraint Programming (CP) [1,2,3] has been successfully applied to numerous combinatorial problems such as scheduling, graph coloring, circuit analysis, or DNA sequencing. Following the success of CP over traditional domains, set variables [4] were also introduced to more declaratively solve a number of different problems. Recently, this also led to the development of a constraint solver over graphs [5,6], as a graph [7,8,9] is composed of a set of vertexes and a set of edges. Developing a framework upon a finite sets computation domain allows one to abstract from many low-level particularities of set operations and focus entirely on graph constraining, consistency checking and propagation.

Graph-based constraint programming can be declaratively used for path and circuit finding problems, to routing, scheduling and allocation problems, among others. CP(Graph) was proposed by G. Dooms et al. [5,6] as a general approach to solve graph-based constraint problems. It provides a key set of basic constraints which represent the framework's core, and higher level constraints for solving path finding and optimization problems, and to enforce graph properties.

In this paper we present GRASPER (GRAph constraint Satisfaction Problem solvER) which is an alternative framework for graph-based constraint solving based on *Cardinal* [10], a finite sets constraint solver with extra inferences developed in Universidade Nova de Lisboa. We present a set of basic constraints which represent the core of our framework and we provide functionality for directed graphs, graph weighting, graph matching, graph path optimization problems and some of the most common graph properties. In addition, we intend to integrate GRASPER in CaSPER [11], a programming environment for the development and integration of constraint solvers, using the Generic Programming [12] methodology.

* Partially supported by PRACTIC - FCT-POSI/SRI/41926/2001.

J. Neves, M. Santos, and J. Machado (Eds.): EPIA 2007, LNAI 4874, pp. 633–644, 2007.
© Springer-Verlag Berlin Heidelberg 2007

This paper is organised as follows. In section 2 we specify the details of our framework, starting with a brief introduction to *Cardinal*, followed by the presentation of our core constraints, other non-trivial ones and the filtering rules used. In section 3 we describe a problem in the context of biochemical networks which we use to test our framework: we present a model for it together with a search strategy to find the solution, and present experimental results, comparing them with the ones obtained with CP(Graph). We conclude in section 4.

2   GRASPER

In this section we start by introducing *Cardinal*, the finite sets constraint solver upon which GRASPER is based and then we present the concepts which build up our framework: core constraints, non-trivial constraints and filtering rules.

2.1   *Cardinal*

Set constraint solving was proposed in [4] and formalized in [13] with ECLiPSe (http://eclipse.crosscoreop.com/eclipse/) library *Conjunto*, specifying set domains by intervals whose lower and upper bounds are known sets ordered by set inclusion. Such bounds are denoted as *glb* (greatest lower bound) and *lub* (least upper bound). The *glb* of a set variable S can be seen as the set of elements that are known to belong to set S, while its *lub* is the set of all elements that can belong to S. Local consistency techniques are then applied using interval reasoning to handle set constraints (e.g. equality, disjointness, containment, together with set operations such as union or intersection). *Conjunto* proved its usefulness in declarativeness and efficiency for NP-complete combinatorial search problems dealing with sets, compared to constraint solving over finite integer domains. Afterwards, *Cardinal* (also in ECLiPSe) [10], improved on *Conjunto* by extending propagation on set functions such as cardinality.

2.2   GRASPER Specification

In this subsection we explain how we defined our framework upon *Cardinal*.

A graph is composed by a set of vertexes and by a set of edges, where each edge connects a pair of the graph's vertexes. So, a possible definition for a graph is to see it as a pair (V, E) where both V and E are finite set variables and where each edge is represented by a pair (X, Y) specifying a directed arc from X towards Y. In our framework we do not constrain the domain of the elements contained in those sets, so the user is free to choose the best representation for the constraint satisfaction problem. The only restriction we impose is that each incidence of an edge in the set of edges must be present in the set of vertexes.

In order to create graph variables we introduce the following constraint:

$graph(G, V, E)$ (G is a compound term of the form $graph(V, E)$)

which is true if G is a graph variable whose set of vertexes is V and whose set of edges is E.

All the basic operations for accessing and modifying the vertexes and edges is supported by *Cardinal*'s primitives, so no additional functionality is needed. Therefore, our framework provides the creation and manipulation of graph variables for constraint satisfaction problems just by providing a single constraint for graph variable creation and delegating to *Cardinal* the underlying core operations on set variables.

While the core constraints of the framework allow basic manipulation of graph variables, it is useful to define some other, more complex, constraints based on the core ones thus providing a more intuitive and declarative set of functions for graph variable manipulation.

We provide a constraint to weigh a graph variable in order to facilitate the modeling of graph optimization or satisfaction problems. The weight of a graph is defined by the weight of the vertexes and of the edges that compose the graph. Therefore, the sum of the weights of both vertexes and edges in the graph variable's *glb* define the lower bound of the graph's weight and, similarly, the sum of the weights of the vertexes and of the edges in the graph variable's *lub* define the upper bound of the graph's weight. The constraint is provided as $weight(G, W_f, W)$, where W_f is a map which associates to each vertex and to each edge a given weight, and can be defined as:

$$weight(graph(V, E), W_f, W) \equiv m = \sum_{v \in glb(V)} W_f(v) + \sum_{e \in glb(E)} W_f(e) \wedge$$
$$M = \sum_{v \in lub(V)} W_f(v) + \sum_{e \in lub(E)} W_f(e) \wedge$$
$$W :: m..M$$

Additionally, we provide a subgraph relation which can be expressed as:

$$subgraph(graph(V_1, E_1), graph(V_2, E_2)) \equiv V_1 \subseteq V_2 \wedge E_1 \subseteq E_2$$

stating that a graph $G_1 = graph(V_1, E_1)$ is a subgraph of a graph $G_2 = graph(V_2, E_2)$ iff $V_1 \subseteq V_2$ and $E_1 \subseteq E_2$.

Obtaining the set of predecessors P of a vertex v in a graph G is performed by the constraint $preds(G, v, P)$ which can be expressed as:

$$preds(graph(V, E), v, P) \equiv P \subseteq V \wedge \forall v' \in V : (v' \in P \equiv (v', v) \in E)$$

Similarly, obtaining the successors S of a vertex v in a graph G is performed by the constraint $succs(G, v, S)$ which can be expressed as:

$$succs(graph(V, E), v, S) \equiv S \subseteq V \wedge \forall v' \in V : (v' \in S \equiv (v, v') \in E)$$

However, this last constraint obtains the set of the immediate successors of a vertex in a graph and we may want to obtain the set of all successors of that vertex, i.e., the set of reachable vertexes of a given initial vertex. Therefore, we provide a $reachables(G, v, R)$ which can be expressed in the following way:

$$reachables(graph(V, E), v, R) \equiv R \subseteq V \wedge$$
$$\forall r \in V : (r \in R \equiv \exists p : p \in paths(graph(V, E), v, r))$$

stating that a set of vertexes each reachable from another if there is a path between this last vertex and each of those vertexes. The rule *paths* represents all possible paths between two given vertexes and $p \in paths(graph(V, E), v, r)$ can be expressed as:

$$p \in paths(graph(V, E), v_0, v_f) \equiv \begin{cases} v_0 \in V \wedge p = \emptyset & , if\ v_0 = v_f \\ \\ \exists v_i \in V : (v_0, v_i) \in E \wedge \\ \exists p' : p' \in paths(graph(V, E), v_i, v_f) \wedge & , if\ v_0 \neq v_f \\ p = cons(v_0, p') \end{cases}$$

Additionally, this last constraint will allow us to build other very useful ones. For example, we can make use of the *reachables*/3 constraint to develop the connectivity property of a graph. By [7], a non-empty graph is said connected if any two vertexes are connected by a path, or in other words, if any two vertexes are reachable from one another. In a connected graph, all vertexes must reach all the other ones, so we define a new constraint *connected(G)* which can be expressed as:

$$connected(graph(V, E)) \equiv \forall v \in V : reachables(graph(V, E), v, R) \wedge R = V$$

Another useful graph property is that of a path: a graph is said to define a path between an initial vertex v_0 and a final vertex v_f if there is a path between those vertexes in the graph and all other vertexes belong to the path and are visited only once. We provide the $path(G, v_0, v_f)$ constraint, which can be expressed in the following way:

$$path(G, v_0, v_f) \equiv quasipath(G, v_0, v_f) \wedge connected(G)$$

This constraint delegates to *quasipath*/3 the task of restricting the vertexes that are or will become part of the graph to be visited only once and delegates to *connected*/1 the task of ensuring that those same nodes belong to the path between v_0 and v_f so as to prevent disjoint cycles from appearing in the graph.

The $quasipath(G, v_0, v_f)$ constraint (for directed graph variables) can be expressed as:

$$quasipath(graph(V, E), v_0, v_f) \equiv \forall v \in V \begin{cases} succs(graph(V, E), S) \wedge \\ \#S = 1 & , if\ v = v_0 \\ \\ preds(graph(V, E), P) \wedge \\ \#P = 1 & , if\ v = v_f \\ \\ preds(graph(V, E), P) \wedge \\ \#P = 1 \wedge \\ succs(graph(V, E), S) \wedge & , otherwise \\ \#S = 1 \end{cases}$$

This constraint, although slighty complex, is very intuitive: it ensures that every vertex that is added to the graph has exactly one predecessor and one sucessor, exceptions being the initial vertex which is only restricted to have one successor and the final vertex which is only restricted to have one predecessor. Therefore, a vertex that is not able to verify these constraints can be safely removed from the set of vertexes.

2.3   Filtering Rules

In this subsection we formalize the filtering rules of the constraints presented earlier, as they are currently implemented. These filtering rules can be seen in higher detail in [14]. For a set variable S, we will denote S' as the new state of the variable (after the filtering) and S as its previous state.

$graph(graph(V, E), V, E)$

- When an edge is added to the set of edges, the vertexes that compose it are added to the set of vertexes:
  $$glb(V') \leftarrow glb(V) \cup \{x : (x, y) \in glb(E) \vee (y, x) \in glb(E)\}$$

- When a vertex is removed from the set of vertexes, all the edges incident on it are removed from the set of edges:
  $$lub(E') \leftarrow lub(E) \cap \{(x, y) : x \in lub(V) \wedge y \in lub(V)\}$$

$weight(graph(V, E), W_f, W)$

Let $min(G)$ and $Max(G)$ be the minimum graph weight and the maximum graph weight of graph G, respectively.

- When an element (vertex or edge) is added to the graph, the graph's weight is updated:
  $$W \geq \sum_{v \in glb(V)} W_f(v) + \sum_{e \in glb(E)} W_f(e)$$

- When an element (vertex or edge) is removed from the graph, the graph's weight is updated:
  $$W \leq \sum_{v \in lub(V)} W_f(v) + \sum_{e \in lub(E)} W_f(e)$$

- When the lower bound of the graph's weight $W :: m..M$ is increased, some elements (vertexes or edges) may be added to the graph:
  $$glb(V') \leftarrow glb(V) \cup \{v : v \in lub(V) \wedge Max(G \backslash \{v\}) < m\}$$
  $$glb(E') \leftarrow glb(E) \cup \{(x, y) : (x, y) \in lub(E) \wedge Max(G \backslash \{(x, y)\}) < m\}$$

- When the upper bound of the graph's weight $W :: m..M$ is decreased, some elements (vertexes or edges) may be removed from the graph:
  $$lub(V') \leftarrow lub(V) \backslash \{v : v \in lub(V) \wedge min(G \cup \{v\}) > M\}$$
  $$lub(E') \leftarrow lub(E) \backslash \{(x, y) : (x, y) \in lub(E) \wedge min(G \cup \{(x, y)\}) > M\}$$

$subgraph(graph(V_1, E_1), graph(V_2, E_2))$

The $V_1 \subseteq V_2$ and $E_1 \subseteq E_2$ *Cardinal* constraints, corresponding to our *subgraph* rule, yield the following filtering rules:

- When a vertex is added to V_1, it is also added to V_2:
$$glb(V_2') \leftarrow glb(V_2) \cup glb(V_1)$$
- When a vertex is removed from V_2 it is also removed from V_1:
$$lub(V_1') \leftarrow lub(V_1) \cap lub(V_2)$$
- When an edge is added to E_1, it is also added to E_2:
$$glb(E_2') \leftarrow glb(E_2') \cup glb(E_1)$$
- When an edge is removed from E_2, it is also removed from E_1:
$$lub(E_1') \leftarrow lub(E_1) \cap lub(E_2)$$

$preds(graph(V, E), v, P)$

- The $P \subseteq V$ constraint of $preds/3$ is managed by *Cardinal*
- When an edge is added to the set of edges, the set of predecessors is updated with the in-vertexes belonging to the edges in $glb(E)$ whose out-vertex is v:
$$glb(P') \leftarrow glb(P) \cup \{x : (x, v) \in glb(E)\}$$

- When an edge is removed from the set of edges, the set of predecessors is limited to the in-vertexes belonging to the edges in $lub(E)$ whose out-vertex is v:
$$lub(P') \leftarrow lub(P) \cap \{x : (x, v) \in lub(E)\}$$

- When a vertex is added to the set of predecessors, the set of edges is updated with the edges that connect each of those nodes in $glb(P)$ to v:
$$glb(E') \leftarrow glb(E) \cup \{(x, v) : x \in glb(P)\}$$

- When a vertex is removed from the set of predecessors, the corresponding edge is removed from the set of edges:
$$lub(E') \leftarrow \{(x, y) : (y = v \wedge x \in lub(P)) \vee (y \neq v \wedge (x, y) \in lub(E))\}$$

$succs(graph(V, E), v, S)$

Similar to $preds(graph(V, E), v, P)$.

$reachables(graph(V, E), v, R)$

- The $R \subseteq V$ constraint $reachables/3$ is managed by *Cardinal*
- When an edge is added to the set of edges, the set of reachable vertexes is updated with all the vertexes in $glb(V)$ that are reachable from v:
$$glb(R') \leftarrow glb(R) \cup \{r : r \in glb(V) \wedge \exists_p\, p \in paths(graph(glb(V), glb(E)), v, r)\}$$
- When a vertex is removed from the set of reachable vertexes, the edge connecting v to it is removed from the set of edges:
$$lub(E') \leftarrow lub(E) \setminus \{(v, r) : r \notin lub(R)\}$$
- When a vertex is added to the set of reachable vertexes, the edge connecting v to it may be added to the set of edges:
$$glb(E') \leftarrow glb(E) \cup \{(v, r) : r \in glb(R) \wedge \nexists(x, r) \in glb(E) : x \neq v\}$$

– When an edge is removed from the set of edges, the set of reachable vertexes is limited to the vertexes in $lub(V)$ that are reachable from v:

$$lub(R') \leftarrow lub(R) \cap \{r : r \in lub(V) \land \exists_p p \in paths(graph(lub(V), lub(E)), v, r)\}$$

connected(graph(V,E))

Let R_v be the set of reachable vertexes of a vertex $v \in glb(V)$. Since this rule makes use of the *reachables*/3 rule it will use its filtering rules. The $R_v = V$ *Cardinal* constraint, included in our *connected*/1 rule can be decomposed into the constraints $R_v \subseteq V$ and $V \subseteq R_v$ for which the filtering rules have been presented previously.

quasipath(graph(V,E), v_0, v_f)

Let P_v and S_v be the set of predecessors and the set of successors, respectively, of a vertex $v \in lub(V)$.

– When a vertex is added to the set of vertexes, the number of predecessors (and successors) is set to 1:

$$\forall v \in glb(V) : \#P_v = 1 \qquad (\forall v \in glb(V) : \#S_v = 1)$$

– When a vertex has no predecessor or successor it is removed from the graph:

$$lub(V') \leftarrow lub(V) \cap \{v : v \in lub(V) \land \#P_v > 0 \land \#S_v > 0\}$$

path(graph(V,E), v_0, v_f):

Since this rule makes use of *connected*/1 and *quasipath*/3 rules, it will use their filtering rules.

3   Results

In this section we describe a problem in biochemical networks, model it, and present obtained results, comparing them with CP(Graph).

3.1   Pathways

Metabolic networks [15,16] are biochemical networks which encode information about molecular compounds and reactions which transform these molecules into substrates and products. A pathway in such a network represents a series of reactions which transform a given molecule into others. In Fig. 1 we present a metabolic network, and in Fig. 2, a possible pathway between the imposed start and finish molecules.

An application for pathway discovery in metabolic networks is the explanation of DNA experiments. An experiment is performed on DNA cells and these mutated cells (called RNA cells) are placed on DNA chips, which contain specific locations for different strands, so when the cells are placed in the chips, the different strands will fit into their specific locations. Once placed, the DNA strands (which encode specific enzymes) are scanned and catalyze a set of reactions. Given this set of reactions the goal is to know which products were active in the cell, given the initial molecule and the final result. Fig. 2 represents a possible pathway between two given nodes regarding the metabolic network of Fig. 1.

A recurrent problem in metabolic networks pathway finding is that many paths take shortcuts, in the sense that they traverse highly connected molecules (act as substrates or products of many reactions) and therefore cannot be considered as belonging to an actual pathway. However there are some metabolic networks for which some of these highly connected molecules act as main intermediaries. In Fig. 1 there are three highly connected compounds, represented by the grid-filled circles.

It is also possible that a path traverses a reaction and its reverse reaction: a reaction from substrates to products and one from products to substrates. Most of the time these reactions are observed in a single direction so we can introduce *exclusive pairs of reactions* to ignore a reaction from the metabolic network when the reverse reaction is known to occur, so that both do not occur simultaneously. Fig. 1 shows the presence of five *exclusive pairs of reactions*, represented by 5 pairs of the ticker arrows.

Additionally, it is possible to have various pathways in a given metabolic experiment and often the interest is not to discover one pathway but to discover a pathway which traverses a given set of intermediate products or substrates, thus introducing the concept of *mandatory molecule*. These *mandatory molecules* are useful, for example, if biologists already know some of the products which are in the pathway but do not know the complete pathway. In Fig. 1 we imposed the existence of a *mandatory molecule*, represented by a diagonal lined-filled circle.

In fact, the pathway represented in Fig. 2 is the shortest pathway obtained from the metabolic network depicted in Fig. 1 that complies with all the above constraints.

3.2 Modeling Pathways

Such network can be represented as a directed bi-partite graph, where the compounds, substrates and products represent one of the partition of the vertexes and the reactions the other partition. The edges link compounds with the set of reactions and these to the substrates and the products. The search of a pathway between two nodes (the original molecule and a final product or substrate) can be easily performed with a breadth-first [17] search algorithm.

Considering the problem of the highly connected molecules, a possible solution is to weight each vertex of the graph, where each vertex's weight is its degree (i.e. the number of edges incident on the vertex) and the solution consists in finding the shortest pathway of the metabolic experiment. This approach allows one to avoid these highly connected molecules whenever it is possible.

The *exclusive pairs of reactions* can also be easily implemented by introducing pairs of *exclusive* vertexes, where as soon as it is known that a given vertex belongs to the graph the other one is instantly removed.

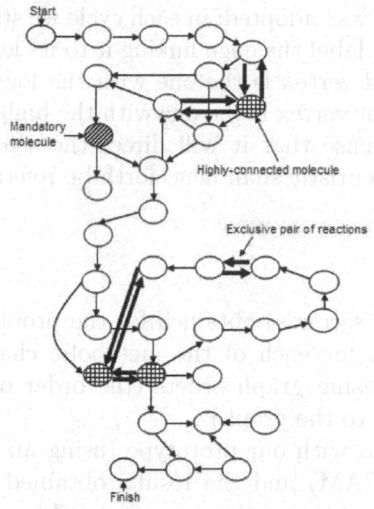

Fig. 1. Metabolic Network

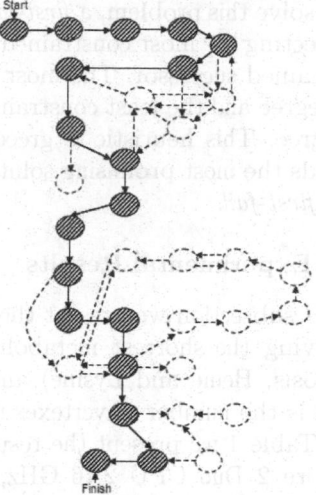

Fig. 2. Metabolic Pathway

Finally, to solve the constraint of *mandatory molecules*, it is sufficient to add the vertexes representing these molecules to the graph thus ensuring that any solution must contain all the specified vertexes. With this mechanism, however, it is not guaranteed that the intended pathway is the shortest pathway between the given initial and final vertexes (e.g. one of the *mandatory* vertexes does not belong to the shortest path), so we cannot rely on breadth-first search again and must find a different search strategy for solving this problem.

Basically, assuming that $G = graph(V, E)$ is the original graph, composed of all the vertexes and edges of the problem, that v_0 and v_f are the initial and the final vertexes, that $Mand = \{v_1, \ldots, v_n\}$ is the set of *mandatory* vertexes, that $Excl = \{(v_{e11}, v_{e12}), \ldots, (v_{em1}, v_{em2})\}$ is the set of *exclusive* pairs of vertexes and that W_f is a map associating each vertex and each edge to its weight, this problem can be easily modeled in GRASPER as:

$$minimize(W) : \begin{aligned} &subgraph(graph(SubV, SubE), G) \land Mand \subseteq SubV \land \\ &\forall(v_{ei1}, v_{ei2}) \in Excl : (v_{ei1} \notin SubV \lor v_{ei2} \notin SubV) \land \\ &path(graph(SubV, SubE), v_0, v_f) \\ &weight(graph(SubV, SubE), W_f, W) \end{aligned}$$

3.3 Search Strategies

The minimization of the graph's weight ensures that one obtains the shortest pathway constrained to contain all the *mandatory vertexes* and not containing both vertexes of any pair of *exclusive* vertexes. However, it may not uniquely determine all the vertexes (the non-mandatory) which belong to that pathway. This must then be achieved by labeling functions which, in graph problems, decide whether or not a given vertex or edge belongs to the graph.

To solve this problem, a *first-fail* heuristic was adopted: in each cycle we start by selecting the most constrained vertex and label the edge linking it to its least constrained successor. The most constrained vertex is the one with the lowest out-degree and the least constrained successor vertex is the one with the highest in-degree. This heuristic is greedy in the sense that it will direct the search towards the most promising solution. This heuristic shall henceforth be referred to as *first-fail*.

3.4 Experimental Results

In this subsection we present the results (in seconds) obtained for the problem of solving the shortest metabolic pathways for each of the metabolic chains (Glycosis, Heme and Lysine) and for increasing graph orders (the order of a graph is the number of vertexes that belong to the graph).

In Table 1 we present the results obtained with our prototype (using an Intel Core 2 Duo CPU 2.16 GHz, 1 Gb of RAM) and the results obtained by CP(Graph)[2] using the same modelling and employing the same *first-fail* heuristic, having been imposed a time limit of 10 minutes. The results for instances that were not solved within the time limit are set to "N.A.".

Table 1. Results obtained for GRASPER and CP(Graph)

Order	GRASPER			CP(Graph)		
	Glycosis	Heme	Lysine	Glycosis	Heme	Lysine
50	0.2	0.2	0.2	0.2	0.2	0.2
100	1.9	1.0	60.8	2.5	0.3	4.7
150	3.1	2.9	106.5	41.7	1.0	264.3
200	6.3	9.5	153.8	55.0	398.8	N.A.
250	13.2	19.0	183.7	127.6	173.3	N.A.
300	98.8	33.0	218.0	2174.4	1520.2	N.A.

In Fig. 3 the speed-up of GRASPER relative to CP(Graph) can be seen, showing that GRASPER presents better results than CP(Graph) for instances of the problem with order of at least 150. The speed-up was calculated as the quotient between CP(Graph) and GRASPER.

CP(Graph) only produces better results for graphs of order 50 and 100 and for the Heme chain of order 150. However, this trend is clearly reversed for higher order instances: results for the Glycosis chain outperform the ones obtained by CP(Graph) from order 150 above, and for the graph of order 300 we achieve almost 35 minutes less; results for graphs of order above 150 are all under 220 seconds managing to decrease the expected time as compared to CP(Graph); finally, for the Lysine chain, we could obtain results for instances of order above 150, for which CP(Graph) presents no results.

---

[2] We could not get access to CP(Graph)'s version implementing this heuristic so we had to use the results presented in [5]

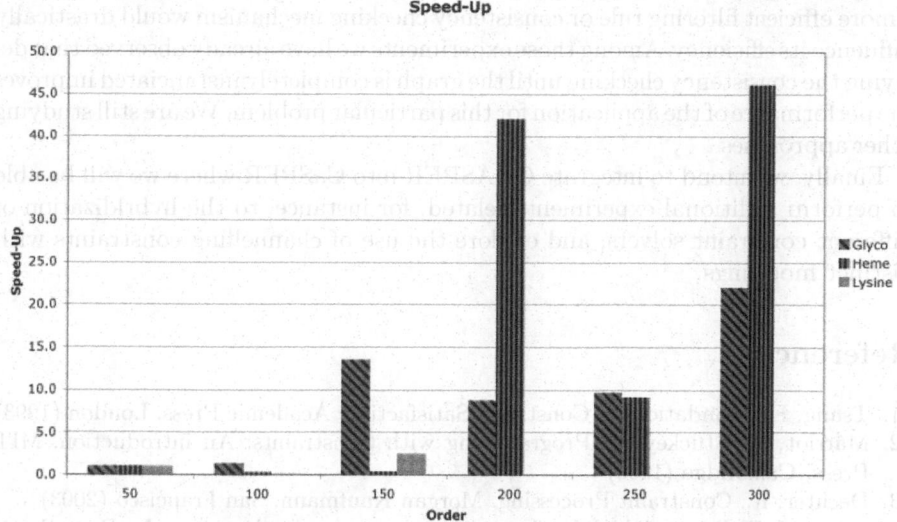

Fig. 3. Grasper's speed-up relative to CP(Graph)

The comparison between these frameworks seems to indicate that GRASPER outperforms CP(Graph) for larger problem instances thus providing a more scalable framework.

4   Conclusions and Future Work

In this paper we presented GRASPER, a new framework for the development of graph-based constraint satisfaction problems. This framework being built upon *Cardinal* [10] allows for a clear and concise manipulation of the elements that constitute a graph, the sets of vertexes and edges, and thus appears as a simple and intuitive interface just by defining a few additional rules for graph creation, manipulation and desirable graph properties.

We tested GRASPER with a problem in the context of biochemical networks (metabolic pathways) and compared results with CP(Graph) [5]. Even though CP(Graph) presented better results for small problem instances, GRASPER clearly outperformed it for larger ones, achieving speed-ups of almost 50. The main differences between these frameworks reside in the underlying set solver and also in the underlying implementation.

More efficient results were published in [6] using a mechanism based on the concept of a shortest path tree and a cost-based filtering [18] mechanism to further constrain search space. We have already started to develop such a mechanism but it is still being improved.

The presented GRASPER's filtering rules still require fine tunning, namely for the *reachables*/3 and the *connected*/1 constraints, for which we are currently trying some different (lighter) rules that are already exhibiting much larger speed-ups in preliminary experiments. Since *connected*/1 is highly dependent on *reachables*/3, a more efficient filtering rule or consistency checking mechanism would drastically influence its efficiency. Among those experiments we have already observed that delaying the consistency checking until the graph is completely instanciated improves the performance of the application for this particular problem. We are still studying other approaches.

Finally, we intend to integrate GRASPER into CaSPER where we will be able to perform additional experiments related, for instance, to the hybridization of different constraint solvers, and explore the use of channelling constraints with distinct modelings.

References

1. Tsang, E.: Foundations of Constraint Satisfaction. Academic Press, London (1993)
2. Marriot, K., Stuckey, P.: Programming with Constraints: An introduction. MIT Press, Cambridge (1998)
3. Dechter, R.: Constraint Processing. Morgan Kaufmann, San Francisco (2003)
4. Puget, J.-F.: Pecos: A high level constraint programming language. In: Proc. Spicis (1992)
5. Dooms, G., Deville, Y., Dupont, P.: CP(Graph): Intoducing a graph computation domain in constraint programming. In: van Beek, P. (ed.) CP 2005. LNCS, vol. 3709, pp. 211–225. Springer, Heidelberg (2005)
6. Dooms, G.: The CP(Graph) Computation Domain in Constraint Programming. PhD thesis, Faculté des Sciences Appliquées, Université Catholique de Louvain (2006)
7. Diestel, R. (ed.): Graph Theory, 3rd edn. Graduate Texts in Mathematics, vol. 173. Springer, Heidelberg (2005)
8. Harary, F.: Graph Theory. Addison-Wesley, Reading (1969)
9. Xu, J.: Theory and Application of Graphs. In: Network Theory and Applications, vol. 10. Kluwer Academic, Dordrecht (2003)
10. Azevedo, F.: Cardinal: A finite sets constraint solver. Constraints journal 12(1), 93–129 (2007)
11. Correia, M., Barahona, P., Azevedo, F.: CaSPER: A programming environment for development and integration of constraint solvers. In: Azevedo, et al. (eds.) BeyondFD 2005. Proc. of the 1st Int. Workshop on Constraint Programming Beyond Finite Integer Domains, pp. 59–73 (2005)
12. Musser, D., Stepanov, A.: Generic programming. In: Gianni, P. (ed.) Symbolic and Algebraic Computation. LNCS, vol. 358, pp. 13–25. Springer, Heidelberg (1989)
13. Gervet, C.: Interval propagation to reason about sets: Definition and implementation of a practical language. Constraints journal 1(3), 191–244 (1997)
14. Viegas, R., Azevedo, F.: GRASPER: a framework for graph CSPs. In: ModRef 2007. 6th Int. Workshop On Constraint Modelling and Reformulation (2007)
15. Mathews, C., Van Holde, K.: Biochemistry, 2nd edn. Benj./Cumm. (1996)
16. Attwood, T., Parry-Smith, D.: Introduction to bioinformatics. Prent. Hall, Englewood Cliffs (1999)
17. Cormen, T., Leiserson, C., Rivest, R., Stein, C.: Introduction to Algorithms, 2nd edn. MIT Press, Cambridge (2001)
18. Sellmann, M.: Cost-based filtering for shorter path constraints. In: Rossi, F. (ed.) CP 2003. LNCS, vol. 2833, pp. 694–708. Springer, Heidelberg (2003)

Chapter 11 - Second Workshop on Text Mining and Applications (TEMA 2007)

Text Segmentation Using Context Overlap

Radim Řehůřek

Faculty of Informatics, Masaryk University
xrehurek@fi.muni.cz

Abstract. In this paper we propose features desirable of linear text segmentation algorithms for the Information Retrieval domain, with emphasis on improving high similarity search of heterogeneous texts. We proceed to describe a robust purely statistical method, based on context overlap exploitation, that exhibits these desired features. Experimental results are presented, along with comparison to other existing algorithms.

1 Introduction

Text segmentation has recently enjoyed increased research focus. Segmentation algorithms aim to split a text document into contiguous blocks, called *segments*, each of which covers a compact topic while consecutive blocks cover different topics. Applications include finding topic boundaries in text transcriptions of audio news, improving text navigation or intrinsic plagiarism (or anomaly) detection. It can also be used to improve Information Retrieval (henceforth IR) performance, which is main target application for the method described in this paper.

To see how text segmentation might improve IR performance, consider standard IR scenario. Here documents are transformed into Vector Space and indexing techniques are employed to allow efficient exact and proximity queries. Given the widely heterogeneous documents that a general IR system may expect, some of these documents may be monothematic and compact, dealing with a single topic. Others can be a mixture of various topics, connected not thematically but rather incidentally (for example, documents containing news agglomerated by date, not topic). Some may cover multiple topics intentionally, such as complex documents involving passages in different languages. The problem here is that once the documents are converted into Vector Space, all structural information is lost. The resulting document vector shifts away from any one topic included in the original document. Still, user queries are typically monothematic and in this way the chance of high similarity match between user query and document vector decreases. This can result in missed hit. Thus having basic retrieval blocks correspond to single topics rather than whole documents seems like a methodologically sound step. It is up to final application to merge and present topical, sub-document hits to the user. It also depends on application to set granularity of topics that we wish to tell apart. Identifying compact document chunks also has applications in intrinsic plagiarism detection, where it helps to reduce number of suspicious passages and subsequent queries.

J. Neves, M. Santos, and J. Machado (Eds.): EPIA 2007, LNAI 4874, pp. 647–658, 2007.
© Springer-Verlag Berlin Heidelberg 2007

1.1 Motivation

There are practical considerations that are important in real-world IR systems. Driven by need to understand system's behaviour (especially unexpected behaviour) and ability to make extensions to the system during development cycle, it is advantageous to keep the system architecture as simple, clear and robust as possible. Based on these concerns, three important properties of text segmentation for IR systems may be identified:

1. Domain independence. As little external knowledge about document content as possible is required for segmentation. For large collections, even semi-automatic techniques (i.e., techniques that require some kind of human intervention during segmentation process) are problematic.
2. Language independence. Although techniques for automatic language detection exist, using separate algorithms for different types of input data is difficult from maintenance point of view. An algorithm should ideally work based solely on document content. Additionally it should be able to deal with the case where input text is not a strictly well-behaved part of a language, but rather a real-world document. An example is robustness towards commonly appearing inserted content such as small ASCII drawings, inserted passages in different languages, short tables, poorly OCR-ed documents and so on.
3. Granularity. Desirable is an option that allows one to set a customizable level of granularity as to what constitutes a "sufficient" topic shift. This allows the system maintainer to set the segmentation policy, based on expected query granularity. There is no point clogging the system with many small segments that are conceptually identical with respect to user interests, just as it's no good to keep widely varying topics in one single document.

In addition to this list, technical aspects such as effectiveness are also important. Segmentation must perform well enough to allow text segmentation of large documents in reasonable time. This limits our choices to algorithms based on readily identifiable surface cues, such as those based on text cohesion and lexical repetition.

2   Existing Methods

Different text segments should more or less correspond to different topics. Paragraph breaks could be indicative of a topic shift. However, simply splitting at new line breaks is problematic. In many documents the new line breaks are not indicative of paragraphs, but rather of the place where text meets right margin of the page. This may be caused by OCR, certain text editors or retransmission through e-mail clients. More sophisticated segmentation algorithms which take into account document content are required.

An early linear segmentation algorithm called TextTiling is due to Hearst [4]. The basic idea is to take note of lexical repetition. A window of fixed length is being gradually slid through the text, and information about word overlap between the left and right part of the window is converted into digital signal. Shape of the post-processed signal is used to determine segment breaks. Complexity of this method is linear in the length of the document and requires virtually no linguistic knowledge or tuning beyond the choice of parameters for window size, step for window slide and segmentation threshold[1]. These features make TextTiling a good candidate for IR purposes.

A more recent approach is that of Choi [1]. His C99 algorithm computes a sentence similarity matrix (again using lexical cohesion), and applies ranking to get rid of unreliable absolute values. Divisive clustering on the rank matrix is applied to obtain final segmentation. Choi's results show that C99 is vastly superior to the previous methods of Hearst, various modifications of Reynar's maximization algorithm [9] and the Segmenter algorithm [5].

3   LSITiling

Our proposed algorithm, called LSITiling, is based on and extends Hearst's Text-Tiling algorithm. It uses the same sliding window to obtain context similarity signal, then analyses this signal to arrive at final segmentation. The difference lies in what constitutes a context. Where TextTiling relies on text surface features of character token units to determine context overlap, LSITiling uses context overlap in a conceptual space. The reasoning behind this enhancement is following: segmentation aims at discerning individual topics. These may be vaguely defined as chunks of text pertaining a single idea. People, however, commonly use different words to express the same idea. Indeed, it is a recommended stylistic guideline to change vocabulary and avoid lexical repetition. TextTiling (as well as other algorithms) go a step in the direction of lexical independence by stemming the text — that is, unifying tokens that share a common word stem. Apart from failing point two of our three point wish-list (language independence), this does not address basic common language phenomena like synonymy.

Our algorithm computes the context similarities in *topic space*. Topic space is a space where each dimension represents a concept present in original data. In a general sense, LSITiling is thus a Vector Space method. Construction of the concept space (also called *embedded* or *latent* space) is done by applying an IR technique called Latent Semantic Indexing (LSI). LSI was chosen for its established role in IR, but other decomposition schemes, such as the recently proposed Non-negative Matrix Factorization [6], are also applicable. LSI is basically a straightforward statistical method which projects data onto a lower dimensional space, where this projection is optimal in the least squares sense. Translated into our text domain, it uses word co-occurrence both of first and higher degrees to derive co-occurrence patterns, called *concepts*, which define dimensions in the new concept space.

In our algorithm, LSI is applied to the domain of text segmentation as follows: let input be a single text document we wish to segment. Next we construct a

---

[1] See Hearst [4] for exhaustive algorithm details.

corpus of *pseudo-documents*. This is a critical step — although previous work by Choi [2] or Foltz [3] also sought to apply LSI to text segmentation, they split the input document into individual sentences and/or paragraphs and used those as corpus of pseudo-documents. Additionally, Foltz added other domain-specific documents such as encyclopaedia entries and book excerpts to the corpus of pseudo-documents to increase quality of the latent model. Here our approach differs in that we consider *overlapping* pieces (or chunks) of the input document. In practice, we choose as pseudo-documents exactly those chunks as considered by the TextTiling algorithm as contexts. All terms are identified (no stemming is performed) and *Term Frequency * Inverse Document Frequency*, or $TF * IDF$, matrix is constructed. On this matrix, Singular Value Decomposition is computed[2]. A number of latent dimensions is chosen and each pseudo-document is projected into the new, reduced latent space.

Dimensionality of the latent space determines how much data variance is retained. Using only the single, most dominant concept would result in each pseudo-document being represented by a single number. Keeping more dimensions keeps more information about the pseudo-document's content, allowing more refined representation in terms of adherence to multiple concepts. As the other extreme, keeping all dimensions means all information present in the data is kept. It is however desirable to keep the number of latent dimensions low. This is to reduce data noise — dimensions with low significance correspond to noise and thus affect performance negatively. Also, for the sake of segmentation, we are only interested in the salient document concepts. Here our approach also differs in that we explore effects of very low dimensionality, as opposed to Choi [2] where the golden IR standard of hundreds of dimensions was used. An interesting algorithmical feature of LSI is that concept spaces are nested. This means that once we have obtained a document representation in for example 100 most significant dimensions, we may work with the document in 10 dimensional space by simply omitting all but its first 10 dimensions. This allows us to produce and store a single latent model, and tune concept granularity by only adjusting vector lengths.

Once the new context representations are obtained, context similarity is computed by standard cosine similarity measure:

$$cos(vec_i, vec_{i+1}) = \frac{vec_i^T \cdot vec_{i+1}}{|vec_i||vec_{i+1}|} \qquad (1)$$

By computing cosine similarity between consecutive pseudo-document vectors we obtain a one-dimensional digital time series, which is smoothed and analyzed for local minima in the same way as described in Hearst [4]. A notable distinction between LSITiling signal and TextTiling signal is signal variance. With LSITiling, most of the signal for overlapping pseudo-documents runs at near exact-match (cosine similarity score of 1.0), with occasional drops near suspected segment boundaries. This allows us to reliably set the segmentation threshold at a high number (such as 1.0), as opposed to TextTiling where the threshold is set

---

[2] The PROPACK [8] package for computing SVD on large, sparse matrices was used.

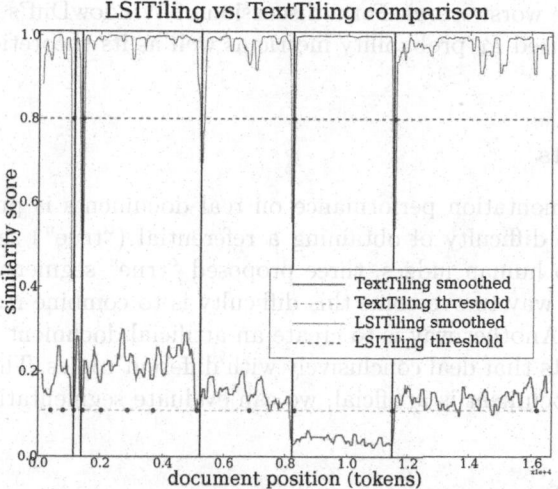

Fig. 1. Comparison of smoothed LSITiling and TextTiling signals on a sample input text of 16,634 tokens. TextTiling threshold is set to signal mean minus half its standard deviation (as proposed by Hearst). The black vertical lines denote true segment boundaries. Compare to blue dots along threshold lines, which denote segment boundaries proposed by respective algorithms. In this example, LSITiling missed one segment boundary. TextTiling missed no boundary but produced seven additional, erroneous boundaries.

according to particular combination of signal mean and variance. See Figure 1 for side-by- side comparison of LSITiling and TextTiling signals over the same sample document. To test general applicability of our new algorithm on more general data, as well as to establish a way to automatically determine its internal parameters, a series of experiments were conducted.

4 Experiments

4.1 Evaluation

Given input document, its true segmentation and a segmentation proposed by some segmentation algorithm, we seek to numerically capture quality (or "closeness") of the two segmentations. This is a common concept in IR evaluation, however traditional IR evaluation measures of precision and recall are not suitable. This is because they penalize "near-misses" too much, whereas we'd like to penalize errors more "softly", depending on distance between true and proposed segment boundary. We chose the WindowDiff metric proposed in [7]. The idea is to run a window of fixed length through input text, taking note of how many times the number of proposed and real segment boundaries within the window disagree. Length of this window is set to half the average true segment length. Resulting error value is scaled to lie between 0.0 and 1.0 (0.0 being the best and 1.0 the worst score). For a discussion of WindowDiff's advantages over the previously used P_k probability metric as well as its theoretical justification, see [7].

4.2   Data Sets

Evaluating segmentation performance on real documents is problematic. This stems from the difficulty of obtaining a referential ("true") segmentation. Indeed, given two human judges, three proposed "true" segmentations are likely to appear. One way to overcome this difficulty is to combine results of multiple human judges. Another way is to create an artificial document by merging several smaller texts that deal conclusively with different topics. This way, although the resulting document is artificial, we can evaluate segmentation quality more precisely.

Table 1. Overall corpora statistics

corpus #1 stats	document length (tokens)	segment length (tokens)	corpus #2 vestats	document length (tokens)	segment length (tokens)
mean	51549	7003	mean	1766	177
stddev	32883	5711	stddev	474	79

We adopted the latter approach. For this purpose, a collection of 31 text *passages* was assembled[3]. These were picked to represent various topics and styles. 24 of these passages were written in English (both by native and non-native speakers), two in German, one in Czech and one in Dutch. Also present were text passages of unidentifiable language, such as an ASCII art picture or text table drawings. These 31 passages thematically covered law, biology, computer science, art, religion and politics. The texts took many different forms, including book excerpts, scientific papers, paper abstracts, text questionnaire, executive summary, part of a stage play, an interview, product overview or end user license. Two of the passages came from poor book OCR, resulting in much garbled content. All passages were chosen with requirements that

- they deal with mutually exclusive topics, so that evaluation of true segment boundaries is possible,
- and that they are real-world documents, the concatenation of which may conceivably appear in a digital collection.

Apart from lack of pair-wise semantic overlap between passages, no other constraints were placed and no text postprocessing was done. Passages had widely varying lengths and formatting, mimicking our requirements placed on segmentation algorithm mentioned above. From these passages, a corpus of 525 documents was assembled. Each document was created by merging one to fourteen

---

[3] All data sets are available from author on request.

passages, which were picked randomly without repetition. Each test document is thus made up from one to fourteen true segments, and the segmentation algorithm then tries to retrieve these segments given their concatenation as its only input. Corpus statistics are summarized in Table 1 under "corpus #1". Average length is 7003 tokens per segment or 51549 tokens per document. Notable is the high passage length deviation, implying text segments of very different lengths.

One objection to constructing our test corpus in this way may be that it produces artificial documents. While this is true, we note that the alternative, tagging a real corpus by hand by human judges, is quite problematic both theoretically and practically. We further note that using artificial corpora is common in segmentation evaluation research and our approach is in this respect comparable to other works in the field.

4.3 Algorithm Comparison

We consider six algorithms in our comparison:

- Hearst. Original C implementation of the TextTiling algorithm [4] with default parameters.
- TextTiling. A slightly modified Python implementation of the same algorithm. The modification is in the way final boundaries are determined — rather than selecting boundaries based on signal mean and variance, a fixed threshold of 0.2 is set. The value of 0.2 was selected for its good performance in preliminary experiments. It was not tuned for this dataset, and thus may not be optimal.
- C99. Choi's 2004 Java implementation of his C99 algorithm[5]. Default parameters were used.
- LSITiling. Proposed algorithm with default parameters. These equal Text-Tiling parameters, plus the dimensionality of the latent space is set to 5 and threshold to 1.0. Again, these parameters were predetermined without looking at the corpus.
- Fixed. An algorithm which inserts a segment boundary after each block of tokens (this block size equals 120 tokens by default).
- Random. A referential algorithm which splits document randomly. The number of segments is selected randomly to lie between one and one plus a thousandth of number of document tokens.

Overall results are summarized in Table 2 and Figure 2 (left). Figure 2 (right) lists average performance over documents which share the same number of true segments. LSITiling consistently outperforms all other algorithms, with the exception of TextTiling for less than 3 true segments. The abysmal performance of the original TextTiling algorithm (called Hearst here) is due to the way it sets its threshold. Upon inspection, it turns out that the segmentation boundaries are

---

[4] As obtained from http://elib.cs.berkeley.edu/src/texttiles, March 2007.
[5] Downloaded from http://www.lingware.co.uk/homepage/
freddy.choi/software/software.htm, March 2007.

Table 2. Experiment results for the first corpus. Numbers show statistical WindowDiff characteristics over all documents scores.

document scores	LSITiling	TextTiling	C99	Random	Hearst	Fixed
WindowDiff mean	0.345	0.374	0.566	0.567	0.950	0.995
WindowDiff stddev	0.186	0.164	0.217	0.275	0.177	0.023

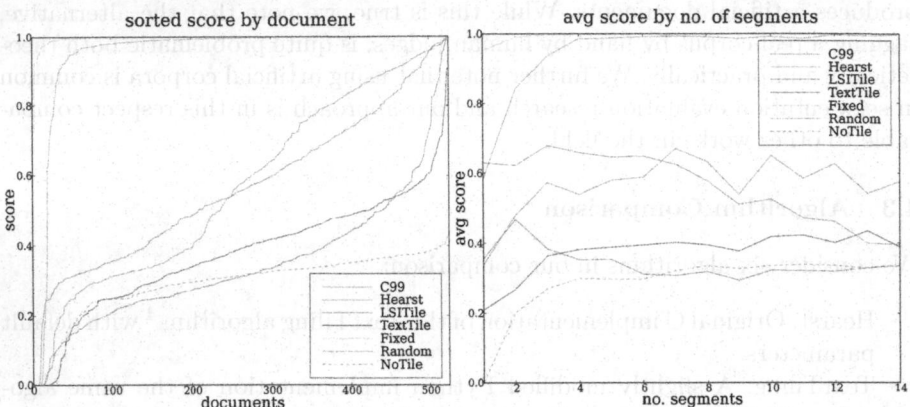

Fig. 2. Left: Experiment results for the first corpus. **Right**: For each number of segments 1–14, scores for all documents with exactly that number of true segments were averaged. This figure compares how different algorithms cope with increasing topical fragmentation of input texts.

being inserted after every couple of sentences, and the *WindowDiff* metric marks this nearly as bad as the Fixed algorithm. The modified TextTiling version with fixed threshold, however, performs much better, and in fact outperforms C99. Poor performance of C99 (which is on par with the Random algorithm here) is surprising. We suspected that a possible reason for such discrepancy between results reported in [1] and our results may lie either in the different metric used (WindowDiff vs. Error Rate) or in the fundamentally different datasets used. For this reason we ran another set of experiments, on a different corpus.

4.4   Second Experiment

The second corpus was created according to Choi's description, with the exception that the British National Corpus (BNC) was used instead of Brown corpus.

Table 3. Experiment results for the second corpus

document scores	C99	LSITiling	TextTiling	Random	Fixed	Hearst
WindowDiff mean	0.175	0.345	0.372	0.378	0.388	0.442
WindowDiff stddev	0.083	0.062	0.038	0.035	0.098	0.092

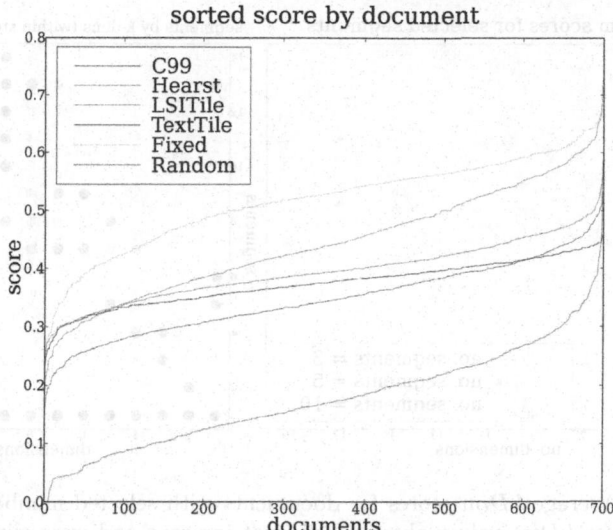

Fig. 3. Experiment results for the second corpus

700 documents were constructed by merging prefixes of ten random BNC documents. The number of true segments was thus fixed to 10, as opposed to 1–14 range from our first experiment. A document prefix here simply means first few sentences extracted from a particular document. The length of the prefix was selected randomly between 3 and 11 sentences for 400 documents, between 3–5 sentences for 100 documents, 6–8 sentences for 100 documents and finally 9–11 sentences for the remaining 100 documents. This is the exact same configuration as used by Choi in [1]. As a general remark, these settings more closely follow the news feed segmentation scenario, with rather short true segments and smaller variation in true segment lengths. Data statistics are summarized in Table 1 under "corpus #2". For this dataset, C99 clearly outperforms all other algorithms by a large margin (Table 3 and Figure 3). It would appear that C99 is well suited for this particular problem, and not so well suited for the IR scenario with widely varying segments of considerable size. On this second corpus, although it lagged far behind C99, LSITiling (still with default parameters) outperformed all other algorithms.

4.5 Algorithm Parameters

The choice of value for the latent space dimensionality parameter is not straightforward. Our next series of tests therefore sought to establish a relationship between the data and this parameter. Ideally, we would like to make setting of this parameter transparent to the user, so that LSITiling requires no additional parameters to these of TextTiling. Remember that in the previous experiments, the semantic space dimensionality value was set to be 5 for all documents. To establish relationship between the number of segments and the dimensionality

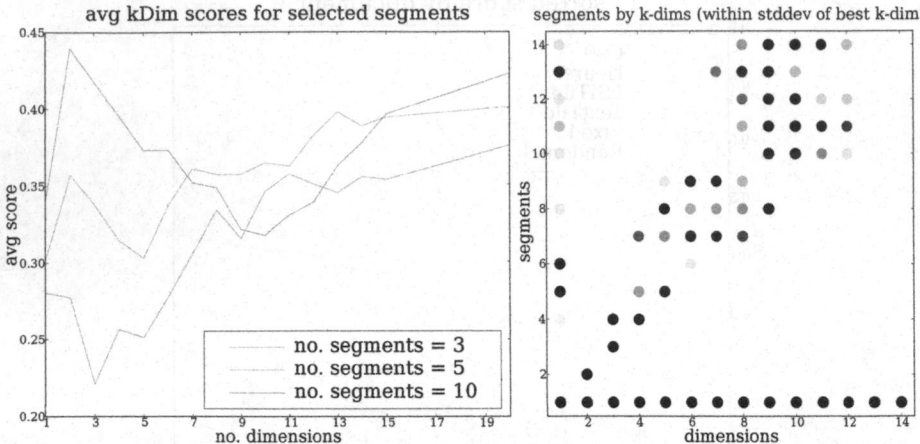

Fig. 4. Left: Average $kDim$ scores for documents with selected number of true segments. Scores for $kDim$ values above 20 did not improve and were clipped from the figure to improve readability. **Right:** Best $kDim$ scores per number of segments. See text for explanation.

(also referred to as $kDim$), we evaluated LSITiling again for each document of the first corpus, with $kDim$ values between 1 and 100. As an example, Fig. 4 (left) shows average scores for all documents that have 3, 5 or 10 true segments. For these documents we force LSITiling to return respectively 3, 5 or 10 proposed segments as well, by splitting the input document at the 3, 5 or 10 gaps with most dissimilar left and right contexts. From this graph we may observe that the optimal $kDim$ value appears to equal the true number of segments. Going through the same procedure for all other segments number (recall that the first corpus has between one and fourteen true segments per document, unlike the second where the number of referential segments is always ten) and plotting the results we obtain Fig. 4 (right). The plots denote all $kDim$ values for which the score was within one standard deviation from the optimal value, with plot intensity decreasing as distance to optimal increases (solid black dots denote best scores). From this graph we may observe three prevailing trends. The horizontal line at true segments equal to 1 is trivial — it simply shows that for one true segment, with LSITiling forced to return one true segment, the segmentation is (trivially) perfect. Second trend is diagonal — it shows that the correlation between number of segments and $kDim$ may in fact be as simple as linear. Third trend is the vertical line at small $kDim$. This is not trivial and shows that basing segmentation on very small fixed number of latent dimensions may be justified. This is in contrast with [2], where the authors use hundreds of latent dimensions in their proposed LSI algorithm. Figure 5 summarizes these results with respect to number of document segments. It plots the original LSITiling result ($kDim$ equal to 5) against results obtained by fixing the number of hypothesized segments to the true number of segments. *LSITiling best* denotes the results obtained by "magically" picking the best $kDim$ for each number of segments. It is

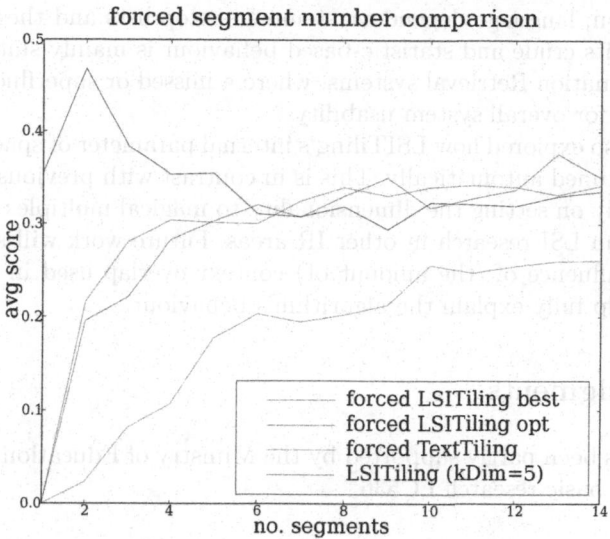

Fig. 5. Comparison of $kDim$ selection strategies

included for illustrative purpose only, because in practice we obviously do not have a referential segmentation to determine what the best $kDim$ is. *LSITiling opt* plots the results obtained by setting $kDim$ to the number of true segments (the diagonal trend). Although not straightforward, it is feasible to determine the number of segments for new, unseen texts. Clusterization techniques (such as the one used in C99) can be employed here, and in particular Spectral Graph Analysis, close in spirit to SVD, seems promising. Also included in the graph for reference are TextTiling scores, where the algorithm is forced to return segmentation with a fixed number of segments in the same way LSITiling was. We can see that for low number of true segments, all forced algorithms perform better than the default LSITiling. This means that default LSITiling overestimates the number of true segments here. However, for higher number of segments, the default LSITiling performs comparably to *LSITiling opt* and forcing it to return a fixed number of segments offers on average no improvement.

5   Conclusion and Future Work

In this paper we have presented a novel algorithm for linear text segmentation based on Latent Semantic Indexing from overlapping document chunks. The algorithm was devised to satisfy Information Retrieval needs for robustness, independence of language, independence of domain and of adjustable segmentation granularity. Its target application is one where heterogeneous texts of widely varying lengths can be expected. Here it exhibits encouraging performance and outperforms other commonly used algorithms. Its robustness draws from design which does not rely on linguistic tools such as sentence or paragraph boundary disambiguation, language dependent stemmers, stop-lists and thesauri. For the same reason, its crude and statistic-based behaviour is mainly suited for refinement of Information Retrieval systems, where a missed or superfluous boundary is not critical for overall system usability.

We have also explored how LSITiling's internal parameter of space dimensionality may be tuned automatically. This is in contrast with previous works in the field which rely on setting the dimensionality to magical multiples of 100, based on results from LSI research in other IR areas. Future work will seek to determine exact influence of (the amount of) context overlap used in LSI training, which will help fully explain the algorithm's behaviour.

Acknowledgments

This work has been partly supported by the Ministry of Education of ČR within the Center of basic research LC536.

References

1. Choi, F.Y.Y.: Advances in domain independent linear text segmentation. In: Proc. of the North American Chapter of the Association of Computational Linguistic, Seattle, pp. 26–33 (2000)
2. Choi, F.Y.Y., Wiemer-Hastings, P., Moore, J.: Latent semantic analysis for text segmentation. In: Proc. of the 6th Conference on Empirical Methods in Natural Language Processing, pp. 109–117 (2001)
3. Foltz, P.W.: Latent Semantic Analysis for text-based research. Behavior Research Methods, Instruments and Computers 28, 197–202 (1996)
4. Hearst, M.A.: Multi-paragraph segmentation of expository text. In: Proc. of the ACL 1994, Las Crces, NM (1994)
5. Min-Yen, K., Klavans, J.L., McKeown, K.R.: Linear segmentation and segment significance. In: Proc. of WVLC-6, Montreal, pp. 197–205 (1998)
6. Lee, D.D., Seung, H.S.: Learning the parts of objects by non-negative matrix factorization. Nature 401(675), 788–791 (1999)
7. Pevzner, L., Hearst, M.: A critique and improvement of an evaluation metric for text segmentation. Computational Linguistics 28(1), 19–36 (2002)
8. http://sun.stanford.edu/~rmunk/PROPACK/ (March 2007)
9. Reynar, J.C.: Topic segmentation: Algorithms and applications. Ph.D. thesis, Computer and Information Science, University of Pennsylvania (1998)

Automatic Extraction of Definitions in Portuguese:
A Rule-Based Approach

Rosa Del Gaudio and António Branco

University of Lisbon
Faculdade de Ciências, Departamento de Informática
NLX - Natural Language and Speech Group
Campo Grande, 1749-016 Lisbon, Portugal
rosa@di.fc.ul.pt, antonio.branco@di.fc.ul.pt

Abstract. In this paper we present a rule-based system for automatic extraction of definitions from Portuguese texts. As input, this system takes text that is previously annotated with morpho-syntactic information, namely on POS and inflection features. It handles three types of definitions, whose connector between *definiendum* and *definiens* is the so-called copula verb "to be", a verb other that one, or punctuation marks. The primary goal of this system is to act as a tool for supporting glossary construction in e-learning management systems. It was tested using a collection of texts that can be taken as learning objects, in three different domains: information society, computer science for non experts, and e-learning. For each one of these domains and for each type of definition typology, evaluation results are presented. On average, we obtain 14% for precision, 86% for recall and 0.33 for F_2 score.

1 Introduction

The aim of this paper is to present a rule-based system for the automatic extraction of definitions from Portuguese texts, and the result of its evaluation against test data made of texts belonging to the domains of computer science, information society and e-learning.

In this work, a *definition* is assumed to be a sentence containing an expression (the *definiendum*) and its definition (the *definiens*). In line with the Aristotelic characterization, there are two types of definitions that typically can be considered, the formal and the semi-formal ones [1]. Formal definitions follow the schema $X = Y + C$, where X is the *definiendum*, " $=$ " is the equivalence relation expressed by some connector, Y is the *Genus*, the class of which X is a subclass, and C represents the characteristics that turn X distinguishable from other subclasses of Y. Semi-formal definitions present a list of characteristics without the *Genus*.

In both types, in case the equivalence relation is expressed by the verb "to be", such definition is classified as a copula definition, as exemplified below:

J. Neves, M. Santos, and J. Machado (Eds.): EPIA 2007, LNAI 4874, pp. 659–670, 2007.

FTP é um protocolo que possibilita a transfêrencia de arquivos de um local para outro pela Internet.
FTP is a protocol that allows the transfer of archives from a place to another through the Internet.

Definitions are not limited to this pattern [2, 3]. It is possible to find definitions expressed by:

- punctuation clues:
  TCP/IP: protocolos utilizados na troca de informações entre computadores.
  TCP/IP: protocols used in the transfer of information between computers.

- linguistic expressions other than the copular verb:
  Uma ontologia pode ser descrita como uma definição formal de objectos.
  An ontology can be described as a formal definition of objects.

- complex syntactic patterns such as apposition, *inter alia*:
  Os Browsers, Navegadores da Web, podem executar som.
  Browsers, tools for navigating the Web, can also reproduce sound.

The definitions taken into account in the present work are not limited to copula definitions. The system is aimed at identifying definitory contexts based on verbs other than "to be" and punctuation marks that act as connectors between the two terms of a definition. Here, we will be calling verb definition to all those definitions that are introduced by a verb other than "to be", and punctuation definitions to the ones introduced by punctuation marks.

The research presented here was carried out within the LT4eL project[1] funded by European Union (FP6) whose main goal is to improve e-learning systems by using multilingual language technology tools and semantic web techniques. In particular, a Learning Management System (LMS) is being improved with new functionalities such us an automatic key-words extractor [4] and a glossary candidate extractor. In this paper, we will focus on the module to extract definition from Portuguese documents.

The reminder of the paper is organized as follows. In Sect. 2 we present the corpus collected in order to develop and test our system. In Sect. 3 the grammars developed to extract definition are described.

In Sect. 4, the results of the evaluation of the grammar, in terms of recall, precision and F2-score, are presented and discussed.

An errors analysis and a discussion on possible alternative methods to evaluate our system are provided in Sec. 5

In Sect. 6, we discuss some related work with special attention to their evaluation results, and finally in Sect. 7 conclusions are presented as well as possible ways to improve the system in future work.

---

[1] www.lt4el.eu

2   The Corpus

The corpus collected in order to develop and test our system is composed by 33 documents covering three different domains: Information Technology for non experts, e-Learning, and Information Society.

Table 1. Corpus domain composition

Domain	tokens
Information SocietyS	92825
Information Technology	90688
e-Learning	91225
Total	274000

Table 1 shows the composition of the corpus.

The documents were preprocessed in order to convert them into a common XML format, conforming to a DTD derived from the XCES DTD for linguistically annotated corpora [5].

The corpus was then automatically annotated with morpho-syntactic information using the LX-Suite [6]. This is a set of tools for the shallow processing of Portuguese with state of the art performance. This pipeline of modules comprises several tools, namely a sentence chunker (99.94% F-score), a tokenizer (99.72%), a POS tagger (98.52%), and nominal and verbal featurizers (99.18%) and lemmatizers (98.73%).

The last step was the manual annotation of definitions. To each definitory context was assigned the information about the type of definition. The definition typology is made of four different classes whose members were tagged with is_def, for copula definitions, verb_def, for verbal non copula definitions, punct_def, for definitions whose connector is a punctuation mark, and finally other_def, for all the remaining definitions. Table 2 displays the distribution of the different types of definitions in the corpus.

The domains of Information Society and Information Technology present a higher number of definitions, in particular of copula definitions. The reason could be that this domain is composed by documents conceived to serve as tutorials for non experts, and have thus a more didactic style. In Sect. 4, we will see how this difference can affect the performance of the system.

Table 2. The distribution of types of definitions in the corpus

Type	Information Society	Information Technology	e-Learning	Total
is_def	80	62	24	166
verb_def	85	93	92	270
punct_def	4	84	18	106
other_def	30	54	23	107
total	199	295	157	651

```
- <s id="s204">
  - <definingText def="m106" def_type1="is_def" id="d01">
      <tok base="o" class="word" ctag="DA" id="t4097" msd="ms" sp="y">O</tok>
    - <markedTerm dt="y" id="m106" kw="y">
        <tok base="tcp" class="word" ctag="PNM" id="t4098" sp="y">TCP</tok>
      </markedTerm>
      <tok base="ser" class="word" ctag="V" id="t4099" msd="pi-3s" sp="y">é</tok>
      <tok base="o" class="word" ctag="DA" id="t4100" msd="ms" sp="y">o</tok>
      <tok base="protocolo" class="word" ctag="CN" id="t4101" msd="ms" sp="y">protocolo</tok>
      <tok base="que" class="word" ctag="CJ" id="t4102" sp="y">que</tok>
      <tok base="dividir" class="word" ctag="V" id="t4103" msd="pi-3s" sp="y">divide</tok>
      <tok base="a" class="word" ctag="DA" id="t4104" msd="fs" sp="y">a</tok>
      <tok base="informação" class="word" ctag="CN" id="t4105" msd="fs" sp="y">informação</tok>
      <tok base="em" class="word" ctag="PREP" id="t4106" sp="y">em</tok>
      <tok base="pacote" class="word" ctag="CN" id="t4107" msd="mp" sp="y">pacotes</tok>
  </definingText>
</s>
```

Fig. 1. The sentence O TCP é um protocolo que divide a informação em pacotes
(The TCP is a protocol that splits information into packets) in final XML
format

In Fig. 1, we present a sample of the final result. Of particular interest for
the development of our grammars are the attribute base, containing the lemma
of each word, the attribute ctag, containing the POS information, and the msd
with the morpho-syntactic information on inflection.

3   The Grammars

The grammars we developed are regular grammars based on the tools lxtrans-
duce, a component of the LTXML2 tool set developed at the University of Edinburgh[2]. It is a transducer which adds or rewrites XML markup on the basis of
the rules provided.

Lxtansduce allows the development of grammars containing a set of rules,
each of which may match part of the input. Grammars are XML documents
conforming to a DTD (lxtransduce.dtd). The XPath-based rules are matched
against the input document. These rules may contain simple regular-expression,
or they may contain references to other rules in sequences or in disjunctions,
hence making it possible to write complex procedures on the basis of simple
rules.

All the grammars we developed present a similar structure and can be divided
in 4 parts. The first part is composed by simple rules for capturing nouns, adjectives, prepositions, etc. The second part by rules that match verbs. The third
part is composed by rules for matching nouns and prepositional phrases. The
last part consist of complex rules that combines the previous ones in order to
match the *definiens* and the *definiendum*.

---

[2] http://www.ltg.ed.ac.uk/~richard/ltxml2/lxtransduce-manual.html

A development corpus, consisting of 75% of the whole 274 000 token corpus, was inspected in order to obtain generalizations helping to concisely delimit lexical and syntactic patterns entering in definitory contexts. This sub-corpus was used also for testing the successive development versions of each grammar.

The held out 25% of the corpus was thus reserved for testing the system.

Three grammars were developed, one for each of the three major types of definitions, namely copula, other verbs, and punctuation definitions.

A sub-grammar for copula definition. Initially we developed a baseline grammar for this type of definition. This grammar marked as definition all that sentences containing the verb "to be" as the main verb of the sentence. In order to improve this grammar with syntactic information, copula definitions manually marked in the developing corpus were gathered. All information was removed except for the information on part-of-speech in order to discover the relevant patterns. Patterns occurring more than three times in the development corpus were implemented in this sub-grammar. Finally the syntactic patterns of all the sentence erroneously marked as definition by the baseline grammar were extracted and analyzed, in order to discover patterns that were common to good and bad definition. We decide not to implement in our grammar patterns whose occurrence was higher in the erroneously marked definitions that in the manually marked ones. We ended up with a grammar composed by 56 rules, 37 simple rules (capturing nouns, adjectives, prepositions, etc), 5 rules to capture the verb and 9 to capture noun and prepositional phrases and 2 rule for capturing the definitory context.

The following rule is a sample of the rules in the copula sub-grammar.

```
<rule name="copula1">
<seq>
<ref name="SERdef"/>
<best> <seq>
<ref name="Art"/>
<ref name="adj|adv|prep|" mult="*"/>
<ref name="Noun" mult="+"/> </seq>
<ref name="tok" mult="*"/>
<end/> </seq> </rule>
```

This is a complex rule that make use of other rules, defined previously in the grammar. This rule matches a sequence composed by the verb "to be" followed by an article and one or more nouns. Between the article and the noun an adjective or an adverb or a preposition can occur. The rule named SERdef matches the verb "to be" only if it occurs in the third person singular or plural of the present or future past or in gerundive or infinitive form.

A sub-grammar for other verbs definition. In order to develop a grammar for this kind of definitions we start immediately to extract lexico-syntactic patterns. In fact it is hard to figure out how a baseline grammar could be implemented. We decided to follow the same methodology used for copula definition.

In a first phase we extracted all the definitions whose connector was a verb other than "to be", and collected all such verbs appearing in the developing corpus, obtaining a list of definitory verbs. This list was improved by adding synonyms. We decided to exclude some verbs initially collected from the final list because their occurrence in the corpus is very high, but their occurrence in definitions is very low. Their introduction in the final list would not improve recall and would have a detrimental effect on the precision score.

In a second phase we divided all the verbs obtained in three different classes: verbs that appear in active form, verbs that appear in passive form and verb that appear in reflexive form. For each class a syntectic rule was wrote. This information was listed in a separate file called lexicon.

The following rule is a sample of how verbs are listed in the lexicon.

```
<lex word="significar"> <cat>act</cat> </lex>
```

In this example the verb *significar* ("to mean") is listed, in his infinitive form that corresponds to the attribute base in the corpus. The tag cat allows to indicate a category for the lexical item. In our grammar, act indicates that the verb occurs in definitions in the active form. A rule was written to match this kind of verbs:

```
<rule name="ActExpr">
<query match="tok[mylex(@base) and (@msd[starts-with(.,'fi-3')]
or @msd[starts-with(.,'pi-3')])]"constraint="mylex(@base)/cat='act'"/>
<ref name="Adv" mult="?"/>
</rule>
```

This rule matches a verb in present and future past (third person singular and plural), but only if the base form is listed in the lexicon and the category is equal to act. Similar rules were developed for verbs that occur in passive and reflexive form.

A sub-grammar for punctuation definition. In this sub-grammar, we take into consideration only those definitions introduced by a colon mark since it is the more frequent pattern in our data. The following rule characterizes this grammar. It marks up sentences that start with a noun phrase followed by a colon.

```
<rule name="punct_def">
<seq> <start/>
<ref name="CompmylexSN" mult="+"/>
<query match="tok[.~'^:']"/>
<ref name="tok" mult="+"/>
<end/> </seq> </rule>
```

4 Results

In this section we report on the results of the three grammar. Further more the results of a fourth grammar are presented. This grammar was obtained by combing the previous three in order to obtain the general performance of our system.

Scores for Recall, Precision and F_2-measure, for developing corpus (dv) and for test (ts) corpus are indicated.

These scores were calculated at the sentence level: a sentence (manually or automatic annotated) is considered a true positive of a definition if it contains a part of a definition. Recall is the proportion of the sentences correctly classified by the system with respect to the sentences (manually annotated) containing a definition. Precision is the proportion of the sentences correctly classified by the system with respect to the sentences automatically annotated.

We opted for an F_2-measure instead of an F_1 one because of the context in which this system is expected to operate. Since the goal is to help the user in the construction of a glossary, it is important that the system retrieves as many definition candidates as possible. The final implementation will allow user to quickly delete or modificate bad definitions.

Table 4 displays the results of the copula grammar. These results can be put in contrast with those obtained with a grammar that provides the performance of the baseline grammar for copula definitions. An improvement of 0.18 in the F_2-measure was obtained.

Table 3. Baseline for copula grammar

	Precision		Recall		F_2	
	dv	ts	dv	ts	dv	ts
IS	0.11	0.12	1.00	0.96	0.27	0.29
IT	0.09	0.26	1.00	0.97	0.22	0.51
e-L	0.04	0.50	0.82	0.83	0.12	0.14
Total	0.09	0.13	0.98	0.95	0.22	0.31

Table 4. Results for copula grammar

	Precision		Recall		F_2	
	dv	ts	dv	ts	dv	ts
IS	0.40	0.33	0.80	0.60	0.60	0.47
IT	0.20	0.51	0.56	0.67	0.40	0.61
e-L	0.13	0.16	0.54	0.75	0.26	0.34
Total	0.30	0.32	0.69	0.66	0.48	0.49

The results obtained with the grammar for other verbs are not as satisfactory as the ones obtained with the copula grammar. This is probably due to the larger diversity of patterns and meaning for each such verb.

As can be seen in Table 2, only 4 definitions of this type occur in the documents of the IS domain and 18 in the e-learning domain. Consequently, this grammar for punctuation definitions ended up by scoring very badly in these documents. Nevertheless, the global evaluation result for this sub-grammar is better than the results obtained with the grammar for other verb definitions.

Finally, Table 7 presents the results obtained by a grammar that combines all the other three sub-grammars described in this work. This table gives the overall performance of the system based on the grammars developed so far, that is, this result represents the performance the end user will face when he will be using the glossary candidate detector.

To obtain the precision and recall score for this grammar, it is not necessary anymore to take into account the type of definition. Any sentence that is correctly tagged as a definitory context (no matter which definition type it receives) will be brought on board.

Table 5. Results for verb grammar

	Precision		Recall		F_2	
	dv	ts	dv	ts	dv	ts
IS	0.13	0.08	0.61	0.78	0.27	0.19
IT	0.13	0.22	0.63	0.66	0.28	0.39
e-L	0.12	0.13	1	0.59	0.28	0.27
Total	0.12	0.14	0.73	0.65	0.27	0.29

Table 6. Results for punctuation grammar

	Precision		Recall		F_2	
	dv	ts	dv	ts	dv	ts
IS	0.00	0.00	0.00	00	0.00	0.00
IT	0.48	0.43	.68	0.60	0.60	0.53
e-L	0.05	0.00	0.58	0.00	0.13	0.00
Total	0.19	0.28	0.64	0.47	0.35	0.38

Table 7. The combined result

	Precision		Recall		F_2	
	dv	ts	dv	ts	dv	ts
IS	0.14	0.14	0.79	0.86	0.31	0.32
IT	0.21	0.33	0.66	0.69	0.38	0.51
e-L	0.11	0.11	0.79	0.59	0.25	0.24
Total	0.15	0.14	0.73	0.86	0.32	0.33

As can be seen, the recall value remains quite high, 86%, while it is clear that for the precision value (14%), there is much room for improvement yet.

In many cases we obtained better results for the test corpus than for the developing one, and this represents an unsuspected result. In order to explain this outcome we analyzed the two sub-corpora separately. When we split the corpus in two parts we just toke in account the size of the two corpora end not the number of definitions in each one. As a matter of fact the developing corpus is characterized by 364 definitory contexts while the training corpus is characterized by 287 definitions. This means that the developing corpus has 55% of all definition instead of 75%, as a consequence the definition density is lower in this corpus than in the test corpus. This result supports our initial hypothesis that the style of a document influence the performance of the system.

5   Error Analysis

As expected, the results obtained with documents from the Information Society and Information Technology domains are better than the results obtained with documents from the e-Learning domain. This confirms our expectation drawn from the style and purpose of the material involved. Documents with a clear educational purpose, like those from IS and IT sub-corpora, are more formal in the structure and are more directed towards explaining concepts, many times via the presentation of the associated definitions. On the other hand, documents with a less educative purpose present less explicit definitions and for this reason it is more difficult to extract definitory contexts from them using basic patterns. More complex patterns and a grammar for deep linguistic processing are likely to be useful in dealing with such documents.

Also worth noting is the fact that though the linguistic annotation tools that were used score at the state of the art level, the above results can be improved with the improvement of the annotation of the corpus. A few errors in the morpho-syntactic annotation were discovered during the development of the grammars that may affect the performance of the grammars.

On the evaluation methodology. Determining the performance of a definition extraction system is not a trivial issue. Many authors have pointed out that a quantitative evaluation as the one we carried out in this work may not be completely appropriate [7]. The first question that arises is about the definition of definitions. If different people are given the same document to mark with definitions, the result may be quite different. Some sentences will be marked by everybody, while others will not. As show in similar studies [8], when different people are asked to mark definitions in a document agreement may be quite low. This can in part explain the low precision we obtained for our system.

Also interesting to note is that a different, perhaps complementary, approach to evaluation is possible. The DEFINDER [9] system for automatic glossary construction was evaluated using a qualitative approach. The definitions extracted by the system were evaluated by end users along three different criteria: readability, usefulness and complexness, taking into account the knowledge of the end user in the domain. This method may be an important complement to the quantitative approach used here.

6   Related Work

In this section, we discuss studies that are related to the work reported here and whose results and methods may be put in contrast with ours.

Note however that comparing the results of different studies against ours is in most cases not easy. Many researches only report recall or precision, or don't specify how these values were obtained (e.g. token level vs. sentence level). The nature of the corpora used (size, composition, structure, etc.) is another sensible aspect that makes comparison more difficult. Also different systems are tuned to finding definitions with different purposes, for instance for relations extraction, or for question answering, etc.

Regarding the methods used for this task, the detection of morpho-syntactic patterns is the most used technique. Since the beginning of 90's Hearst [10] proposed a method to identify a set of lexico-syntactic patterns (e.g. such NP as NP, NP and other NP, NP especially NP, etc) to extract hyponym relations from large corpora and extend WordNet with them. This method was extended in recent years to cover other types of relations[11]. In particular, Malaise and colleagues [2] developed a system for the extraction of definitory expressions containing hyperonym and synonym relations from French corpora. They used a training corpus with documents from the domain of anthropology and a test corpus from the domain of dietetics. The evaluation of the system using a corpus of a different domain makes results more interesting as this puts the system under a more stressing performance. Nevertheless, it is not clear what is the nature and purpose of the documents making this corpora, namely if they are consumer-oriented, technical, scientific papers, etc. These authors used lexical-syntactic markers and patterns to detect at the same time definitions and relations. For the two different, hyponym and synonym, relations, they obtained, respectively, 4% and 36% of recall, and 61% and 66% of precision.

Saggion [12] combines probabilistic technics with shallow linguistic analysis. In a first stage are collected 20 candidate texts with a probabilistic document retrieval system using as input the definition question expanded with a list of related terms. In a second stage the candidate sentences are analyzed in order to match the definition patterns. Regarding the results obtained at the TREC QA 2003 competition, he reports F_5 score, where the recall is 5 times more important than precision. His system obtained a F-score of 0.236, where the best score in the same competition was of 0.555 and the median was of 0.192. The main difference between this task and our work resides in the fact that we do not know beforehand the expressions that should receive definitions. This lack of information makes the task more difficult because it not possible to use the term as a clue for extracting its definitions.

DEFINDER [9], an automatic definition extraction system combines shallow natural language processing with deep grammatical analysis. Furthermore it makes use of cue-phrase and structural indicators that introduce definitions and the defined term. In terms of quantitative evaluation, this system presents 87% precision and 75% recall. This very high values are probably due to the nature of the corpus composed by consumer oriented medical jornal article.

More recently machine learning techniques were combined with patterns recognition in order to improve the general results. In particular [13] used a maximum entropy classifier to extract definition and syntactic features to distinguish actual definitions from other sentences.

Turning more specifically to the Portuguese language, there is only one publication in this area. Pinto and Oliveira [14] present a study on the extraction of definitions with a corpus from a medical domain. They first extract the relevant terms and then extract definitions for each term. The comparison of results is not feasible because they report results for each term. Recall and precision range between 0% and 100%.

By using the same methodology for Dutch as the one used here, Westerhout and Monachesi [8] obtained 0.26 of precision and 0.72 of recall, for copula definitions, and 0.44 of precision and a 0.56 of recall, for other verbs definition. This means that their system outperforms ours in precision though not in recall.

7   Conclusions and Future Work

In this work, we presented preliminary results of a rule-based system for the extraction of definitions from corpora. The practical objective of this system is to support the creation of glossaries in e-learning environments, and it is part of a larger project aiming at improving e-learning management systems with human language technology.

The better results were obtained with the system running over documents that are tutorials on information technology, where it scored a recall of 69%

and a precision of 33%. For less educational oriented documents, 59% and 11%, respectively, was obtained.

We also studied its performance on different types of definitions. The better results were obtained with copula definitions, with 67% of recall and 51% of precision, in the Information Technology domain.

Compared to work and results reported in other publications concerning related research, our results seem thus very promising. Nevertheless, further strategies should be explored to improve the performance of the grammar, in particular its precision.

In general, we will seek to take advantage of a module that allows deep linguistic analysis, able to deal with anaphora and apposition, for instance. At present, we know that a grammar for deep linguistic processing of Portuguese is being developed [15]. We plan to integrate this grammar in our system.

Regarding punctuation definition, the pattern in the actual grammar can also be extended. At present, the pattern can recognize sentences composed by a simple noun followed by a colon plus the definition. Other rules with patterns involving brackets, quotation marks, and dashes will be integrated.

Finally, in this work we ignored an entire class of definitions that we called "other definition", which represents 16% of all definitions in our corpus. These definitions are introduced by lexical clues such as *that is, in other words*, etc. This class also contains definitions spanning over several sentences, where the terms to be defined appear in the first sentence, which is then characterized by a list of features, each one of them conveyed by expressions occurring in different sentences. These patterns need thus also to be taken into account in future efforts to improve the grammar and its results reported here.

References

[1] Meyer, I.: Extracting knowledge-rich contexts for terminography. In: Bourigault, D. (ed.) Recent Advances in Computational Terminology, pp. 279–302. John Benjamins, Amsterdam (2001)

[2] Malais, V., Zweigenbaum, P., Bachimont, B.: Detecting semantic relations between terms in definitions. In: CompuTerm 2004. CompuTerm Workshop at Coling 2004, 3rd edn., pp. 55–62 (2004)

[3] Alarcn, R., Sierra, G.: El rol de las predicaziones verbales en la extraccin automtica de conceptos. Estudios de Linguistica Aplicada 22(38), 129–144 (2003)

[4] Lemnitzer, L., Degórski, L.: Language technology for elearning – implementing a keyword extractor. In: EDEN Research Workshop Research into online distance education and eLearning. Making the Difference, Castelldefels, Spain (2006)

[5] NIKS: Xml, corpus encoding standard, document xces 0.2. Technical report, Department of Computer Science, Vassar College and Equipe Langue ed Dialogue, New York, USA and LORIA/CNRS, Vandouvre-les-Nancy, France (2002)

[6] Silva, J.R.: Shallow processing of Portuguese: From sentence chunking to nominal lemmatization. Master's thesis, Universidade de Lisboa, Faculdade de Ciências (2007)

[7] Przepiórkowski, A., adn Miroslav Spousta, L.D., Simov, K., Osenova, P., Lemnitzer, L., Kubon, V., Wójtowicz, B.: Towards the automatic extraction of definitions in Slavic. In: Piskorski, J., Pouliquen, B., Steinberger, R., Tanev, H. (eds.) Proceedings ofo the BSNLP workshop at ACL 2007, Prague (2007)

[8] Westerhout, E., Monachesi, P.: Extraction of Dutch definitory contexts for elearning purposes. In: CLIN proceedings 2007 (2007)

[9] Klavans, J., Muresan, S.: Evaluation of the DEFINDER system for fully automatic glossary construction. In: AMIA 2001. Proceedings of the American Medical Informatics Association Symposium (2001)

[10] Hearst, M.A.: Automatic acquisition of hyponyms from large text corpora. In: Proceedings of the 14th conference on Computational linguistics, Morristown, NJ, USA, Association for Computational Linguistics, pp. 539–545 (1992)

[11] Person, J.: The expression of definitions in specialised text: a corpus-based analysis. In: Gellerstam, M., Jaborg, J., Malgren, S.G., Noren, K., Rogstrom, L., Papmehl, C. (eds.) EURALEX 1996. 7th Internation Congress on Lexicography, Goteborg, Sweden, pp. 817–824 (1996)

[12] Saggion, H.: Identifying definitions in text collections for question answering. In: LREC 2004 (2004)

[13] Fahmi, I., Bouma, G.: Learning to identify definitions using syntactic feature. In: Basili, R., Moschitti, A. (eds.) Proceedings of the EACL workshop on Learning Structured Information in Natural Language Applications, Trento, Italy (2006)

[14] Pinto, A.S., Oliveira, D.: Extracção de definições no Corpógrafo. Technical report, Faculdade de Letras da Universidade do Porto (2004)

[15] Branco, A., Costa, F.: LXGRAM – deep linguistic processing of Portuguese with HSPG. Technical report, Departement of Informatics, University of Lisbon (2005)

N-Grams and Morphological Normalization in Text Classification: A Comparison on a Croatian-English Parallel Corpus

Artur Šilić[1], Jean-Hugues Chauchat[2],
Bojana Dalbelo Bašić[1], and Annie Morin[3]

[1] University of Zagreb, Department of Electronics, Microelectronics, Computer and Intelligent Systems, KTLab, Unska 3, 10000 Zagreb, Croatia
[2] Université de Lyon 2, Faculté de Sciences Economique et de Gestion, Laboratoire Eric, 5 avenue Pierre Mendès France, 69676 Bron Cedex, France
[3] Université de Rennes 1, IRISA, 35042 Rennes Cedex, France
artur.silic@fer.hr, jean-hugues.chauchat@univ-lyon2.fr,
bojana.dalbelo@fer.hr, annie.morin@inria.fr

Abstract. In this paper we compare n-grams and morphological normalization, two inherently different text-preprocessing methods, used for text classification on a Croatian-English parallel corpus. Our approach to comparing different text preprocessing techniques is based on measuring computational performance (execution time and memory consumption), as well as classification performance. We show that although n-grams achieve classifier performance comparable to traditional word-based feature extraction and can act as a substitute for morphological normalization, they are computationally much more demanding.

1 Introduction

There are two criteria for the construction of an automated text classification system: classification performance and computational performance. Computational performance consists of memory consumption, learning time, and classification time of a new document. Short learning time is important in the case of batch learning since new examples can be introduced very often.

Usually, there are two steps in the construction of an automated text classification system. In the first step, the texts are being preprocessed into a representation more suitable for the learning algorithm. There are various levels of text representations [9] such as word fragments, words, phrases, meanings, and concepts. In this work we show that different text representations are influenced differently by the language of the text. In the second step, the learning algorithm is applied on the preprocessed data. In this work we focus on the first step.

The two basic levels of text representation include extraction of word fragments and extraction of words. More sophisticated text representations do not

J. Neves, M. Santos, and J. Machado (Eds.): EPIA 2007, LNAI 4874, pp. 671–682, 2007.
© Springer-Verlag Berlin Heidelberg 2007

yield significantly better results as shown in [3], [12]. In this paper we compare different text representations for text classification in two languages: Croatian, which has a rich morphology, and English, which has simple morphology. The comparison includes the above mentioned criteria: classification performance and computational performance.

For this comparison we use a Croatian-English parallel corpus consisting of 3580 text documents, each written in both languages. Since the corpus is parallel we can fairly compare the performance of the selected methods for text representation.

In our experiments we choose SVM (Support Vector Machine) as the learning algorithm used for classification and χ^2 test as the score used for feature selection. Many authors claim that these methods are efficient and robust in text classification [9], [18], [24].

Feature extraction methods used in our experiments are described in Section 2. Experiment settings are given in Section 3 followed by results and discussion in Section 4. Conclusion is given in Section 5.

2   Feature Extraction

In text classification the text is usually represented as a vector of weighted features. The differences in text representations come from the definition of a *feature*. This work explores two feature extraction methods with their variations.

2.1   Bag-of-Words and Morphological Normalization

Bag-of-words is the most widely used model for text representation. The method consists of counting word occurrences in the text. This produces a vector in the space of features, which correspond to the words found in the text.

The morphology of the language implies that a word can have many forms, for example "wash", "washed", "washing", "washer", etc. Although not having completely the same meaning, such words are semantically very close regarding other words in the text. To strengthen this semantic relationship among different word forms, morphological normalization can be applied.

Inflectional normalization converts a word form into its normalized linguistic form called the *lemma*, hence it is called lemmatization. Only the effects of syntax are removed, but the part-of-speech remains unchanged. For example, "eats" and "eaten" would be inflectionally normalized to "eat", but "eater" would remain the same because the part-of-speech is different. In contrast to this, derivational normalization eliminates the effect of derivational morphology by converting many lemmas into a single representative lemma. In the example above "eater" would be derivationally normalized to "eat". Inflectional and derivational normalization are usually performed with a dictionary that can be automatically or manually built [21]. Stemming is a morphological normalization technique that finds the roots of the words by removing affixes (usually suffixes).

Regarding feature space, both inflectional and a combination of derivational and inflectional morphological normalization are a mixture of feature conflation and feature inflation. Feature conflation, for the reason that many different word forms are mapped onto the same dimension, as shown by the examples above.

In the case of homographs, different words have one of many morphological forms equal. For example, in Croatian language the word "para" can have three possible meanings: nominative of the noun "steam", third person singular of the verb "to rip", or genitive form of the noun "money" (colloquial). This is considered feature inflation, because one word form will be mapped onto three dimensions in the feature space. Our results show that for Croatian the dominant effect of a combination of inflectional and derivational morphological normalization is feature conflation (see Table 1). Morphological normalization by stemming is pure feature conflation.

If no morphological normalization is applied, the word frequencies in the vector representing a text are dispersed among many dimensions making specific terms less confident. The impact on classification performance depends on the language of the text [13]. Morphologically richer languages may benefit much more from morphological normalization than morphologically simple languages. In [17] it is shown that information retrieval techniques on some European languages benefit from stemming.

Note that morphological normalization and word segmentation are language-dependent. Obviously, morphological normalization techniques such as lemmatization or stemming need language-specific dictionaries or stripping rules. Non-trivial word segmentation examples include Germanic languages [11] or Arab languages that use word compounds and Chinese or Japanese languages that have no whitespaces between words.

Stopwords are functional words that have little semantic content. They are frequently found in texts and are regarded as a noise for classification, so they are filtered out before transforming the texts into vectors.

2.2 Character n-Grams

In this work we subsume that an n-gram is a sequence of n consecutive characters that appear in the text. For example, the string "new shipment" yields the following set of 3-grams: "new", "ew ", "w s", " sh", "shi", "hip", "ipm", "pme", "men", "ent".

N-grams have first been mentioned by Shannon in 1948 [19] in a context of theory of communication. Since then, n-grams have found their use in many applications such as spell checking, string searching, prediction, and speech recognition [14]. Information retrieval systems based on n-grams have been studied in [14]. Text classification using n-grams has been explored in [2], [4], [23]. As noted in [11], most fuzzy string matching algorithms are based on character n-gram techniques. An automatic keyword proposal algorithm based on n-grams has been presented in [7].

Advantages of letter based n-grams. There are several advantages of using n-grams in information retrieval tasks. Decomposing words to n-grams captures the stems of the words automatically. For example, the words "industry", "industries", and "industrial" are different, but their first five 3-grams match ("ind",

Table 1. Number of unique features in Croatian and English data sets (stopwords removed), for exact definitons of feature extraction methods see Subsection 3.2

	Croatian set	English set
2-grams	922	930
3-grams	10959	10309
4-grams	65624	55699
5-grams	234892	187687
6-grams	612663	474134
words	101951	38083
words-i	48548	-
words-id	44350	-
words-p	-	28015

"ndu", "dus", "ust", "str"). N-gram extraction on a large corpus will yield a large number of possible n-grams, but only the important ones that form the roots of the words will have significant frequency values in vectors representing the texts. If needed, feature selection techniques can select the most significant n-grams, i.e. those that are specific to some classes out of the whole class set.

Secondly, n-gram-based techniques are language-independent. Several authors note that stopword removal or morphological normalization is unnecessary when representing text with n-grams [6], [8], [14]. No morphological normalization is needed because associations among similar words are achieved by increasing the feature values for the intersecting n-grams of different words. No compound word analysis is needed when representing a text with n-grams because two separated words yield the same n-grams as their compound with an addition of space-including n-grams that will not be discriminative. For similar reasons n-grams can be applied to texts written in languages with no whitespaces. Therefore, no dictionaries or language-specific techniques are needed when working with n-grams. This is an excellent property of any text-processing technique.

Thirdly, n-grams are capable of dealing with noise in the texts such as OCR-generated (Optical Character Recognition) errors, or orthographic errors done by humans [2]. For example, in a bag-of-words model words "bicycle" and "bycycle" generate two distinct feature values, but in an n-gram model these words generate an intersecting set of feature values therefore keeping a part of the association between the words. This property is especially important for short texts (such as Internet posts or email messages) since a small number of words increases the importance of each word occurrence in the text.

Shortcomings of letter based n-grams. In English, the number of unique 4-grams converges to the number of unique words found in a large corpus. For n-grams of length five and longer the total number of unique n-grams is larger than the total number of word forms found in text. This has been noted in [14]. We obtained similar results on our corpus in English as well as in Croatian (Table 1).

As the main result of this work, we will show that although effective in terms of F1 and feature space size, 3-grams and 4-grams yield much larger document-feature matrices regarding memory size thus slowing down the classifier learning, and the classification of a new document.

3 Experiments

3.1 Experimental Settings

The experiments are done on a Croatian-English parallel corpus. The source is Croatia Weekly (CW) [22], a newspaper that was published in English and translated to Croatian from 1998 to 2000 by the Croatian Institute for Information and Culture. Each article is assigned to one of four classes: politics, economy, nature, and events. The corpus includes 3580 articles from 113 issues of Croatia Weekly. The article distribution is shown in Table 2.

We use the SVM classification method that was introduced by Vladimir Vapnik and his colleagues. SVM has shown to be robust and efficient for the task of text classification because of the following reasons: high dimensional input space, data sparsity, and linear separability of the classification problems [10]. We use a linear kernel since text classification problems are usually lineary separable [10]. Learning parameter is set to C=1.

Feature selection is a feature space reduction method that attempts to select the more discriminative features extracted from documents in order to improve classification quality and reduce computational complexity [24]. Each feature is assigned a numeric score based on the occurrence of features within the different document classes.

In this work choice of scoring method is the χ^2 test. The χ^2 measures the lack of independence between features and classes. For binary classification, χ^2 is commonly defined as [24]:

$$\chi^2(t) = \frac{N(AD - CB)^2}{(A + C)(B + D)(A + B)(C + D)},$$

where N is the total number of documents. Definitions for A, B, C, and D are given by the contingency table 3.

Table 2. Distribution of articles in the CW corpus

Class	Number of documents
Politics	2033
Events	892
Economy	404
Nature	251
Total	3580

Table 3. A is the number of documents within the class c containing the feature t, B is the number of documents out of the class c containing the feature t, C is the number of documents within a class c not containing the feature t, and D is the number of documents out of the class c not containing the feature t.

$$
\begin{array}{c|cc}
 & c & \bar{c} \\
\hline
t & A & B \\
\bar{t} & C & D
\end{array}
$$

There are many other tests available as summarized in [18] but the χ^2 is often claimed to be one of the best tests for feature selection [24]. The χ^2 gives a similar result as Information Gain because it is numerically equivalent as shown by Benzécri in 1979 [1]. In our experiment, when doing feature selection, sum is used as a global function among the categories [18].

For the term-weightening scheme, we use *TF-IDF* normalization of the document vectors is performed [18]. The tests are done using 5-fold crossvalidation. All letters in the text are lowercased.

While working with n-grams all the punctuation signs are replaced by spaces and all consecutive white space characters are replaced by a single space. Early tests showed that this modification does not influence classification performance. Stopwords are removed while working with the bag-of-words model.

The framework used to execute the experiments is TMT (Text Mining Tools) [20]. In TMT vectors representing the text documents are stored in a sparse form because usually the number of non-zero elements in vectors is substantially smaller than the total number of dimensions [10]. For example, in Croatian CW set plain bag-of-words generates vectors with an average of 169 non-zero elements out of 101951 possible elements per document.

3.2 Feature Extraction Methods

The feature extraction methods include extraction of character n-grams and extraction of words with or without morphological normalization. These methods are compared using different numbers of features selected from the CW corpus.

For the n-grams method we perform *2-gram*, *3-gram*, and *4-gram* feature extraction. For the bag-of-words method on Croatian documents we perform experiments with the following settings: no normalization (*words*), inflectional normalization (*words-i*) and a combination of inflectional and derivational normalization (*words-id*). We use an automatically generated morphological normalization dictionary [21]. On the English documents we perform experiments with the following settings: no normalization (*words*) or normalization using the Porter's stemming algorithm (*words-p*) [16].

3.3 Performance Measures

Classification performance is measured with the usual microaveraged and macroaveraged F1 measure [18]. Computational performance is measured in terms of the number of non-zero elements in sparse document-feature matrices which is equivalent to memory allocation size for these matrices. The same approach to evaluation was already used in [15]. Next, execution time needed for the construction of a classifier and execution time needed for the classification of a new document are measured. These times are presented in milliseconds per document. The configuration used to run the experiments is a PC with AMD Sempron 2800 on 1.76 GHz with 960 MB of RAM. Used operating system is Windows XP Professional Edition.

4   Results

The plotted curves in Figs. 1 and 2 are an average of 5 folds. The average standard deviation is less than 1% and the maximum standard deviation among the folds is 2.5%.

4.1   Classification Performance

Classification performance is presented in Fig. 1. For both languages *3-grams* and *4-grams* are able to achieve similar result as *words*, whereas *2-grams* perform significantly worse.

For the Croatian documents, morphological normalization improves the classification quality of *words* from 1% up to 2%. We present only *words-id*, as *words-i* give almost the same result (*words-id* improve performance by less than 0,5% on feature set sizes from 500 above; this is also confimed by [13]). In terms of the F1 measure, *words* are worse than *3-grams* or *4-grams* on an equal feature set size up to the size of about 45k features. At higher feature set sizes, *words* achieve quite good results similar to *words-id*. This is because although the Croatian morphology allows many words to have more than 30 forms, not all of these forms are used often. Besides that, if a corpus and the documents are large enough, all word forms will be covered much better and the classification with *words* will outperform *3-grams* and *4-grams*. The *3-grams* achieve better

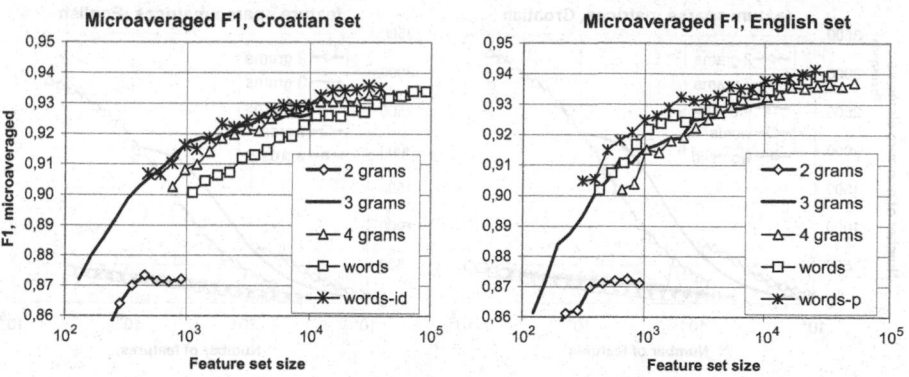

Fig. 1. Microaveraged F1 plotted against the feature set size

results than *4-grams* on low to middle sized feature sets. That is because *3-grams* are shorter thus yielding more intersecting features among documents at lower numbers of features.

For the English documents, *words* perform better than the n-grams for all feature set sizes. Normalized set *words-p* performs just slightly better than the *words*, but not as significantly as in Croatian. This is due to morphology richness [13]. N-gram performance is comparable to *words* and *words-p*.

The figures show that n-grams are less sensitive to language morphology than the bag-of-words approach. On the English set, the classification quality obtained by n-grams is close to the classification quality obtained by *words*, but on the Croatian corpus n-grams are closer to the morphological normalizing method.

4.2   Computational Efficiency

Fig. 2 shows a great difference between *n-grams* and *words*. N-gram feature extraction produces much denser document-feature matrices for the same number of selected features. This is due to the fact that each letter in the text begins an n-gram (except the n-1 last ones), so the number of n-grams in a text is much greater than the number of words. Many n-grams will appear in many documents, more frequently than the actual words appear in the documents. The number of non-zero elements in document-feature sparse matrices differ up to the factor of seven between *words* and *4-grams*. When comparing *n-gram* curves for Croatian and English, we can see that Croatian documents need more memory for n-gram representation.

When using morphological normalization with the bag-of-words model somewhat denser document-feature matrices are generated in comparison to the non-normalized bag-of-words model. This corresponds to the notion of feature conflation since many different word forms can occur in the text to be recognized as a single feature. This increases the probability of finding a feature in the raw text thus generating denser document-feature matrices.

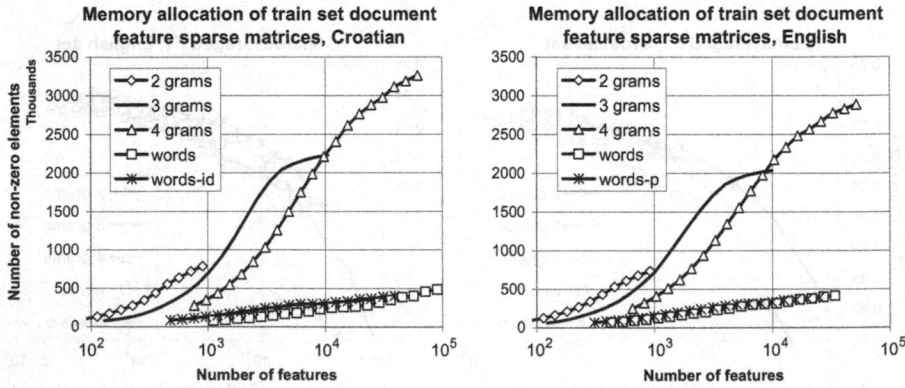

Fig. 2. Number of non-zero elements in the document-feature sparse matrix for the training sets

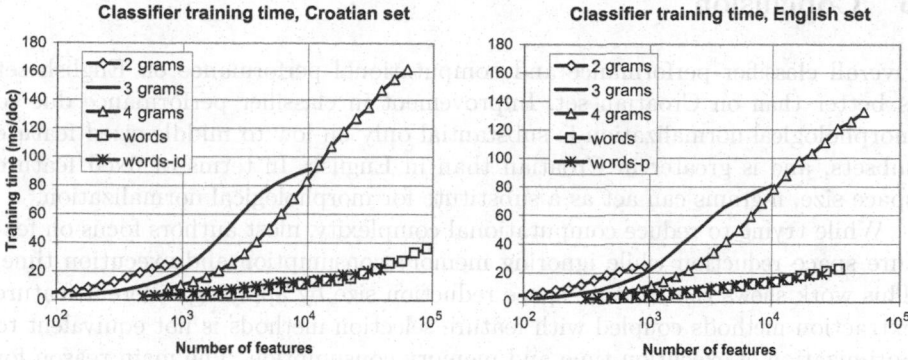

Fig. 3. Milliseconds per document to train a classifier on different feature set sizes

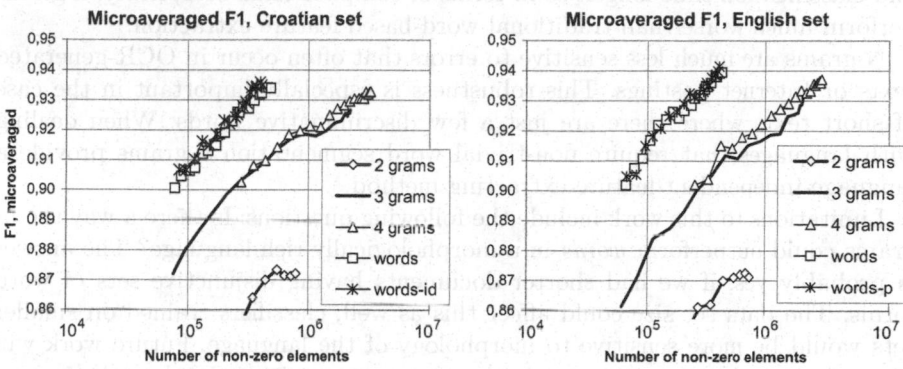

Fig. 4. Microaveraged F1 plotted against the number of non-zero elements in the document-feature sparse matrices of training sets

Fig. 3 shows the time needed to construct an SVM classifier from a document-feature matrix. It is clear that the classifier construction times are proportionally related to memory size needed to represent the document-feature matrices of the training document set in sparse format (Fig. 2). The memory sizes of document-feature matrices of the test sets are proportional to those of the train sets. Likewise, classification times of new documents are proportional to the classifier construction times.

On Fig. 4 we can see the F1 performance in respect to the memory usage of the document-feature matrices of the training set. The difference between n-grams and bag-of-words is clear; n-grams simply need more memory in order to achieve the same result as the bag-of-words model regardless of morphological normalization. Since the times needed to construct a classifier or to classify a new document are proportional to document-feature matrix memory sizes, Fig. 4 would be similar if the x-axis was chosen to be time.

5   Conclusion

Overall classifier performance and computational performance on English set is better than on Croatian set. Improvement in classifier performance due to morphological normalization is substantial only on low to middle sized feature subsets, and is greater in Croatian than in English. In terms of fixed feature space size, n-grams can act as a substitute for morphological normalization.

While trying to reduce computational complexity, most authors focus on feature space reduction while ignoring memory consumption and execution time. This work shows that feature space reduction size by applying different feature extraction methods coupled with feature selection methods is not equivalent to optimization of execution time and memory consumption. The main reason for this dissonance is that the structure of the produced data can be different; n-grams produce much denser matrices thus making classifier construction time and classification time longer, so in terms of computational complexity, n-grams perform much worse than traditional word-based feature extraction.

N-grams are much less sensitive to errors that often occur in OCR-generated texts or Internet postings. This robustness is especially important in the case of short texts where there are just a few discriminative words. When dealing with languages that require non-trivial word segmentation n-grams provide a language-independent feature extracting method.

Limitations to this work include the following questions: Is there a way the n-grams could outperform *words* in a morphologically rich language? The answer is probably yes, if we had shorter documents having disjunctive sets of word forms. The data set size could affect this as well; classifiers trained on smaller sets would be more sensitive to morphology of the language. Future work will try to answer these questions and compare n-grams with string kernels [5].

Acknowledgements. This research has been jointly supported by the French Ministry of Foreign Affairs through the Egide cooperation program under the joint project "Knowledge discovery in textual data and visualisation", and the Croatian Ministry of Science, Education and Sports under the grants No. 036-1300646-1986 and No. 098-0982560-2563. The authors would like to thank Marko Tadić for making the Croatian-English parallel corpus available to them.

References

1. Benzécri, J.-P.: L'Analyse des Données, T1 = la Taxinomie. In: Dunod (ed.), 3rd edn. (1979)
2. Cavnar, W.B., Trenkle, J.M.: N-Gram-Based Text Categorization. In: Proc. of the Third Annual Symposium on Document Analysis and Information Retrieval, pp. 161–175 (1994)
3. Dumais, S.T., Platt, J., Heckerman, D., Sahami, M.: Inductive Learning Algorithms and Representations for Text Categorization. In: Proc. of the Seventh International Conference on Information and Knowledge Management, pp. 148–155 (1998)

4. Dunning, T.: Statistical Identification of Languages. Comp. Res. Lab. Technical Report, MCCS, pp. 94–273 (1994)
5. Fortuna, B., Mladenić, D.: Using String Kernels for Classification of Slovenian Web Documents. In: Proc. of the 29th Annual Conference of the Gesellschaft für Klassifikation e.V. University of Magdeburg (2005)
6. Ekmekçioglu, F.Ç., Lynch, M.F., Willett, P.: Stemming and N-gram Matching for Term Conflation in Turkish Texts. J. Inf. Res. 7(1), 2–6 (1996)
7. Jalam, R., Chauchat, J.-H.: Pourquoi les N-grammes Permettent de Classer des Textes? Recherche de Mots-clefs Pertinents à l'Aide des N-grammes Caractéristiques. In: Proc. of the 6es Journées internationales d'Analyse statistique des Données Textuelles, pp. 77–84 (2002)
8. Jalam, R.: Apprentissage Automatique et Catégorisation de Textes Multilingues Ph.D. thesis, Université Lumière Lyon 2 (2003)
9. Joachims, T.: Learning to Classify Text Using Support Vector Machines. Kluwer Academic Publishers, Dordrecht (2002)
10. Joachims, T.: Text Categorization With Support Vector Machines: Learning with Many Relevant Features. In: Nédellec, C., Rouveirol, C. (eds.) Machine Learning: ECML 1998. LNCS, vol. 1398, pp. 137–142. Springer, Heidelberg (1998)
11. Kraaij, W.: Variations on Language Modeling for Information Retrieval. Ph.D. thesis, University of Twente (2004)
12. Lewis, D.D.: An Evaluation of Phrasal and Clustered Representations on a Text Categorization Task. In: Proc. of the 15th Annual International ACM SIGIR Conference on Research and Development in Information Retrieval, pp. 37–50 (1992)
13. Malenica, M., Šmuc, T., Šnajder, J., Dalbelo Bašić, B.: Language Morphology Offset: Text Classification on a Croatian-English Parallel Corpus. J. Inf. Proc. Man (to appear, 2007)
14. Miller, E., Shen, D., Liu, J., Nicholas, C.: Performance and Scalability of a Large-Scale N-gram Based Information Retrieval System. J. Dig. Inf. 1(5), 1–25 (2000)
15. Mladenić, D., Brank, J., Grobelnik, M., Milic-Frayling, N.: Feature Selection Using Linear Classifier Weights: Interaction with Classification Models. In: Proc. of the 27th Annual International ACM SIGIR Conference on Research and Development in Information Retrieval (2004)
16. Porter, M.F.: An Algorithm for Suffix Stripping. Program 14(3), 130–137 (1980)
17. Savoy, J.: Light Stemming Approaches for the French, Portuguese, German and Hungarian Languages. In: Proc. of the 2006 ACM Symposium on Applied Computing, pp. 1031–1035 (2006)
18. Sebastiani, F.: Machine Learning in Automated Text Categorization. J. ACM Comp. Surv. 1, 1–47 (2002)
19. Shannon, C.: The Mathematical Theory of Communication. J. Bell Sys. Tech. 27, 379–423, 623–656 (1948)
20. Šilić, A., Šarić, F., Dalbelo Bašić, B., Šnajder, J.: TMT: Object-oriented Library for Text Classification. In: Proc. of the 30th International Conference on Information Technology Interfaces, pp. 559–564 (2007)
21. Šnajder, J.: Rule-Based Automatic Acquisition of Large-Coverage Morphological Lexicons for Information Retrieval. Technical Report MZOS 2003-082, Department of Electronics, Microelectronics, Computer and Intelligent Systems, FER, University of Zagreb (2006)
22. Tadić, M.: Building the Croatian-English Parallel Corpus. In: Proc. of the Third International Conference On Language Resources And Evaluation, vol. 1, pp. 523–530 (2000)

23. Teytaud, O., Jalam, R.: Kernel based text categorization. In: Proc. of the International Joint Conference on Neural Networks, vol. 3, pp. 1891–1896 (2001)
24. Yang, Y., Pedersen, J.O.: A Comparative Study on Feature Selection in Text Categorization. In: Proc. of the 14h International Conference on Machine Learning, pp. 412–420 (1997)

Detection of Strange and Wrong Automatic Part-of-Speech Tagging

Vitor Rocio[1,3], Joaquim Silva[2,3], and Gabriel Lopes[2,3]

[1] DCET Universidade Aberta, 1250-100 Lisboa, Portugal
vjr@univ-ab.pt
[2] DI/FCT Universidade Nova de Lisboa, Quinta da Torre, 2829-516 Caparica,
Portugal
jfs@di.fct.unl.pt   gpl@di.fct.unl.pt
[3] CITI Centro de Informática e Tecnologias de Informação - FCT/UNL

Abstract. Automatic morphosyntactic tagging of corpora is usually imperfect. Wrong or strange tagging may be automatically repeated following some patterns. It is usually hard to manually detect all these errors, as corpora may contain millions of tags. This paper presents an approach to detect sequences of part-of-speech tags that have an internal cohesiveness in corpora. Some sequences match to syntactic chunks or correct sequences, but some are strange or incorrect, usually due to systematically wrong tagging. The amount of time spent in separating incorrect bigrams and trigrams from correct ones is very small, but it allows us to detect 70% of all tagging errors in the corpus.

1   Introduction

Annotated corpora are vital resources for modern computational linguistics. Usually, millions of words are required in order to ensure statistical significance on the results obtained. Manual annotation is therefore an enormous and costly task that rarely is adopted in CL projects. Instead, automatic and semi-automatic systems (e.g. POS-taggers) are built and they guarantee high precision and recall of results. However, automatic systems are not perfect and a percentage of errors still exist. Therefore, those errors need to be manually detected and corrected. In particular, automatic annotators are prone to systematic errors that derive from flaws in their conception, erroneous machine learning, even software bugs. In some cases, it is less costly to correct *a posteriori* the output of the automatic annotator than to correct or refine the annotator itself, given the complexity associated to such specific cases. In this paper, we address the problem of correcting systematic errors in automatically POS-tagged corpora, with the aid of the computer.

Previous work on automatic POS-tagging detection and correction by Dickinson and Meurers [1] use variation n-grams to detect possible errors. In their method, an occurrence of an n-gram is considered erroneously tagged if in most other occurrences in the corpus, it is tagged in a different way. The method is aimed at corpora that have only occasional errors, and their goal is to achieve

J. Neves, M. Santos, and J. Machado (Eds.): EPIA 2007, LNAI 4874, pp. 683–690, 2007.

a "gold standard" (near 100% precision) tagged corpus. However, this approach doesn't work for systematic errors caused by automatic annotators, since these errors are not an exception to the rule, rather becoming the rule themselves.

Another approach by Květoň and Oliva [2] uses the notion of "impossible n-grams" (n-grams that don't occur in a given language) in order to detect erroneously tagged text. However, their assumption of an "error-free and representative" corpus from which to extract the n-grams is difficult to achieve in practice, leading to a manual revision of the n-grams extracted. The problem is that, for n larger than 2, the amount of impossible n-grams makes this task unfeasible.

Our approach overcomes these problems by extracting cohesive sequences of tags from the very corpus we intend to correct. We use an algorithm, LIPxtractor [8] that has been successfully used for multi-word unit extraction and multi-letter unit extraction, and it was generalized for the extraction of multi-element units (MEUs) of text [6].

From the multi-tag sequences (tag n-grams) extracted, we manually (with the aid of an example selector) selected those that don't make sense linguistically and promoted them to the category of likely errors. The results obtained are encouraging in the sense that a large percentage of errors was found in minimal time. The rest of the paper is organized as follows: section 2 presents the n-gram extraction tools in detail, while section 3 explains the manual method used to detect errors; the obtained results are discussed in section 4 and in section 5 we draw conclusions and point to future work, including improvements to our approach.

2 Extracting Multi-element Units of Text From Corpora

Three tools working together, are used for extracting MEUs from any corpus: the LocalMaxs algorithm, the Symmetric Conditional Probability (SCP) statistical measure and the Fair Dispersion Point Normalization (FDPN). A full explanation of these tools is given in [6]. However, a brief description is presented here for paper self-containment.

For a simple explanation, first we will use words instead of tags. Thus, let us consider that an n-gram is a string of words in any text[1]. For example the word *president* is an 1-gram; the string *President of the Republic* is a 4-gram. LocalMaxs is based on the idea that each n-gram has a kind of "glue" or cohesion sticking the words together within the n-gram. Different n-grams usually have different cohesion values. One can intuitively accept that there is a strong cohesion within the n-gram (*Giscard d'Estaing*) i.e. between the words *Giscard* and *d'Estaing*. However, one cannot say that there is a strong cohesion within the n-gram (*or uninterrupted*) or within the (*of two*). Now, considering morphosyntactic tags instead of words, one can intuitively accept that, for example in English, there is a strong cohesion within the n-gram (_DET _N), that is the sequence *Determinant* followed by *Noum*. But one would not expect a strong

---

[1] We use the notation $(w_1 \ldots w_n)$ or $w_1 \ldots w_n$ to refer an n-gram of length n.

cohesion for the tag sequence ($\_DET$ $\_V$), that is *Determinant* followed by *Verb*, as it is a rare or even an incorrect tag sequence. So, the $SCP(.)$ cohesion value of a generic bigram $(x\ y)$ is obtained by

$$SCP(x\ y) = p(x|y) \cdot p(y|x) = \frac{p(x\ y)}{p(y)} \cdot \frac{p(x\ y)}{p(x)} = \frac{p(x\ y)^2}{p(x) \cdot p(y)} \tag{1}$$

where $p(x\ y)$, $p(x)$ and $p(y)$ are the probabilities of occurrence of bigram $(x\ y)$ and unigrams x and y in the corpus; $p(x|y)$ stands for the conditional probability of occurrence of x in the first (left) position of a bigram in the text, given that y appears in the second (right) position of the same bigram. Similarly $p(y|x)$ stands for the probability of occurrence of y in the second (right) position of a bigram, given that x appears in the first (left) position of the same bigram.

However, in order to measure the cohesion value of each n-gram of any size in the corpus, the FDPN concept is applied to the $SCP(.)$ measure and a new cohesion measure, $SCP\_f(.)$, is obtained.

$$SCP\_f(w_1 \ldots w_n) = \frac{p(w_1 \ldots w_n)^2}{\frac{1}{n-1} \sum_{i=1}^{n-1} p(w_1 \ldots w_i) \cdot p(w_{i+1} \ldots w_n)} \tag{2}$$

where $p(w_1 \ldots w_n)$ is the probability of the n-gram $w_1 \ldots w_n$ in the corpus. So, any n-gram of any length is "transformed" in a pseudo-bigram that reflects the *average cohesion* between each two adjacent contiguous sub-n-gram of the original n-gram. Now it is possible to compare cohesions from n-grams of different sizes.

2.1 LocalMaxs Algorithm

LocalMaxs ([7] and [6]) is a language independent algorithm to extract cohesive n-grams of text elements (words, tags or characters).

Definition 1. *Let* $W = w_1 \ldots w_n$ *be an n-gram and* $g(.)$ *a cohesion generic function. And let:* $\Omega_{n-1}(W)$ *be the set of* $g(.)$ *values for all contiguous* $(n-1)$-*grams contained in the n-gram* W; $\Omega_{n+1}(W)$ *be the set of* $g(.)$ *values for all contiguous* $(n+1)$-*grams which contain the n-gram* W, *and let* $len(W)$ *be the length (number of elements) of n-gram* W. *We say that*

W *is a MEU if and only if,*

$$for\ \forall x \in \Omega_{n-1}(W), \forall y \in \Omega_{n+1}(W)$$

$$(len(W) = 2 \wedge g(W) > y)\ \vee$$

$$(len(W) > 2 \wedge g(W) > \frac{x+y}{2})\ .$$

Then, for n-grams with $n \geq 3$, LocalMaxs algorithm elects every n-gram whose cohesion value is greater than the average of two maxima: the greatest cohesion value found in the contiguous $(n-1)$-grams contained in the n-gram, and the greatest cohesion found in the contiguous $(n+1)$-grams containing the n-gram. LiPXtractor is the MEUs extractor that uses $SCP\_f(.)$ cohesion function as $g(.)$ in LocalMaxs algorithm.

3 Detecting Error Sequences

The process to select tag n-grams that likely point to errors is manual. A previous characterization of errors in the corpus (legal texts from the Portuguese General Attorney) showed that most of them are caused by unknown words or forms in the original text. These unknown words/forms are of several types:

- genuine words for which no knowledge exists in our lexicon. For example: "conceitual" (conceptual), an adjective; "115°" (115th) an ordinal; "finalístico" (finalistic) an adjective; fielmente (faithfully) an adverb.
- genuine morphological variants that do not exist in the lexicon (mainly irregular forms). For example: "constituído" (constituted) a past participle verb; expresso (expressed) a past participle verb; "contrapôs" (opposed) a 3rd person singular verb.
- corrupted forms due to diacritics encoding. For example: "consequ ncia" (consequence) a noun where an ê is missing; "circunst ncias" (circumstances) where an â is missing.
- components of hyphenated words that were separated, namely verbs with clitics attached. For example: "verifica-se" (one verifies); "sublinhe-se" (let us underline).
- upper/lower case variations;
- abbreviations and acronyms. For example: "Art." (Article) a noun; "Ob." (Observation); "cit." (citation) a noun; "cfr." (according to) a preposition.
- foreign words. For example "donnés" (data) and "audelà" (beyond), both in French; "ratione legis" (according to the law).
- previously unprocessed contractions. For example: "nesse"="em"+"esse" (in that); "àquilo"="a"+"aquilo" (to that).

There is also a number of errors caused by ambiguous words. For instance, the sequence "é certo que" (it is certain that) is wrongly tagged as _VSER _PIND _PR (verb to be, indefinite pronoun, relative pronoun), but it should be _VSER _ADJ _CONJSUB (verb to be, adjective, subordinative conjunction).

The relatively small number of tag n-grams (up to tetragrams) extracted by LIPxtractor allowed us to manually check and select the ones that likely point to tagging errors. The task was aided by an example viewer that showed the occurrences of sequences of words corresponding to the original tags, in order to allow us to better judge on the validity of the n-grams.

Our method has two steps:

1. The first step is to separate suspect tag n-grams from linguistically sound ones; grammar knowledge is used, as well as our knowledge of observed error patterns. For instance, unknown words are often tagged as proper names, adverbs, ordinal numbers, etc., which may not be the correct tags.
2. The suspect tag n-grams are then analysed one by one, by inspecting the sequences of words in the corpus that were tagged with those n-grams. In this step, it is to important to collect those cases where a wrong pattern can be easily corrected by a substitution rule.

With the information collected by this method, we noticed that the subsequent correction process can be successfully automated.

4   Results

We applied the LIPxtractor tool to the Portuguese General Attorney corpus, consisting of legal texts. The corpus has 1.5 million words, from which 18,675 were manually tagged. The whole corpus was automatically tagged by a neural network based tagger [3], using a part of the hand tagged texts as training set. The tag set has 33 tags.

LIPxtractor extracted a set of n-grams (up to octograms) that are considered MEUs of POS-tags. The graph on figure 1 shows the number of sequences extracted for each n-gram dimension. As expected, data sparseness causes a larger number of n-grams of higher dimension to be extracted.

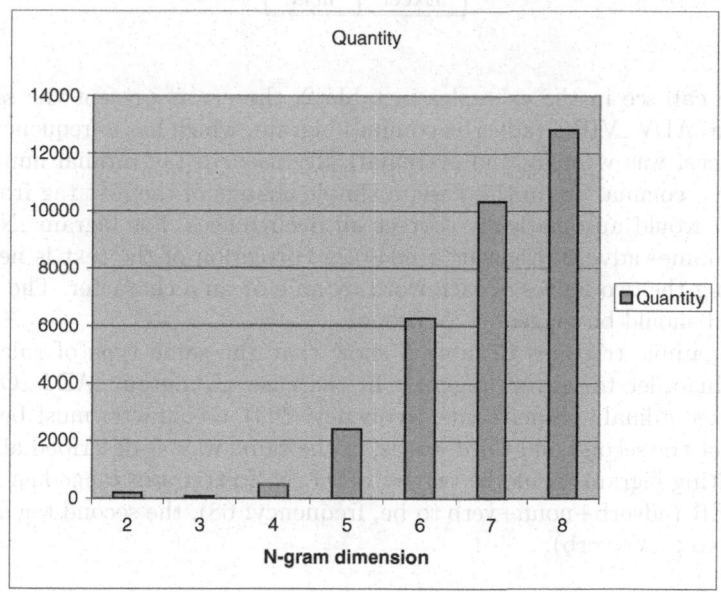

Fig. 1. Number of sequences extracted

We manually checked the bigrams, trigrams and tetragrams extracted, according to the method described in section 3, discriminating between good (sound) n-grams and likely errors. The results are shown in table 1. The analysis of tetragrams suggested that most errors were already detected in bigrams and trigrams. Longer n-grams tend to have the same type of errors, often including the same errors that occur in smaller n-grams. So, in this paper, we have only addressed errors occurring in n-grams up to tetragrams.

Table 1. Number of n-grams analysed

	Extracted	Wrong/strange
Bigrams	185	54
Trigrams	68	32
Tetragrams	485	247

Table 2. Examples of wrongly tagged bigrams

_ADV	_VIRG
115	,
281	,
43	,

_NP	_ADV
jurisprud	ncia
C	mara
Excel	ncia

As we can see in the examples in table 2, the errors present the same pattern. The _ADV _VIRG (adverb+comma) bigram, which has a frequency of 9218 occurrences, was wrongly and systematically used to tag ordinal numbers followed by a comma. So, in this case, a simple change of the first tag from _ADV to _ORD would automatically correct all occurrences. For bigram _NP _ADV (proper name+adverb, frequency: 6849), a correction of the text is needed, by connecting the two halves of each word by an ê or an â character. The resulting new word should be tagged as _N (noun).

The example trigrams in table 3 show that the same type of rules can be applied in order to correct tagging. In the case of trigram _ADJ _ORD _NP (adjective+ordinal+proper name, frequency: 293), a character must be inserted to connect the second and third words, in the same way as described above, and the resulting bigram should be tagged _ADJ _N. In trigrams tagged as _ADV _N _VINFSER (adverb+noun+verb to be, frequency: 68), the second tag should be changed to a _V (verb).

Table 3. Examples of wrongly tagged trigrams

_ADJ	_ORD	_NP
maior	import	ncia
especial	relev	ncia
eventuais	consequ	ncias

_ADV	_N	_VINFSER
não	poderão	ser
só	possam	ser
só	poderão	ser

Table 4. Examples of wrongly tagged tetragrams

_ART	_VINFSER	_V	_VINF
a	inquirida	respondeu	estar
o	expropriado	pode	obter
os	expropriados	deixaram	consumar

_VSER	_ART	_PREP	_VINF
é	o	de	proteger
é	a	de	suportar
foi	a	de	rejeitar

Finally, both tag tetragrams shown in table 4 are corrected by changing one of the tags: in the first case (frequency: 4), the second tag should be changed from _VINFSER (infinitive verb to be) to _N, and in the second case (frequency: 38), the second tag should be changed from _ART (article) to _PD (demonstrative pronoun).

We observe that, as the dimension of n-grams increase, the number of error occurrences detected per n-gram dramatically decreases.

An evaluation of the precision of our approach was carried on the manually tagged subset of the corpus. A sample of 200 errors obtained by comparing both manually and automatically tagged subset revealed that 70% of tagging errors were detected by our approach. The remaining errors are caused by lexical ambiguity, that is, not easily resolved by the immediately surrounding context. For instance, _V _ART (verb+article) and _V _PREP (verb+preposition) are both valid tag sequences, however the Portuguese word "a" can be an article or a preposition, and can be tagged either way. For these cases, a deeper analysis is needed.

5   Conclusions and Future Work

We proposed in this paper a method for detecting systematic errors in automatically POS-tagged corpora. The method is based on the assumption that those errors lead to tag n-grams that, despite having internal cohesiveness, are linguistically wrong or at least unusual.

With this method, we aimed at studying the n-grams having strong internal cohesiveness, since this set of n-grams includes both correct tag sequences and systematic errors in tag n-grams. In order to obtain these patterns, we used the LIPxtractor tool. The number of bigrams, trigrams and tetragrams extracted was sufficiently small to allow us, in a very short time, to manually detect (and correct) a large percentage of errors. The number of larger n-grams is greater than smaller ones. So, the task of analysing all of them one by one would be very time-consuming. Fortunately, these larger n-grams mostly contained the same type of error cases already included in smaller n-grams. Then, when the smaller n-grams are corrected, most of the larger n-grams errors will automatically disappear. This is very important for the sake of effort economy.

The experiences we made showed that, in many error cases, there is an error tag pattern which can be easily substituted by the correct sequences of tags. This is very promising in terms of a future tag corrector, since the number automatic corrections must be as large as possible.

Another direction that we would like to pursue is the successive application of the method, in order to improve precision. That is, after a round of corpus correction, we can extract a new set of n-grams that will reveal new errors, and continue until there are no more errors we can correct.

As we pointed out, there are always errors that escape this kind of methods, and more knowledge (syntactic, semantic, pragmatic) is needed. The use of a syntactic parser with fault finding capabilities [5,4] is a possibility not only to detect these errors but also to automatically propose possible corrections.

References

1. Dickinson, M., Meurers, W.D.: Detecting Errors in Part-of-Speech Annotation. In: EACL 2003. Proceedings of the 10th Conference of the European Chapter of the Association for Computational Linguistics (2003), http://ling.osu.edu/~dickinso/papers/dickinson-meurers-03.html
2. Kveton, P., Oliva, K.: (Semi-) Automatic Detection of Errors in PoS-Tagged Corpora. In: Proceedings of the 19th International Conference on Computational Linguistics (COLING) (2002), http://acl.ldc.upenn.edu/C/C02/C02-1021.pdf
3. Marques, N.C., Lopes, G.P.: Tagging With Small Training Corpora. In: Hoffmann, F., Adams, N., Fisher, D., Guimarães, G., Hand, D.J. (eds.) IDA 2001. LNCS, vol. 2189, pp. 63–72. Springer, Heidelberg (2001)
4. Rocio, V., Lopes, G.P., de la Clergerie, E.: Tabulation for multi-purpose partial parsing. Grammars 4(1), 41–65 (2001)
5. Rocio, V.: Syntactic Infra-structure for fault finding and fault overcoming. PhD thesis. FCT/UNL (2002)
6. Silva, J.F., Dias, G., Guilloré, S., Lopes, G.P.: Using LocalMaxs Algorithm for the Extraction of Contiguous and Non-contiguous Multiword Lexical Units. In: Barahona, P., Alferes, J.J. (eds.) EPIA 1999. LNCS (LNAI), vol. 1695, pp. 113–132. Springer, Heidelberg (1999)
7. Silva, J.F., Lopes, G.P.: A Local Maxima Method and a Fair Dispersion Normalization for Extracting Multiword Units. In: Proceedings of the 6th Meeting on the Mathematics of Language, Orlando, pp. 369–381 (1999)
8. Silva, J.F., Lopes, G.P., Mexia, J.T.: A Statistical Approach for Multilingual Document Clustering and Topic Extraction from Clusters. Pliska Studia Mathematica Bulgarica 15, 207–228 (2004)

New Techniques for Relevant Word Ranking and Extraction

João Ventura and Joaquim Ferreira da Silva

DI/FCT Universidade Nova de Lisboa, Quinta da Torre, 2829-516 Caparica, Portugal
joao_ventura@netvisao.pt, jfs@di.fct.unl.pt

Abstract. In this paper we first propose two new metrics to rank the relevance of words in a text. The metrics presented are purely statistic and language independent and are based in the analysis of each word's neighborhood. Typically, a relevant word is more strongly connected to some of its neighbors in despite of others. We also present a new technique based on the syllable analysis and show that despite it can be a metric by itself, it can also improve the quality of the proposed methods as also greatly improve the quality of other proposed methods (such as *Tf-idf*). Finally, based on the rankings previously obtained and using another neighborhood analysis, we present a new method to decide about the relevance of words on a yes/no basis.

1 Introduction

In this paper, our goal is to identify and extract words (unigrams) which are considered relevant either in a certain text, or in a corpus of documents or even in a representative corpus of a language, using statistical methods and techniques which are independent from the language, context and word frequency. More specifically, through the study of the statistical properties of a certain corpus, and without having access to external references, we aim to obtain a ranking of that corpus' words ordered by relevance, and from that same ranking, we want to be able to say which words are really relevant and which ones are not. Our basic assumption is that a relevant word must be strongly connected to some words that occur in its neighborhood in despite of others. In a simpler way, a relevant word is less promiscuous in its relation with the words in its neighborhood; some words co-occur more with that relevant word and some words co-occur less. On the other hand, through the analysis of the syllables, we also assume that a relevant word usually has a certain number of syllables, not too small (usually function words) neither too long (usually rare words that might have origin in orthographical errors or words having too specific meanings). After obtaining the ranking, we use a technique that we've denominated "Islands" where we aim to obtain a boolean classification about the relevance of words. Basically, in order to be considered relevant, a word must have greater or equal relevance than the words that occur in its neighborhood. The knowledge of relevant words on texts is very important and has many possible applications. Thus, they may provide greater efficiency of search engines and they may improve precision in document

J. Neves, M. Santos, and J. Machado (Eds.): EPIA 2007, LNAI 4874, pp. 691–702, 2007.

clustering, as relevant unigrams are fundamental in this process. In fact, the relevant unigrams of a document are usually part of the important topics in it. So, this unigrams may be used to complete the documents' "ID-card" proposed in [3], where only multiword relevant expressions are extracted from corpora; relevant unigrams are missing. Other applications are easily predicted.

Some techniques to extract relevant word have already been suggested. Luhn [1] suggested that with a simple analysis to the frequency of occurrence of the words, it could be noted that the words with a very high frequency and the words with a very low frequency could be considered as very common or very rare, and then not very relevant. However, this analysis is too simplistic because there are words that are frequent and relevant, and words in the intermediate frequencies that are not relevant at all. Another problem of this approach would be to find the frequency thresholds. In [2] and [4] other approaches are also suggested. Basically those authors assume that because a relevant word is usually the main subject on local contexts, it should occur more often in some areas and less frequently in others giving raise to clusters. Such approach is very intuitive, however it is very punitive to relevant words that are relatively frequent, because they tend to exist in the entire document, not forming significative or evident clusters.

This paper is structured as follows. In section 2 we will present two new metrics to classify each unigram's relevance. In this section we will also present a technique based on the syllable analysis that improves the results obtained and that could be considered an autonomous metric by itself. In section 3 we will present the "Islands" method that allows one to decide if a word is or is not relevant based on the ranking scores. In section 4, other approaches currently used to extract relevant words are analyzed, and we will comment the results obtained by those approaches after the application of the syllables and Islands methods. Finally, in section 5 we will do the conclusions.

2   New Ranking Methodologies

In this section we propose a new approach to extract relevant unigrams using two new metrics. We also present the syllable technique and show how to apply that technique to improve results.

2.1   New Metrics to Relevance Ranking

Although the notion of relevance is a pretty clear concept, there is no consensus about the frontier between the relevant and the irrelevant words. The reader will certainly find strong relevance in the words "republic" and "London" and will certainly find irrelevance in the words "since" or "and". But the problem comes in some words like "read", "finish" or "next", because usually there is no consensus about their semantic value. So, there is a fuzzy area about the words' relevance. In our result's evaluations we preferred to consider as non-relevant any word that couldn't guarantee the consensus about its relevance, in a hypothetical evaluation. This criterion has inevitably affected the results presented in this paper, which might be considered conservative if the reader was not aware of that criterion.

Starting from a corpus composed of several documents, from which we want to understand which words are more relevant and which are less relevant, the underlying idea is that relevant words have a special preference to relate with a small group of other words. In this way, it is possible to use a metric that measures the importance of a word in a corpus based on the study of the relation that that word has with the words that follow it. We have denominated that measure the successor's score of a word w, that is $Sc_{\mathrm{suc}}(w)$.

$$Sc_{\mathrm{suc}}(w) = \sqrt{\frac{1}{\|\mathcal{Y}\| - 1} \sum_{y_i \in \mathcal{Y}} \left(\frac{p(w, y_i) - p(w, .)}{p(w, .)}\right)^2}. \tag{1}$$

In (1), $\mathcal{Y}$ is the set of distinct words in the corpus, and $\|\mathcal{Y}\|$ stands for the size of that set; $p(w, y_i)$ means the probability of y_i to be a successor of word w; $p(w, .)$ gives the average probability of the successors of w, which is given by:

$$p(w, .) = \frac{1}{\|\mathcal{Y}\|} \sum_{y_i \in \mathcal{Y}} p(w, y_i) \qquad p(w, y_i) = \frac{f(w, y_i)}{N}. \tag{2}$$

N stands for the number of words occurred in the corpus and $f(w, y_i)$ is the frequency of bigram (w, y_i) in the same corpus.

Resuming the mathematical formalism, $Sc_{\mathrm{suc}}(w)$ is given by a standard deviation "normalized" by the average probability of the successors of w. It measures the variation of the current word's *preference* to appear before the rest of the words in the corpus. The higher values will appear for the words that have more diversified frequency on bigrams where it appears with its successors, and the lowest values will appear in the words that have less variations of frequency on bigrams where it appears with its successors. Similarly, we may measure the *preference* that a word has to the words the precede it using the following metric that we've denominated predecessor's score, that is $Sc_{\mathrm{pre}}(w)$.

$$Sc_{\mathrm{pre}}(w) = \sqrt{\frac{1}{\|\mathcal{Y}\| - 1} \sum_{y_i \in \mathcal{Y}} \left(\frac{p(y_i, w) - p(., w)}{p(., w)}\right)^2}. \tag{3}$$

The meanings of $p(y_i, w)$ and $p(., w)$ are obvious.

So, using both (1) and (3), through the arithmetic average, we will obtain the metric that allows us to classify the relevance of a word based on the predecessors and successors scores. This metric is simply denominated $Sc(w)$:

$$Sc(w) = \frac{Sc_{\mathrm{pre}}(w) + Sc_{\mathrm{suc}}(w)}{2}. \tag{4}$$

Table 1 shows some examples of $Sc(.)$ values and ranking positions for the words of a corpus made from the EurLex corpus on European legislation in force, about

Table 1. Some examples of $Sc(.)$ values and ranking positions for words in an English corpus

Word	$Sc(.)$	Ranking position
pharmacopoeia	135.17	48
oryctolagus	134.80	64
embryonic	132.67	76
and	10.82	6,696
the	19.34	6,677
of	24.15	6,627

several topics (http://europa.eu.int/eur-lex/). It has about half million words; and there are 18,172 distinct ones. We have only studied the words that occurred at least 3 times. As one can see, the more common words like "the", "and" and "of" are positioned lower in the ranking while words with semantic value are positioned upper in the list.

On the other hand, observing some characteristics of the unigrams, we have verified that usually the words considered relevant have some interesting characteristics about the number of predecessors and successors. For instance, with a Portuguese corpus of half million words (also from European Union documents), it could be noted that the relevant word "comissão" (*commission*) occurred 1,909 times in the corpus, with 41 distinct predecessors and 530 distinct successors. Also, the relevant word "Europa" (*Europe*) occurred 466 times in the corpus, with 29 distinct predecessors and 171 distinct successors. In both cases, most of the predecessors are articles or prepositions such as "a", "na" e "da" (*the, on* and *of*). In fact, function words (articles, prepositions, etc.) show no special preference to a small set of words: one may say that *they populate all the corpus*.

The morphosyntactic sequence <article> <name> <verb> is very frequent in the case of Latin languages such as Portuguese, Spanish, Italian and French, among others; for example, "a comissão lançou" (*the commission presented*), "a comissão considera" (*the commission considers*). Then, given that there are more verbs than articles, the number of successors must be larger than the number of predecessors. Following this reasoning, we propose another statistic metric that measures the importance of a word based on the quotient between the number of its distinct successors and the number of its distinct predecessors. We have called it SPQ (Successor-Predecessor Quotient).

$$SPQ(w) = \frac{Nsuc(w)}{Npred(w)} . \tag{5}$$

$Nsuc(w)$ and $Npred(w)$ give the number of successors of the word w and the number of predecessors of w.

Although both presented metrics (Sc and SPQ) measure the relevance of words, in a language-independent basis, when we tested SPQ, the results were better for the Portuguese corpus than for the English one. However, assuming this, it may be preferably to use this metric if one is working only with Latin languages (see both English and Portuguese results in section 2.2).

Thus, we can build word rankings ordered by relevance. The words that occur in the upper areas of the ranking are considered more relevant and the words that occur in the lower areas are considered less relevant. However, it is not easy to separate the really relevant from the really irrelevant words; i.e., we just have a list saying what is more relevant than what, but not specifying what is relevant and what is not.

2.2   Syllable Analysis

By the examples of table 1, one can find that from 6 words, 3 are relevant and 3 are not. It is easy to conclude that the relevant words ("oryctolagus", "pharma-copoeia" e "embryonic") are, in fact, larger than the non-relevant ("of", "the" e "and"). We could build a metric in order to favor larger words as they tend to be more relevant, but, as we will see, it is preferable to consider the number of syllables instead of the length of the words. For instance, the probability of occurrence of the definite article "the" in oral or textual speeches, is identical to its Portuguese counterpart article "o". However, there is a 3-to-1 relation about the number of characters, while the number of syllables is identical in both languages (one syllable). Thus, a metric based on the length of words, would value the word "the" 3 times more relevant than the word "o", which wouldn't be correct. Using a metric based on the number of syllables, that distortion would not occur.

Figure 1 shows the distribution of the average frequency of occurrence of words of each syllable group, for the English corpus we have mentioned; the values are normalized such that its sum is 1. By the values used to build this graphic, as we may estimate by figure 1, the average frequency of occurrence of the words having 1 syllable is 6.9 times the average frequency of occurrence of the words having 2 syllables, and 15.99 times the average frequency of occurrence of the words of the 3 syllables group — $0.7549/0.1094 = 6.9$ and $0.7549/0.0472 = 15.99$ —.

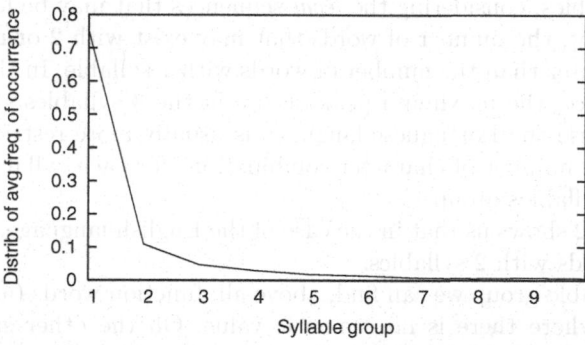

Fig. 1. Normalized distribution of the average frequency of occurrence of words of each syllable group, for an English corpus

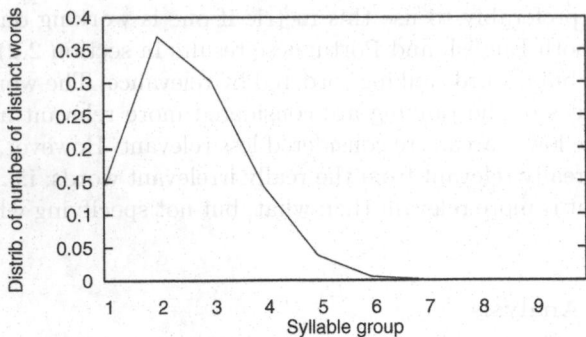

Fig. 2. Normalized distribution of the number of distinct words for each syllable group, for an English corpus

Thus, the average frequency of occurrence of the words in each syllable group decreases with the increasing of the number of syllables. This phenomenon is certainly related to the economy of speech. It is necessary that the words that occur more often are the ones easier to say, otherwise the discourses would be too long. The words having 1 syllable are usually articles and other function words like "and", "the", "of" and "or" (in Portuguese "e", "o", "de" and "ou"); because they occur more frequently in texts, they are easier and faster to pronounce.

Figure 2 shows the distribution of the number of distinct words for each syllable group, for the same English corpus; the values are normalized such that its sum is 1. By the values used to build this graphic, as we may estimate by figure 2, the number of distinct words having 2 syllables is 2.59 times the number of distinct words having 1 syllable, and 1.27 times the number of distinct words of the 3 syllables group. The peak on the 2 syllables group allows us to say that this is the *most popular group*.

The interpretation of this curve is beyond the domain of this paper, but without a secure certainty, we believe that this distribution is probably connected to the number of distinct words that may be formed preferably with the least number of syllables, considering the *legal* sequences that may be formed in each language. In fact, the number of words that may exist with 2 or more syllables is certainly greater than the number of words with 1 syllable. In the Portuguese case, for instance, the maximum peak occurs in the 3 syllables group. This is, probably, because the Portuguese language is usually more restrictive concerning the possible number of character combinations for each syllable, needing to occupy the 3 syllables group.

Thus, figure 2 shows us that in the case of the English language, there is more diversity of words with 2 syllables.

In the 1-syllable group we can find, above all, function words (articles, prepositions, etc.), where there is no semantic value. On the other side, very rare words, with many syllables, have semantic contents which are too specific to be considered relevant and broad simultaneously.

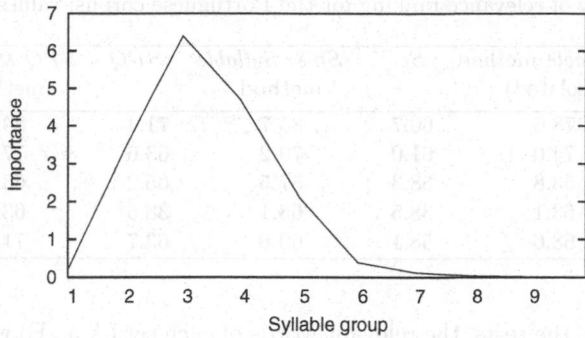

Fig. 3. Importance of each syllable group, for an English corpus

Figure 3 shows a graphic that represents the importance of each syllable group. For each syllable group, importance is determined by the corresponding value used in the graphic of the figure 2 (the Normalized distribution of the number of distinct words) divided by the corresponding value used in the graphic of the figure 1 (Normalized distribution of the average frequency of occurrence of words). If the distribution in figure 3 was used to classify the words on a text, 3-syllables words would be considered the most important, followed by the words with 2 and 4 syllables, etc. The words with 7, 8, 9 and 10 syllables (the rarest) would be considered less relevant. In section 2.2 we will show some results using this method.

Tables 2 and 3 show the quality/efficiency of the relevance ranking in several tests to the English and Portuguese corpora, respectively.

"A" test was made to the 100 most frequent words of the corpus. "B" test was made to 200 random words from the 1000 most frequent ones. For "C" test, 300 random words were taken from the 3000 most frequent ones. "D" test used 200 random words that occur at least 2 times in the corpus. "E" test considers all the previous test groups. The reader might think that the "D" test would be sufficient to evaluate the performance of the metrics, because it uses a frequency-independent sample. However, the other tests add information about the efficiency of each metric in specific areas of frequency of occurrence.

Table 2. Quality of relevance ranking for the English corpus; values in percentage

Test	Syllable method (isolated)	Sc	Sc & Syllable method	SPQ	SPQ & Syllable method
A	73.3	56.6	80.0	56.7	73.3
B	65.4	48.1	63.0	53.1	65.4
C	60.3	48.4	65.1	54.0	68.3
D	69.4	47.9	70.1	46.5	70.8
E	66.6	49.7	69.8	59.1	71.1

Table 3. Quality of relevance ranking for the Portuguese corpus; values in percentage

Test	*Syllable* method (isolated)	*Sc*	*Sc* & *Syllable* method	*SPQ*	*SPQ* & *Syllable* method
A	78.6	60.7	85.7	71.4	89.3
B	74.0	61.0	79.2	63.6	77.9
C	53.8	58.3	57.5	65.2	63.6
D	63.1	38.5	63.1	38.5	63.1
E	68.6	58.1	69.0	63.7	71.3

In order to do the tests, the relevant words of each set (A ... E) were identified and counted. Then, ranking quality was measured by the following criterion: if all those relevant words were in the top of the test ranking, there was 100% quality. But if, for instance, there were 30 relevant words, but only 25 of them were in the first 30 ranking positions, then we had 25/30 = 83.3% quality.

Each column in tables 2 and 3 represents a metric. The first column represents the syllable method isolated, where the evaluation of the words is made accordingly with the distribution in (3). As it can be seen, the syllable method isolated has good efficiency values. The values of Sc column follow the Sc metric defined in (4), while the SPQ column follow the SPQ metric defined in (5). Both metrics present respectable results, however the SPQ method is more efficient in Portuguese than in English. Both columns are followed by the same methods with the application of the syllable method. Basically, for each word, the values of the metrics (Sc and SPQ) were multiplied by the importance of the syllable group of the word (see figure 3). Tables 2 and 3 show that syllable technique improves the ranking quality in all tests. SC & *syllable* and SPQ & *syllable* methods show good results even for the most frequent relevant words group (tests "A" and "B").

3   The Islands Method

Although it is possible to obtain relevance rankings using the metrics previously presented, in certain situations it is necessary to obtain the information about the veracity or falsity of the words' relevance, i.e., it becomes necessary to know if a certain word is relevant or not. On a first analysis one could consider that all the words on top of the ranking are relevant, and that the words on the bottom are not. But the problem is to define the frontier that separates the relevant from the non-relevant words. As far as we know, there is no such method. We present a method that we have designated "Islands" that allows us to extract relevant words from a text based on a relevance ranking.

3.1   Method's Definition

Our first assumption was: in order to be considered relevant, a word must be more important than words occurring in its neighborhood. Thus, using the relevance rankings previously obtained, we are able to verify if a word is more relevant than another, by verifying if the score of each word is higher than the score of its neighbors.

After some research, current approach states that the score of a relevant word must be higher than 90% of the average score of its neighbors; but this is a weighted average based on the frequency of co-occurrence. Basically it measures the weight that each neighbor has on the word, considering the frequency of the bigram the word forms with each neighbor. Equation (6) measures the weighted average of the predecessors of a word.

$$Avg_{\text{pre}}(w) = \sum_{y_i \in \{\text{predecs of } w\}} p(y_i, w) \cdot r(y_i) \ . \tag{6}$$

$$Avg_{\text{suc}}(w) = \sum_{y_i \in \{\text{succecs of } w\}} p(w, y_i) \cdot r(y_i) \ , \tag{7}$$

where $p(y_i, w)$ means the probability of occurrence of bigram (y_i, w); $r(y_i)$ is the relevance value given by a generic $r(.)$ metric. Similarly, (7) gives the weighted average of the successors of a word.

Thus, accordingly to the Islands criterion, a word w is considered relevant *if and only if* $r(w) \cdot 0.9 \geq \max(Avg_{\text{pre}}(w), Avg_{\text{suc}}(w))$.

3.2 Results of the Islands Method

Tables 4 and 5 show precision and recall values for the Islands method: the generic relevance metric $r(.)$ in (6) and (7), and in the Islands criterion definition, was instantiated with each method shown in this tables.

Precision and recall values are a little better for Portuguese than English. However, we believe that this is due to the personal criteria used to evaluate

Table 4. Precision and recall for the Islands criterion, for the English corpus; values in percentage

	Syllable method (isolated)	*Sc*	*Sc & Syllable* method	*SPQ*	*SPQ & Syllable* method
Precision	68.2	61.1	69.2	63.6	71.6
Recall	82.2	76.8	77.0	48.4	65.7

Table 5. Precision and recall for the Islands criterion, for the Portuguese corpus; values in percentage

	Syllable method (isolated)	*Sc*	*Sc & Syllable* method	*SPQ*	*SPQ & Syllable* method
Precision	76.4	70.6	77.0	75.6	82.0
Recall	78.1	85.8	75.3	64.9	72.1

the results in these particular corpora, as the Sc metric does not follow any language-specific morphosyntactic sequence.

4   Related Work

In this section we shall analyze other approaches proposed by other authors. As we'll see, the inclusion of the syllable method greatly improves the results of those approaches. All methods listed here were implemented and the results can be seen in section 4.4.

4.1   The *Tf-idf*

Tf-idf [5] is a widely used statistic metric that measures how important a word is in a document in relation to other documents. Generally speaking, a word is more important in a document if it occurs more often in that document. However, if that word occurs in other documents, its importance decreases. So, it has a local importance. Because our corpora are based in single isolated documents, we have adapted *Tf-idf* in the following manner: the ranking score for *Tf-idf* based ranking is given by the maximum *Tf-idf* value of all documents of the corpus.

4.2   Zhou et al. Method

Zhou et al. method [2] is based on a search for clusters formed by relevant words on texts. The authors of that paper assume that relevant words tend to form clusters in certain areas of the texts while in other areas the frequency of occurrence should be very low or null. However, as we can see in subsection 4.4, this method is quite punitive for frequent words. On the other hand, in [2] the authors do not present results in terms of efficiency of their method as they do not provide an answer to determine the relevancy of a word. This method has some similarities with *Tf-idf*.

4.3   Other Approaches

Other approaches have been suggested like the elimination of the first N frequent words and the less frequent ones (the common and rare). However this empirical rule is too simplistic and prone to errors. Those methods were not implemented.

4.4   Syllable and Island Method Application

Tables 6 and 7 show the quality of the ranking scores based on each method implemented in this section. Results for the same methods are also shown after the syllable method application. Tables 8 and 9 present results for the Islands criterion instantiating each related work method.   By tables 6 and 7 we can see that Zhou et al. method is very punitive for the most frequent relevant words (tests "A"). However, both Zhou et al. and *Tf-idf* methods presented better results when combined with the syllable method.

Table 6. Quality of relevance ranking for related work methods, for the English corpus; values in percentage

Test	Tf-idf	Tf-idf & Syllable method	Zhou et al.	Zhou et al. & Syllable method
A	56.7	70.0	46.7	80.0
B	61.7	70.4	62.7	69.1
C	59.5	71.4	59.5	69.8
D	68.8	75.7	56.3	72.9
E	65.0	74.1	62.0	72.2

Table 7. Quality of relevance ranking for related work methods, for the Portuguese corpus; values in percentage

Test	Tf-idf	Tf-idf & Syllable method	Zhou et al.	Zhou et al. & Syllable method
A	46.4	78.6	25.0	85.7
B	54.5	76.6	58.4	77.9
C	63.6	62.1	66.7	58.3
D	47.7	60.0	35.4	60.0
E	56.8	68.0	58.4	69.3

Table 8. Precision and recall for the Islands criterion applied to the related work methods, for the English corpus; values in percentage

	Tf-idf	Tf-idf & Syllable method	Zhou et al.	Zhou et al. & Syllable method
Precision	73.6	81.5	66.7	71.5
Recall	47.1	55.4	75.4	76.8

Table 9. Precision and recall for the Islands criterion applied to the related work methods, for the Portuguese corpus; values in percentage

	Tf-idf	Tf-idf & Syllable method	Zhou et al.	Zhou et al. & Syllable method
Precision	80.0	83.5	70.1	78.9
Recall	59.5	65.8	79.1	77.4

5   Conclusions

In this paper we proposed two new metrics (Sc and SPQ) to obtain relevance rankings of the words in corpora. We have showed that some related work approaches excluded frequent relevant words. On the other hand, we have presented the syllable method that is a new approach, and beside the fact that it can be an autonomous metric by itself, it can also improve the quality of other methods. We believe that further research should be done in the syllables area. Finally, using the Islands criterion, we have presented another approach (unique, as far as we know) to decide what are the relevant words in corpora. We had average values of 75% for precision and recall for Islands criterion when instanciated with the combined method, Sc & Syllable, that we proposed; values that can be considered quite reasonable. SPQ method can sometimes overcome those values, although it is better in Latin languages corpus.

Although approaches such as $Tf\text{-}idf$ and the one proposed by Zhou et al. in [2] present modest precision and recall values, when we add the influence of the syllables presented in this paper, they get higher values (when applied as ranking metrics and when used with the Islands criterion).

References

1. Luhn, H.P.: The Automatic Creation of Literature Abstracts. IBM Journal of Research and Development 2, 159–165 (1958)
2. Zhou, H., Slater, G.W.: A metric to search for relevant words. Physica A: Statistical Mechanics and its Applications 329(1-2), 309–327
3. Silva, J.F., Mexia, J.T., Coelho, C.A., Lopes, G.P.: Multilingual document clustering, topic extraction and data transformation. In: Brazdil, P.B., Jorge, A.M. (eds.) EPIA 2001. LNCS (LNAI), vol. 2258, pp. 74–87. Springer, Heidelberg (2001)
4. Ortuño, M., Carpena, P., Bernaola-Galván, P., Muñoz, E., Somoza, A.M.: Europhys. Lett. 57(5), 759–764 (2002)
5. Salton, G., Buckley, C.: Term-weighing approaches in automatic text retrieval. Information Processing & Management 24(5), 513–523

Author Index

Abelha, António 160
Aerts, Marc 183
Alberto, Carlos 323
Alferes, José Júlio 3
Alves, Victor 332
Analide, Cesar 160
Antunes, Bruno 357
Antunes, Luis 605
Azevedo, Francisco 633

Bădică, Amelia 43
Bădică, Costin 43
Balsa, João 605
Banzhaf, Wolfgang 223
Bazzan, Ana L.C. 195
Belo, Orlando 383
Bianchi, Reinaldo A.C. 520
Bousmalis, Konstantinos 247
Branco, António 659
Brandão, Rui 437
Brazdil, Pavel 87
Brito, Luis 160
Brito, Pedro Quelhas 437
Brito, Robison Cris 347
Bula-Cruz, José 309

Cavique, Luís 406
Celiberto Jr., Luiz A. 520
Certo, João 542
Chauchat, Jean-Hugues 671
Chen, Huowang 485
Chen, Sushing 463
Chiang, Mark Ming-Tso 395
Coelho, Helder 605
Corchado, J.M. 53
Correia, Luís 235
Corrente, Gustavo A. 499
Cortez, Paulo 124
Costa, Ricardo 323
Cruz, Filipe 593

da Rocha Costa, Antônio Carlos 580
Dalbelo Bašić, Bojana 671
Del Gaudio, Rosa 659
Dias, Jorge 530

Díaz, F. 53
Dimuro, Graçaliz Pereira 580
Duarte, Pedro 593

Endisch, Christian 15

Fdez-Riverola, F. 53
Ferreira da Silva, Joaquim 691
Ferreira, Eugénio C. 473
Freddo, Ademir Roberto 347
Frommberger, Lutz 508

Gago, Pedro 415
Gama, João 112, 133
Gimenez-Lugo, Gustavo 347
Glez-Peña, D. 53
Gomes, Nuno 296
Gomes, Paulo 357
Graça, Ana 621
Grilo, Carlos 235

Hackl, Christoph 15
Hayes, Gillian M. 247
He, Li 463
Heymer, Mourylise 332
Hitzler, Pascal 3

Ilić, Manoela 29

Jorge, Alípio 426, 437

Kennedy, James 259
Klügl, Franziska 195
Knorr, Matthias 3
Kriksciuniene, Dalia 371

Lau, Nuno 542
Leite, João 29
Leite, Rui 87
Li, Shutao 485
Lima, Luís 323
Lopes, Gabriel 683
Lopes, Gonçalo 73
Lopes, Manuel C. 555
Loureiro, Jorge 383
Lucas, Joel P. 426
Lynce, Inês 621

Machado, Amauri A.   426
Machado, José   160, 309, 323
Maia, Paulo   473
Marques-Silva, João   621
Marreiros, Filipe   332
Marreiros, Goreti   309
Matsuura, Jackson   520
Melo, Francisco S.   555
Mendes, Rui   259, 473
Méndez, J.R.   53
Mirkin, Boris   395
Moons, Elke   183
Morin, Annie   671
Munteanu, Cristian   209

Nagel, Kai   195
Nelas, Luís   332
Neto, João Pedro   568
Neves, António J.R.   499
Neves, José   160, 309, 323, 332
Novais, Paulo   309, 323
Nunes, Francisco   296

Oliveira, Arlindo L.   621

Pereira, António   593
Pereira, Cristiano   296
Pereira, Fernando   426
Pereira, Luís Moniz   63, 73, 99
Pereira Rodrigues, Pedro   133
Pernas, Ana M.   426
Pfaffmann, Jeffrey O.   247
Pillay, Nelishia   223
Pinho, Armando J.   499
Pinto, José P.   473
Popescu, Elvira   43

Quintela, Hélder   124

Rafael, Pedro   568
Ramos, Carlos   285, 296, 309

Rebelo, Carmen   437
Řehuřek, Radim   647
Reis, Luís Paulo   542, 593
Rett, Jörg   530
Revett, Kenneth   145
Rocha, Isabel   473
Rocha, Miguel   473
Rocio, Vitor   683
Rosa, Agostinho   209

Sakalauskas, Virgilijus   371
Santos, Henrique M.D.   145
Santos, Manuel Filipe   124, 415
Santos, Ricardo   309
Saptawijaya, Ari   99
Schröder, Dierk   15
Sebastião, Raquel   112
Seco, Nuno   357
Shen, Bairong   463
Shukla, Pradyumn Kumar   271
Šilić, Artur   671
Silva, Álvaro   415
Silva, Joaquim   683
Soares, Carlos   437

Tacla, Cesar Augusto   347
Tang, Binhua   463
Tenreiro de Magalhães, Sérgio   145
Torgo, Luis   449

Valente, Pedro   593
Ventura, João   691
Viegas, Ruben   633

Wang, Shulin   485
Wets, Geert   183
Wheeler, Gregory   170

Zhang, Dingxing   485

Lecture Notes in Artificial Intelligence (LNAI)

Vol. 4874: J. Neves, M.F. Santos, J.M. Machado (Eds.), Progress in Artificial Intelligence. XVIII, 704 pages. 2007.

Vol. 4830: M.A. Orgun, J. Thornton (Eds.), AI 2007: Advances in Artificial Intelligence. XIX, 841 pages. 2007.

Vol. 4828: M. Randall, H.A. Abbass, J. Wiles (Eds.), Progress in Artificial Life. XII, 402 pages. 2007.

Vol. 4827: A. Gelbukh, Á.F. Kuri Morales (Eds.), MICAI 2007: Advances in Artificial Intelligence. XXIV, 1234 pages. 2007.

Vol. 4798: Z. Zhang, J.H. Siekmann (Eds.), Knowledge Science and Engineering and Management. XVI, 669 pages. 2007.

Vol. 4795: F. Schilder, G. Katz, J. Pustejovsky (Eds.), Annotating, Extracting and Reasoning about Time and Events. VII, 141 pages. 2007.

Vol. 4790: N. Dershowitz, A. Voronkov (Eds.), Logic for Programming, Artificial Intelligence, and Reasoning. XIII, 562 pages. 2007.

Vol. 4788: D. Borrajo, L. Castillo, J.M. Corchado (Eds.), Current Topics in Artificial Intelligence. XI, 280 pages. 2007.

Vol. 4775: A. Esposito, M. Faundez-Zanuy, E. Keller, M. Marinaro (Eds.), Verbal and Nonverbal Communication Behaviours. XII, 325 pages. 2007.

Vol. 4772: H. Prade, V.S. Subrahmanian (Eds.), Scalable Uncertainty Management. X, 277 pages. 2007.

Vol. 4766: N. Maudet, S. Parsons, I. Rahwan (Eds.), Argumentation in Multi-Agent Systems. XII, 211 pages. 2007.

Vol. 4755: V. Corruble, M. Takeda, E. Suzuki (Eds.), Discovery Science. XI, 298 pages. 2007.

Vol. 4754: M. Hutter, R.A. Servedio, E. Takimoto (Eds.), Algorithmic Learning Theory. XI, 403 pages. 2007.

Vol. 4737: B. Berendt, A. Hotho, D. Mladenic, G. Semeraro (Eds.), From Web to Social Web: Discovering and Deploying User and Content Profiles. XI, 161 pages. 2007.

Vol. 4733: R. Basili, M.T. Pazienza (Eds.), AI*IA 2007: Artificial Intelligence and Human-Oriented Computing. XVII, 858 pages. 2007.

Vol. 4724: K. Mellouli (Ed.), Symbolic and Quantitative Approaches to Reasoning with Uncertainty. XV, 914 pages. 2007.

Vol. 4722: C. Pelachaud, J.-C. Martin, E. André, G. Chollet, K. Karpouzis, D. Pelé (Eds.), Intelligent Virtual Agents. XV, 425 pages. 2007.

Vol. 4720: B. Konev, F. Wolter (Eds.), Frontiers of Combining Systems. X, 283 pages. 2007.

Vol. 4702: J.N. Kok, J. Koronacki, R. Lopez de Mantaras, S. Matwin, D. Mladenič, A. Skowron (Eds.), Knowledge Discovery in Databases: PKDD 2007. XXIV, 640 pages. 2007.

Vol. 4701: J.N. Kok, J. Koronacki, R. Lopez de Mantaras, S. Matwin, D. Mladenič, A. Skowron (Eds.), Machine Learning: ECML 2007. XXII, 809 pages. 2007.

Vol. 4696: H.-D. Burkhard, G. Lindemann, R. Verbrugge, L.Z. Varga (Eds.), Multi-Agent Systems and Applications V. XIII, 350 pages. 2007.

Vol. 4694: B. Apolloni, R.J. Howlett, L. Jain (Eds.), Knowledge-Based Intelligent Information and Engineering Systems, Part III. XXIX, 1126 pages. 2007.

Vol. 4693: B. Apolloni, R.J. Howlett, L. Jain (Eds.), Knowledge-Based Intelligent Information and Engineering Systems, Part II. XXXII, 1380 pages. 2007.

Vol. 4692: B. Apolloni, R.J. Howlett, L. Jain (Eds.), Knowledge-Based Intelligent Information and Engineering Systems, Part I. LV, 882 pages. 2007.

Vol. 4687: P. Petta, J.P. Müller, M. Klusch, M. Georgeff (Eds.), Multiagent System Technologies. X, 207 pages. 2007.

Vol. 4682: D.-S. Huang, L. Heutte, M. Loog (Eds.), Advanced Intelligent Computing Theories and Applications. XXVII, 1373 pages. 2007.

Vol. 4676: M. Klusch, K.V. Hindriks, M.P. Papazoglou, L. Sterling (Eds.), Cooperative Information Agents XI. XI, 361 pages. 2007.

Vol. 4667: J. Hertzberg, M. Beetz, R. Englert (Eds.), KI 2007: Advances in Artificial Intelligence. IX, 516 pages. 2007.

Vol. 4660: S. Džeroski, L. Todorovski (Eds.), Computational Discovery of Scientific Knowledge. X, 327 pages. 2007.

Vol. 4659: V. Mařík, V. Vyatkin, A.W. Colombo (Eds.), Holonic and Multi-Agent Systems for Manufacturing. VIII, 456 pages. 2007.

Vol. 4651: F. Azevedo, P. Barahona, F. Fages, F. Rossi (Eds.), Recent Advances in Constraints. VIII, 185 pages. 2007.

Vol. 4648: F. Almeida e Costa, L.M. Rocha, E. Costa, I. Harvey, A. Coutinho (Eds.), Advances in Artificial Life. XVIII, 1215 pages. 2007.

Vol. 4635: B. Kokinov, D.C. Richardson, T.R. Roth-Berghofer, L. Vieu (Eds.), Modeling and Using Context. XIV, 574 pages. 2007.

Vol. 4632: R. Alhajj, H. Gao, X. Li, J. Li, O.R. Zaïane (Eds.), Advanced Data Mining and Applications. XV, 634 pages. 2007.

Vol. 4629: V. Matoušek, P. Mautner (Eds.), Text, Speech and Dialogue. XVII, 663 pages. 2007.

Vol. 4626: R.O. Weber, M.M. Richter (Eds.), Case-Based Reasoning Research and Development. XIII, 534 pages. 2007.

Vol. 4617: V. Torra, Y. Narukawa, Y. Yoshida (Eds.), Modeling Decisions for Artificial Intelligence. XII, 502 pages. 2007.

Vol. 4612: I. Miguel, W. Ruml (Eds.), Abstraction, Reformulation, and Approximation. XI, 418 pages. 2007.

Vol. 4604: U. Priss, S. Polovina, R. Hill (Eds.), Conceptual Structures: Knowledge Architectures for Smart Applications. XII, 514 pages. 2007.

Vol. 4603: F. Pfenning (Ed.), Automated Deduction – CADE-21. XII, 522 pages. 2007.

Vol. 4597: P. Perner (Ed.), Advances in Data Mining. XI, 353 pages. 2007.

Vol. 4594: R. Bellazzi, A. Abu-Hanna, J. Hunter (Eds.), Artificial Intelligence in Medicine. XVI, 509 pages. 2007.

Vol. 4585: M. Kryszkiewicz, J.F. Peters, H. Rybinski, A. Skowron (Eds.), Rough Sets and Intelligent Systems Paradigms. XIX, 836 pages. 2007.

Vol. 4578: F. Masulli, S. Mitra, G. Pasi (Eds.), Applications of Fuzzy Sets Theory. XVIII, 693 pages. 2007.

Vol. 4573: M. Kauers, M. Kerber, R. Miner, W. Windsteiger (Eds.), Towards Mechanized Mathematical Assistants. XIII, 407 pages. 2007.

Vol. 4571: P. Perner (Ed.), Machine Learning and Data Mining in Pattern Recognition. XIV, 913 pages. 2007.

Vol. 4570: H.G. Okuno, M. Ali (Eds.), New Trends in Applied Artificial Intelligence. XXI, 1194 pages. 2007.

Vol. 4565: D.D. Schmorrow, L.M. Reeves (Eds.), Foundations of Augmented Cognition. XIX, 450 pages. 2007.

Vol. 4562: D. Harris (Ed.), Engineering Psychology and Cognitive Ergonomics. XXIII, 879 pages. 2007.

Vol. 4548: N. Olivetti (Ed.), Automated Reasoning with Analytic Tableaux and Related Methods. X, 245 pages. 2007.

Vol. 4539: N.H. Bshouty, C. Gentile (Eds.), Learning Theory. XII, 634 pages. 2007.

Vol. 4529: P. Melin, O. Castillo, L.T. Aguilar, J. Kacprzyk, W. Pedrycz (Eds.), Foundations of Fuzzy Logic and Soft Computing. XIX, 830 pages. 2007.

Vol. 4520: M.V. Butz, O. Sigaud, G. Pezzulo, G. Baldassarre (Eds.), Anticipatory Behavior in Adaptive Learning Systems. X, 379 pages. 2007.

Vol. 4511: C. Conati, K. McCoy, G. Paliouras (Eds.), User Modeling 2007. XVI, 487 pages. 2007.

Vol. 4509: Z. Kobti, D. Wu (Eds.), Advances in Artificial Intelligence. XII, 552 pages. 2007.

Vol. 4496: N.T. Nguyen, A. Grzech, R.J. Howlett, L.C. Jain (Eds.), Agent and Multi-Agent Systems: Technologies and Applications. XXI, 1046 pages. 2007.

Vol. 4483: C. Baral, G. Brewka, J. Schlipf (Eds.), Logic Programming and Nonmonotonic Reasoning. IX, 327 pages. 2007.

Vol. 4482: A. An, J. Stefanowski, S. Ramanna, C.J. Butz, W. Pedrycz, G. Wang (Eds.), Rough Sets, Fuzzy Sets, Data Mining and Granular Computing. XIV, 585 pages. 2007.

Vol. 4481: J. Yao, P. Lingras, W.-Z. Wu, M. Szczuka, N.J. Cercone, D. Ślęzak (Eds.), Rough Sets and Knowledge Technology. XIV, 576 pages. 2007.

Vol. 4476: V. Gorodetsky, C. Zhang, V.A. Skormin, L. Cao (Eds.), Autonomous Intelligent Systems: Multi-Agents and Data Mining. XIII, 323 pages. 2007.

Vol. 4460: S. Aguzzoli, A. Ciabattoni, B. Gerla, C. Manara, V. Marra (Eds.), Algebraic and Proof-theoretic Aspects of Non-classical Logics. VIII, 309 pages. 2007.

Vol. 4457: G.M.P. O'Hare, A. Ricci, M.J. O'Grady, O. Dikenelli (Eds.), Engineering Societies in the Agents World VII. XI, 401 pages. 2007.

Vol. 4456: Y. Wang, Y.-m. Cheung, H. Liu (Eds.), Computational Intelligence and Security. XXIII, 1118 pages. 2007.

Vol. 4455: S. Muggleton, R. Otero, A. Tamaddoni-Nezhad (Eds.), Inductive Logic Programming. XII, 456 pages. 2007.

Vol. 4452: M. Fasli, O. Shehory (Eds.), Agent-Mediated Electronic Commerce. VIII, 249 pages. 2007.

Vol. 4451: T.S. Huang, A. Nijholt, M. Pantic, A. Pentland (Eds.), Artifical Intelligence for Human Computing. XVI, 359 pages. 2007.

Vol. 4442: L. Antunes, K. Takadama (Eds.), Multi-Agent-Based Simulation VII. X, 189 pages. 2007.

Vol. 4441: C. Müller (Ed.), Speaker Classification II. X, 309 pages. 2007.

Vol. 4438: L. Maicher, A. Sigel, L.M. Garshol (Eds.), Leveraging the Semantics of Topic Maps. X, 257 pages. 2007.

Vol. 4434: G. Lakemeyer, E. Sklar, D.G. Sorrenti, T. Takahashi (Eds.), RoboCup 2006: Robot Soccer World Cup X. XIII, 566 pages. 2007.

Vol. 4429: R. Lu, J.H. Siekmann, C. Ullrich (Eds.), Cognitive Systems. X, 161 pages. 2007.

Vol. 4428: S. Edelkamp, A. Lomuscio (Eds.), Model Checking and Artificial Intelligence. IX, 185 pages. 2007.

Vol. 4426: Z.-H. Zhou, H. Li, Q. Yang (Eds.), Advances in Knowledge Discovery and Data Mining. XXV, 1161 pages. 2007.

Vol. 4411: R.H. Bordini, M. Dastani, J. Dix, A.E.F. Seghrouchni (Eds.), Programming Multi-Agent Systems. XIV, 249 pages. 2007.

Vol. 4410: A. Branco (Ed.), Anaphora: Analysis, Algorithms and Applications. X, 191 pages. 2007.

Vol. 4399: T. Kovacs, X. Llorà, K. Takadama, P.L. Lanzi, W. Stolzmann, S.W. Wilson (Eds.), Learning Classifier Systems. XII, 345 pages. 2007.

Vol. 4390: S.O. Kuznetsov, S. Schmidt (Eds.), Formal Concept Analysis. X, 329 pages. 2007.

Vol. 4389: D. Weyns, H. Van Dyke Parunak, F. Michel (Eds.), Environments for Multi-Agent Systems III. X, 273 pages. 2007.